Ricci Solitons in Dimensions 4 and Higher

Mathematical
Surveys
and
Monographs

Volume 293

Ricci Solitons in Dimensions 4 and Higher

Bennett Chow
Brett Kotschwar
Ovidiu Munteanu

AMS
American Mathematical Society
Providence, Rhode Island

2020 *Mathematics Subject Classification.* Primary 53E20, 53E10, 53E30, 58J05, 58J35, 57K30, 57M50, 57K40, 57R55, 53C21.

For additional information and updates on this book, visit
www.ams.org/bookpages/surv-293

Library of Congress Cataloging-in-Publication Data

Cataloging-in-Publication Data has been applied for by the AMS.
See http://www.loc.gov/publish/cip/.

♾ The paper used in this book is acid-free and falls within the guidelines
established to ensure permanence and durability.
Visit the AMS home page at `https://www.ams.org/`

10 9 8 7 6 5 4 3 2 1   30 29 28 27 26 25

In memory of Richard S. Hamilton (1943–2024),
to whom this book is dedicated.

Contents

Foreword

All my mathematical life I have followed and used differential geometry and its attendant PDEs. I have had amazing tutors, including S.T. Yau, Rick Schoen, Richard Hamilton, Cliff Taubes, and Ben Chow. I am much more a consumer than a producer of differential geometry. When set against my personal skills, the subject has an NP flavor. Given the correct clues (witness, as they say in computer science) the arguments flow beautifully and do make sense. But how does one think them up in the first place? Which quantities control which others and which ones are amenable to monotonicity estimates? To me these mysteries are hidden, and I have always relied on guides. I am grateful to those who have explained and taught me what they could of the subject. Certainly, low-dimensional topologists and physicists can understand very little of the worlds they seek to explore without access to this school of geometry. In the last decade, I would say that any wall separating elliptic and parabolic methods has come down. Now it is conventional wisdom in geometry to consider evolving your manifold and submanifolds as an integral part of geometric analysis. Of course, the premiere evolution equation is Ricci flow. This book is an essential technical introduction to those who would study 4-manifolds geometrically. The Thurston–Hamilton–Perelman story, to my mind, has been one of the two most profound developments to touch topology in the last 50 years. (The other has been the invention of topological quantum field theory.) One naturally asks: "What is next?" "Ricci Solitons in Dimension 4 and Higher" boldly proposes that Ricci flow under the leadership of a new generation of mathematicians can tackle the structure of smooth 4-manifolds. It is an amazing thought. The authors explain how the famous 11/8-conjecture is a plausible target and that work of Richard Bamler already constitutes the start of a program. (The conjecture is that among closed spin 4-manifolds, the density of the signature in the second Betti number never exceeds 8/11ths, as exemplified by the K3 surface.) They also, sagely, point out that the slightly more famous problem: the smooth 4D Poincaré conjecture (S4DPC) has no known starting point within Ricci flow. This to me is a second great challenge: Conceive of a program for approaching S4DPC using Ricci flow, or related equations. The book explains a host of technical developments in marvelous detail and invites the reader to think on the largest possible scale. It is quite possible that the next great advance in 4-manifolds will come from a reader of this book.

Michael Hartley Freedman

Preface

Hamilton and Perelman's proof of Thurston's Geometrization Conjecture is one of the most spectacular mathematical achievements of the past several decades — a far-reaching result which not only subsumes the Poincaré Conjecture but provides a classification for all closed 3-manifolds. The key mechanism in their proof is a program of Ricci flow with surgery, in which, by a sequence of careful and timely interventions, the flow may be continued past any singularities which would otherwise halt the flow under the natural evolution of the space. This program is made in part possible by a thorough understanding of 2- and 3-dimensional *Ricci solitons* — special solutions to the equation which model the geometry nearby a developing singularity.

Currently, there is hope that the Ricci flow may prove useful in the exploration of various conjectures related to the topology of smooth 4-manifolds, an optimism partly rooted in the success already achieved by the theory of the anti-self-dual and Seiberg–Witten equations in this direction. Any potential program of Ricci flow with surgery in dimension 4 will, however, require a reasonably detailed understanding of the kinds of singularities one may encounter along the flow, and therefore a correspondingly detailed understanding of Ricci solitons in that dimension, given the central role they play in the description of these singularities. This is a difficult problem: whereas those Ricci solitons in dimensions 2 and 3 which arise as singularity models are completely classified, starting in dimension 4 the theory becomes considerably more complex and the possibilities for such spaces substantially more varied.

The purpose of this book is to present an organized collection of results for Ricci solitons in dimensions 4 and higher and to advertise some related open questions whose resolutions would help to advance the theory. Our primary focus will be to present a coherent theory in dimension 4, where applications to topology appear most promising.

Acknowledgments

The authors would like to offer their special thanks to AMS publisher Sergei Gelfand for his role in making the publication of this book possible and to AMS book acquisitions editor Ina Mette for her splendid work and invaluable guidance in shepherding this volume to its completion. We would like to thank Auden Boatman, Savannah Cooper, and Christine Thivierge for their assistance with contracts and communications regarding the publication of this book. We would like to thank Arlene O'Sean for her wonderful copy editing of parts of the book.

The authors would also like to thank the editors of the AMS Mathematical Surveys and Monographs series and the anonymous reviewers for their careful reading of the manuscript. Their suggestions greatly improved the organization and exposition of the book. They are also thankful for the help and expertise of their friends, colleagues, and students throughout the writing process.

Finally, the authors would like to acknowledge their heartfelt gratitude to Mike Freedman for generously writing a foreword to this volume and for his encouragement and support of the project. We are deeply grateful to Richard Bamler for numerous very helpful suggestions and corrections.

During the preparation of this volume, Bennett was partially supported by the Institute for Advanced Study (IAS) Director's Discretionary Fund. Bennett would like to express his deepest gratitude to Aaron Naber, who made this opportunity possible.

During the preparation of this volume, Brett was partially supported by Simons Foundation grant #709004.

During the preparation of this volume, Ovidiu was partially supported by the NSF grant DMS-1811845 and by a Simons Foundation grant.

Bennett Chow
University of California San Diego

Brett Kotschwar
Arizona State University

Ovidiu Munteanu
University of Connecticut

Introduction

The Ricci flow was first introduced more than forty years ago by Richard Hamilton. In the decades since, the theory of singularity formation under the Ricci flow has undergone several revolutions. From the seminal and visionary works of Hamilton to the dramatic breakthroughs of Perelman, the major contributions of Brendle, and the recent fundamental advances of Bamler, these revolutions — all supported and driven forward by the work of many other authors — have substantially improved our understanding of the longtime behavior of the flow. Through each of these revolutions, Ricci solitons – the self-similar solutions which are generalized fixed points of the equation – have consistently occupied a place at the heart of the theory, both as models for the singular regions themselves and as the critical cases of inequalities and monotonicity formulas valid for all solutions.

In this book, we continue the study of Ricci solitons begun in the previous volume [**Cho23**] by one of the authors. Our exposition will, of necessity, be an incomplete snapshot of the current state of this vast and still changing area of research and one that raises as many questions as it answers. Our aim is to present some select and broadly representative results for Riemannian shrinking and steady solitons and to describe their relevance to the study of the singularity formation of the Ricci flow. In the Kähler and expanding cases, we will for the most part content ourselves to survey relevant results, leaving proofs and more detailed discussion for a planned later volume.

We will focus on results for Ricci solitons in dimensions greater than three. Although (as explained in the first volume [**Cho23**]) there is now a complete classification of those 2- and 3-dimensional Ricci solitons which arise as singularity models (a classification made possible by certain simplifying features of the problem which are automatic in — but also specific to — those dimensions), the higher-dimensional case is considerably more complex. It is currently unclear whether classifying all shrinking and steady soliton singularity models in dimension 4 is a realistic goal. However, as we will see in this book, significant progress in understanding their structure remains possible.

As an indication of the novel difficulties one encounters when crossing from three dimensions to four, consider that in dimension 4 one must contend with the existence of nontrivial Einstein manifolds (i.e., those without constant sectional curvature). Any Einstein manifold is in particular an example of a gradient Ricci soliton, taking the potential function to be constant. However, the notorious complexity of the class of Einstein manifolds should not deter one from hoping to say *anything* about Ricci solitons: indeed, the solitons of most interest to the analysis of singularities are those with a nonconstant potential function, and this extra bit of structure adds a substantial amount of rigidity.[1] The complementary nature of the Einstein and Ricci soliton equations will be a recurring theme throughout the text.

With our sights set instead, then, on obtaining qualitative and quantitative statements constraining the geometric behavior and large-scale structure of solitons, and on coarse classifications by type rather than an exhaustive taxonomy, we will progress from a review of basic examples and foundational concepts to a sequence of more specialized and essentially independent topics. We will be particularly concerned with what can be said for complete noncompact Ricci solitons, the case of greatest significance to singularity analysis.[2] For these solitons, we will present estimates on the growth and decay of curvature, bounds on the number of topological ends, descriptions of their asymptotic geometry, and results on their uniqueness subject to constraints on their asymptotic geometry.

Finally, one feature of Bamler's theory of singularity development in dimension 4 is its emphasis on singularities modeled on *smooth orbifolds with isolated singularities*. This appears to be the appropriate (and unavoidable) setting for studying any potential formulation of Ricci flow with surgery, even when the initial data is a smooth closed Riemannian 4-manifold. Many results on Ricci solitons extend in a straightforward manner to the category of such orbifolds, and we will make an effort to state the results in this book for that category when possible (see Appendix A and the references therein for some background). Unless stated otherwise, however, we assume that all objects and morphisms belong to the smooth (C^∞) category.

This book is organized as follows:

Chapter 1 provides background and motivation from the singularity analysis of Ricci flow. We provide an overview of examples of Ricci solitons as well as a survey of aspects of the general theory with an emphasis on results that are proved in this book. A simplified outline suggesting a possible direction for applying Ricci flow to 4-dimensional smooth topology is presented.

Chapter 2 discusses curvature estimates for 4-dimensional shrinking gradient Ricci solitons and their relation to asymptotically conical and asymptotically cylindrical solitons.

Chapter 3 introduces the concept of counting topological ends of Ricci solitons, illustrating the relationship between function theory and weighted manifolds.

Chapter 4 takes an in-depth look at asymptotically conical shrinking gradient Ricci solitons and proves that they satisfy a holographic principle, in the sense that the geometry of the asymptotic slice uniquely determines the global geometry of the shrinking soliton.

Chapter 5 discusses curvature and volume estimates for steady solitons, which detail the interplay between the Riemann, Ricci, and scalar curvatures of solitons as in Chapter 2.

Chapter 6 discusses aspects of the asymptotic geometry at infinity of steady Ricci solitons that can be proved using Bamler's theory. This includes tangent flows at infinity, volume lower bounds, and curvature lower bounds.

[1]Furthermore, one may speculate that, for topological applications, one would only need to understand those shrinking Ricci solitons which are stable in some sense (including Einstein metrics with positive scalar curvature).

[2]It is known that any compact singularity model must be a shrinking soliton and hence yields finite time extinction for the Ricci flow.

Appendix A provides a review of orbifolds with isolated singularities.

At the end of each chapter is a short collection of open problems on Ricci solitons in relation to the topics of the chapter. For each problem regarding shrinking or steady solitons, the reader may add if they so wish the hypotheses that the Ricci soliton is both 4-dimensional and a singularity model.

For the most part, we will use the same notation as the companion text [**Cho23**] by one of the authors. However, this book is largely independent of that text (hereafter referred to as Volume 1).

Notation and Symbols

Here we list some of the notation and symbols used throughout the book.

0^n	the origin in $\mathbb{R}^n$
$\vec{0}$	the origin in a tangent space
$\measuredangle$	angle
$\alpha * \beta$	a contraction of the tensor product of tensors α and β
$A \setminus B$	the set difference between sets A and B
$X^\flat$	dual 1-form to the vector field X
$\alpha^\natural$	dual vector field to the 1-form α
$\alpha(g)$	aperture of a metric g
Area	area of a surface or volume of a hypersurface
ACR	asymptotic (Riemann) curvature ratio
ASCR	asymptotic scalar curvature ratio
AVR	asymptotic volume ratio
$B(x,r)$	open ball of radius r centered at x
$B_g(x,r)$	open ball with respect to the metric g
$B(x,t,r)$	open ball $B_{g(t)}(x,r)$ with respect to a Ricci flow $g(t)$
$B_{ijk\ell}$	the curvature quadratic: $-\sum_{p,q=1}^n R_{ipj}^q R_{\ell qk}^p$
Γ_{ij}^k	Christoffel symbols
$\mathrm{CA}(g)$	cone angle at infinity of g
$C_0\Sigma$, $\mathcal{C}^\Sigma$	cone on Σ
$\mathfrak{C}$	correspondence between metric flows
∇	covariant derivative
∇^k	k-th covariant derivative
$\mathrm{Cut}(p)$	cut locus of $p \in M^n$
$d(x,y)$	Riemannian distance between x and y
$d_{W_p}^g$	p-Wasserstein distance
$d_{\mathbb{F}}^J$	$\mathbb{F}$-distance
$d_{\mathbb{F}}^{\mathfrak{C},J}$	$\mathbb{F}$-distance within a correspondence $\mathfrak{C}$
dA	area form (volume form in dimension 2)
$\mathrm{diam}(M^n)$	diameter of M^n
$\mathrm{Diff}(M^n)$	diffeomorphism group of M^n
div	divergence

div_f	f-divergence
D_p°	injectivity domain of $\exp_p$
$\frac{d_-}{dt}(t)$	lim inf of backward difference quotients
$d\nu_{x_0,t_0;t}$	conjugate heat kernel measure based at (x_0, t_0)
$\{e_i\}_{i=1}^n$	orthonormal frame (field)
χ	Euler characterisic
$\exp_p$	exponential map at $p \in M^n$
f	potential function of a gradient Ricci soliton
f_t	potential of the conjugate heat kernel
f_{Gau}	potential function for Gaussian soliton
$\mathcal{F}$	Perelman's $\mathcal{F}$-energy
g	Riemannian metric
g_∞	limit Riemannian metric
$g(t)$	1-parameter family of metrics, often a solution to Ricci flow
g_{Bry}	metric of Bryant soliton
$g_{\mathbb{E}}$	Euclidean metric on $\mathbb{R}^n$
$g_{\mathbb{S}^n}$	standard metric on $\mathbb{S}^n$
g_Σ	cigar soliton metric
γ_V	geodesic with initial velocity V
grad or ∇	gradient
Γ	a finite subgroup of $\mathrm{O}(4)$
$\square$	heat operator
$\square^*$	conjugate heat operator
∇^2	Hessian or double covariant derivative
H_f	f-mean curvature of a hypersurface
H_n	the dimensional constant $\frac{(n-1)\pi^2}{2} + 4$
id_X	identity map of a set X
$\int$	integral
$\oint$	contour integral
inj	injectivity radius
$\mathrm{Isom}(M^n, g)$	isometry group of (M^n, g)
J	almost complex structure
J	Jacobian of the exponential map
J_f	f-Jacobian of the exponential map
K	Gauss curvature of a surface
$K(x, t; y, s)$	(conjugate) heat kernel
$K_{\mathbb{E}}(x, t; y, s)$	Euclidean (conjugate) heat kernel
ker	kernel
K_{orb}	orbital sectional curvature of a rotationally symmetric metric
K_{rad}	radial sectional curvature of a rotationally symmetric metric
ℓ	reduced distance

L	length
$\mathcal{L}$	Lie derivative
Δ	Laplacian
Δ_f	f-Laplacian
Δ_X	X-Laplacian $= \Delta - X \cdot \nabla$
Δ_Σ	Laplacian of a hypersurface Σ
Δ_L	Lichnerowicz Laplacian
$\Delta_{L,f}$	f-Lichnerowicz Laplacian
$\Lambda^k T^* M$	vector bundle of k-forms on M^n
Σ_c	level set of a function
M^n	manifold
M^n_∞	limit manifold
$\tilde{M}^n$	(usually universal) cover of M^n
$\mathrm{Met}(M^n)$	space of Riemannian metrics on M^n
M_{ij}, P_{ijk}	components of Hamilton's matrix Harnack quadratic form
μ	Perelman's μ-invariant
μ_∞	asymptotic Nash entropy
$\cdot^\perp$	normal component or projection
$N(M^n)$	number of ends of M^n
$\mathcal{N}[g, f, \tau]$	Nash entropy
$\mathcal{N}_{x_0,t_0}(\tau)$	pointed Nash entropy based at (x_0, t_0)
$(\nu_{x;s})_{s\in I, s\leq \mathbf{t}(x)}$	conjugate heat kernel based at $x \in \mathcal{X}$
$n\omega_n$	volume of the $(n-1)$-dimensional Euclidean unit sphere
ω_n	volume of the n-dimensional Euclidean unit ball
$\{\omega^i\}_{i=1}^n$	orthonormal coframe (field)
ω_i^j	connection 1-form
Ω_i^j	curvature 2-form
$\mathcal{O}$	orbifold
$M^-(\mathbf{z}_0)$	ODE unstable manifold at $\mathbf{z}_0$
$W^-(\mathbf{z}_0)$	ODE unstable set at $\mathbf{z}_0$
∂_t	$\frac{\partial}{\partial t}$ partial time-derivative
Ψ	the error function
$Q(\mathrm{Rm})$	quadratic of Rm arising in Hamilton's evolution equation
$\mathbb{R}_+$	set of strictly positive real numbers
$\mathbb{R}^n$	n-dimensional Euclidean space
R	scalar curvature
Ric	Ricci tensor
Rm	Riemann curvature operator
$R_{ijk\ell}$	components of Rm
R_f	f-scalar curvature
Ric_f	f-Ricci tensor

$r_R(x)$	scalar curvature scale of x
$\mathcal{R}^\infty$	set of regular points
II	second fundamental form of a hypersurface
$\mathfrak{so}(n)$	special orthogonal group
$\mathbb{S}^n$	n-dimensional unit sphere
$\mathbb{S}^n(r)$	n-dimensional sphere of radius r
$S_r(p)$	geodesic sphere of radius r centered at p
$\mathcal{S}_p^{n-1}$	the unit sphere in $T_p\mathcal{M}$
sect	sectional curvature
Sym	symmetization (of a 2-tensor)
$S^2(T^*M)$	bundle of symmetric 2-tensors on M^n
$\mathcal{S}^\infty$	set of singular points
$\mathbb{S}^3/\Gamma$	spherical space form
T_pM	tangent space at $p \in M^n$
$\cdot^\top$	tangential component or projection
$\mathbf{t}$	time function of a metric flow
tr	trace
$\overset{\circ}{\alpha}$	trace-free part of a 2-tensor α
Var	variance
Vol	volume
Vol_f	f-volume
$d\mu$	Riemannian volume form or measure
$d\sigma$	volume form or measure on a hypersurface
W	Lambert W function
$\mathcal{W}$	Perelman's $\mathcal{W}$-entropy
$\mathcal{X}$	(spacetime of a) metric flow
$\mathcal{X}^\lambda$	λ-rescaling of a metric flow
$\mathcal{X}_t$	time slice of a metric flow
$\mathcal{X} - t_0$	time shift of a metric flow
$\mathcal{X}\|_{I'}$	restriction of a metric flow to I'
$(\mathcal{X}, (\mu_t)_{t\in I'})$	metric flow pair
$\mathcal{X}^\infty$	tangent flow at infinity
z_t	H_n-center at time t

CHAPTER 1

Ricci Solitons in the Context of Ricci Flow

We will begin our study of 4-dimensional Ricci solitons with a high-level discussion of the role that solitons play in the analysis of singularities of the Ricci flow. After a brief review of some foundational results for solitons, we will examine the central role they play in the program of Ricci-flow-with-surgery in 3-dimensions developed by Hamilton and Perelman to prove the Poincaré Conjecture and Thurston's Geometrization Conjecture.

This book is motivated by the hope that a similar program of weak Ricci flow or flow-with-surgery might yet be developed in dimension 4 with the power to advance the understanding of smooth 4-manifold topology (see Bamler [**Bam21b**]). Such a program would follow the rough progression of ideas

$$\text{Singularity analysis} \longrightarrow \text{Surgery theory} \longrightarrow \text{Topological applications}$$

in which the study of Ricci solitons lies at the very center of the first step.

Following our review of solitons in 3-dimensional singularity analysis, we will present a cross-sectional survey of examples and results for solitons in general dimensions. This survey is intended to provide a broad (though, of necessity, still incomplete) overview of the current state of our understanding of Ricci solitons and to provide context for the selected and more detailed investigations we will pursue in subsequent chapters. We will conclude the chapter with an open-ended and speculative discussion about the potential applications of a program of Ricci flow-with-surgery to 4-manifold topology.

1.1. Ricci solitons in Ricci flow

Let us begin with a review of definitions and other foundational results.

1.1.1. Ricci solitons.

We start with the definition of a Ricci soliton on a Riemannian orbifold.

Definition 1.1 (Ricci soliton). Let $(\mathcal{O}^n, g)$ be a Riemannian orbifold with isolated singularities,[1] let X be a smooth vector field on $\mathcal{O}^n$, and let λ be a real number. The **Ricci soliton** equation for the quadruple $(\mathcal{O}^n, g, X, \lambda)$ is

$$\operatorname{Ric} + \frac{1}{2}\mathcal{L}_X g = \frac{\lambda}{2} g, \tag{1.1}$$

where λ is normalized to equal 1, 0, or -1 for a **shrinking**, **steady**, or **expanding** Ricci soliton, respectively (see Hamilton [**Ham88**, §10], [**Ham93b**, §3], [**Ham93a**, §1], [**Ham95b**, §20] and Friedan [**Fri80**, equation (10)]).

[1] See Definitions A.3 and A.12 for the definition of a Riemannian orbifold with isolated singularities.

In the definition above, Ric denotes the Ricci tensor of g and $\mathcal{L}_X$ denotes the Lie derivative with respect to X.

DEFINITION 1.2 (Gradient Ricci soliton). If $X = \nabla f$ is the gradient of some smooth function $f : \mathcal{O}^n \to \mathbb{R}$, then we have the quadruple $(\mathcal{O}^n, g, f, \lambda)$ and we obtain the **gradient Ricci soliton** equation (cf. (I-2.24)):

$$\mathrm{Ric}_f := \mathrm{Ric} + \nabla^2 f = \frac{\lambda}{2} g, \tag{1.2}$$

where $\nabla^2 f$ denotes the Hessian of f.

The symmetric 2-tensor Ric_f is called the **Bakry–Emery tensor** or the **f-Ricci tensor**; see [**BE85**] and [**Per02**].

Shrinking gradient Ricci solitons first entered the study of the Ricci flow through Hamilton's surface monotonicity formula [**Ham88**, §9]. A link between shrinking solitons and Hamilton's surface entropy was later observed in [**Cho91a**]. Hamilton discovered remarkable connections between steady solitons, differential Harnack estimates, and Type II singularity formation [**Ham93a, Ham93b, Ham95b**]. This link was interpreted in the space-time setting by [**CC95, CC96**]. Perelman's entropy and reduced volume monotonicity formulas [**Per02**] established deep connections between shrinking solitons and monotonicity in all dimensions. In particular, the shrinking soliton equation itself naturally results from parabolically rescaling the Ricci flow. In a similar vein, monotonicity has been used to study blow-ups of solutions to a plethora of geometric equations.

Virtually all Ricci solitons which arise in singularity analysis are gradient Ricci solitons. For example, it is a result of Naber [**Nab10**] that every complete smooth shrinking soliton (M^n, g, X, λ) with bounded curvature is gradient (at least in the sense that $X = \nabla f + K$ for some Killing vector K). In this book, we will focus almost exclusively on gradient Ricci solitons, and will generally use the terms 'soliton' and 'Ricci soliton' as synonyms for "gradient Ricci soliton." We will also frequently use the informal alternatives **shrinker**, **steady**, and **expander** to describe shrinking, steady, and expanding gradient solitons.

1.1.2. The Ricci flow equation.

Most of the basic analytic theory for the Ricci flow carries over to the orbifold setting.

DEFINITION 1.3 (Hamilton's Ricci flow [**Ham82**]). A **Ricci flow** on an orbifold $\mathcal{O}^n$ with isolated singularities and time interval I is a 1-parameter family of Riemannian metrics $g_t := g(t)$, $t \in I$, satisfying the tensorial equation

$$\partial_t g_t = -2\mathrm{Ric}_{g_t} \tag{1.3}$$

at all points in the space-time $\mathcal{O}^n \times I$.

We say that a Ricci flow $(\mathcal{O}^n, (g_t)_{t\in I})$ is **closed** if $\mathcal{O}^n$ is a closed orbifold, and **complete** if the Riemannian metrics g_t are complete for each $t \in I$.

A Ricci flow is said to have **bounded curvature** if

$$\sup_{x\in\mathcal{O}^n} |\mathrm{Rm}|(x,t) < \infty \tag{1.4}$$

for each $t \in I$. That is, the Riemann curvature tensor is bounded on each time slice by a function of time. However, this function need not be uniform in time.

The Ricci flow is often interpreted as a heat equation for Riemannian metrics (see [**Ham95b**, §1] or the book [**CK04**, Chapter 3, §5]), and heuristically one expects the Ricci flow to tend to smooth out the metric over small time scales.

1.1.3. Short-time existence of Ricci flows.
In his original paper, [**Ham82**, §4-6] Hamilton proved that any compact Riemannian manifold may be evolved in a unique way by the Ricci flow for a short time. A simplified proof was provided shortly thereafter by DeTurck [**DeT83**, **DeT03**]. Later Shi [**Shi89**] and Chen and Zhu [**CZ06**] extended these results to the class of complete manifolds with bounded curvature. See Schoen [**Sch91**, §2] and Chapter 3 of [**CK04**] for expositions. Hamilton's result carries over more or less directly to orbifolds: see Kleiner and Lott [**KL14**, §5.2] or §A.4 in Appendix A.

THEOREM 1.4 (Short-time existence and uniqueness). *If $(\mathcal{O}^n, g_0)$ is a closed Riemannian orbifold with isolated singularities, then there exist $\varepsilon > 0$ and a unique Ricci flow $g(t)$, $t \in [0, \varepsilon)$, on $\mathcal{O}^n$ with $g(0) = g_0$.*

This result is in keeping with the heuristic picture of the Ricci flow as a heat equation for Riemannian metrics, however, as (1.3) is a nonlinear and only weakly parabolic system, some device is required to make the question amenable to standard parabolic theory.

1.1.4. Long-time existence theorems.
Let $\mathcal{O}^n$ be a closed smooth orbifold with isolated singularities.

DEFINITION 1.5 (Singular Ricci flow). We say that a finite-time Ricci flow $(g_t)_{t\in[0,T)}$ on $\mathcal{O}^n$ **goes singular at the time** $0 < T < \infty$ if it cannot be continued as a Ricci flow past time T. We call T the **singularity time** and refer to $(\mathcal{O}^n, (g_t)_{t\in[0,T)})$ as a **singular Ricci flow**.

According to the long-time existence theorem of Hamilton [**Ham82**, Theorem 14.1] (which extends to orbifold Ricci flows), if $(\mathcal{O}^n, (g_t)_{t\in[0,T)})$ is a singular Ricci flow, then

$$\sup_{\mathcal{O}^n\times[0,T)} |\mathrm{Rm}| = \infty.$$

In fact, by a refinement due to Šešum [**Š05**, Theorem 2], we actually have

$$\sup_{\mathcal{O}^n\times[0,T)} |\mathrm{Ric}| = \infty.$$

Chen and Wang have made further extensions; see [**CW13**] for details.

1.1.5. Cheeger–Gromov–Hamilton compactness theorem.
Compactness theorems are fundamental tools for understanding Ricci flow singularity formation. The standard compactness theorem in the subject is a Cheeger–Gromov-type result due to Hamilton [**Ham95a**] which requires completeness, bounded curvature,[2] and noncollapsing at the base point. By a **pointed** manifold or orbifold, we mean a manifold or orbifold with a distinguished point, called the base point. A **pointed complete Ricci flow** is a Ricci flow with a base point, where the metric is complete at each time.

[2] Or at least bounded curvature at bounded distances.

THEOREM 1.6 (CGH compactness theorem). *Let $(\mathcal{O}_i^n, g_i(t), p_i)_{t\in I_i}$ be a sequence of pointed complete Ricci flows on orbifolds with isolated singularities.*[3] *Suppose that the curvatures are uniformly bounded (independent of i) and the volumes of the unit balls centered at (p_i, t_0) are uniformly bounded from below for some $t_0 \in \bigcap_i I_i$, and $I_\infty := (\lim_{i\to\infty} I_i)^\circ \cup \{t_0\}$ exists. Then a subsequence converges in the smooth pointed Cheeger–Gromov–Hamilton* (**CGH**) *sense on I_∞ to an n-dimensional pointed complete Ricci flow on an orbifold with isolated singularities. See Figures* 1.1 *and* 1.2.

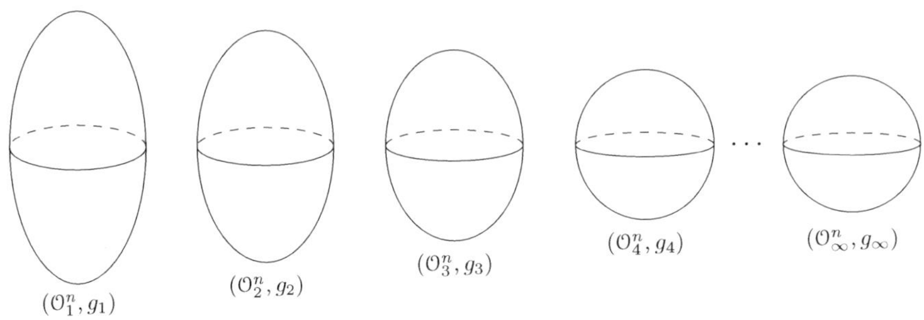

FIGURE 1.1. A sequence of Riemannian orbifolds converging to a closed Riemannian orbifold. In general, if the limit $\mathcal{O}_\infty^n$ is closed, then the orbifolds $\mathcal{O}_i^n$ in the sequence are closed for i sufficiently large.

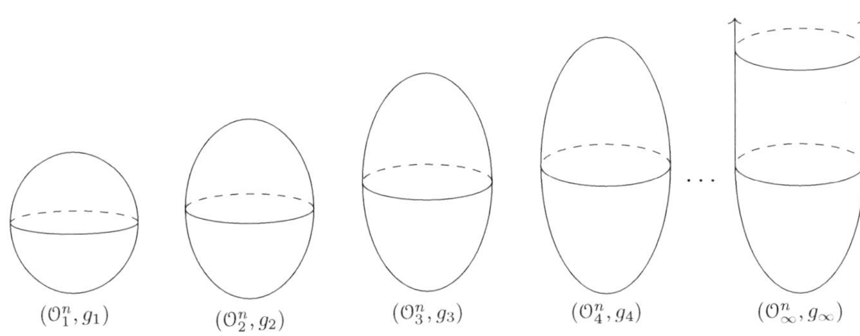

FIGURE 1.2. A sequence of closed Riemannian orbifolds converging to a complete noncompact Riemannian orbifold.

For a proof in the smooth manifold category, see e.g. [**CCG$^+$07**, Chapters 3 and 4] for an exposition and the references therein. Their exposition also includes the definition of smooth pointed Cheeger–Gromov–Hamilton convergence of Ricci flows, as well as smooth pointed Cheeger–Gromov convergence of Riemannian manifolds. For the case of orbifolds, see P. Lu [**Lu01**, Theorem 3.2] (cf. Kleiner and Lott [**KL14**, Proposition 4.1] and [**CCG$^+$07**, Theorems 3.24 and 3.25]).

[3]Here, we allow p_i to be a singular point of $\mathcal{O}_i^n$.

1.1.6. Ancient and eternal Ricci flows.

DEFINITION 1.7 (Infinite-time solutions to the Ricci flow). We say that a Ricci flow $(\mathcal{O}^n, (g_t)_{t\in I})$ is **ancient** if $(-\infty, a] \subset I$ for some $a \in \mathbb{R}$, **eternal** if $I = \mathbb{R}$, and **immortal** if $[a, \infty) \subset I$ for some $a \in \mathbb{R}$; see [**Ham95b**, §19]. See Figure 1.3.

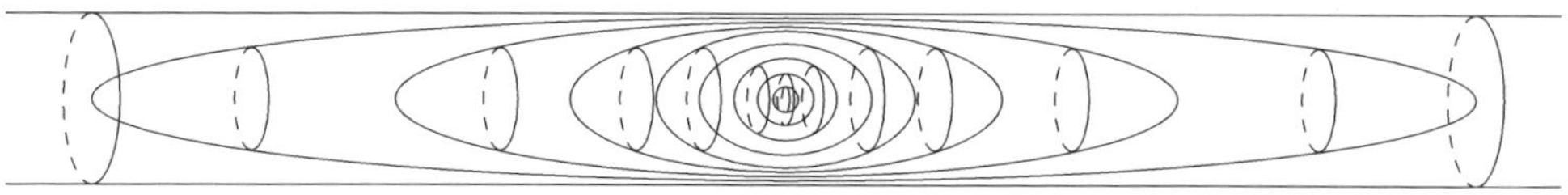

FIGURE 1.3. The King–Rosenau ancient solution [**Kin93, Ros95**] to the Ricci flow on a 2-sphere. The backward asymptotic limit is a cylinder.

According to a result of Zhang [**Zha09b**], if the manifold (M^n, g) which underlies a gradient Ricci soliton (M^n, g, f) is complete, then the gradient ∇f of f is a complete vector field. Thus, the self-similar solutions associated to complete gradient shrinkers are special cases of ancient solutions, and those associated to complete steadies and expanders are special cases of eternal and immortal solutions, respectively.

Since we expect the Ricci flow to tend to improve metrics over time, we expect that solutions which exist on infinite time intervals ought to have nice properties. In particular, we expect ancient solutions (shrinkers and steadies included) to be *extremely* nice – and more rigid than immortal solutions or solutions defined only on finite time intervals – given that they have benefited from the ameliorating effects of the equation for an infinite amount of time.

1.1.7. No local collapsing and singularity models.

DEFINITION 1.8 (Singularity model). A **(classical) singularity model** of a singular Ricci flow on a closed orbifold with isolated singularities is a complete Ricci flow that is the pointed CGH limit of a sequence of rescalings of the Ricci flow based at a sequence of space-time points where the curvatures tend to infinity (see Definition 1.10 in Volume 1 [**Cho23**]).

That is, a singularity model of a closed Ricci flow $(\mathcal{O}^n, (g_t)_{t\in[0,T)})$, where $0 < T < \infty$, is a complete ancient solution

$$(\mathcal{O}^n_\infty, g_\infty(t), p_\infty), \quad t \in (-\infty, 0],$$

which is the pointed CGH limit of $(\mathcal{O}_i, g_i(t), p_i)$, where

$$g_i(t) = \lambda_i g(t_i + t/\lambda_i)$$

$\lambda_i \to \infty$, and $K_i = |\mathrm{Rm}|(p_i, t_i) \to \infty$. Typically, one chooses λ_i to equal K_i.

DEFINITION 1.9 (Noncollapsed). Let $\kappa, \rho > 0$. We say that a Riemannian orbifold $(\mathcal{O}^n, g)$ with isolated singularities is **κ-noncollapsed up to the scale ρ** if for any $p \in \mathcal{O}^n$ and $r \in (0, \rho)$ with the scalar curvature satisfying $R \leq r^{-2}$ in $B(p, r)$, we have that

$$\mathrm{Vol}\,(B(p,r)) \geq \kappa r^n. \tag{1.5}$$

If $(\mathcal{O}^n, g)$ is κ-noncollapsed up to the scale ρ for all $\rho \in (0, \infty)$, then we say that $(\mathcal{O}^n, g)$ is **κ-noncollapsed on all scales**. See Figure 1.4.

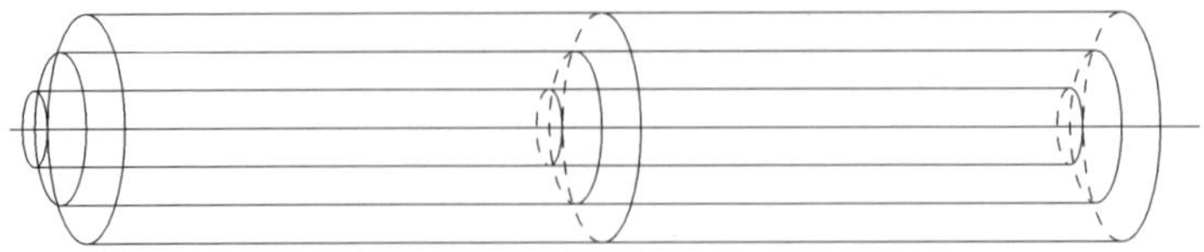

FIGURE 1.4. A sequence of flat finite 2-dimensional cylinders $\mathbb{S}^1(\varepsilon_i) \times [-\pi, \pi]$, $\varepsilon_i \to 0$, collapsing to a line. If $\mathbb{S}^1$ is replaced by $\mathbb{S}^{n-1}$, where $n \geq 3$, then the sequence is not collapsing; note that in this case the curvatures tend to infinity.

For the following result, which holds because the distance function, the curvature, and the volume all converge under the limit, we refer the reader to, for example, the expository Lemma 6.48 in [**CCG+07**] (note that the case of orbifolds is no different).

LEMMA 1.10 (Preservation of noncollapsing under CG-convergence). *Suppose that* $(\mathcal{O}_i^n, g_i)$ *are* κ*-noncollapsed up to the scale* ρ *for all* i. *If* $(\mathcal{O}_i^n, g_i, p_i)$ *converges in the smooth pointed Cheeger–Gromov (CG) sense to* $(\mathcal{O}_\infty^n, g_\infty, p_\infty)$, *then the limit* $(\mathcal{O}_\infty, g_\infty)$ *is* κ*-noncollapsed up to the scale* ρ.

If, on the other hand, $(\mathcal{O}_i^n, g_i)$ *is* κ*-noncollapsed up to the scale* ρ_i, *where* $\rho_i \to \infty$, *then* $(\mathcal{O}_\infty, g_\infty)$ *is* κ*-noncollapsed on all scales.*

The fundamental result underlying singularity analysis is the following (Perelman [**Per02**, §4] and cf. [**Per02**, §8], [**Top06**, §8.3], and the earlier works in [**Ham95c**, 1.1 Main Theorem], and [**Ham95b**, §15, §23, §25]).

THEOREM 1.11 (Perelman's No Local Collapsing Theorem). *For any closed Riemannian orbifold with isolated singularities* $(\mathcal{O}^n, g_0)$, $\rho > 0$, *and* $T \in (0, \infty)$ *there exists* $\kappa > 0$ *depending only on* g_0, ρ, *and* T *such that for any solution* $(\mathcal{O}^n, g(t))$, $t \in [0, T)$, *to the Ricci flow with* $g(0) = g_0$, *each metric* $g(t)$ *is* κ*-noncollapsed up to the scale* ρ.

Let $(\mathcal{O}^n, (g_t)_{t \in [0,T)})$ be a Ricci flow on a closed orbifold with isolated singularities that goes singular at time T. Let (x_i, t_i) be a sequence of space-time points with $t_i \to T$ and let $K_i \to \infty$. Define the rescaled solutions

$$g_i(t) := K_i g(t_i + K_i^{-1} t), \quad t \in \big[-K_i t_i, K_i(T - t_i)\big). \tag{1.6}$$

Note that $-K_i t_i \to -\infty$ so that these solutions 'tend to ancient'. Suppose that there exists a sequence $r_i \to \infty$ and a continuous function $C : [0, \infty) \to \mathbb{R}$ such that

$$\sup_{(x,t) \in B_{g_i(0)}(x_i, r) \times [-r, 0]} |\mathrm{Rm}_{g_i(t)}(x)| \leq C(r) \quad \text{for all } r \in [0, r_i]. \tag{1.7}$$

COROLLARY 1.12 (Existence of singularity models). *Under the hypotheses in the paragraph above, there exists a subsequence such that the sequence of Ricci flows* $\big(\mathcal{O}^n, g_i(t), (x_i, 0)\big)$ *converges in the smooth pointed Cheeger–Gromov sense to a complete ancient Ricci flow* $\big(\mathcal{O}^n, g_\infty(t), (x_\infty, 0)\big)$, $t \in (-\infty, 0]$, *on an orbifold with isolated singularities.*

PROOF. By Perelman's no local collapsing theorem and the curvature bound assumption for the sequence, we can apply the Cheeger–Gromov–Hamilton compactness theorem for Ricci flows (Theorem 1.6) to obtain subconvergence to a complete limit. In general, this limit depends on the subsequence. □

As a consequence, we have the following (compare with the expository [**CCG$^+$07**, Theorem 6.68]).

COROLLARY 1.13 (Singularity models are noncollapsed). *Every singularity model is κ-noncollapsed on all scales.*

*In particular, if a noncompact singularity model has zero scalar curvature, i.e., $R \equiv 0$, then it has **Euclidean volume growth**; that is, there exists $\kappa > 0$ such that* $\mathrm{Vol}(B(p,r)) \geq \kappa r^n$ *for all* $r \in (0,\infty)$.

PROOF. This is because $(\mathcal{O}^n, (g_t)_{t\in[0,T)})$ being κ-noncollapsed up to the scale $\rho := 1$ implies that $(\mathcal{O}^n, g_i(t))$, where $g_i(t) = \lambda_i g(t_i + \lambda_i^{-1} t)$, is κ-noncollapsed up to the scale $\sqrt{\lambda_i} \to \infty$ (where κ is independent i). □

Perelman [**Per02**, **Per03a**] also proved that all 3-dimensional singularity models have bounded curvature (see [**CCG$^+$15**, Theorem 28.20] for an exposition).

Bamler's [**Bam20b**] more general definition of singularity models, as well as his much more general existence theorem, will be introduced in Chapter 6.

1.1.8. Compact singularity models are shrinking solitons.

Regarding compact singularity models, in any dimension, Zhang [**Zha07**] proved the following result, which highlights the most desirable outcome of singularity formation: finite-time extinction with convergence to a compact shrinker.

THEOREM 1.14 (Compact singularity models are shrinking Ricci solitons). *Let $(\mathcal{O}^n_\infty, g_\infty(t))$ be a singularity model whose underlying orbifold $\mathcal{O}^n_\infty$ is closed. Then there exists a function $f_\infty : \mathcal{O}^n_\infty \to \mathbb{R}$ such that $(\mathcal{O}^n_\infty, g_\infty(0), f_\infty)$ is a shrinking gradient Ricci soliton, and the underlying orbifold $\mathcal{O}^n_\infty$ is diffeomorphic to the underlying orbifold $\mathcal{O}^n$ of the Ricci flow solution of which $(\mathcal{O}^n_\infty, g_\infty(t))$ is a singularity model.*

Note that neither the King–Rosenau ancient solution on S^2, pictured in Figure 1.3, nor Perelman's ancient solution [**Per03a**, §1.4] on S^n, $n \geq 3$, is a singularity model since both are defined on a closed manifold, but neither is a shrinking gradient Ricci soliton. Alternatively, the King–Rosenau ancient solution can also be ruled out by Perelman's no local collapsing theorem.

1.1.9. The convergence theorem in two dimensions.

We now specialize our attention to convergence theorems in two dimensions. Recall that a gradient Ricci soliton is a self-similar solution to the Ricci flow; see Proposition 1.73 below.

As has been discussed in Chapter 1 of Volume 1, we have convergence to a soliton for the Ricci flow on 2-orbifolds, due to Hamilton [**Ham88**], Wu [**Wu91**], one of the authors [**Cho91b**], and Wu and one of the authors [**CW91**]. Recall that the definition of a smooth Riemannian orbifold with isolated singularities, including the distinction between good and bad, is provided in Appendix A.

THEOREM 1.15 (Convergence theorem for Ricci flow on 2-orbifolds). *Let $(\mathcal{O}^2, g_0)$ be a 2-dimensional smooth closed Riemannian orbifold with isolated singularities. Then there exists a unique solution $g(t)$ to the normalized Ricci flow $\partial_t g = (r-R)g$ with $g(0) = g_0$ defined for all $t \in [0,\infty)$, where r denotes the average scalar curvature.*

(1) *If $\mathcal{O}^2$ is a good orbifold, then $g(t)$ converges in C^∞ exponentially fast to a constant curvature metric in the same conformal class as g_0.*

(2) *If $\mathcal{O}^2$ is a bad orbifold, then after pulling back by a suitable family of diffeomorphisms, $g(t)$ converges in C^∞ exponentially fast to a shrinking gradient Ricci soliton metric on $\mathcal{O}^2$. See Figure* 1.5.

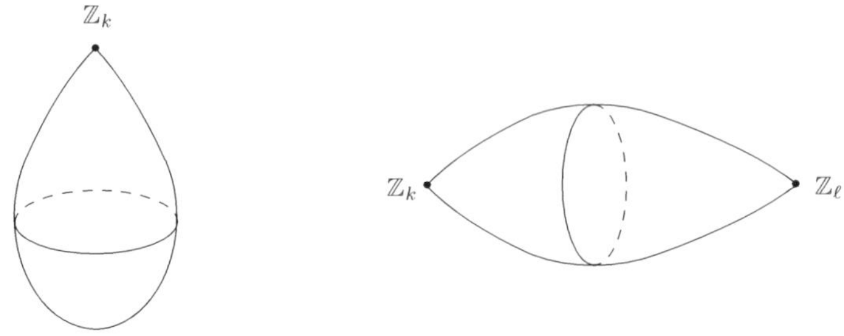

FIGURE 1.5. The teardrop (left) and American football (right) shrinking solitons are the asymptotic geometric limits of the normalized Ricci flow on bad 2-orbifolds. The orbifold points are labelled by their isotropy groups.

By Theorem 1.15, in dimension 2, the only singularity models of the Ricci flow on an oriented closed smooth surface or on a 2-dimensional good orbifold are the shrinking round 2-sphere $\mathbb{S}^2$ and its quotients $\mathbb{S}^2/\mathbb{Z}_k$, where $k \geq 2$. For a 2-dimensional closed bad orbifold, which must be a teardrop or an American football (see Example A.4), the only singularity models are shrinking solitons, which exhibit rotational symmetry and are positively curved (see [**Ham88**, §10], [**Wu91**, §4], [**CLT06**, Lemma 0.1]). In particular, steady and expanding solitons are not singularity models in dimension 2.

For a survey of the Ricci flow on noncompact surfaces, see Isenberg, Mazzeo, and Šešum [**IMŠ11**]. See Topping and Yin [**TY24**], and the references therein, for the Ricci flow on surfaces, including those with weak initial data.

1.1.10. Sphere theorems.

A classical direction in Riemannian geometry is comprised of topological and differentiable sphere theorems. In particular, there is the 1/4-pinching topological sphere theorem of Rauch, Klingenberg, and Berger. Using minimal surfaces, Micallef and Moore proved a topological sphere theorem under the hypotheses of simple-connectivity and positive isotropic curvature. Under strong pinching hypotheses, differentiable sphere theorems were proved by Sugimoto, Shiohama, and Karcher, by Ruh, and by Grove, Karcher, and Ruh.

Using Ricci flow, Hamilton's classification of 3-manifolds with positive Ricci curvature and 4-manifolds with positive curvature operator may be considered as differentiable sphere theorems. Under strong pointwise-pinching hypotheses, differentiable sphere theorems were proved in higher dimensions by Huisken, Margerin, and Nishikawa. Remarkable further developments are the differentiable sphere theorems of Böhm and Wilking under the hypothesis of positive curvature operator and Brendle and Schoen under a hypothesis weaker than pointwise 1/4-pinching, thus proving an improved version of the longstanding differentiable-sphere-theorem conjecture. All of the results above may be interpreted as finite-time extinction to the most ideal type of shrinking soliton, namely, constant positive sectional curvature manifolds. For excellent accounts of these developments, see the survey article by Brendle and Schoen [**BS11**] and the book by Brendle [**Bre10**].

1.2. 3-Dimensional Ricci flow and geometrization

In this section, we review the central role that Ricci solitons play in the proof of the Geometrization Conjecture. Throughout our discussion, all 3-manifolds will be orientable.

1.2.1. Ricci flow origins and development.

The Ricci flow was introduced by Richard Hamilton in his seminal 1982 paper [**Ham82**] on 3-manifolds with positive Ricci curvature. Hamilton's work on the equation over the next two decades, including [**Ham86, Ham88, Ham93a, Ham93b, Ham95a, Ham95b, Ham97, Ham99**], showcased its power in solving problems at the intersection of geometry and topology and in the process developed a new subfield of geometric analysis. Within his papers, Hamilton laid out a detailed program to attack Thurston's Geometrization Conjecture [**Thu82, Thu97**], and with it the Poincaré Conjecture, a program which he carried out to a large extent.

1.2.2. Perelman's breakthrough.

Hamilton's program for 3-dimensional Ricci flow came to fruition in the spectacular work of Perelman in the early 2000's. In the papers [**Per02, Per03a, Per03b**], Perelman overcame the remaining obstacles in the program with a variety of ingenious innovations, and completed the proof of:

THEOREM 1.16 (Perelman–Hamilton theorem). *The following are true:*

(1) (Poincaré conjecture) *Any closed smooth 3-manifold that is homotopy equivalent to S^3 (equivalently, simply-connected) is diffeomorphic to S^3.*
(2) (Thurston geometrization conjecture) *Any closed smooth 3-manifold admits a decomposition into a finite number of* ***geometric pieces****, each of which admits a canonical metric, namely, one of Thurston's eight model geometries* [**Thu97**].

The theory of Ricci flow with surgery developed in the work of Hamilton and Perelman relies on a deep understanding of singularity formation and, specifically, of the structure of noncollapsed 3-dimensional ancient solutions. This, in turn, relies on a classification of 3-dimensional shrinking gradient Ricci solitons. In Perelman's theory, the asymptotic behaviors of noncompact ancient solutions are characterized by those of the *asymptotic soliton* associated to such a solution (see [**Per02**, 11.2], [**Per03a**, 1.1]).

On the other hand, fundamental to 3-manifold topology is the *prime* (*sphere*) *decomposition* (see Kneser [**Kne29**] and Milnor [**Mil62**]). The fact that the topological prime decomposition can be achieved geometrically through Ricci flow is directly related to the fact that (per their classification) the only orientable 3-dimensional *shrinking gradient Ricci solitons* are

$$\mathbb{S}^3/\Gamma,\ \mathbb{S}^2\times\mathbb{R},\ (\mathbb{S}^2\times\mathbb{R})/\mathbb{Z}_2,\ \text{and } \mathbb{R}^3. \tag{1.8}$$

Here, Γ is a finite subgroup of $\mathrm{O}(4)$ and $\mathbb{S}^3/\Gamma$ denotes a **spherical space form**, i.e., a closed Riemannian 3-manifold with constant positive sectional curvature.

The product $\mathbb{S}^2 \times \mathbb{R}$ denotes a round cylinder, and $(\mathbb{S}^2 \times \mathbb{R})/\mathbb{Z}_2$ denotes its $\mathbb{Z}_2$-quotient, where the $\mathbb{Z}_2$-action is generated by the antipodal map $(x, y) \mapsto (-x, -y)$ for $x \in \mathbb{S}^2$ and $y \in \mathbb{R}$. We remark that $(\mathbb{S}^2 \times \mathbb{R})/\mathbb{Z}_2$ is diffeomorphic to the nontrivial line bundle over $\mathbb{R}P^2$, which we also denote by L; these 3-manifolds are also diffeomorphic to $\mathbb{R}P^3 \setminus \overline{B}^3$. Note, on the other hand, that $\mathbb{R}P^2 \times \mathbb{R}$ is non-orientable and hence does not appear in the list (1.8).[4]

Associated to the classification of 3-dimensional shrinking solitons in (1.8) (and more generally the classification of 3-dimensional singularity models) is the fact that, for finite times, 3-dimensional Ricci flow with surgery only effects neckpinches (including degenerate ones) geometrically and topologically this corresponds to 2-surgeries, consistent with the prime decomposition.

Another fundamental part of 3-manifold topology is the *toral decomposition*, also known as the JSJ decomposition in honor of Jaco, Shalen, and Johannson (see [**JS79**] and [**Joh79**]). In the Hamilton–Perelman theory [**Ham99**], [**Per03a**, §6-7], the toral decomposition is geometrically manifested in the infinite time behavior of Ricci flow. In particular, the boundary components of the hyperbolic pieces comprising the noncollapsed regions of an immortal 3-dimensional Ricci flow with surgery are incompressible tori (see [**Ham99**, §11-12], [**Per03a**, §7.3]).[5]

In the context of Hamilton's program described above, and with the aid of Perelman's extension of Cheeger and Gromov's collapsing theory (the thin part of a **thick-thin decomposition**),[6] Perelman was able to infer Thurston's Geometrization Conjecture from the qualitative behavior of the Ricci flow with surgery. The Poincaré Conjecture is a special case of this. See, e.g., Kleiner and Lott [**KL08**] and Morgan and Tian [**MT07, MT14**] for complete expositions of the Perelman–Hamilton proof.

1.2.3. The works of Bamler, Brendle, Kleiner, and Lott.

Following Hamilton's 1982–1999 and Perelman's 2002–2003 works, there have been a number of other major breakthroughs in 3-dimensional Ricci flow. These include the following results:

Brendle's proof of Perelman's conjecture [**Per02**] on the classification of 3-dimensional noncollapsed ancient solutions.

THEOREM 1.17 (Perelman's conjecture on ancient solutions [**Bre13, Bre20**]). *Any 3-dimensional singularity model is isometric to one of the following:*

$$\mathbb{S}^3/\Gamma,\ \mathbb{S}^2 \times \mathbb{R},\ (\mathbb{S}^2 \times \mathbb{R})/\mathbb{Z}_2, \quad \textit{or the Bryant soliton.} \tag{1.9}$$

This result not only characterizes noncollapsed 3-dimensional ancient solutions as shrinking or steady solitons, but completely determines these solitons. Here, in dimensions 3 and above, the **Bryant soliton** is the (unique up to scaling) nonflat complete rotationally symmetric steady gradient Ricci soliton; see [**Bry**] and Theorem 1.76 below. See Figure 1.6.

[4]For a discussion, including the classification, of the geometric quotients of $\mathbb{S}^2 \times \mathbb{R}$, see the section titled "The geometry of $\mathbb{S}^2 \times \mathbb{R}$" in Scott [**Sco83**, p. 457ff].

[5]A torus embedded in a 3-manifold is **incompressible** if the inclusion map induces an injection of the fundamental group of the torus into the fundamental group of the 3-manifold.

[6]See Perelman [**Per03a**, §7.4] and Shioya and Yamaguchi [**SY05**]. Cheeger and Gromov's collapsing theory was developed in [**CG86, CG90, CFG92**]. For manifolds of negative sectional curvature, this is the Margulis lemma.

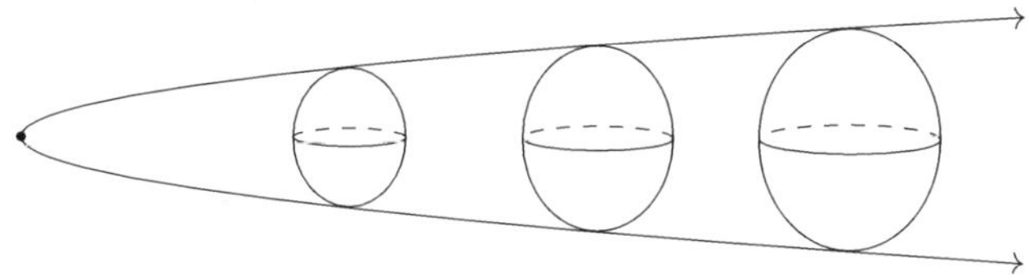

FIGURE 1.6. The Bryant steady GRS is rotationally symmetric and opens up like a paraboloid.

Bamler's proof of Perelman's conjecture [**Per03a**] on the finiteness of the number of surgeries and immortal solutions being Type III.

THEOREM 1.18 (Perelman's conjecture on finiteness of number of surgeries [**Bam18a, Bam18b, Bam18c, Bam18d, Bam18e**]). *Given any closed orientable 3-dimensional Riemannian manifold (M^3, g_0), there exists a Ricci flow $(M_t^3, g(t))$ with only finitely many surgeries and initial condition (M^3, g_0) that either goes extinct in finite time or exists for all time, where, in the latter case, $|\mathrm{Rm}| \leq C/t$, i.e., this immortal solution is Type III.*

Bamler and Kleiner's proof of the **generalized Smale conjecture.**

THEOREM 1.19 (Generalized Smale conjecture [**BK22, BK23, BK19, BK24**]). *For a spherical space form, the inclusion of its isometry group into its diffeomorphism group is a homotopy equivalence.*

The original Smale conjecture (for S^3) was proved by Hatcher [**Hat83**]. Various special cases of the generalized Smale conjecture were proved by Hong, Kalliongis, McCullough, and Rubinstein [**HKMR12**].

Additionally, subsequent work of Kleiner and Lott [**KL17, KL20**] has established the existence of a weak Ricci flow on 3-manifolds, i.e., a Ricci flow with surgery on 3-manifolds with a surgery parameter effectively equal to zero, as well as provided a proof of Thurston's orbifold theorem ([**CHK00, BLP05**]) via Ricci flow with surgery, extending the Hamilton–Perelman theory [**KL14**].[7]

For excellent surveys of 3-dimensional Ricci flow, with a focus on work subsequent to Hamilton and Perelman, see Bamler [**Bam22**] and Brendle [**Bre22**].

1.2.4. Ricci flow singularity formation in dimension 3.

In this subsection, we give a high-level overview of central themes in the geometric analysis of singularities of the 3-dimensional Ricci flow, which is a key component of the Hamilton–Perelman theory. Recall that a finite-time solution $(M^n, g(t))$, $t \in [0, T)$, is said to go singular at a time $T \in (0, \infty)$ if $\sup_{M^n \times [0,T)} |\mathrm{Rm}| = \infty$. Recall also that if $g(t)$ does not become singular at time T, it may be smoothly continued beyond time T.

1.2.4.1. *The microscope problem in singularity analysis.*

A cornerstone of the study of singularity formation under the Ricci flow is the method of *microscopic examination*, a technique with precedents in minimal surface theory. Its applications to the Ricci flow were pioneered by Hamilton ([**Ham95b**, §16,§22,§26], [**Ham97**, §5]) and further developed by Perelman ([**Per02**, §10-12], [**Per03a**, §4-5]) and Bamler ([**Bam20a, Bam23, Bam20b**]).

---

[7]Roughly speaking, Thurston's orbifold theorem provides a characterization of when a 3-dimensional orbifold with a 1-dimensional singular locus admits a geometric structure.

The technique involves basing a solution that goes singular at a sequence of space-time points and then applying increasingly large rescaling factors, a process we metaphorically describe as *focusing*. With the aid of compactness theorems, one may in certain cases extract limits from these sequences – the *singularity models* described above – which capture various aspects of the solution's behavior near the singularity.

The *microscope problem* is the problem of selecting suitable sequences of space-time points about which to focus these microscopes in order to capture the best glimpse of the forming singularity. The Perelman–Hamilton proof of the Geometrization Conjecture includes in some sense a successful resolution of this microscope problem in dimension 3, with a classification of 3-dimensional singularity models sufficient enough to produce a well-defined flow with surgery. 3-Dimensional singularity models were later fully classified by Brendle [**Bre20**].

In essence, 3-dimensional singularity models have themselves been geometrized, leading to the understanding of the geometry of high curvature regions in 3-dimensional Ricci flows, as encapsulated in the *canonical neighborhood theorem* (see [**Per02**, §12]), which in turn made possible a surgery procedure to continue the Ricci flow past a finite time singularity.

In dimension 4, one would like to geometrize singularity models as much as possible. The potential classification of shrinking and steady solitons should be intimately tied to this pursuit. Toward this end, a key question is: *What is a suitable canonical neighborhood theorem in dimension* 4*?*

While the term "microscope problem" may not be explicitly referenced in the subsequent chapters, the techniques used to address this problem will be nevertheless woven into the fabric of our broader discussion as we explore the relationship between the Ricci flow and Ricci solitons in dimension 4.

1.2.4.2. *The Hamilton–Ivey curvature estimate.*

We now recall the crucial *Hamilton-Ivey estimate* (see [**Ham95b**, Theorem 24.4], [**Ham99**, Theorem 4.1], cf. [**Ive93**, Theorem 2]) which implies that 3-dimensional singularity models have non-negative curvature. This imposes substantial constraints on their topology and geometry.

THEOREM 1.20. *Let* $(M^3, g(t))$*,* $t \in [0, T)$*, be a solution of the Ricci flow on a closed* 3*-manifold. Suppose that* $\lambda_1(\mathrm{Rm})(x, 0) \geq -A$ *for all* $x \in M^3$*, where* A *is a positive real number and* λ_1 *denotes the smallest eigenvalue. Then, at any point* $(x, t) \in M^3 \times [0, T)$ *where* $\lambda_1(\mathrm{Rm})(x, t) < 0$*, we have*

$$R \geq |\lambda_1(\mathrm{Rm})| \cdot \big(\ln |\lambda_1(\mathrm{Rm})| + \ln(A^{-1} + t) - 3\big). \tag{1.10}$$

In particular, without any assumption on $g(0)$*, the result holds for* $A^{-1} := 0$ *and provided that* $t > 0$*.*

Inequality (1.10) implies that there exists a function $\Phi : \mathbb{R} \to \mathbb{R}$ such that

$$\lambda_1(\mathrm{Rm}) \geq -\Phi(R) \tag{1.11}$$

on all of $M^3 \times [0, T)$, where Φ is positive and non-decreasing, and for $u > 0$ we have that $u \mapsto \frac{\Phi(u)}{u}$ decreases to zero as $u \to \infty$ (see [**KL08**, Definition 52.1 and Appendix B], cf. [**Per02**, §12.1]). Note that Φ depends on the choice of A. When a solution $g(t)$ satisfies (1.11) we say that it has **Φ-almost non-negative curvature**.

The Hamilton–Ivey estimate has been localized by Chen, Xu, and Zhang [**CXZ13**, Theorem 1.3]. For simplicity, in the following we state their result in the special case where $R > 0$ and for the consequential result for complete noncompact solutions.[8]

THEOREM 1.21 ([**CXZ13**, Corollary 1.5]). *Let $(M^3, g(t))$ be a 3-dimensional complete noncompact solution to the Ricci flow on $[0, t_0]$ with positive scalar curvature. Then, at any point (x, t_0) with $\lambda_1(\mathrm{Rm})(x, t_0) < 0$, we have*[9]

$$R \geq |\lambda_1(\mathrm{Rm})|\big(\ln|\lambda_1(\mathrm{Rm})| + \ln t_0 - 3\big). \tag{1.12}$$

In particular, if $g(t)$, $t \in (-\infty, 0]$, is a 3-dimensional complete ancient solution, then $g(t)$ has non-negative sectional curvature for all t.

We conclude from Theorem 1.21 that 3-dimensional singularity models must have non-negative sectional curvature. This last fact also follows from Theorem 1.20 and the fact that singularity models are pointed Cheeger–Gromov limits of rescalings of some singular Ricci flow on a closed 3-manifold. Armed with this curvature restriction, we can begin to think about classifying singularity models. This has been successfully done.

1.2.4.3. *Classification of 3-dimensional singularity models.*

The implementation of an effective surgery procedure for a 3-dimensional Ricci flow (see [**Per03a**]) requires a qualitatively precise understanding of the high curvature regions of singular Ricci flows. The idea is that, with the help of a robust compactness theorem (see [**Per02**, 11.7]), one can show that neighborhoods in high curvature regions are *close to* 3-dimensional oriented κ-noncollapsed complete ancient solutions with bounded curvatures (see [**Per02**, 12.1]), called **κ-solutions**.

Within this theoretical framework, these κ-solutions are singularity models. The Hamilton–Ivey curvature estimate (see Theorem 1.21) implies that 3-dimensional ancient solutions have non-negative sectional curvature. 3-Dimensional oriented κ-solution singularity models have been completely classified. Recall that an exhaustive list is given by (1.9) as:[10]

$$\mathbb{S}^3/\Gamma,\ \mathbb{S}^2 \times \mathbb{R},\ (\mathbb{S}^2 \times \mathbb{R})/\mathbb{Z}_2, \quad \text{and the Bryant soliton} \tag{1.13}$$

by the works of Hamilton [**Ham82**], [**Ham95b**, §26], Perelman [**Per02, Per03a**], and Brendle [**Bre13**], [**Bre20**, Theorems 1.1 and 1.2].[11] Note that all of the singularity models mentioned above are either shrinking Ricci solitons or, specifically in the case of the Bryant soliton, a steady gradient Ricci soliton.

We remark that 3-dimensional singularity models, apart from $\mathbb{S}^3/\Gamma$, are κ_0-noncollapsed with a universal constant $\kappa_0 > 0$.[12] This was originally proved by

---

[8]Any non-Ricci-flat complete ancient solution to the Ricci flow must have positive scalar curvature by a result of Chen [**Che09**]. In other words, if $R = 0$ at some point, then the ancient solution is Ricci-flat.

[9]The original Hamilton–Ivey estimate implies this when M^3 is compact.

[10]Note that Perelman's Type II non-round ancient solutions on $\mathbb{S}^3$ and $\mathbb{R}P^3$ are not singularity models (see Theorem 1.14). Also, compact quotients of $\mathbb{S}^2 \times \mathbb{R}$ are not singularity models because they are not noncollapsed.

[11]Brendle's result appeared after Perelman's work and can be used to simplify the analysis of 3-dimensional high curvature regions. See Bamler and Kleiner [**BK21**] for an alternative proof of the rotational symmetry.

[12]Note that if $|\Gamma_i| \to \infty$ as $i \to \infty$, then $\mathbb{S}^3/\Gamma_i$ is a collapsing sequence of Riemannian 3-manifolds.

Perelman [**Per03a**, §1.5]. By the classification in (1.9), we have that κ_0 may be chosen to be the minimum of the largest value of κ for $(\mathbb{S}^2 \times \mathbb{R})/\mathbb{Z}_2$ and the largest value of κ for the Bryant soliton. We call κ_0 the **universal noncollapsing constant**.

Having classified singularity models, one can understand the geometry and topology of high curvature regions in a 3-dimensional singular Ricci flow; see [**Per02**, §12].

1.3. Examples of noncollapsed Ricci solitons in dimension 4

For Ricci flow singularity analysis in any dimension, it is crucial to gain a comprehensive understanding of solitons. The qualitative study of the geometry, topology, and analysis of gradient Ricci solitons is guided by examples. In this section, we present a terse outline of what is known for 4-dimensional noncollapsed shrinking and steady solitons, organized according to whether the soliton is shrinking or steady and/or compact or noncompact. For the noncompact cases, we organize the discussion according to the geometry at infinity.

In the next subsection, we will discuss how the examples in this section fit into the broader landscape of all 4-dimensional noncollapsed shrinking and steady Ricci solitons. There we will also survey additional geometric and topological properties of 4-dimensional Ricci solitons. Our aim is to give readers a birds-eye view of the 4-dimensional Ricci soliton landscape before we begin its detailed study. The subsequent chapters are intended to serve as a more detailed resource for further study and research.

1.3.1. The Gaussian soliton.

Let Γ be a finite subgroup of $\mathrm{O}(n)$ acting freely on $\mathbb{S}^{n-1}$, where the action is induced by the standard action on $\mathbb{R}^n$. Then the gradient Ricci soliton $(\mathbb{R}^n/\Gamma, g, f)$ is called the **(quotient) Gaussian soliton**, where the metric g on the orbifold $\mathbb{R}^n/\Gamma$ is induced by the Euclidean metric $g_{\mathbb{E}}$ and where $f : \mathbb{R}^n/\Gamma \to \mathbb{R}$ is induced by the Gaussian potential function $x \mapsto \frac{\lambda|x|^2}{4}$ on $\mathbb{R}^n$, where $\lambda = 1$, 0, or -1. If a Gaussian soliton has $\Gamma = 1$, then we call it the **Euclidean Gaussian soliton**. See Figure 1.7.

In the following, we assume that the solitons are not Euclidean Gaussian solitons, which, in and of themselves, are not singularity models.[13]

1.3.2. Examples of shrinking solitons.

We divide our discussion into the compact and noncompact cases.

1.3.2.1. *Compact shrinking solitons.*

Compact shrinking gradient Ricci solitons may either be Einstein or nontrivial shrinking gradient Ricci solitons. By a shrinking gradient Ricci soliton being **nontrivial**, we mean that its potential function is not constant.

1.3.2.1.a. *Einstein metrics with positive scalar curvature.*

Any 4-dimensional Einstein metric with positive scalar curvature is a shrinking gradient Ricci soliton. The simplest examples are the standard metrics on $\mathbb{S}^4$ and

[13] On the other hand, orbifold quotients of the Gaussian soliton can be singularity models. For example, $\mathbb{R}^4/\mathbb{Z}_2$ is a singularity model; see Appleton [**App23**].

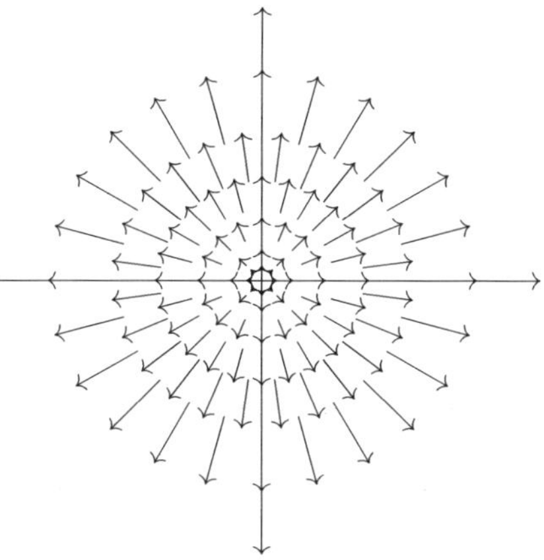

FIGURE 1.7. The gradient of the potential function f of the Gaussian shrinker $(\mathbb{R}^2, g_{\mathbb{E}}, f)$, where $f = \frac{|x|^2}{4}$. The reference metric at time t is $\phi_t^* g_{\mathbb{E}} = \mathrm{e}^{t/2} g_{\mathbb{E}}$. Since the Ricci flow is static, $g(t) \equiv g_{\mathbb{E}}$, we have that $g(t)$ is shrinking relative to the reference metric $\phi_t^* g(0)$.

$\mathbb{C}P^2$. By the **Hitchin–Thorpe inequality**, if a closed oriented 4-manifold M^4 admits an Einstein metric, then

$$|\tau(M^4)| \leq \frac{2}{3}\chi(M^4),$$

where τ and χ denote the signature and Euler characteristic, respectively; see [**Hit74**], [**Tho69**]. Equality holds if and only if the Einstein metric is **hyper-Kähler**, in particular **Ricci-flat**, i.e., $\mathrm{Ric} \equiv 0$.[14] In particular, if M^4 is simply-connected, then it is diffeomorphic to the $K3$ surface, which satisfies $\tau(M^4) = -16$ and $\chi(M^4) = 24$.

By a theorem of Tian [**Tia90**], a closed Kähler surface M^4 with positive first Chern class, a.k.a. a **Fano surface**, admits a Kähler–Einstein metric with positive scalar curvature if and only if the Lie algebra $\mathfrak{h}(M^4)$ of holomorphic vector fields is reductive; i.e., it is the complexification of a compact real subalgebra (Matsushima [**Mat57**] proved the necessity of this condition).

On the other hand, del Pezzo [**dP87**] proved that a complex surface is a Fano surface if and only if it is biholomorphic to $\mathbb{C}P^2$, $\mathbb{C}P^1 \times \mathbb{C}P^1$, or $\mathbb{C}P^2$ blown up at k points in general position, where $1 \leq k \leq 8$. Each of these is a Kähler surface; see §1.3.2.2.d below.[15] In the latter case, the underlying smooth manifold is diffeomorphic to $\mathbb{C}P^2 \# k\overline{\mathbb{C}P^2}$, where $1 \leq k \leq 8$. As a consequence, Tian proved that any Fano surface admits a Kähler–Einstein metric with positive scalar curvature unless it is biholomorphic to $\mathbb{C}P^2$ blown up at 1 or 2 points.

---

[14] A Riemannian manifold (M^{4k}, g) is *hyper-Kähler* if it admits three integrable almost complex structures I, J, K, each of which is Kähler with respect to g, and satisfies $IJK = -\mathrm{Id}$. Equivalently, the holonomy group is contained in $\mathrm{Sp}(k)$.

[15] Since $1 \leq k \leq 8$, this means that no 3 points lie on the same line and no 6 points lie on the same conic (curve of degree 2).

We remark that the $M^4 \cong j\mathbb{C}P^2\#k\overline{\mathbb{C}P^2}$ satisfies $\chi(M^4) = j + k + 2$ and $\tau(M^4) = j - k$, so that by the Hitchin–Thorpe inequality (these 4-manifolds are not hyper-Kähler),

$$-\frac{2}{3}(j+k+2) < j - k < \frac{2}{3}(j+k+2),$$

which is equivalent to

$$\frac{j-4}{5} < k < 5j + 4.$$

For example, if $j = 1$, so that $M^4 \cong \mathbb{C}P^2\#k\overline{\mathbb{C}P^2}$, then we have $k < 9$.

1.3.2.1.b. *Non-Einstein compact shrinking solitons.*

The closed 4-manifold $\mathbb{C}P^2\#\overline{\mathbb{C}P^2}$ admits a Kähler non-Einstein shrinking gradient Ricci soliton by the independent works of Koiso [**Koi90**] and Cao [**Cao96**]. Wang and Zhu [**WZ04**] proved the existence of Kähler non-Einstein shrinking gradient Ricci solitons on $\mathbb{C}P^2\#2\overline{\mathbb{C}P^2}$. Hence, all Fano surfaces admit Kähler shrinking gradient Ricci soliton structures.

1.3.2.2. *Noncompact shrinking Ricci solitons.*

We categorize noncompact shrinking solitons according to their geometry at infinity.

1.3.2.2.a. *Splitting shrinking solitons.*

If a 4-dimensional complete shrinking gradient Ricci soliton splits as a product with an $\mathbb{R}$ factor, then by the classification of 3-dimensional shrinking gradient Ricci solitons, it must be $(\mathbb{S}^3/\Gamma) \times \mathbb{R}$, $\mathbb{S}^2 \times \mathbb{R}^2$ or $((\mathbb{S}^2 \times \mathbb{R})/\mathbb{Z}_2) \times \mathbb{R}$.

1.3.2.2.b. *Asymptotically conical shrinking solitons.*

For a complex m-dimensional Kähler manifold, we define **complex rotational symmetry** as having its isometry group containing $\mathrm{U}(m)$, in which case we say that the metric is $\mathrm{U}(m)$-symmetric. Let M_ℓ^{2m} denote the total space of the complex line bundle $\mathcal{O}(\ell)$ over $\mathbb{C}P^{m-1}$ with degree ℓ; that is, its first Chern class satisfies $c_1(\mathcal{O}(\ell))[\Sigma] = \ell$, where Σ is a positively-oriented generator of $H_2(\mathbb{C}P^{m-1};\mathbb{Z})$. The following result proven by Feldman, Ilmanen, and Knopf [**FIK03**, Theorem 1.5] holds for asymptotically conical shrinking gradient Kähler–Ricci solitons.

THEOREM 1.22. *For any integers $m \geq 2$ and $0 < k < m$, there exists a unique $\mathrm{U}(m)$-symmetric asymptotically conical shrinking gradient Kähler–Ricci soliton on the complex line bundle M_{-k}^{2m}. Such a soliton is called an **FIK shrinking soliton**.*

In particular, in real dimension 4, i.e., $m = 2$, this states that there exists a unique $\mathrm{U}(2)$-symmetric asymptotically conical shrinking gradient Kähler–Ricci soliton. This soliton, which is defined on the complex line bundle over $\mathbb{C}P^1$ with first Chern class -1 (equivalently, on $\mathbb{C}^2$ blown up at one point),[16] is the only known nontrivial 4-dimensional asymptotically conical shrinking gradient Ricci soliton.

REMARK 1.23. Given a complex manifold M^m of complex dimension m, i.e., real dimension $2m$, let $\mathrm{Bl}_p(M^m)$ denote the **blowup** of M^m at a point $p \in M^m$. This is a complex manifold, where the point p is replaced by a copy of $\mathbb{C}P^{m-1}$ in a complex analytic way compatible with the complex structure of M^m. Moreover, the blowup of a Kähler manifold is a Kähler manifold. See Griffiths and Harris [**GH94**] or Voisin [**Voi07**, Proposition 3.24].

[16]See Remark 1.23 below for the definition of blowup.

The asymptotic cone of the FIK shrinker has cross-section isometric to a Berger 3-sphere. Its Ricci curvature is indefinite. The Riemann curvature curvature operator has quadratic decay and has a quartically decaying negative lower bound.

In the Kähler category, Conlon, Deruelle, and Sun [**CDS24**] proved the following uniqueness result:

THEOREM 1.24. *The FIK shrinking soliton is the unique shrinking gradient Ricci soliton* (M^4, g, f) *that is Kähler and has scalar curvature tending to zero at infinity.*

1.3.2.2.c. *Asymptotically* 3*-cylindrical shrinking solitons.*

The only known 4-dimensional shrinking soliton that is strongly asymptotically 3-cylindrical is the cylinder $(\mathbb{S}^3/\Gamma) \times \mathbb{R}$.

1.3.2.2.d. *Asymptotically* 2*-cylindrical shrinking solitons.*

The shrinking gradient Ricci solitons $\mathbb{S}^2 \times \mathbb{R}^2$ and $((\mathbb{S}^2 \times \mathbb{R})/\mathbb{Z}_2) \times \mathbb{R}$ are strongly asymptotically 2-cylindrical.

Bamler, Cifarelli, Conlon, and Deruelle [**BCCD24**] proved the following existence result for shrinking solitons. Their shrinking gradient Ricci soliton, called the **BCCD shrinking soliton**, is weakly asymptotically 2-cylindrical.

THEOREM 1.25 (BCCD shrinking soliton). *There exists a Kähler shrinking gradient Ricci soliton on* $\mathrm{Bl}_p(\mathbb{C}P^1 \times \mathbb{C})$ *with bounded curvature which is invariant under the standard* 2*-torus action on* $\mathrm{Bl}_p(\mathbb{C}P^1 \times \mathbb{C})$*. It arises a singularity model of the Kähler–Ricci flow and is unique up to automorphism.*

The underlying space of the BCCD shrinking soliton is diffeomorphic to $\mathbb{C}P^1 \times \mathbb{C}$ blown up at one point. It is smoothly asymptotic at infinity to the product of S^2 with a 2-dimensional cone $C(S^1_r)$, where S^1_r is the circle of radius $r \in (0,1)$. Initially, one observes that the shrinker is asymptotic an S^2 fibration over a 2-dimensional cone $C(S^1_r)$. Metrically, one can describe this asymptotic manifold by taking the product $S^2 \times C(S^1_r)$, cutting $C(S^1_r)$ along a ray γ emanating from its vertex, and producing two rays, γ_+ and γ_-, on either side of the cut; here, the S^2 factor has constant scalar curvature 1, i.e., it has radius $\sqrt{2}$. The product regions $S^2 \times \gamma_+$ and $S^2 \times \gamma_-$ are then glued together by rotating the S^2 factor by a fixed angle, which can then be shown to be zero (see [**Bam**]). In particular, the BCCD shrinking soliton has quadratic volume growth and its asymptotic cone is $C(S^1_r)$. Note that, for any sequence of points $x_i \to \infty$ in the BCCD shrinking soliton, the pointed manifolds consisting of the soliton based at the x_i converge to $S^2 \times \mathbb{R}^2$.

Cifarelli, Conlon, and Deruelle [**CCD22**] proved the following uniqueness theorem for the BCCD soliton:

THEOREM 1.26. *If a complete noncompact Kähler shrinking gradient Ricci soliton* (M^4, g, f) *has bounded scalar curvature* R*, and* R *admits a positive lower bound along some integral curve of* ∇f*, then* (M^4, g, f) *must be either* $\mathbb{C}P^1 \times \mathbb{C}$ *or the BCCD soliton.*

Together with Theorem 1.24, it follows that any complete noncompact Kähler shrinking Ricci soliton (M^4, g, f) with bounded scalar curvature must be one of the following: the Gaussian on $\mathbb{C}^2$, the product soliton $\mathbb{C}P^1 \times \mathbb{C}$, the FIK shrinking soliton, or the BCCD shrinking soliton. The boundedness assumption for scalar curvature has been recently removed by Li and Wang in [**LW23**].

1.3.2.2.e. *Shrinking solitons with 2-cylinder limits.*

Inspired by Lai's steady solitons (see §1.3.3.2.d.i below), Bamler [**Bam**] asks whether there exists a 4-dimensional shrinking gradient Ricci soliton with the asymptotic limits (at spatial infinity) of this shrinking gradient Ricci soliton being either $\mathbb{S}^2 \times \mathbb{R}^2$ or the product of the 3-dimensional Bryant soliton with $\mathbb{R}$.

1.3.3. Examples of steady solitons.

Again we divide our discussion into the compact and noncompact cases.

1.3.3.1. *Compact steady solitons.*

Any compact steady soliton must be Ricci-flat. Besides the flat examples, the only known 4-dimensional compact simply-connected Ricci-flat manifolds are $K3$ surfaces with Ricci-flat Kähler metrics, called **Calabi–Yau surfaces** [**Yau78**].

1.3.3.2. *Noncompact steady solitons.*

We categorize noncompact steady solitons according to their geometry at infinity.

1.3.3.2.a. *Splitting steady solitons.*

Any noncollapsed 4-dimensional complete steady gradient Ricci soliton that splits as a product with an $\mathbb{R}$ factor must have its 3-dimensional factor be isometric to the Bryant soliton. This follows from Brendle's uniqueness theorem [**Bre13**].

1.3.3.2.b. *Ricci-flat asymptotically locally Euclidean 4-manifolds.*

Consider a 4-dimensional complete steady gradient Ricci soliton with a constant potential function. Such a soliton must be Ricci-flat. Further assuming that the steady gradient Ricci soliton is a singularity model, Perelman's no local collapsing theorem implies that it has Euclidean volume growth; see Corollary 1.13. In other words, $\mathrm{Vol}\,(B(o,r)) \geq cr^4$ for all $r > 0$ and where $c > 0$. Using that we are in dimension 4, by a fundamental theorem of Cheeger and Naber [**CN15**, Corollary 8.85], our steady gradient Ricci soliton must be a Ricci-flat asymptotically locally Euclidean (ALE) 4-manifold; see §2.7.2 below for the definition of ALE manifold. If we further assume that these Riemannian 4-manifolds are hyper-Kähler, then Kronheimer [**Kro89**] has classified them.[17] These last hyper-Kähler examples are the only known Ricci-flat ALE 4-manifolds.

Appleton [**App23**] has proved that the Eguchi–Hanson Ricci-flat ALE 4-manifold is the singularity model of a 4-dimensional Ricci flow. For a related question, see Open Problem 1.9 below.

1.3.3.2.c. *Asymptotically 3-cylindrical steady solitons.*

The 4-dimensional Bryant soliton is asymptotic, after rescaling, to the 3-cylinder $\mathbb{S}^3 \times \mathbb{R}$. Furthermore, the Bryant steady soliton is rotationally symmetric, and in particular, is cohomogeneity-one.

It is natural that canonical geometric objects, such as Ricci solitons, exhibit symmetries. One of the strongest forms of symmetry is rotational symmetry. In the Riemannian case, **rotational symmetry** of a real n-dimensional manifold manifests itself as an isometry group containing $\mathrm{O}(n)$. In this case we say that the Riemannian metric is $\mathrm{O}(n)$-symmetric. Note that the orbit space of the action of the isometry group on the manifold is 1-dimensional. If we only require this last condition, then we say that the metric has **cohomogeneity one**.

[17] We will also call these 4-manifolds **Kronheimer spaces**.

In even dimensions at least 4, Appleton [**App17**] has constructed steady gradient Ricci solitons that are noncollapsed, cohomogeneity-one, non-Kähler, and non-Ricci-flat. In dimension 4, the statement is as follows. Let M_k^4 be the total space of the complex line bundle $\mathcal{O}(-k)$ over $\mathbb{C}P^1$ with degree $-k$. Topologically, M_k^4 is diffeomorphic to $(\mathbb{S}^3/\mathbb{Z}_k) \times (0, \infty)$ by adding a 2-sphere at the origin. Appleton proved the following result:

THEOREM 1.27. *On each manifold M_k^4, $k \geq 3$, there exists a κ-noncollapsed* U(2)*-symmetric complete steady gradient Ricci soliton which is asymptotic at infinity to the $\mathbb{Z}_k$-quotient of the* 4*-dimensional Bryant soliton.*

Thus, the 4-dimensional cohomogeneity-one steady gradient Ricci solitons of Appleton are asymptotic, after rescaling, to the quotient 3-cylinder $(\mathbb{S}^3/\mathbb{Z}_k) \times \mathbb{R}$, where $k \geq 3$. The $k = 1$ case corresponds to the Taub-Bolt metric[18] and the $k = 2$ case corresponds to the Eguchi–Hanson metric (see [**But21**, §4.2-4.3]).

To date, these and the Bryant soliton are the only known steady solitons that are asymptotic to $(\mathbb{S}^3/\Gamma) \times \mathbb{R}$ at infinity.

Buttsworth [**But21**] proved the existence of a 1-parameter family of SU(2)-symmetric complete steady gradient Ricci solitons on M_4^4 that are not U(2)-symmetric. These steady gradient Ricci solitons are expected to be collapsed (see the remark about the $b = 0$ case in [**But21**, §7]). Buttsworth also simplified the proof of the existence of Appleton's steady gradient Ricci solitons.

1.3.3.2.d. *Steady solitons with* 2*-cylinder asymptotic limits.*

The asymptotic limits of such steady gradient Ricci solitons are not unique in general.

1.3.3.2.d.i. *Lai's steady solitons.*

Lai [**Lai24**] proved the following result.

THEOREM 1.28. *For every dimension $n \geq 4$ and every real number $\beta > 1$ there exists an $\mathrm{O}(n-1) \times \mathbb{Z}_2$-symmetric noncollapsed steady gradient Ricci soliton (M^n, g, f) with $\mathrm{Rm} > 0$ such that $n - 1$ of the eigenvalues of* Ric *at the origin 0^n are equal to β times the n-th eigenvalue of* Ric.

The underlying manifold M^n is diffeomorphic to $\mathbb{R}^n$. In particular, in dimension 4 we have $\mathrm{O}(3) \times \mathbb{Z}_2$-symmetric solitons. Some asymptotic limits of these solitons are the product of the 3-dimensional Bryant soliton with $\mathbb{R}$, while others are $\mathbb{S}^2 \times \mathbb{R}^2$; see Figure 1.8. Answering a conjecture of Hamilton and Cao, Lai [**Lai25**] proved that in dimension 3 any steady soliton is O(2)-symmetric.

CONJECTURE 1.29 (Lai). *In dimension* 4*, the only noncollapsed steady gradient Ricci solitons with positive curvature operator are the Bryant soliton and the family of $\mathrm{O}(3) \times \mathbb{Z}_2$-symmetric steady gradient Ricci solitons constructed in* [**Lai24**]. *Moreover, the asymptotic cone of each such steady gradient Ricci soliton is a ray.*

Ma, Mahmoudian, and Šešum [**MMŠ**] give precise asymptotics for 4-dimensional O(3)-symmetric noncollapsed steady gradient Ricci solitons with bounded nonnegative curvature operator and curvature decaying faster than $d^{-1/3}$.

[18] See Page [**Pag78**] for the Taub-Bolt metric.

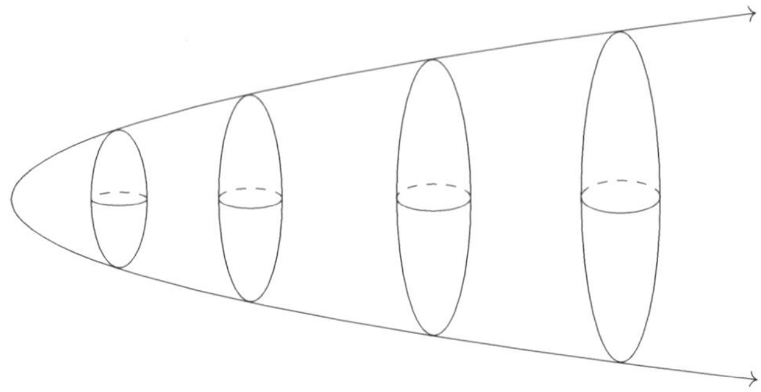

FIGURE 1.8. Lai's 4-dimensional steady soliton. The slices are ovaloids that elongate and should asymptotically approach Perelman's ancient solution.

1.4. An overview of the theory of 4-dimensional Ricci solitons

In this section, we will give a high-level (and necessarily incomplete) outline of what is known today about the geometry and topology of 4-dimensional noncollapsed Ricci solitons. Many of the results that we state will be proved in this book. The aim of these investigations is to advance our understanding of singularity formation for the 4-dimensional Ricci flow and to obtain a description of solitons which is sufficiently precise to assist in a potential formulation of a program of flow-with-surgery.

We will mostly limit our focus to complete noncompact shrinking and steady solitons as these have the greatest relevance to finite-time singularity formation: shrinking and steady solitons model a solution *entering* a finite-time singularity, whereas expanding solitons are the models for the *exit* of a solution of Ricci flow with surgery from a finite-time singularity, as well as for some types of infinite-time singularities. We focus on the noncompact case because, in the event of a noncompact singularity model, the goal is to continue the Ricci flow via surgery or a weak solution. In contrast, for compact singularity models, finite-time extinction occurs for a connected component of the Ricci flow.

In subsequent chapters, we will present the details for the proofs of some of the results we mention below. It is our hope to revisit other of the results, including those for expanding solitons, in a future volume.

The overview in this section is somewhat terse, and the reader who is interested in diving straight into the details of the proofs of the results may proceed directly to Chapter 2.

1.4.1. Theory of 4-dimensional shrinking solitons.

The study of shrinking solitons has expanded considerably over the past couple of decades. Much of this investigation has focused on either bounds on geometric invariants such as curvature or volume or on constraints on topological invariants such as the fundamental group and the number of ends. Another direction of study has focused on the asymptotic geometry of noncompact shrinking solitons, and in particular on those with asymptotically conical or asymptotically cylindrical ends.

1.4.1.1. *Topology of shrinking solitons.*

Wylie [**Wyl08**] proved that any shrinking Ricci soliton must have finite fundamental group; see §1.6.12 below.

Kähler shrinking solitons can have only one end; see §3.3 below. However, in the Riemannian case, even in dimension 4, it is unknown if shrinking solitons must have only a finite number of ends.

1.4.1.2. *Curvature estimates for shrinking solitons.*

We consider lower and upper curvature bounds.

1.4.1.2.a. *Lower scalar curvature bounds for shrinking solitons.*

If a shrinking gradient Ricci soliton (M^n, g, f) is not the Gaussian shrinking soliton and if $p \in M^n$, then there exists a positive constant c such that

$$C_n|\mathrm{Rm}| \geq R \geq c(d_p + 1)^{-2},$$

where $d_p := d(\cdot, p)$ denotes the distance to p and where C_n is a constant depending only on n; see Theorem 4.31 in Volume 1.

1.4.1.2.b. *Upper curvature bounds for shrinking solitons.*

For any shrinking gradient Ricci soliton (M^n, g, f) and for any point $p \in M^n$, there exists a constant $C > 0$ such that

$$0 < R \leq f \leq C d_p^2.$$

In dimension 4, with the additional assumption of being a singularity model, we have that

$$|\mathrm{Rm}| \leq C d_p^2; \tag{1.14}$$

see §2.7 below.

1.4.1.3. *Volume estimates for shrinking solitons.*

We consider upper and lower volume estimates.

1.4.1.3.a. *Upper volume bounds for shrinking solitons.*

Cao and Zhou [**CZ10**] proved that for any shrinking gradient Ricci soliton (M^n, g, f) and $p \in M^n$, there exists a constant $C(n,p) > 0$ such that $\mathrm{Vol}\,(B(p,r)) \leq C(n,p)r^n$ for all $r > 0$. That is, by definition, g has **at most Euclidean volume growth**; for an exposition, see §4.2 in Volume 1. This upper estimate has been improved by Li and Wang [**LW24a**] to $\mathrm{Vol}\,(B(p,r)) \leq C(n)r^n$ for all $r > 0$.

Furthermore, Wang and Wang [**WW25**] proved that shrinking gradient Ricci solitons must have asymptotic volume ratio (defined in (2.57) below) less than or equal to 1; see §2.8.

1.4.1.3.b. *Lower volume bounds for shrinking solitons.*

Assuming noncollapsing and by the quadratic in distance scalar curvature upper bound, for any shrinking gradient Ricci soliton (M^n, g, f) and any point $p \in M^n$ there exists a positive constant c such that

$$\mathrm{Vol}\left(B(x, d_p^{-1}(x))\right) \geq c(1 + d_p(x))^{-n}.$$

1.4.1.4. *Shrinking solitons with bounded scalar curvature.*

For any 4-dimensional shrinking gradient Ricci soliton with bounded scalar curvature, we have that

$$|\mathrm{Rm}| \leq CR \leq C;$$

see §2.2 below.

Li and Wang [**LW23**] proved that any complex 2-dimensional Kähler shrinking gradient Ricci soliton must have bounded scalar curvature, and hence bounded sectional curvature.

We divide the study of Riemannian shrinking solitons with bounded scalar curvature into those with scalar curvature decay and those without.

1.4.1.4.a. *Shrinking solitons with scalar curvature decay.*

If a shrinking gradient Ricci soliton (M^4, g, f) satisfies $R \to 0$, then $|\mathrm{Rm}| \leq C d_p^{-2}$; see §2.4 below.

1.4.1.4.a.i. *Asymptotically conical shrinking solitons.*

The condition that $|\mathrm{Rm}| \leq C d_p^{-2}$ implies that the shrinking gradient Ricci soliton is asymptotically conical. In this case, the isometry class of the asymptotic cone determines the isometry class of the shrinking gradient Ricci soliton; see Chapter 4.

1.4.1.4.b. *Shrinking solitons without scalar curvature decay.*

Suppose that (M^4, g, f) is a 4-dimensional shrinking gradient Ricci soliton with $R \leq C$ and $R \not\to 0$. In this case, results are only known if we have a positive lower bound for the scalar curvature on a topological end. Suppose that E is an end of M^4 on which $R \geq c > 0$. Then either:

(i) the end E is smoothly asymptotic at infinity to

$$(\mathbb{S}^3/\Gamma) \times \mathbb{R},$$

in which case the end is **asymptotically 3-cylindrical**, or

(ii) for any sequence of points $\{x_i\}$ in the end E tending to infinity on a fixed integral curve γ to the gradient of the potential function f, the associated unrescaled pointed sequence (M^4, g, x_i) converges smoothly to

$$\mathbb{S}^2 \times \mathbb{R}^2 \quad \text{or its } \mathbb{Z}_2 \text{ quotient},$$

in which case the shrinking gradient Ricci soliton **has a 2-cylinder asymptotic limit**; this limit depends only on γ. For all of this, see §2.5 below.

1.4.1.4.c. *Lower sectional curvature bounds for shrinking solitons.*

Any 4-dimensional shrinking gradient Ricci soliton with bounded scalar curvature must have asymptotically non-negative curvature operator in the following quantitative sense. If (M^4, g, f) is a shrinking gradient Ricci soliton with bounded scalar curvature and $p \in M^4$, then there exists a constant $C > 0$ such that the curvature operator satisfies

$$\mathrm{Rm} \geq -\frac{C}{\ln d_p}\,\mathrm{id}$$

on the complement of $B(p, C)$; see §2.6 below.

1.4.1.5. *Shrinking solitons with unbounded curvature.*

Currently, it is unclear if the curvature bound $|\mathrm{Rm}| \leq C d_p^2$ for 4-dimensional shrinking soliton singularity models given by (1.14) is qualitatively sharp. In any dimension, the scalar curvature has at most linear growth on a net in the following sense. For any $\varepsilon > 0$, there exists a constant $C(\varepsilon, n)$ such that for any $x \in M^n$ there exists $y \in B(x, \varepsilon)$ such that

$$R(y) \leq C(\varepsilon, n)\,(d_o(y) + 1), \tag{1.15}$$

where $o \in M^n$ is a minimum point of f; see §2.7.10. Furthermore, the scalar curvature cannot have quadratic growth along any geodesic ray; see §2.7.11 for the statement and proof. However, so far, there does not exist an example of a 4-dimensional shrinking gradient Ricci soliton with unbounded curvature.

1.4.2. Theory of 4-dimensional steady solitons.

The topological and geometric study of steady solitons has been, so far, focused on understanding the following topics: the number of topological ends, the curvature and volume estimates, and the asymptotic geometry at infinity.

1.4.2.1. *Topology of steady solitons.*

By a result of Bamler [**Bam21a**], any singularity model of the Ricci flow must have a finite fundamental group (see Yokota [**Yok09, Yok12**] for an earlier partial result). In particular, any steady Ricci soliton singularity model must have a finite fundamental group.

Any steady gradient Ricci soliton either has exactly one *topological end* or splits as the product of $\mathbb{R}$ with a compact Ricci-flat manifold; see §3.2 below.

Regarding the definition of a topological end, let M^n be a noncompact smooth manifold. For each compact subset Ω, define $\mathcal{E}(\Omega)$ to be the set of connected unbounded components E of $M^n \setminus \Omega$, where each such component is called an **end with respect to** Ω. Let $N_\Omega(M^n) := |\mathcal{E}(\Omega)|$ denote the number of unbounded components of $M^n \setminus \Omega$. Given a compact subset $\Omega' \supset \Omega$, the map $\phi : \mathcal{E}(\Omega') \to \mathcal{E}(\Omega)$ is defined by taking $\phi(E')$ to be the connected component of $M^n \setminus \Omega$ containing E'. Since ϕ is a surjection, $N_\Omega(M^n) \leq N_{\Omega'}(M^n)$.

DEFINITION 1.30 (Cf. [**Cho23**, §3.4.2]). The **number of (topological) ends of** M^n is defined by

$$N(M^n) := \lim_{i\to\infty} N_{\Omega_i}(M^n) \in [1, \infty], \tag{1.16}$$

where $\{\Omega_i\}$ is a compact exhaustion of M^n. We have that $N(M^n)$ is independent of the choice of $\{\Omega_i\}$. If $N(M^n) = 1$, then we say that M^n is **connected at infinity**.

We say that Ω is **end-complementary** if $\phi : \mathcal{E}(\Omega') \to \mathcal{E}(\Omega)$ is a bijection for all $\Omega' \supset \Omega$. If Ω is end-complementary, then $N_\Omega(M^n) = N(M^n)$ and each element of $\mathcal{E}(\Omega)$ is called a **topological end**. If $N(M^n) < \infty$, then there exists Ω such that $N_\Omega(M^n) = N(M^n)$ and any such Ω is end-complementary.

1.4.2.2. *Curvature estimates for steady solitons.*

For steady solitons, we consider norm bounds for the curvature and positive lower bounds for the scalar curvature.

1.4.2.2.a. *Curvature norm bounds for steady solitons.*

The scalar curvature of any steady gradient Ricci soliton is non-negative and bounded. Unless this steady solition is Ricci-flat, it has positive scalar curvature. Any 4-dimensional steady gradient Ricci soliton singularity model must have bounded sectional curvatures; see §5.4 below. Moreover, in this case, we have the estimate

$$|\mathrm{Rm}| \leq CR;$$

see §5.5 below.

1.4.2.2.b. *Scalar curvature lower bounds for steady solitons.*

For any steady gradient Ricci soliton (M^n, g, f) with the potential function bounded from above and $p \in M^n$, there exists a positive constant c such that $R \geq c\mathrm{e}^{-d_p}$; see §5.3 below.

For any 4-dimensional steady gradient Ricci soliton singularity model, there exists a positive constant $c(p)$ such that $R \geq c(1+d_p)^{-2}$; see §6.5 below.

1.4.2.3. *Volume estimates for steady solitons.*

For any steady gradient Ricci soliton singularity model, there exists a positive constant c such that

$$cr^{\frac{n+1}{2}} \leq \mathrm{Vol}\,(B(p,r)) \leq C(n)r^n;$$

see §6.6 below.

1.4.2.4. *Asymptotic geometry of steady solitons.*

The classification of tangent flows (at infinity) of 4-dimensional steady gradient Ricci soliton singularity models is provided in §6.4 below; we refer to that section for the definition of a tangent flow. Roughly speaking, the tangent flow of a steady gradient Ricci soliton can be thought of as a core, or central, limit at infinity that consists of a shrinking gradient Ricci soliton. The only possibilities for these tangent flows are $\mathbb{R}^4/\Gamma$, $(\mathbb{S}^3/\Gamma)\times\mathbb{R}$, and $\mathbb{S}^2\times\mathbb{R}^2$ or its $\mathbb{Z}_2$-quotient.

1.4.2.4.a. *Steady solitons with tangent flow* $\mathbb{R}^4/\Gamma$.

In this case, the steady gradient Ricci soliton must be a Ricci-flat asymptotically locally Euclidean 4-manifold; see §6.4 below.

1.4.2.4.b. *Steady solitons with tangent flow* $(\mathbb{S}^3/\Gamma)\times\mathbb{R}$.

In this case, the steady gradient Ricci soliton must have positive curvature operator outside a compact subset, with linear curvature decay; see §6.4.3.2 below.

1.4.2.4.c. *Steady solitons with tangent flow* $\mathbb{S}^2\times\mathbb{R}^2$ *or its* $\mathbb{Z}_2$*-quotient.*

This is the most difficult case, about which not much is known. Firstly, we have the examples of $\mathbb{S}^2\times\mathbb{R}^2$ and $((\mathbb{S}^2\times\mathbb{R})/\mathbb{Z}_2)\times\mathbb{R}$ themselves. Secondly, Lai's $\mathrm{O}(3)\times\mathbb{Z}_2$-symmetric steady gradient Ricci solitons have tangent flow $\mathbb{S}^2\times\mathbb{R}^2$. Besides this tangent flow, other asymptotic limits of her solitons are the Cartesian product of the 3-dimensional Bryant soliton with $\mathbb{R}$.

Regarding steady solitons with tangent flow $\mathbb{S}^2\times\mathbb{R}^2$, we have the following:

CONJECTURE 1.31 (Bamler). *Suppose that (M^4,g,X) is a 4-dimensional steady (gradient) soliton with tangent flow $S^2\times\mathbb{R}^2$. Suppose that (M^4,g,X) admits a proper isometric $\mathbb{R}$ or $\mathbb{Z}$-action. Then (M^4,g) must be isometric to the cartesian product of a 3-dimensional Bryant soliton and $\mathbb{R}$.*

1.4.2.4.d. *Remarks on Kähler steady solitons.*

In the Kähler case, Cao [**Cao96**] constructed $\mathrm{U}(m)$-symmetric Kähler steady gradient Ricci solitons on $\mathbb{C}^m$ and on the canonical line bundle of $\mathbb{C}P^{m-1}$ for all $m\geq 2$.

Apostolov and Cifarelli [**AC23**] constructed for any $\ell\geq 1$ and positive integers $d_1,\ldots,d_\ell$ summing to m, $\mathrm{U}(d_1)\times\cdots\times\mathrm{U}(d_\ell)$-symmetric Kähler steady gradient Ricci solitons on $\mathbb{C}^m$. When $m=2$, they obtained a $\mathrm{U}(1)\times\mathrm{U}(1)$ -symmetric, but not $\mathrm{U}(2)$-symmetric, Kähler steady gradient Ricci soliton on $\mathbb{C}^2$ with positive sectional curvature.

Chan, Conlon, and Lai [**CCL24**] proved the existence of m-complex dimensional Kähler steady gradient Ricci solitons with positive sectional curvature that are $\mathrm{U}(m-1)\times\mathrm{U}(1)$-symmetric, but not $\mathrm{U}(m)$-symmetric, for all $m\geq 2$.

1.4.3. Theory of 4-dimensional expanding solitons.

In this subsection we highlight some of the results regarding expanding Ricci solitons.

1.4.3.1. *Rotationally symmetric expanding solitons and their uniqueness.* For all dimensions $n \geq 3$, Bryant [**Bry**] proved the following existence theorem for expanding solitons with the highest degree of symmetry, namely rotational symmetry.

THEOREM 1.32 (Rotationally symmetric, asymptotically conical expanding solitons). *There exists a 1-parameter family of rotationally symmetric complete expanding gradient Ricci solitons* $(\mathbb{R}^n, g_c, f_c)$, $c \in (0,1)$, *with positive curvature operator* $\mathrm{Rm}_{g_c} > 0$ *and which are asymptotic to the cone* $(C_0\mathbb{S}^{n-1}, dr^2 + (cr)^2 g_{\mathbb{S}^{n-1}})$, *i.e., to the Riemannian cone over* $(\mathbb{S}^{n-1}, c^2 g_{\mathbb{S}^{n-1}})$. *As* $c \to 1$, $(\mathbb{R}^n, g_c, 0^n)$ *converges in the pointed* C^∞ *Cheeger–Gromov sense to Euclidean* n*-space* $(\mathbb{R}^n, g_{\mathrm{Euc}}, 0^n)$.

Chodosh [**Cho14**, Theorem 1.2] proved the following uniqueness theorem for expanding solitons asymptotic to a rotationally symmetric cone.

THEOREM 1.33. *Let* (M^n, g, f), $n \geq 3$, *be a complete expanding gradient Ricci soliton with positive sectional curvature and asymptotic to the Riemannian cone over* $(\mathbb{S}^{n-1}, c^2 g_{\mathbb{S}^{n-1}})$, *where* $c \in (0,1)$. *Then* (M^n, f, g) *is one of the rotationally symmetric expanding Bryant solitons.*

Imposing the condition of positive Ricci curvature has a significant impact on the asymptotic geometry of expanding solitons. Firstly, if $\mathrm{Ric} \geq 0$, then

$$\nabla^2 f = -\frac{1}{2} g - \mathrm{Ric} \leq -\frac{1}{2} g, \tag{1.17}$$

so that f is a uniformly concave function on M^n. Furthermore, we have the following result of Hamilton (for an exposition, see Proposition 9.46 of [**CLN06**]).

THEOREM 1.34. *If* (M^n, g, f) *is a complete noncompact expanding gradient Ricci soliton with* $\mathrm{Ric} > 0$, *then the asymptotic volume ratio of* g *is positive:*

$$\mathrm{AVR}(g) = \limsup_{r \to \infty} \frac{\mathrm{Vol}\,(B\,(p, r))}{\omega_n r^n} > 0. \tag{1.18}$$

1.4.3.2. *Asymptotically conical expanding solitons.*

Chen and Deruelle [**CD15**, Theorem 1.2] proved that expanders with quadratic curvature decay are asymptotic to unique metric cones.

THEOREM 1.35. *If* (M^n, g, f), $n \geq 3$, *is a complete expanding gradient Ricci soliton satisfying* $|\mathrm{Rm}|(x) d(x,p)^2 \leq C$ *for some constant* $C > 0$ *and all* $x \in M^n$, *then the metric* g *is asymptotic to a unique metric cone over a Riemannian manifold* (Σ^{n-1}, g_Σ), *where the metric* g_Σ *is* $C^{1,\alpha}$, $\alpha \in (0,1)$.

Given λ_0 and Λ_k, for $k \geq 0$, let $\mathfrak{M}$ denote the class of complete pointed expanding gradient Ricci solitons (M^n, g, f, p) satisfying the following conditions:

(1) $\mathrm{Ric} \geq 0$,
(2) p is the unique critical point of f,
(3) $\sup_{M^n} |\mathrm{Rm}| \leq \lambda_0$,
(4) $\limsup_{d_p \to \infty} d_p^{k+2} |\nabla^k \mathrm{Rm}| \leq \Lambda_k$ for all $k \geq 0$,

where d_p denotes the distance to p. Deruelle [**Der17a**] proved the following compactness theorem for expanders.

THEOREM 1.36. *Given any constants λ_0 and Λ_k, for $k \geq 0$, the space $\mathfrak{M}$ is compact with respect to the pointed C^∞ topology. Moreover, the asymptotic cones of any convergent sequence subconverges with respect to the Gromov–Hausdorff distance.*[19]

Deruelle [**Der16**, Theorem 1.3] proved the following theorem regarding the existence of a unique asymptotically conical expanding soliton for any given asymptotic cone whose slice has positive curvature operator.

THEOREM 1.37. *Let (Σ^{n-1}, g_Σ) be a simply-connected closed Riemannian manifold with its curvature operator satisfying* $\mathrm{Rm} > \mathrm{id}$. *Then there exists a unique (up to isometry) complete expanding gradient Ricci soliton (M^n, g, f) with positive curvature operator such that the metric g is asymptotic to the Riemannian cone $(C_0\Sigma, dr^2 + r^2 g_\Sigma)$. For this unique expanding soliton, the gradient of the potential function ∇f is asymptotic to the radial vector field $r\partial_r/2$.*

In [**Der16**, Theorem 1.4], Deruelle proved the following result regarding the existence of 1-parameter families of asymptotically conical expanding solitons connecting to some Bryant expanding soliton.

THEOREM 1.38. *Let $(\mathbb{R}^n, g, f)$ be any nonflat complete expanding gradient Ricci soliton with* $\mathrm{Rm} \geq 0$ *and asymptotic to the Riemannian cone over a closed manifold (Σ^{n-1}, g_Σ). Then there exists $c = c(\mathrm{Vol}(g_\Sigma), n) \in (0,1)$ and a 1-parameter family of expanding gradient Ricci solitons $(\mathbb{R}^n, g(s), f(s))$, $s \in [0,1]$, satisfying:*

(1) *$g(s)$ is asymptotically conical for $s \in [0,1]$,*
(2) *$(\mathbb{R}^n, g(0), f(0)) = (\mathbb{R}^n, g, f)$ and $(\mathbb{R}^n, g(1), f(1)) = (\mathbb{R}^n, g_c, f_c)$, a Bryant expanding soliton,*
(3) $\mathrm{Rm}_{g(s)} > 0$ *for all $s \in (0,1]$.*

REMARK 1.39. There is no obstruction to obtaining an expanding Ricci soliton asymptotic to a Kähler cone; cf. [**CDS24**, Corollary B]. One only requires an algebraic-geometric condition to be satisfied, namely the existence of a smooth canonical model of the cone (which is the analogue of having negative first Chern class).

1.4.3.3. *Bamler and Chen's existence theorem.*
Bamler and Chen [**BC23**, Corollary 1.6] proved the following result.

THEOREM 1.40. *Let C be any 4-dimensional Riemannian cone with $R \geq 0$ and whose link (a.k.a. cross-section) is diffeomorphic to a spherical space form $\mathbb{S}^3/\Gamma$. Then there exists an expanding gradient Ricci soliton that is asymptotic to C, which is diffeomorphic to $\mathbb{R}^4/\Gamma$, and has non-negative scalar curvature.*

Nienhaus and Wink [**NW24**] proved the existence of an expanding gradient Ricci soliton that is asymptotic to the cone over a standard product $\mathbb{S}^2 \times \mathbb{S}^1$.

1.4.3.4. *Curvature estimates for expanding solitons.*
Let (M^4, g, f) be a 4-dimensional expanding gradient Ricci soliton. If the potential function f is proper and if the scalar curvature is bounded, then the

[19] See Burago, Burago, and Ivanov [**BBI01**] for the definition of Gromov–Hausdorff distance.

Riemann curvature tensor is bounded. If, moreover, the scalar curvature converges to zero at infinity, then the Riemann curvature tensor also converges, in norm, to zero at infinity; see Chan [**Cha23**].

Any expanding Ricci soliton with non-negative Ricci curvature has proper potential and bounded scalar curvature. Cao and Liu [**CL22**] proved that 4-dimensional expanding gradient Ricci solitons with non-negative Ricci curvature satisfy some curvature bounds of the form

$$|\mathrm{Rm}| \leq \frac{C(a)}{1-a} R^a,$$

for any constant $0 \leq a < 1$.

1.4.4. Summary of shrinking and steady solitons.

According to our current knowledge, the landscape of 4-dimensional shrinking and steady solitons looks roughly as follows.

1.4.4.1. *Shrinking solitons.*

A list of complete noncompact 4-dimensional shrinking gradient Ricci solitons is as follows:

(1) Asymptotically conical shrinking gradient Ricci solitons, where the only known example is the FIK shrinking soliton.
(2) $(\mathbb{S}^3/\Gamma) \times \mathbb{R}$. There are no other known examples sharing its asymptotics.
(3) $\mathbb{S}^2 \times \mathbb{R}^2$, $((\mathbb{S}^2 \times \mathbb{R})/\mathbb{Z}_2) \times \mathbb{R}$, and the BCCD shrinking soliton. These are the only known examples whose asymptotic limits are unique and equal to $\mathbb{S}^2 \times \mathbb{R}^2$.

1.4.4.2. *Steady solitons.*

A list of noncollapsed complete noncompact 4-dimensional steady gradient Ricci solitons is as follows:

(1) Ricci-flat ALE 4-manifolds. All known examples are hyper-Kähler.
(2) The Cartesian product of the 3-dimensional Bryant soliton and $\mathbb{R}$. These are the only possible splitting steady gradient Ricci solitons.
(3) The 4-dimensional Bryant and Appleton solitons. These are the only known steady gradient Ricci solitons asymptotic to $(\mathbb{S}^3/\Gamma) \times \mathbb{R}$.
(4) Lai's $\mathrm{O}(3) \times \mathbb{Z}_2$-symmetric steady gradient Ricci solitons. These are the only known steady gradient Ricci solitons with asymptotic limits consisting of $\mathbb{S}^2 \times \mathbb{R}^2$ and the 3-dimensional Bryant soliton Cartesian product with $\mathbb{R}$.

1.5. Potential applications to smooth 4-manifold topology

In this section, we give a heuristic discussion of the role that Ricci solitons might potentially play in a hypothetical program of Ricci flow with surgery. First we review some aspects of 4-manifold topology. For detailed presentations of 4-manifold topology in the topological and smooth categories, see the books by Kirby [**Kir89**], Freedman and Quinn [**FQ90**], Behrens, Kalmár, Kim, Powell, and Ray [**BKK$^+$21**], Gompf and Stipsicz [**GS99**], Freed and Uhlenbeck [**FU91**], Donaldson and Kronheimer [**DK90**], and Scorpan [**Sco05**].

1.5.1. Topological 4-manifolds.

Let $\mathcal{M}^4$ be a closed oriented topological 4-manifold. Then the top-dimensional integral homology group $H_4(\mathcal{M}^4;\mathbb{Z}) \cong \mathbb{Z}$ is infinite cyclic. The orientation on $\mathcal{M}^4$ naturally defines a generator $[\mathcal{M}^4]$ of $H_4(\mathcal{M}^4;\mathbb{Z})$, known as the fundamental class of $\mathcal{M}^4$; see Hatcher [**Hat02**, p. 236].

The **cohomology intersection form** $Q_{\mathcal{M}} : H^2(\mathcal{M}^4;\mathbb{Z}) \times H^2(\mathcal{M}^4;\mathbb{Z}) \to \mathbb{Z}$ is defined by

$$Q_{\mathcal{M}}(a,b) := \langle a \smile b, [\mathcal{M}^4] \rangle =: a \cdot b,$$

where $\smile$ denotes the cup product; see [**GS99**, Definition 1.2.1]. We have $Q_{\mathcal{M}}(a,b) = 0$ if a or b is torsion. By Poincaré duality, this implies that $Q_{\mathcal{M}}$ descends to a pairing on the second integral homology group modulo torsion $H_2(\mathcal{M}^4;\mathbb{Z})_{\mathrm{tf}}$.[20] The bilinear form $Q_{\mathcal{M}}$ over $\mathbb{Z}$ is symmetric and unimodular. By unimodular, we mean that the matrix $\left(Q_{\mathcal{M}}\right)_{ij} = Q_{\mathcal{M}}(e_i, e_j)$, where $\{e_i\}$ is a basis for $H_2(\mathcal{M}^4;\mathbb{Z})_{\mathrm{tf}}$, has determinant ± 1, which is equivalent to $\left(Q_{\mathcal{M}}\right)_{ij}$ being invertible over $\mathbb{Z}$. We have that $Q_{\overline{\mathcal{M}}} = -Q_{\mathcal{M}}$, where $\overline{\mathcal{M}}$ denotes $\mathcal{M}$ with the reversed orientation; see [**GS99**, Remark 1.2.2(c)].

Let $\mathcal{M}_1^4$ and $\mathcal{M}_2^4$ be two oriented topological 4-manifolds. By a Mayer–Vietoris argument, one can prove that the homology groups of the connected sum of $\mathcal{M}_1$ and $\mathcal{M}_2$ satisfy

$$H_k(\mathcal{M}_1 \# \mathcal{M}_2;\mathbb{Z}) \cong H_k(\mathcal{M}_1;\mathbb{Z}) \oplus H_k(\mathcal{M}_2;\mathbb{Z}) \quad \text{for } k = 1,2,3; \tag{1.19}$$

see [**Hat02**, Exercise 6(a) on p. 257]. Moreover, the intersection form of the connected sum is given by

$$Q_{\mathcal{M}_1 \# \mathcal{M}_2} = Q_{\mathcal{M}_1} \oplus Q_{\mathcal{M}_2}; \tag{1.20}$$

see [**Sco05**, §3.2], [**GS99**, Exercise 1.3.5(a)].

The intersection form of $Q_{\mathcal{M}}$ of an oriented topological 4-manifold $\mathcal{M}^4$ is **even** if for all $a \in H^2(\mathcal{M}^4;\mathbb{Z})_{\mathrm{tf}}$, $Q_{\mathcal{M}}(a,a)$ is an even integer. Otherwise, we say that $Q_{\mathcal{M}}$ is **odd**.

We will call an integral symmetric bilinear form a **lattice**. So the intersection form of a 4-manifold is a unimodular lattice.

EXAMPLE 1.41 (The E_8 lattice). The unimodular lattice

$$E_8 := \begin{bmatrix} 2 & 1 & 0 & 0 & 0 & 0 & 0 & 0 \\ 1 & 2 & 1 & 0 & 0 & 0 & 0 & 0 \\ 0 & 1 & 2 & 1 & 0 & 0 & 0 & 0 \\ 0 & 0 & 1 & 2 & 1 & 0 & 0 & 0 \\ 0 & 0 & 0 & 1 & 2 & 1 & 0 & 1 \\ 0 & 0 & 0 & 0 & 1 & 2 & 1 & 0 \\ 0 & 0 & 0 & 0 & 0 & 1 & 2 & 0 \\ 0 & 0 & 0 & 0 & 1 & 0 & 0 & 2 \end{bmatrix}$$

is positive-definite and even; cf. [**GS99**, p. 13].

The **signature** $\sigma(\mathcal{M}^4)$ of $\mathcal{M}^4$ is the difference:

$$\sigma(\mathcal{M}^4) = b_2^+(\mathcal{M}^4) - b_2^-(\mathcal{M}^4), \tag{1.21}$$

where $b_2^+(\mathcal{M}^4)$ and $b_2^-(\mathcal{M}^4)$ denote the dimensions of the maximal positive- and

[20]The subscript "tf" stands for torsion-free.

negative-definite subspaces of $H_2(\mathcal{M}^4;\mathbb{Z})_{\mathrm{tf}}$ under the form $Q_{\mathcal{M}}$. Note that $b_2^+(\mathcal{M}^4)+b_2^-(\mathcal{M}^4) = b_2(\mathcal{M}^4)$ is the second Betti number, which is equal to the dimension of $H_2(\mathcal{M}^4;\mathbb{Z})_{\mathrm{tf}}$. More generally, the signature of a unimodular lattice is defined in the same way.

For a proof of the following result of van der Blij [**vdB59**], see Milnor and Husemoller's book [**MH73**, Chapter II, Theorem 5.1].

PROPOSITION 1.42. *The signature of any even unimodular lattice is divisible by* 8.

We have the following result of Mordell [**Mor38**] (cf. [**Ser73**, p. 55]).

PROPOSITION 1.43. *Any even positive-definite unimodular lattice of rank* 8 *is isomorphic to* E_8.

The unimodular lattice

$$H := \begin{bmatrix} 0 & 1 \\ 1 & 0 \end{bmatrix}, \tag{1.22}$$

which is indefinite, is the intersection form of $S^2 \times S^2$. H is called the **hyperbolic lattice**.

In the indefinite case, we have the following classification of Serre [**Ser73**, Chapter V, Theorems 5 and 6]; see [**MH73**, Chapter II, Theorem 5.3].

THEOREM 1.44. *If* Q_1 *and* Q_2 *are two indefinite unimodular integral symmetric bilinear forms, then* Q_1 *and* Q_2 *are isomorphic if and only if* Q_1 *and* Q_2 *have the same rank, signature, and parity.*

If Q *is an indefinite unimodular integral symmetric bilinear form, then:*

(1) *If* Q *is odd, then*

$$Q = m\,[+1] \oplus n\,[-1]\,.$$

That is, up to equivalence, the matrix (Q_{ij}) *is diagonal with* m *entries equal to* 1 *and* n *entries equal to* -1.

(2) *If* Q *is even, then*

$$Q = \pm m E_8 \oplus nH.$$

The intersection form of the $K3$ complex surface $\mathcal{M}_{K3}$ is

$$Q_{K3} = 2(-E_8) \oplus 3H\,; \tag{1.23}$$

see [**GS99**, p. 23] Hence, the intersection form of the connected sum of k copies of the $K3$ surface and ℓ copies of $S^2 \times S^2$ is

$$2k(-E_8) \oplus (3k+\ell)H. \tag{1.24}$$

EXAMPLE 1.45 (Signatures and second Betti numbers). By (1.22), (1.23), (1.24), we have:

(1) $S^2 \times S^2$ satisfies $b_2^-(S^2 \times S^2) = 1$, $b_2^+(S^2 \times S^2) = 1$, $b_2(S^2 \times S^2) = 2$, and $\sigma(S^2 \times S^2) = 0$.
(2) The $K3$ complex surface $\mathcal{M}_{K3}$ satisfies $b_2^-(\mathcal{M}_{K3}) = 19$, $b_2^+(\mathcal{M}_{K3}) = 3$, $b_2(\mathcal{M}_{K3}) = 22$, and $\sigma(\mathcal{M}_{K3}) = -16$.
(3) For the connected sum of k $K3$ surfaces and ℓ $S^2 \times S^2$'s, we have $b_2^- = 19k+\ell$, $b_2^+ = 3k+\ell$, $\sigma = -16k$, and $b_2 = 22k+2\ell$.

With this foundation, Freedman provided a classification of simply-connected topological 4-manifolds:

THEOREM 1.46 ([**Fre82**, Theorem 1.5]). *Let Q be a unimodular lattice. Then there exists a closed simply-connected topological 4-manifold $\mathcal{M}^4$ with Q as its intersection form. Moreover, two closed simply-connected topological 4-manifolds are homeomorphic if and only if their intersection forms are isomorphic and their Kirby–Siebenmann invariants κ are the same.*[21]

An important corollary of Freedman's theorem is the resolution of the 4-dimensional Poincaré conjecture:

COROLLARY 1.47 ([**Fre82**, Theorem 1.6]). *If a closed topological 4-manifold is homotopy equivalent to S^4, then it is homeomorphic to S^4.*

In dimensions 5 and higher, Smale [**Sma61**] proved the topological Poincaré conjecture that if a closed topological n-manifold is homotopy equivalent to S^n, where $n \geq 5$, then it is homeomorphic to S^n. In dimensions 7 and higher, there exist exotic spheres.

1.5.2. Distinguishing between the smooth and topological categories for 4-manifolds.

Let M^n be a smooth manifold. Associated with the smooth manifold M^n is its **underlying topological manifold**, which we denote by $|M^n|$, obtained by forgetting the smooth structure while retaining the topological structure. Two smooth manifolds can have homeomorphic underlying topological manifolds but may not be diffeomorphic.

DEFINITION 1.48. Two smooth manifolds M^n and N^n are said to have the same **diffeomorphism type** if there exists a bijective map $f : M^n \to N^n$ such that both f and its inverse f^{-1} are smooth. In this case, f is said to be a **diffeomorphism** and M^n and N^n are said to be **diffeomorphic**.

The **smooth 4-dimensional Poincaré conjecture** states:

CONJECTURE 1.49. *Every smooth 4-manifold whose underlying topological manifold is homeomorphic to the 4-sphere S^4 is diffeomorphic to S^4.*

Conjecture 1.49 asserts that there is a unique diffeomorphism type of smooth 4-manifolds whose underlying topological manifold is homeomorphic to S^4.

In general, smooth structures on topological 4-manifolds do not have to be unique. Originally, the existence of a smooth 4-manifold not diffeomorphic to $\mathbb{R}^4$ but whose underlying topological manifold is homeomorphic to $\mathbb{R}^4$ was observed by Kirby as a consequence of the works of Freedman [**Fre82**] and Donaldson [**Don83**]. Such manifolds are called **exotic** $\mathbb{R}^4$'s. In particular, by performing surgeries on the $K3$ complex surface to remove the three H's from its intersection form $2(-E_8) \oplus 3H$, and by combining the results of Freedman and Donaldson, one can infer the existence of an exotic $\mathbb{R}^4$; see Freed and Uhlenbeck [**FU91**]. Later, Gompf [**Gom85**] found a countably infinite set of diffeomorphism types of smooth 4-manifolds whose underlying topological manifolds are homeomorphic to $\mathbb{R}^4$. This result was further

[21]The Kirby–Siebenmann invariant, $\mathrm{ks}(\mathcal{M}^4)$, is a topological invariant that takes values in $\mathbb{Z}_2$. It is defined as $\mathrm{ks}(\mathcal{M}^4) = 0$ if $\mathcal{M}^4$ admits a smooth structure and $\mathrm{ks}(\mathcal{M}^4) = 1$ if it does not.

improved by Taubes as follows:

THEOREM 1.50 ([**Tau87**]). *There are uncountably many diffeomorphism types of smooth* 4*-manifolds whose underlying topological manifolds are homeomorphic to* $\mathbb{R}^4$.

REMARK 1.51. Freedman and Taylor [**FT86**, Corollary B] proved that there exists an exotic $\mathbb{R}^4$, called a **universal exotic** $\mathbb{R}^4$, into which every exotic $\mathbb{R}^4$ is smoothly embeddable.

REMARK 1.52. In contrast, if $n \neq 4$, then any smooth n-manifold whose underlying topological manifold is homeomorphic to $\mathbb{R}^n$ must be diffeomorphic to $\mathbb{R}^n$; see Stallings [**Sta62**, Corollary 5.2].

There exist compact topological 4-manifolds which admit inequivalent smooth structures:

THEOREM 1.53 (Donaldson [**Don87**]). *There exist two smooth, closed, non-diffeomorphic* 4*-manifolds whose underlying topological manifolds are homeomorphic.*

In fact, there exist compact topological 4-manifolds which admit an infinite number of inequivalent smooth structures:

THEOREM 1.54 (Okonek and Van de Ven [**OVdV86**], Friedman and Morgan [**FM88a, FM88b**]). *There exist infinitely many smooth, closed, non-diffeomorphic* 4*-manifolds whose underlying topological manifolds are all homeomorphic.*

REMARK 1.55. For any closed topological n-manifold, where $n \geq 5$, there at most a finite number of inequivalent smooth structures; see Kervaire and Milnor [**KM63**] and Kirby and Siebenmann [**KS77**].

The problem of understanding the set of distinct smooth structures on a topological 4-manifold is called the **botany problem**; see Fintushel and Stern [**FS09**].

1.5.3. Smoothability of topological 4-manifolds.

We say that a smooth 4-manifold M^4 is **spin** if its second Stiefel–Whitney class vanishes: $w_2(M^4) = 0$.[22] This implies that the intersection form Q_M of M^4 is even.

We say that a topological 4-manifold $\mathcal{M}^4$ is **smoothable** if there exists a smooth 4-manifold M^4 such that $|M^4| = \mathcal{M}^4$; that is, the underlying topological manifold of M^4 is $\mathcal{M}^4$.

A longstanding problem in 4-manifold topology, known as the **geography problem**, is to classify those unimodular lattices that can arise as the intersection forms of closed simply-connected *smooth* 4-manifolds; see [**FS09**].

First, we have the following obstruction to smoothability due to Rokhlin:

THEOREM 1.56 ([**Rok52**]). *If a closed oriented topological* 4*-manifold is smoothable and spin, then its signature must be divisible by* 16.

For example, there does not exist a closed oriented simply-connected smooth 4-manifold whose intersection form is isomorphic to E_8, although by Freedman's theorem there exists such a topological 4-manifold; see [**Fre82**, Theorem 1.7].

[22] For a discussion of spin orbifolds, see Belgun, Ginoux, and Rademacher [**BGR07**].

Second, the geography problem for closed simply-connected 4-manifolds with positive-definite intersection form was answered by Donaldson:

THEOREM 1.57 ([**Don83**]). *If M^4 is a closed oriented simply-connected smooth 4-manifold with positive-definite intersection form Q_M, then Q_M is diagonalizable over the integers to the identity matrix.*

On the other hand, the unimodular lattice $k[1]$ can be realized as the intersection form of the closed oriented simply-connected smooth manifold $\#^k \mathbb{C}P^2$.

EXAMPLE 1.58 (Non-smoothable 4-manifold with signature 16). By Donaldson's theorem, the unimodular lattice $2E_8 := E_8 \oplus E_8$, which is positive-definite with rank and signature equal to 16, is not the intersection form of a closed oriented simply-connected smooth 4-manifold.

On the other hand, if a unimodular lattice Q is indefinite and odd, then by Theorem 1.44(1), $Q = m[+1] \oplus n[-1]$, so that $(\#^m \mathbb{C}P^2)\#(\#^n \overline{\mathbb{C}P^2})$ realizes Q as its intersection form.

Thus, the geography problem is reduced to the case of *even indefinite* unimodular lattices. Furthermore, by Theorems 1.44(2) and 1.56, we may assume that the unimodular lattice Q is of the form

$$Q = -2mE_8 \oplus nH, \quad \text{where } m \geq 0 \text{ and } n \geq 1. \tag{1.25}$$

If $n \geq 3m$, then the connected sum of m copies of the $K3$ surface and $n - 3m \geq 0$ copies of $S^2 \times S^2$ realizes Q in (1.25) as its intersection form. In fact, any topological 4-manifold $\mathcal{M}^4$ with intersection form Q is smoothable because, by Freedman's theorem (see [**Fre82**, Theorem 1.5]), it is homeomorphic to the connected sum of m copies of the $K3$ surface and $n - 3m \geq 0$ copies of $S^2 \times S^2$. Observe that, in this case, $b_2(\mathcal{M}^4) = 16m + 2n$ and $|\sigma(\mathcal{M}^4)| = 16m$. Hence, the inequality that $n \geq 3m$ is equivalent to $b_2(\mathcal{M}^4) \geq \frac{11}{8}|\sigma(\mathcal{M}^4)|$.

The following conjecture of Matsumoto [**Mat82**], asserts that the converse is true. This would resolve the remaining open cases of the geography problem.

CONJECTURE 1.59 (The 11/8-conjecture). *Any smooth closed 4-manifold M^4 with indefinite intersection form that is even (and spin, if M^4 is not simply-connected) satisfies*

$$b_2(M^4) \geq \frac{11}{8}|\sigma(M^4)|. \tag{1.26}$$

In this conjecture, the manifold M^4 is not assumed to be simply-connected. Seminal progress toward resolving this conjecture was made by Furuta [**Fur01**] using Seiberg–Witten theory. In particular, he proved that

$$b_2(M^4) \geq \frac{10}{8}|\sigma(M^4)| + 2.$$

This result was later improved by Hopkins, Lin, Shi, and Xu [**HLSX22**] to

$$b_2(M^4) \geq \frac{10}{8}|\sigma(M^4)| + 4$$

provided that M^4 is simply connected and not homeomorphic to S^4, $S^2 \times S^2$, or the $K3$ surface.

1.5.4. Extending the notion of topological surgery.

In this subsection, we discuss some topological surgeries that could arise in a hypothetical Ricci flow with surgery in dimension 4. This includes traditional k-surgeries, quotient k-surgeries, smooth blowdowns, and other potential topological modifications. In particular, a 4-dimensional Ricci flow with surgery is expected to introduce geometric surgeries that produce topologically novel transformations. In §1.5.5, we will briefly indicate how Ricci flow with surgery might realize these surgeries as topological transformations.

1.5.4.1. *Traditional topological surgeries.*

The definition of topological surgery, which we consider in the category of smooth orbifolds with isolated singularities, was introduced by Milnor [**Mil61**] and Wallace [**Wal60**].[23]

In this subsection, $\mathcal{O}^n$ denotes a smooth oriented n-dimensional orbifold with isolated singularities. Let S^k denote the standard smooth k-dimensional sphere, and let B^ℓ denote the open ℓ-dimensional Euclidean ball. A map $\phi : S^k \times B^{n-k} \to \mathcal{O}^n$ is called a **smooth embedding** if it is a smooth diffeomorphism onto its image and if ϕ extends to a smooth map $\overline{\phi} : S^k \times \overline{B}^{n-k} \to \mathcal{O}^n$ with the restriction of $\overline{\phi}$ to the boundary $S^k \times S^{n-k-1}$ also a smooth embedding, where $\overline{B}^{n-k}$ denotes the closed $(n-k)$-ball.

DEFINITION 1.60 (k-Surgery). A **k-surgery** on a smooth oriented n-dimensional orbifold $\mathcal{O}^n$ with isolated singularities, where $0 \leq k \leq n-1$, removes a smoothly embedded $S^k \times B^{n-k}$ from the regular set of the orbifold, where the induced orientation on $S^k \times B^{n-k}$ is the standard orientation,[24] and replaces it by $B^{k+1} \times S^{n-k-1}$. The replacement (or 'gluing') occurs along their shared smooth boundary $S^k \times S^{n-k-1}$.[25]

Surgery provides the foundation for the h-cobordism theorem in the smooth category and the s-cobordism theorem in the topological category; see Milnor [**Mil65**] and Wall [**Wal16**].

DEFINITION 1.61. We say that closed oriented smooth 4-manifolds M_1^4 and M_2^4 are **oriented cobordant** if there exists a compact oriented smooth 5-manifold W^5 such that $\partial W = \overline{M_1^4} \sqcup M_2^4$.

We have that M_1^4 and M_2^4 are oriented cobordant if and only if there exists a finite sequence of surgeries that transforms M_1^4 to M_2^4, where each 0-surgery is orientation preserving; see [**GS99**, Exercise 5.2.2(d)].

1.5.4.2. *Topological quotient* 3*-surgeries.*

In view of potential Ricci flow singularity formation, it is natural to extend the topological notion of a 3-surgery (associated with a Ricci flow neckpinch) as follows:

DEFINITION 1.62 (Quotient 3-Surgery). Let $\mathcal{O}^4$ be a smooth oriented 4-dimensional orbifold with isolated singularities. A **quotient 3-surgery** on $\mathcal{O}^4$ removes a smoothly embedded $(S^3/\Gamma) \times B^1$ from $\mathcal{O}^4$ and replaces it by $(B^4/\Gamma) \times S^0$, where $S^0 = \{\pm 1\}$. The gluing occurs along their shared smooth boundary $(S^3/\Gamma) \times S^0$.

---

[23] For the relationship between surgery and Morse theory, see Milnor [**Mil65**].

[24] In other words, we have an orientation-preserving embedding of $S^k \times B^{n-k}$ into the regular set of $\mathcal{O}^n$.

[25] Such a surgery was originally termed a *surgery of type* $(k+1, n-k)$ by Milnor.

In particular, a quotient 3-surgery removes a quotient 3-cylinder and replaces it by two quotient 4-balls, and hence introduces two orbifold points with isotropy groups isomorphic to Γ; see Figures 1.9 and 1.10 for quotient 3-surgery in the context of Ricci flow, which will be discussed in §1.5.5.2.a below. When $\Gamma = 1$, we obtain a traditional 3-surgery.

FIGURE 1.9. Left: An embedded S^3/Γ in a quotient neck region. Right: Neckpinch at the singularity time.

FIGURE 1.10. Left: Removed $(S^3/\Gamma) \times B^1$. Right: Attached $B^4/\Gamma \times S^0$. This introduces two orbifold points with Γ isotropy group.

1.5.4.3. *Topological prophylactic 2-surgery.*

Given the potential complexity with which the $S^2 \times \mathbb{R}^2$ singularity model might form (discussed in §1.5.5.2.b, 1.5.5.2.e, 1.5.5.2.f below), it is reasonable to anticipate that extending the definition of 2-surgery as follows could prove beneficial.

DEFINITION 1.63 (Prophylactic 2-surgery). Let Σ^2 be a smooth compact surface with nonempty boundary (which is a disjoint union of circles). A **prophylactic 2-surgery**[26] on $\mathcal{O}^4$ removes a smoothly embedded $S^2 \times (\Sigma^2)^\circ$ from $\mathcal{O}^4$ and replaces it by $B^3 \times \partial\Sigma$; see Figures 1.11 and 1.12.

If Σ^2 is a closed 2-ball, then we obtain a traditional 2-surgery.

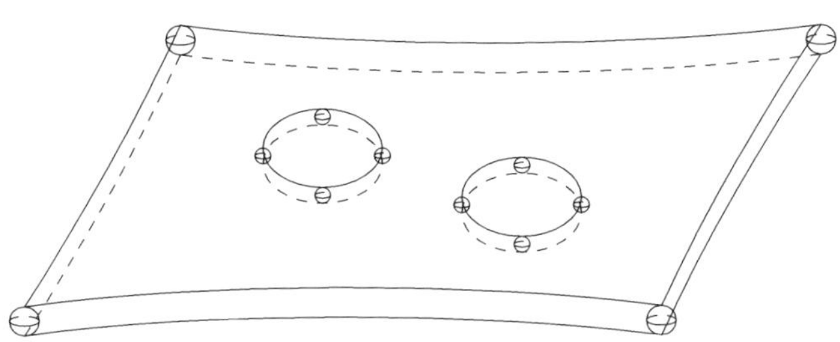

FIGURE 1.11. $S^2 \times (\Sigma^2)^\circ$ is removed, where $\partial\Sigma$ consists of two circles.

1.5.4.4. *Smooth blowdown.*

In view of the formation of the FIK shrinking soliton singularity model, we consider the following:

DEFINITION 1.64 (Smooth blowdown). A **smooth blowdown** on $\mathcal{O}^4$ removes a smoothly embedded $\overline{\mathbb{C}P^2} \setminus \overline{B}^4$ and replaces it by B^4.

The topological effect of a smooth blowdown is to remove a $\overline{\mathbb{C}P^2}$ topological summand; see Figure 1.13.

[26]We would like to thank Mike Freedman for suggesting this colorful terminology.

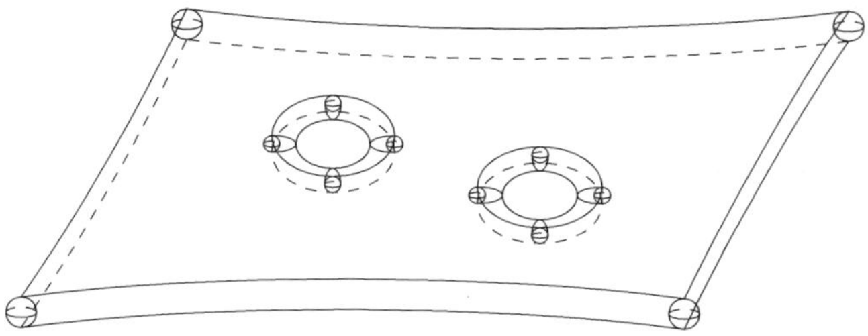

FIGURE 1.12. Attached $B^3 \times \partial\Sigma$.

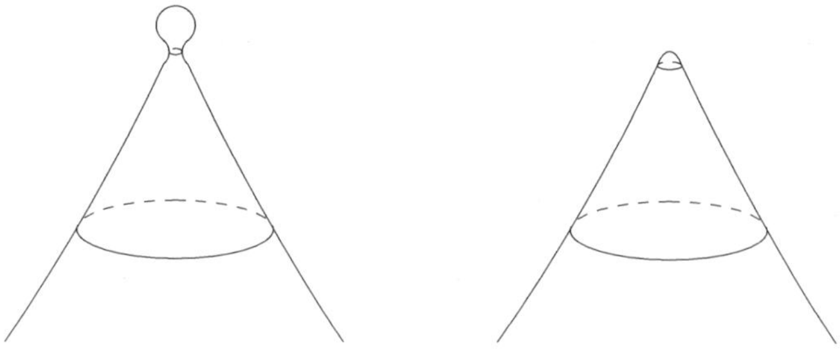

FIGURE 1.13. A smooth blowdown. Left: Pre-surgery. Right: Post-surgery.

1.5.4.5. *Topological surgeries resolving conical singularities.*

Geometrically, surgeries should exist which correspond to the formation of asymptotically conical singularity models, combined with the Bamler–Chen resolution via expanding solitons as given in Theorem 1.40. Topologically, the general framework is as follows:

Let $\mathcal{N}^4$ be an open smooth 4-orbifold with isolated singularities embedded in $\mathcal{O}^4$ with boundary $\partial\mathcal{N}$ diffeomorphic to the connected sum

$$\#_{i=1}^k (S^3/\Gamma_i)\#(\#^\ell(S^2 \times S^1)).$$

We assume that $\mathcal{N}^4$ admits a complete Riemannian metric with positive scalar curvature that is asymptotically conical. To any compact 4-orbifold $\mathcal{P}^4$ with $\partial\mathcal{P} = \partial\mathcal{N}$, we can associate a surgery on $\mathcal{O}^4$ by removing the smoothly embedded $\mathcal{N}^4$ and replacing it with $\mathcal{P}^4$.

1.5.5. Hypothetical 4-dimensional Ricci flow with surgery.

In this subsection, we discuss some topological-geometric surgeries that hypothetical Ricci flow with surgery in dimension 4 might produce. The following discussion on Ricci flow with surgery in four dimensions is heuristic and included here to suggest to the reader that such a theory may eventually be developed with Bamler's higher-dimensional theory as a foundation. While we hope it captures some useful intuition, it should be taken with a grain of salt, as our treatment here remains informal.

1.5.5.1. *Contrasting topological monotonicity for Ricci flow with surgery in dimensions* 3 *and* 4.

We first observe some essential differences between the potential behavior of hypothetical 4-dimensional Ricci flow with surgery and the established behavior of

3-dimensional Ricci flow with surgery. From a topological perspective, surgeries in 3-dimensional Ricci flow result only in connected sum decompositions or the removal of $S^2 \times S^1$ or $\mathbb{R}P^3$ topological summands. This fact, which we interpret as a form of topological monotonicity, is instrumental in the proof of the Geometrization Conjecture. In contrast, in dimension 4, Ricci flow with surgery does not appear to satisfy topological monotonicity, except perhaps in certain limited senses that we touched on above. By "topological monotonicity," we refer to the informal idea that after Ricci flow surgery, the resulting 4-orbifold is topologically simpler, or at least not more complex, than the 4-orbifold just before surgery.

As an example of this lack of topological monotonicity for hypothetical 4-dimensional Ricci flow with surgery, consider the following situation. Given any 2-sphere K embedded in a closed 4-orbifold $\mathcal{O}^4$ with a topologically trivial normal bundle, a.k.a. **2-knot**, there should exist an initial Riemannian metric g_0 such that the 2-sphere K contracts to a point under the Ricci flow.[27] A potential Ricci flow with surgery resolving such a singularity would topologically effect a 2-surgery on this 2-knot in $\mathcal{O}^4$. For example, if the initial 4-orbifold $\mathcal{O}^4$ is a manifold diffeomorphic to S^4, then the 4-manifold which the 2-surgery produces has its fundamental group isomorphic to the 2-knot group $\pi(K) \cong \pi_1(S^4 \setminus K)$.[28]

Thus, 4-dimensional Ricci flow with surgery can increase the topological complexity of a 4-orbifold, at least from the perspective of the fundamental group. In contrast, 1-knots in Riemannian 3-manifolds have no intrinsic curvature, and under the 3-dimensional Ricci flow, they do not contract to points. Hence, Ricci flow with surgery on 3-manifolds does not produce 1-surgeries.

1.5.5.2. *How geometric-topological surgeries might arise in* 4*-dimensional Ricci flow.*

In this subsection, we itemize some hypothetical 4-dimensional Ricci flows with surgery on closed 4-manifolds and, more generally, closed 4-orbifolds with isolated singularities.

1.5.5.2.a. *Geometric quotient* 3*-surgeries.*

Corresponding to the topological quotient 3-surgeries that we discussed in §1.5.4.2, there should be **quotient neckpinch** singularities, which are analogous to the neckpinch singularity. Let $\mathcal{O}^4$ be a 4-orbifold with isolated singularities, and let $\mathcal{N}$ be an open subset of $\mathcal{O}^4$ diffeomorphic to $(S^3/\Gamma) \times B^1$, called a topological quotient neck. Such an $\mathcal{N}$ exists whenever there is an embedded S^3/Γ in $\mathcal{O}^4$. Endow $\mathcal{O}^4$ with a smooth Riemannian metric g_0 such that its restriction to $\mathcal{N}$ forms a long neck, which is very thin in the middle, and where the geometry of g_0 is well-controlled throughout $\mathcal{O}^4$. If g_0 is chosen appropriately, the Ricci flow $g(t)$ with initial metric g_0 *should* develop a quotient neckpinch singularity at some finite time; see Figure 1.9.

More generally, we expect singularity formation where the neck pinches simultaneously along a subset of B^1, which could be complex, potentially even a Cantor-type set. In such a region, all associated singularity models are of the form $(S^3/\Gamma) \times \mathbb{R}$. One expects that Perelman's methods, which have been successful in

[27] As of the time of writing, this expectation remains unproven. But its verification should be independent of the development of any geometrization program.

[28] Since the collection of 2-knots in S^4 includes **spun knots** (see Artin [**Art25**], Epstein [**Eps60**], Zeeman [**Zee65**]), the set of knot groups for 2-knots in S^4 includes all knot groups of 1-knots in S^3.

dimension 3, should effect a quotient 3-surgery and address the issue of continuing the Ricci flow past these singularities. Note that such a surgery produces two orbifold points with isotropy group Γ; see Figure 1.10. So, even if one starts with the Ricci flow on a smooth closed 4-*manifold*, the Ricci flow with surgery can produce an underlying space that is a 4-*orbifold* with isolated singularities. Observe also that this surgery either increases the number of connected components of $\mathcal{O}^4$ by one or leaves the number of components unchanged.[29]

EXAMPLE 1.65 (Cohomogeneity-one Q_8-neckpinch). Suppose that a cohomogeneity-one Ricci flow on S^4 forms a Q_8-neckpinch singularity, as conjectured in Open Problem 1.5. An associated Ricci flow will produce a geometric surgery at the singular time with the following topological effect: Cut along a regular orbit diffeomorphic to S^3/Q_8 to obtain normal disk bundles over two Veronese embeddings of $\mathbb{R}P^2$ into S^4. Attach a B^4/Q_8 to each of these two resulting smooth 4-manifolds with boundary S^3/Q_8. Denote each of the two underlying smooth 4-orbifolds by $\mathcal{O}^4$. Open Problem 2.7 asks if they each admit shrinking soliton structures; see Figure 1.14.

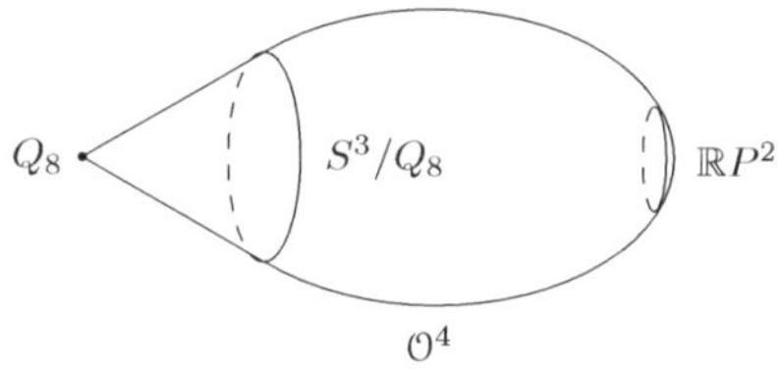

FIGURE 1.14. The orbifold $\mathcal{O}^4$.

1.5.5.2.b. *Geometric prophylactic 2-surgery.*

Ricci flow should also produce the prophylactic 2-surgeries discussed in §1.5.4.3 that generalize traditional 2-surgeries on 4-manifolds. Consider $\mathcal{O}^4$ with a smoothly embedded $S^2 \times \Sigma^2$, where Σ^2 is a connected compact surface with boundary $\partial\Sigma$. Endow $\mathcal{O}^4$ with a Riemannian metric g_0 such that its restriction to $S^2 \times \Sigma^2$ is a warped product of S^2 over Σ^2, where the 2-spheres are uniformly small and very small over some subset of Σ^2. One can imagine the Ricci flow forming a singularity where, at the singularity time, the 2-sphere fibers pinch to points simultaneously along some other subset of Σ^2. Here, we assume that the only singularity models near such a singularity region are isometric to $\mathbb{S}^2 \times \mathbb{R}^2$.

We speculate that there exists a finite collection of disjoint connected compact subsurfaces $\{\Sigma_i\}$ of Σ^2 such that performing a geometric prophylactic 2-surgery along the surface $\bigsqcup_i \Sigma_i^2$ with boundary is appropriate, where each $S^2 \times \Sigma_i^2$ is removed and replaced by $B^3 \times \partial\Sigma_i$. According to Bamler [**Bam**], there could be more complicated singularity formations along these lines, involving the coalescence of subsurfaces, which should also be tractable via suitable surgery.

However, in general, the mere existence of an $\mathbb{S}^2 \times \mathbb{R}^2$ singularity model may indicate the simultaneous presence of qualitatively different types of singularity

---

[29] An example of the latter case is when a component of $\mathcal{O}^4$ is $(S^3/\Gamma) \times S^1$, and the surgery produces the suspension of S^3/Γ, thereby preserving the number of components of $\mathcal{O}^4$.

models (for example, where $\mathbb{S}^2 \times \mathbb{R}^2$ is a limit at spatial infinity of another singularity model). Presumably, addressing such issues is necessary for the development of a successful 4-dimensional surgery theory. See §1.5.5.2.f below.

1.5.5.2.c. *Smooth flowdown.*

Isenberg, Knopf, and Šešum [**IKŠ19**] proved the existence of non-Kähler Ricci flows forming the FIK shrinker as a singularity model.

More generally, for each instance of an FIK singularity model formation, there should be an associated continuation of the flow beyond the singularity, where a subset diffeomorphic to the FIK is removed and the U(2)-invariant expanding soliton is grafted in to resolve the singularity.

Since the underlying space of the FIK shrinker is $\mathbb{C}^2$ blown up at one point and since the U(2)-invariant expander is defined on $\mathbb{C}^2$, we have that, topologically, the resulting surgery is a smooth blowdown, as discussed in §1.5.4.4.

As an aside, any 4-manifold forming an FIK singularity model under Ricci flow contains a $\overline{\mathbb{C}P^2}$ topological summand and hence cannot be spin.

1.5.5.2.d. *Geometric surgeries resolving conical singularities.*

More generally, Stolarski [**Sto22**] has shown that in any dimension, every asymptotically conical shrinking gradient Ricci soliton occurs as a singularity model.

In dimension 4, consider the formation of an asymptotically conical singularity model $\mathcal{S}$. Its asymptotic cone $C(N^3, h)$ has its link (N^3, h) a closed 3-manifold with positive scalar curvature. By Perelman's theorem [**Per03a**], the link N^3 is diffeomorphic to a connected sum of spherical space forms and $S^2 \times S^1$s:

$$N^3 \cong \left(\#_{i=1}^k (S^3/\Gamma_i) \right) \# \left(\#^\ell (S^2 \times S^1) \right)$$

for some non-negative integers k, ℓ with $k + \ell \geq 1$ and where each Γ_i is a finite subgroup of SO(4) acting freely on S^3 by isometries. By a standard result, N^3 is cobordant to the disjoint union

$$N^3 \overset{\text{cob}}{\sim} \bigsqcup_{i=1}^{k} (S^3/\Gamma_i) \sqcup (S^2 \times S^1)^{\sqcup k}.$$

In turn, each $S^2 \times S^1$ is cobordant to the empty set $\emptyset$, with a choice of cobordism being either $B^3 \times S^1$ or $S^2 \times B^2$. Similarly, for each S^3/Γ_i, we may bound it by the 4-dimensional orbifold B^4/Γ_i. Thus, we may consider the asymptotic slice N^3 of the singularity model $\mathcal{S}$ as being *orbifold cobordant* to the empty set $\emptyset$, where the (4-dimensional) cobordism has k singular points, with associated local groups isomorphic to Γ_i, $i = 1, \ldots, k$.

To address the challenge of formulating 4-dimensional Ricci flow with surgery for asymptotically conical singularity formation, Bamler and Chen [**BC23**] considered the case where (N^3, h) has the property that N^3 is diffeomorphic to a spherical space form in addition to h having positive scalar curvature. They proved the existence of an expanding gradient Ricci soliton on a 4-orbifold with isolated singularities that is asymptotic to the cone $C(N^3, h)$. They conjecture that their result should be extendable to the general case where (N^3, h) is any closed oriented 3-manifold with positive scalar curvature. A positive resolution of their conjecture would provide the framework needed to address conical singularity formation in 4-dimensional Ricci flow with surgery.

1.5.5.2.e. *Geometric surgery resolving a quotient 3-cylindrical steady soliton singularity model.*

Suppose that a 4-dimensional Ricci flow $(\mathcal{O}^4, g(t))$ forms a quotient 3-cylindrical steady soliton singularity model $(M^4_\infty, g_\infty(t))$. For example, for $\Gamma \cong 1$, there is the Bryant soliton, which can form as a singularity model, and for $\Gamma \cong \mathbb{Z}_k$, where $k \geq 3$, there are the Appleton cohomogeneity-one steady solitons (see Theorem 1.27), which may possibly form. Then the complement of some compact subset of M^4_∞ is covered by quotient ε-necks. One should be able to continue the flow beyond the singularity by performing a surgery: removing a suitable ε-neck, which is diffeomorphic to $(S^3/\Gamma) \times B^1$, and replacing it by two caps, each diffeomorphic to B^4/Γ.

We remark that if M^4_∞ is spin and $\Gamma \cong \mathbb{Z}_k$, then k is even. Moreover, by attaching $B^4/\mathbb{Z}_k$ to M^4_∞, we obtain a closed 4-orbifold $\mathcal{O}^4_\infty$ with a single orbifold point with isotropy group $\mathbb{Z}_k$. Using Theorem 6.77, one can show that $\mathcal{O}^4_\infty$ admits a smooth Riemannian metric with positive scalar curvature.

1.5.5.2.f. *The problem of formulating geometric surgeries for additional singularity models.*

In general, the prototypical Ricci flow singularity models are shrinking and steady Ricci solitons. In the previous paragraphs, we have considered 4-dimensional Ricci flow singularity formations where the associated singularity models are $(\mathbb{S}^3/\Gamma) \times \mathbb{R}$, $\mathbb{S}^2 \times \mathbb{R}^2$, $(\mathbb{S}^2 \times \mathbb{R}^2)/\mathbb{Z}_2$, the FIK shrinking soliton, a general 4-dimensional asymptotically conical shrinking soliton, the BCCD shrinking soliton, and a quotient 3-cylindrical steady soliton.

In the penultimate case above, we have, as singularity models, both the BCCD shrinker itself and $\mathbb{S}^2 \times \mathbb{R}^2$ occurring as limits at infinity of the BCCD shrinker. Furthermore, in the last case, we have, as singularity models, both a quotient 3-cylindrical steady soliton and the $(\mathbb{S}^3/\Gamma) \times \mathbb{R}$ shrinking soliton as a blowdown at spatial infinity of the steady soliton. Note that any 4-manifold forming a BCCD singularity model under Ricci flow contains a $\overline{\mathbb{C}P^2}$ topological summand and hence cannot be spin.

However, given a singularity formation in a region of the Ricci flow, nonuniqueness of the associated singularity models may arise in more complex ways. These cases will be the most challenging to address when formulating Ricci flow with surgery. We highlight some difficulties here, without speculating too much on the potential structure of Ricci flow with surgery in such situations.

An example of a challenging case to formulate Ricci flow with surgery is where a singularity has the product of $\mathbb{R}$ and the 3-dimensional Bryant soliton Σ^3 as one of its associated singularity models. In order to perform surgeries in such a case, it would be helpful to obtain a strong *canonical neighborhood theorem* for 4-dimensional Ricci flow.[30] Note that, in such a case, $\mathbb{S}^2 \times \mathbb{R}^2$ is an associated singularity model of the singularity formation, obtained by blowdown. Examples of such conjectured singularity models are Lai's 4-dimensional $\mathrm{O}(3) \times \mathbb{Z}_2$-symmetric, but not $\mathrm{O}(4)$-symmetric, steady gradient Ricci solitons. These examples are still weakly asymptotic to $\Sigma^3 \times \mathbb{R}$, so Ricci flow surgery can be understood using the $\Sigma^3 \times \mathbb{R}$ model.

[30] By the work of Hamilton, Perelman, and Brendle, a strong canonical neighborhood theorem holds in dimension 3.

This leads to the following important question: If a singularity model has $\mathbb{S}^2 \times \mathbb{R}^2$ as a *non-unique* limit at infinity, must it be weakly asymptotic to $\Sigma^3 \times \mathbb{R}$?

Note that if we consider the special case of steady gradient Ricci solitons under the restrictive condition that they strongly dimension reduce, then the answer to this question is yes; see Deng, Ma, and one of the authors [**CDM22**].

1.5.5.2.g. *Summary of hypothetical Ricci flow surgeries.*

To summarize, topologically, Ricci flow in dimension 4 is expected to perform quotient 3-surgeries, 2-surgeries along surfaces with boundaries, smooth blowdowns (which do not occur in the spin case), and possibly novel surgeries as described above, but not 1-surgeries (cf. Hamilton [**Ham95b**, §3] for this last fact).[31]

1.5.6. Ricci flow surgeries on 4-manifolds and topological invariants. Given a hypothetical Ricci flow with surgery defined on a finite time interval $[t_1, t_2]$, let $\mathcal{O}_t^4$ denote the 4-dimensional smooth orbifold with isolated singularities at time t. Here, we allow the Ricci flow with surgery to instantaneously switch to a new smooth metric at a finite number of times. We say that $\mathcal{O}_{t_2}^4$ is **flowbordant** to $\mathcal{O}_{t_1}^4$, and we call the space-time of the Ricci flow a **flowbordism**. Note that the relation of being flowbordant is reflexive and transitive, but not symmetric. In this language, an objective is to understand the topological effects of flowbordisms on the underlying orbifolds.

1.5.6.1. *The signature to second-Betti-number ratio under hypothetical Ricci flow with surgery: First difficulties.*

Recall that the 11/8-conjecture says that the **signature-to-second-Betti-number ratio** satisfies

$$\mathcal{R} := \frac{|\sigma|}{b_2} \leq \frac{8}{11}, \tag{1.27}$$

where b_2 denotes the second Betti number and σ denotes the signature. Ideally, under Ricci flow, if $\mathcal{R} \leq 8/11$ holds post-surgery, one might hope that this would imply that $\mathcal{R} \leq 8/11$ holds pre-surgery as well. A sufficient condition for this implication is the monotonicity inequality:

$$\text{If } \mathcal{O}_2^4 \text{ is flowbordant to } \mathcal{O}_1^4, \text{ then } \mathcal{R}(\mathcal{O}_2^4) \geq \mathcal{R}(\mathcal{O}_1^4).$$

However, the situation is not quite this simple. In particular, let M^4 be a closed spin 4-manifold. Whenever there exists a 4-manifold N^4 embedded in M^4 that is diffeomorphic to the underlying smooth manifold of a Ricci-flat ALE 4-manifold $\mathcal{A}^4$, there is the possibility that $\mathcal{A}^4$ forms as a singularity model for some Ricci flow on M^4 that goes singular at some finite time $T > 0$.

In this case, Ricci flow with surgery should produce at time T the topological change of replacing N^4 by $\mathbb{R}^4/\Gamma$, where $\mathbb{R}^4/\Gamma$ models the unique end of $\mathcal{A}^4$. Hence, if there is a finite collection of 4-manifolds $\{N_i^4\}$ disjointly embedded in M^4 that are diffeomorphic to the underlying smooth manifolds of Ricci-flat ALE 4-manifolds $\mathcal{A}_i^4$, then some Ricci flow with surgery on M^4 might produce the topological change of replacing each N_i^4 by $\mathbb{R}^4/\Gamma_i$, where $\mathbb{R}^4/\Gamma_i$ models the end of $\mathcal{A}_i^4$.

This results in a 4-orbifold $\mathcal{O}^4$ with isolated singularities. In such a case, it is possible that the signature-to-second-Betti-number ratio is not monotone under

[31] Hamilton [**Ham95b**, §3] wrote: "In dimension 4, we expect the Ricci flow could perform surgeries $S^3 \times B^1 \to B^4 \times S^0$ and $S^2 \times B^2 \to B^3 \times S^1$, but not the reverse; this gives hope the Ricci flow may provide topological information on 4-manifolds."

Ricci flow with surgery. In particular, it is conceivable that Ricci flow with surgery, starting from a topological K3 surface, could produce the orbifold $T^4/\mathbb{Z}_2$, which has $\tau = 0$, $b_2 = 6$, and hence $\mathcal{R} = 0$, where 16 Eguchi–Hanson Ricci-flat ALEs have formed as singularity models (effectively, bubbled off). Thus, if we choose not to modify the definition of signature or Betti number for orbifolds, then we could not have monotonicity of the ratio $\mathcal{R}$ under Ricci flow with surgery.

We remark that analogous to this are the finite diameter degenerations of the $K3$ surface limiting to hyper-Kähler orbifolds with isolated singularities; see Bando, Kasue, and Nakajima [**BKN89**], Anderson [**And89**], and Kobayashi and Todorov [**KT87**].

1.5.6.2. *Monotonicity of the signature to second-Betti-number ratio under quotient 3-surgeries and prophylactic 2-surgeries.*

Examples of situations where we do have monotonicity of the ratio are the neck-pinch surgeries and the hole punch surgeries, corresponding to 3- and 2-cylinder singularity models formation. Recall that two oriented manifolds are oriented cobordant if and only they are related by some sequence of k-surgeries with the condition that the 0-surgeries preserve the orientations. The signature of a closed oriented 4-manifold is an oriented cobordism invariant and hence invariant under k-surgeries; this is a result of Thom (see e.g. Milnor and Stasheff [**MS74**, Lemma 19.3]). On the other hand, with regard to the denominator b_2 of $\mathcal{R}$, the second Betti number is non-increasing under 2-surgeries and 3-surgeries. This last fact follows from the Mayer–Vietoris sequence for homology groups.

1.5.6.2.a. *The ratio under quotient 3-surgeries.*

Regarding quotient 3-surgeries, *suppose* that the embedded $(S^3/\Gamma) \times B^1$ separates $\mathcal{O}^4$ into two components $\mathcal{O}_1^4$ and $\mathcal{O}_2^4$. Although the additivity of the intersection forms formula (1.20) may not hold, as in the case $\Gamma = 1$, since S^3/Γ is a rational homology sphere, we have that $b_2(\mathcal{O}^4) = b_2(\mathcal{O}_1^4) + b_2(\mathcal{O}_2^4)$ and $\sigma(\mathcal{O}^4) = \sigma(\mathcal{O}_1^4) + \sigma(\mathcal{O}_2^4)$. The same formulas hold if we attach B^4/Γ to each of $\mathcal{O}_i^4$, $i = 1, 2$. Note that if $\frac{|\sigma(\mathcal{O}_i^4)|}{b_2(\mathcal{O}_i^4)} \leq \frac{8}{11}$ for $i = 1, 2$, then

$$\frac{|\sigma(\mathcal{O}^4)|}{b_2(\mathcal{O}^4)} \leq \frac{|\sigma(\mathcal{O}_1^4)| + |\sigma(\mathcal{O}_2^4)|}{b_2(\mathcal{O}_1^4) + b_2(\mathcal{O}_2^4)} \leq \frac{8}{11}.$$

Situations where quotient a 3-surgery will be performed are when there exists a singularity model which is asymptotic to a cylinder $(S^3/\Gamma) \times \mathbb{R}$. For example, when an Appleton steady soliton is a singularity model, in which case $\Gamma = \mathbb{Z}_k$, where $k \geq 3$.

1.5.6.2.b. *The ratio under prophylactic 2-surgeries.*

Under a prophylactic 2-surgery, the second Betti number is non-increasing and the signature is preserved.[32] To see this, suppose that $\mathcal{O}^4 = (\Sigma^2 \times S^2) \cup \mathcal{W}^4$, where $\partial \mathcal{W} = \partial \Sigma \times S^2 = \partial(\Sigma^2 \times S^2)$. By Mayer–Vietoris, with respect to any coefficient ring, which we take to be $\mathbb{Q}$, we have the long exact sequence

$$\cdots \to H_2(\partial\Sigma \times S^2) \cong H_2(\partial \mathcal{W}) \overset{(i_*, j_*)}{\longrightarrow} H_2(\Sigma^2 \times S^2) \oplus H_2(\mathcal{W}) \to H_2(\mathcal{O}^4) \to \cdots,$$

where the map $i_* : H_2(\partial\Sigma \times S^2) \to H_2(\Sigma^2 \times S^2)$ is an isomorphism. Hence

$$b_2(\mathcal{O}^4) \geq b_2(\mathcal{W}^4). \tag{1.28}$$

[32]We would like to thank Feng Luo for showing us the arguments in this paragraph.

Moreover, let $\widetilde{\mathcal{O}}^4 = (\partial\Sigma \times B^3) \cup \mathcal{W}$, where $\partial\mathcal{W} = \partial(\partial\Sigma \times B^3)$. By Mayer–Vietoris,

$$\cdots \to H_2(\partial\Sigma \times B^3) \oplus H_2(\mathcal{W}^4) \to H_2(\widetilde{\mathcal{O}}^4) \to H_1(\partial\Sigma \times S^2) \xrightarrow{(i_*,j_*)} H_1(\partial\Sigma \times B^3) \oplus H_1(\mathcal{W}^4) \to \cdots,$$

where the map $i_* : H_1(\partial\Sigma \times S^2) \to H_1(\partial\Sigma \times B^3)$ is an isomorphism and $H_2(\partial\Sigma \times B^3) = 0$. Hence $b_2(\mathcal{W}^4) \geq b_2(\widetilde{\mathcal{O}}^4)$. We conclude that $b_2(\mathcal{O}^4) \geq b_2(\widetilde{\mathcal{O}}^4)$. Furthermore, $\widetilde{\mathcal{O}}^4$ is cobordant to $\mathcal{O}^4$. Hence, the signature is preserved under a prophylactic 2-surgery.

1.5.6.3. *Positive scalar curvature and the Lichnerowicz vanishing theorem: Implications for compact and noncompact singularity models.*

If a closed spin Riemannian 4-manifold (M^4, g) satisfies $R > 0$, then $\widehat{A}(M^4) = 0$ by the Lichnerowicz vanishing theorem [**Lic63**], where $\widehat{A}$ denotes the **A-hat genus** (see Hirzebruch [**Hir95**, p. 197], Lawson and Michelsohn [**LM89**, p. 138 and Chapter III, §11]). Since $\widehat{A}(M^4) = \frac{1}{8}\sigma(M^4)$, a fact specific to dimension 4, this implies that the signature $\sigma(M^4) = 0$.

Recall from Theorem 1.14 that if a singularity model of a connected component of a Ricci flow has its underlying manifold being compact, then the singularity model must be a shrinking gradient Ricci soliton. Such a singularity model must have $R > 0$ and hence, if it is spin, $\sigma = 0$ by the Lichnerowicz vanishing theorem. Bamler has highlighted the importance of this consequence for the 4-dimensional Ricci flow.

It will be interesting to understand in what way this fact generalizes to orbifolds with isolated singularities; perhaps this is in the literature. For the Atiyah–Singer index theorem on orbifolds, see Kawasaki [**Kaw81**]; for discussions of elliptic genera, including the $\widehat{A}$-genus, on orbifolds, see Borisov and Libgober [**BL03, BL02**] and Dong, Liu, and Ma [**DLM02**].

Moreover, any complete noncompact non-Ricci-flat singularity model of the Ricci flow must satisfy $R > 0$. Furthermore, the geometry and topology at spatial infinity of noncompact singularity models are anticipated to be well-controlled. Consequently, it is promising that the Lichnerowicz vanishing theorem may extend in some form to noncompact singularity models, potentially yielding control of the $\widehat{A}$-genus and signature.

1.5.6.4. *Tracking signature-to-second-Betti-number ratio under Ricci-flat ALE bubbling.*

As a possible scenario, in this subsection we consider the case where only Ricci-flat ALE singularity models bubble and the final orbifold satisfies the Hitchin–Thorpe inequality. Let $\mathcal{O}^4$ be an orbifold with a finite number of orbifold points p_i and corresponding isotropy groups $\Gamma_i \subset \mathrm{SO}(4)$, $1 \leq i \leq k$. Suppose that, for each i, there exists a Ricci-flat ALE 4-manifold (M_i^4, g_i) asymptotic to $\mathbb{R}^4/\Gamma_i$; by a result of Kronheimer, this is the case if $\Gamma_i \subset \mathrm{SU}(2) \subset \mathrm{SO}(4)$.

By applying the Hitchin–Thorpe inequality to this ALE (see (2.215)–(2.216) below), we obtain

$$(1.29) \qquad \left|\sigma(M_i^4) + \eta(\mathbb{S}^3/\Gamma_i)\right| \leq \frac{2}{3}\left(\chi(M_i^4) - \frac{1}{|\Gamma_i|}\right) = \frac{2}{3}\left(b_2(M_i^4) + 1 - \frac{1}{|\Gamma_i|}\right),$$

where equality holds if and only if $W(g_i)_\pm \equiv 0$, i.e., g_i is (anti-)self-dual. In this case, the condition of being hyper-Kähler is equivalent to g_i being anti-self-dual.

For example, the Eguchi–Hanson hyper-Kähler ALE $\mathbb{EH}^4$ is asymptotic to $\mathbb{R}/\Gamma$, where $\Gamma = \mathbb{Z}_2$, and satisfies $\sigma(\mathbb{EH}^4) = -1$, $\eta(\mathbb{S}^3/\mathbb{Z}_2) = \eta(\mathbb{R}P^3) = 0$, and $b_2(\mathbb{EH}^4) = 1$, where equality holds in (1.29).

Before going further, first consider the example of the Kummer construction of the $K3$ surface. Namely, we have the $\mathbb{Z}_2$ action on T^4 generated by the involution $\mathbf{x} \to -\mathbf{x}$. The $K3$ surface is diffeomorphic to the resolution of the orbifold $T^4/\mathbb{Z}_2$, which has 16 orbifold points with isotropy group $\mathbb{Z}_2$, by attaching 16 Eguchi–Hanson Ricci-flat ALE 4-manifolds, each asymptotic at infinity to $\mathbb{R}^4/\mathbb{Z}_2$. There is the possibility of the existence of an initial metric on the $K3$ surface that forms k Eguchi–Hanson singularity models, where $1 \leq k \leq 16$. In this case, at the singular time, the underlying orbifold is diffeomorphic to $T^4/\mathbb{Z}_2$ with $16 - k$ Eguchi–Hanson ALEs attached, or equivalently, a $K3$ surface with k Eguchi–Hanson ALEs replaced by k $B^4/\mathbb{Z}_2$s.

More generally, consider a 2-complex dimensional Kähler orbifold $\mathcal{G}^4$ with isolated singularities, endowed with a hyper-Kähler (Ricci-flat) metric. The underlying complex manifold of such a 4-orbifold is called a **generalized $K3$ surface**. The isotropy groups of the orbifold points of $\mathcal{G}^4$ are finite subgroups of SU(2). Such a generalized $K3$ surface $\mathcal{G}^4$ can occur as a limit of a finite diameter degeneration of hyper-Kähler metrics on the $K3$ surface. Let p_i, $1 \leq i \leq k$, be the orbifold points of $\mathcal{G}^4$, and let $\Gamma_i \subset \mathrm{SU}(2)$ be the corresponding isotropy groups. There is the possibility that, for a suitable initial metric on the $K3$ surface, the Ricci flow bubbles off at the singular time the corresponding hyper-Kähler ALEs $\mathcal{A}_i$, which are asymptotic at infinity to $\mathbb{R}^4/\Gamma_i$.

Note that

$$\mathcal{G}^4 \setminus \bigcup_i \{p_i\} \sim \mathcal{G}^4 \setminus \bigcup_i B_i^4/\Gamma_i \cong K3 \setminus \bigcup_i \mathcal{A}_i,$$

where $\sim$ denotes homotopy equivalent and where $B_i^4 := B_\varepsilon(p_i)$, with $\varepsilon > 0$ small. We have the equality case of the Hitchin–Thorpe inequality for Einstein 4-orbifolds with boundary:[33]

$$(1.30)\quad \left|\sigma\left(\mathcal{G}^4 \setminus \bigcup_i B_i^4/\Gamma_i\right) - \sum_i \eta(S^3/\Gamma_i)\right| \leq \frac{2}{3}\left(b_2\left(\mathcal{G}^4 \setminus \bigcup_i B_i^4/\Gamma_i\right) + 2 + \sum_i \left(\frac{1}{|\Gamma_i|} - 1\right)\right).$$

For example, for $\mathcal{G}^4 = T^4/\mathbb{Z}_2$ and $\mathcal{N}^4 := \mathcal{G}^4 \setminus \bigcup_{i=1}^{16} B_i^4/\Gamma_i$, we have $\sigma(\mathcal{N}^4) = 0$, $\eta(S^3/\mathbb{Z}_2) = 0$, $b_2(\mathcal{N}^4) = 6$, and $|\Gamma_i| = 2$.

On the other hand, *suppose* that $\mathcal{P}^4$ is a 4-orbifold with isolated singularities and corresponding isotropy groups Γ_i, $1 \leq i \leq k$, that satisfies the Hitchin–Thorpe inequality:

$$(1.31)\quad \left|\sigma\left(\mathcal{P}^4 \setminus \bigcup_i B_i^4/\Gamma_i\right) - \sum_i \eta(S^3/\Gamma_i)\right| \leq \frac{2}{3}\left(b_2\left(\mathcal{P}^4 \setminus \bigcup_i B_i^4/\Gamma_i\right) + 2 + \sum_i \left(\frac{1}{|\Gamma_i|} - 1\right)\right).$$

[33] This is proved in essentially the same way as the Hitchin–Thorpe inequality for Ricci-flat ALE 4-manifolds (2.215), where here we take $\varepsilon \to 0$.

For example, (1.31) holds if $\mathcal{P}^4$ admits an Einstein metric. Further suppose that $\Gamma_i \subset \mathrm{SO}(4)$ and $\mathcal{A}_i$ are Ricci-flat ALE 4-manifolds asymptotic to $\mathbb{R}^4/\Gamma_i$.[34] Define

$$M^4 := \Big(\mathcal{P}^4 \setminus \bigcup_i B^4/\Gamma_i\Big) \cup \bigcup_i \mathcal{A}_i.$$

Now further suppose that the orbifold $\mathcal{P}^4$ arises from a Ricci flow originating on the manifold M^4, i.e., the Ricci flow with surgery produces bubbling of the Ricci-flat ALEs $\mathcal{A}_i$ to obtain $\mathcal{P}$. We also assume that the intersection form of M^4 is even and indefinite. By (1.29) and (1.31), it follows that

$$|\sigma(M^4)| \leq \frac{2}{3}\chi(M^4) = \frac{2}{3}(b_2(M^4) + 2), \tag{1.32}$$

i.e., any such M^4 satisfies the Hitchin–Thorpe inequality. The upshot of this discussion is that, for a smooth closed initial 4-manifold M^4, if Ricci flow with surgery produces only the bubbling of Ricci-flat ALEs, and the resulting orbifold satisfies the Hitchin–Thorpe inequality (e.g., by admitting an Einstein metric), then M^4 must satisfy the Hitchin–Thorpe inequality as well.

Observe that (1.32) implies the 11/8 conjecture for the manifold M^4. Indeed, by Theorem 1.56, we may assume $|\sigma(M^4)| \geq 16$. Combining this with (1.32) gives $b_2(M^4) \geq 22$, which, together with (1.32), implies (1.26).

A moral of the discussion above is that the Hitchin–Thorpe inequality bounds the signature of Einstein manifolds, and in particular it bounds the signature of Ricci-flat ALE bubbles, which yields a favorable inequality direction regarding Ricci-flat ALE bubbling under Ricci flow.

1.5.6.5. *Bamler's thick-thin decomposition for immortal solutions to the Ricci flow.*

In the spin case, recall that if a Ricci flow has finite-time extinction, then each connected component of the Ricci flow that has no orbifold points (i.e., is a manifold) will have vanishing signature. What can one say about the components that contain orbifold points?[35]

Otherwise, the Ricci with surgery will exist for all time. In this case, Bamler [**Bam20b**, §2.8] has proved a thick-thin decomposition result applicable to dimension 4 and analogous to the thick-thin decompositions of Hamilton [**Ham99**] and Perelman [**Per03a**] in dimension 3. In Bamler's theory, the 3-dimensional hyperbolic pieces of Hamilton and Perelman are replaced by 4-dimensional Einstein orbifolds with isolated singularities and finite volume.

This, invariably, will be a key component to any Ricci flow approach to the 11/8-conjecture. A natural question arises: How does the Hitchin–Thorpe inequality extend to such finite-volume Einstein 4-orbifolds? For some works on the Hitchin–Thorpe inequality for noncompact Einstein 4-manifolds, see Kotschick [**Kot98**], and Dai and Wei [**DW07**], and the references therein.

1.5.7. Type I singularities.

We end this subsection by mentioning some results that are known about singularity formation of the Ricci flow in the Type I case.

[34] If $\Gamma_i \subset \mathrm{SU}(2)$, then by Kronheimer's theorem, such Ricci-flat ALE 4-manifolds $\mathcal{A}_i$ exist with the additional property that they are hyper-Kähler.

[35] Keep in mind that if an Appleton steady soliton is coned off, its signature satisfies $|\sigma| = 1$. Does this orbifold admit a shrinking soliton structure?

Šešum [**Š06**] proved that compact singularity models associated to Type I singularities are shrinking gradient Ricci solitons; see Theorem 1.14 for a generalization of this. Naber [**Nab10**] proved that natural rescalings of Type I singularity models are shrinking gradient Ricci solitons. Enders, Müller, and Topping [**EMT11**] proved that the rescalings can be chosen so that the shrinking solitons are nonflat.

Recently, Conlon, Hallgren, and Ma [**CHM25**] proved that any singularity of the Kähler–Ricci flow on a closed complex surface is Type I, and therefore the associated singularity model is the FIK shrinker.

1.6. Foundational formulas and results

Having surveyed the role of Ricci solitons in Ricci flow, this section compiles a selection of standard formulas and results related to gradient Ricci solitons, as presented in Volume 1 of [**Cho23**]. These technical tools will be instrumental in studying solitons in greater depth.

When referencing a formula like (1.1) in Volume 1, we will denote it as (I-1.1). Similarly, a result like Theorem 1.1 in Volume 1 will be cited as Theorem I-1.1. We will use the Einstein summation convention throughout this book and generally we do not bother to raise indices.

1.6.1. Scalar formulas on gradient Ricci solitons.

Let $(\mathcal{O}^n, g, f, \lambda)$ be a gradient Ricci soliton on an orbifold with isolated singularities, so that

$$\operatorname{Ric}_f = \operatorname{Ric} + \nabla^2 f = \frac{\lambda}{2} g \tag{1.33}$$

by (1.2), where $\lambda = 1$, 0, or -1. By tracing this soliton equation, we obtain (cf. (I-2.39)):

$$R + \Delta f = \frac{n\lambda}{2}. \tag{1.34}$$

If $\lambda = \pm 1$, then we normalize the potential function by (cf. (I-2.45))

$$R + |\nabla f|^2 = \lambda f. \tag{1.35}$$

On the other hand, if $\lambda = 0$ and g is not Ricci-flat, then by Theorem 2.23 in Volume 1 we have

$$R > 0 \tag{1.36}$$

and we may normalize the metric so that (cf. (I-2.46))

$$R + |\nabla f|^2 = 1. \tag{1.37}$$

The **f-Laplacian** is the elliptic operator (cf. (I-2.50))

$$\Delta_f := \Delta - \nabla f \cdot \nabla. \tag{1.38}$$

The f-Laplacian of the potential function is given as follows. For a shrinking gradient Ricci soliton (cf. (I-2.54)),

$$\Delta_f f = \frac{n}{2} - f. \tag{1.39}$$

For a (normalized) non-Ricci-flat steady gradient Ricci soliton (cf. (I-2.56)),

$$\Delta_f f = -1. \tag{1.40}$$

For an expanding soliton (cf. (I-2.58)),

$$\Delta_f f = f - \frac{n}{2}. \tag{1.41}$$

By taking the divergence of (1.46) below, we have that the f-Laplacian of the scalar curvature is given by (cf. (I-2.60))

$$\Delta_f R = -2\,|\mathrm{Ric}|^2 + \lambda R. \tag{1.42}$$

1.6.2. The Bochner and f-Bochner formulas.

Given a Riemannian orbifold $\mathcal{O}^n$ with isolated singularities and a function u on $\mathcal{O}^n$, the fundamental Bochner formula is (see e.g. [**CC24**, Proposition 11.3]; cf. (I-1.9))

$$\frac{1}{2}\Delta|\nabla u|^2 = \langle \nabla u, \nabla \Delta u\rangle + |\nabla^2 u|^2 + \mathrm{Ric}(\nabla u, \nabla u). \tag{1.43}$$

Generalizing this, the f-Bochner formula says that for any function f on $\mathcal{O}^n$ we have (cf. (I-4.32))

$$\frac{1}{2}\Delta_f|\nabla u|^2 = |\nabla^2 u|^2 + \langle \nabla u, \nabla \Delta_f u\rangle + \mathrm{Ric}_f(\nabla u, \nabla u). \tag{1.44}$$

In particular, for any soliton we have that

$$\frac{1}{2}\Delta_f|\nabla u|^2 = |\nabla^2 u|^2 + \langle \nabla u, \nabla \Delta_f u\rangle + \frac{\lambda}{2}|\nabla u|^2. \tag{1.45}$$

As we will see, beginning in Chapter 2, the Bochner and f-Bochner formulas will be applied to $u = f$, the potential function of a Ricci soliton, as well as to f-harmonic functions u; i.e., functions satisfying $\Delta_f u = 0$.

1.6.3. Tensor formulas on Ricci solitons.

In this subsection, we recall some tensor formulas that Hamilton has derived for gradient Ricci solitons [**Ham93b**, §3], [**Ham95b**, §20].

By taking the divergence of (1.2), we have that (cf. (I-2.40))

$$2\,\mathrm{Ric}\,(\nabla f) = \nabla R. \tag{1.46}$$

We remark that substituting (1.2) back into the preceding formula yields equations (1.35) and (1.37). We also observe that the preceding formula is the trace of the following tensor formula (cf. (I-4.99)):

$$R_{ijk\ell}\nabla_\ell f = \nabla_i R_{jk} - \nabla_j R_{ik}. \tag{1.47}$$

Here, the components of the Riemann curvature tensor are defined by

$$R_{ijk\ell} := \left\langle \mathrm{Rm}\left(\frac{\partial}{\partial x^i}, \frac{\partial}{\partial x^j}\right)\frac{\partial}{\partial x^k}, \frac{\partial}{\partial x^\ell}\right\rangle. \tag{1.48}$$

By taking the divergence of (1.47), we obtain (cf. (I-4.104))

$$\begin{aligned} \Delta R_{ij} - \frac{1}{2}\nabla_i\nabla_j R + 2R_{kij\ell}R_{k\ell} - R_{ik}R_{jk} - \frac{\lambda}{2}R_{ij} &= (\nabla_k R_{ji} - \nabla_j R_{ki})\nabla_k f \\ &= R_{kij\ell}\nabla_k f \nabla_\ell f, \end{aligned} \tag{1.49}$$

where ΔR_{ij} denotes $(\Delta \mathrm{Ric})_{ij}$. Similarly, denote $\Delta_f R_{ij} := (\Delta_f \mathrm{Ric})_{ij}$. Using (1.49), we compute that

$$\begin{aligned}\Delta_f R_{ij} + 2R_{kij\ell}R_{k\ell} - \frac{\lambda}{2}R_{ij} &= \frac{1}{2}\nabla_i\nabla_j R + R_{ik}R_{jk} - \nabla_j R_{ki}\nabla_k f \\ &= \frac{\lambda}{2}R_{ij},\end{aligned}$$

where we used $\frac{1}{2}\nabla_i\nabla_j R = \nabla_j(R_{ik}\nabla_k f)$ and the soliton equation for the last equality. Therefore,

$$\Delta_f R_{ij} = \lambda R_{ij} - 2R_{kij\ell}R_{k\ell}. \tag{1.50}$$

We also write this equation in the rough form (cf. (I-4.82))

$$\Delta_f \mathrm{Ric} = \lambda \mathrm{Ric} - 2\,\mathrm{Rm} * \mathrm{Ric}, \tag{1.51}$$

where in general $*$ denotes the product with contractions of tensors and where the expression on the right-hand side satisfies $|\mathrm{Rm} * \mathrm{Ric}| \leq |\mathrm{Rm}|\,|\mathrm{Ric}|$.

For any soliton, we have (cf. (I-4.114))

$$\Delta_f R_{ijk\ell} - \lambda R_{ijk\ell} = 2(-B_{ijk\ell} + B_{ij\ell k} - B_{ikj\ell} + B_{i\ell jk}), \tag{1.52}$$

where $B_{ijk\ell} := -R_{ipjq}R_{kp\ell q}$ and $\Delta_f R_{ijk\ell} := (\Delta_f \mathrm{Rm})_{ijk\ell}$. We also write this in the abbreviated form

$$\Delta_f \mathrm{Rm} = \lambda \mathrm{Rm} + \mathrm{Rm} * \mathrm{Rm}, \tag{1.53}$$

where $|\mathrm{Rm} * \mathrm{Rm}| \leq 8|\mathrm{Rm}|^2$.

1.6.4. Scalar curvature and potential function estimates for Ricci solitons.

Firstly, we recall the scalar curvature lower bounds. By Theorem I-2.14, we have Chen's **[Che09]** result that for a complete shrinking or steady Ricci soliton we have

$$R \geq 0. \tag{1.54}$$

For a shrinking gradient Ricci soliton, if $R = 0$ at some point, then it is the Gaussian soliton. For a steady Ricci soliton, if $R = 0$ at some point, then it is Ricci-flat. For a shrinking gradient Ricci soliton, this further implies that $0 \leq |\nabla f|^2 \leq f$ by (1.35). For a complete expanding soliton, we have (see Pigola, Rimoldi, and Setti **[PRS11]**)

$$R \geq -\frac{n}{2}. \tag{1.55}$$

Thus, by (1.35), for an expanding soliton, we have

$$|\nabla f|^2 \leq -f + \frac{n}{2}. \tag{1.56}$$

Furthermore, for any shrinking gradient Ricci soliton which is not the Gaussian shrinker, we have by Theorem I-2.14 that

$$R > 0. \tag{1.57}$$

For any expanding soliton, either g is Einstein or $R > -\frac{n}{2}$.

Secondly, we recall the potential function upper and lower bounds and a scalar curvature growth estimate. For a shrinking gradient Ricci soliton, the minimum of f on $\mathcal{O}^n$ is attained; see Theorem I-4.3. Let o be a minimum point of f. By applying the elliptic maximum principle to (1.39), we obtain

$$f(o) \leq \frac{n}{2}. \tag{1.58}$$

Since $R \geq 0$, (1.35) yields $|\nabla f|^2 \leq f$. Integrating this last inequality along minimal geodesics emanating from the minimum point o yields in turn that (cf. (I-2.64))[36]

$$f(x) \leq \frac{1}{4}\left(d(x,o) + \sqrt{2n}\right)^2. \tag{1.59}$$

That is, the potential function has a quadratic (in distance) upper bound. Moreover, by (1.35), this yields a quadratic upper bound for the scalar curvature:

$$R(x) \leq \frac{1}{4}\left(d(x,o) + \sqrt{2n}\right)^2. \tag{1.60}$$

We have the following fundamental lower bound of Cao and Zhou [**CZ10**], where the improved constant was observed by Haslhofer and Müller [**HM11**] (cf. (I-4.7) and Theorem I-4.3)

$$f(x) \geq \frac{1}{4}\left((d(x,o) - 5n)_+\right)^2, \tag{1.61}$$

where $u_+ := \max\{u, 0\}$ denotes the non-negative part of a real number u. Let $\gamma : [0, r_x] \to M^n$ be a minimal geodesic from o to x, where $r_x := d(x,o)$. The proof of (1.61) uses the stability inequality:

$$(n-1)\int_0^{r_x} (\zeta')^2(s)\,ds \geq \int_0^{r_x} \zeta^2(s)\operatorname{Ric}(\gamma'(s), \gamma'(s))\,ds, \tag{1.62}$$

where $\zeta(0) = \zeta(r_x) = 0$, $\zeta(1) = \zeta(r_x - 1) = 1$, and $\zeta(s)$ is linear in between. Define $f(s) := f(\gamma(s))$. By the shrinking gradient Ricci soliton equation and integration by parts, we calculate that

$$\begin{aligned} 2(n-1) &\geq \int_0^{r_x} \zeta^2(s)\left(\frac{1}{2} - f''(s)\right)ds \\ &= \frac{1}{2}r_x - \frac{2}{3} + 2\int_0^1 sf'(s)\,ds - 2\int_{r_x-1}^{r_x} (r_x - s)f'(s)\,ds. \end{aligned} \tag{1.63}$$

Since

$$|f'(s)| \leq |\nabla f|(\gamma(s)) \leq \sqrt{f}(\gamma(s))$$

and since $\sqrt{f}$ has Lipschitz constant $\frac{1}{2}$, this yields

$$\begin{aligned} 2(n-1) \geq \frac{1}{2}r_x - \frac{2}{3} &- 2\int_0^1 s\left(\sqrt{f}(o) + \frac{s}{2}\right)ds \\ &- 2\int_{r_x-1}^{r_x} (r_x - s)\left(\sqrt{f}(x) + \frac{r_x - s}{2}\right)ds. \end{aligned}$$

We obtain

$$2(n-1) \geq \frac{1}{2}d(x,o) - \frac{4}{3} - \sqrt{f}(o) - \sqrt{f}(x). \tag{1.64}$$

This implies Cao and Zhou's estimate (1.61).

---

[36]Minimal geodesics on orbifolds with isolated singularities cannot contain singular points except at their endpoints.

1.6.5. Shi's local curvature derivative estimates for Ricci flow. It is often useful to apply general results from the Ricci flow theory to the study of Ricci solitons. W.-X. Shi's [**Shi89**] local derivative of curvature estimates for solutions to the Ricci flow are as follows (cf. Theorem I-1.6).

THEOREM 1.66 (Local derivative of curvature estimates). *For any positive integers n and m, and for any positive constant c, there exists a constant $C \in (0, \infty)$ such that if for some K, $(U^n, g(t), p)$, $t \in [0, T]$, $0 < T \leq cK^{-1}$, is a (not necessarily complete) solution to the Ricci flow on an orbifold with isolated singularities such that the closed ball $\overline{B}_{g(0)}(p, cK^{-\frac{1}{2}})$ is compact and satisfies*

$$|\mathrm{Rm}|(x,t) \leq K \quad \text{in } B_{g(0)}(p, cK^{-\frac{1}{2}}) \times [0,T], \tag{1.65}$$

then

$$|\nabla^m \mathrm{Rm}|(x,t) \leq \frac{CK}{t^{\frac{m}{2}}} \quad \text{in } B_{g(0)}(p, \tfrac{1}{2}cK^{-\frac{1}{2}}) \times (0,T]. \tag{1.66}$$

In particular,

$$|\nabla^m \mathrm{Rm}|(p, cK^{-1}) \leq Cc^{-\frac{m}{2}} K^{1+\frac{m}{2}}. \tag{1.67}$$

1.6.6. 2-Dimensional Ricci solitons.

In dimension 2, Ricci solitons have essentially been classified. In this subsection we survey these classification results.

Gutperle, Headrick, Minwalla, and Schomerus [**GHMS03**] proved the existence of 2-dimensional expanding solitons (cf. Theorem I-3.4):

THEOREM 1.67. *For each $\alpha \in (0, 2\pi)$ there exists a complete 2-dimensional rotationally symmetric expanding soliton (M^2, g, f), where M^2 is diffeomorphic to $\mathbb{R}^2$, g has positive, exponentially decaying curvature, and is asymptotically conical with cone angle α.*

Hamilton's [**Ham88**] classification of Ricci solitons on closed surfaces is given by (cf. Theorem I-3.5):

THEOREM 1.68. *If (M^2, g, X) is a Ricci soliton on a closed surface, then g has constant curvature. Hence:*

1. *A shrinking Ricci soliton must be a constant positive curvature metric on the 2-sphere or the real projective plane.*
2. *A steady Ricci soliton must be a flat metric on the torus or the Klein bottle.*
3. *An expanding Ricci soliton must be a constant negative curvature metric on a surface of genus at least 2.*

Recall that the **cigar soliton** is the steady gradient soliton on $\mathbb{R}^2$ with metric and potential function defined by

$$g_\Sigma(x,y) := \frac{4g_{\mathrm{Euc}}}{1+x^2+y^2} \quad \text{and} \quad f_\Sigma(x,y) := -\ln\left(1+x^2+y^2\right), \tag{1.68}$$

respectively, see §3.1.2 of Volume 1.

2-Dimensional steady solitons are classified by Hamilton [**Ham88**, §10], [**Ham95b**, Theorem 26.3], (cf. Theorem I-3.11) as follows.

THEOREM 1.69. *If (M^2, g, f) is a complete steady gradient Ricci soliton, then it is either the cigar soliton or flat.*

2-Dimensional shrinking solitons are classified by Hamilton [**Ham88**], [**Ham95b**, Theorem 26.1] as (cf. Theorem I-3.12):

THEOREM 1.70. *If (M^2, g, f) is a complete nonflat shrinking gradient Ricci soliton, then M^2 is closed; hence g has constant positive curvature. Moreover, the only complete noncompact shrinking gradient Ricci soliton is the Euclidean plane.*

2-Dimensional expanding solitons are classified by Bernstein and Mettler [**BM15**] and by Ramos [**Ram13**] as (cf. Theorem I-3.14):

THEOREM 1.71. *If (M^2, g, f) is a complete expanding soliton not of constant curvature, then g and f are rotationally symmetric and either:*

(i) *g has positive curvature and has one end, which is asymptotic to a flat Riemannian cone with cone angle in $(0, 2\pi)$. These are the expanding gradient Ricci solitons on $\mathbb{R}^2$ given by Theorem* 1.67.

(ii) *g has negative curvature and has one end, which is asymptotic to a flat Riemannian cone with cone angle in $(2\pi, \infty)$.*

(iii) *g has negative curvature and has two ends, where one end is asymptotic to a hyperbolic cusp and the other end is asymptotic to a flat Riemannian cone with cone angle in $(0, \infty)$. There is an "infinite cone angle" version of this described in the proof below.*

The classification of 2-dimensional expanding Ricci solitons has been extended to the non-gradient case by Peachey and Topping [**PT25**].

1.6.7. Volume growth of shrinking solitons.

Volume is an utmost fundamental invariant of Riemannian manifolds. In all dimensions, Cao and Zhou [**CZ10**] proved that shrinking solitons have at most Euclidean volume growth. Wang and one of the authors [**MW14**] proved the following sharper version (cf. Theorem I-4.8):

THEOREM 1.72. *For any shrinking gradient Ricci soliton and $p \in \mathcal{O}^n$,*

$$\mathrm{Vol}\left(B(p,r)\right) \leq \omega_n \mathrm{e}^{f(p)} r^n \quad \textit{for } r > 0. \tag{1.69}$$

In particular, if $o \in \mathcal{O}^n$ is a minimum point of f, then

$$\mathrm{Vol}\left(B(o,r)\right) \leq \omega_n \mathrm{e}^{n/2} r^n. \tag{1.70}$$

1.6.8. The canonical form of a Ricci soliton.

The Ricci flow associated to a soliton is given by the following (cf. Proposition I-4.19).

PROPOSITION 1.73. *If $(\mathcal{O}^n, g, f, \lambda)$ is a complete gradient Ricci soliton, then there exist a solution $g(t)$ of the Ricci flow with $g(0) = g$, diffeomorphisms $\varphi_t : \mathcal{O}^n \to \mathcal{O}^n$ with $\varphi_0 = \mathrm{id}$, and $f(t) : \mathcal{O}^n \to \mathbb{R}$ with $f(0) = f$, all of the above defined for all t with $\tau(t) := 1 - \lambda t > 0$, such that the following hold:*

(1) $\frac{\partial}{\partial t}\varphi_t(x) = \frac{1}{\tau(t)}(\nabla_g f)(\varphi_t(x))$,

(2) $g(t) = \tau(t)\, \varphi_t^* g$,

(3) $f(t) = f \circ \varphi_t$.

Moreover,

$$\operatorname{Ric}_{g(t)} + \nabla^2_{g(t)} f(t) = \frac{\lambda}{2\tau(t)} g(t) \tag{1.71}$$

and

$$\frac{\partial f}{\partial t}(t) = |\nabla f(t)|^2_{g(t)}. \tag{1.72}$$

We call $(\mathcal{O}^n, g(t), f(t), \frac{\lambda}{\tau(t)})$ *the* ***canonical form*** *or* ***evolving soliton associated to*** $(\mathcal{O}^n, g, f, \lambda)$.

Note that since

$$R_{g(t)} + \Delta_{g(t)} f(t) = \frac{n\lambda}{2\tau(t)}, \tag{1.73}$$

we may rewrite (1.72) as

$$\frac{\partial f}{\partial t} = -\Delta_{g(t)} f(t) + |\nabla f(t)|^2_{g(t)} - R_{g(t)} + \frac{n\lambda}{2\tau(t)}. \tag{1.74}$$

Moreover, since

$$R_{g(t)} + |\nabla f(t)|^2_{g(t)} = \frac{f(t)}{\tau(t)} \tag{1.75}$$

(cf. (I-4.85)), we also have that

$$\tau(t)\left(\frac{\partial}{\partial t} - \Delta_{g(t)}\right) f(t) = -\tau(t)\left(2\Delta_{g(t)} f(t) - |\nabla f(t)|^2_{g(t)} + R_{g(t)}\right) + \frac{n\lambda}{2}. \tag{1.76}$$

1.6.9. 3-Dimensional gradient Ricci solitons.

According to Hamilton [**Ham95b**], 3-dimensional solitons can be categorized by (cf. Theorem I-5.7):

THEOREM 1.74. *Let* (M^3, g, f, λ) *be a 3-dimensional complete gradient Ricci soliton. Then it has one of the following properties:*

(1) *The metric* g *is flat. In the case of a shrinking gradient Ricci soliton or an expanding gradient Ricci soliton, the soliton must be a Gaussian soliton.*
(2) *The metric* g *has positive sectional curvature. Such shrinking gradient Ricci solitons must be (compact) spherical space forms. Such steady and expanding gradient Ricci solitons must be noncompact.*
(3) *The universal covering soliton* $(\tilde{M}^3, \tilde{g}, \tilde{f}, \lambda)$ *splits as the product of the 1-dimensional Gaussian soliton and a nonflat 2-dimensional gradient Ricci soliton. In the case of a shrinking gradient Ricci soliton, this is a round* $\mathbb{S}^2$ *or its* $\mathbb{Z}_2$*-quotient* $\mathbb{RP}^2$*. In the case of a steady gradient Ricci soliton, this is the cigar soliton.*

By the works of Hamilton [**Ham95b**], Perelman [**Per03a**], Cao, Chen, and Zhu [**CCZ08**], Ni and Wallach [**NW08**], and Petersen and Wylie [**PW10**], 3-dimensional shrinking solitons have been classified as follows (cf. Corollary I-5.8):

COROLLARY 1.75. *Any 3-dimensional complete shrinking gradient Ricci soliton* (M^3, g, f) *is either a quotient of a round 3-sphere, a quotient of a round*

$\mathbb{S}^2 \times \mathbb{R}$, or the Gaussian shrinking soliton. If M^3 is orientable, then the only nontrivial quotient of $\mathbb{S}^2 \times \mathbb{R}$ is by the $\mathbb{Z}_2$-action generated by the antipodal map on each factor.

1.6.10. The Bryant soliton.

The Bryant steady soliton, first mentioned in §1.2.4.3, is specified as follows; see [**Bry**]; cf. Theorem I-6.1.

THEOREM 1.76. *For each $n \geq 3$ there exists a complete nonflat rotationally symmetric steady gradient Ricci soliton (M^n, g, f) called the Bryant soliton. Up to scaling, this is the unique such soliton. For the Bryant soliton, M^n is diffeomorphic to $\mathbb{R}^n$, the metric*

$$g = dr^2 + w^2(r) g_{\mathbb{S}^{n-1}}$$

*has positive curvature operator everywhere, and $R_{\max} = R(0^n) = 1$, where 0^n denotes the origin in $\mathbb{R}^n$. As $r \to \infty$, we have that the **warping function** w and its first two derivatives satisfy*

$$w(r) = \mathrm{O}(r^{1/2}), \quad w'(r) = \mathrm{O}(r^{-1/2}), \quad -w''(r) = \mathrm{O}(r^{-3/2}),$$

and the orbital and radial sectional curvatures satisfy

$$K_{\mathrm{orb}} = \mathrm{O}(r^{-1}) \quad \text{and} \quad K_{\mathrm{rad}} = \mathrm{O}(r^{-2}).$$

1.6.11. Brendle's classification.

Brendle's [**Bre13**] classification of noncollapsed 3-dimensional steady solitons is given by (cf. Theorem I-8.1):

THEOREM 1.77. *Any 3-dimensional κ-noncollapsed nonflat steady gradient Ricci soliton must be the Bryant soliton.*

For further examples of Ricci solitons, see Dancer and Wang [**DW09, DW11**], Yang [**Yan12**], and Angenent and Knopf [**AK22**].

1.6.12. The fundamental groups of shrinking solitons are finite. We have the following result of Wylie [**Wyl08**], which applies to shrinking Ricci solitons in all dimensions.

THEOREM 1.78. *If (M^n, g) is a complete Riemannian manifold and if X is a vector field satisfying the Bakry–Émery tensor lower bound:*

$$\operatorname{Ric} + \frac{1}{2}\mathcal{L}_X g \geq \frac{1}{2} g, \tag{1.77}$$

then M^n has finite fundamental group. In particular, any shrinking Ricci soliton has finite fundamental group.

PROOF. Let $p, q \in M^n$ satisfy $d_g(p,q) \geq 2$, and let $\gamma : [0, d_g(p,q)] \to M^n$ be a unit-speed minimal geodesic joining p to q. Define, for $y \in M^n$,

$$\mathrm{S}_g(y) := \max\left\{0, \sup \operatorname{Ric}_g(V,V) : V \in T_x M,\ |V| = 1,\ x \in B(y,1)\right\}.$$

Observe that

$$\frac{1}{2}(\mathcal{L}_X g)\left(\gamma'(r), \gamma'(r)\right) = \left\langle \nabla_{\gamma'(r)} X, \gamma'(r) \right\rangle = \gamma'(r)\left(\langle X, \gamma'(r)\rangle\right). \tag{1.78}$$

Since γ is a minimal geodesic, by the second variation of the arc-length formula (see Lemma 2.20 in Volume 1), we obtain that

$$
\begin{aligned}
(1.79) \qquad 2(n-1)+\frac{2}{3}\left(\mathrm{S}_g(p)+\mathrm{S}_g(q)\right) &\geq \int_0^{d(p,q)} \operatorname{Ric}\left(\gamma'(r),\gamma'(r)\right) dr \\
&\geq \int_0^{d(p,q)} \left(\frac{1}{2}-\frac{1}{2}(\mathcal{L}_X g)\left(\gamma'(r),\gamma'(r)\right)\right) dr \\
&= \frac{1}{2} d_g(p,q)+\langle X,\gamma'\rangle(p)-\langle X,\gamma'\rangle(q),
\end{aligned}
$$

where we also used (1.78).

Now let $(\widetilde{M}^n,\widetilde{g})$ be the Riemannian universal covering of (M^n,g) and let $\widetilde{X}$ be the lift of X to $\widetilde{M}^n$. Fix $\widetilde{p}\in\widetilde{M}^n$ and let $\phi:\widetilde{M}^n\to\widetilde{M}^n$ be a deck transformation. By basic covering space theory, ϕ corresponds to an element of $\pi_1(M^n)$. Observe that $|\widetilde{X}(\phi(\widetilde{p}))|=|\widetilde{X}(\widetilde{p})|$ and $\mathrm{S}_{\widetilde{g}}(\phi(\widetilde{p}))=\mathrm{S}_{\widetilde{g}}(\widetilde{p})$ on $\widetilde{M}^n$. Hence, by (1.79) applied to $\widetilde{p}$ and $\phi(\widetilde{p})$, for all $\phi\in\pi_1(M^n)$ such that $d_{\widetilde{g}}(\widetilde{p},\phi(\widetilde{p}))\geq 2$ we have

$$
(1.80) \qquad d_{\widetilde{g}}(\widetilde{p},\phi(\widetilde{p}))\leq 4(n-1)+\frac{8}{3}\mathrm{S}_{\widetilde{g}}(\widetilde{p})+4\left|\widetilde{X}(\widetilde{p})\right|.
$$

This implies that $\pi_1(M^n)$ is finite. □

Recently, Esparza [**Esp25**] proved that any shrinking gradient Kähler–Ricci soliton is simply connected.

See the surveys of H.-D. Cao [**Cao10, Cao11**] and X. Cao and H. Tran [**CT24**] for these and other related results on Ricci solitons.

1.7. Open problems on 4-dimensional Ricci flow

While it seems reasonable to expect additional important advances in our understanding of 3-dimensional Ricci flow, the focus of most current research has shifted to higher dimensions, particularly to dimension 4, where, with dimension reduction, one has a hope of leveraging all of the successes of the 3-dimensional theory.

A fundamental question at the time of writing is:

Can Ricci flow be used to study 4-dimensional smooth topology?

The crucial issue is to understand the geometry and topology of singularity models of the Ricci flow on oriented closed orbifolds with isolated singularities. With a suitably detailed understanding of these singularity models, one could hope to formulate a program of Ricci flow through singularities.

Recent research, especially the new compactness and partial regularity theory developed for the Ricci flow in all dimensions by Richard Bamler [**Bam20a, Bam23, Bam20b**], suggests that there is some hope for a geometric-based surgery theory for smooth 4-dimensional orbifolds with isolated singularities in dimension 4 (see [**Bam21b**]).[37] In fact, from Bamler's work, there emerges the possibility that the Ricci flow might yield a geometrization-type theorem for all smooth closed

[37] We highlight a few of Bamler's results in Chapter 6.

4-dimensional orbifolds with isolated singularities.[38] Of course, even if this possibility were eventually realized to some extent, it might still remain challenging to infer from it extensive topological conclusions about smooth 4-manifolds. Nevertheless, the number of potential topological applications should only continue to increase with the growth of our understanding of 4-dimensional Ricci flow. We discuss some potential applications below.

1.7.1. Ricci flow as an approach to the 11/8-conjecture.

Bamler [**Bam20a, Bam23, Bam20b, Bam21b**] has developed an extensive theory to understand singularity formation of the Ricci flow in dimensions 4 and higher. In particular, his results are the strongest in dimension 4 and form a basis for approaching the 11/8-conjecture; the idea of a Ricci flow approach was initially suggested by Tian, Freedman, and others.

OPEN PROBLEM 1.1 (Tian, Freedman, Bamler). *Prove the* 11/8*-conjecture using Ricci flow. It is possible that under a suitable formulation of* 4*-dimensional Ricci flow with surgery, the signature to the second Betti number ratio is monotone. In any case, the topological invariants one needs to track under Ricci flow should be much easier to deal with as compared to the fundamental group for example.*

It is possible that under a suitable formulation of 4-dimensional Ricci flow with surgery, the signature to the second Betti number ratio is monotone. In any case, the topological invariants one needs to track under Ricci flow should be much easier to deal with as compared to the fundamental group for example.

1.7.2. Other potential applications to smooth 4-dimensional topology.

Some other well-known and generally formulated problems regarding the application of Ricci flow to smooth 4-dimensional topology are the following:

OPEN PROBLEM 1.2 (4-Dimensional Geometrization Conjecture). *Can one prove a geometrization-type theorem for closed* 4*-manifolds using the Ricci flow? This is an open-ended question.*

Should one aim to understand the Ricci flow for all initial data, or focus on some formulation for generic initial data? If the latter is not possible, the former seems to be the only option.

The formulation of a singular Ricci flow solution by Bamler, Kleiner, and Lott in dimension 3 *may be a template for dimension* 4*; see* [**BK22**] *and the references therein. However, unlike dimension* 3*, one might not expect uniqueness and continuous dependence on initial metrics in dimension* 4.

OPEN PROBLEM 1.3 (4-Dimensional smooth Poincaré conjecture). *Can one formulate a Ricci flow approach to the* 4*-dimensional smooth Poincaré conjecture?*

This seems too optimistic. For example, for any 2*-knot in* $\mathbb{S}^4$*, Ricci flow with surgery should effect a* 2*-surgery on that* 2*-knot for a suitable initial metric; see*

[38] Even if one is only interested in *smooth* closed 4-manifolds, it is important to note that certain scenarios, including when embedded spherical space forms $\mathbb{S}^3/\Gamma$ undergo pinching, can bring us – via Ricci flow with surgery – into the realm of 4-dimensional orbifolds with isolated singularities.

Problem 1.6. *Note, for the special case of spun* 2*-knots in* $\mathbb{S}^4$, *their knot groups (i.e., fundamental groups of their complements) are isomorphic to the knot groups of the* 1*-knots in* $\mathbb{S}^3$ *that they spin.*

OPEN PROBLEM 1.4. *For any embedded* $\mathbb{S}^3/\Gamma$ *in a closed* 4*-manifold* M, *does there exist an initial metric* g_0 *for which the Ricci flow contracts the embedded* $\mathbb{S}^3/\Gamma$ *to a point at its singularity time* $T \in (0, \infty)$ *and such that an associated singularity model is* $(\mathbb{S}^3/\Gamma) \times \mathbb{R}$*?*

OPEN PROBLEM 1.5. *As a special case of the previous problem, can one prove the existence of a cohomogeneity-one* Q_8*-neckpinch on* $\mathbb{S}^4$ *in the following setting? Consider* $\mathbb{S}^4$ *as the unit sphere in the space of real trace-free symmetric* 3×3 *matrices, with the matrix norm. Consider* $\mathrm{SO}(3)$ *acting on* $\mathbb{S}^4$ *by conjugation. There are exactly two singular orbits of this action, each diffeomorphic to* $\mathbb{R}P^2$ *and represented by Veronese embeddings; see* [**Law80**, p. 20]. *The regular orbits are each diffeomorphic to* $\mathbb{S}^3/Q_8$, *where* $Q_8 = \{\pm 1, \pm i, \pm j, \pm k\}$ *is the order* 8 *subgroup of the unit quarternions.*

Does there exist an initial $\mathrm{SO}(3)$*-symmetric metric on* $\mathbb{S}^4$ *such that the Ricci flow starting from that metric forms a* Q_8*-neckpinch, or alternatively, a degenerate* Q_8*-neckpinch? Note that in the latter case, the resulting steady gradient Ricci soliton singularity model should be cohomogeneity-one and on a plane bundle over* $\mathbb{R}P^2$ *(the normal bundle of the Veronese embedding, which is orientable) that is the* $\mathbb{Z}_2$*-quotient of an Appleton soliton on* M_4^4*; see* [**App17**].

OPEN PROBLEM 1.6. *Let* M^4 *be a closed* 4*-manifold and suppose that* Σ *is an embedded* $\mathbb{S}^2$ *with trivial normal bundle. For example, if* M^4 *is homeomorphic to* $\mathbb{S}^4$, *then any embedded* $\mathbb{S}^2$ *must have trivial normal bundle. Does there exist a metric* g_0 *for which the Ricci flow contracts* Σ *to a point at its singularity time* $T \in (0, \infty)$ *and such that an associated singularity model is* $\mathbb{S}^2 \times \mathbb{R}^2$*?*

OPEN PROBLEM 1.7. *Is there an orbifold version of Theorem* 1.78 *that shrinking gradient Ricci solitons must have finite fundamental group?*

OPEN PROBLEM 1.8. *Is the product of* $\mathbb{R}$ *and the* 3*-dimensional Bryant steady soliton in some sense not a generic* 4*-dimensional singularity model?*

This question is motivated by the viewpoint that such a singularity model may be less desirable, as its formation could complicate the formulation of a 4-dimensional Ricci flow with surgery theory. On the other hand, even though the 4*-dimensional Bryant steady soliton may not form generically as a singularity model, it should not cause problems for formulating a Ricci flow with surgery or singular Ricci flow theory.*

One may start in dimension 3 *and try to show that the* 3*-dimensional Bryant steady soliton is not a generic* 3*-dimensional singularity model. Is the linearized Ricci flow helpful?*

OPEN PROBLEM 1.9. *Can hyper-Kähler Ricci-flat ALE* 4*-manifolds besides the Eguchi–Hanson Ricci-flat ALE occur as singularity models of the Ricci flow?*

OPEN PROBLEM 1.10. *Must any ALE Ricci flat* 4*-manifold be hyper-Kähler?*

This is a well-known open problem; see Lock and Viaclovsky [**LV16**]. *Since ALE Ricci flat* 4*-manifolds may potentially occur as singularity models, one would like to know as much as possible about the topology of such spaces.*

OPEN PROBLEM 1.11. *Can an exotic $\mathbb{R}^4$ admit a soliton structure? One hopes not. One can specialize this question based on type: shrinking soliton, steady soliton, or expanding soliton. In the shrinking soliton case, one can ask this question under the additional assumption of being asymptotically conical.*

OPEN PROBLEM 1.12 (Bamler and Chen; see [**BC23**, Question 1.1]). *For any 4-dimensional Riemannian cone with $R \geq 0$, does there exist an expanding gradient Ricci soliton with non-negative scalar curvature that is asymptotic to this cone?*

CHAPTER 2

Curvature Estimates for 4-Dimensional Shrinking Solitons

This chapter is devoted to the analytic study of the geometry of 4-dimensional *shrinking* gradient Ricci solitons. The results below belong to a broader effort to classify — or at least understand, sufficiently well — those 4-dimensional shrinking gradient Ricci solitons which arise as singularity models, in order to formulate a program of Ricci flow with surgery in that dimension.

We focus on two themes that are crucial to understanding gradient Ricci solitons: the behavior of the Riemann curvature tensor and the geometry at infinity. In particular, we bound the curvature growth and decay and we discuss under what hypotheses the shrinking gradient Ricci soliton is asymptotically conical and under what hypotheses it asymptotically splits off a line.

A general belief is that the geometry and topology of a Ricci soliton 'simplifies' at spatial infinity. This leads us to study the asymptotic geometry at infinity of Ricci solitons. A suitable understanding of the asymptotic geometry may yield constraints, rigidity, and symmetry for the global geometry of the Ricci soliton. This approach has been particularly successful in dimensions 2 and 3, as we have surveyed in the previous chapter.

In dimension 4 we show that $|\mathrm{Rm}|$ can be estimated by R. To accomplish this, we exploit the interplay between the metric, curvature, and potential function for a soliton. The process is fairly intricate and involves various applications of the maximum principle.

We will assume everywhere that the shrinking gradient Ricci solitons are complete noncompact and that we are not on the Gaussian shrinker, so $R > 0$ according to (1.57).

2.1. Scalar curvature and bounds on Riemann curvature

One of the most basic questions one can ask about the geometry of a complete noncompact Riemannian manifold is whether the Riemann curvature tensor is bounded. Bounding the norm of the Riemann curvature tensor by the scalar curvature presents a strategic approach in the study of solitons, thereby allowing us to control a stronger curvature invariant by a weaker one. Such a bound is not only insightful for constraining the curvature, but it also helps us divide the study of solitons into the cases where the scalar curvature is either bounded or unbounded, as well as the case where it tends to zero at infinity. In situations where the scalar curvature is indeed bounded, this bound furnishes us with additional information, which we will extend and exploit, for example in the case where we have scalar curvature decay.

In this section we prove the following result of Wang and one of the authors [**MW15a**].

THEOREM 2.1 (R bound implies $|\mathrm{Rm}|$ bound). *Any complete noncompact 4-dimensional shrinking gradient Ricci soliton with bounded scalar curvature must have bounded Riemann curvature tensor.*

Roughly speaking, the structure of the proof involves two key steps: first, we bound the Ricci curvature by its trace, which is the scalar curvature; then, we bound the Riemann curvature by its trace, the Ricci curvature. This method can be regarded as a bootstrapping process.

Let (M^n, g, f) be a shrinking gradient Ricci soliton. Recall that f (more aptly, $2\sqrt{f}$) may be considered as a proxy for the distance function. Denote

$$\begin{aligned} D(t) &:= \{x \in M^n : f(x) \leq t\}, \\ \Sigma(t) &:= \partial D(t) = \{x \in M^n : f(x) = t\} \end{aligned} \tag{2.1}$$

for $t \in \mathbb{R}$, which are the sublevel and level sets of the potential function, respectively. We may think of $D(t)$ and $\Sigma(t)$ as analogues of geodesic balls and geodesic spheres, respectively. In certain respects, they are more well-adapted to the Ricci soliton structure.

We henceforth assume that (M^4, g) has bounded scalar curvature and let

$$A := \sup_{M^n} R < \infty.$$

Recall from (1.35) that the scalar curvature is normalized by

$$R + |\nabla f|^2 = f.$$

Since $R > 0$, it follows that

$$|\nabla f|^2 \geq \frac{1}{2} f + 1 > 0 \quad \text{on } M^4 \setminus D(t_1), \tag{2.2}$$

where $t_1 := 2(A+1)$. In particular, the level sets $\Sigma(t)$ of f are smooth closed hypersurfaces of M^4 for all $t \geq t_1$. Let

$$\nu_x := \frac{\nabla f}{|\nabla f|}(x) \in T_x^{\perp} \Sigma(t), \tag{2.3}$$

which is a choice of the unit normal vector field to $\Sigma(t)$.

Throughout this section, unless otherwise stated, C denotes a universal constant, whose value may change from line to line.

The Riemann curvature tensor of a general 4-dimensional Riemannian manifold in the level set directions may be expressed in terms of Rm in the normal direction and Ric using the following result:

LEMMA 2.2 (Rm in the level set directions). *Let (M^4, g, f) be a 4-dimensional Riemannian manifold and let Σ be a hypersurface in M^4 with unit normal field*

ν. For any $x \in \Sigma$ and vectors $X, Y, Z, W \in T_x\Sigma$, we have

$$\begin{aligned}
\mathrm{Rm}(X,Y,Z,W) &= \mathrm{Ric}(X,W)g(Y,Z) - \mathrm{Ric}(X,Z)g(Y,W) \\
&\quad + \mathrm{Ric}(Y,Z)g(X,W) - \mathrm{Ric}(Y,W)g(X,Z) \\
&\quad - \frac{1}{2}R\big(g(X,W)g(Y,Z) - g(X,Z)g(Y,W)\big) \\
&\quad - \mathrm{Rm}(X,\nu,\nu,W)g(Y,Z) + \mathrm{Rm}(X,\nu,\nu,Z)g(Y,W) \\
&\quad - \mathrm{Rm}(Y,\nu,\nu,Z)g(X,W) + \mathrm{Rm}(Y,\nu,\nu,W)g(X,Z) \\
&\quad + \mathrm{Ric}(\nu,\nu)\big(g(X,W)g(Y,Z) - g(X,Z)g(Y,W)\big). \qquad (2.4)
\end{aligned}$$

PROOF. Since both sides of identity (2.4) are algebraic curvature tensors in $X, Y, Z, W \in T_x\Sigma$, it suffices to prove the identity for $X = W = e_1$ and $Y = Z = e_2$, where $\{e_1, e_2, e_3\}$ is any orthonormal frame tangent to Σ.[1] That is, the lemma will follow from proving that

$$R_{1221} = R_{11} + R_{22} - \frac{1}{2}R - \big(K(e_1,\nu) + K(e_2,\nu)\big) + \mathrm{Ric}\,(\nu,\nu)\,,$$

where $K(e_i,\nu)$ denotes the sectional curvature of the plane spanned by e_i and ν. Our sign convention is such that R_{1221} denotes the sectional curvature of the plane spanned by e_1 and e_2, see (1.48).

However, since we are in dimension 4, for any orthonormal frame $\{e_i\}_{i=1}^4$ it is easy to see by a direct computation that

$$R_{1221} = \frac{1}{2}(R_{11} + R_{22} - R_{33} - R_{44}) + R_{3443}.$$

Applying this for the frame $\{e_1, e_2, e_3, e_4 = \nu\}$ proves the desired result. □

2.1.1. Bounding Riemann curvature by Ricci curvature and its covariant derivative.

The idea is that we can bound the Riemann curvature by the Ricci curvature and its covariant derivative by applying the previous lemma together with the derived equations for gradient Ricci solitons.

PROPOSITION 2.3 ($|\mathrm{Rm}|$ bounded by $|\mathrm{Ric}|$ and $|\nabla \mathrm{Ric}|$). *Let (M^4, g, f) be a complete noncompact 4-dimensional gradient Ricci soliton. Then, for a universal constant $C > 0$,*

$$|\mathrm{Rm}| \leq C\left(|\mathrm{Ric}| + \frac{|\nabla \mathrm{Ric}|}{|\nabla f|}\right) \qquad (2.5)$$

on $M^4 \setminus D(t_1)$.

PROOF. According to Lemma 2.2,

$$|\mathrm{Rm}| \leq C\left(|\mathrm{Ric}| + |\mathrm{Rm}(\nu)|\right). \qquad (2.6)$$

In the equation above, the term

$$|\mathrm{Rm}(\nu)| := \sup_{U,V,W} |\mathrm{Rm}(U,V,W,\nu)|$$

represents the supremum over unit tangent vectors U, V, W on M^4, where $\nu = \frac{\nabla f}{|\nabla f|}$.

[1]This is for the same reason that the curvature tensor is determined by the sectional curvatures and the inner product; see Cheeger and Ebin [**CE08**, pp. 13–14].

We now observe that by

$$R_{ijkl}\nabla_l f = \nabla_i R_{jk} - \nabla_j R_{ik}$$

(see (1.47)), which relates the Riemann curvature tensor to the covariant derivative of the Ricci tensor, we have the estimate

$$|\mathrm{Rm}\,(\nu)| \leq 2\frac{|\nabla \mathrm{Ric}|}{|\nabla f|}. \tag{2.7}$$

By applying this to (2.6), we conclude that

$$|\mathrm{Rm}| \leq C\left(|\mathrm{Ric}| + \frac{|\nabla \mathrm{Ric}|}{|\nabla f|}\right)$$

as claimed. □

Let $\Delta_f := \Delta - \nabla f \cdot \nabla$ denote the f-Laplacian as defined by (1.38). We now record the following useful estimate for solutions to elliptic inequalities with a quadratic right-hand side.[2] Let $o \in M^n$ be a minimum point of f if M^n is a shrinking gradient Ricci soliton, and any arbitrary point if M^n is a steady soliton.

LEMMA 2.4 (Upper estimate for quadratic right-hand side). *Let (M^n, g, f) be a shrinking or steady gradient Ricci soliton. Assume that u is a function on M^n satisfying the differential inequality*

$$\Delta_f u \geq \frac{1}{C_0}u^2 - C_1 \tag{2.8}$$

on $M^n \setminus B(o, r_0)$, for some positive constants C_0, C_1, and for some $r_0 > 1$. Then there exists a constant $C = C(n) > 0$ such that

$$u \leq CC_0 + \sqrt{C_1 C_0} \quad \text{on } M^n \setminus B(o, 2r_0). \tag{2.9}$$

PROOF. For arbitrary $x \in M^n \setminus B(o, 2r_0)$ we let

$$T := \frac{d(o,x)}{4}.$$

Let ϕ be a smooth non-negative function defined on the real line so that $\phi(t) = 1$ for $0 \leq t \leq T$ and $\phi(t) = 0$ for $t \geq 2T$. We may choose ϕ so that

$$|\phi'|^2(t) + |\phi''|(t) \leq \frac{C}{T^2}.$$

We use $\phi(z) := \phi(d(x,z))$ as a cut-off function with support in the ball $B(x, 2T) \subset M^n \setminus B(o, r_0)$. According to [**WW09**], cf. (4.31) in Volume 1, the distance function $r(z) := d(x,z)$ satisfies, in barrier sense,

$$\Delta_f r \leq \frac{n-1}{r} + a \quad \text{on } B(x, 2T)$$

provided $\mathrm{Ric}_f \geq 0$ and $|\nabla f| \leq a$ on $B(x, 2T)$. As $B(x, 2T) \subset B(o, 6T)$, we have that $|\nabla f| \leq 3T + C(n)$ if M^n is a shrinking gradient Ricci soliton; see (1.59). Moreover, any steady soliton is assumed to be normalized such that $|\nabla f| \leq 1$; see (1.37). In conclusion,

$$\Delta_f r \leq C(n)T \quad \text{on } B(x, 2T) \setminus B(x, T).$$

[2]This form of the maximum principle, which is now ubiquitous, was first used in geometric analysis by Yau [**Yau75**] to prove a Liouville theorem for harmonic functions.

Therefore, the cut-off ϕ satisfies

$$|\nabla\phi| \le \frac{C}{T} \quad \text{and} \quad \Delta_f \phi \ge -C, \tag{2.10}$$

for C depending only on the dimension n. Define the function

$$G := u\phi^2,$$

which has support in $B(x, 2T)$. Moreover,

$$\begin{aligned}\phi^2\Delta_f G &= \phi^4(\Delta_f u) + \phi^2 u(\Delta_f \phi^2) + 2\phi^2\langle\nabla u, \nabla\phi^2\rangle \\ &= \phi^4(\Delta_f u) + 2\phi(\Delta_f\phi)G - 6|\nabla\phi|^2 G + 2\left\langle\nabla G, \nabla\phi^2\right\rangle.\end{aligned}$$

As u satisfies the inequality

$$\Delta_f u \ge \frac{1}{C_0}u^2 - C_1,$$

it follows that

$$\phi^2\Delta_f G \ge \frac{1}{C_0}G^2 - C_1 + 2\phi(\Delta_f\phi)G - 6|\nabla\phi|^2 G + 2\left\langle\nabla G, \nabla\phi^2\right\rangle. \tag{2.11}$$

We apply the maximum principle to this elliptic inequality. At the maximum point x_0 of G we may assume that $G(x_0) > 0$. Then (2.10) and (2.11) imply that

$$0 \ge \frac{1}{C_0}G^2(x_0) - CG(x_0) - C_1,$$

from which it follows that

$$u(x) \le G(x_0) \le CC_0 + \sqrt{C_0C_1}.$$

Consequently, $u \le CC_0 + \sqrt{C_0C_1}$ on $M^n \setminus B(o, 2r_0)$. This proves the result. □

Lemma 2.4 reduces proving many apriori estimates to finding suitable quantities that satisfy elliptic inequalities of the form (2.8).

2.1.2. Ricci curvature bounds from scalar curvature bounds.

We now prove a bound on the Ricci curvature assuming a bound on the scalar curvature. The proof of this bound relies on establishing an elliptic inequality for the ratio of the Ricci curvature norm over the square root of the scalar curvature. We accomplish this by employing our previous estimate of the Riemann curvature by the Ricci curvature and its covariant derivative.

PROPOSITION 2.5 (Bound for $|\mathrm{Ric}|$). *Let (M^4, g, f) be a complete noncompact 4-dimensional shrinking gradient Ricci soliton with bounded scalar curvature $R \le A$. Then there exists a universal constant $C > 0$ such that*

$$|\mathrm{Ric}| \le C(A+1) \quad \text{on } M^4 \setminus D(t_3), \tag{2.12}$$

where $t_3 := C(A+1)$.

PROOF. Recall from (1.51) that

$$\Delta_f \mathrm{Ric} = \mathrm{Ric} - 2\mathrm{Rm} * \mathrm{Ric},$$

where $|\mathrm{Rm} * \mathrm{Ric}| \leq |\mathrm{Rm}||\mathrm{Ric}|$. By Proposition 2.3 this implies that

$$\begin{aligned} \Delta_f |\mathrm{Ric}|^2 &\geq 2|\nabla \mathrm{Ric}|^2 + 2|\mathrm{Ric}|^2 - 4|\mathrm{Rm}||\mathrm{Ric}|^2 \\ &\geq 2|\nabla \mathrm{Ric}|^2 + 2|\mathrm{Ric}|^2 - C|\mathrm{Ric}|^3 \\ &\quad - \frac{C}{|\nabla f|}|\nabla \mathrm{Ric}||\mathrm{Ric}|^2. \end{aligned} \tag{2.13}$$

By direct computation we obtain[3]

$$\begin{aligned} \Delta_f\left(|\mathrm{Ric}|^2 R^{-\frac{1}{2}}\right) &= R^{-\frac{1}{2}}\Delta_f\left(|\mathrm{Ric}|^2\right) + |\mathrm{Ric}|^2 \Delta_f\left(R^{-\frac{1}{2}}\right) \\ &\quad + 2\left\langle \nabla(R^{-\frac{1}{2}}), \nabla|\mathrm{Ric}|^2\right\rangle. \end{aligned}$$

As the scalar curvature satisfies

$$\Delta_f R = R - 2|\mathrm{Ric}|^2,$$

it follows that

$$\begin{aligned} \Delta_f\left(R^{-\frac{1}{2}}\right) &= -\frac{1}{2}(\Delta_f R)R^{-\frac{3}{2}} + \frac{3}{4}|\nabla R|^2 R^{-\frac{5}{2}} \\ &= -\frac{1}{2}R^{-\frac{1}{2}} + |\mathrm{Ric}|^2 R^{-\frac{3}{2}} + \frac{3}{4}|\nabla R|^2 R^{-\frac{5}{2}}. \end{aligned}$$

Furthermore, we have

$$\begin{aligned} 2\left\langle \nabla(R^{-\frac{1}{2}}), \nabla|\mathrm{Ric}|^2\right\rangle &\geq -2|\nabla \mathrm{Ric}||\nabla R|R^{-\frac{3}{2}}|\mathrm{Ric}| \\ &\geq -\frac{3}{4}|\nabla R|^2 R^{-\frac{5}{2}}|\mathrm{Ric}|^2 - \frac{4}{3}|\nabla \mathrm{Ric}|^2 R^{-\frac{1}{2}}. \end{aligned}$$

Hence, it follows from (2.13) that

$$\begin{aligned} \Delta_f\left(|\mathrm{Ric}|^2 R^{-\frac{1}{2}}\right) &\geq \frac{2}{3}|\nabla \mathrm{Ric}|^2 R^{-\frac{1}{2}} + \frac{3}{2}|\mathrm{Ric}|^2 R^{-\frac{1}{2}} - C|\mathrm{Ric}|^3 R^{-\frac{1}{2}} \\ &\quad - \frac{C}{|\nabla f|}|\nabla \mathrm{Ric}||\mathrm{Ric}|^2 R^{-\frac{1}{2}} + |\mathrm{Ric}|^4 R^{-\frac{3}{2}}. \end{aligned}$$

The following inequality holds:

$$\frac{C}{|\nabla f|}|\nabla \mathrm{Ric}||\mathrm{Ric}|^2 R^{-\frac{1}{2}} \leq \frac{2}{3}|\nabla \mathrm{Ric}|^2 R^{-\frac{1}{2}} + \frac{C}{|\nabla f|^2}|\mathrm{Ric}|^4 R^{-\frac{1}{2}}.$$

Consequently, we obtain from above that

$$\Delta_f\left(|\mathrm{Ric}|^2 R^{-\frac{1}{2}}\right) \geq |\mathrm{Ric}|^4 R^{-\frac{3}{2}} - C|\mathrm{Ric}|^3 R^{-\frac{1}{2}} - \frac{C}{|\nabla f|^2}|\mathrm{Ric}|^4 R^{-\frac{1}{2}}.$$

Moreover, for some $t_2 \geq C(A+1)$ we have

$$\frac{C}{|\nabla f|^2}|\mathrm{Ric}|^4 R^{-\frac{1}{2}} = \frac{CR}{|\nabla f|^2}|\mathrm{Ric}|^4 R^{-\frac{3}{2}} \leq \frac{1}{4}|\mathrm{Ric}|^4 R^{-\frac{3}{2}}$$

on $M^4 \setminus D(t_2)$. Finally, we also have that

$$C|\mathrm{Ric}|^3 R^{-\frac{1}{2}} \leq \frac{1}{4}|\mathrm{Ric}|^4 R^{-\frac{3}{2}} + CR^{\frac{5}{2}}. \tag{2.14}$$

---

[3]In the context of 3-dimensional Ricci flow, Hamilton [**Ham82**] considered the similar, but scale-invariant, quantity $|\mathrm{Ric}|^2 R^{-2}$.

In conclusion, we obtain the following inequality

$$\begin{aligned}\Delta_f\left(|\mathrm{Ric}|^2 R^{-\frac{1}{2}}\right) &\geq \frac{1}{2}|\mathrm{Ric}|^4 R^{-\frac{3}{2}} - CR^{\frac{5}{2}}\\ &\geq \frac{1}{2\sqrt{A}}|\mathrm{Ric}|^4 R^{-1} - CA^{\frac{5}{2}}.\end{aligned}$$

In other words, the function

$$u := \frac{|\mathrm{Ric}|^2}{\sqrt{R}}$$

satisfies

$$\Delta_f u \geq \frac{1}{2\sqrt{A}}u^2 - CA^2\sqrt{A} \quad \text{on } M^4 \setminus D(t_2).$$

By Lemma 2.4, we conclude that $u \leq C(A+1)\sqrt{A}$ on $M^4 \setminus D(4t_2)$; therefore

$$|\mathrm{Ric}|^2 \leq CA(A+1) \quad \text{on } M^4 \setminus D(t_3),$$

where $t_3 := 4t_2$. This proves the proposition. □

2.1.3. Riemann curvature bounds from scalar curvature bounds.

We now prove the main result of this section that the Riemann curvature tensor of a 4-dimensional shrinking gradient Ricci soliton with bounded scalar curvature is bounded. Here, we establish an elliptic inequality for the Riemann curvature norm plus a constant times the Ricci curvature norm squared. This is possible by our previous estimates.

THEOREM 2.6 (Bound for $|\mathrm{Rm}|$). *Let (M^4, g, f) be a complete noncompact 4-dimensional shrinking gradient Ricci soliton with bounded scalar curvature $R \leq A$. Then there exists a universal constant $C > 0$ such that*

$$|\mathrm{Rm}| \leq C(A+1)$$

on $M^4 \setminus D(t_0)$, where $t_0 := C(A+1)$.

PROOF. Recall from (1.53) that

$$\Delta_f \mathrm{Rm} = \mathrm{Rm} - \mathrm{Rm} * \mathrm{Rm},$$

where $|\mathrm{Rm} * \mathrm{Rm}| \leq 8|\mathrm{Rm}|^2$. Consequently, we have[4]

$$\Delta_f |\mathrm{Rm}|^2 \geq 2|\nabla \mathrm{Rm}|^2 - 16|\mathrm{Rm}|^3. \tag{2.15}$$

Applying the Kato inequality ($|\nabla|\alpha|| \leq |\nabla \alpha|$ for any tensor α), one sees that

$$\Delta_f |\mathrm{Rm}| \geq -8|\mathrm{Rm}|^2. \tag{2.16}$$

By Proposition 2.3 and Proposition 2.5 we have

$$|\mathrm{Rm}|^2 \leq C(A+1)^2 + \frac{C}{|\nabla f|^2}|\nabla \mathrm{Ric}|^2$$

on $M^4 \setminus D(t_3)$. Plugging this into (2.16), we conclude that

$$\Delta_f |\mathrm{Rm}| \geq |\mathrm{Rm}|^2 - C(A+1)^2 - \frac{C}{|\nabla f|^2}|\nabla \mathrm{Ric}|^2. \tag{2.17}$$

[4]We remark that this inequality, as well as the one following it, which we will use later in the book, also holds for steady solitons.

On the other hand, by (2.13), we know that on $M^4 \setminus D(t_3)$,

$$\Delta_f |\mathrm{Ric}|^2 \geq 2|\nabla \mathrm{Ric}|^2 - C|\mathrm{Ric}|^3 - \frac{C}{|\nabla f|}|\nabla \mathrm{Ric}||\mathrm{Ric}|^2.$$

Applying Proposition 2.5 we obtain the inequality

$$\begin{aligned}\frac{C}{|\nabla f|}|\nabla \mathrm{Ric}||\mathrm{Ric}|^2 &\leq |\nabla \mathrm{Ric}|^2 + \frac{C}{|\nabla f|^2}|\mathrm{Ric}|^4 \\ &\leq |\nabla \mathrm{Ric}|^2 + \frac{C}{f}(A+1)^4.\end{aligned}$$

Since $f \geq t_3 \geq C(A+1)$, it follows that

$$\Delta_f |\mathrm{Ric}|^2 \geq |\nabla \mathrm{Ric}|^2 - C(A+1)^3. \tag{2.18}$$

Therefore, by combining (2.17) and (2.18), we obtain

$$\begin{aligned}\Delta_f \left(|\mathrm{Rm}| + \frac{1}{A+1}|\mathrm{Ric}|^2\right) &\geq |\mathrm{Rm}|^2 - C(A+1)^2 \\ &\quad + \left(\frac{1}{A+1} - \frac{C}{|\nabla f|^2}\right)|\nabla \mathrm{Ric}|^2 \\ &\geq |\mathrm{Rm}|^2 - C(A+1)^2 \\ &\geq \frac{1}{2}\left(|\mathrm{Rm}| + \frac{1}{A+1}|\mathrm{Ric}|^2\right)^2 - C(A+1)^2.\end{aligned}$$

In other words, the function

$$v := |\mathrm{Rm}| + \frac{1}{A+1}|\mathrm{Ric}|^2$$

satisfies the following differential inequality on $M^4 \setminus D(t_3)$:

$$\Delta_f v \geq \frac{1}{2}v^2 - C(A+1)^2$$

for some universal constant C. By Lemma 2.4 we conclude that $v \leq C(A+1)$ on $M^4 \setminus D(4t_3)$. This proves the theorem. □

Applying a standard Shi-type argument (see [**Shi89**]), we obtain derivative estimates for the Riemann curvature tensor. We will require the following formula for the covariant derivatives of the curvature.

LEMMA 2.7. *For any gradient Ricci soliton (M^n, g, f, λ) and any integer $m \geq 0$,*

$$\Delta_f (\nabla^m \mathrm{Rm}) = \left(\frac{m}{2}+1\right)\lambda\, \nabla^m \mathrm{Rm} + \sum_{k=0}^{m} \nabla^k \mathrm{Rm} * \nabla^{m-k}\mathrm{Rm}. \tag{2.19}$$

PROOF. Let $\{e_i\}_{i=1}^n$ be a local orthonormal frame. In this frame,

$$\Delta_f (\nabla^m \mathrm{Rm}) = \nabla_i \nabla_i (\nabla^m \mathrm{Rm}) - \nabla_i (\nabla^m \mathrm{Rm}) \nabla_i f, \tag{2.20}$$

where $\nabla_i := \nabla_{e_i}$. For simplicity, we denote

$$\nabla^m \mathrm{Rm} := \nabla_j \left(\nabla^{m-1}\mathrm{Rm}\right).$$

Applying the Ricci identities it follows that

$$\begin{aligned}\nabla_i \nabla^m \mathrm{Rm} &= \nabla_i \nabla_j \left(\nabla^{m-1}\mathrm{Rm}\right) \\ &= \nabla_j \nabla_i \left(\nabla^{m-1}\mathrm{Rm}\right) + \mathrm{Rm} * \nabla^{m-1}\mathrm{Rm}.\end{aligned}$$

Taking a covariant derivative of this identity, we obtain

$$\nabla_i \nabla_i \left(\nabla^m \mathrm{Rm}\right) = \nabla_i \nabla_j \nabla_i \left(\nabla^{m-1}\mathrm{Rm}\right) + \mathrm{Rm} * \nabla^m \mathrm{Rm} + \nabla \mathrm{Rm} * \nabla^{m-1}\mathrm{Rm}.$$

Applying the Ricci identities on the first term on the right side implies that

$$\nabla_i \nabla_j \nabla_i \left(\nabla^{m-1}\mathrm{Rm}\right) = \nabla_j \nabla_i \nabla_i \left(\nabla^{m-1}\mathrm{Rm}\right) + \mathrm{Rm} * \nabla^m \mathrm{Rm}.$$

In conclusion, we have proved that

$$\begin{aligned}\nabla_i \nabla_i \left(\nabla^m \mathrm{Rm}\right) &= \nabla_j \nabla_i \nabla_i \left(\nabla^{m-1}\mathrm{Rm}\right) \\ &\quad + \mathrm{Rm} * \nabla^m \mathrm{Rm} + \nabla \mathrm{Rm} * \nabla^{m-1}\mathrm{Rm}.\end{aligned} \tag{2.21}$$

Proceeding similarly, we have

$$\begin{aligned}\nabla_i \left(\nabla^m \mathrm{Rm}\right) \nabla_i f &= \nabla_i \nabla_j \left(\nabla^{m-1}\mathrm{Rm}\right) \nabla_i f \\ &= \nabla_j \nabla_i \left(\nabla^{m-1}\mathrm{Rm}\right) \nabla_i f + \mathrm{Rm}(\nabla f) * \nabla^{m-1}\mathrm{Rm}.\end{aligned}$$

Applying (1.47) it follows that

$$\begin{aligned}\nabla_i \left(\nabla^m \mathrm{Rm}\right) \nabla_i f &= \nabla_j \left(\nabla_i \nabla^{m-1}\mathrm{Rm}\right) \nabla_i f + \nabla \mathrm{Rm} * \nabla^{m-1}\mathrm{Rm} \\ &= \nabla_j \left(\nabla_i \left(\nabla^{m-1}\mathrm{Rm}\right) \nabla_i f\right) - \nabla^2_{ij} f \, \nabla_i \nabla^{m-1}\mathrm{Rm} \\ &\quad + \nabla \mathrm{Rm} * \nabla^{m-1}\mathrm{Rm},\end{aligned}$$

where $\nabla^2_{ij} f := \nabla^2_{e_i, e_j} f$. Since $\nabla^2_{ij} f = -R_{ij} + \frac{\lambda}{2} g_{ij}$, we get that

$$\begin{aligned}\nabla_i \left(\nabla^m \mathrm{Rm}\right) \nabla_i f &= \nabla_j \left(\nabla_i \left(\nabla^{m-1}\mathrm{Rm}\right) \nabla_i f\right) - \frac{\lambda}{2} \nabla^m \mathrm{Rm} \\ &\quad + \mathrm{Rm} * \nabla^m \mathrm{Rm} + \nabla \mathrm{Rm} * \nabla^{m-1}\mathrm{Rm}.\end{aligned}$$

Plugging this identity and (2.21) into (2.20) implies that

$$\begin{aligned}\Delta_f \left(\nabla^m \mathrm{Rm}\right) &= \nabla \left(\Delta_f \nabla^{m-1}\mathrm{Rm}\right) + \frac{\lambda}{2} \nabla^m \mathrm{Rm} \\ &\quad + \mathrm{Rm} * \nabla^m \mathrm{Rm} + \nabla \mathrm{Rm} * \nabla^{m-1}\mathrm{Rm}.\end{aligned}$$

This immediately implies (2.19) by an induction argument. □

We are now ready to prove the Shi-type estimates for curvature.

THEOREM 2.8 (Derivative of curvature estimates). *Let $\left(M^4, g, f\right)$ be a complete noncompact 4-dimensional shrinking gradient Ricci soliton with bounded scalar curvature $R \leq A$. Then there exists a constant $C(k) > 0$, depending only on k, such that*

$$\left|\nabla^k \mathrm{Rm}\right| \leq C(k) \left(A+1\right)^{\frac{k}{2}+1}$$

on $M^4 \setminus D(t_k)$, where $t_k := C(k)(A+1)$.

PROOF. We use induction on k. The case $k=0$ was proved in Theorem 2.6. We assume that for $j = 0, 1, \ldots, k-1$ we have

$$\left|\nabla^j \mathrm{Rm}\right| \leq C(j) \left(A+1\right)^{\frac{j}{2}+1} \tag{2.22}$$

on $M^4 \setminus D(t_j)$.

According to Lemma 2.7 we have

$$\Delta_f \left(\nabla^k \mathrm{Rm}\right) = \left(\frac{k}{2}+1\right) \nabla^k \mathrm{Rm} + \sum_{j=0}^{k} \nabla^j \mathrm{Rm} * \nabla^{k-j} \mathrm{Rm}.$$

Therefore, it follows that

$$\begin{aligned}\Delta_f \left|\nabla^k \mathrm{Rm}\right|^2 \geq & \, 2|\nabla^{k+1}\mathrm{Rm}|^2 - C|\mathrm{Rm}||\nabla^k \mathrm{Rm}|^2 \\ & - C|\nabla^k \mathrm{Rm}| \sum_{j=1}^{k-1} |\nabla^j \mathrm{Rm}||\nabla^{k-j}\mathrm{Rm}|.\end{aligned}$$

Applying (2.22), we get

$$\begin{aligned}\Delta_f \left|\nabla^k \mathrm{Rm}\right|^2 \geq & \, 2|\nabla^{k+1}\mathrm{Rm}|^2 - C\left(A+1\right)|\nabla^k \mathrm{Rm}|^2 \\ & - C\left(A+1\right)^{\frac{k}{2}+2}|\nabla^k \mathrm{Rm}|\end{aligned} \tag{2.23}$$

on $M^4 \setminus D(t_{k-1})$. Using the Kato inequality, this implies

$$\Delta_f |\nabla^k \mathrm{Rm}| \geq -C\left(A+1\right)|\nabla^k \mathrm{Rm}| - C\left(A+1\right)^{\frac{k}{2}+2}. \tag{2.24}$$

By (2.23), applied for $k-1$ derivatives, and using (2.22) we have

$$\begin{aligned}\Delta_f \left|\nabla^{k-1} \mathrm{Rm}\right|^2 \geq & \, 2|\nabla^{k}\mathrm{Rm}|^2 - C\left(A+1\right)|\nabla^{k-1} \mathrm{Rm}|^2 \\ & - C\left(A+1\right)^{\frac{k-1}{2}+2}|\nabla^{k-1} \mathrm{Rm}| \\ \geq & \, 2|\nabla^k \mathrm{Rm}|^2 - C\left(A+1\right)^{k+2}.\end{aligned} \tag{2.25}$$

By (2.24) and (2.25), we obtain

$$\begin{aligned}& \Delta_f \left(|\nabla^k \mathrm{Rm}| + \frac{1}{(A+1)^{\frac{k}{2}}}|\nabla^{k-1}\mathrm{Rm}|^2\right) \\ & \geq \frac{2}{(A+1)^{\frac{k}{2}}}|\nabla^k \mathrm{Rm}|^2 - C\left(A+1\right)|\nabla^k \mathrm{Rm}| - C\left(A+1\right)^{\frac{k}{2}+2} \\ & \geq \frac{1}{(A+1)^{\frac{k}{2}}}\left(|\nabla^k \mathrm{Rm}| + \frac{1}{(A+1)^{\frac{k}{2}}}|\nabla^{k-1}\mathrm{Rm}|^2\right)^2 - C\left(A+1\right)^{\frac{k}{2}+2},\end{aligned}$$

where in the last line we have used (2.22). This implies that the function

$$w := |\nabla^k \mathrm{Rm}| + \frac{1}{(A+1)^{\frac{k}{2}}}|\nabla^{k-1}\mathrm{Rm}|^2$$

satisfies

$$\Delta_f w \geq \frac{1}{(A+1)^{\frac{k}{2}}} w^2 - C\left(A+1\right)^{\frac{k}{2}+2}$$

on $M^4 \setminus D(t_{k-1})$. Applying Lemma 2.4 it follows that $w \leq C\left(A+1\right)^{\frac{k}{2}+1}$ on $M^4 \setminus D(4t_{k-1})$. Consequently,

$$|\nabla^k \mathrm{Rm}| \leq C\left(A+1\right)^{\frac{k}{2}+1}$$

on $M^4 \setminus D\left(t_k\right)$, where $t_k := 4t_{k-1}$. □

REMARK 2.9. While we focus here on 4-dimensional shrinking solitons with bounded scalar curvature, the methods are flexible enough to handle certain unbounded cases. For instance, Cao, Ribeiro, and Zhou [**CRZ21**] established that if a shrinking gradient Ricci soliton (M^4, g, f) satisfies

$$R \leq A + \varepsilon f$$

for some constants $A > 0$ and $0 \leq \varepsilon < 1$, then

$$|\mathrm{Rm}| \leq c_0 + c_1 \varepsilon f^2$$

for some constants $c_0, c_1 > 0$. Certain Shi-type estimates for the derivatives of the curvature can be obtained similarly; see Remark 4 in [**CRZ21**].

2.2. Riemann curvature norm is bounded by scalar curvature

In this section, we establish that for 4-dimensional shrinking gradient Ricci solitons, the norm of the Riemann curvature tensor is bounded by a constant multiple of the scalar curvature. In other words, everywhere on a 4-dimensional shrinker, the Riemann curvature scale is comparable to the scalar curvature scale. This improves the estimate in Theorem 2.1 of the previous section. To lay the groundwork for this result, we initially discuss some general considerations applicable in arbitrary dimensions.

2.2.1. Estimates for f-subharmonic functions.

In the following result, we illustrate how to establish bounds for functions that are f-subharmonic outside a compact set and exhibit sub-quadratic exponential growth.[5]

We continue to denote with $D(t)$ and $\Sigma(t)$ the sublevel and level sets of the potential f, see (2.1).

LEMMA 2.10 (Maximum principle for f-subharmonic functions). *Suppose that (M^n, g, f) is a complete noncompact n-dimensional shrinking gradient Ricci soliton. For $t_0 > C(n)$ large enough, assume that $u : M^n \setminus D(t_0) \to \mathbb{R}$ satisfies*

$$\Delta_f u \geq 0 \quad \text{and} \quad \limsup_{x \to \infty} \frac{u(x)}{\rho(x)} = 0,$$

where

$$\rho := \frac{e^f}{f^{\frac{n}{2}}}. \tag{2.26}$$

Then

$$u \leq \sup_{\Sigma(t_0)} |u| \quad \text{on } M^n \setminus D(t_0).$$

[5]Note, for example, that by (1.39), the function $-f$ is f-subharmonic outside the compact set $\left\{f \leq \frac{n}{2}\right\}$.

PROOF. By $\Delta_f(f) = \frac{n}{2} - f$ and $|\nabla f|^2 = f - R$, it follows that

$$\begin{aligned}\Delta_f \rho &= \rho\left(1 - \frac{n}{2f}\right)\left(\frac{n}{2} - f\right) + \rho\left(\left(1 - \frac{n}{2f}\right)^2 + \frac{n}{2f^2}\right)|\nabla f|^2 \\ &= \frac{n}{2f}\rho - \left(\left(1 - \frac{n}{2f}\right)^2 + \frac{n}{2f^2}\right)\rho R \\ &\leq \frac{n}{2f}\rho.\end{aligned}$$

Similarly, the function

$$\sigma := \frac{e^f}{f^{\frac{n+1}{2}}}$$

satisfies

$$\begin{aligned}\Delta_f \sigma &= \sigma\left(1 - \frac{n+1}{2f}\right)\left(\frac{n}{2} - f\right) + \sigma\left(\left(1 - \frac{n+1}{2f}\right)^2 + \frac{n+1}{2f^2}\right)|\nabla f|^2 \\ &= -\left(\frac{1}{2} - \frac{3(n+1)}{4f}\right)\sigma - \left(\left(1 - \frac{n+1}{2f}\right)^2 + \frac{n+1}{2f^2}\right)\sigma R \\ &\leq -\frac{1}{4}\sigma,\end{aligned}$$

by assuming $t_0 > 3(n+1)$. In conclusion,

$$\Delta_f\left(\frac{e^f}{f^{\frac{n}{2}}} + \frac{e^f}{f^{\frac{n+1}{2}}}\right) \leq 0 \qquad \text{on } M^n \setminus D(t_0)$$

if additionally $t_0 > 4n^2$.

Fix $\varepsilon > 0$ and let

$$v := u - \sup_{\Sigma(t_0)} |u| - \varepsilon\left(\frac{e^f}{f^{\frac{n}{2}}} + \frac{e^f}{f^{\frac{n+1}{2}}}\right).$$

Then v satisfies

$$\Delta_f v \geq 0 \quad \text{on } M^n \setminus D(t_0)$$

and

$$\limsup_{x\to\infty} v(x) \leq 0 \quad \text{and} \quad v \leq 0 \quad \text{on } \Sigma(t_0).$$

The maximum principle implies that $v \leq 0$ on $M^n \setminus D(t_0)$. Since $\varepsilon > 0$ is arbitrary, this proves that

$$u \leq \sup_{\Sigma(t_0)} |u| \quad \text{on } M^n \setminus D(t_0).$$

□

We obtain the following useful consequence, which will be frequently utilized in subsequent sections.

LEMMA 2.11 (Elliptic decay estimate). *Let (M^n, g, f) be a complete noncompact n-dimensional shrinking gradient Ricci soliton. For $\delta > 0$, assume that $t_0 > C(n, \delta)$ is large enough and that $u : M^n \setminus D(t_0) \to \mathbb{R}$ satisfies*

$$\Delta_f u \geq \delta u \quad \textit{and} \quad \limsup_{x\to\infty} \frac{u(x)}{\rho(x)} = 0,$$

where ρ is defined by (2.26). *Then*

$$u \leq 2\left(\frac{t_0}{f}\right)^{\delta} \sup_{\Sigma(t_0)} |u| \quad \textit{on } M^n \setminus D(t_0).$$

PROOF. We have that

$$\begin{aligned}\Delta_f(f^{-\delta}) &= \delta\left(f - \frac{n}{2}\right) f^{-\delta-1} + \delta(\delta+1)|\nabla f|^2 f^{-\delta-2} \\ &\leq \delta f^{-\delta} + \delta^2 f^{-\delta-1}.\end{aligned}$$

Similarly,

$$\begin{aligned}\Delta_f(f^{-\delta-1}) &= (\delta+1)\left(f - \frac{n}{2}\right) f^{-\delta-2} + (\delta+1)(\delta+2)|\nabla f|^2 f^{-\delta-3} \\ &\geq (\delta+1)\left(f - \frac{n}{2}\right) f^{-\delta-2} \\ &\geq \left(\delta + \frac{1}{2}\right) f^{-\delta-1},\end{aligned}$$

assuming that $t_0 > n(\delta+1)$. This implies that

$$\begin{aligned}\Delta_f\left(f^{-\delta} - 2\delta^2 f^{-\delta-1}\right) &\leq \left(\delta f^{-\delta} + \delta^2 f^{-\delta-1}\right) - 2\delta^2\left(\delta + \frac{1}{2}\right) f^{-\delta-1} \\ &= \delta\left(f^{-\delta} - 2\delta^2 f^{-\delta-1}\right).\end{aligned}$$

Therefore, the function

$$\eta := f^{-\delta} - 2\delta^2 f^{-\delta-1}$$

satisfies

$$\Delta_f \eta \leq \delta\eta \quad \text{on } M^n \setminus D(t_0).$$

Consequently, we have that

$$w := u - \frac{\sup_{\Sigma(t_0)} |u|}{\eta(t_0)} \eta$$

satisfies

$$\Delta_f w \geq \delta w \quad \text{on } M^n \setminus D(t_0)$$

and

$$w \leq 0 \quad \text{on } \Sigma(t_0).$$

In particular, $w_+ = \max\{w, 0\}$ satisfies

$$\Delta_f w_+ \geq 0 \quad \text{on } M^n \setminus D(t_0)$$

in the weak sense and $w_+ = 0$ on $\Sigma(t_0)$. By Lemma 2.10 we conclude that $w_+ = 0$; that is,

$$w \leq 0 \quad \text{on } M^n \setminus D(t_0).$$

We may assume $t_0 > 4\delta^2$, so that

$$\eta(t_0) \geq \frac{1}{2t_0^{\delta}}.$$

Then we have obtained that

$$u \leq \sup_{\Sigma(t_0)} |u| \frac{\eta}{\eta(t_0)} \leq 2\left(\frac{t_0}{f}\right)^{\delta} \sup_{\Sigma(t_0)} |u| \quad \text{on } M^n \setminus D(t_0).$$

□

2.2.2. Bounding Riemann curvature by the scalar curvature. Recall that Lu, Yang, and one of the authors [**CLY11**] (cf. Theorem I-4.31) established the following lower bound of the scalar curvature:

$$R \geq \frac{1}{2}\Big(\inf_{\Sigma(t_0)} R\Big)\frac{t_0}{f} \quad \text{on } M^n \setminus D(t_0). \tag{2.27}$$

We are ready to prove the main result of this section, which is a bound of the norm of the Riemann curvature tensor by a constant multiple of the scalar curvature; see Wang and one of the authors [**MW15a**].

THEOREM 2.12 ($|\mathrm{Rm}|$ is bounded by a constant times R). *Let (M^4, g, f) be a complete noncompact 4-dimensional shrinking gradient Ricci soliton with bounded scalar curvature $R \leq A$. Then there exists a universal constant $C_0 > 0$ such that*

$$|\mathrm{Rm}| \leq \frac{C_0(A+1)}{\inf_{\Sigma(t_0)} R} R$$

on $M^4 \setminus D(t_0)$, where $t_0 = C_0(A+1)$.

PROOF. Recall from (1.53) that

$$\Delta_f \mathrm{Rm} = \mathrm{Rm} - \mathrm{Rm} * \mathrm{Rm}.$$

Consequently, we have

$$\Delta_f |\mathrm{Rm}|^2 \geq 2|\nabla \mathrm{Rm}|^2 + 2|\mathrm{Rm}|^2 - C|\mathrm{Rm}|^3. \tag{2.28}$$

In particular, by Theorem 2.6 we obtain that

$$\Delta_f |\mathrm{Rm}|^2 \geq 2|\nabla \mathrm{Rm}|^2 - C(A+1)|\mathrm{Rm}|^2 \tag{2.29}$$

on $M^4 \setminus D(t_0)$. Moreover, by (2.28) and the Kato inequality, it follows that

$$\Delta_f |\mathrm{Rm}| \geq |\mathrm{Rm}| - C|\mathrm{Rm}|^2. \tag{2.30}$$

Hence, for any constant $\alpha > 0$, (2.29) and (2.30) imply that the function

$$u := |\mathrm{Rm}| + \frac{1}{A+1}|\mathrm{Rm}|^2 - \alpha R$$

satisfies

$$\Delta_f u \geq u + \frac{2}{A+1}|\nabla \mathrm{Rm}|^2 + 2\alpha|\mathrm{Ric}|^2 - C|\mathrm{Rm}|^2$$

on $M^4 \setminus D(t_0)$, by additionally using that $\Delta_f R = R - 2|\mathrm{Ric}|^2$. According to Proposition 2.3 we have that

$$|\mathrm{Rm}|^2 \leq C|\mathrm{Ric}|^2 + C\frac{|\nabla \mathrm{Ric}|^2}{|\nabla f|^2}.$$

Hence, for $\alpha \geq C$ large enough, this proves that

$$\frac{2}{A+1}|\nabla \mathrm{Rm}|^2 + 2\alpha|\mathrm{Ric}|^2 \geq C|\mathrm{Rm}|^2 \quad \text{on } M^4 \setminus D(t_0).$$

In conclusion, for α as above, u satisfies

$$\Delta_f u \geq u \quad \text{on } M^4 \setminus D(t_0).$$

Lemma 2.11 implies that

$$u \leq 2\left(\sup_{\Sigma(t_0)} |u|\right)\frac{t_0}{f} \quad \text{on } M^4 \setminus D(t_0).$$

Since by (2.27) and Theorem 2.6 we have that

$$2\left(\sup_{\Sigma(t_0)}|u|\right)\frac{t_0}{f}\leq\frac{C(A+1)}{\inf_{\Sigma(t_0)}R}R,$$

this proves that

$$|\mathrm{Rm}|\leq\frac{C(A+1)}{\inf_{\Sigma(t_0)}R}\,R$$

on $M^4\setminus D(t_0)$, as claimed in the theorem. □

From here on, unless otherwise stated, C denotes a constant with dependency given by

$$C=C\left(\sup_{M^4}R,\ \sup_{D(t_0)}|\mathrm{Rm}|,\ \left(\inf_{D(t_0)}R\right)^{-1}\right),\tag{2.31}$$

for t_0 specified in Theorem 2.12. This constant may change from line to line.

Theorem 2.12 therefore implies that

$$|\mathrm{Rm}|\leq CR\quad\text{on } M^4.\tag{2.32}$$

Moreover, the argument in Theorem 2.8 implies global (i.e., on M^4) Shi-type estimates for curvature.

COROLLARY 2.13. *Assume that (M^4,g,f) is a complete noncompact 4-dimensional shrinking gradient Ricci soliton with bounded scalar curvature. For each $k\geq 0$, there exists a constant $C_k>0$ with dependency specified in (2.31) and additionally depending on k, such that*

$$|\nabla^k\mathrm{Rm}|\leq C_k\quad\text{on } M^4.\tag{2.33}$$

PROOF. The case $k=0$ follows from (2.32). Assume by induction that (2.33) holds for $j=0,1,..,k-1$, therefore

$$|\nabla^j\mathrm{Rm}|\leq C_{k-1}\quad\text{on } M^4\tag{2.34}$$

for any $0\leq j\leq k-1$. According to Lemma 2.7 we have

$$\begin{aligned}\Delta_f|\nabla^k\mathrm{Rm}|^2\geq{}&2|\nabla^{k+1}\mathrm{Rm}|^2-C|\mathrm{Rm}||\nabla^k\mathrm{Rm}|^2\\&-C|\nabla^k\mathrm{Rm}|\sum_{j=1}^{k-1}|\nabla^j\mathrm{Rm}||\nabla^{k-j}\mathrm{Rm}|,\end{aligned}\tag{2.35}$$

where $C>0$ is a universal constant. By (2.34) and the Kato inequality, this implies that

$$\Delta_f|\nabla^k\mathrm{Rm}|\geq -C_k\left(|\nabla^k\mathrm{Rm}|+1\right),$$

for some $C_k>0$ depending on k and as in (2.31). Moreover, the inequality corresponding to (2.35) for $k-1$ derivatives yields

$$\Delta_f|\nabla^{k-1}\mathrm{Rm}|^2\geq 2|\nabla^k\mathrm{Rm}|^2-C_k.$$

Combining these two inequalities we conclude that the function

$$v := |\nabla^k \mathrm{Rm}| + |\nabla^{k-1}\mathrm{Rm}|^2$$

satisfies

$$\Delta_f v \geq v^2 - C_k \quad \text{on } M^4.$$

A cut-off argument as in Lemma 2.4, applied on $B(o, r)$ where o is a minimum point of f and $r > 1$ arbitrary, proves the result. □

2.3. Scale-invariant derivative estimates of Riemann curvature

In this section, our objective is to establish scale-invariant derivative bounds for the Riemann curvature in terms of the scalar curvature. The estimates we seek are of the form

$$|\nabla^k \mathrm{Rm}|^2 \leq C_k R^{k+2}, \tag{2.36}$$

where C_k is as specified in (2.31) and additionally depends on k. Our strategy to prove (2.36) is similar to that of Theorem 2.12, which corresponds to the case $k = 0$. In this sense, it will be useful to evaluate the weighted Laplacian of $|\nabla^k \mathrm{Rm}|^2 R^{-a}$, for $a \geq 0$ to be specified later.

LEMMA 2.14. *Let (M^n, g, f) be a gradient shrinking Ricci soliton. For any integer $k \geq 0$ and any constant $a \geq 0$ we have*

$$\begin{aligned}\Delta_f \left(|\nabla^k \mathrm{Rm}|^2 R^{-a}\right) \geq\ & (k+2-a)\, |\nabla^k \mathrm{Rm}|^2 R^{-a} \\ & + a\,(1-a)\, |\nabla \ln R|^2\, |\nabla^k \mathrm{Rm}|^2 R^{-a} \\ & - C |\nabla^k \mathrm{Rm}| R^{-a} \sum_{j=0}^{k} |\nabla^j \mathrm{Rm}| |\nabla^{k-j}\mathrm{Rm}|,\end{aligned}$$

where $C > 0$ is a universal constant.

PROOF. According to Lemma 2.7 with $\lambda = 1$, we have the following identity:

$$\Delta_f \left(\nabla^k \mathrm{Rm}\right) = \left(\frac{k}{2} + 1\right) \nabla^k \mathrm{Rm} + \sum_{j=0}^{k} \nabla^j \mathrm{Rm} * \nabla^{k-j}\mathrm{Rm}.$$

This implies that

$$\begin{aligned}\Delta_f \left|\nabla^k \mathrm{Rm}\right|^2 \geq\ & 2\left|\nabla^{k+1}\mathrm{Rm}\right|^2 + (k+2)\left|\nabla^k \mathrm{Rm}\right|^2 \\ & - C\left|\nabla^k \mathrm{Rm}\right| \sum_{j=0}^{k} \left|\nabla^j \mathrm{Rm}\right| \left|\nabla^{k-j}\mathrm{Rm}\right|,\end{aligned}$$

for a universal constant $C > 0$. Note also that

$$\begin{aligned}\Delta_f (R^{-a}) &= -a\,(\Delta_f R)\, R^{-a-1} + a\,(a+1)\,|\nabla R|^2\, R^{-a-2} \\ &= -aR^{-a} + 2a\,|\mathrm{Ric}|^2\, R^{-a-1} + a\,(a+1)\,|\nabla R|^2\, R^{-a-2}.\end{aligned}$$

For any $k \geq 0$ and any $a > 0$, we have that

$$\begin{aligned}\Delta_f \left(\left|\nabla^k \mathrm{Rm}\right|^2 R^{-a}\right) =\ & \left(\Delta_f \left|\nabla^k \mathrm{Rm}\right|^2\right) R^{-a} + \left|\nabla^k \mathrm{Rm}\right|^2 \Delta_f(R^{-a}) \\ & + 2\left\langle \nabla \left|\nabla^k \mathrm{Rm}\right|^2, \nabla(R^{-a})\right\rangle.\end{aligned}$$

Consequently, it follows from above that

$$\begin{aligned}\Delta_f\left(\left|\nabla^k \mathrm{Rm}\right|^2 R^{-a}\right) \geq\; & 2\left|\nabla^{k+1}\mathrm{Rm}\right|^2 R^{-a}+(k+2)\left|\nabla^k \mathrm{Rm}\right|^2 R^{-a} \\ & - C\left(\left|\nabla^k \mathrm{Rm}\right| R^{-a}\right)\sum_{j=0}^{k}\left|\nabla^j \mathrm{Rm}\right|\left|\nabla^{k-j}\mathrm{Rm}\right| \\ & + \left|\nabla^k \mathrm{Rm}\right|^2\left(-aR^{-a}+2a\left|\mathrm{Ric}\right|^2 R^{-a-1}\right) \\ & + a(a+1)\left|\nabla \ln R\right|^2\left|\nabla^k \mathrm{Rm}\right|^2 R^{-a} \\ & + 2\left\langle \nabla\left|\nabla^k \mathrm{Rm}\right|^2, \nabla(R^{-a})\right\rangle. \end{aligned} \tag{2.37}$$

Using the Kato inequality $\left|\nabla\left|\nabla^k \mathrm{Rm}\right|\right| \leq \left|\nabla^{k+1}\mathrm{Rm}\right|$ we obtain that

$$\begin{aligned} 2\left\langle \nabla\left|\nabla^k \mathrm{Rm}\right|^2, \nabla(R^{-a})\right\rangle &\geq -4a\left|\nabla^{k+1}\mathrm{Rm}\right|\left|\nabla^k \mathrm{Rm}\right|\left|\nabla R\right| R^{-a-1} \\ &\geq -2\left|\nabla^{k+1}\mathrm{Rm}\right|^2 R^{-a} \\ &\quad - 2a^2\left|\nabla \ln R\right|^2\left|\nabla^k \mathrm{Rm}\right|^2 R^{-a}. \end{aligned}$$

Plugging this into (2.37), it follows that

$$\begin{aligned}\Delta_f\left(|\nabla^k \mathrm{Rm}|^2 R^{-a}\right) \geq\; & (k+2-a)\,|\nabla^k \mathrm{Rm}|^2 R^{-a} \\ & + a(1-a)\left|\nabla \ln R\right|^2 |\nabla^k \mathrm{Rm}|^2 R^{-a} \\ & - C|\nabla^k \mathrm{Rm}| R^{-a}\sum_{j=0}^{k}|\nabla^j \mathrm{Rm}||\nabla^{k-j}\mathrm{Rm}| \end{aligned}$$

for a universal constant $C>0$. This proves the result. □

We first establish (2.36) for $k=1$. Our strategy involves a bootstrap method, where we bound $|\nabla \mathrm{Rm}|^2 R^{-a}$ for increasing powers of a, until we reach the optimal $a=3$.

PROPOSITION 2.15 ($|\nabla \mathrm{Rm}|$ bound by $R^{3/2}$). *Let $\left(M^4, g, f\right)$ be a complete non-compact 4-dimensional shrinking gradient Ricci soliton with bounded scalar curvature. Then there exists $C>0$ such that*

$$|\nabla \mathrm{Rm}|^2 \leq C R^3 \quad \text{on } M^4.$$

PROOF. Applying Lemma 2.14 for $k=1$ implies that

$$\begin{aligned}\Delta_f\left(|\nabla \mathrm{Rm}|^2 R^{-a}\right) \geq\; & (3-a)\,|\nabla \mathrm{Rm}|^2 R^{-a} \\ & + a(1-a)\left|\nabla \ln R\right|^2 |\nabla \mathrm{Rm}|^2 R^{-a} \\ & - C|\mathrm{Rm}||\nabla \mathrm{Rm}|^2 R^{-a}. \end{aligned}$$

Using that $|\mathrm{Rm}| \leq CR$ by (2.32), it follows that

$$\begin{aligned}\Delta_f\left(|\nabla \mathrm{Rm}|^2 R^{-a}\right) \geq\; & (3-a)\,|\nabla \mathrm{Rm}|^2 R^{-a} \\ & + a(1-a)\left|\nabla \ln R\right|^2 |\nabla \mathrm{Rm}|^2 R^{-a} \\ & - C|\nabla \mathrm{Rm}|^2 R^{-a+1}. \end{aligned} \tag{2.38}$$

STEP 1: $|\nabla \mathrm{Rm}|^2$ *bound by* R^2.

Choosing $a = 1$ in (2.38), we obtain

$$\Delta_f \left(|\nabla \mathrm{Rm}|^2 R^{-1}\right) \geq 2|\nabla \mathrm{Rm}|^2 R^{-1} - C|\nabla \mathrm{Rm}|^2.$$

By (2.33) and Young's inequality, we have that

$$\begin{aligned} C|\nabla \mathrm{Rm}|^2 &\leq C|\nabla \mathrm{Rm}|^{\frac{4}{3}} \\ &\leq \left(|\nabla \mathrm{Rm}|^{\frac{4}{3}} R^{-\frac{2}{3}}\right)^{\frac{3}{2}} + C\left(R^{\frac{2}{3}}\right)^3 \\ &= |\nabla \mathrm{Rm}|^2 R^{-1} + CR^2. \end{aligned}$$

Hence, we obtain the following:

$$\Delta_f \left(|\nabla \mathrm{Rm}|^2 R^{-1}\right) \geq |\nabla \mathrm{Rm}|^2 R^{-1} - CR^2.$$

Proceeding as in Theorem 2.12, we note that for any $\alpha > 0$ we have

$$\Delta_f \left(|\nabla \mathrm{Rm}|^2 R^{-1} - \alpha R\right) \geq \left(|\nabla \mathrm{Rm}|^2 R^{-1} - \alpha R\right) + 2\alpha|\mathrm{Ric}|^2 - CR^2.$$

Note that $|\mathrm{Ric}|^2 \geq \frac{1}{4}R^2$. Hence, choosing α so that $2\alpha|\mathrm{Ric}|^2 \geq CR^2$, this proves that the function

$$u := |\nabla \mathrm{Rm}|^2 R^{-1} - \alpha R$$

satisfies

$$\Delta_f u \geq u \quad \text{on } M^4. \tag{2.39}$$

Lemma 2.11 and (2.27) imply that

$$u \leq 2\left(\sup_{\Sigma(t_0)} |u|\right) \frac{t_0}{f} \leq CR$$

on $M^4 \setminus D(t_0)$. Hence, $|\nabla \mathrm{Rm}| \leq CR$ on $M^4 \setminus D(t_0)$. By (2.33) and (2.31), this proves that

$$|\nabla \mathrm{Rm}| \leq CR \quad \text{on } M^4. \tag{2.40}$$

STEP 2: $|\nabla \mathrm{Rm}|^2$ *bound by* $R^{2+\varepsilon}$.

For $\frac{1}{2} > \varepsilon > 0$ to be specified later, we let $a = 1 + \varepsilon$ in (2.38) to obtain that

$$\begin{aligned} \Delta_f \left(|\nabla \mathrm{Rm}|^2 R^{-1-\varepsilon}\right) \geq \frac{3}{2}|\nabla \mathrm{Rm}|^2 R^{-1-\varepsilon} &- C\varepsilon|\nabla \ln R|^2 |\nabla \mathrm{Rm}|^2 R^{-1-\varepsilon} \\ &- C|\nabla \mathrm{Rm}|^2 R^{-\varepsilon}. \end{aligned}$$

Applying (2.40) and Young's inequality, we get that

$$\begin{aligned} C|\nabla \mathrm{Rm}|^2 R^{-\varepsilon} &\leq C|\nabla \mathrm{Rm}| R^{1-\varepsilon} \\ &\leq \frac{1}{4}|\nabla \mathrm{Rm}|^2 R^{-1-\varepsilon} + CR^{3-\varepsilon}. \end{aligned}$$

Furthermore, as $|\nabla \ln R| \leq C$ by (2.40), it follows that

$$\Delta_f \left(|\nabla \mathrm{Rm}|^2 R^{-1-\varepsilon}\right) \geq \left(\frac{5}{4} - C\varepsilon\right)|\nabla \mathrm{Rm}|^2 R^{-1-\varepsilon} - CR^2.$$

We choose $\varepsilon > 0$ small enough so that $\frac{5}{4} - C\varepsilon \geq 1$. We have therefore proved that

$$\Delta_f \left(|\nabla \mathrm{Rm}|^2 R^{-1-\varepsilon}\right) \geq |\nabla \mathrm{Rm}|^2 R^{-1-\varepsilon} - CR^2.$$

For $\alpha > 0$ sufficiently large it follows as in (2.39) that

$$\begin{aligned}\Delta_f\left(|\nabla \mathrm{Rm}|^2 R^{-1-\varepsilon} - \alpha R\right) &\geq \left(|\nabla \mathrm{Rm}|^2 R^{-1-\varepsilon} - \alpha R\right) + 2\alpha|\mathrm{Ric}|^2 - CR^2\\ &\geq |\nabla \mathrm{Rm}|^2 R^{-1-\varepsilon} - \alpha R.\end{aligned}$$

By Lemma 2.11, it follows that

$$|\nabla \mathrm{Rm}|^2 R^{-1-\varepsilon} - \alpha R \leq \frac{C}{f}.$$

Applying (2.27), we conclude that

$$|\nabla \mathrm{Rm}|^2 \leq CR^{2+\varepsilon} \quad \text{on } M^4. \tag{2.41}$$

STEP 3: $|\nabla \mathrm{Rm}|^2$ *bound by* $R^{\frac{5}{2}}$.

We now choose $a = \frac{3}{2}$ in (2.38) to obtain that

$$\begin{aligned}\Delta_f\left(|\nabla \mathrm{Rm}|^2 R^{-\frac{3}{2}}\right) \geq \frac{3}{2}|\nabla \mathrm{Rm}|^2 R^{-\frac{3}{2}} &- C|\nabla \ln R|^2 |\nabla \mathrm{Rm}|^2 R^{-\frac{3}{2}}\\ &- C|\nabla \mathrm{Rm}|^2 R^{-\frac{1}{2}}.\end{aligned}$$

By (2.41) it follows that

$$|\nabla \ln R|^2 |\nabla \mathrm{Rm}|^2 R^{-\frac{3}{2}} + |\nabla \mathrm{Rm}|^2 R^{-\frac{1}{2}} \leq C|\nabla \mathrm{Rm}|^2 R^{-\frac{3}{2}+\varepsilon}.$$

Hence, this proves the following estimate:

$$\Delta_f\left(|\nabla \mathrm{Rm}|^2 R^{-\frac{3}{2}}\right) \geq \frac{3}{2}|\nabla \mathrm{Rm}|^2 R^{-\frac{3}{2}} - C|\nabla \mathrm{Rm}|^2 R^{-\frac{3}{2}+\varepsilon}. \tag{2.42}$$

By (2.40) and Young's inequality, we know that

$$\begin{aligned}C|\nabla \mathrm{Rm}|^2 R^{-\frac{3}{2}+\varepsilon} &\leq C|\nabla \mathrm{Rm}|^{2-\frac{4}{3}\varepsilon} R^{-\frac{3}{2}+\frac{7}{3}\varepsilon}\\ &\leq \frac{1}{4}\left(|\nabla \mathrm{Rm}|^{2-\frac{4}{3}\varepsilon} R^{-\frac{3}{2}+\varepsilon}\right)^{\frac{1}{1-\frac{2\varepsilon}{3}}} + C\left(R^{\frac{4}{3}\varepsilon}\right)^{\frac{3}{2\varepsilon}}\\ &= \frac{1}{4}|\nabla \mathrm{Rm}|^2 R^{-\frac{3}{2}} + CR^2.\end{aligned}$$

By (2.42), this implies that

$$\Delta_f\left(|\nabla \mathrm{Rm}|^2 R^{-\frac{3}{2}}\right) \geq \frac{5}{4}|\nabla \mathrm{Rm}|^2 R^{-\frac{3}{2}} - CR^2. \tag{2.43}$$

For $\alpha > 0$ sufficiently large so that $2\alpha|\mathrm{Ric}|^2 \geq CR^2$, this implies that

$$\begin{aligned}\Delta_f\left(|\nabla \mathrm{Rm}|^2 R^{-\frac{3}{2}} - \alpha R\right) &\geq \left(|\nabla \mathrm{Rm}|^2 R^{-\frac{3}{2}} - \alpha R\right) + \left(2\alpha|\mathrm{Ric}|^2 - CR^2\right)\\ &\geq |\nabla \mathrm{Rm}|^2 R^{-\frac{3}{2}} - \alpha R.\end{aligned}$$

Applying Lemma 2.11 and (2.27) it follows that

$$|\nabla \mathrm{Rm}|^2 \leq CR^{\frac{5}{2}} \quad \text{on } M^4. \tag{2.44}$$

STEP 4: $|\nabla \mathrm{Rm}|^2$ *bound by* R^3.

Letting $a = 2$ in (2.38) implies that

$$\begin{aligned}\Delta_f\left(|\nabla \mathrm{Rm}|^2 R^{-2}\right) \geq |\nabla \mathrm{Rm}|^2 R^{-2} &- C|\nabla \ln R|^2 |\nabla \mathrm{Rm}|^2 R^{-2}\\ &- C|\nabla \mathrm{Rm}|^2 R^{-1}.\end{aligned}$$

By (2.44) we know that $|\nabla \ln R|^2 \leq CR^{\frac{1}{2}}$. Thus, it follows that

$$\Delta_f \left(|\nabla \mathrm{Rm}|^2 R^{-2}\right) \geq |\nabla \mathrm{Rm}|^2 R^{-2} - C|\nabla \mathrm{Rm}|^2 R^{-\frac{3}{2}}. \tag{2.45}$$

By (2.45) and (2.43), we find that

$$\begin{aligned}\Delta_f \left(|\nabla \mathrm{Rm}|^2 R^{-2} + \gamma|\nabla \mathrm{Rm}|^2 R^{-\frac{3}{2}}\right) \geq &\left(|\nabla \mathrm{Rm}|^2 R^{-2} + \gamma|\nabla \mathrm{Rm}|^2 R^{-\frac{3}{2}}\right)\\ &+ \left(\frac{\gamma}{4} - C\right)|\nabla \mathrm{Rm}|^2 R^{-\frac{3}{2}} - \gamma C R^2\end{aligned}$$

for any $\gamma > 0$. We choose γ so that $\frac{\gamma}{4} - C = 0$, to obtain that

$$\begin{aligned}\Delta_f \left(|\nabla \mathrm{Rm}|^2 R^{-2} + \gamma|\nabla \mathrm{Rm}|^2 R^{-\frac{3}{2}}\right) \geq &\left(|\nabla \mathrm{Rm}|^2 R^{-2} + \gamma|\nabla \mathrm{Rm}|^2 R^{-\frac{3}{2}}\right)\\ &- CR^2.\end{aligned}$$

The function

$$v := |\nabla \mathrm{Rm}|^2 R^{-2} + \gamma|\nabla \mathrm{Rm}|^2 R^{-\frac{3}{2}} - \alpha R$$

therefore satisfies

$$\Delta_f v \geq v + \left(2\alpha|\mathrm{Ric}|^2 - CR^2\right).$$

Hence, choosing $\alpha > 0$ large enough so that $2\alpha|\mathrm{Ric}|^2 - CR^2 \geq 0$, we conclude that

$$\Delta_f v \geq v \quad \text{on } M^4.$$

Lemma 2.11 and (2.27) then imply that

$$v \leq CR \quad \text{on } M^4.$$

In conclusion,

$$|\nabla \mathrm{Rm}|^2 R^{-2} \leq CR \quad \text{on } M^4,$$

which proves the result. □

We are now ready to prove the main result of this section, due to Wang and one of the authors [**MW19**], which is a sharp bound on the higher derivatives of curvature.

THEOREM 2.16 (Scaled derivative of curvature estimates). *Let (M^4, g, f) be a complete noncompact 4-dimensional shrinking gradient Ricci soliton with bounded scalar curvature. Then, for any $k \geq 0$ there exists $C_k > 0$ such that*

$$\left|\nabla^k \mathrm{Rm}\right|^2 \leq C_k R^{k+2} \quad \text{on } M^4.$$

PROOF. We use induction on $k \geq 0$. For $k = 0$ the result holds by Theorem 2.12 and for $k = 1$ the result holds by Proposition 2.15. We assume that

$$\left|\nabla^j \mathrm{Rm}\right|^2 \leq C_j R^{j+2} \quad \text{on } M^4, \tag{2.46}$$

for all $0 \leq j \leq k-1$, and we prove that

$$\left|\nabla^k \mathrm{Rm}\right|^2 \leq C_k R^{k+2} \quad \text{on } M^4. \tag{2.47}$$

According to Lemma 2.14,

$$\begin{aligned}\Delta_f \left(|\nabla^k \mathrm{Rm}|^2 R^{-a}\right) \geq\ &(k+2-a)\,|\nabla^k \mathrm{Rm}|^2 R^{-a}\\ &+ a\,(1-a)\,|\nabla \ln R|^2\,|\nabla^k \mathrm{Rm}|^2 R^{-a}\\ &- C\left(|\nabla^k \mathrm{Rm}| R^{-a}\right)\sum_{j=0}^{k} |\nabla^j \mathrm{Rm}||\nabla^{k-j} \mathrm{Rm}|.\end{aligned} \tag{2.48}$$

By the induction hypothesis (2.46), we have that

$$\sum_{j=1}^{k-1} |\nabla^j \mathrm{Rm}||\nabla^{k-j}\mathrm{Rm}| \leq C_k R^{\frac{k}{2}+2}.$$

Together with the estimate $|\mathrm{Rm}| \leq CR$, which was proved in (2.32), this implies that

$$|\nabla^k \mathrm{Rm}|R^{-a} \sum_{j=0}^{k} |\nabla^j \mathrm{Rm}||\nabla^{k-j}\mathrm{Rm}| \leq C_k |\nabla^k \mathrm{Rm}| R^{-a+\frac{k}{2}+2} + C|\nabla^k \mathrm{Rm}|^2 R^{-a+1}.$$

By the inequality

$$2|\nabla^k \mathrm{Rm}|R^{-a+\frac{k}{2}+2} \leq |\nabla^k \mathrm{Rm}|^2 R^{-a+1} + R^{k+3-a},$$

we therefore get that

$$|\nabla^k \mathrm{Rm}|R^{-a} \sum_{j=0}^{k} |\nabla^j \mathrm{Rm}||\nabla^{k-j}\mathrm{Rm}| \leq C_k |\nabla^k \mathrm{Rm}|^2 R^{-a+1} + C_k R^{k+3-a}.$$

Moreover, since $|\nabla \ln R|^2 \leq CR$ by Proposition 2.15, it follows that

$$|\nabla \ln R|^2 \, |\nabla^k \mathrm{Rm}|^2 R^{-a} \leq C|\nabla^k \mathrm{Rm}|^2 R^{-a+1}.$$

In conclusion, we have proved that

$$a\,(1-a)\,|\nabla \ln R|^2 \, |\nabla^k \mathrm{Rm}|^2 R^{-a} - C\left(|\nabla^k \mathrm{Rm}|R^{-a}\right) \sum_{j=0}^{k} |\nabla^j \mathrm{Rm}||\nabla^{k-j}\mathrm{Rm}| \geq -C_k |\nabla^k \mathrm{Rm}|^2 R^{-a+1} - C_k R^{k+3-a}.$$

Plugging this into (2.48) implies that

$$\Delta_f\left(|\nabla^k \mathrm{Rm}|^2 R^{-a}\right) \geq (k+2-a)\,|\nabla^k \mathrm{Rm}|^2 R^{-a} - C_k |\nabla^k \mathrm{Rm}|^2 R^{-a+1} - C_k R^{k+3-a}. \tag{2.49}$$

We first apply (2.49) for $a = k+1$ to obtain that

$$\Delta_f\left(|\nabla^k \mathrm{Rm}|^2 R^{-k-1}\right) \geq |\nabla^k \mathrm{Rm}|^2 R^{-k-1} - C_k |\nabla^k \mathrm{Rm}|^2 R^{-k} - C_k R^2. \tag{2.50}$$

Applying now (2.49) for $a = k$, it follows that

$$\Delta_f\left(|\nabla^k \mathrm{Rm}|^2 R^{-k}\right) \geq 2|\nabla^k \mathrm{Rm}|^2 R^{-k} - C_k |\nabla^k \mathrm{Rm}|^2 R^{-k+1} - C_k R^3. \tag{2.51}$$

According to (2.33), we have $|\nabla^k \mathrm{Rm}| \leq C_k$. Using this and applying Young's inequality, it follows that

$$\begin{aligned} C_k |\nabla^k \mathrm{Rm}|^2 R^{-k+1} &\leq C_k |\nabla^k \mathrm{Rm}|^{\frac{2(k+1)}{k+2}} R^{-k+1} \\ &\leq \frac{1}{2}\left(|\nabla^k \mathrm{Rm}|^{\frac{2(k+1)}{k+2}} R^{-\frac{k(k+1)}{k+2}}\right)^{\frac{k+2}{k+1}} + C_k \left(R^{\frac{2}{k+2}}\right)^{k+2} \\ &= \frac{1}{2}|\nabla^k \mathrm{Rm}|^2 R^{-k} + C_k R^2. \end{aligned}$$

Hence, (2.51) becomes

$$\Delta_f\left(|\nabla^k \mathrm{Rm}|^2 R^{-k}\right) \geq \frac{3}{2}|\nabla^k \mathrm{Rm}|^2 R^{-k} - C_k R^2. \tag{2.52}$$

By (2.50) and (2.52), we infer that the function

$$u := |\nabla^k \mathrm{Rm}|^2 R^{-k-1} + \gamma |\nabla^k \mathrm{Rm}|^2 R^{-k}$$

satisfies

$$\Delta_f u \geq u + \left(\frac{\gamma}{2} - C_k\right)|\nabla^k \mathrm{Rm}|^2 R^{-k} - C_k(\gamma+1)R^2$$

for any $\gamma > 0$. We choose γ so that $\frac{\gamma}{2} - C_k = 0$, which then gives us that

$$\Delta_f u \geq u - C_k R^2.$$

As $\Delta_f R = R - 2|\mathrm{Ric}|^2$, this proves that the function

$$\begin{aligned} w &:= u - \alpha R \\ &= |\nabla^k \mathrm{Rm}|^2 R^{-k-1} + \gamma |\nabla^k \mathrm{Rm}|^2 R^{-k} - \alpha R \end{aligned}$$

satisfies

$$\Delta_f w \geq w + \left(2\alpha |\mathrm{Ric}|^2 - C_k R^2\right) \quad \text{on } M^4.$$

Choosing $\alpha = 2C_k$ implies that $2\alpha|\mathrm{Ric}|^2 - C_k R^2 \geq 0$. Therefore,

$$\Delta_f w \geq w \quad \text{on } M^4.$$

By Lemma 2.11 and (2.27), we conclude that $w \leq C_k R$ on M^4, which in turn implies that

$$|\nabla^k \mathrm{Rm}|^2 \leq C_k R^{k+2} \quad \text{on } M^4.$$

This proves (2.47) and completes the induction step. □

Theorem 2.16 allows us to establish the following decay estimate for the scalar curvature. Assume that (M^4, g, f) is a 4-dimensional shrinking gradient Ricci soliton with bounded scalar curvature. For $t \geq t_0$, define ϕ_t as the solution to the following ODE initial value problem:

$$\begin{aligned} \frac{d\phi_t}{dt} &= \frac{\nabla f}{|\nabla f|^2}(\phi_t), \\ \phi_{t_0} &= \mathrm{Id} \quad \text{on } \overline{M^4 \setminus D(t_0)}. \end{aligned} \tag{2.53}$$

For $x_0 \in \Sigma(t_0)$ fixed, let $R(t) := R(\phi_t(x_0))$, where $t \geq t_0$. Then

$$\begin{aligned} \frac{dR}{dt} &= \frac{\langle \nabla R, \nabla f\rangle}{|\nabla f|^2} \\ &= \frac{\Delta R - R + 2|\mathrm{Ric}|^2}{t - R}. \end{aligned}$$

Using Theorem 2.12 and the estimate $|\Delta R| \leq C R^2$ from Theorem 2.16, we get

$$\left|\frac{dR}{dt} + \frac{R}{t}\right| \leq C_0 \frac{R^2}{t}$$

for some constant $C_0 > 0$ as specified in (2.31). This can be rewritten as

$$\left|\frac{d}{dt}\left(\frac{1}{tR}\right)\right| \leq \frac{C_0}{t^2}.$$

Integrating in t gives

$$\left|\frac{1}{t_2 R(t_2)} - \frac{1}{t_1 R(t_1)}\right| \le C_0 \left(\frac{1}{t_1} - \frac{1}{t_2}\right) \tag{2.54}$$

for any t_1 and t_2 with $t_0 < t_1 < t_2$. In particular, if $R(t_1) \le \frac{1}{2C_0}$, then (2.54) implies that

$$\frac{1}{tR(t)} \ge \frac{1}{t_1}\left(\frac{1}{R(t_1)} - C_0\right) \ge \frac{C_0}{t_1}$$

for all $t_1 < t$. In conclusion, we obtain the following decay estimate of scalar curvature along the integral curves of the potential.

COROLLARY 2.17 (Smallness of R at a point implies R decays quadratically along its integral curve of ∇f). *Assume that (M^4, g, f) is a complete noncompact 4-dimensional shrinking Ricci soliton with bounded scalar curvature. For $t_0 > 0$ large enough, let ϕ_t be the solution of* (2.53). *Then there exists a constant $C_0 > 0$ such that if $x_0 \in \Sigma(t_0)$ satisfies*

$$R(\phi_{t_1}(x_0)) \le \frac{1}{2C_0} \tag{2.55}$$

for some $t_1 \ge t_0$, then

$$R(\phi_t(x_0)) \le \frac{1}{C_0}\frac{t_1}{t} \quad \textit{for all } t \ge t_1. \tag{2.56}$$

The curvature bounds obtained in Theorems 2.6 and 2.12 have some corresponding statements for steady gradient Ricci solitons, which will be presented in Chapter 5.

Related curvature estimates have also been considered for expanding Ricci solitons, leading to results of a similar, but not fully analogous, nature. We will not discuss the details here, but refer the interested reader to Cao and Liu [**CL22**] and Chan [**Cha23**].

2.4. 4-Dimensional shrinkers with scalar curvature tending to zero

Theorems 2.12 and 2.16 show that the scalar curvature of a complete 4-dimensional shrinking gradient Ricci soliton controls the growth of the entire curvature tensor. This has some interesting consequences for the ends of shrinkers on which $R(x)$ decays to zero as $x \to \infty$. For example, it follows from these estimates that, if the scalar curvature tends to zero along integral curves of the gradient of the potential function, the shrinker must be smoothly asymptotic to a cone at infinity. We will give a precise statement and proof of this result below in Theorem 4.11, as part of a more detailed study of asymptotically conical shrinking gradient Ricci solitons in Chapter 4.

For now, we will derive another result in this direction, and show that a shrinking gradient Ricci soliton satisfying the weaker condition

$$\liminf_{x \to \infty} R(x) = 0$$

must have positive asymptotic volume ratio. Here, recall that the **asymptotic volume ratio** of an n-dimensional Riemannian manifold (M^n, g) is defined by

$$\mathrm{AVR}(g) = \limsup_{r \to \infty} \frac{\mathrm{Vol}(B(p,r))}{\omega_n r^n}, \tag{2.57}$$

where ω_n is the volume of the unit Euclidean n-ball. By Theorem I-4.14, the limit in (2.57) always exists for shrinking gradient Ricci solitons, so we may replace "lim sup" by "lim" in the definition above in this case.

PROPOSITION 2.18 (Scalar curvature infimum decay implies AVR > 0). *Suppose that $\left(M^4, g, f\right)$ is a complete noncompact 4-dimensional shrinking gradient Ricci soliton with bounded scalar curvature R and*

$$\liminf_{x\to\infty} R(x) = 0.$$

Then we have

$$\mathrm{AVR}(g) > 0.$$

PROOF. Since R converges to zero along a sequence of points, there exists $x_0 \in \Sigma(t_0)$ and $x_1 = \phi_{t_1}(x_0) \in \Sigma(t_1)$ such that $R(x_1) \leq \frac{1}{4C_0}$, where C_0 is the constant in (2.55). Therefore, there exists a ball $B(x_1, \varepsilon_1)$ such that

$$R(x) \leq \frac{1}{2C_0}$$

for all $x \in B(x_1, \varepsilon_1) \cap \Sigma(t_1)$. As $\Sigma(t_0)$ is diffeomorphic to $\Sigma(t_1)$ via the diffeomorphism ϕ_t defined in (2.53), it follows that there exists $\varepsilon > 0$ such that

$$\phi_{t_1}\big(B(x_0, \varepsilon) \cap \Sigma(t_0)\big) \subset B(x_1, \varepsilon_1) \cap \Sigma(t_1).$$

This proves that

$$R(\phi_{t_1}(y)) \leq \frac{1}{2C_0}$$

for all $y \in B(x_0, \varepsilon) \cap \Sigma(t_0)$. By (2.56), we conclude that

$$R(\phi_t(y)) \leq \frac{1}{C_0}\frac{t_1}{t}$$

for all $y \in B(x_0, \varepsilon) \cap \Sigma(t_0)$ and all $t > t_1$. For $t \geq t_0$ we let

$$\Sigma_0(t) := \phi_t\big(B(x_0, \varepsilon) \cap \Sigma(t_0)\big) \subset \Sigma(t).$$

We have that

$$R(z) \leq \frac{C}{t} \quad \text{for all } z \in \Sigma_0(t). \tag{2.58}$$

By the first variation of area formula (see, for example, [**CC24**, Lemma 12.23]), we know that

$$\frac{d}{dt}\mathrm{Area}(\Sigma_0(t)) = \int_{\Sigma_0(t)} \frac{1}{|\nabla f|^2}\left(\sum_{a=1}^{3} \nabla^2_{aa} f\right) da, \tag{2.59}$$

where $\{e_a\}_{a=1}^{3}$ is an orthonormal frame tangential to $\Sigma(t)$. However,

$$\begin{aligned}\sum_{a=1}^{3} \nabla^2_{aa} f &= \frac{3}{2} - \sum_{a=1}^{3} \mathrm{Ric}(e_a, e_a) \\ &= \frac{3}{2} - R + \mathrm{Ric}(\nu, \nu),\end{aligned}$$

where $\nu = \frac{\nabla f}{|\nabla f|}$. Theorem 2.16 shows that

$$\begin{aligned}
|\mathrm{Ric}(\nu,\nu)| &= \frac{1}{|\nabla f|^2}|\mathrm{Ric}(\nabla f,\nabla f)| \\
&= \frac{1}{2}\frac{1}{|\nabla f|^2}|\langle \nabla R, \nabla f\rangle| \\
&= \frac{1}{2}\frac{1}{|\nabla f|^2}\left|\Delta R - R + 2|\mathrm{Ric}|^2\right| \\
&\leq \frac{C}{f}R.
\end{aligned}$$

By (2.58) we therefore obtain that

$$|\mathrm{Ric}(\nu,\nu)| \leq \frac{C}{t^2} \quad \text{on } \Sigma_0(t).$$

Consequently, on $\Sigma_0(t)$ we have that

$$\begin{aligned}
\frac{1}{|\nabla f|^2}\sum_{a=1}^{3}\nabla^2_{aa}f &= \frac{1}{t-R}\left(\frac{3}{2} - R + \mathrm{Ric}(\nu,\nu)\right) \\
&\geq \frac{3}{2t} - \frac{C}{t^2}.
\end{aligned}$$

In conclusion, (2.59) implies that

$$\frac{d}{dt}\ln \mathrm{Area}(\Sigma_0(t)) \geq \frac{3}{2t} - \frac{C}{t^2}$$

for all $t \geq t_0$. Integrating over t, we get that

$$\mathrm{Area}(\Sigma_0(t)) \geq \frac{1}{C}\mathrm{Area}(\Sigma_0(t_0))\, t^{\frac{3}{2}} \tag{2.60}$$

for all $t \geq t_0$. Recall that by (1.59) and (1.61)

$$\frac{1}{4}r^2(x) - C_0 r(x) \leq f(x) \leq \frac{1}{4}r^2(x) + C_0 r(x),$$

where $r(x) = d(o,x)$ is the distance to a minimum point $o \in M^4$ of f. Therefore, if $x \in D(4t)$, then $\frac{1}{4}r^2(x) - C_0 r(x) \leq 4t$, and hence

$$r(x) \leq 4\sqrt{t} + 3C_0.$$

Similarly, if $x \in M^4 \setminus D(t)$, then $\frac{1}{4}r^2(x) + C_0 r(x) > t$, and hence

$$r(x) \geq 2\sqrt{t} - 3C_0.$$

This implies that

$$D(4t) \setminus D(t) \subset B\left(o, 4\sqrt{t} + 3C_0\right) \setminus B\left(o, 2\sqrt{t} - 3C_0\right).$$

By (2.60) and the co-area formula (see, for example, Lin and Yang's book [**LY02**]), we have that

$$\mathrm{Vol}(D(4t) \setminus D(t)) = \int_t^{4t}\left(\int_{\Sigma(s)}\frac{1}{|\nabla f|}da\right)ds \geq \frac{1}{C}\mathrm{Area}(\Sigma_0(t_0))\, t^2,$$

which implies that

$$\operatorname{Vol}\left(B\left(o,2s+3C_0\right)\setminus B\left(o,s-3C_0\right)\right)\geq\frac{1}{C}\operatorname{Area}\left(\Sigma_0\left(t_0\right)\right)s^4$$

for arbitrary $s=2\sqrt{t}\geq 2\sqrt{t_0}$. This proves the result. □

2.5. Asymptotically cylindrical shrinking solitons

As we have remarked in the introduction to this chapter, the hope is that 4-dimensional Ricci solitons asymptotically split off lines at their infinities. In this case, rescaled limits at the spatial infinities of the 4-dimensional solitons are hoped to be products of $\mathbb{R}$ and 3-dimensional noncollapsed ancient solutions. Now, 3-dimensional nonflat noncollapsed ancient solutions are either $\mathbb{S}^3/\Gamma$ or in turn themselves have $\mathbb{S}^2\times\mathbb{R}$ limits (in particular, for the 3-dimensional Bryant soliton and Perelman's ancient solutions on $\mathbb{S}^3$ and $\mathbb{R}P^3$). Thus, the expectation is that all 4-dimensional shrinking and steady gradient Ricci solitons have limits at their infinities that are isometric to $\mathbb{R}^4$ (in the asymptotically conical case), $(\mathbb{S}^3/\Gamma)\times\mathbb{R}$, or $\mathbb{S}^2\times\mathbb{R}^2$.

The question we address in this section is: what can we say if a glimpse at infinity is equal to $(\mathbb{S}^3/\Gamma)\times\mathbb{R}$ or $\mathbb{S}^2\times\mathbb{R}^2$? More precisely, what can we say about the geometry at infinity of 4-dimensional shrinkers for which there exists some sequence of points tending to infinity about which the rescalings converge to $(\mathbb{S}^3/\Gamma)\times\mathbb{R}$ or $\mathbb{S}^2\times\mathbb{R}^2$?

2.5.1. Curvature pinching estimates for shrinking solitons with bounded scalar curvature.

We begin with the following curvature equation, which is inspired by Hamilton's work on the Ricci flow on 3-manifolds with positive Ricci curvature [**Ham82**]. We continue to denote with $D(t)$ and $\Sigma(t)$ the sublevel and level sets of the potential f, see (2.1). Unless otherwise specified, $C>0$ denotes a constant with dependency specified in (2.31), whose value may change from line to line.

LEMMA 2.19 (Hamilton-type Bochner formula). *Assume $\left(M^4,g,f\right)$ is a 4-dimensional shrinking gradient Ricci soliton with bounded scalar curvature. Then there exists $C>0$ such that*

$$\frac{1}{2}\Delta_F\left(\frac{|\mathrm{Ric}|^2}{R^2}\right)\geq\frac{P}{R^3}-\frac{CR}{f}\tag{2.61}$$

holds on $M^4\setminus D\left(t_0\right)$, where $F:=f-2\ln R$ and P is the non-negative quartic curvature quantity defined by

$$\begin{aligned}P:=&\left(\lambda_1+\lambda_2-\lambda_3\right)^2\left(\lambda_1-\lambda_2\right)^2+\left(\lambda_2+\lambda_3-\lambda_1\right)^2\left(\lambda_2-\lambda_3\right)^2\\&+\left(\lambda_1+\lambda_3-\lambda_2\right)^2\left(\lambda_1-\lambda_3\right)^2,\end{aligned}\tag{2.62}$$

where $\lambda_1\leq\lambda_2\leq\lambda_3$ denote the eigenvalues of the Ricci tensor restricted to the tangent space of the level set $\Sigma\left(t\right)$ for $t\geq t_0$.

PROOF. The equation (see (1.50))

$$\Delta_f R_{ij}=R_{ij}-2R_{kij\ell}R_{k\ell}$$

implies the following identity:

$$\frac{1}{2}\Delta_f\left|\mathrm{Ric}\right|^2=\left|\nabla\mathrm{Ric}\right|^2+\left|\mathrm{Ric}\right|^2-2\,\mathrm{Rm}(\mathrm{Ric}\odot\mathrm{Ric}),\tag{2.63}$$

where

$$\mathrm{Rm}(\mathrm{Ric}\odot\mathrm{Ric}) := R_{kij\ell}R_{ij}R_{k\ell}.$$

We proceed to simplify this formula on a 4-dimensional shrinker by applying Lemma 2.2. Denote by $\{e_1, e_2, e_3\}$ an orthonormal frame tangent to $\Sigma(t)$ and let $e_4 := \nu = \frac{\nabla f}{|\nabla f|}$, the normal vector to $\Sigma(t)$. In what follows, indices $a, b, c \in \{1,2,3\}$ and $i, j, k \in \{1,2,3,4\}$. We assume everywhere that $t \geq t_0$ is large enough. Let

$$\begin{aligned} |\mathrm{Ric}_\Sigma|^2 &:= \sum_{a,b} |R_{ab}|^2, \\ \mathrm{Ric}_\Sigma^3 &:= \sum_{a,b,c} R_{ab}R_{bc}R_{ac}, \\ R_\Sigma &:= R - R_{44}. \end{aligned} \tag{2.64}$$

CLAIM 2.20. *There exists $C > 0$ such that*

$$\left|\mathrm{Rm}(\mathrm{Ric}\odot\mathrm{Ric}) - \frac{5}{2}|\mathrm{Ric}_\Sigma|^2 R_\Sigma + 2\mathrm{Ric}_\Sigma^3 + \frac{1}{2}R_\Sigma^3\right| \leq \frac{CR^3}{f} \quad \text{on } M^4 \setminus D(t_0).$$

PROOF OF CLAIM 2.20. We have that

$$\begin{aligned} \mathrm{Rm}(\mathrm{Ric}\odot\mathrm{Ric}) = {} & R_{abcd}R_{bc}R_{ad} + 4R_{abc4}R_{bc}R_{a4} + 2R_{4bc4}R_{bc}R_{44} \\ & - 2R_{a44b}R_{a4}R_{b4}. \end{aligned}$$

Moreover,

$$R_{abc4}R_{bc}R_{a4} = R_{ijk4}R_{jk}R_{i4} - R_{4bc4}R_{bc}R_{44} + R_{a44b}R_{a4}R_{b4},$$

which implies that

$$\begin{aligned} \mathrm{Rm}(\mathrm{Ric}\odot\mathrm{Ric}) &= R_{abcd}R_{bc}R_{ad} + 4R_{ijk4}R_{jk}R_{i4} \\ &\quad - 2R_{4bc4}R_{bc}R_{44} + 2R_{4bc4}R_{b4}R_{c4} \\ &= R_{abcd}R_{bc}R_{ad} + 4R_{ijk4}R_{jk}R_{i4} \\ &\quad - 2R_{4ij4}R_{ij}R_{44} + 2R_{4ij4}R_{i4}R_{j4}. \end{aligned} \tag{2.65}$$

According to Lemma 2.2,

$$\begin{aligned} R_{abcd} = {} & R_{ad}g_{bc} - R_{ac}g_{bd} + R_{bc}g_{ad} - R_{bd}g_{ac} \\ & - \frac{1}{2}R\left(g_{ad}g_{bc} - g_{ac}g_{bd}\right) \\ & - R_{a44d}g_{bc} + R_{a44c}g_{bd} - R_{b44c}g_{ad} + R_{b44d}g_{ac} \\ & + R_{44}\left(g_{ad}g_{bc} - g_{ac}g_{bd}\right). \end{aligned}$$

Therefore, it follows that

$$\begin{aligned} R_{abcd}R_{bc}R_{ad} = {} & 2|\mathrm{Ric}_\Sigma|^2 R_\Sigma - 2\mathrm{Ric}_\Sigma^3 - \frac{1}{2}RR_\Sigma^2 + \frac{1}{2}R|\mathrm{Ric}_\Sigma|^2 \\ & + 2R_{4ij4}R_{ib}R_{jb} - 2R_\Sigma R_{4ij4}R_{ij} \\ & + R_{44}R_\Sigma^2 - R_{44}|\mathrm{Ric}_\Sigma|^2. \end{aligned}$$

We rewrite this as

$$\begin{aligned}R_{abcd}R_{bc}R_{ad} &= \frac{5}{2}\left|\mathrm{Ric}_\Sigma\right|^2 R_\Sigma - 2\mathrm{Ric}_\Sigma^3 - \frac{1}{2}R_\Sigma^3 \\ &\quad + 2R_{4ij4}R_{ik}R_{jk} - 2RR_{4ij4}R_{ij} \\ &\quad - 2R_{4ij4}R_{i4}R_{j4} + 2R_{4ij4}R_{ij}R_{44} \\ &\quad + \frac{1}{2}R_{44}R_\Sigma^2 - \frac{1}{2}R_{44}\left|\mathrm{Ric}_\Sigma\right|^2.\end{aligned}$$

In conclusion, we have that

$$\mathrm{Rm}(\mathrm{Ric}\odot\mathrm{Ric}) = \frac{5}{2}\left|\mathrm{Ric}_\Sigma\right|^2 R_\Sigma - 2\mathrm{Ric}_\Sigma^3 - \frac{1}{2}R_\Sigma^3 + \Gamma, \tag{2.66}$$

where

$$\begin{aligned}\Gamma &:= 2R_{4ij4}R_{ik}R_{jk} - 2RR_{4ij4}R_{ij} + 4R_{ijk4}R_{jk}R_{i4} \\ &\quad + \frac{1}{2}R_{44}R_\Sigma^2 - \frac{1}{2}R_{44}\left|\mathrm{Ric}_\Sigma\right|^2.\end{aligned} \tag{2.67}$$

It remains to prove that $|\Gamma| \leq \frac{CR^3}{f}$. We have that

$$\begin{aligned}R_{4ij4} &= R(\nu, e_i, e_j, \nu) \\ &= \frac{1}{|\nabla f|^2}R_{kij\ell}\nabla_k f\nabla_\ell f \\ &= \frac{1}{|\nabla f|^2}\left(\nabla_k R_{ij} - \nabla_i R_{jk}\right)\nabla_k f.\end{aligned}$$

We note that

$$\nabla_k f\nabla_k R_{ij} = \Delta R_{ij} - R_{ij} + 2R_{pijq}R_{pq}$$

and that

$$\begin{aligned}\nabla_k f\nabla_i R_{jk} &= \nabla_i\left(R_{jk}\nabla_k f\right) - R_{jk}\nabla_{ik}^2 f \\ &= \frac{1}{2}\nabla_{ij}^2 R - \frac{1}{2}R_{ij} + R_{ik}R_{jk}.\end{aligned}$$

By Theorem 2.16, we therefore obtain that

$$|R_{4ij4}| \leq \frac{CR}{f}. \tag{2.68}$$

Similarly, we have

$$|R_{ijk4}| \leq \frac{CR^{\frac{3}{2}}}{\sqrt{f}}. \tag{2.69}$$

By (2.68) and (2.69) it follows that

$$|\Gamma| \leq \frac{CR^3}{f}. \tag{2.70}$$

This proves Claim 2.20. □

Using Claim 2.20 and (2.63), it follows that

$$\frac{1}{2}\Delta_f\left|\mathrm{Ric}\right|^2 = \left|\nabla\mathrm{Ric}\right|^2 + \left|\mathrm{Ric}\right|^2 - 5\left|\mathrm{Ric}_\Sigma\right|^2 R_\Sigma + 4\mathrm{Ric}_\Sigma^3 + R_\Sigma^3 - 2\Gamma,$$

for Γ specified in (2.67). By

$$\frac{1}{2}\Delta_f\left(|\mathrm{Ric}|^2R^{-2}\right)=\left(\frac{1}{2}\Delta_f|\mathrm{Ric}|^2\right)R^{-2}+\left(\frac{1}{2}\Delta_f(R^{-2})\right)|\mathrm{Ric}|^2$$
$$+\left\langle\nabla|\mathrm{Ric}|^2,\nabla(R^{-2})\right\rangle$$

and

$$\frac{1}{2}\Delta_f(R^{-2})=-\left(R-2|\mathrm{Ric}|^2\right)R^{-3}+3|\nabla R|^2R^{-4},$$

we therefore obtain that

$$\begin{aligned}\frac{1}{2}\Delta_f\left(|\mathrm{Ric}|^2R^{-2}\right)&=|\nabla\mathrm{Ric}|^2R^{-2}+3|\nabla R|^2R^{-4}|\mathrm{Ric}|^2\\&\quad+\left\langle\nabla|\mathrm{Ric}|^2,\nabla(R^{-2})\right\rangle\\&\quad+\left(|\mathrm{Ric}|^2-5|\mathrm{Ric}_\Sigma|^2R_\Sigma+4\mathrm{Ric}_\Sigma^3+R_\Sigma^3\right)R^{-2}\\&\quad-\left(R-2|\mathrm{Ric}|^2\right)|\mathrm{Ric}|^2R^{-3}-2\Gamma R^{-2}.\end{aligned}$$

Note that

$$\begin{aligned}&|\nabla\mathrm{Ric}|^2R^{-2}+3|\nabla R|^2R^{-4}|\mathrm{Ric}|^2+\left\langle\nabla|\mathrm{Ric}|^2,\nabla(R^{-2})\right\rangle\\&\quad=|R\nabla\mathrm{Ric}-\nabla R\otimes\mathrm{Ric}|^2R^{-4}-\left\langle\nabla\left(|\mathrm{Ric}|^2R^{-2}\right),\nabla\ln R\right\rangle\\&\quad\geq-\left\langle\nabla\left(|\mathrm{Ric}|^2R^{-2}\right),\nabla\ln R\right\rangle.\end{aligned}$$

Hence, as we recall that

$$F=f-2\ln R,$$

we arrive at

$$\begin{aligned}\frac{1}{2}\Delta_F\left(|\mathrm{Ric}|^2R^{-2}\right)\geq&\left(-5R|\mathrm{Ric}_\Sigma|^2R_\Sigma+4R\mathrm{Ric}_\Sigma^3+RR_\Sigma^3+2|\mathrm{Ric}|^4\right)R^{-3}\\&-2\Gamma R^{-2}.\end{aligned}$$

Moreover, we have

$$\begin{aligned}R&=R_\Sigma+R_{44},\\|\mathrm{Ric}|^2&=|\mathrm{Ric}_\Sigma|^2+2|R_{i4}|^2-|R_{44}|^2.\end{aligned}\tag{2.71}$$

Hence, we obtain that (cf. Lemma 10.3 in Hamilton [**Ham82**])

$$\begin{aligned}\frac{1}{2}\Delta_F\left(|\mathrm{Ric}|^2R^{-2}\right)\geq&\left(-5|\mathrm{Ric}_\Sigma|^2R_\Sigma^2+R_\Sigma^4+4R_\Sigma\mathrm{Ric}_\Sigma^3+2|\mathrm{Ric}_\Sigma|^4\right)R^{-3}\\&-\Gamma_0,\end{aligned}$$

where

$$\begin{aligned}\Gamma_0:=\,&2\Gamma R^{-2}+5R_{44}|\mathrm{Ric}_\Sigma|^2R_\Sigma R^{-3}-4R_{44}\mathrm{Ric}_\Sigma^3R^{-3}-R_{44}R_\Sigma^3R^{-3}\\&-8|R_{i4}|^4R^{-3}-2|R_{44}|^4R^{-3}+8|R_{i4}|^2|R_{44}|^2R^{-3}\\&-8|\mathrm{Ric}_\Sigma|^2|R_{i4}|^2R^{-3}+4|\mathrm{Ric}_\Sigma|^2|R_{44}|^2R^{-3}.\end{aligned}$$

By (2.68), (2.69), and (2.70) we know that

$$|\Gamma_0|\leq\frac{CR}{f}.$$

Finally, denote by $\lambda_1 \leq \lambda_2 \leq \lambda_3$ the eigenvalues of $\mathrm{Ric}|_{T_x\Sigma}$ at a point $x \in \Sigma(t)$. Then (2.64) implies that

$$\begin{aligned}\mathrm{Ric}_\Sigma^3 &= \lambda_1^3 + \lambda_2^3 + \lambda_3^3,\\ |\mathrm{Ric}_\Sigma|^2 &= \lambda_1^2 + \lambda_2^2 + \lambda_3^2,\\ R_\Sigma &= \lambda_1 + \lambda_2 + \lambda_3.\end{aligned}$$

By a direct computation we obtain the following (cf. [**CK04**, Lemma 6.34][6]):

$$\begin{aligned}&-5\,|\mathrm{Ric}_\Sigma|^2 R_\Sigma^2 + R_\Sigma^4 + 4R_\Sigma \mathrm{Ric}_\Sigma^3 + 2\,|\mathrm{Ric}_\Sigma|^4\\ &= 2(\lambda_1^4 + \lambda_2^4 + \lambda_3^4 + \lambda_1^2\lambda_2\lambda_3 + \lambda_1\lambda_2^2\lambda_3 + \lambda_1\lambda_2\lambda_3^2\\ &\quad - \lambda_1^3\lambda_2 - \lambda_1\lambda_2^3 - \lambda_1^3\lambda_3 - \lambda_1\lambda_3^3 - \lambda_2^3\lambda_3 - \lambda_2\lambda_3^3)\\ &= (\lambda_1 + \lambda_2 - \lambda_3)^2(\lambda_1 - \lambda_2)^2 + (\lambda_2 + \lambda_3 - \lambda_1)^2(\lambda_2 - \lambda_3)^2\\ &\quad + (\lambda_1 + \lambda_3 - \lambda_2)^2(\lambda_1 - \lambda_3)^2.\end{aligned}$$

In conclusion,

$$\frac{1}{2}\Delta_F\left(|\mathrm{Ric}|^2 R^{-2}\right) \geq PR^{-3} - \frac{CR}{f},$$

for

$$\begin{aligned}P = {} &(\lambda_1 + \lambda_2 - \lambda_3)^2(\lambda_1 - \lambda_2)^2 + (\lambda_2 + \lambda_3 - \lambda_1)^2(\lambda_2 - \lambda_3)^2\\ &+ (\lambda_1 + \lambda_3 - \lambda_2)^2(\lambda_1 - \lambda_3)^2.\end{aligned}$$

This completes the proof of the lemma. □

For future reference, let us record that

$$|\mathrm{Ric}_\Sigma|^2 = \sum_{a,b=1}^{3} |R_{ab}|^2 = \lambda_1^2 + \lambda_2^2 + \lambda_3^2, \tag{2.72a}$$

$$R_\Sigma = R - R_{44} = \lambda_1 + \lambda_2 + \lambda_3, \tag{2.72b}$$

where λ_1, λ_2, and λ_3 are the eigenvalues of the Ricci curvature restricted to $T\Sigma(t)$.

Let $\{e_1, e_2, e_3\}$ be an orthonormal frame tangent to $\Sigma(t)$ for $t \geq t_0$ and let $e_4 := \frac{\nabla f}{|\nabla f|}$ be the normal vector to $\Sigma(t)$. The indices $a, b, c \in \{1, 2, 3\}$ and $i, j, k \in \{1, 2, 3, 4\}$.

As Lemma 2.10 does not seem to apply to (2.61) directly, we take a different approach that transforms (2.61) into an equation on each hypersurface $\Sigma := \Sigma(t)$. To achieve this, we remark that for any function w on $M^4 \setminus D(t_0)$ the gradient and the Laplacian on the submanifold Σ are given by

$$\begin{aligned}\nabla_\Sigma w &= \nabla w - \frac{\langle \nabla w, \nabla f\rangle}{|\nabla f|^2}\nabla f,\\ \Delta_\Sigma w &= \Delta w - \nabla_{44}^2 w - H\nabla_4 w\\ &= \Delta w - \frac{\nabla^2 w(\nabla f, \nabla f)}{|\nabla f|^2} - \frac{\langle \nabla w, \nabla f\rangle}{|\nabla f|^2}\left(\frac{3}{2} - R_\Sigma\right),\end{aligned} \tag{2.73}$$

[6]In [**CK04**] the Riemann curvature operator Rm is diagonalized with eigenvalues ordered as $\lambda \geq \mu \geq \nu$. These correspond to the eigenvalues of the Ricci tensor via the relations $\lambda_3 = \frac{\lambda+\mu}{2}$, $\lambda_2 = \frac{\lambda+\nu}{2}$, and $\lambda_1 = \frac{\mu+\nu}{2}$.

where we used that the mean curvature H of Σ is given by

$$H = \frac{1}{|\nabla f|} \sum_{a=1}^{3} \nabla^2_{aa} f = \frac{1}{|\nabla f|} \left(\frac{3}{2} - R_\Sigma \right).$$

Define the operator $\mathcal{L}$ on functions $w : M^4 \setminus D(t_0) \to \mathbb{R}$ by

$$\mathcal{L}w := \Delta_\Sigma w - \langle \nabla f, \nabla w \rangle + 2 \langle \nabla_\Sigma \ln R, \nabla_\Sigma w \rangle . \tag{2.74}$$

Using (2.73), we see that $\Delta_F w$ and $\mathcal{L}w$ are related by the formula

$$\begin{aligned} \Delta_F w = \mathcal{L}w &+ \frac{\nabla^2 w(\nabla f, \nabla f)}{|\nabla f|^2} + \frac{\langle \nabla w, \nabla f \rangle}{|\nabla f|^2} \left(\frac{3}{2} - R_\Sigma \right) \\ &+ \frac{2}{|\nabla f|^2} \langle \nabla \ln R, \nabla f \rangle \langle \nabla w, \nabla f \rangle . \end{aligned} \tag{2.75}$$

We will apply (2.75) for $w = \frac{|\mathrm{Ric}|^2}{R^2}$ to obtain the following estimate of $\mathcal{L}w$:

LEMMA 2.21 (Bochner formula along level sets of f). *Let (M^4, g, f) be a 4-dimensional shrinking gradient Ricci soliton with bounded scalar curvature. Then the function*

$$w := \frac{|\mathrm{Ric}|^2}{R^2}$$

satisfies

$$\mathcal{L}w \geq \frac{2P}{R^3} - \frac{CR}{f} \quad \textit{on } M^4 \setminus D(t_0),$$

where P is defined by (2.62) and where $\mathcal{L}$ is defined by (2.74).

PROOF. Our goal is to prove that each of the last three terms on the right side of (2.75) can be estimated by $\frac{CR}{f}$. Let

$$u := |\mathrm{Ric}|^2 \quad \text{and} \quad v := R^2,$$

so that $w = \frac{u}{v}$. The first term in (2.75) that we estimate is $\frac{\nabla^2 w(\nabla f, \nabla f)}{|\nabla f|^2}$, and in this sense we apply the formula

$$\begin{aligned} \nabla^2 w(\nabla f, \nabla f) = &\frac{\nabla^2 u(\nabla f, \nabla f)}{v} - \frac{u \nabla^2 v(\nabla f, \nabla f)}{v^2} \\ &+ 2\frac{u \langle \nabla v, \nabla f \rangle^2}{v^3} - 2\frac{\langle \nabla u, \nabla f \rangle \langle \nabla v, \nabla f \rangle}{v^2}. \end{aligned} \tag{2.76}$$

Our goal is to prove that $\left|\nabla^2 w(\nabla f, \nabla f)\right| \leq CR$. We will estimate the difference of the first two terms and that of the last two terms on the right-hand side of (2.76) separately.

STEP 1: *Bounding the last two terms on the right side of* (2.76).

First, by (2.63) it follows that

$$\langle \nabla u, \nabla f \rangle + 2u = \Delta |\mathrm{Ric}|^2 - 2 |\nabla \mathrm{Ric}|^2 + 4 \, \mathrm{Rm}(\mathrm{Ric} \odot \mathrm{Ric}).$$

According to Theorem 2.16, we therefore have

$$|\langle \nabla u, \nabla f \rangle + 2u| \leq CR^3. \tag{2.77}$$

Moreover,

$$\langle \nabla v, \nabla f \rangle + 2v = 2R\Delta R + 4R |\mathrm{Ric}|^2 .$$

Therefore, we have

$$|\langle\nabla v, \nabla f\rangle + 2v| \leq CR^3. \tag{2.78}$$

In conclusion, we find that

$$\left|\frac{u\langle\nabla v, \nabla f\rangle^2}{v^3} - \frac{\langle\nabla u, \nabla f\rangle\langle\nabla v, \nabla f\rangle}{v^2}\right| \leq CR. \tag{2.79}$$

Furthermore, as

$$\langle\nabla w, \nabla f\rangle = \frac{\langle\nabla u, \nabla f\rangle}{v} - \frac{\langle\nabla v, \nabla f\rangle u}{v^2},$$

it follows from (2.77) and (2.78) that

$$|\langle\nabla w, \nabla f\rangle| \leq CR. \tag{2.80}$$

STEP 2: *Bounding the first two terms on the right side of* (2.76).

We now proceed to estimate

$$\left|\frac{\nabla^2 u(\nabla f, \nabla f)}{v} - \frac{u\nabla^2 v(\nabla f, \nabla f)}{v^2}\right|$$

using similar, but more intricate, computations. First, note that

$$\begin{aligned}\nabla^2 v(\nabla f, \nabla f) &= 2\langle\nabla R, \nabla f\rangle^2 + 2R\nabla^2 R(\nabla f, \nabla f)\\ &= 2\langle\nabla R, \nabla f\rangle^2 + 2R\langle\nabla\langle\nabla R, \nabla f\rangle, \nabla f\rangle\\ &\quad - 2R\nabla^2 f(\nabla R, \nabla f).\end{aligned}$$

We have that

$$\langle\nabla\langle\nabla R, \nabla f\rangle, \nabla f\rangle = \left\langle\nabla\left(\Delta R - R + 2|\mathrm{Ric}|^2\right), \nabla f\right\rangle. \tag{2.81}$$

Using Lemma 2.7 and Theorem 2.16, we see that

$$\left|\Delta_f\left(\nabla^2\mathrm{Rm}\right)\right| \leq CR^2, \tag{2.82}$$

which in particular yields the inequality

$$|\Delta_f(\Delta R)| \leq CR^2.$$

By Theorem 2.16, we therefore have that

$$|\langle\nabla(\Delta R), \nabla f\rangle| \leq CR^2.$$

From $\Delta_f R = R - 2|\mathrm{Ric}|^2$ we know that

$$|\langle\nabla R, \nabla f\rangle + R| \leq CR^2. \tag{2.83}$$

Therefore, (2.81) implies that

$$|\langle\nabla\langle\nabla R, \nabla f\rangle, \nabla f\rangle - R| \leq CR^2. \tag{2.84}$$

Finally, by $\nabla^2 f = \frac{1}{2}g - \mathrm{Ric}$ and $\mathrm{Ric}(\nabla f) = \frac{1}{2}\nabla R$, it follows that

$$\begin{aligned}2R\nabla^2 f(\nabla R, \nabla f) &= R\langle\nabla R, \nabla f\rangle - 2R\mathrm{Ric}(\nabla R, \nabla f)\\ &= -R^2 + R\Delta R + 2R|\mathrm{Ric}|^2 - R|\nabla R|^2.\end{aligned}$$

Theorem 2.16 implies that

$$\left|2R\nabla^2 f(\nabla R, \nabla f) + R^2\right| \leq CR^3. \tag{2.85}$$

In conclusion, by (2.83), (2.84), and (2.85), we have established that

$$\left|\nabla^2 v(\nabla f, \nabla f) - 5v\right| \leq CR^3. \tag{2.86}$$

Similarly, we have

$$\begin{aligned}\nabla^2 u(\nabla f, \nabla f) &= \langle \nabla \langle \nabla u, \nabla f \rangle, \nabla f \rangle - \nabla^2 f(\nabla u, \nabla f) \\ &= - \langle \nabla (\Delta_f u), \nabla f \rangle + \langle \nabla(\Delta u), \nabla f \rangle - \frac{1}{2} \langle \nabla u, \nabla f \rangle \\ &\quad + \frac{1}{2} \langle \nabla u, \nabla R \rangle,\end{aligned}$$

where in the second line we used that $\nabla^2 f = \frac{1}{2} g - \mathrm{Ric}$ and $\mathrm{Ric}(\nabla f) = \frac{1}{2} \nabla R$. According to (2.63),

$$\begin{aligned}\langle \nabla (\Delta_f u), \nabla f \rangle = 2 \left\langle \nabla |\nabla \mathrm{Ric}|^2, \nabla f \right\rangle + 2 \left\langle \nabla |\mathrm{Ric}|^2, \nabla f \right\rangle \\ - 4 \langle \nabla (\mathrm{Rm}(\mathrm{Ric} \odot \mathrm{Ric})), \nabla f \rangle.\end{aligned}$$

Applying Theorem 2.16 similarly to the above, we obtain that

$$\begin{aligned}\left| \left\langle \nabla |\nabla \mathrm{Ric}|^2, \nabla f \right\rangle \right| &\leq C R^3, \\ |\langle \nabla (\mathrm{Rm}(\mathrm{Ric} \odot \mathrm{Ric})), \nabla f \rangle| &\leq C R^3.\end{aligned}$$

Moreover, by (2.77), we have

$$\left| \left\langle \nabla |\mathrm{Ric}|^2, \nabla f \right\rangle + 2u \right| \leq C R^3. \tag{2.87}$$

This proves that

$$|\langle \nabla (\Delta_f u), \nabla f \rangle + 4u| \leq C R^3. \tag{2.88}$$

Tracing (2.82) we deduce that

$$|\Delta_f (\Delta \mathrm{Ric})| \leq C R^2,$$

which, by Theorem 2.16, implies that

$$|\langle \nabla (\Delta \mathrm{Ric}), \nabla f \rangle| \leq C R^2.$$

In turn, this implies that

$$|\langle \nabla(\Delta u), \nabla f \rangle| \leq C R^3. \tag{2.89}$$

Finally, Theorem 2.16 implies that

$$|\langle \nabla u, \nabla R \rangle| \leq C R^3.$$

Consequently, by (2.87), (2.88), and (2.89), it follows that

$$\left| \nabla^2 u(\nabla f, \nabla f) - 5u \right| \leq C R^3. \tag{2.90}$$

From (2.86) and (2.90) we infer that

$$\left| \frac{\nabla^2 u(\nabla f, \nabla f)}{v} - \frac{u \nabla^2 v(\nabla f, \nabla f)}{v^2} \right| \leq C R. \tag{2.91}$$

STEP 3: *Concluding the proof.*

Plugging (2.79) and (2.91) into (2.76) proves that

$$\left| \nabla^2 w(\nabla f, \nabla f) \right| \leq C R.$$

The last two terms in (2.75) can be estimated by

$$2\left|\langle\nabla\ln R,\nabla f\rangle\langle\nabla w,\nabla f\rangle\right|+\left|\langle\nabla w,\nabla f\rangle\left(\frac{3}{2}-R_{\Sigma}\right)\right|\leq C\left|\langle\nabla w,\nabla f\rangle\right|,$$

using that

$$\begin{aligned}\left|\langle\nabla R,\nabla f\rangle+R\right|&\leq|\Delta R|+2\left|\mathrm{Ric}\right|^2\\&\leq CR^2.\end{aligned}$$

Hence, in view of (2.80), this implies that

$$2\left|\langle\nabla\ln R,\nabla f\rangle\langle\nabla w,\nabla f\rangle\right|+\left|\langle\nabla w,\nabla f\rangle\left(\frac{3}{2}-R_{\Sigma}\right)\right|\leq CR.$$

Summing up, by (2.75), this yields

$$\Delta_F w\leq\mathcal{L}w+\frac{CR}{f}.$$

On the other hand, by (2.61) we have

$$\Delta_F w\geq 2PR^{-3}-\frac{CR}{f}.$$

In conclusion,

$$\mathcal{L}w\geq 2PR^{-3}-\frac{CR}{f}.$$

This proves the lemma. □

We will derive several consequences of Lemma 2.21. We start with a curvature pinching estimate that holds for any 4-dimensional shrinking gradient Ricci soliton with bounded scalar curvature.

THEOREM 2.22 (Ricci curvature pinching estimate). *Let (M^4,g,f) be a complete noncompact 4-dimensional shrinking gradient Ricci soliton with bounded scalar curvature. If $R>0$, then there exists a constant $C>0$ such that*

$$\frac{\left|\mathrm{Ric}\right|^2}{R^2}\leq\frac{1}{2}+\frac{C}{R\ln f}\quad\textit{on } M^4\setminus D\left(t_0\right).\tag{2.92}$$

PROOF. We apply the maximum principle along with Lemma 2.21. To this end, we first compute the operator $\mathcal{L}$, as defined in (2.74), for a function analogous to the right-hand side of (2.92). Specifically, define the function

$$\rho:=\frac{1}{R\ln^2 f}+\frac{1}{\ln f}.\tag{2.93}$$

CLAIM 2.23. *There exists a constant $C>0$, with dependency as specified in (2.31), such that*

$$\mathcal{L}\rho\leq-\frac{1}{2R\ln^2 f}+\frac{C}{\ln^2 f}\quad\textit{on } M^4\setminus D(t_0).\tag{2.94}$$

PROOF OF CLAIM 2.23. Let us note that for $F = f - 2\ln R$, we have

$$\begin{aligned}\Delta_F(R^{-1}) &= \Delta_f(R^{-1}) + 2\left\langle \nabla \ln R, \nabla(R^{-1})\right\rangle \\ &= -\left(\Delta_f R\right) R^{-2} \\ &= -R^{-1} + 2\left|\mathrm{Ric}\right|^2 R^{-2} \\ &\leq -R^{-1} + C.\end{aligned} \tag{2.95}$$

By (2.75), it follows that

$$\begin{aligned}\Delta_F(R^{-1}) = \mathcal{L}(R^{-1}) &+ \frac{\nabla^2\left(R^{-1}\right)(\nabla f, \nabla f)}{|\nabla f|^2} + \frac{\langle \nabla(R^{-1}), \nabla f\rangle}{|\nabla f|^2}\left(\frac{3}{2} - R_\Sigma\right) \\ &+ \frac{2}{|\nabla f|^2}\langle \nabla \ln R, \nabla f\rangle \left\langle \nabla(R^{-1}), \nabla f\right\rangle.\end{aligned}$$

We have that

$$\begin{aligned}\nabla^2(R^{-1})(\nabla f, \nabla f) &= \left\langle \nabla\left\langle \nabla(R^{-1}), \nabla f\right\rangle, \nabla f\right\rangle - \nabla^2 f\left(\nabla(R^{-1}), \nabla f\right) \\ &= -\left\langle \nabla\left\langle \nabla R, \nabla f\right\rangle, \nabla f\right\rangle R^{-2} + 2\left\langle \nabla R, \nabla f\right\rangle^2 R^{-3} \\ &\quad + \nabla^2 f(\nabla R, \nabla f) R^{-2}.\end{aligned}$$

By (2.84) we know that

$$\left|\left\langle \nabla\left\langle \nabla R, \nabla f\right\rangle, \nabla f\right\rangle R^{-2} - R^{-1}\right| \leq C.$$

From the identity

$$\nabla^2 f(\nabla R, \nabla f) R^{-2} = \frac{1}{2}\left\langle \nabla R, \nabla f\right\rangle R^{-2} - \frac{1}{2}\left|\nabla R\right|^2 R^{-2}$$

and from (2.83)

$$\left|\left\langle \nabla R, \nabla f\right\rangle + R\right| \leq C R^2,$$

we similarly infer that

$$\left|\nabla^2 f(\nabla R, \nabla f) R^{-2} + \frac{1}{2} R^{-1}\right| \leq C.$$

The above estimates imply that

$$\left|\nabla^2(R^{-1})(\nabla f, \nabla f) - \frac{1}{2} R^{-1}\right| \leq C.$$

By applying (2.27) we conclude that

$$\frac{\left|\nabla^2\left(R^{-1}\right)(\nabla f, \nabla f)\right|}{|\nabla f|^2} \leq C.$$

Moreover, let us note that by (2.83),

$$\begin{aligned}\frac{1}{|\nabla f|^2}\left|\langle \nabla \ln R, \nabla f\rangle\right|\left|\left\langle \nabla(R^{-1}), \nabla f\right\rangle\right| &\leq \frac{\langle \nabla R, \nabla f\rangle^2}{R^3\left|\nabla f\right|^2} \\ &\leq C.\end{aligned}$$

We have therefore proved that

$$\Delta_F(R^{-1}) \geq \mathcal{L}(R^{-1}) - C,$$

for $\mathcal{L}$ defined by (2.74). By (2.95), this implies the following:

$$\mathcal{L}(R^{-1}) \leq -R^{-1} + C.$$

Consequently, it follows that ρ given by (2.93) satisfies

$$\begin{aligned}\mathcal{L}\rho &= \left(\mathcal{L}(R^{-1})\right)(\ln f)^{-2} - \left\langle \nabla f, \nabla (\ln f)^{-2}\right\rangle R^{-1} - \langle \nabla f, \nabla(\ln f)^{-1}\rangle \\ &\leq -R^{-1}(\ln f)^{-2} + C(\ln f)^{-2} + 2R^{-1}(\ln f)^{-3} \\ &\leq -\frac{1}{2}R^{-1}(\ln f)^{-2} + C(\ln f)^{-2}.\end{aligned}$$

This proves Claim 2.23. □

Recall that by Lemma 2.21 the function

$$w = \frac{|\mathrm{Ric}|^2}{R^2}$$

satisfies

$$\mathcal{L}w \geq \frac{2P}{R^3} - \frac{CR}{f} \quad \text{on } M^4 \setminus D(t_0), \tag{2.96}$$

for P defined by (2.62). Claim 2.23 and (2.96) imply that there exists a constant $C_0 > 0$ such that the function

$$\begin{aligned} G &:= \left(\frac{|\mathrm{Ric}|^2}{R^2} - \frac{1}{2}\right) - \frac{\Upsilon}{\ln f} - \frac{\Upsilon}{R\ln^2 f} \\ &= \left(w - \frac{1}{2}\right) - \Upsilon\rho \end{aligned}$$

satisfies

$$\mathcal{L}G \geq \frac{2P}{R^3} + \frac{\Upsilon}{2R\ln^2 f} - \frac{C_0\Upsilon}{\ln^2 f} \quad \text{on } M^4 \setminus D(t_0), \tag{2.97}$$

for any constant $\Upsilon \geq 1$. We may assume that

$$C_0 \geq (100A)^3,$$

where $A \geq 1$ is a constant such that $R \leq A$ on M^4. Since the Ricci curvature in the direction $\nu = \frac{\nabla f}{|\nabla f|}$ satisfies

$$|\mathrm{Ric}(\nu)| = \frac{1}{2}\frac{|\nabla R|}{|\nabla f|} \leq \frac{C}{\sqrt{f}},$$

there exists $t_1 \geq t_0$ such that

$$|\mathrm{Ric}(\nu)| \leq \frac{1}{100C_0}\frac{1}{\ln^2 f} \quad \text{on } M^4 \setminus D(t_1). \tag{2.98}$$

Moreover, we may assume that t_1 is large enough so that

$$\ln t_1 \geq \mathrm{e}^{C_0}. \tag{2.99}$$

Set

$$\Upsilon := (\ln t_1) \sup_{\Sigma(t_1)} \frac{|\mathrm{Ric}|^2}{R^2} \geq 1. \tag{2.100}$$

Then for $T > t_1$ arbitrary, assume by contradiction that

$$G(x_0) := \max_{\overline{D(T)\setminus D(t_1)}} G > 0. \tag{2.101}$$

Since $G<0$ on $\Sigma(t_1)$, we may assume henceforth that G achieves its maximum at $x_0\in\Sigma(t)$ for some $t\in(t_1,T]$. By the maximum principle and the fact that $\Sigma(t)=\{f=t\}$ it follows that

$$\begin{aligned}\Delta_\Sigma G(x_0)&\le 0,\\ \langle\nabla f,\nabla G\rangle(x_0)&\ge 0,\\ \langle\nabla_\Sigma \ln R,\nabla_\Sigma G\rangle(x_0)&=0.\end{aligned}$$

Hence, this proves that

$$\mathcal{L}G(x_0)\le 0. \tag{2.102}$$

By (2.97) we obtain that

$$\frac{2P}{R^3}+\frac{\Upsilon}{2R\ln^2 f}\le\frac{C_0\Upsilon}{\ln^2 f}\quad\text{at}\quad x_0, \tag{2.103}$$

for P specified in (2.62) and Υ defined in (2.100). In particular, as $P\ge 0$, it follows that

$$R(x_0)\ge\frac{1}{2C_0}. \tag{2.104}$$

As $R\le A$ on M^4 and $C_0\ge(100A)^3$, by (2.103) we also obtain that

$$\sqrt{P}\le\frac{1}{100}\frac{C_0\sqrt{\Upsilon}}{\ln f}\quad\text{at}\quad x_0. \tag{2.105}$$

Let $\lambda_1\le\lambda_2\le\lambda_3$ be the eigenvalues of the Ricci tensor restricted to $T\Sigma(t)$. We aim to prove that $|\lambda_1|(x_0)$ and $(\lambda_3-\lambda_2)(x_0)$ are small, specifically on the scale of $\frac{C_0^2\sqrt{\Upsilon}}{\ln f}$, which will lead to a contradiction with (2.101).

CLAIM 2.24. *We have that*

$$\lambda_3-\lambda_2\le\frac{1}{10}\frac{C_0^2\sqrt{\Upsilon}}{\ln f}\quad\textit{at}\quad x_0. \tag{2.106}$$

PROOF OF CLAIM 2.24. Assume by contradiction that

$$\lambda_3-\lambda_2>\frac{1}{10}\frac{C_0^2\sqrt{\Upsilon}}{\ln f}\quad\text{at}\quad x_0. \tag{2.107}$$

According to (2.105) and (2.62),

$$|\lambda_3+\lambda_2-\lambda_1|(\lambda_3-\lambda_2)\le\frac{1}{100}\frac{C_0\sqrt{\Upsilon}}{\ln f}\quad\text{at}\quad x_0.$$

Then (2.107) implies that

$$\lambda_3\le\lambda_3+\lambda_2-\lambda_1\le\frac{1}{10C_0}\quad\text{at}\quad x_0.$$

In view of (2.98), this implies that

$$R\le 3\lambda_3+R_{44}<\frac{1}{2C_0}\quad\text{at}\quad x_0,$$

which provides a contradiction to (2.104). This proves Claim 2.24. □

CLAIM 2.25. *We have that*

$$\lambda_3-\lambda_1\ge\frac{1}{100C_0}\quad\textit{at}\quad x_0. \tag{2.108}$$

Proof of Claim 2.25. Assume by contradiction that

$$\lambda_3 - \lambda_1 < \frac{1}{100C_0} \quad \text{at} \quad x_0. \tag{2.109}$$

Setting

$$\epsilon := \frac{1}{100C_0},$$

we then obtain by (2.104) and (2.98) that

$$\begin{aligned}\frac{1}{2C_0} &\leq R(x_0)\\ &\leq 3\lambda_1(x_0) + 2\epsilon + R_{44}(x_0)\\ &\leq 3\lambda_1(x_0) + \frac{1}{8C_0}.\end{aligned}$$

This implies that

$$\lambda_1(x_0) \geq \frac{1}{8C_0}. \tag{2.110}$$

According to (2.101) we have that

$$2|\mathrm{Ric}|^2 \geq R^2 \quad \text{at} \quad x_0.$$

By (2.71), (2.72a), and (2.72b) we infer that

$$2\left(\lambda_1^2 + \lambda_2^2 + \lambda_3^2\right) + 4|R_{i4}|^2 \geq \left(\lambda_1 + \lambda_2 + \lambda_3 + R_{44}\right)^2. \tag{2.111}$$

However, (2.98) and (2.109) imply that

$$2\left(\lambda_1^2 + \lambda_2^2 + \lambda_3^2\right) + 4|R_{i4}|^2 \leq 2\left(\lambda_1^2 + 2\left(\lambda_1 + \epsilon\right)^2\right) + \epsilon^2.$$

Together with the inequality $(\lambda_1 + \epsilon)^2 \leq \frac{3}{2}\lambda_1^2 + 3\epsilon^2$, this yields

$$2\left(\lambda_1^2 + \lambda_2^2 + \lambda_3^2\right) + 4|R_{i4}|^2 \leq 8\lambda_1^2 + 13\epsilon^2.$$

On the other hand, (2.110) and (2.98) imply that

$$\left(\lambda_1 + \lambda_2 + \lambda_3 + R_{44}\right)^2 \geq \left(3\lambda_1 - \epsilon\right)^2.$$

By (2.111) we conclude that

$$\left(3\lambda_1 - \epsilon\right)^2 \leq 8\lambda_1^2 + 13\epsilon^2.$$

This contradicts the inequality $\lambda_1 \geq 10\epsilon$, which results from (2.110) and from $\epsilon = \frac{1}{100C_0}$. This proves Claim 2.25. □

Claim 2.26. *We have that*

$$|\lambda_1| \leq \frac{2C_0^2\sqrt{\Upsilon}}{\ln f} \quad \textit{at} \quad x_0. \tag{2.112}$$

Proof of Claim 2.26. According to (2.105) and (2.62) we have that at x_0,

$$\left(\lambda_3 - \lambda_1\right)|\lambda_3 + \lambda_1 - \lambda_2| \leq \frac{1}{100}\frac{C_0\sqrt{\Upsilon}}{\ln f}.$$

Using (2.108) it follows that at x_0,

$$|\lambda_3 + \lambda_1 - \lambda_2| \leq \frac{C_0^2\sqrt{\Upsilon}}{\ln f}.$$

Moreover, (2.106) implies that

$$|\lambda_3 - \lambda_2| \leq \frac{1}{10} \frac{C_0^2 \sqrt{\Upsilon}}{\ln f} \qquad \text{at} \quad x_0.$$

Adding these two inequalities implies that

$$|\lambda_1| \leq \frac{2C_0^2 \sqrt{\Upsilon}}{\ln f} \quad \text{at} \quad x_0.$$

This proves Claim 2.26. □

We now use Claims 2.24 and 2.26 to reach a contradiction with (2.101). Indeed, according to (2.71) we have that

$$\begin{aligned} 2|\mathrm{Ric}|^2 - R^2 &= 2\left(\lambda_1^2 + \lambda_2^2 + \lambda_3^2\right) + 4|R_{i4}|^2 - 2|R_{44}|^2 \\ &\quad - \left(\lambda_1 + \lambda_2 + \lambda_3 + R_{44}\right)^2 \\ &= \lambda_1^2 + \left(\lambda_3 - \lambda_2\right)^2 - 2\lambda_1\left(\lambda_2 + \lambda_3\right) \\ &\quad + 4|R_{i4}|^2 - |R_{44}|^2 - 2RR_{44}. \end{aligned}$$

By (2.98) and (2.112) it follows that

$$\begin{aligned} \lambda_1^2 - 2\lambda_1(\lambda_2 + \lambda_3) &= 3\lambda_1^2 - 2\lambda_1 R + 2\lambda_1 R_{44} \\ &\leq \frac{12C_0^4 \Upsilon}{\ln^2 f} + \frac{C_0^3 \sqrt{\Upsilon}}{\ln f}, \end{aligned}$$

by additionally using that $R \leq A$ and $C_0 \geq (100A)^3$. Moreover, (2.98) implies that

$$4|R_{i4}|^2 - |R_{44}|^2 - 2RR_{44} \leq \frac{1}{\ln^2 f}.$$

Therefore, by additionally using (2.106), we obtain that

$$\begin{aligned} 2|\mathrm{Ric}|^2 - R^2 &\leq \frac{13C_0^4 \Upsilon}{\ln^2 f} + \frac{C_0^3 \sqrt{\Upsilon}}{\ln f} \\ &\leq C_0^5 \left(\frac{\Upsilon}{\ln^2 f} + \frac{\sqrt{\Upsilon}}{\ln f}\right). \end{aligned}$$

Since $R \geq \frac{1}{2C_0}$ by (2.104), this implies that

$$\frac{|\mathrm{Ric}|^2}{R^2} - \frac{1}{2} \leq 2C_0^7 \left(\frac{\Upsilon}{\ln^2 f} + \frac{\sqrt{\Upsilon}}{\ln f}\right).$$

Consequently, we have that at x_0,

$$\begin{aligned} G &\leq \left(\frac{|\mathrm{Ric}|^2}{R^2} - \frac{1}{2}\right) - \frac{\Upsilon}{\ln f} \\ &\leq 2C_0^7 \left(\frac{\Upsilon}{\ln^2 f} + \frac{\sqrt{\Upsilon}}{\ln f}\right) - \frac{\Upsilon}{\ln f}. \end{aligned} \tag{2.113}$$

Since $|\mathrm{Ric}|^2 \geq \frac{1}{4}R^2$, it follows from (2.100) that $\Upsilon \geq \frac{1}{4}\ln t_1$. By (2.99) and (2.113) it follows that $G(x_0) < 0$, which is a contradiction to (2.101). Hence, we have established that

$$\frac{|\mathrm{Ric}|^2}{R^2}-\frac{1}{2}\leq\frac{\Upsilon}{\ln f}+\frac{\Upsilon}{R\ln^2 f}\quad\text{on } D(T)\setminus D(t_1)$$

for any $T>t_1$. This proves the result. □

2.5.2. Shrinking solitons with 3-cylinder limits at infinity.

We now prove the following pinching result, which says that the symmetry of a far-enough level set propagates to infinity.[7]

THEOREM 2.27 (3-cylindrical Ricci pinching propagates to infinity). *Let (M^4,g,f) be a complete noncompact 4-dimensional shrinking gradient Ricci soliton with bounded scalar curvature and let E be an end of M^4. There exists a large enough constant $C_0>0$ with the following property: if $R>0$ and*

$$\frac{|\mathrm{Ric}|^2}{R^2}<\frac{1}{3}+\varepsilon\quad\text{on } \Sigma(T_0)\cap E,$$

for some $0<\varepsilon<\frac{1}{100}$ and for some $T_0\geq\frac{C_0}{\varepsilon}$, then we have the same inequality

$$\frac{|\mathrm{Ric}|^2}{R^2}<\frac{1}{3}+\varepsilon\quad\text{on } E\setminus D(T_0).$$

PROOF. According to Lemma 2.21 the function

$$w:=\frac{|\mathrm{Ric}|^2}{R^2}$$

satisfies

$$\mathcal{L}w\geq\frac{2P}{R^3}-\frac{C_1R}{f}\quad\text{on } M^4\setminus D(t_0)\tag{2.114}$$

for some constant $C_1>0$. We may assume that $C_1>1$. Moreover, the Ricci curvature in the direction $\nu=\frac{\nabla f}{|\nabla f|}$ satisfies

$$|\mathrm{Ric}(\nu)|=\frac{1}{2}\frac{|\nabla R|}{|\nabla f|}\leq\frac{C_2R}{\sqrt{f}}\quad\text{on } M^4\setminus D(t_0)\tag{2.115}$$

for some constant $C_2>1$.

For Ric_Σ and R_Σ defined in (2.72a) and (2.72b), it follows from (2.71) that

$$\left(|\mathrm{Ric}|^2-\frac{1}{3}R^2\right)-\left(|\mathrm{Ric}_\Sigma|^2-\frac{1}{3}R_\Sigma^2\right)=2|R_{i4}|^2-\frac{2}{3}|R_{44}|^2-\frac{2}{3}RR_{44}.$$

According to (2.83),

$$|R_{44}|=\frac{1}{2}\frac{|\langle\nabla R,\nabla f\rangle|}{|\nabla f|^2}\leq\frac{CR}{f},$$

whereas Theorem 2.16 implies that

$$|R_{i4}|^2=\frac{1}{4}\frac{|\nabla_iR|^2}{|\nabla f|^2}\leq\frac{CR^3}{f}.$$

Hence, it follows that

$$\left|\left(|\mathrm{Ric}|^2-\frac{1}{3}R^2\right)-\left(|\mathrm{Ric}_\Sigma|^2-\frac{1}{3}R_\Sigma^2\right)\right|\leq\frac{CR^2}{f}.\tag{2.116}$$

[7] By far-enough level set we mean that f is sufficiently large.

Moreover, we have

$$|\mathrm{Ric}_\Sigma|^2 - \frac{1}{3}R_\Sigma^2 = \frac{1}{3}\left((\lambda_3-\lambda_1)^2 + (\lambda_3-\lambda_2)^2 + (\lambda_2-\lambda_1)^2\right), \tag{2.117}$$

where $\lambda_1 \le \lambda_2 \le \lambda_3$ are the eigenvalues of the Ricci tensor restricted to $\Sigma(t)\cap E$. As

$$(\lambda_3-\lambda_1)^2 \le (\lambda_3-\lambda_1)^2 + (\lambda_3-\lambda_2)^2 + (\lambda_2-\lambda_1)^2 \le 3(\lambda_3-\lambda_1)^2,$$

we see from (2.116) and (2.117) that

$$\left(|\mathrm{Ric}|^2 - \frac{1}{3}R^2\right) - \frac{C_3R^2}{f} \le (\lambda_3-\lambda_1)^2 \le 3\left(|\mathrm{Ric}|^2 - \frac{1}{3}R^2\right) + \frac{C_3R^2}{f} \tag{2.118}$$

for some constant $C_3 > 1$. Let us now define

$$C_0 := 100\,(C_1+C_2+C_3)^2, \tag{2.119}$$

where C_1, C_2, and C_3 are the constants in (2.114), (2.115), and (2.118). We aim to prove that if

$$\frac{|\mathrm{Ric}|^2}{R^2} < \frac{1}{3} + \varepsilon \quad \text{on } \Sigma(T_0)\cap E,$$

for some $0 < \varepsilon < \frac{1}{100}$ and for some $T_0 \ge \frac{C_0}{\varepsilon}$, then we have the same inequality

$$\frac{|\mathrm{Ric}|^2}{R^2} < \frac{1}{3} + \varepsilon \quad \text{on } E\setminus D(T_0). \tag{2.120}$$

Assume by contradiction that (2.120) is not true. Then there exists a first $T > T_0$ such that

$$\sup_{\Sigma(T)\cap E} \frac{|\mathrm{Ric}|^2}{R^2} = \frac{1}{3} + \varepsilon. \tag{2.121}$$

Let $x_0 \in \Sigma(T)\cap E$ be such that

$$\begin{aligned}\frac{|\mathrm{Ric}|^2}{R^2}(x_0) &= \sup_{\Sigma(T)\cap E} \frac{|\mathrm{Ric}|^2}{R^2}\\ &= \frac{1}{3} + \varepsilon.\end{aligned}$$

In other words, w achieves its maximal value on $D(T)\setminus D(T_0)$ at $x_0 \in \Sigma(T)$ and $w(x_0) = \frac{1}{3} + \varepsilon$. By (2.121) and (2.118) it follows that at x_0,

$$(\lambda_3-\lambda_1)^2 \le 3\varepsilon R^2 + \frac{C_3R^2}{T}.$$

Since T satisfies $T \ge T_0 \ge \frac{C_0}{\varepsilon}$ for C_0 specified in (2.119), it follows that $\frac{C_3}{T} \le \varepsilon$. Therefore, we obtain that

$$0 \le \lambda_3 - \lambda_1 \le 2\sqrt{\varepsilon}R \quad \text{at} \quad x_0.$$

As $R = (\lambda_1+\lambda_2+\lambda_3) + R_{44}$, we furthermore obtain that at x_0,

$$\begin{aligned}\left|\lambda_1 - \frac{1}{3}R\right| &\le \frac{1}{3}\left(|\lambda_3-\lambda_1| + |\lambda_2-\lambda_1| + |R_{44}|\right)\\ &\le \frac{4\sqrt{\varepsilon}}{3}R + \frac{1}{3}\frac{C_2R}{\sqrt{T}}\\ &\le \sqrt{2\varepsilon}R,\end{aligned} \tag{2.122}$$

where in the second line we used (2.115) and in the last line we used that $T_0 \geq \frac{C_0}{\varepsilon}$. Since $\varepsilon < \frac{1}{100}$, we infer from (2.122) that

$$\lambda_1 \geq \left(\frac{1}{3} - \sqrt{2\varepsilon}\right) R \geq \frac{1}{6} R \quad \text{at} \quad x_0. \tag{2.123}$$

According to the maximum principle and (2.121), the function $w = \frac{|\mathrm{Ric}|^2}{R^2}$ satisfies

$$\begin{aligned} \Delta_\Sigma w\,(x_0) &\leq 0, \\ \langle \nabla f, \nabla w \rangle\,(x_0) &\geq 0, \\ \langle \nabla_\Sigma \ln R, \nabla_\Sigma w \rangle\,(x_0) &= 0. \end{aligned}$$

Therefore, $\mathcal{L}w(x_0) \leq 0$. Hence, (2.114) implies that

$$P \leq \frac{C_1}{T} R^4 \qquad \text{at} \quad x_0. \tag{2.124}$$

By (2.62) and (2.124), we have in particular that

$$(\lambda_3 - \lambda_1)(\lambda_1 + \lambda_3 - \lambda_2) \leq \frac{\sqrt{C_1}}{\sqrt{T}} R^2 \quad \text{at} \quad x_0.$$

By (2.123) we know that

$$\lambda_1 + \lambda_3 - \lambda_2 \geq \lambda_1 \geq \frac{1}{6} R \quad \text{at} \quad x_0.$$

We obtain that

$$0 \leq \lambda_3 - \lambda_1 \leq \frac{6\sqrt{C_1}}{\sqrt{T}} R \quad \text{at} \quad x_0. \tag{2.125}$$

In conclusion, by (2.118) and (2.125), it follows that

$$\begin{aligned} \sup_{\Sigma(T)\cap E} \frac{|\mathrm{Ric}|^2}{R^2} &= \frac{|\mathrm{Ric}|^2}{R^2}(x_0) \\ &\leq \frac{1}{3} + \frac{(\lambda_3 - \lambda_1)^2}{R^2}(x_0) + \frac{C_3}{T} \\ &\leq \frac{1}{3} + \frac{36C_1 + C_3}{T}. \end{aligned}$$

Therefore, by (2.119) it follows that

$$\sup_{\Sigma(T)\cap E} \frac{|\mathrm{Ric}|^2}{R^2} < \frac{1}{3} + \varepsilon.$$

This contradicts (2.121). In conclusion,

$$\frac{|\mathrm{Ric}|^2}{R^2} < \frac{1}{3} + \varepsilon \quad \text{on } E \setminus D(T_0).$$

This proves the result. □

DEFINITION 2.28 (C^k-asymptotically $(n-1)$-cylindrical end). A Riemannian manifold (M^n, g) is said to be **C^k-asymptotic along an end $E \subset M^n$ to the cylinder $L = \left(\mathbb{R} \times N^{n-1}, g_c\right)$**, where g_c is the product metric, if there exists a diffeomorphism $\Phi : L = (0, \infty) \times N^{n-1} \to E \setminus \Omega$ such that $\rho_\lambda^* \Phi^* g \to g_c$ as $\lambda \to \infty$ in $C^k_{loc}(L, g_c)$, where Ω is a compact subset of M^n and $\rho_\lambda : L \to L$ is the translation given by $\rho_\lambda(r, \sigma) := (\lambda + r, \sigma)$ for $r \in \mathbb{R}$ and $\sigma \in N^{n-1}$.

We say that an end is C^∞**-asymptotically** $(n-1)$**-cylindrical** if it is C^k-asymptotically $(n-1)$-cylindrical for all k. We obtain the following consequence.

THEOREM 2.29 (Asymptotically 3-cylindrical shrinkers). *Let (M^4, g, f) be a complete noncompact 4-dimensional shrinking gradient Ricci soliton with bounded scalar curvature. Assume that there exists a sequence $x_k \to \infty$ along an end E such that (M^4, g, x_k) converges in the smooth pointed Cheeger–Gromov sense to $(\mathbb{R} \times (\mathbb{S}^3/\Gamma), g_c, x_\infty)$, where g_c is the standard cylinder metric. Then (M^4, g) is C^∞-asymptotic along E to $(\mathbb{R} \times (\mathbb{S}^3/\Gamma), g_c)$.*

PROOF. Let $\{x_k\}$ be such a sequence. Define

$$f_k(x) := \frac{f(x) - f(x_k)}{|\nabla f|(x_k)}$$

for k large enough so that $|\nabla f|(x_k) > 0$. By Theorem 2.6, which establishes that (M^4, g) has bounded curvature, the sequence f_k has the following properties:

$$f_k(x_k) = 0, \quad |\nabla f_k|(x_k) = 1, \quad \text{and} \quad |\nabla^2 f_k|(x) \leq \frac{C}{|\nabla f|(x_k)}$$

for all $x \in M^4 \setminus D(t_0)$. Hence f_k converges, up to a subsequence, in $C^{1,\alpha}$-sense to a function f_∞ on $(\mathbb{R} \times (\mathbb{S}^3/\Gamma), g_c, x_\infty)$ satisfying

$$f_\infty(x_\infty) = 0, \quad |\nabla f_\infty|(x_\infty) = 1, \quad \text{and} \quad \nabla^2 f_\infty \equiv 0.$$

This means that f_∞ must be linear on $\mathbb{R} \times (\mathbb{S}^3/\Gamma)$, and we can normalize it as $f_\infty(t) = t$. In particular, for $a < b$ fixed, the level subset $(\{a < f_k < b\}, g, x_k)$ must converge smoothly to $((a,b) \times (\mathbb{S}^3/\Gamma), g_c, x_\infty)$. Hence, for k large enough, the scalar curvature is uniformly bounded from below by a positive constant on the level sets $\Sigma(f(x_k))$.

In fact, Corollary 2.17 implies that the scalar curvature of (M^4, g) is uniformly bounded from below by a positive constant along the end E. For, otherwise, by (2.56) it converges to zero uniformly along an integral curve ϕ_t of the potential. This contradicts the fact that ϕ_t must intersect the associated almost cylindrical regions which have uniformly positive R.

The proof of the theorem relies on Theorem 2.27 which, for R bounded below by a positive constant, can be improved to polynomial decay at infinity. In turn, this implies convergence the Riemann curvature tensor.

STEP 1: $|\mathrm{Ric}|^2 - \frac{1}{3}R^2$ *decays polynomially.*

By Lemma 2.21 we have

$$\mathcal{L}\left(\frac{|\mathrm{Ric}|^2}{R^2}\right) \geq \frac{2P}{R^3} - \frac{CR}{f}, \tag{2.126}$$

where P is defined by (2.62) and where $\mathcal{L}$ is defined by (2.74). As (M^4, g, x_k) converges to $(\mathbb{R} \times (\mathbb{S}^3/\Gamma), g_c, x_\infty)$, we may assume there exists a level set $\Sigma(T_0)$ with $T_0 \geq \frac{C_0}{\varepsilon}$ such that

$$\frac{|\mathrm{Ric}|^2}{R^2} < \frac{1}{3} + \varepsilon \quad \text{on } \Sigma(T_0) \cap E. \tag{2.127}$$

Now, (2.127) and Theorem 2.27 imply that

$$\frac{|\mathrm{Ric}|^2}{R^2} < \frac{1}{3} + \varepsilon \quad \text{on } E \setminus D(T_0).$$

By (2.116) and (2.117), we obtain as in (2.118) that

$$\left(|\mathrm{Ric}|^2 - \frac{1}{3}R^2\right) - \frac{CR^2}{f} \leq (\lambda_3 - \lambda_1)^2 \leq 3\left(|\mathrm{Ric}|^2 - \frac{1}{3}R^2\right) + \frac{CR^2}{f}. \tag{2.128}$$

In particular, this yields

$$|\lambda_3 - \lambda_1| \leq 2\sqrt{\varepsilon}R;$$

hence we obtain as in (2.123) that $\lambda_1 \geq \frac{1}{6}R$. Consequently,

$$\lambda_1 + \lambda_3 - \lambda_2 \geq \frac{1}{6}R$$

at any $x \in \Sigma(t) \cap E$ with $t \geq T_0$. In particular, we obtain

$$\begin{aligned} 2P &\geq 2(\lambda_1 + \lambda_3 - \lambda_2)^2 (\lambda_3 - \lambda_1)^2 \\ &\geq \frac{1}{18}R^2 (\lambda_3 - \lambda_1)^2. \end{aligned} \tag{2.129}$$

Since the scalar curvature is uniformly bounded below by a positive constant on E, there exists a constant $\sigma > 0$ such that

$$\frac{1}{18}R \geq \sigma \quad \text{on } E. \tag{2.130}$$

Consequently, by (2.129) and (2.128) it follows that

$$\begin{aligned} 2P &\geq \sigma R (\lambda_3 - \lambda_1)^2 \\ &\geq \sigma R \left(|\mathrm{Ric}|^2 - \frac{1}{3}R^2\right) - \frac{C}{f}. \end{aligned}$$

Therefore, we get from (2.126) that the function

$$G := \frac{|\mathrm{Ric}|^2}{R^2} - \frac{1}{3}$$

satisfies

$$\mathcal{L}G \geq \sigma G - \frac{C_1}{f} \quad \text{on } E \setminus D(T_0), \tag{2.131}$$

for some constant $C_1 > 0$. We may assume that $\ln T_0 \geq C_1$. Let

$$\Upsilon := T_0 \sup_{\Sigma(T_0)\cap E} \frac{|\mathrm{Ric}|^2}{R^2}. \tag{2.132}$$

For C_1 the constant in (2.131), it follows that

$$\begin{aligned} \mathcal{L}\left(\frac{\Upsilon}{f^\sigma} - \frac{4C_1}{f}\right) &= -\left\langle \nabla f, \nabla\left(\frac{\Upsilon}{f^\sigma} - \frac{4C_1}{f}\right)\right\rangle \\ &= \Upsilon\sigma\frac{|\nabla f|^2}{f^{\sigma+1}} - \frac{4C_1|\nabla f|^2}{f^2} \\ &\leq \frac{\Upsilon\sigma}{f^\sigma} - \frac{2C_1}{f}. \end{aligned}$$

Without loss of generality we may assume that $\sigma \leq \frac{1}{4}$. By (2.131), we conclude that the function

$$\tilde{G} := G - \left(\frac{\Upsilon}{f^\sigma} - \frac{4C_1}{f} \right)$$

satisfies

$$\mathcal{L}\tilde{G} \geq \sigma \tilde{G} \quad \text{on } E \setminus D(T_0).$$

We claim that

$$\tilde{G} \leq 0 \quad \text{on } E \setminus D(T_0).$$

Indeed, by (2.132) we have that $\tilde{G} < 0$ on $\Sigma(T_0) \cap E$. For $T \geq T_0$ arbitrary, suppose that $\tilde{G}$ achieves its maximum on $\left(\overline{D(T)} \setminus D(T_0)\right) \cap E$ at some $x_0 \in \Sigma(t) \cap E$, where $T_0 < t \leq T$. Applying the maximum principle as in (2.102) then implies that $\mathcal{L}\tilde{G}(x_0) \leq 0$. As $\mathcal{L}\tilde{G} \geq \sigma \tilde{G}$, it follows that $\tilde{G}(x_0) \leq 0$, and thus $\tilde{G} \leq 0$ on $(D(T) \setminus D(T_0)) \cap E$. Since $T \geq T_0$ is arbitrary, we obtain that $\tilde{G} \leq 0$ on $E \setminus D(T_0)$.

In conclusion, we have proved that

$$|\mathrm{Ric}|^2 - \frac{1}{3}R^2 \leq \frac{C}{f^\sigma} \quad \text{on } E. \tag{2.133}$$

STEP 2: *Decay of* $|\nabla^k \mathrm{Ric}|$ *for* $k \geq 1$.

Define the tensor

$$U := \mathrm{Ric} - \frac{1}{3}Rg + \frac{1}{3}R\frac{\nabla f \otimes \nabla f}{|\nabla f|^2}.$$

By $\mathrm{Ric}(\nabla f, \nabla f) = \frac{1}{2}\langle \nabla R, \nabla f \rangle$, it follows that

$$|U|^2 = |\mathrm{Ric}|^2 - \frac{1}{3}R^2 + \frac{1}{3}R\frac{\langle \nabla R, \nabla f \rangle}{|\nabla f|^2}.$$

Hence, (2.133) implies that

$$|U|^2 \leq \frac{C}{f^\sigma} \quad \text{on } E. \tag{2.134}$$

For $x \in \Sigma(t) \cap E$ and

$$\theta := t^{-\frac{\sigma}{4}},$$

let ϕ be a cutoff function with support in $B(x, \theta)$ such that $\phi = 1$ on $B\left(x, \frac{\theta}{2}\right)$ and $|\nabla \phi| \leq \frac{C}{\theta}$. Theorem 2.16 implies that

$$|\nabla U| + |\nabla^2 U| \leq C.$$

Integrating by parts, we have

$$\begin{aligned} \int_{B(x,\theta)} |\nabla U|^2 \phi^2 dv &= -\int_{B(x,\theta)} \langle \Delta U, U \rangle \phi^2 dv - \frac{1}{2}\int_{B(x,\theta)} \langle \nabla |U|^2, \nabla \phi^2 \rangle dv \\ &\leq \int_{B(x,\theta)} \left(|\Delta U| + \frac{C}{\theta} |\nabla U| \right) |U| \, dv \\ &\leq C\left(1 + \frac{1}{\theta}\right) \int_{B(x,\theta)} |U| \, dv. \end{aligned} \tag{2.135}$$

It follows from (2.134) and $\theta = t^{-\frac{\sigma}{4}}$ that

$$\int_{B(x,\theta)} |\nabla U|^2 \phi^2 dv \leq C t^{-\frac{\sigma}{4}} \mathrm{Vol}\left(B(x, \theta)\right).$$

As $\left(M^4,g\right)$ has bounded curvature, the volume comparison theorem implies that $\operatorname{Vol}\left(B\left(x,\theta\right)\right)\leq C\operatorname{Vol}\left(B\left(x,\frac{\theta}{2}\right)\right)$. We conclude from the above that

$$\inf_{B\left(x,\frac{\theta}{2}\right)}\left|\nabla U\right|^2\leq Ct^{-\frac{\sigma}{4}}.$$

Since $\left|\nabla\left|\nabla U\right|\right|\leq\left|\nabla^2 U\right|\leq C$, this leads to

$$\begin{aligned}\left|\nabla U\right|(x)&\leq\inf_{B\left(x,\frac{\theta}{2}\right)}\left|\nabla U\right|+C\theta\\&\leq Ct^{-\frac{\sigma}{8}}\\&=Cf\left(x\right)^{-\frac{\sigma}{8}}.\end{aligned}$$

Consequently, we obtain that

$$\left|\nabla U\right|\leq\frac{C}{f^{\frac{\sigma}{8}}}\quad\text{on }E,\tag{2.136}$$

which, in a local orthonormal frame $\{e_i\}$, means that

$$\left|\nabla_k\left(R_{ij}-\frac{1}{3}Rg_{ij}+\frac{1}{3}R\frac{\nabla_i f\nabla_j f}{\left|\nabla f\right|^2}\right)\right|\leq\frac{C}{f^{\frac{\sigma}{8}}}.$$

By

$$\begin{aligned}\nabla_k\left(\frac{\nabla_i f\nabla_j f}{\left|\nabla f\right|^2}\right)&=\frac{1}{\left|\nabla f\right|^2}\left(\nabla^2_{ik}f\nabla_j f+\nabla^2_{jk}f\nabla_i f\right)\\&\quad-\frac{1}{\left|\nabla f\right|^4}\left(\nabla_k f\nabla_i f\nabla_j f-\nabla_k R\nabla_i f\nabla_j f\right),\end{aligned}$$

we infer that

$$\left|\nabla\left(\frac{\nabla_i f\nabla_j f}{\left|\nabla f\right|^2}\right)\right|\leq\frac{C}{\left|\nabla f\right|}.$$

Hence, (2.136) proves that

$$\left|\nabla_k R_{ij}-\frac{1}{3}\left(\nabla_k R\right)g_{ij}+\frac{1}{3}\left(\nabla_k R\right)\frac{\nabla_i f\nabla_j f}{\left|\nabla f\right|^2}\right|\leq\frac{C}{f^{\frac{\sigma}{8}}}\quad\text{on }E.\tag{2.137}$$

We take $i=k$ in (2.137) and sum up, noting that

$$\left|\frac{\left\langle\nabla R,\nabla f\right\rangle\nabla_j f}{\left|\nabla f\right|^2}\right|\leq\frac{C}{\left|\nabla f\right|}.$$

It follows that

$$\left|\nabla R\right|\leq\frac{C}{f^{\frac{\sigma}{8}}}\quad\text{on }E,$$

which when plugged into (2.137) yields

$$\left|\nabla\operatorname{Ric}\right|\leq\frac{C}{f^{\frac{\sigma}{8}}}\quad\text{on }E.$$

We interpolate again as in (2.135) to obtain decay for higher derivatives of the Ricci curvature. We have

$$\left|\nabla^2\operatorname{Ric}\right|\leq\frac{C}{f^{\frac{\sigma}{32}}}\quad\text{on }E.$$

The proof is the same as that of (2.135), with U replaced by $\nabla \mathrm{Ric}$ and θ adjusted accordingly to $\theta = t^{-\frac{\sigma}{16}}$. In particular, there exists $\delta := \frac{\sigma}{32} > 0$ such that

$$|\Delta R| \leq \frac{C}{f^{\delta}} \quad \text{on } E. \tag{2.138}$$

STEP 3: $\left|R - \frac{3}{2}\right|$ *decays polynomially.*

We now consider ϕ_t defined to be the solution of the ODE initial value problem

$$\begin{aligned} \frac{d\phi_t}{dt} &= \frac{\nabla f}{|\nabla f|^2}, \\ \phi_{t_0} &= \mathrm{id} \text{ on } \Sigma(t_0) \cap E. \end{aligned} \tag{2.139}$$

For a fixed $x \in \Sigma(t_0) \cap E$, we denote $R(t) := R(\phi_t(x))$, where $t \geq t_0$. Then

$$\begin{aligned} \frac{dR}{dt} &= \frac{\langle \nabla R, \nabla f \rangle}{|\nabla f|^2} \\ &= \frac{\Delta R - R + 2|\mathrm{Ric}|^2}{t - R}. \end{aligned}$$

Using (2.134) and (2.138), it follows that

$$\left| t\frac{dR}{dt} - \frac{2}{3}R^2 + R \right| \leq Ct^{-\delta}.$$

Then the function

$$\rho(t) := \frac{1}{t}\left(R(t) - \frac{3}{2}\right)$$

satisfies

$$\left| \frac{d\rho}{dt} - \frac{2}{3}\rho^2 \right| \leq Ct^{-(2+\delta)}. \tag{2.140}$$

In particular,

$$\frac{d\rho}{dt} \geq \frac{2}{3}\rho^2 - Ct^{-(2+\delta)} \geq -Ct^{-(2+\delta)}.$$

Integrating this over t and using that $\rho(t) \to 0$ as $t \to \infty$, we obtain that

$$\rho(t) \leq C_2 t^{-(1+\delta)}, \tag{2.141}$$

for some constant $C_2 > 0$. Define the function

$$\xi(t) := C_2 t^{-(1+\delta)} - \rho(t) \geq 0.$$

According to (2.140) we have

$$-\rho'(t) + \frac{2}{3}\rho^2(t) \geq -Ct^{-(2+\delta)}.$$

It then follows that

$$\begin{aligned} \xi'(t) + \frac{2}{3}\xi^2(t) &= -\rho'(t) + \frac{2}{3}\rho^2(t) + \frac{4}{3}C_2 t^{-(1+\delta)}\xi(t) \\ &\quad - (1+\delta)C_2 t^{-(2+\delta)} - \frac{2}{3}(C_2)^2 t^{-2(1+\delta)} \\ &\geq -Ct^{-(2+\delta)}. \end{aligned} \tag{2.142}$$

Recall that by (2.130), we have

$$R \geq 18\sigma > 0.$$

By

$$\xi(t) = C_2 t^{-(1+\delta)} - \rho$$
$$= C_2 t^{-(1+\delta)} + \frac{3}{2} t^{-1} - t^{-1} R(t)$$

we therefore obtain that

$$0 \leq \xi(t) \leq \frac{3}{2}(1-\sigma) t^{-1} \quad \text{for all } t \geq T_0. \tag{2.143}$$

Therefore, (2.142) and (2.143) yield that

$$\xi'(t) + (1-\sigma) t^{-1} \xi(t) \geq \xi'(t) + \frac{2}{3}\xi^2(t) \geq -Ct^{-(2+\delta)}.$$

Rewrite this as

$$\left(t^{1-\sigma}\xi(t)\right)' \geq -Ct^{-1-\delta-\sigma} \quad \text{for all } t \geq T_0.$$

Integrating over t, we obtain that

$$T^{1-\sigma}\xi(T) - t^{1-\sigma}\xi(t) \geq -\frac{C}{\delta+\sigma} t^{-\delta-\sigma} + \frac{C}{\delta+\sigma} T^{-\delta-\sigma}.$$

Letting $T \to \infty$ and using (2.143), it follows that

$$\xi(t) \leq \frac{C}{\delta+\sigma} t^{-(1+\delta)}, \quad \text{for all } t \geq T_0.$$

In particular,

$$\rho(t) \geq -\frac{C}{\delta+\sigma} t^{-(1+\delta)}.$$

Combined with (2.141), this implies that

$$\left| R - \frac{3}{2} \right| \leq \frac{C}{f^\delta} \quad \text{on } E,$$

for some $\delta > 0$ and $C > 0$. Together with (2.134), this proves that

$$\left| R_{ij} - \frac{1}{2} g_{ij} + \frac{1}{2} \frac{\nabla_i f \nabla_j f}{|\nabla f|^2} \right| \leq \frac{C}{f^\delta} \quad \text{on } E. \tag{2.144}$$

STEP 4: *Convergence of the Riemann curvature.*

Using Lemma 2.2, it follows that the tensor

$$V_{ijk\ell} := R_{ijk\ell} - \frac{1}{4}\left(g_{i\ell} g_{jk} - g_{ik} g_{j\ell}\right)$$
$$+ \frac{1}{4}\left(g_{i\ell} \frac{\nabla_j f \nabla_k f}{|\nabla f|^2} - g_{j\ell} \frac{\nabla_i f \nabla_k f}{|\nabla f|^2} + g_{jk} \frac{\nabla_i f \nabla_\ell f}{|\nabla f|^2} - g_{ik} \frac{\nabla_j f \nabla_\ell f}{|\nabla f|^2} \right)$$

satisfies

$$|V| \leq \frac{C}{f^\delta} \quad \text{on } E. \tag{2.145}$$

Interpolating as in (2.135), we also obtain decay estimates for all derivatives of V, which in particular translates to

$$\left|\nabla^k \mathrm{Rm}\right| \leq \frac{C_k}{f^{\delta_k}} \quad \text{on } E, \text{ for all } k \geq 1. \tag{2.146}$$

STEP 5: *Completion of the proof of Theorem* 2.29.

We now conclude that M^4 is C^∞-asymptotic to $\mathbb{R} \times (\mathbb{S}^3/\Gamma)$ along E. Indeed, for ϕ_t defined in (2.139), consider $\widetilde{g}(t) = \phi_t^*(g)$ on $\Sigma(t_0) \cap E$, the pullback of the metric g on $\Sigma(t) \cap E$. Then

$$\begin{aligned}\frac{d}{dt}\widetilde{g}_{ab}(t) &= 2\,\phi_t^*\left(\frac{\nabla^2_{ab} f}{|\nabla f|^2}\right)\\ &= \frac{1}{|\nabla f|^2}\phi_t^*\left(g_{ab} - 2R_{ab}\right).\end{aligned}$$

Hence, by (2.144),

$$-Ct^{-1-\delta}\widetilde{g}_{ab}(t) \leq \frac{d}{dt}\widetilde{g}_{ab}(t) \leq Ct^{-1-\delta}\widetilde{g}_{ab}(t).$$

Integrating over t implies that

$$\left|\widetilde{g}_{ab}(t) - g^\infty_{ab}\right| \leq Ct^{-\delta},$$

where g^∞_{ab} is the round metric on $\Sigma(t_0) \cap E$. Note that (2.146) implies decay estimates for $\left|\partial^k \widetilde{g}_{ab}\right|$ for all k. It follows that M^4 is smoothly asymptotic to $\mathbb{R} \times (\mathbb{S}^3/\Gamma)$ along E. □

2.5.3. Shrinkers with 2-sided positive scalar curvature bounds. We obtain the following result about the asymptotic structure of shrinking gradient Ricci solitons with 2-sided positive scalar curvature bounds, cf. [**MW19**].

THEOREM 2.30 (2-sided R bounds imply weakly 2-cylindrical or strongly 3-cylindrical). *Suppose that (M^4, g, f) is a complete noncompact 4-dimensional shrinking gradient Ricci soliton with bounded scalar curvature R. If R is bounded from below by a positive constant on an end E of M^4, then either:*

(1) *E is C^∞-asymptotic to the round cylinder $\mathbb{R} \times (\mathbb{S}^3/\Gamma)$, or*
(2) *E is* ***weakly asymptotically 2-cylindrical*** *in the sense that for any sequence $x_i \in E$ going to infinity along an integral curve of ∇f, (M^4, g, x_i) converges in the smooth pointed Cheeger–Gromov sense to $\mathbb{R}^2 \times \mathbb{S}^2$ or its $\mathbb{Z}_2$ quotient. Moreover, the limit is uniquely determined by the integral curve.*

PROOF. Since R is bounded, (M^4, g) has bounded curvature by Theorem 2.6. Recall that the associated canonical form $(M^4, g(t))$ is an ancient solution to the Ricci flow defined on $(-\infty, 0)$, where

$$g(t) := (-t)\,\phi_t^* g$$

and ϕ_t is the family of diffeomorphisms defined by the ODE initial value problem

$$\begin{aligned}\frac{d\phi}{dt} &= \frac{\nabla f}{(-t)},\\ \phi_{-1} &= \mathrm{id}.\end{aligned}$$

For any sequence $\tau_i \to 0$, consider the sequence of rescaled flows $(M^4, g_i(t))$ for $t < 0$, where

$$g_i(t) := \frac{1}{\tau_i} g(\tau_i t).$$

According to Theorem 1.5 in [**Nab10**] regarding Type I solutions to the Ricci flow, for any $x_0 \in \Sigma(t_0)$, a subsequence of $\left(M^4, g_i(t), (x_0, -1)\right)$ converges smoothly to a gradient shrinking Ricci soliton $\left(M^4_\infty, g_\infty(t), (x_\infty, -1)\right)$.

Now for any sequence $x_i \in E$ going to infinity along an integral curve of ∇f, one may write $x_i := \phi_{-\tau_i}(x_0)$ for some point x_0 and $\tau_i \to 0$. However, as $g_i(-1) = \phi^*_{-\tau_i} g$, we see that a subsequence of $\left(M^4, g, x_i\right)$ converges in the smooth pointed Cheeger–Gromov sense to $\left(M^4_\infty, g_\infty(-1), x_\infty\right)$.

Since $x_i \to \infty$, invoking Proposition 5.1 in Naber [**Nab10**], we conclude that $\left(M^4_\infty, g_\infty(-1)\right)$ splits as $\left(\mathbb{R}, ds^2\right) \times \left(N^3, h\right)$, where $\left(N^3, h\right)$ is a normalized 3-dimensional gradient shrinking Ricci soliton. By the classification of 3-dimensional shrinking gradient Ricci solitons, $\left(N^3, h\right)$ is isometric to a quotient of either $\mathbb{S}^3$ or $\mathbb{R} \times \mathbb{S}^2$.

If the quotient of $\mathbb{S}^3$ ever occurs, then Theorem 2.29 implies that E is C^∞-asymptotic to $\mathbb{R} \times (\mathbb{S}^3/\Gamma)$, so we may assume that N^3 is never isometric to a quotient of $\mathbb{S}^3$. In this case, for any sequence $x_i \in E$ going to infinity along an integral curve of ∇f, a subsequence of (M^4, g, x_i) converges smoothly to $\mathbb{R} \times N^3$, where N^3 is either $\mathbb{R} \times \mathbb{S}^2$ or its $\mathbb{Z}_2$ quotient. However, by Corollary 5.1 and Remark 5.1 in Naber [**Nab10**], N^3 is uniquely determined by the integral curve. □

If (M^4, g, f) is weakly asymptotically 2-cylindrical, then the limit of (M^4, g, x_i) in Theorem 2.30 may depend on the integral curve of ∇f that contains the sequence $\{x_i\}$. As pointed out by Naber [**Nab10**, Remark 7.1], consider the case when M^4 is the $\mathbb{Z}_2$ quotient of $\mathbb{R}^2 \times \mathbb{S}^2$ i.e.,

$$M^4 = \mathbb{R} \times \left(\mathbb{R} \times \mathbb{S}^2\right)/\mathbb{Z}_2,$$

where $\mathbb{Z}_2$ acts on $\mathbb{R} \times \mathbb{S}^2$ by the antipodal map $(t, \theta) \to (-t, -\theta)$ for $t \in \mathbb{R}$ and $\theta \in \mathbb{S}^2$. If the sequence $\{x_i\}$ tends to infinity along the first $\mathbb{R}$-factor, then the limit remains the $\mathbb{Z}_2$ quotient of $\mathbb{R}^2 \times \mathbb{S}^2$. However, if the sequence $\{x_i\}$ tends to infinity along the second $\mathbb{R}$-factor, the limit becomes $\mathbb{R}^2 \times \mathbb{S}^2$ (without the $\mathbb{Z}_2$ quotient).

2.5.4. Volume dichotomy for shrinking solitons with bounded scalar curvatures.

According to Proposition 2.18, if the scalar curvature of a 4-dimensional shrinker converges to zero along some sequence going to infinity, then $\left(M^4, g\right)$ has positive AVR; that is

$$\operatorname{Vol}\left(B\left(o, r\right)\right) \geq \frac{1}{C} r^4 \quad \text{for all } r \geq 1$$

and some constant $C > 0$. Here $o \in M^4$ is a minimum point of f.

We henceforth consider the complementary case and assume the 2-sided bound that $C \geq R \geq \frac{1}{C}$ on M^4 and we study the volume growth of geodesic balls in M^4. Certainly, if $\left(M^4, g, f\right)$ is C^∞-asymptotic to $\mathbb{R} \times (\mathbb{S}^3/\Gamma)$ along some end E, then (E, g) has linear volume growth; i.e.,

$$\operatorname{Vol}\left(B\left(o, r\right) \cap E\right) \leq Cr \quad \text{for all } r \geq 1.$$

The remaining case is when $\left(M^4, g, x_i\right)$ converges to $\mathbb{R}^2 \times \mathbb{S}^2$ or its $\mathbb{Z}_2$ quotient along each integral curve of f contained by some end $E \subset M^4$, which we address below.

We will make use of the following lemma. Recall that $D(t)$ and $\Sigma(t)$ denote the sublevel and level sets of f as in (2.1).

LEMMA 2.31. *Let (M^4, g, f) be a 4-dimensional shrinking gradient Ricci soliton with bounded scalar curvature R. Then for any $t \geq t_0$,*

$$\frac{d}{dt}\left(\int_{\Sigma(t)} R\,|\nabla f|\,da - \int_{\Sigma(t)} \frac{\langle \nabla R, \nabla f\rangle}{|\nabla f|} da\right) = \int_{\Sigma(t)} \frac{R}{|\nabla f|} da + \int_{\Sigma(t)} \frac{2\,|\mathrm{Ric}|^2 - R^2}{|\nabla f|} da.$$

PROOF. Integrating by parts and using that the unit normal to $\Sigma(t)$ is $\nu = \frac{\nabla f}{|\nabla f|}$, we have that

$$\begin{aligned}
&\int_{\Sigma(t)} \frac{\langle \nabla R, \nabla f\rangle}{|\nabla f|} da - \int_{\Sigma(t_0)} \frac{\langle \nabla R, \nabla f\rangle}{|\nabla f|} da \\
&= \int_{D(t)\setminus D(t_0)} \Delta R\, dv \\
&= \int_{D(t)\setminus D(t_0)} \langle \nabla f, \nabla R\rangle\, dv + \int_{D(t)\setminus D(t_0)} \left(R - 2\,|\mathrm{Ric}|^2\right) dv \\
&= -\int_{D(t)\setminus D(t_0)} R\Delta f\, dv + \int_{\Sigma(t)} R\,|\nabla f|\,da - \int_{\Sigma(t_0)} R\,|\nabla f|\,da \\
&\quad + \int_{D(t)\setminus D(t_0)} \left(R - 2\,|\mathrm{Ric}|^2\right) dv.
\end{aligned}$$

This proves that

$$\begin{aligned}
\int_{\Sigma(t)} R\,|\nabla f|\,da - \int_{\Sigma(t)} \frac{\langle \nabla R, \nabla f\rangle}{|\nabla f|} da = &\int_{\Sigma(t_0)} R\,|\nabla f|\,da \\
&- \int_{\Sigma(t_0)} \frac{\langle \nabla R, \nabla f\rangle}{|\nabla f|} da \\
&+ \int_{D(t)\setminus D(t_0)} \left(2\,|\mathrm{Ric}|^2 - R^2 + R\right) dv.
\end{aligned}$$

Taking a derivative in t implies the desired result. □

We now establish the following integral estimate for the scalar curvature.

PROPOSITION 2.32 (Integral of R over level sets of f). *Let (M^4, g, f) be a complete noncompact 4-dimensional shrinking gradient Ricci soliton with bounded scalar curvature R. There exist constants $C_0 > 0$ and $C_1 > 0$ so that for any $t \geq t_0$*

$$\int_{\Sigma(t)} R\, da \leq C_0 \sqrt{t}\,(\ln t)^{C_1}.$$

PROOF. By Lemma 2.31 we have that

$$\frac{d}{dt}\left(\int_{\Sigma(t)} R\,|\nabla f|\,da - \int_{\Sigma(t)} \frac{\langle \nabla R, \nabla f\rangle}{|\nabla f|} da\right) = \int_{\Sigma(t)} \frac{R}{|\nabla f|} da + \int_{\Sigma(t)} \frac{2\,|\mathrm{Ric}|^2 - R^2}{|\nabla f|} da.$$

According to Theorem 2.22 we have

$$2\left|\mathrm{Ric}\right|^2 - R^2 \leq \frac{CR}{\ln t} \quad \text{on } \Sigma(t).$$

Therefore

$$\frac{d}{dt}\left(\int_{\Sigma(t)} R\left|\nabla f\right| da - \int_{\Sigma(t)} \frac{\langle \nabla R, \nabla f\rangle}{\left|\nabla f\right|} da\right) \leq \left(1+\frac{C}{\ln t}\right)\int_{\Sigma(t)} \frac{R}{\left|\nabla f\right|} da.$$

Moreover, we have

$$\frac{1}{\left|\nabla f\right|^2} \leq \frac{1}{f}\left(1+\frac{C}{f}\right).$$

Thus, it follows that

$$\int_{\Sigma(t)} \frac{R}{\left|\nabla f\right|} da \leq \frac{1}{t}\left(1+\frac{C}{t}\right)\int_{\Sigma(t)} R\left|\nabla f\right| da.$$

In conclusion, there exists $C_1 > 0$ such that

$$\frac{d}{dt}\left(\int_{\Sigma(t)} R\left|\nabla f\right| da - \int_{\Sigma(t)} \frac{\langle \nabla R, \nabla f\rangle}{\left|\nabla f\right|} da\right) \leq \frac{1}{t}\left(1+\frac{C_1}{\ln t}\right)\int_{\Sigma(t)} R\left|\nabla f\right| da.$$

Denote by

$$u(t) := \int_{\Sigma(t)} R\left|\nabla f\right| da \quad \text{and} \quad v(t) := \int_{\Sigma(t)} \frac{\langle \nabla R, \nabla f\rangle}{\left|\nabla f\right|} da.$$

The above differential inequality can be written as

$$u'(t) - \frac{1}{t}\left(1+\frac{C_1}{\ln t}\right)u(t) \leq v'(t).$$

We rewrite this as

$$\frac{d}{dt}\left(\frac{u(t)}{t\left(\ln t\right)^{C_1}}\right) \leq \frac{v'(t)}{t\left(\ln t\right)^{C_1}}.$$

Integrating over t, it follows that

$$\frac{u(t)}{t\left(\ln t\right)^{C_1}} \leq \frac{u(t_0)}{t_0\left(\ln t_0\right)^{C_1}} + \int_{t_0}^{t} \frac{v'(r)}{r\left(\ln r\right)^{C_1}} dr. \tag{2.147}$$

Integrating by parts, we get

$$\begin{aligned}\int_{t_0}^{t} \frac{v'(r)}{r\left(\ln r\right)^{C_1}} dr = &\ \frac{v(t)}{t\left(\ln t\right)^{C_1}} - \frac{v(t_0)}{t_0\left(\ln t_0\right)^{C_1}} \\ &+ \int_{t_0}^{t} \frac{1}{r^2\left(\ln r\right)^{C_1}}\left(1+\frac{C_1}{\ln r}\right)v(r)\, dr.\end{aligned} \tag{2.148}$$

Note that by Theorem 2.16, we have

$$\begin{aligned} v(r) &= \int_{\Sigma(r)} \frac{\langle \nabla R, \nabla f \rangle}{|\nabla f|} da \\ &= \int_{\Sigma(r)} \frac{1}{|\nabla f|} \left(\Delta R - R + |\mathrm{Ric}|^2 \right) da \\ &\leq \frac{C}{r} \int_{\Sigma(r)} R |\nabla f| \, da \\ &= \frac{C}{r} u(r). \end{aligned}$$

Hence, (2.148) implies that

$$\int_{t_0}^{t} \frac{v'(r)}{r (\ln r)^{C_1}} dr \leq C + \frac{C u(t)}{t^2 (\ln t)^{C_1}} + C \int_{t_0}^{t} \frac{u(r)}{r^3 (\ln r)^{C_1}} dr.$$

By (2.147), we get

$$\frac{u(t)}{t (\ln t)^{C_1}} \leq C + C \int_{t_0}^{t} \frac{u(r)}{r^3 (\ln r)^{C_1}} dr. \tag{2.149}$$

This implies that the function

$$w(t) := \int_{t_0}^{t} \frac{u(r)}{r^3 (\ln r)^{C_1}} dr$$

satisfies

$$t^2 w'(t) \leq C (1 + w(t)).$$

We integrate this inequality over t to obtain that $w(t) \leq C$. Plugging this into (2.149) implies that

$$\int_{\Sigma(t)} R |\nabla f| \, da = u(t) \leq C_0 t (\ln t)^{C_1}$$

for some constant $C_0 > 0$, which proves the result. □

We obtain the following dichotomy for the volume growth rate of 4-dimensional shrinkers; see Wang and one of the authors [**MW**].

THEOREM 2.33 (Volume dichotomy for shrinkers). *Let (M^4, g, f) be a complete noncompact 4-dimensional shrinking gradient Ricci soliton with bounded scalar curvature. Then either (M^4, g) has positive asymptotic volume ratio, i.e.,*

$$\mathrm{Vol}(B(o, r)) \geq \frac{1}{C} r^4 \quad \textit{for all } r \geq 1,$$

or there exist constants $C_0, C_1 > 0$ such that

$$\mathrm{Vol}(B(o, r)) \leq C_0 r^2 (\ln r)^{C_1} \quad \textit{for all } r \geq 2.$$

PROOF. By the first variation of area formula (2.59), we have

$$\frac{d}{dt} \mathrm{Area}(\Sigma(t)) = \int_{\Sigma(t)} \frac{1}{|\nabla f|^2} \left(\sum_{a=1}^{3} \nabla^2_{aa} f \right) da,$$

where $\nabla^2_{aa} f := \nabla^2 f(e_a, e_e)$ and $\{e_a\}_{a=1,2,3}$ is an orthonormal frame on $T\Sigma(t)$. Since

$$\sum_{a=1}^{3} \nabla^2_{aa} f = \frac{3}{2} - R + \frac{1}{2} \frac{\langle \nabla R, \nabla f \rangle}{|\nabla f|^2},$$

it follows that

$$\frac{d}{dt} \text{Area}\left(\Sigma\left(t\right)\right) \geq \frac{1}{t} \int_{\Sigma(t)} \left(\frac{3}{2} - R\right) da - \frac{C}{t^2} \text{Area}\left(\Sigma\left(t\right)\right).$$

Using Proposition 2.32, we get

$$\frac{d}{dt} \text{Area}\left(\Sigma\left(t\right)\right) \geq \left(\frac{3}{2t} - \frac{C}{t^2}\right) \text{Area}\left(\Sigma\left(t\right)\right) - \frac{C_0}{\sqrt{t}} \left(\ln t\right)^{C_1}.$$

This shows that

$$\frac{d}{dt} \left(\frac{\mathrm{e}^{-\frac{C}{t}}}{t\sqrt{t}} \text{Area}\left(\Sigma\left(t\right)\right) \right) \geq -\frac{C_0}{t^2} \left(\ln t\right)^{C_1}$$

for all $t \geq t_0$. Integrating in $t \geq s$, and using that

$$\int_s^t \frac{1}{r^2} \left(\ln r\right)^{C_1} dr \leq \frac{2}{s} \left(\ln s\right)^{C_1} \quad \text{for all } s \geq t_0,$$

it follows that

$$\frac{\mathrm{e}^{-\frac{C}{t}}}{t\sqrt{t}} \text{Area}\left(\Sigma\left(t\right)\right) \geq \frac{\mathrm{e}^{-\frac{C}{s}}}{s\sqrt{s}} \text{Area}\left(\Sigma\left(s\right)\right) - \frac{2C_0}{s} \left(\ln s\right)^{C_1}.$$

In particular, we get

$$\frac{1}{t\sqrt{t}} \text{Area}\left(\Sigma\left(t\right)\right) \geq \frac{1}{2} \frac{1}{s\sqrt{s}} \left(\text{Area}\left(\Sigma\left(s\right)\right) - 5C_0 \sqrt{s} \left(\ln s\right)^{C_1} \right). \tag{2.150}$$

Suppose that there exists $s \geq t_0$ such that

$$\text{Area}\left(\Sigma\left(s\right)\right) \geq 6C_0 \sqrt{s} \left(\ln s\right)^{C_1}. \tag{2.151}$$

Then (2.150) implies that

$$\text{Area}\left(\Sigma\left(t\right)\right) \geq \frac{1}{C} t^{\frac{3}{2}} \text{ for all } t \geq s.$$

This is the same as the inequality (2.60) in the proof of Proposition 2.18. The same argument as given there proves that

$$\text{Vol}\left(B\left(o, r\right)\right) \geq \frac{1}{C} r^4 \quad \text{for all } r \geq r_0.$$

On the other hand, if (2.151) fails for all $s \geq t_0$, then

$$\text{Area}\left(\Sigma\left(t\right)\right) \leq C\sqrt{t} \left(\ln t\right)^{C_1} \quad \text{for all } t \geq t_0.$$

The co-area formula implies that

$$\begin{aligned} \text{Vol}(D(t)) &= \text{Vol}(D(t_0)) + \int_{t_0}^t \left(\int_{\Sigma(r)} \frac{1}{|\nabla f|} da \right) dr \\ &\leq C + C \int_{t_0}^t \frac{1}{\sqrt{r}} \text{Area}(\Sigma(r)) dr \\ &\leq Ct \left(\ln t\right)^{C_1} \end{aligned}$$

for all $t \geq t_0$. As in Proposition 2.18, this implies that

$$\mathrm{Vol}\,(B\,(o,r)) \leq Cr^2\,(\ln r)^{C_1} \quad \text{for all } r \geq r_0.$$

This proves the result. □

2.5.5. Rigidity theorem for shrinkers close to cylindrical.

If a shrinking gradient Ricci soliton is sufficiently close to a cylinder in the sense that both the metric and the potential function are close to that of the cylinder, then it must be isometric to the cylinder.

Let $(\mathbb{S}^k \times \mathbb{R}^{n-k}, \overline{g})$ denote the standard cylinder shrinking gradient Ricci soliton, so that $R_{\overline{g}} = \frac{k}{2}$ and the potential function is given by $\overline{f}(x,y) := \frac{|y|^2}{4}$ for $x \in \mathbb{S}^k$ and $y \in \mathbb{R}^{n-k}$. Let $\overline{o}$ denote a fixed point of the form $(x_0, 0)$, where $x_0 \in \mathbb{S}^k$.

Colding and Minicozzi [**CM21**, Theorem 0.2] and Li and Wang [**LW24b**, Theorem 1.1] proved rigidity results which include the following as a special case:

THEOREM 2.34 (Shrinking soliton rigidity). *For any $n \geq 3$ there exists $\varepsilon(n) > 0$ such that if a complete pointed shrinking gradient Ricci soliton (M^n, g, f, o) is $\varepsilon(n)$-close to $(\mathbb{S}^k \times \mathbb{R}^{n-k}, \overline{g}, \overline{f}, \overline{o})$, then (M^n, g) must be isometric to $(\mathbb{S}^k \times \mathbb{R}^{n-k}, \overline{g})$.*

By ε-**close** we mean that balls centered at the base points of radius $1/\varepsilon$ are metrically ε-close in $C^{1/\varepsilon}$-norm and that under the diffeomorphism used to obtain this metric closeness the same is true for the potential functions. See the aforementioned papers for the proofs of the results that contain the theorem above.

This result leads to the following question: To what extent can we remove the potential function closeness in the hypothesis of the theorem? For example, this is not possible when $n = 4$ and $k = 2$ by the existence of the BCCD shrinking gradient Ricci soliton discussed in §1.3.2.2.d. However, this might possibly be true when $n = 4$ and $k = 3$ since there are no counterexamples.

2.6. A decaying lower bound on sectional curvature

In this section we will prove, for any 4-dimensional shrinking gradient Ricci soliton with bounded scalar curvature, that the Riemann curvature operator has a negative lower bound that is logarithmically decaying in the distance function; this is Theorem 2.38 below. We can think of this result as an analogue of the Hamilton–Ivey estimate (Theorem 1.21) for the 3-dimensional Ricci flow. A consequence of Theorem 2.38 is that limits at infinity of the shrinking gradient Ricci soliton have non-negative curvature operator provided that they exist. In particular, this result suggests that the geometries at the infinities of 4-dimensional shrinkers are relatively simple.

We note that a logarithmically decaying bound is already available for the Ricci curvature by Theorem 2.22. Indeed, let $\lambda_1 \leq \lambda_2 \leq \lambda_3$ be the eigenvalues of the Ricci tensor restricted to the tangent space of the level set $\Sigma\,(t)$ of f. By estimates similar to (2.116), we have that

$$\begin{aligned}\frac{|\mathrm{Ric}|^2}{R^2} - \frac{1}{2} &\geq \frac{1}{2R^2}\left(2|\mathrm{Ric}_\Sigma|^2 - R_\Sigma^2\right) - \frac{C}{f} \\ &= \frac{1}{2R^2}\left((\lambda_1 + \lambda_3 - \lambda_2)^2 - 4\lambda_1\lambda_3\right) - \frac{C}{f} \\ &\geq -\frac{4\lambda_3\lambda_1}{R^2} - \frac{C}{f}.\end{aligned}$$

Together with Theorem 2.22, which says that

$$\frac{|\mathrm{Ric}|^2}{R^2} - \frac{1}{2} \leq \frac{C}{R \ln f},$$

this implies that

$$\lambda_3 \lambda_1 \geq -\frac{CR}{\ln f}.$$

From

$$R = \lambda_1 + \lambda_2 + \lambda_3 + R_{44} \leq 3\lambda_3 + \frac{CR}{f}$$

we deduce that

$$\lambda_3 \geq \frac{1}{4} R > 0 \quad \text{on } M^4 \setminus D(t_0).$$

Consequently, $\lambda_1 \geq -\frac{C}{\ln f}$, which implies that

$$\mathrm{Ric} \geq -\frac{C}{\ln f}.$$

Our goal is to establish a similar lower bound for the sectional curvatures. We begin with a tensor computation similar to the scalar computation of Lemma 2.19. Related tensorial computations were carried out by Eminenti, La Nave, and Mantegazza in [**ELNM08**] under the additional assumption that the gradient shrinking Ricci soliton is locally conformally flat, i.e., its Weyl tensor vanishes.

LEMMA 2.35 (Bochner formula for eigenvalues of Ric). *Let $\left(M^4, g, f\right)$ be a 4-dimensional shrinking gradient Ricci soliton with bounded scalar curvature. Let $\lambda_1 \leq \lambda_2 \leq \lambda_3$ denote the eigenvalues of the Ricci tensor restricted to the tangent space of the level set $\Sigma(t)$ of f, where $t \geq t_0$. Then there exists a constant $C > 0$ such that the inequality*

$$\Delta_F \left(\frac{\lambda_3}{R} \right) \geq \frac{Q}{R^2} - \frac{C}{|\nabla f|} \tag{2.152}$$

holds in the barrier sense (cf. Definition 2.17 in Volume 1) on the super-level set $M^4 \setminus D(t_0)$ of f, where $F = f - 2\ln R$ and where

$$Q := \left(\lambda_1 + \lambda_3 - \lambda_2\right)^2 \left(\lambda_3 - \lambda_1\right) + \left(\lambda_3 + \lambda_2 - \lambda_1\right)^2 \left(\lambda_3 - \lambda_2\right). \tag{2.153}$$

PROOF. Recall from (1.50) that

$$\Delta_f R_{ij} = R_{ij} - 2R_{kij\ell} R_{k\ell}.$$

We also compute that

$$\begin{aligned} \Delta_f \left(\frac{1}{R} \right) &= -\frac{1}{R^2} \Delta_f R + \frac{2}{R^3} |\nabla R|^2 \\ &= -\frac{1}{R^2} \left(R - 2 |\mathrm{Ric}|^2 \right) + \frac{2}{R^3} |\nabla R|^2 . \end{aligned}$$

Hence, it follows that

$$\begin{aligned}\Delta_f\left(\frac{R_{ij}}{R}\right) &= \frac{1}{R}\Delta_f R_{ij} + \left(\Delta_f \frac{1}{R}\right) R_{ij} + 2\left\langle \nabla R_{ij}, \nabla \frac{1}{R}\right\rangle \\ &= \frac{1}{R}\left(R_{ij} - 2R_{kij\ell}R_{k\ell}\right) - \frac{1}{R^2}\left(R - 2\left|\mathrm{Ric}\right|^2\right) R_{ij} \\ &\quad + 2\left\langle \nabla R_{ij}, \nabla \frac{1}{R}\right\rangle + \frac{2}{R^3}\left|\nabla R\right|^2 R_{ij} \\ &= \frac{2}{R^2}\left|\mathrm{Ric}\right|^2 R_{ij} - \frac{2}{R}R_{kij\ell}R_{k\ell} - 2\left\langle \nabla \ln R, \nabla\left(\frac{R_{ij}}{R}\right)\right\rangle,\end{aligned}$$

where $\Delta_f\left(\frac{R_{ij}}{R}\right) := \left(\Delta_f \frac{\mathrm{Ric}}{R}\right)_{ij}$. Therefore, we obtain the following equation:

$$\Delta_F\left(\frac{R_{ij}}{R}\right) = \frac{2}{R^2}\left|\mathrm{Ric}\right|^2 R_{ij} - \frac{2}{R}R_{kij\ell}R_{k\ell}. \tag{2.154}$$

Let $x_0 \in M^4 \setminus D(t_0)$ be an arbitrary point. Assume $x_0 \in \Sigma(r)$ for some $r \geq t_0$ and let $\{e_1, e_2, e_3\}$ be unit eigenvectors of $\mathrm{Ric}|_{T_{x_0}\Sigma(r)}$ with eigenvalues $\lambda_1 \leq \lambda_2 \leq \lambda_3$. Then $\{e_1, e_2, e_3, e_4\}$ is an orthonormal frame at x_0, where $e_4 = \nu = \frac{\nabla f}{|\nabla f|}$.

We extend e_3 through parallel transport in a neighborhood of x_0, and denote by Z the resulting vector field. Note that $Z(x)$ may not be tangential to the level sets of f for $x \neq x_0$. However, Z satisfies

$$\langle Z, \nabla f\rangle (x_0) = 0, \quad \nabla Z(x_0) = 0, \quad \text{and} \quad \nabla^2_{V,V} Z(x_0) = 0 \tag{2.155}$$

for any unit vector $V \in T_{x_0}M$, where $\nabla^2 Z$ denotes the double covariant derivative of Z. Denote by

$$X := Z^T = Z - \frac{\langle Z, \nabla f\rangle}{|\nabla f|^2}\nabla f \tag{2.156}$$

the tangential component of Z.

CLAIM 2.36. *The vector field X in (2.156) satisfies*

$$|\nabla X| + |\nabla_{\nabla f} X| + \left|\nabla^2_{V,V} X\right| \leq \frac{C}{|\nabla f|} \quad \text{at} \quad x_0, \tag{2.157}$$

for any unit vector $V \in T_{x_0}M$.

PROOF OF CLAIM 2.36. Let $\{\tilde{e}_1, \tilde{e}_2, \tilde{e}_3, \tilde{e}_4\}$ be an arbitrary parallel orthonormal frame at x_0. We denote with ∇_j the covariant derivative with respect to $\tilde{e}_j$. We have, for any $j \in \{1, 2, 3, 4\}$, that

$$\begin{aligned}\nabla_j X &= \nabla_j Z - \nabla_j\left(\frac{\langle Z, \nabla f\rangle}{|\nabla f|^2}\nabla f\right) \\ &= \nabla_j Z - \frac{\nabla_j \langle Z, \nabla f\rangle}{|\nabla f|^2}\nabla f + \frac{\nabla_j |\nabla f|^2}{|\nabla f|^4}\langle Z, \nabla f\rangle \nabla f - \frac{\langle Z, \nabla f\rangle}{|\nabla f|^2}\nabla_j \nabla f.\end{aligned}$$

By (2.155) and the estimate $\left|\nabla^2 f(Z, \tilde{e}_j)\right|(x_0) \leq C$ we get that

$$|\nabla X| \leq \frac{C}{|\nabla f|} \quad \text{at} \quad x_0. \tag{2.158}$$

Since $|X|(x_0) = 1$, we see that X is nonzero in a neighborhood of x_0. Moreover, (2.155) implies that at x_0,

$$\begin{aligned}\nabla_{\nabla f} X &= -\frac{\nabla^2 f(Z, \nabla f)}{|\nabla f|^2}\nabla f \\ &= \frac{\mathrm{Ric}(Z, \nabla f)}{|\nabla f|^2}\nabla f \\ &= \frac{\langle \nabla R, Z\rangle}{2|\nabla f|^2}\nabla f.\end{aligned}$$

Therefore, it follows that

$$|\nabla_{\nabla f} X| \le \frac{C}{|\nabla f|} \quad \text{at} \quad x_0. \tag{2.159}$$

We similarly compute the double covariant derivative of X at x_0 and find for any fixed $j \in \{1, 2, 3, 4\}$ that

$$\begin{aligned}\nabla^2_{jj} X = &-\nabla_j\left(\frac{\nabla_j \langle Z, \nabla f\rangle}{|\nabla f|^2}\nabla f\right) + \nabla_j\left(\frac{\nabla_j |\nabla f|^2}{|\nabla f|^4}\langle Z, \nabla f\rangle \nabla f\right) \\ &- \nabla_j\left(\frac{\langle Z, \nabla f\rangle}{|\nabla f|^2}\nabla_j \nabla f\right),\end{aligned}$$

where, due to $\tilde{e}_j$ being parallel, we have that $\nabla^2_{jj} X := \nabla^2_{\tilde{e}_j, \tilde{e}_j} X = \nabla_j (\nabla_j X)$. The first term on the right side can be expanded as

$$\begin{aligned}-\nabla_j\left(\frac{\nabla_j \langle Z, \nabla f\rangle}{|\nabla f|^2}\nabla f\right) = &-\frac{\nabla^2_{jj}\langle Z, \nabla f\rangle}{|\nabla f|^2}\nabla f - \frac{\nabla_j \langle Z, \nabla f\rangle}{|\nabla f|^2}\nabla_j \nabla f \\ &+ \frac{(\nabla_j \langle Z, \nabla f\rangle)(\nabla_j |\nabla f|^2)}{|\nabla f|^4}\nabla f.\end{aligned}$$

Applying (2.155) and the estimate $|\nabla^2 f| + |\nabla^3 f| \le C$ we find that

$$\left|\frac{\nabla^2_{jj}\langle Z, \nabla f\rangle}{|\nabla f|^2}\nabla f\right| \le \frac{|\nabla^3 f||Z|}{|\nabla f|} \le \frac{C}{|\nabla f|} \quad \text{at} \quad x_0.$$

Similar arguments imply that at x_0,

$$\left|\frac{\nabla_j \langle Z, \nabla f\rangle}{|\nabla f|^2}\nabla_j \nabla f\right| \le \frac{|\nabla^2 f|^2 |Z|}{|\nabla f|^2} \le \frac{C}{|\nabla f|^2}$$

and

$$\left|\frac{(\nabla_j \langle Z, \nabla f\rangle)(\nabla_j |\nabla f|^2)}{|\nabla f|^4}\nabla f\right| \le \frac{|\nabla^2 f|^2 |Z|}{|\nabla f|^2} \le \frac{C}{|\nabla f|^2}.$$

Therefore, this proves that

$$\left|\nabla_j\left(\frac{\nabla_j \langle Z, \nabla f\rangle}{|\nabla f|^2}\nabla f\right)\right| \le \frac{C}{|\nabla f|} \quad \text{at} \quad x_0.$$

Proceeding similarly with remaining two terms in the expression of $\nabla^2_{jj} X$ implies that

$$|\nabla^2_{jj} X| \le \frac{C}{|\nabla f|} \quad \text{at} \quad x_0. \tag{2.160}$$

Since the frame $\{\tilde{e}_j\}$ is arbitrary, Claim 2.36 now follows from (2.158), (2.159), and (2.160). □

For X defined in (2.156), let

$$\psi := \frac{\mathrm{Ric}\,(X,X)\,/\,|X|^2}{R}. \tag{2.161}$$

Clearly, since $X \in T\Sigma\,(t)$ and $\lambda_1 \leq \lambda_2 \leq \lambda_3$ are the eigenvalues of Ric restricted to $T\Sigma(t)$, it follows that

$$\psi \leq \frac{\lambda_3}{R}$$

in a neighborhood of x_0. Moreover, as $X(x_0) = e_3$ is an eigenvector of Ric with eigenvalue λ_3, we have that

$$\psi\,(x_0) = \frac{\lambda_3}{R}\,(x_0)\,.$$

To finish the proof, we now check that

$$\Delta_F \psi \geq \frac{Q}{R^2} - \frac{C}{|\nabla f|} \quad \text{at} \quad x_0.$$

We extend the orthonormal frame $\{e_i\}$ in a neighborhood of x_0 through parallel transport and continue to denote the resulting orthonormal frame with $\{e_i\}$; note that $e_3 = Z$ defined above and satisfying (2.155). Express X as

$$X = \sum_{i=1}^{4} X_i e_i,$$

where, for convenience, we will not raise indices for contravariant tensors. We have by (2.161) that

$$\Delta_F \psi =. \Delta_F \left(\frac{R_{ij}}{R}\right) \frac{X_i X_j}{|X|^2} + \Delta_F \left(\frac{X_i X_j}{|X|^2}\right) \frac{R_{ij}}{R} + 2\left\langle \nabla \left(\frac{R_{ij}}{R}\right), \nabla \left(\frac{X_i X_j}{|X|^2}\right)\right\rangle.$$

By (2.157), we have that

$$\left|\Delta_F \left(\frac{X_i X_j}{|X|^2}\right)\right| \leq \frac{C}{|\nabla f|} \quad \text{at} \quad x_0,$$

and

$$\left|\left\langle \nabla \left(\frac{R_{ij}}{R}\right), \nabla \left(\frac{X_i X_j}{|X|^2}\right)\right\rangle\right| \leq \frac{C}{|\nabla f|} \quad \text{at} \quad x_0.$$

Consequently, we obtain at x_0 that

$$\begin{aligned} \Delta_F \psi &\geq \Delta_F \left(\frac{R_{ij}}{R}\right) \frac{X_i X_j}{|X|^2} - \frac{C}{|\nabla f|} \\ &= \frac{2}{R^2}\,|\mathrm{Ric}|^2\, R_{33} - \frac{2}{R} R_{k33\ell} R_{k\ell} - \frac{C}{|\nabla f|}, \end{aligned} \tag{2.162}$$

where in the last line we have applied (2.154) and used that $X\,(x_0) = e_3$. Since

$$R_{k33l} R_{kl} = R_{a33b} R_{ab} + 2R_{a334} R_{a4} + R_{4334} R_{44}$$

and

$$|R_{ijk4}| \leq \frac{2|\nabla \mathrm{Ric}|}{|\nabla f|} \leq \frac{C R^{\frac{3}{2}}}{|\nabla f|},$$

it follows that

$$R_{k33\ell}R_{k\ell} \leq R_{a33b}R_{ab} + \frac{CR^2}{f}.$$

According to Lemma 2.2, the following holds at x_0:

$$\begin{aligned} R_{a33b}R_{ab} &= \lambda_1 R_{1331} + \lambda_2 R_{2332} \\ &\leq \lambda_1 \left(\lambda_1 + \lambda_3 - \frac{1}{2}R\right) + \lambda_2 \left(\lambda_2 + \lambda_3 - \frac{1}{2}R\right) + \frac{CR^2}{f}. \end{aligned}$$

In conclusion,

$$2R_{k33l}R_{kl} \leq \lambda_1^2 + \lambda_2^2 + \lambda_1\lambda_3 + \lambda_2\lambda_3 - 2\lambda_1\lambda_2 + \frac{CR^2}{f}.$$

Hence, this implies that

$$\begin{aligned} 2\left|\mathrm{Ric}\right|^2 R_{33} - 2RR_{k33\ell}R_{k\ell} &\geq 2\lambda_3 \left|\mathrm{Ric}_\Sigma\right|^2 - 2R_\Sigma R_{k33\ell}R_{k\ell} - \frac{CR^3}{f} \\ &\geq 2\lambda_3^3 - \lambda_1^3 - \lambda_2^3 + \lambda_1^2\lambda_2 + \lambda_2^2\lambda_1 \\ &\quad - \lambda_3^2\lambda_1 - \lambda_3^2\lambda_2 - \frac{CR^3}{f} \\ &= \left(\lambda_1 + \lambda_3 - \lambda_2\right)^2 \left(\lambda_3 - \lambda_1\right) \\ &\quad + \left(\lambda_3 + \lambda_2 - \lambda_1\right)^2 \left(\lambda_3 - \lambda_2\right) - \frac{CR^3}{f}. \end{aligned}$$

Plugging this into (2.162) implies that

$$\Delta_F \psi \geq \frac{Q}{R^2} - \frac{C}{|\nabla f|} \quad \text{at} \quad x_0,$$

where

$$Q = \left(\lambda_1 + \lambda_3 - \lambda_2\right)^2 \left(\lambda_3 - \lambda_1\right) + \left(\lambda_3 + \lambda_2 - \lambda_1\right)^2 \left(\lambda_3 - \lambda_2\right).$$

This proves the result. □

We now express (2.152) as an equation on $\Sigma = \Sigma(t)$, by an argument analogous to that used to prove Lemma 2.21. According to (2.73), for an arbitrary function w, the gradient and the Laplacian of w on $\Sigma(t)$ are expressed by

$$\begin{aligned} \nabla_\Sigma w &= \nabla w - \frac{\langle \nabla w, \nabla f\rangle}{|\nabla f|}, \\ \Delta_\Sigma w &= \Delta w - \frac{\nabla^2 w(\nabla f, \nabla f)}{|\nabla f|^2} \\ &\quad - \frac{\langle \nabla w, \nabla f\rangle}{|\nabla f|^2}\left(\frac{3}{2} - R + \frac{1}{2}\frac{\langle \nabla R, \nabla f\rangle}{|\nabla f|^2}\right). \end{aligned} \tag{2.163}$$

As in (2.74), we denote

$$\mathcal{L}w = \Delta_\Sigma w - \langle \nabla f, \nabla w\rangle + 2\left\langle \nabla_\Sigma w, \nabla_\Sigma \ln R\right\rangle.$$

LEMMA 2.37. *Let $\left(M^4, g, f\right)$ be a 4-dimensional shrinking gradient Ricci soliton with bounded scalar curvature. Let $\lambda_1 \leq \lambda_2 \leq \lambda_3$ be the eigenvalues of the*

Ricci curvature restricted to the tangent space of $\Sigma(t)$, *where* $t \geq t_0$. *There exists a constant* $C > 0$ *such that the inequality*

$$\mathcal{L}\left(\frac{\lambda_3}{R}\right) \geq \frac{Q}{R^2} - \frac{C}{|\nabla f|}$$

holds in the barrier sense on $M^4 \setminus D(t_0)$, *where* Q *is defined by* (2.153).

PROOF. According to (2.163), the function

$$\psi = \frac{\operatorname{Ric}(X,X)/|X|^2}{R}$$

defined in (2.161) satisfies (cf. (2.75))

$$\begin{aligned}\Delta_F \psi = \mathcal{L}\psi &+ \frac{\nabla^2\psi(\nabla f, \nabla f)}{|\nabla f|^2} + \frac{\langle \nabla\psi, \nabla f\rangle}{|\nabla f|^2}\left(\frac{3}{2} - R_\Sigma\right) \\ &+ \frac{2}{|\nabla f|^2}\langle \nabla\psi, \nabla f\rangle \langle \nabla \ln R, \nabla f\rangle .\end{aligned}$$

By (2.157) and Theorem 2.16, we immediately get

$$|\langle \nabla\psi, \nabla f\rangle| \leq C.$$

Moreover, (2.157) implies that

$$|\nabla^2_{\nabla f, \nabla f} X| \leq C|\nabla f|.$$

Furthermore, Theorem 2.16 and Lemma 2.7 imply that

$$|\nabla^2_{\nabla f, \nabla f}\operatorname{Ric}| \leq C,$$

see the proof of Lemma 2.21 for similar estimates. We obtain that

$$\left|\nabla^2\psi(\nabla f, \nabla f)\right| \leq C|\nabla f|.$$

Concluding, we have proved that

$$\Delta_F \psi \leq \mathcal{L}\psi + \frac{C}{|\nabla f|}.$$

Together with Lemma 2.35, this yields

$$\mathcal{L}\psi \geq \frac{Q}{R^2} - \frac{C}{|\nabla f|}.$$

This proves the result. □

We now prove the following curvature lower bound due to Wang and one of the authors [**MW15a**]. We denote by $r(x) = d(o,x)$ the distance from x to a minimum point $o \in M^4$ of the potential f.

THEOREM 2.38 (Curvature operator lower bound for shrinkers). *Let* (M^4, g, f) *be a complete noncompact 4-dimensional shrinking gradient Ricci soliton with bounded scalar curvature. Then the curvature operator satisfies the lower bound*

$$\operatorname{Rm} \geq -\frac{C}{\ln r}\operatorname{id}_{\wedge^2(T^*M)} \quad \text{on } M^4 \setminus B(o, r_0).$$

PROOF. Recall that by Lemma 2.2,

$$R_{1221} = \frac{1}{2}(R - 2R_{33}) - R_{44} + R_{3443}$$

for any orthonormal frame $\{e_i\}_{i=1}^4$ such that $e_4 = \frac{\nabla f}{|\nabla f|}$. By (2.68),

$$|R_{44}| + |R_{3443}| \leq \frac{C}{f}.$$

Consequently, to prove the theorem we will establish that

$$R - 2\lambda_3 \geq -\frac{C}{\ln t} \quad \text{on } \Sigma(t), \tag{2.164}$$

where λ_3 is the largest eigenvalue of the Ricci tensor restricted to $T\Sigma(t)$.

Recall that, by (2.94), the function

$$\rho = \frac{1}{R \ln^2 f} + \frac{1}{\ln f},$$

defined by (2.93), satisfies

$$\mathcal{L}\rho \leq -\frac{1}{2R \ln^2 f} + \frac{C}{\ln^2 f}. \tag{2.165}$$

For $\Upsilon \geq 1$ arbitrary define the function

$$\begin{aligned} G &:= \frac{R - 2\lambda_3}{R} + \frac{\Upsilon}{\ln f} + \frac{\Upsilon}{R \ln^2 f} \\ &= -\frac{2\lambda_3}{R} + 1 + \Upsilon\rho. \end{aligned}$$

By Lemma 2.37 and (2.165), there exists a constant $C_0 > 0$ so that G satisfies in the barrier sense

$$\mathcal{L}G \leq -\frac{2Q}{R^2} - \frac{1}{2}\frac{\Upsilon}{R \ln^2 f} + \frac{C_0 \Upsilon}{\ln^2 f} \tag{2.166}$$

for any $\Upsilon \geq 1$, where Q is defined by (2.153). Since R is bounded above, there exists a constant $A \geq 1$ such that $R \leq A$ on M^4. Moreover, we may assume that $C_0 > 0$ in (2.166) is large enough such that

$$C_0 \geq \max\left\{100A^2, \sup_{M^4} \frac{|\mathrm{Ric}|}{R}\right\}.$$

According to (2.98), there exists $t_1 > 1$ large enough so that

$$\ln t_1 \geq \mathrm{e}^{C_0}, \tag{2.167}$$

and

$$|\mathrm{Ric}(\nu)| \leq \frac{1}{100C_0} \frac{1}{\ln^2 f} \quad \text{on } M^4 \setminus D(t_1). \tag{2.168}$$

Now let

$$\Upsilon := (\ln t_1) \sup_{\Sigma(t_1)} \frac{2\,|\mathrm{Ric}|}{R}. \tag{2.169}$$

Let us note that

$$\ln t_1 \leq \Upsilon \leq 2C_0 \ln t_1. \tag{2.170}$$

Assume by contradiction that there exists $T > t_1$ such that

$$G(x_0) := \min_{D(T)\setminus D(t_1)} G < 0. \tag{2.171}$$

By (2.169), we see that $G > 0$ on $\Sigma(t_1)$; hence $x_0 \in \Sigma(t)$ for some $t > t_1$. Following the argument in (2.102), but applying it to a minimum point x_0 instead of a maximum, we obtain $\mathcal{L}G(x_0) \geq 0$. By (2.166), it follows that

$$\frac{2Q}{R^2} + \frac{1}{2}\frac{\Upsilon}{R\ln^2 f} \leq \frac{C_0\Upsilon}{\ln^2 f} \quad \text{at} \quad x_0. \tag{2.172}$$

We will see that (2.172) and (2.171) imply upper and lower bounds of $|\lambda_1|$, which lead to a contradiction. We start with the lower bound.

CLAIM 2.39. *We have that*

$$|\lambda_1| \geq \frac{1}{4C_0}\frac{\Upsilon}{\ln f} \quad \textit{at} \quad x_0. \tag{2.173}$$

PROOF OF CLAIM 2.39. As $Q \geq 0$, we see from (2.172) that

$$R(x_0) \geq \frac{1}{2C_0}. \tag{2.174}$$

Note that by (2.168),

$$\begin{aligned} R &= \lambda_1 + \lambda_2 + \lambda_3 + R_{44} \\ &\leq 3\lambda_3 + \frac{1}{8C_0}. \end{aligned}$$

Hence, by (2.174), we conclude that

$$\lambda_3(x_0) \geq \frac{1}{4}R(x_0) \geq \frac{1}{8C_0}. \tag{2.175}$$

On the other hand, (2.172) and the scalar curvature bound $R \leq A \leq \frac{\sqrt{C_0}}{10}$ imply that

$$(\lambda_3 + \lambda_2 - \lambda_1)^2 (\lambda_3 - \lambda_2) \leq \frac{C_0^2\Upsilon}{100\ln^2 f} \quad \text{at} \quad x_0. \tag{2.176}$$

Assume by contradiction that

$$\lambda_3 - \lambda_2 \geq \frac{C_0^4\Upsilon}{\ln^2 f} \quad \text{at} \quad x_0.$$

Then (2.176) would imply that

$$\lambda_3 + \lambda_2 - \lambda_1 \leq \frac{1}{10C_0} \quad \text{at} \quad x_0.$$

In particular, as $\lambda_1 \leq \lambda_2 \leq \lambda_3$, we get that

$$\lambda_3(x_0) \leq \frac{1}{10C_0}.$$

This contradicts (2.175). Therefore, we obtain that

$$\lambda_3 - \lambda_2 \leq \frac{C_0^4\Upsilon}{\ln^2 f} \quad \text{at} \quad x_0. \tag{2.177}$$

Consequently, (2.174) implies that

$$\frac{\lambda_2-\lambda_3}{R} \geq -\frac{C_0^4\Upsilon}{R\ln^2 f} \geq -\frac{2C_0^5\Upsilon}{\ln^2 f} \quad \text{at} \quad x_0.$$

Hence, using (2.168), (2.170), and (2.167), we conclude that

$$\frac{\lambda_2-\lambda_3}{R}+\frac{R_{44}}{R}+\frac{1}{2}\frac{\Upsilon}{\ln f} \geq 0 \quad \text{at} \quad x_0.$$

Therefore, it follows that at x_0,

$$\begin{aligned} G &= \frac{\lambda_1+\lambda_2-\lambda_3+R_{44}}{R}+\frac{\Upsilon}{\ln f}+\frac{\Upsilon}{R\ln^2 f} \\ &\geq \frac{\lambda_1}{R}+\frac{1}{2}\frac{\Upsilon}{\ln f}. \end{aligned}$$

Since $G(x_0)<0$, this proves that

$$\lambda_1 \leq -\frac{1}{2}\frac{\Upsilon}{\ln f}R \quad \text{at} \quad x_0. \tag{2.178}$$

Together with (2.174), this implies that

$$|\lambda_1| \geq \frac{\Upsilon R}{2\ln f} \geq \frac{1}{4C_0}\frac{\Upsilon}{\ln f} \quad \text{at} \quad x_0.$$

This proves Claim 2.39. □

We now establish an upper bound for $|\lambda_1|$.

CLAIM 2.40. *We have that*

$$|\lambda_1| \leq \frac{C_0^2\sqrt{\Upsilon}}{\ln f}+\frac{C_0^4\Upsilon}{\ln^2 f} \quad \textit{at} \quad x_0. \tag{2.179}$$

PROOF OF CLAIM 2.40. By (2.175) and (2.178), we obtain in particular that

$$\lambda_3-\lambda_1 \geq \frac{1}{8C_0} \quad \text{at} \quad x_0.$$

However, (2.172) and $R \leq \frac{\sqrt{C_0}}{10}$ imply that

$$(\lambda_1+\lambda_3-\lambda_2)^2(\lambda_3-\lambda_1) \leq \frac{C_0^2\Upsilon}{100\ln^2 f} \quad \text{at} \quad x_0.$$

We conclude from the above two inequalities that

$$|\lambda_1+\lambda_3-\lambda_2| \leq \frac{C_0^2\sqrt{\Upsilon}}{\ln f} \quad \text{at} \quad x_0.$$

According to (2.177),

$$|\lambda_2-\lambda_3| \leq \frac{C_0^4\Upsilon}{\ln^2 f} \quad \text{at} \quad x_0.$$

Adding these two inequalities, we obtain that

$$|\lambda_1| \leq \frac{C_0^2\sqrt{\Upsilon}}{\ln f}+\frac{C_0^4\Upsilon}{\ln^2 f} \quad \text{at} \quad x_0.$$

This proves Claim 2.40. □

From (2.179) and (2.173) we conclude that

$$1 \leq \frac{4C_0^3}{\sqrt{\Upsilon}} + \frac{4C_0^5}{\ln f} \quad \text{at} \quad x_0.$$

However, (2.170) and (2.167) imply that

$$\frac{4C_0^3}{\sqrt{\Upsilon}} + \frac{4C_0^5}{\ln f} \leq \frac{5C_0^5}{\sqrt{\ln t_1}} < 1.$$

This is a contradiction. In conclusion, $G > 0$ on $M^4 \setminus D(t_1)$, which proves (2.164). This in turn completes the proof of Theorem 2.38, concerning the logarithmic decay of the curvature lower bound. □

2.7. Curvature estimates for shrinking soliton singularity models

As previously noted, a fundamental question is to understand what bounds hold for the curvature tensor norms of arbitrary complete shrinking gradient Ricci solitons. In this section we provide an answer in dimension 4, specifically in the case of singularity models. Namely, we prove the following result due to Freedman, Shin, Zhang, and one of the authors [**CFSZ20**, **CFSZ25**].

THEOREM 2.41 (Quadratic bound of $|\mathrm{Rm}|$ for shrinkers). *For any 4-dimensional shrinking gradient Ricci soliton singularity model (M^4, g, f) and point $p \in M^4$, there exists a constant C (which depends on the soliton singularity model and p) such that*

$$|\mathrm{Rm}|(x) \leq C(d(x,p)^2 + 1). \tag{2.180}$$

We will prove this theorem in detail below. The rough idea of the proof is as follows. Recall from (1.60) that the scalar curvature R, which is positive, has a quadratic upper bound for any shrinking gradient Ricci soliton. We want to replace R by $|\mathrm{Rm}|$ in this bound for the case where the shrinker is both 4-dimensional and a singularity model of the Ricci flow.

To prove this theorem, suppose for a contradiction that there exists a sequence of points in M^4 which violates the quadratic curvature bound. By point selection, we may assume that the rescalings about some sequence of points converge to a complete limit (M_∞^4, g_∞) which is a Ricci-flat asymptotically locally Euclidean (ALE) 4-manifold. Since the shrinking gradient Ricci soliton is also a singularity model of a Ricci flow $(\overline{M}^4, \overline{g}(t))$ on a closed 4-manifold, there would also be an unbounded number of disjoint copies of this Ricci-flat ALE manifold contained in $\overline{M}^4$. We can then obtain a contradiction by topological and geometric methods and by the compactness of the manifold $\overline{M}^4$. More generally, the idea of ALE exclusion originates with Anderson [**And89**, Lemma 6.3], [**And08**, Proposition 3.10], with a modification by Bamler (see [**Zha19**, §6]).

On the analytic side, two key ingredients to implement this strategy are Perelman's no local collapsing theorem and a deep result of Cheeger and Naber [**CN15**], which we recall below. On the topological and geometric sides, the result relies on homological arguments and index theorems.

2.7.1. No local collapsing on all scales.

By Perelman's no local collapsing theorem [**Per02**] (see [**CCG**$^+$**07**, Theorem 6.74] for an exposition), any singularity model (M^n, g) of the Ricci flow is κ-noncollapsed on all scales in the following sense: There exists $\kappa > 0$ such that if $x \in M^n$ and $r \in (0, \infty)$ have the property that $R \leq r^{-2}$ in $B(x, r)$, then

$$\text{Vol}(B(x,r)) \geq \kappa r^n. \tag{2.181}$$

In particular, if a singularity model (M^n, g) has zero scalar curvature, i.e., $R \equiv 0$ on M^n, then

$$\text{Vol}(B(x,r)) \geq \kappa r^n \tag{2.182}$$

for all $x \in M^n$ and $r > 0$. By definition, this says that (M^n, g) has Euclidean volume growth.

2.7.2. Asymptotically locally Euclidean spaces.

We are interested in complete noncompact Riemannian 4-manifolds that have Euclidean volume growth and asymptotically (at spatial infinity) look locally like Euclidean space. It is natural to consider such spaces modeled on the quotient of $\mathbb{R}^4$ by a finite subgroup of SO(4).

DEFINITION 2.42 (ALE space). A 4-dimensional oriented complete noncompact Riemannian manifold (M^4, g) is **asymptotically locally Euclidean** (**ALE**) if there exists $\tau > 0$, a compact subset K of M^4, a finite subgroup Γ of SO(4) acting freely on $\mathbb{S}^3$, and an orientation-preserving diffeomorphism

$$\Phi : (\mathbb{R}^4 \setminus B(0, C))/\Gamma \to M^4 \setminus K,$$

where C is some positive real number, such that the pullback $\widetilde{h}$ to the cover $\mathbb{R}^4 \setminus B(0, C)$ of the metric $h := \Phi^* g$ on $(\mathbb{R}^4 \setminus B(0, C))/\Gamma$ satisfies with respect to Euclidean coordinates:

$$\left|\partial_I \big(\widetilde{h}_{jk} - \delta_{jk}\big)\right| = \mathrm{O}(|x|^{-\tau - |I|}) \quad \text{as } |x| \to \infty, \tag{2.183}$$

for each multi-index $I = (i_1, \ldots, i_m)$, where $1 \leq i_1, \ldots, i_m \leq 4$, $m \in \mathbb{Z}_{\geq 0}$, $|I| := m$, and $\partial_I := \partial_{i_1} \cdots \partial_{i_m}$.

The parameter τ from equation (2.183) is defined to be the **decay rate**. By Bando, Kasue, and Nakajima [**BKN89**, Theorem 1.5], if an ALE 4-manifold is Ricci flat, then there exists a diffeomorphism Φ with associated decay rate $\tau = 4$.

Now let (M^4, g) be a complete noncompact Ricci-flat Riemannian 4-manifold (this is a special case of being a steady soliton). Suppose in addition that (M^4, g) is a singularity model of the Ricci flow. By Perelman's no local collapsing theorem, there exists a positive real number κ such that

$$\text{Vol}(B(x,r)) \geq \kappa r^4 \tag{2.184}$$

for all $x \in M^4$ and $r > 0$. By a result of Cheeger and Naber [**CN15**, Corollary 8.85], this implies that (M^4, g) is an asymptotically locally Euclidean space.

2.7.3. Point selection.

In general, given a sequence of points in a Riemannian manifold that tend to spatial infinity, one can consider the associated sequence of pointed Riemannian manifolds (all sharing the same underlying smooth manifold). In this sequence, we rescale the Riemannian metric according to its curvatures at these points. The goal is to select suitable sequences of points and to extract a convergent subsequence from the corresponding sequence of pointed manifolds.

To elaborate further, given a point x_0 in a Riemannian manifold (M^n, g), its curvature scale is $f(x_0) := |\mathrm{Rm}|(x_0)$. The corresponding distance scale is $\frac{1}{f(x_0)^{1/2}}$. One can always select points in M^n so that the curvature is locally bounded in terms of the curvature scale at a point.

More generally, in the setting of metric spaces, Perelman's point selection procedure is the following:

LEMMA 2.43 (Perelman's point selection on general metric spaces). *Let (X, d) be a metric space and $f : X \to \mathbb{R}$ be any locally bounded non-negative function. Let $y_0 \in X$, $\alpha > 0$, and $A_0 \in \mathbb{R}^+$. Then there exists a point $x_0 \in B\big(y_0, \frac{1}{1-4^{-\alpha}} \frac{A_0}{f(y_0)^\alpha}\big)$ such that $f(x_0) \geq f(y_0)$ and*

$$f(x) \leq 4f(x_0) \quad \text{in } B\big(x_0, \tfrac{A_0}{f(x_0)^\alpha}\big). \tag{2.185}$$

PROOF. If (2.185) holds for $x_0 = y_0$, then we are done; so we assume otherwise. Then there exists $y_1 \in B\big(y_0, \frac{A_0}{f(y_0)^\alpha}\big)$ such that $f(y_1) > 4f(y_0)$. As long as (2.185) does not hold for $x_0 = y_{j-1}$, we may inductively continue in this way to obtain $y_j \in B\big(y_{j-1}, \frac{A_0}{f(y_{j-1})^\alpha}\big)$ such that $f(y_j) > 4f(y_{j-1})$. This implies that

$$f(y_j) > 4^j f(y_0) \tag{2.186}$$

for all such j. We also have that

$$d(y_j, y_{j-1}) < \frac{A_0}{f(y_{j-1})^\alpha} < 4^{-(j-1)\alpha} \frac{A_0}{f(y_0)^\alpha}.$$

Since

$$d(y_k, y_0) \leq \sum_{j=1}^{k} d(y_j, y_{j-1}) < \frac{A_0}{f(y_0)^\alpha} \sum_{j=1}^{k} 4^{-(j-1)\alpha} < \frac{1}{1-4^{-\alpha}} \frac{A_0}{f(y_0)^\alpha},$$

this implies that $y_k \in B\big(y_0, \frac{1}{1-4^{-\alpha}} \frac{A_0}{f(y_0)^\alpha}\big)$ for all $k \geq 1$. Since f is locally bounded, we have a contradiction to (2.186) unless (2.185) holds for $x_0 = y_{k_0}$ at some finite step k_0. The lemma follows. □

Although $A_0 > 0$ above is fixed, for a sequence we may let it tend to infinity. Specifically, for finding suitable sequences of points on Riemannian manifolds, Perelman's point selection procedure is the following:

COROLLARY 2.44 (Perelman's point selection on Riemannian manifolds). *Let (M^n, g) be a complete Riemannian manifold, let $\{y_i\}$ be a sequence of points in M^n, and let $A_i \to \infty$. Then, for each i, there exists a point $x_i \in B\big(y_i, 2A_i|\mathrm{Rm}|^{-\frac{1}{2}}(y_i)\big)$ such that $|\mathrm{Rm}|(x_i) \geq |\mathrm{Rm}|(y_i)$ and the rescaled metrics $g_i := |\mathrm{Rm}|(x_i)g$ satisfy*

$$|\mathrm{Rm}|_{g_i}(x) \leq 4 \quad \text{in } B_{g_i}(x_i, A_i). \tag{2.187}$$

In particular, if the sequence $\{g_i\}$ is noncollapsing and satisfies uniform derivative of curvature bounds, then the rescaled manifolds (M^n, g_i, x_i) subconverge to a complete Riemannian manifold $(M^n_\infty, g_\infty, x_\infty)$ in the pointed C^∞ Cheeger–Gromov sense.

PROOF. By Lemma 2.43 with $f := |\mathrm{Rm}|$, $y_0 = y_i$, $A_0 = A_i$, and $\alpha = \frac{1}{2}$, we have that for each i there exists $x_i \in B\big(y_i, 2A_i|\mathrm{Rm}|^{-\frac{1}{2}}(y_i)\big)$ such that $|\mathrm{Rm}|(x) \leq 4|\mathrm{Rm}|(x_i)$ in $B\big(x_i, A_i|\mathrm{Rm}|^{-\frac{1}{2}}(x_i)\big)$. By the definition of g_i, this implies (2.187). Since $A_i \to \infty$, when the sequence $\{g_i\}$ is noncollapsing, the convergence to a complete limit follows from the pointed C^∞ Cheeger–Gromov compactness theorem; see, for example, [**CCG$^+$07**, Chapters 3 and 4]. □

For sequences of solutions to the Ricci flow, derivative of curvature bounds are typically obtained as a direct consequence of Shi's local derivative estimates (Theorem 1.66). Furthermore, for singularity models, since they are noncollapsed on all scales, the rescaled metrics g_i are noncollapsed.

2.7.4. Ricci-flat limits of shrinker singularity models.

By Perelman's point selection procedure, if a 4-dimensional shrinking gradient Ricci soliton singularity model does not satisfy a quadratic curvature bound, then we can extract a Ricci-flat ALE limit. This fact, which is we use later to obtain a contradiction, is the content of the following:

PROPOSITION 2.45 (ALE limits in the contradiction proof of Theorem 2.41). *Suppose Theorem* 2.41 *is false. Then there exists a sequence of points* $\{x_i\}$ *in* (M^4, g, f) *with the following property: Based at these points, the metric rescalings* (M^4, g_i, x_i) *converge in the pointed* C^∞ *Cheeger–Gromov sense to a Ricci-flat ALE* 4*-manifold* $(M^4_\infty, g_\infty(t))$, *which is also a singularity model.*

PROOF. By hypothesis, there exists a sequence of points $\{y_i\}$ in M^4 with $|\mathrm{Rm}|(y_i)\big(d(y_i, o) + 1\big)^{-2} \to \infty$. By Corollary 2.44 with

$$A_i = \frac{1}{3}|\mathrm{Rm}|^{\frac{1}{2}}(y_i), \tag{2.188}$$

there exists a sequence of points $\{x_i\}$ such that for each $i \geq 1$ we have that $x_i \in B\big(y_i, \frac{2}{3}\big)$,

$$Q_i := |\mathrm{Rm}|(x_i) \geq |\mathrm{Rm}|(y_i),$$

and

$$|\mathrm{Rm}|(x) \leq 4Q_i \quad \text{in } B\big(x_i, A_i Q_i^{-\frac{1}{2}}\big). \tag{2.189}$$

We will rescale the metric g about the points x_i.

Consider the canonical form $(M^4, g(t), f(t))$ associated to (M^4, g, f) defined by Proposition 1.73. We now show that there are uniform scaled curvature bounds in parabolic cylinders based at $(x_i, 0)$. We have that $g(t) = (1-t)\varphi_t^* g$ is a solution to the Ricci flow and $f(t) = f \circ \varphi_t$, where $\frac{\partial}{\partial t}\varphi_t(x) = \frac{1}{1-t}(\nabla f)(\varphi_t(x))$. Using this, we estimate for $t \leq 0$ that

$$\begin{aligned}
-\frac{d}{dt}d(\varphi_t(x), o) &\leq \left|\frac{\partial}{\partial t}\varphi_t(x)\right| \\
&\leq \frac{1}{1-t}|\nabla f|(\varphi_t(x)) \\
&\leq \frac{1}{1-t}f^{1/2}(\varphi_t(x)) \\
&\leq d(\varphi_t(x), o) + C.
\end{aligned}$$

Let $x \in B\big(x_i, \frac{3}{4}A_iQ_i^{-\frac{1}{2}}\big)$ and $\bar{t} \in [-Q_i^{-1}, 0]$. We therefore have that

$$d(\varphi_t(x), o) + C \leq \mathrm{e}^{-t}\left(d(x_i, o) + \frac{3}{4}A_iQ_i^{-\frac{1}{2}} + C\right) \leq C\big(d(x_i, o) + C\big)$$

for $t \in [\bar{t}, 0]$, and

$$\begin{aligned} d(\varphi_{\bar{t}}(x), x) &\leq \int_{\bar{t}}^{0} \left|\frac{\partial}{\partial t}\varphi_t(x)\right| dt \\ &\leq \int_{\bar{t}}^{0} \big(d(\varphi_t(x), o) + C\big) dt \\ &\leq CQ_i^{-1}\big(d(x_i, o) + C\big). \end{aligned} \tag{2.190}$$

Since $Q_i^{-\frac{1}{2}}\big(d(x_i, o) + C\big) \leq 2|\mathrm{Rm}|(y_i)^{-\frac{1}{2}}\big(d(y_i, o) + 1\big) \to 0$ (assume that i is sufficiently large), we have that

$$\varphi_{\bar{t}}(x) \in B\big(x_i, A_iQ_i^{-\frac{1}{2}}\big).$$

Thus

$$|\mathrm{Rm}|(x, \bar{t}) = |\mathrm{Rm}|(\varphi_{\bar{t}}(x), 0) \leq 4Q_i \tag{2.191}$$

for $x \in B\big(x_i, \frac{3}{4}A_iQ_i^{-\frac{1}{2}}\big)$ and $\bar{t} \in [-Q_i^{-1}, 0]$.

Now consider the sequence of rescaled shrinking self-similar solutions to the Ricci flow:

$$g_i(t) := Q_i g(Q_i^{-1}t).$$

By the curvature bound (2.191), we have Shi's local derivative of curvature estimates (Theorem 1.66). Hence we may apply Perelman's no local collapsing theorem and Hamilton's Cheeger–Gromov compactness theorem for solutions to the Ricci flow (Theorem 1.6) to obtain that there exists a subsequence such that $(M^4, g_i(t), x_i)$ converges in the pointed C^∞ Cheeger–Gromov sense to a complete pointed solution to the Ricci flow $\big(M_\infty^4, g_\infty(t), x_\infty\big)$, $t \in \big[-\frac{1}{2}, 0\big]$, satisfying

$$|\mathrm{Rm}_{g_\infty}| \leq 4 \quad \text{and} \quad |\mathrm{Rm}_{g_\infty}|(x_\infty, 0) = 1.$$

In general, suppose that $\mathcal{N}$ is the pointed C^∞ Cheeger–Gromov limit of a space $\mathcal{M}$ based at a sequence of points in $\mathcal{M}$. If $\mathcal{P}$ is the pointed C^∞ Cheeger–Gromov limit of $\mathcal{N}$ based at a sequence of points in $\mathcal{N}$, then $\mathcal{P}$ is the pointed C^∞ Cheeger–Gromov limit of $\mathcal{M}$ based at a different sequence of points in $\mathcal{M}$. In view of this general fact, we conclude that $(M_\infty^4, g_\infty(t))$ is also a singularity model.

Since (M^4, g, f) is a shrinking gradient Ricci soliton, we have that $0 \leq R \leq f$. On the other hand, $0 \leq Q_i^{-1}f(x_i) \leq Q_i^{-1}(d(x_i, o) + C)^2 \to 0$. Thus,

$$R_{g_i}(x_i, 0) = Q_i^{-1}R(x_i) \leq Q_i^{-1}f(x_i) \to 0.$$

This implies that $R_{g_\infty}(x_\infty, 0) = 0$. Since $R_{g_\infty}(x, t) \geq 0$ and since $g_\infty(t)$ is a solution to the Ricci flow, by the strong maximum principle we have that $R_{g_\infty(t)} \equiv 0$. Furthermore, by this and the equation $\frac{\partial R}{\partial t} = \Delta R + 2|\mathrm{Ric}|^2$ for $g_\infty(t)$, we conclude that $\mathrm{Ric}_{g_\infty(t)} \equiv 0$. Since $(M_\infty^4, g_\infty(t))$ is a singularity model and by a result of Cheeger and Naber (see §2.7.2), we conclude that $(M_\infty^4, g_\infty(t))$ is an ALE 4-manifold. □

2.7.5. ALE exclusion principle.

Through primarily topological approaches, albeit with an essential geometric component, it is possible to establish the following result.

THEOREM 2.46 (ALE exclusion principle). *If $\overline{M}^4$ is a closed 4-manifold and if M_∞^4 is a Ricci-flat ALE 4-manifold, then there can exist at most a bounded number of disjoint copies of M_∞^4 in $\overline{M}^4$. See Figure 2.1.*

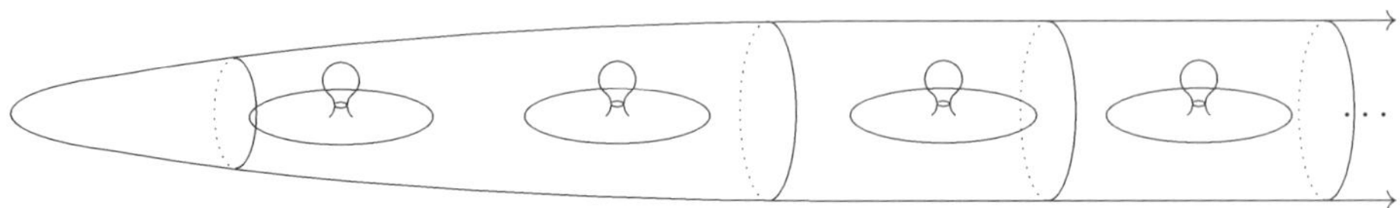

FIGURE 2.1. An infinite number of disjoint embedded ALEs in a 4-dimensional singularity model implies an unbounded number of disjoint embedded ALEs M_∞^4 in the associated singular solution to the Ricci flow on a closed 4-manifold $\overline{M}^4$. A contradiction is derived.

First, we give a rough sketch of the proof of Theorem 2.46; the full proof will be given in §2.7.7 below. Suppose for a contradiction that there exist an unbounded number of disjoint copies of a Ricci-flat ALE M_∞^4 in a closed 4-manifold $\overline{M}^4$. Since M_∞^4 is a Ricci-flat ALE 4-manifold, it has a single topological end diffeomorphic to $\mathbb{S}^3/\Gamma \times (0, \infty)$, where $\mathbb{S}^3/\Gamma$ is a spherical space form.[8]

By a homological argument, we will show that $H_2(M_\infty^4; \mathbb{Z}) \cong 0$ and that $H_1(\partial M_\infty; \mathbb{Z}) \cong H_1(\mathbb{S}^3/\Gamma; \mathbb{Z})$ is a **direct double**; i.e., it is isomorphic to $A \oplus A$ for some finite abelian group A; see Proposition 2.48 and its improvement, Corollary 2.54, below. This significantly limits the possible admissible groups Γ.

A contradiction, and thus Theorem 2.46, now follows from Bamler's modification of Anderson's argument in Zhang [**Zha19**, §6]. Specifically, since $H_1(\mathbb{S}^3/\Gamma, \mathbb{Z}) \cong A \oplus A$, by Lemma 2.52 below, which is proved by index theorems, either Γ would be the binary dihedral group $2D_n$ with n even or Γ would be the binary icosahedral group $2I$ of order 120.[9] Finally, by Lemma 2.55 below, which is proved using index theorems and eta invariant formulas, this would imply that $b_2(M_\infty^4) \geq 1$, which contradicts the fact that $H_2(M_\infty^4; \mathbb{Z}) \cong 0$.

2.7.6. Proof of Theorem 2.41 on the quadratic curvature bound. Using the preceding ALE exclusion principle (Theorem 2.46), we can now give a proof of the main Theorem 2.41 of this section, which establishes the quadratic curvature bound.

Suppose for a contradiction that Theorem 2.41 were false. Then, by Proposition 2.45 there would exist a sequence of points $\{x_i\}$ in (M^4, g, f) about whose metric rescalings converge in the pointed C^∞ Cheeger–Gromov sense to a Ricci-flat ALE 4-manifold M_∞^4. This would imply that there exists an unbounded number of disjoint copies of M_∞^4 in the closed 4-manifold $\overline{M}^4$ as well as in shrinker M^4.

---

[8]If M_∞^4 had more than one topological end, it would contain a geodesic line and hence split off an $\mathbb{R}$ factor by the Cheeger–Gromoll splitting theorem, which is a contradiction.

[9]The definitions of these binary polyhedral groups are recalled in Definition 2.51 below.

(Geometrically, the copies of M^4_∞ in M^4 are contained in balls of radii 1 centered at the x_i.) By Theorem 2.46, we obtain a contradiction. $\square$

2.7.7. Proof of Theorem 2.46 on ALE exclusion.

2.7.7.1. *$H_1(\partial M_\infty; \mathbb{Z}_p)$ is a direct double for p prime.*

By truncating the Ricci-flat ALE 4-manifold M^4_∞, we may replace it by a compact 4-manifold with boundary ∂M_∞ diffeomorphic to $\mathbb{S}^3/\Gamma$. We will need the following homological result for M^4_∞ and its boundary to eventually obtain a contradiction to the assumption that Theorem 2.46 is false.

LEMMA 2.47 (The map i_* on homology is surjective). *Suppose that Theorem 2.46 is false; that is, suppose that there are an unbounded number of disjoint copies of M^4_∞ in a closed 4-manifold $\overline{M}^4$. Then, for any homology coefficient ring R, the homomorphism*

$$i_* : H_m(\partial M_\infty; R) \to H_m(M^4_\infty; R),$$

induced by the inclusion map $i : \partial M_\infty \hookrightarrow M^4_\infty$, is surjective for $0 \leq m \leq 3$.

PROOF. In the following, all homology groups are taken with coefficients in the ring R. The case $m = 0$ is obvious since M^4_∞ is connected. So we assume that $1 \leq m \leq 3$. Consider one embedded copy of M^4_∞ in the closed 4-manifold $\overline{M}^4$. Let $(M^4_\infty)^\circ$ denote the interior of M^4_∞. In $\overline{M}^4$, we have the pair of compact submanifolds with boundary, M^4_∞ and $\overline{M}^4 \setminus (M^4_\infty)^\circ$, which have intersection equal to ∂M_∞. By the Mayer–Vietoris sequence for the subspaces M^4_∞ and $\overline{M}^4 \setminus (M^4_\infty)^\circ$, which satisfy $M^4_\infty \cup \big(\overline{M}^4 \setminus (M^4_\infty)^\circ\big) = \overline{M}^4$ and $M^4_\infty \cap \big(\overline{M}^4 \setminus (M^4_\infty)^\circ\big) = \partial M_\infty$ (see [**Hat02**, p. 149]), we have the long exact homology sequence

$$\begin{aligned}(2.192)\qquad \cdots \longrightarrow H_m(\partial M_\infty) &\xrightarrow{(i_*, j_*)} H_m(M^4_\infty) \oplus H_m\big(\overline{M}^4 \setminus (M^4_\infty)^\circ\big) \\ &\xrightarrow{k_* - l_*} H_m(\overline{M}^4) \xrightarrow{\partial_*} H_{m-1}(\partial M_\infty) \longrightarrow \cdots,\end{aligned}$$

where $j : \partial M_\infty \hookrightarrow \overline{M}^4 \setminus (M^4_\infty)^\circ$, $k : M^4_\infty \hookrightarrow \overline{M}^4$, and $l : \overline{M}^4 \setminus (M^4_\infty)^\circ \hookrightarrow \overline{M}^4$ are the inclusion maps.

Suppose, for a contradiction, that $i_* : H_m(\partial M_\infty) \to H_m(M^4_\infty)$ is not surjective. Then there exists a homology class $\alpha \in H_m(M^4_\infty)$ which is not contained in the image of i_*. Thus, by the exactness of the homology sequence (2.192), for any such α and any $\beta \in H_m\big(\overline{M}^4 \setminus (M^4_\infty)^\circ\big)$, we have

$$k_*(\alpha) - l_*(\beta) \neq 0. \qquad (2.193)$$

Firstly, we claim that this implies that the cokernel of i_*,

$$\operatorname{coker}(i_*) = H_m(M^4_\infty)/i_*(H_m(\partial M_\infty)) \neq 0,$$

injects into the quotient $H_m(\overline{M}^4)/l_*\big(H_m\big(\overline{M}^4 \setminus (M^4_\infty)^\circ\big)\big)$ by the map induced by $k_* : H_m(M^4_\infty) \to H_m(\overline{M}^4)$. Indeed, observe that this induced map is well-defined since

$$k_*(i_*(H_m(\partial M_\infty))) = l_*(j_*(H_m(\partial M_\infty)) \subset l_*\big(H_m\big(\overline{M}^4 \setminus (M^4_\infty)^\circ\big)\big).$$

Suppose that there were a nonzero coset $\alpha\, i_*(H_m(\partial M_\infty))$ which maps to zero in the quotient. Then we would have that $\alpha \in H_m(M^4_\infty)$ is not in the image of i_* and that $k_*(\alpha) \in l_*\big(H_m(\overline{M}^4 \setminus (M^4_\infty)^\circ)\big)$. This contradicts (2.193), and hence proves that $\operatorname{coker}(i_*)$ injects into $H_m(\overline{M}^4)/l_*\big(H_m(\overline{M}^4 \setminus (M^4_\infty)^\circ)\big)$.

Secondly, by hypothesis, the closed 4-manifold $\overline{M}^4$ contains a finite collection of disjoint embedded copies of M^4_∞, denoted by $\{M_p\}_{p=1}^I$, for $I \in \mathbb{N}$ arbitrarily large. Let $M^I := \bigcup_{p=1}^I M_p$, so that $\partial M^I = \bigcup_{p=1}^I \partial M_p$. Let $i^I : \partial M^I \hookrightarrow M^I$ and $l^I : \overline{M}^4 - (M^I)^\circ \hookrightarrow \overline{M}^4$ denote the inclusion maps. Since, in the argument above, we may replace M^4_∞ by M^I (we did not assume that M^4_∞ is connected), we have that

$$\operatorname{coker}(i^I_*) = H_m(M^I)/i^I_*(H_m(\partial M^I)) = \bigoplus_{p=1}^{I} \operatorname{coker}(i_{p*})$$

injects into $H_m(\overline{M}^4)/l^I_*(H_m(\overline{M}^4 - (M^I)^\circ)$, where $i_p : \partial M_p \hookrightarrow M_p$ is the inclusion map. Note that the isomorphism class of $\operatorname{coker}(i_{p*})$ is independent of p. In this way, we obtain an injection from modules of increasing rank into $H_m(\overline{M}^4)/l^I_*\big(H_m(\overline{M}^4 - (M^I)^\circ)\big)$. This contradicts that the homology groups of $\overline{M}^4$ are finitely generated, a property derived from $\overline{M}^4$ being a closed manifold. This completes the proof of Lemma 2.47. ☐

The preceding lemma says that, under our contradiction hypothesis, every homology class in M^4_∞ can be represented by a cycle in its boundary $\partial M_\infty \cong \mathbb{S}^3/\Gamma$, for dimensions $k = 0, 1, 2, 3$ and any coefficient ring. This has the following implication on the first homology group of ∂M_∞ with $\mathbb{Z}_p$ coefficients.

Proposition 2.48 (Direct double property for $H_1(\partial M_\infty; \mathbb{Z}_p)$). *Suppose that Theorem* 2.46 *is false. Then*

$$H_1(\partial M_\infty; \mathbb{Z}) \otimes \mathbb{Z}_p \cong \big(H_1(M^4_\infty; \mathbb{Z}) \oplus H_1(M^4_\infty; \mathbb{Z})\big) \otimes \mathbb{Z}_p \tag{2.194}$$

for all prime numbers p.

Proof. By Lemma 2.47 with the homology coefficient ring $R = \mathbb{Z}$ and by the homology long exact sequence for the pair $(M^4_\infty, \partial M_\infty)$, we have the short exact sequence

$$0 \to H_{k+1}(M^4_\infty, \partial M_\infty; \mathbb{Z}) \to H_k(\partial M_\infty; \mathbb{Z}) \to H_k(M^4_\infty; \mathbb{Z}) \to 0 \tag{2.195-k}$$

for $k = 1, 2, 3$.

Lemma 2.49. $H_2(M^4_\infty; \mathbb{Z}) \cong 0$ *and* $H_3(M^4_\infty; \mathbb{Z}) \cong 0$.

Proof of Lemma 2.49. We have that $H_1(\partial M_\infty; \mathbb{Z}) \cong H_1(\mathbb{S}^3/\Gamma; \mathbb{Z})$ is isomorphic to the abelianization of Γ and in particular it is a finite abelian group; i.e., is torsion.[10] In particular, the first Betti number $b_1(\partial M_\infty)$ is zero. As a consequence of (2.195-k) with $k = 1$, $H_1(M^4_\infty; \mathbb{Z})$ is also torsion.

Firstly, by Poincaré duality (see [**Hat02**, Theorem 3.30 on p. 241]) and by the cohomology universal coefficient theorem (see [**Hat02**, Corollary 3.3 on p. 196]), we have that

$$H_2(\partial M_\infty; \mathbb{Z}) \cong H^1(\partial M_\infty; \mathbb{Z}) \cong \mathbb{Z}^{b_1(\partial M_\infty)} \oplus \operatorname{Tor}(H_0(\partial M_\infty; \mathbb{Z})) \cong 0,$$

[10]An abelian group **is torsion** if every element has finite order. For homology groups of closed manifolds, since they are finitely generated, being torsion is equivalent to being finite.

where $\mathrm{Tor}(G)$ denotes the torsion subgroup of G. Now, by (2.195-k) with $k = 2$, we have that $H_2(M^4_\infty;\mathbb{Z}) \cong 0$. This proves the first assertion of the lemma.

Secondly, $H_3(\partial M_\infty;\mathbb{Z}) \cong \mathbb{Z}$ since ∂M_∞ is orientable. By (2.195-k) with $k = 3$, we have that $H_3(\partial M_\infty;\mathbb{Z})$ maps surjectively onto $H_3(M^4_\infty;\mathbb{Z})$. On the other hand, the map $H_4(M^4_\infty, \partial M_\infty;\mathbb{Z}) \to H_3(\partial M_\infty;\mathbb{Z})$ is an isomorphism (the boundary map sends a fundamental class of M^4_∞ to a fundamental class of ∂M_∞; see, for example, May [**May99**, Chapter 21, §4]), which implies that $H_3(\partial M_\infty;\mathbb{Z}) \to H_3(M^4_\infty;\mathbb{Z})$ is the zero map. We conclude that $H_3(M^4_\infty;\mathbb{Z}) \cong 0$. This completes the proof of Lemma 2.49. □

Next, we prove that the first homology direct double property with $\mathbb{Z}_p$ coefficients, where p is prime. The proposition will quickly follow from this.

LEMMA 2.50. *For each prime p, we have*

$$H_1(\partial M_\infty;\mathbb{Z}_p) = H_1(M^4_\infty;\mathbb{Z}_p) \oplus H_1(M^4_\infty;\mathbb{Z}_p). \tag{2.196}$$

PROOF OF LEMMA 2.50. By Lemma 2.47 with $R = \mathbb{Z}_p$, we have the short exact sequence

$$0 \to H_2(M^4_\infty, \partial M_\infty;\mathbb{Z}_p) \to H_1(\partial M_\infty;\mathbb{Z}_p) \to H_1(M^4_\infty;\mathbb{Z}_p) \to 0.$$

Since $\mathbb{Z}_p$ is a field, this short exact sequence splits. Hence the lemma will follow from showing that

$$H_2(M^4_\infty, \partial M_\infty;\mathbb{Z}_p) \cong H_1(M^4_\infty;\mathbb{Z}_p). \tag{2.197}$$

By Lefschetz duality (see [**Hat02**, Theorem 3.43 on p. 254, with $A = \emptyset$]),

$$H_2(M^4_\infty, \partial M_\infty;\mathbb{Z}_p) \cong H^2(M^4_\infty;\mathbb{Z}_p). \tag{2.198}$$

Moreover, by the cohomology universal coefficient theorem (see [**Hat02**, Theorem 3.2 on p. 195]),

(2.199)

$$0 \to \mathrm{Ext}^1_{\mathbb{Z}}(H_1(M^4_\infty;\mathbb{Z}),\mathbb{Z}_p) \to H^2(M^4_\infty;\mathbb{Z}_p) \to \mathrm{Hom}_{\mathbb{Z}}(H_2(M^4_\infty;\mathbb{Z}),\mathbb{Z}_p) \to 0$$

is a short exact sequence, where Ext and Hom are the Ext and Hom functors, respectively. Since $\mathrm{Hom}_{\mathbb{Z}}(H_2(M^4_\infty;\mathbb{Z}),\mathbb{Z}_p) \cong 0$ by Lemma 2.49 and using (2.199), we obtain

$$H^2(M^4_\infty;\mathbb{Z}_p) \cong \mathrm{Ext}^1_{\mathbb{Z}}(H_1(M^4_\infty;\mathbb{Z}),\mathbb{Z}_p). \tag{2.200}$$

We have the following general formulas for $\mathrm{Ext}^1_{\mathbb{Z}}$ (see [**Hat02**, Chapter 3]):

$$\mathrm{Ext}^1_{\mathbb{Z}}(\mathbb{Z},\mathbb{Z}_p) \cong 0 \quad \text{and} \quad \mathrm{Ext}^1_{\mathbb{Z}}(\text{torsion},\mathbb{Z}_p) \cong \mathrm{Hom}_{\mathbb{Z}}(\text{torsion},\mathbb{Z}_p). \tag{2.201}$$

By (2.201), since $H_1(M^4_\infty;\mathbb{Z})$ is torsion, and since $\mathrm{Ext}^1_{\mathbb{Z}}$ commutes with direct sums, we have that

$$\begin{aligned}\mathrm{Ext}^1_{\mathbb{Z}}(H_1(M^4_\infty;\mathbb{Z}),\mathbb{Z}_p) &\cong \mathrm{Hom}_{\mathbb{Z}}(\mathrm{Tor}(H_1(M^4_\infty;\mathbb{Z})),\mathbb{Z}_p)\\ &\cong \mathrm{Hom}_{\mathbb{Z}}(H_1(M^4_\infty;\mathbb{Z}),\mathbb{Z}_p)\\ &\cong H_1(M^4_\infty;\mathbb{Z}_p),\end{aligned}$$

where the last isomorphism is by the homology universal coefficient theorem (see (2.202) below). By combining (2.198), (2.200), and this formula, we obtain

$$H_2(M^4_\infty, \partial M_\infty;\mathbb{Z}_p) \cong H_1(M^4_\infty;\mathbb{Z}_p),$$

which is (2.197). Lemma 2.50 follows. □

Now, by the homology universal coefficient theorem (see [**Hat02**, Theorem 3A.3 on p. 264]), we have for both $X = M^4_\infty$ and $X = \partial M_\infty$ that

$$0 \to H_1(X;\mathbb{Z}) \otimes \mathbb{Z}_p \to H_1(X;\mathbb{Z}_p) \to \operatorname{Tor}(H_0(X;\mathbb{Z}),\mathbb{Z}_p) \cong 0 \to 0,$$

where Tor denotes the Tor functor (note that $\operatorname{Tor}(\mathbb{Z}, A) \cong 0$ for any abelian group A; see [**Hat02**, Proposition 3A.5(3) on p. 265]). Hence

$$\operatorname{Hom}_{\mathbb{Z}}\left(H_1(X;\mathbb{Z}),\mathbb{Z}_p\right) \cong H_1(X;\mathbb{Z}) \otimes \mathbb{Z}_p \cong H_1(X;\mathbb{Z}_p). \tag{2.202}$$

Thus Lemma 2.50 implies that

$$H_1(\partial M_\infty;\mathbb{Z}) \otimes \mathbb{Z}_p \cong \left(H_1(M^4_\infty;\mathbb{Z}) \oplus H_1(M^4_\infty;\mathbb{Z})\right) \otimes \mathbb{Z}_p \tag{2.203}$$

for all prime numbers p. This completes the proof of Proposition 2.48. □

2.7.7.2. *The cardinality of $H_1(\partial M_\infty;\mathbb{Z})$ is a perfect square.*

We claim that

$$|H_1(\partial M_\infty;\mathbb{Z})| = |H_1(M^4_\infty;\mathbb{Z})|^2, \tag{2.204}$$

where $|G|$ denotes the order of a group G. To see (2.204), we argue as follows: By (2.195-k) with $k = 1$, we have that

$$\left|H_1(\partial M_\infty;\mathbb{Z})\right| = \left|H_2(M^4_\infty,\partial M_\infty;\mathbb{Z})\right| \left|H_1(M^4_\infty;\mathbb{Z})\right|.$$

However, by Lefschetz duality,

$$H_2(M^4_\infty,\partial M_\infty;\mathbb{Z}) \cong H^2(M^4_\infty;\mathbb{Z}) \cong \operatorname{Ext}^1_{\mathbb{Z}}(H_1(M^4_\infty;\mathbb{Z}),\mathbb{Z}) \cong H_1(M^4_\infty;\mathbb{Z})$$

with the last isomorphism holding since $H_1(M^4_\infty;\mathbb{Z})$ is torsion; equation (2.204) follows.

2.7.7.3. *Restrictions on the spherical space form groups.*

In view of a Ricci-flat ALE 4-manifold being diffeomorphic to the interior of a compact 4-manifold with boundary diffeomorphic to a spherical space form S^3/Γ, we consider the binary polyhedral groups.

Definition 2.51 (Binary polyhedral groups).

(1) The **binary dihedral group** $2D_{n,k}$ of order $4n$ is isomorphic to the finite group generated by x and y with relations $xyx^{-1} = y^{-1}$, $x^{2^k} = y^n$, which we denote by

$$\langle x, y \mid xyx^{-1} = y^{-1},\, x^{2^k} = y^n\rangle \tag{2.205}$$

and where k, n are positive integers with $n \geq 2$. We define $2D_n := 2D_{n,1}$.

(2) The **binary tetrahedral group** $2T_k$ is the finite group with the presentation

$$\left\langle x, y, z \,\middle|\, (xy)^2 = x^2 = y^2,\, zxz^{-1} = y,\, zyz^{-1} = xy,\, z^{3^k} = 1\right\rangle, \tag{2.206}$$

where k is a positive integer.

(3) The **binary octahedral group** $2O$ of order 48 is isomorphic to the finite group generated by x and y with relations $(xy)^2 = x^3 = y^4$, which we denote by

$$\langle x, y \mid (xy)^2 = x^3 = y^4\rangle. \tag{2.207}$$

The condition on S^3/Γ provided by Proposition 2.48 turns out to be restrictive. In particular, we have the following result; see e.g. Zhang [**Zha19**, Lemma 6.3], based on unpublished notes of Bamler.

LEMMA 2.52 (Spherical space forms whose first $\mathbb{Z}_p$-coefficient homology are direct doubles). *If a spherical space form $\mathbb{S}^3/\Gamma$ satisfies*

$$H_1(\mathbb{S}^3/\Gamma;\mathbb{Z})\otimes\mathbb{Z}_p \cong H_1(\mathbb{S}^3/\Gamma;\mathbb{Z}_p)\cong A\oplus A, \tag{2.208}$$

for all primes p, then Γ is isomorphic to either:

(1) *a binary dihedral group $2D_{n,k}$, where n is even, or*
(2) *the binary icosahedral group $2I$ of order* 120.

PROOF. 3-Dimensional spherical space forms $\mathbb{S}^3/\Gamma$, in terms of the finite subgroup Γ of SO(4), have been classified in 5 categories as follows (see e.g. Wolf [**Wol11**, §7.5]):

(1) (Cyclic) Let $k>1$ and $\ell>0$ be coprime integers. The **lens space** $L(k;\ell)$ is $\mathbb{S}^3/\Gamma$, where the subgroup $\Gamma<\mathrm{SU}(2)<\mathrm{SO}(4)$ is generated by the element

$$\begin{pmatrix} e^{2\pi i/k} & 0\\ 0 & e^{2\pi i\ell/k}\end{pmatrix},$$

using the identification $\mathbb{R}^4\cong\mathbb{C}^2$ so that

$$e^{2\pi i\theta}=\begin{pmatrix}\cos(2\pi\theta) & \sin(2\pi\theta)\\ -\sin(2\pi\theta) & \cos(2\pi\theta)\end{pmatrix}.$$

The fundamental group of $L(k;\ell)$ is given by $\pi_1(L(k;\ell))\cong\mathbb{Z}_k:=\mathbb{Z}/k\mathbb{Z}$. Since this group is cyclic and in particular abelian, we have that $H_1(L(k;\ell);\mathbb{Z})\cong\mathbb{Z}_k$, which contradicts (2.208) for p a prime divisor of k.

(2) (Dihedral) In this case, we have that

$$\Gamma\cong\mathbb{Z}_m\times 2D_{n,k},$$

where $2D_{n,k}$ is given by (2.205) and m is a positive integer coprime to $2n$. In this case, we say that $\mathbb{S}^3/\Gamma$ is a **prism manifold**. We have that the abelianization of $\Gamma\cong\pi_1(\mathbb{S}^3/\Gamma)$ is

$$H_1(\mathbb{S}^3/\Gamma;\mathbb{Z})\cong\mathbb{Z}_m\oplus\langle x,y\,|\,y=y^{-1},\ x^{2^k}=y^n\rangle.$$

If n is even, then

$$H_1(\mathbb{S}^3/\Gamma;\mathbb{Z})\cong\mathbb{Z}_m\oplus\mathbb{Z}_2\oplus\mathbb{Z}_{2^k},$$

and if n is odd, then

$$H_1(\mathbb{S}^3/\Gamma;\mathbb{Z})\cong\mathbb{Z}_m\oplus\mathbb{Z}_{2^{k+1}}.$$

Since m and 2 are coprime, (2.208) implies that $m=1$ and n is even. Hence

$$\Gamma\cong 2D_{n,k},$$

where n is even, and $H_1(\mathbb{S}^3/\Gamma;\mathbb{Z})\cong\mathbb{Z}_2\oplus\mathbb{Z}_{2^k}$.

(3) (Tetrahedral) In this case, we have that

$$\Gamma\cong\mathbb{Z}_m\times 2T_k,$$

where $2T_k$ is given by (2.206) and m is a positive integer coprime to 6. In this case we say that $\mathbb{S}^3/\Gamma$ is a **tetrahedral manifold**. The abelianization of Γ is

$$\begin{aligned}H_1(\mathbb{S}^3/\Gamma;\mathbb{Z})&\cong\mathbb{Z}_m\oplus\langle x,y,z\,|\,x^2=y^2=1,\ x=y,\ y=xy,\ z^{3^k}=1\rangle\\ &\cong\mathbb{Z}_m\oplus\langle x,z\,|\,x^2=1,\ x=x^2,\ z^{3^k}=1\rangle\\ &\cong\mathbb{Z}_m\oplus\mathbb{Z}_{3^k}.\end{aligned}$$

Since m and 3 are coprime, this contradicts (2.208) for $p=3$.

(4) (Octahedral) In this case, we have that

$$\Gamma \cong \mathbb{Z}_m \times 2O,$$

where $2O$ is given by (2.207) and m is a positive integer coprime to 6. Here, we say that $\mathbb{S}^3/\Gamma$ is an **octahedral manifold**. We have $H_1(\mathbb{S}^3/\Gamma;\mathbb{Z}) \cong \mathbb{Z}_m \oplus \mathbb{Z}_2$, which contradicts (2.208) for $p = 2$ since m and 2 are coprime.

(5) (Icosahedral) The **binary icosahedral group** $2I$ of order 120 is isomorphic to $\langle x, y \mid (xy)^2 = x^3 = y^5 \rangle$. In this case we have that

$$\Gamma \cong \mathbb{Z}_m \times 2I,$$

where m is a positive integer coprime to 30. Here, we say that $\mathbb{S}^3/\Gamma$ is an **icosahedral manifold**. We have $H_1(\mathbb{S}^3/\Gamma;\mathbb{Z}) \cong \mathbb{Z}_m$. If $m > 1$, then this contradicts (2.208) for p a prime divisor of m. Thus $m = 1$ and

$$\Gamma \cong 2I$$

is the binary icosahedral group, so that $\mathbb{S}^3/\Gamma$ is the **Poincaré homology sphere**, with $H_1(\mathbb{S}^3/\Gamma;\mathbb{Z}) \cong 0$. □

2.7.7.4. *The direct double property for* $H_1(\partial M_\infty;\mathbb{Z})$.

Freedman, using a stabilization argument, proved that can improve Proposition 2.48 to the following result; see [**CFSZ25**, Proposition 2] for the details of the proof of a slight generalization of this result.

PROPOSITION 2.53. *Let M^4 be a compact 4-manifold with non-empty, connected boundary ∂M diffeomorphic to a spherical space form. Let $i : \partial M \to M^4$ denote the inclusion map. Assume for all coefficient rings R, $i_* : H_*(\partial M; R) \to H_*(M^4; R)$ is onto for $* = 0, 1, 2, 3$. Let N^4 be a compact 4-manifold with or without boundary. Assume, for all $n \in \mathbb{N}$, there exist embeddings $e_n : \bigsqcup_{n\text{copies}} M^4 \hookrightarrow N^4$ of the disjoint unions of n M^4's. Then $H_1(\partial M;\mathbb{Z})$ is isomorphic to $A \oplus A$ for some finite abelian group A.*

By Lemma 2.47, for all coefficient rings R, $i_* : H_*(\partial; R) \to H_*(M; R)$ is onto for $* = 0, 1, 2, 3$, and $H_1(\partial, \mathbb{Z}) = \tau$ is torsion. Thus, Proposition 2.53 yields the following:

COROLLARY 2.54. *Under the same hypotheses as Proposition 2.48, we have that*

$$H_1(\partial M_\infty;\mathbb{Z}) \cong H_1(\mathbb{S}^3/\Gamma;\mathbb{Z})$$

is a direct double; i.e., $H_1(\partial M_\infty;\mathbb{Z})$ is isomorphic to $A \oplus A$ for some finite abelian group A.

In particular, under the hypotheses of Proposition 2.48, we can improve the conclusion of Lemma 2.52 to obtain that Γ is isomorphic to either:

(1) a binary dihedral group $2D_n := 2D_{n,1}$, where n is even, or
(2) the binary icosahedral group $2I$ of order 120.

Indeed, if $\Gamma \cong 2D_{n,k}$, then $H_1(\mathbb{S}^3/\Gamma;\mathbb{Z}) \cong \mathbb{Z}_2 \oplus \mathbb{Z}_{2^k}$, which is a direct double if and only if $k = 1$.

2.7.7.5. *By Hitchin–Thorpe, the Ricci-flat ALE limits have* $b_2 \geq 1$.

Since $H_2(M^4_\infty; \mathbb{Z}) \cong 0$ by Lemma 2.49, the following result (see [**Zha19**, Lemma 6.5]) yields a contradiction and hence finishes the proof of Theorem 2.46.

LEMMA 2.55 (Ricci-flat ALE limit has $b_2 \geq 1$). *Let* (M^4_∞, g) *be a Ricci-flat ALE 4-manifold asymptotic to* $(\mathbb{R}^4 \setminus B(0,C))/\Gamma$*, where* Γ *is isomorphic to either a binary dihedral group* $2D_n$*,* $n \in \mathbb{N}$*, or the binary icosahedral group* $2I$ *of order* 120*. Then*

$$b_2(M^4_\infty) \geq 1. \tag{2.209}$$

PROOF. The **Chern–Gauss–Bonnet formula** [**Che44**] says that for a closed oriented Riemannian manifold (M^4, g), its Euler characteristic $\chi(M^4)$ satisfies

$$\chi(M^4) = \frac{1}{8\pi^2} \int_{M^4} \left(\frac{R^2}{24} + |W_+|^2 + |W_-|^2 - \frac{|\overset{\circ}{\mathrm{Ric}}|^2}{2} \right) dv, \tag{2.210}$$

where W_+ and W_- are the self-dual and anti-self-dual parts of the Weyl tensor and where $\overset{\circ}{\mathrm{Ric}} = \mathrm{Ric} - \frac{1}{4}Rg$ is the trace-free part of the Ricci tensor.

If (M^4, g) is an asymptotically locally Euclidean 4-manifold asymptotic to $(\mathbb{R}^4 \setminus B(0,C))/\Gamma$, then the Chern–Gauss–Bonnet formula says that

$$\chi(M^4) - \frac{1}{|\Gamma|} = \frac{1}{8\pi^2} \int_{M^4} \left(\frac{R^2}{24} + |W_+|^2 + |W_-|^2 - \frac{|\overset{\circ}{\mathrm{Ric}}|^2}{2} \right) dv. \tag{2.211}$$

See §2.7.8 below for the proof of this. In particular, if (M^4, g) is a Ricci-flat ALE 4-manifold asymptotic to $(\mathbb{R}^4 \setminus B(0,C))/\Gamma$, then we obtain the **Chern–Gauss–Bonnet formula for ALE spaces**:

$$\chi(M^4) - \frac{1}{|\Gamma|} = \frac{1}{8\pi^2} \int_{M^4} \left(|W_+|^2 + |W_-|^2 \right) dv. \tag{2.212}$$

On the other hand, the **Hirzebruch signature theorem** [**Hir95**] states that for any closed oriented Riemannian manifold (M^4, g), its signature $\sigma(M^4) := b_2^+(M^4) - b_2^-(M^4)$ satisfies

$$\sigma(M^4) = \frac{1}{12\pi^2} \int_M \left(|W_+|^2 - |W_-|^2 \right) dv. \tag{2.213}$$

This result for closed manifolds was extended by Atiyah, Patodi, and Singer [**APS75a, APS75b**] to manifolds with boundary as a special case of their index theorem. As a consequence of their formula, we have that if (M^4, g) is an ALE 4-manifold asymptotic to $(\mathbb{R}^4 \setminus B(0,C))/\Gamma$, then we obtain the **signature theorem for ALE spaces**:

$$\sigma(M^4) + \eta(\mathbb{S}^3/\Gamma) = \frac{1}{12\pi^2} \int_M \left(|W_+|^2 - |W_-|^2 \right) dv, \tag{2.214}$$

where η is the **eta invariant** associated to the signature operator. Thus we have the Hitchin–Thorpe inequality for Ricci-flat ALE 4-manifolds:

$$\left| \sigma(M^4) + \eta(\mathbb{S}^3/\Gamma) \right| \leq \frac{2}{3} \left(\chi(M^4) - \frac{1}{|\Gamma|} \right). \tag{2.215}$$

The equality cases are characterized as follows:

$$\sigma(M^4) + \eta(\mathbb{S}^3/\Gamma) = \frac{2}{3}\left(\chi(M^4) - \frac{1}{|\Gamma|}\right) \quad \text{if and only if} \quad W_- \equiv 0, \tag{2.216}$$

$$-(\sigma(M^4) + \eta(\mathbb{S}^3/\Gamma)) = \frac{2}{3}\left(\chi(M^4) - \frac{1}{|\Gamma|}\right) \quad \text{if and only if} \quad W_+ \equiv 0.$$

Now, as in our hypothesis, let (M^4_∞, g) be a Ricci-flat ALE 4-manifold asymptotic to $(\mathbb{R}^4 \setminus B(0,C))/\Gamma$, where the group Γ is isomorphic to $2D_n$ or $2I$. Suppose for a contradiction that $b_2(M^4_\infty) = 0$. Then

$$\sigma(M^4_\infty) = 0 \quad \text{and} \quad \chi(M^4_\infty) = 1 - b_1(M^4_\infty) - b_3(M^4_\infty) \leq 1.$$

Therefore (2.215) yields

$$\eta(\mathbb{S}^3/\Gamma) \leq \frac{2}{3}\left(1 - \frac{1}{|\Gamma|}\right) < \frac{2}{3}. \tag{2.217}$$

On the other hand (see Nakajima [**Nak90**, top of p. 393] and Tsuboi [**Tsu81**, (4.2)]; see also Remark 2.56 below), we have that

$$\eta(\mathbb{S}^3/2D_n) = \frac{2(2n+2)^2 - 8(2n+2) + 9}{6((2n+2)-2)} = \frac{8n^2+1}{12n} > \frac{2}{3} \tag{2.218}$$

and

$$\eta(\mathbb{S}^3/2I) = \frac{361}{180} > \frac{2}{3}. \tag{2.219}$$

This yields a contradiction and hence the lemma is proved. □

As remarked above, this completes the proof of Theorem 2.46.

REMARK 2.56. When $\Gamma \subset SU(2)$, the signature-operator eta invariant calculation of the spherical space form S^3/Γ follows from (2.216) and the calculations of the Euler characteristics and signatures of the corresponding hyper-Kähler (anti-self-dual) ALE 4-manifolds. In particular, these spaces are simply-connected, satisfy $b_1 = b_3 = b_4 = 0$, and have negative-definite intersection form and hence satisfy $\chi = -\sigma + 1$. For example:

(1) For $\Gamma = \mathbb{Z}_k$, we have $|\Gamma| = k$, $\chi = k$, $\sigma = -k+1$, which yields $\eta(\mathbb{S}^3/\mathbb{Z}_k) = \frac{(k-1)(k-2)}{3k}$. For example, $\eta(\mathbb{S}^3) = \eta(\mathbb{R}P^3) = 0$.
(2) Equation (2.218) follows from the facts that when $\Gamma = 2D_n$, we have $|\Gamma| = 8n$, $\chi = 2n+3$, and $\sigma = -2n-2$.
(3) For $\Gamma = 2T$, we have $|\Gamma| = 24$, $\chi = 7$, $\sigma = -6$, which yields $\eta(\mathbb{S}^3/2T) = \frac{49}{36}$.
(4) For $\Gamma = 2O$, we have $|\Gamma| = 48$, $\chi = 8$, $\sigma = -7$, which yields $\eta(\mathbb{S}^3/2O) = \frac{121}{72}$.
(5) Equation (2.219) follows from the facts that when $\Gamma = 2I$, we have $|\Gamma| = 120$, $\chi = 9$, and $\sigma = -8$.

Alternate notations for $2T$, $2O$, and $2I$ are T^*, O^*, and I^*, respectively.

2.7.8. Proof of the Chern–Gauss–Bonnet formula for ALE 4-manifolds.

In this section, for completeness, we provide a derivation of formula (2.211) for a (not necessarily Ricci-flat) ALE 4-manifold (M^4, g). Denote the Chern–Gauss–Bonnet integrand of a metric g by

$$\text{CGB}(g) := \left(\frac{R^2}{24} + |W_+|^2 + |W_-|^2 - \frac{|\overset{\circ}{\text{Ric}}|^2}{2} \right) dv. \tag{2.220}$$

Fix a point $o \in M^4$ and let ε be a small positive real number. Consider $(B_{\varepsilon g}(o, \varepsilon), \varepsilon g)$, which is a rescaling of the ball $(B_g(o, \varepsilon^{-1}), g)$ since the metric scales like distance squared. Since (M^4, g) is asymptotically locally Euclidean and since the Chern–Gauss–Bonnet integral is scale invariant, we have that

$$\int_{B_{\varepsilon g}(o,\varepsilon)} \text{CGB}(\varepsilon g) \to \int_{M^4} \text{CGB}(g) \tag{2.221}$$

as $\varepsilon \to 0$.

On the other hand, consider the unit 4-sphere $(\mathbb{S}^4, g_{\mathbb{S}^4})$ and the isometric action of Γ on $(\mathbb{S}^4, g_{\mathbb{S}^4})$ given by the restriction of the action of Γ on $\mathbb{R}^5 = \mathbb{R}^4 \times \mathbb{R}$. The resulting quotient space $(\mathbb{S}^4/\Gamma, \overline{g})$, with $\overline{g}$ as the induced metric, forms a 4-dimensional Riemannian orbifold. This orbifold contains two singular points, both of which have isotropy groups isomorphic to Γ. Topologically, $\mathbb{S}^4/\Gamma$ is the suspension of $\mathbb{S}^3/\Gamma$.

By the Chern–Gauss–Bonnet formula for orbifolds, the orbifold Euler characteristic χ_{orb} of $\mathbb{S}^4/\Gamma$ is given by

$$\chi_{\text{orb}}(\mathbb{S}^4/\Gamma) = \frac{1}{|\Gamma|}\chi(\mathbb{S}^4) = \frac{2}{|\Gamma|} = \frac{1}{8\pi^2}\int_{\mathbb{S}^4/\Gamma} \text{CGB}(\overline{g}). \tag{2.222}$$

Let $\overline{n}$ denote one of the singular points of $\mathbb{S}^4/\Gamma$. Since (M^4, g) is an asymptotically locally Euclidean space, we can now glue the ball $(B_{\varepsilon g}(o, \varepsilon), \varepsilon g)$ and $((\mathbb{S}^4/\Gamma) \setminus B_{\overline{g}}(\overline{n}, \varepsilon/2), \overline{g})$ together along an annulus without changing significantly the metrics εg and $\overline{g}$ to obtain a Riemannian orbifold which we denote by $(\mathcal{O}^4, g_{\text{orb}})$. By the Chern–Gauss–Bonnet formula for orbifolds, we have that

$$\chi_{\text{orb}}(\mathcal{O}^4) = \chi(M^4) + \frac{1}{|\Gamma|} = \frac{1}{8\pi^2}\int_{\mathcal{O}^4} \text{CGB}(g_{\text{orb}}). \tag{2.223}$$

On the other hand,

$$\int_{B_{\varepsilon g}(o,\varepsilon)} \text{CGB}(\varepsilon g) + \int_{\mathbb{S}^4/\Gamma} \text{CGB}(\overline{g}) \to \int_{\mathcal{O}^4} \text{CGB}(g_{\text{orb}}) \tag{2.224}$$

as $\varepsilon \to 0$. We conclude from this that

$$\chi(M^4) - \frac{1}{|\Gamma|} = \frac{1}{8\pi^2}\int_{M^4} \text{CGB}(g), \tag{2.225}$$

which is the Chern–Gauss–Bonnet formula for Ricci-flat ALE 4-manifolds (2.211).

This completes our discussion of the proof of the quadratic curvature bound for 4-dimensional shrinking gradient Ricci soliton singularity models.

2.7.9. Steady soliton limits of shrinkers with quadratic curvature growth.

In this subsection and the next two, we present a few additional results about the geometry at infinity of shrinking solitons.

By Theorem 2.41, any 4-dimensional shrinking gradient Ricci soliton singularity model has at most quadratic curvature growth rate. We now consider the case in which there exists a sequence of points tending to infinity with the curvature growth rate being exactly quadratic. For the following result, cf. [**CFSZ20**, Proposition 10].

PROPOSITION 2.57 (Limits of rescalings based at quadratic growth rate points). *Let (M^4, g, f) be a shrinking gradient Ricci soliton singularity model and let $o \in M^4$. Suppose that a sequence of points $x_i \to \infty$ has the property that there exists a constant $c > 0$ such that*[11]

$$|\mathrm{Rm}|(x_i) \geq c d(x_i, o)^2 \quad \textit{for all } i.$$

Let $Q_i := |\mathrm{Rm}|(x_i)$ and let $g_i := Q_i g$. Then (M^4, g_i, x_i) subconverges in the pointed C^∞ Cheeger–Gromov sense to a complete pointed Riemannian manifold $(M^4_\infty, g_\infty, x_\infty)$ with bounded curvature. Moreover, there exists a function f_∞ such that $(M^4_\infty, g_\infty, f_\infty)$ is a non-Ricci-flat steady gradient Ricci soliton.

PROOF. By Theorem 2.41, there exists a constant $C > 0$ such that

$$|\mathrm{Rm}|(x) \leq C d(x, o)^2 \quad \text{for all } x \in M^4 \setminus B(o, 1). \tag{2.226}$$

We may assume without loss of generality that $x_i \in M^4 \setminus B(o, 2)$ for all i. Hence,

$$|\mathrm{Rm}|(x) \leq 3C d(x_i, o)^2 \leq \frac{3C}{c} Q_i \quad \text{for all } x \in B(x_i, 1). \tag{2.227}$$

Equivalently,

$$|\mathrm{Rm}_{g_i}|(x) \leq \frac{3C}{c} \quad \text{for all } x \in B_{g_i}(x_i, Q_i^{1/2}). \tag{2.228}$$

Now let $(M^4, g(t), f(t))$ be the canonical form associated to (M^4, g, f) as given by Proposition 1.73, which is a solution to the Ricci flow with $g(0) = g$. Let $x \in B(x_i, \frac{1}{2})$ and $t \in [-Q_i^{-1}, 0]$. By (2.190), we have that

$$d(\varphi_t(x), x) \leq C Q_i^{-1} \big(d(x_i, o) + C\big) \leq C Q_i^{-\frac{1}{2}},$$

where the diffeomorphisms $\varphi_t : M^4 \to M^4$ are as in Proposition 1.73 and where the last inequality follows from $d(x_i, o) \leq c^{-1/2} Q_i^{1/2}$. Thus, $\varphi_t(x) \in B(x_i, 1)$ provided that i is sufficiently large, which we henceforth assume. We conclude that

$$|\mathrm{Rm}|(x, t) = |\mathrm{Rm}|(\varphi_t(x), 0) \leq \frac{3C}{c} Q_i. \tag{2.229}$$

In other words, for the rescaled solutions $g_i(t) := Q_i g(Q_i^{-1} t)$, we have

$$|\mathrm{Rm}_{g_i}|(x, t) \leq \frac{3C}{c} \quad \text{for all } (x, t) \in B_{g_i}\left(x_i, \tfrac{1}{2} Q_i^{1/2}\right) \times [-1, 0]. \tag{2.230}$$

[11] A necessary and sufficient condition for the existence of such a sequence $\{x_i\}$ is that $\limsup_{x \to \infty} R(x) d(x, o)^{-2} > 0$.

We may now argue exactly as in the proof of Proposition 2.45 to obtain that, after passing to a subsequence, the pointed sequence $(M^4, g_i(t), x_i)$ converges in the pointed C^∞ Cheeger–Gromov sense to a complete pointed solution $(M^4_\infty, g_\infty(t), x_\infty)$, $t \in \left[-\frac{1}{2}, 0\right]$, to the Ricci flow satisfying

$$|\mathrm{Rm}_{g_\infty}| \leq \frac{3C}{c} \quad \text{and} \quad |\mathrm{Rm}_{g_\infty}|(x_\infty, 0) = 1.$$

Consider the sequence of Riemannian metrics $g_i := Q_i g = Q_i g(0)$, so that (M^4, g_i, x_i) converges in the pointed C^∞ Cheeger–Gromov sense to $(M^4_\infty, g_\infty, x_\infty)$, where $g_\infty := g_\infty(0)$. Define the potential functions $f_i(x) := f(x) - f(x_i)$, so that $f_i(x_i) = 0$. Since (M^4, g, f) is a shrinking gradient Ricci soliton, we have that

$$\mathrm{Ric}_{g_i} + \nabla^2_{g_i} f_i = \frac{1}{2} Q_i^{-1} g_i \tag{2.231}$$

and

$$R_{g_i} + |\nabla f_i|^2_{g_i} = Q_i^{-1} f = Q_i^{-1} f_i + Q_i^{-1} f(x_i). \tag{2.232}$$

Using that $Q_i \geq c f(x_i)$, we see that for each $A < \infty$, the potential functions satisfy the following derivative estimates for all $x \in B_{g_i}(x_i, A)$:

(1)

$$|\nabla f_i|^2_{g_i}(x) = Q_i^{-1} |\nabla f|^2_g(x) \leq Q_i^{-1} f(x) \leq C(A), \tag{2.233}$$

(2)

$$|\nabla^2 f_i|^2_{g_i}(x) = \left| \mathrm{Ric}_{g_i} - \frac{1}{2} Q_i^{-1} g_i \right|_{g_i} \leq C(A), \tag{2.234}$$

(3)

$$|\nabla^k f_i|^2_{g_i}(x) = \left|\nabla^{k-2} \mathrm{Ric}_{g_i}\right|_{g_i} \leq C_k(A) \quad \text{for all } k \geq 3. \tag{2.235}$$

Since $f(x_i) = 0$, by passing to a subsequence, we have that f_i converges to a smooth function f_∞ in C^∞ with respect to the pointed C^∞ Cheeger–Gromov of g_i to g_∞. Moreover, by passing to a subsequence, we may assume that $Q_i^{-1} f(x_i) \to c_0$, where $c_0 > 0$. Thus, by taking the limit as $i \to \infty$ of (2.231) and (2.232), we obtain

$$\mathrm{Ric}_{g_\infty} + \nabla^2_{g_\infty} f_\infty = 0 \tag{2.236}$$

and

$$R_{g_\infty} + |\nabla f_\infty|^2_{g_\infty} = c_0 > 0, \tag{2.237}$$

respectively. Thus, $(M^4_\infty, g_\infty, f_\infty)$ is a steady soliton.

Finally, suppose for a contradiction that $\mathrm{Ric}_{g_\infty} \equiv 0$. Then

$$\nabla^2_{g_\infty} f_\infty = 0 \quad \text{and} \quad |\nabla_{g_\infty} f_{g_\infty}|^2_{g_\infty} = c_0 > 0. \tag{2.238}$$

This implies that g_∞ splits as the Riemannian product of $\mathbb{R}$ and a Ricci-flat 3-manifold, which in turn must be flat. This is a contradiction. Therefore we have proved that $(M^4_\infty, g_\infty, f_\infty)$ is a non-Ricci-flat steady soliton. □

2.7.10. At most linear growth for R on a net.

Using the methods of the proof of (1.61), one can prove the following estimate (cf. [**Cho25**]).

PROPOSITION 2.58. *Let (M^n, g, f) be a complete noncompact shrinking gradient Ricci soliton and suppose that f attains its minimum at $o \in M^n$. For any $\varepsilon \in (0, 1]$, there exists a constant $C(n)$ such that for any $x \in M^n$ there exists $y \in B(x, \varepsilon)$ such that*

$$R(y) \leq C(n)\varepsilon^{-1} (d(y, o) + 1). \tag{2.239}$$

By an ε-**net** we mean a subset $\mathcal{N}$ of M^n with the property that for every $x \in M^n$ there exists $y \in \mathcal{N}$ such that $d(x, y) < \varepsilon$. So this proposition shows that for any $\varepsilon > 0$ there exists an ε-net on which the scalar curvature grows at most linearly.

PROOF OF PROPOSITION 2.58. Let $x \in M^n$, denote $r_x := d(x, o)$, and let $\gamma : [0, r_x] \to M^n$ be a minimal geodesic from o to x. Define $\zeta : [0, r_x] \to [0, 1]$ so that $\zeta(0) = \zeta(r_x) = 0$, $\zeta(\varepsilon) = \zeta(r_x - \varepsilon) = 1$, and so that ζ is linear in between. By the stability inequality (1.62) for minimal geodesics, we know that

$$2\varepsilon^{-2} \int_{r_x - \varepsilon}^{r_x} (r_x - s) f'(s)\, ds \geq \frac{r_x}{2} - \frac{2}{3}\varepsilon + 2\varepsilon^{-2} \int_0^{\varepsilon} s f'(s)\, ds - 2(n-1)\varepsilon^{-1};$$

cf. (1.63). Thus, for $\varepsilon > 0$ sufficiently small, we have that

$$2\varepsilon^{-2} \int_{r_x - \varepsilon}^{r_x} (r_x - s)|\nabla f|(\gamma(s))\, ds \geq \frac{r_x}{2} - 2n\varepsilon^{-1}. \tag{2.240}$$

By the mean value theorem for integrals, there exists $s_1 \in [r_x - \varepsilon, r_x]$ such that

$$|\nabla f|(\gamma(s_1)) \geq \frac{1}{2}(r_x - 4n\varepsilon^{-1}). \tag{2.241}$$

By (2.241) and (1.59), we conclude that for all $x \in M^n$ with $d(x, o) \geq 4n\varepsilon^{-1}$,

$$\begin{aligned} R(\gamma(s_1)) &= (f - |\nabla f|^2)(\gamma(s_1)) \\ &\leq \frac{1}{4}\left(s_1 + \sqrt{2n}\right)^2 - \frac{1}{4}(s_1 - 4n\varepsilon^{-1})^2 \\ &= \left(\sqrt{n/2} + 2n\varepsilon^{-1}\right) s_1 + \frac{n}{2} - 4n^2\varepsilon^{-2} \\ &\leq 3n\varepsilon^{-1} s_1. \end{aligned} \tag{2.242}$$

On the other hand, for $x \in M^n$ satisfying $d(x, o) < 4n\varepsilon^{-1}$, by

$$R(x) \leq f(x) \leq \frac{1}{4}\left(d(x, o) + \sqrt{2n}\right)^2$$

we have that

$$R(x) \leq \frac{1}{4}\left(4n\varepsilon^{-1} + \sqrt{2n}\right)\left(d(x, o) + \sqrt{2n}\right). \tag{2.243}$$

This completes the proof of the proposition. □

2.7.11. Non-quadratic scalar curvature growth along rays.
The following result says that the curvature growth rate along minimizing geodesic rays, of a shrinking gradient Ricci soliton, must be subquadratic.

PROPOSITION 2.59. *For any complete noncompact shrinking gradient Ricci soliton (M^n, g, f), there does not exist a minimizing geodesic ray $\gamma : [0, \infty) \to M^n$ such that $R(\gamma(s)) \geq cs^2$ for all $s \in [0, \infty)$, where $c > 0$.*

PROOF. Suppose, for a contradiction, that the proposition is not true. Denote $f(s) := f(\gamma(s))$. By (1.63), we have for all $\overline{s} \geq 1$ that

$$2(n-1) \geq \frac{1}{2}\overline{s} - \frac{2}{3} + 2\int_0^1 sf'(s)ds - 2\int_{\overline{s}-1}^{\overline{s}} (\overline{s} - s)f'(s)ds. \tag{2.244}$$

On the other hand, using the assumption that $R(\gamma(s)) \geq cs^2$, we have that

$$\begin{aligned} |f'(s)| &\leq |\nabla f|(\gamma(s)) \\ &= \sqrt{f - R}(\gamma(s)) \\ &\leq \left(\left(\sqrt{f}(\gamma(0)) + \frac{s}{2} \right)^2 - cs^2 \right)^{\frac{1}{2}}. \end{aligned}$$

By substituting this inequality into (2.244), we obtain a contradiction for $\overline{s}$ sufficiently large. □

As a final remark, we note that Propositions 2.58 and 2.59 are both consistent with the expected behavior of the hypothetical 4-dimensional Bamler shrinking solitons mentioned in §1.3.2.2.e.

2.8. The asymptotic volume ratio of shrinking solitons

Let (M^n, g, f) be a complete noncompact shrinking gradient Ricci soliton. Recall from (2.57) that the limit

$$\mathrm{AVR}(g) = \lim_{r\to\infty} \frac{\mathrm{Vol}\,(B\,(p, r))}{\omega_n r^n} \tag{2.245}$$

exists and is called the asymptotic volume ratio.

We have the following result of Wang and Wang [**WW25**, Theorem 2.8].

THEOREM 2.60. *If (M^n, g, f) is a complete noncompact shrinking gradient Ricci soliton, then* $\mathrm{AVR}(g) \leq 1$.

PROOF. Let $\rho := 2\sqrt{f}$ and following [**CZ10**], define

$$\mathrm{V}(r) := \int_{\{\rho<r\}} dv = \mathrm{Vol}\,\{\rho < r\}, \tag{2.246a}$$

$$\mathrm{R}(r) := \int_{\{\rho<r\}} R\,dv. \tag{2.246b}$$

For $\tau > 0$, define

$$h(\tau) := (4\pi\tau)^{-\frac{n}{2}} \int_M \mathrm{e}^{-\frac{f}{\tau}} dv. \tag{2.247}$$

STEP 1: *The function $\tau \mapsto h(\tau)$ is non-increasing for $\tau \geq 1$.*

We compute that

$$h'(\tau) = (4\pi\tau)^{-\frac{n}{2}} \int_M \left(\frac{f}{\tau^2} - \frac{n}{2\tau}\right) \mathrm{e}^{-\frac{f}{\tau}} dv \tag{2.248}$$
$$= (4\pi\tau)^{-\frac{n}{2}} \int_M \left(\frac{f}{\tau^2} - \frac{1}{\tau}\Delta f - \frac{1}{\tau}R\right) \mathrm{e}^{-\frac{f}{\tau}} dv$$
$$= (4\pi\tau)^{-\frac{n}{2}} \int_M \left(\frac{f - |\nabla f|^2}{\tau^2} - \frac{1}{\tau}R\right) \mathrm{e}^{-\frac{f}{\tau}} dv$$
$$= (4\pi\tau)^{-\frac{n}{2}} \left(\frac{1}{\tau^2} - \frac{1}{\tau}\right) \int_M R\,\mathrm{e}^{-\frac{f}{\tau}} dv,$$

where we integrated by parts to obtain the third equality. Since $R \geq 0$, we conclude for all $\tau \geq 1$ that

$$h'(\tau) \leq 0. \tag{2.249}$$

This proves that $h(\tau)$ is non-increasing for $\tau \geq 1$.

STEP 2:

$$\lim_{\tau \to \infty} h(\tau) = \mathrm{AVR}(g). \tag{2.250}$$

Let $\varepsilon > 0$. Since

$$\lim_{r \to \infty} \frac{\mathrm{V}(r)}{r^n} = \omega_n \,\mathrm{AVR}(g),$$

there exists $\overline{r} > 1$ such that

$$\frac{\mathrm{V}(r)}{r^n} \geq \omega_n(\mathrm{AVR}(g) - \varepsilon) \quad \text{for all } r \geq \overline{r}. \tag{2.251}$$

We compute using the co-area formula and integrating by parts that

$$(4\pi\tau)^{\frac{n}{2}} h(\tau) \geq \int_{\{\rho \geq \overline{r}\}} \mathrm{e}^{-\frac{f}{\tau}} dv$$
$$= \int_{\overline{r}}^{\infty} \mathrm{e}^{-\frac{r^2}{4\tau}} \mathrm{V}'(r)\, dr$$
$$= \mathrm{V}(r)\mathrm{e}^{-\frac{r^2}{4\tau}}\Big|_{\overline{r}}^{\infty} + \int_{\overline{r}}^{\infty} \frac{r}{2\tau} \mathrm{e}^{-\frac{r^2}{4\tau}} \mathrm{V}(r)\, dr$$
$$= -\mathrm{V}(\overline{r})\mathrm{e}^{-\frac{\overline{r}^2}{4\tau}} + \int_{\overline{r}}^{\infty} \frac{r^{n+1}}{2\tau} \mathrm{e}^{-\frac{r^2}{4\tau}} \frac{\mathrm{V}(r)}{r^n}\, dr.$$

Therefore, in view of (2.251), we have that

$$(4\pi\tau)^{\frac{n}{2}} h(\tau) \geq -\mathrm{V}(\overline{r})\mathrm{e}^{-\frac{\overline{r}^2}{4\tau}} + \omega_n(\mathrm{AVR}(g) - \varepsilon) \int_{\overline{r}}^{\infty} \frac{r^{n+1}}{2\tau} \mathrm{e}^{-\frac{r^2}{4\tau}} dr.$$

Regarding the last term on the right-hand side, we make the substitution $s = \frac{r^2}{4\tau}$ to see that

$$\int_{\overline{r}}^{\infty} \frac{r^{n+1}}{2\tau} \mathrm{e}^{-\frac{r^2}{4\tau}} dr = (4\tau)^{\frac{n}{2}} \int_{\frac{\overline{r}^2}{4\tau}}^{\infty} s^{\frac{n}{2}} \mathrm{e}^{-s} ds \tag{2.252}$$
$$= (4\tau)^{\frac{n}{2}} \left(\frac{\pi^{\frac{n}{2}}}{\omega_n} - \int_0^{\frac{\overline{r}^2}{4\tau}} s^{\frac{n}{2}} \mathrm{e}^{-s} ds\right).$$

Now, by

$$\int_0^{\frac{\overline{r}^2}{4\tau}} s^{\frac{n}{2}} \mathrm{e}^{-s} ds \leq \int_0^{\frac{\overline{r}^2}{4\tau}} s^{\frac{n}{2}} ds = \frac{2}{n+2} \frac{\overline{r}^{n+2}}{(4\tau)^{\frac{n}{2}+1}}$$

it follows that

$$\int_{\overline{r}}^{\infty} \frac{r^{n+1}}{2\tau} \mathrm{e}^{-\frac{r^2}{4\tau}} dr \geq \frac{(4\pi\tau)^{\frac{n}{2}}}{\omega_n} - \frac{1}{n+2} \frac{\overline{r}^{n+2}}{2\tau}.$$

Thus, we obtain

$$(4\pi\tau)^{\frac{n}{2}} h(\tau) \geq -\mathrm{V}(\overline{r}) \mathrm{e}^{-\frac{\overline{r}^2}{4\tau}} + (\mathrm{AVR}(g) - \varepsilon) \left((4\pi\tau)^{\frac{n}{2}} - \frac{\omega_n}{n+2} \frac{\overline{r}^{n+2}}{2\tau} \right),$$

so that

$$h(\tau) \geq (\mathrm{AVR}(g) - \varepsilon) - \frac{\mathrm{V}(\overline{r}) \mathrm{e}^{-\frac{\overline{r}^2}{4\tau}}}{(4\pi\tau)^{\frac{n}{2}}} - (\mathrm{AVR}(g) - \varepsilon) \frac{\omega_n}{n+2} \frac{\overline{r}^{n+2}}{2\tau (4\pi\tau)^{\frac{n}{2}}}.$$

By taking the limit $\tau \to \infty$, we obtain that

$$\lim_{\tau \to \infty} h(\tau) \geq \mathrm{AVR}(g) - \varepsilon.$$

Since this inequality holds for all $\varepsilon > 0$, we conclude that

$$\lim_{\tau \to \infty} h(\tau) \geq \mathrm{AVR}(g). \tag{2.253}$$

We now establish the reverse inequality. Let $\varepsilon > 0$. Then there exists $\overline{r} > 1$ such that

$$\frac{\mathrm{V}(r)}{r^n} \leq \omega_n (\mathrm{AVR}(g) + \varepsilon) \quad \text{for all } r \geq \overline{r}. \tag{2.254}$$

Estimating $\mathrm{e}^{-\frac{f}{\tau}} \leq 1$ on $\{\rho < \overline{r}\}$, we note that

$$\begin{aligned}
(4\pi\tau)^{\frac{n}{2}} h(\tau) &\leq \mathrm{V}(\overline{r}) + \int_{\{\rho \geq \overline{r}\}} \mathrm{e}^{-\frac{f}{\tau}} dv \\
&= \mathrm{V}(\overline{r}) + \int_{\overline{r}}^{\infty} \mathrm{e}^{-\frac{r^2}{4\tau}} \mathrm{V}'(r) dr \\
&= \mathrm{V}(\overline{r}) \left(1 - \mathrm{e}^{-\frac{\overline{r}^2}{4\tau}}\right) + \int_{\overline{r}}^{\infty} \frac{r^{n+1}}{2\tau} \mathrm{e}^{-\frac{r^2}{4\tau}} \frac{\mathrm{V}(r)}{r^n} dr,
\end{aligned} \tag{2.255}$$

where the last line follows after integration by parts. Making use of (2.254) in the last integral, it follows that

$$\int_{\overline{r}}^{\infty} \frac{r^{n+1}}{2\tau} \mathrm{e}^{-\frac{r^2}{4\tau}} \frac{\mathrm{V}(r)}{r^n} dr \leq \omega_n (\mathrm{AVR}(g) + \varepsilon) \int_{\overline{r}}^{\infty} \frac{r^{n+1}}{2\tau} \mathrm{e}^{-\frac{r^2}{4\tau}} dr.$$

However, (2.252) implies that

$$\begin{aligned}
\int_{\overline{r}}^{\infty} \frac{r^{n+1}}{2\tau} \mathrm{e}^{-\frac{r^2}{4\tau}} dr &= (4\tau)^{\frac{n}{2}} \left(\frac{\pi^{\frac{n}{2}}}{\omega_n} - \int_0^{\frac{\overline{r}^2}{4\tau}} s^{\frac{n}{2}} \mathrm{e}^{-s} ds \right) \\
&\leq \frac{(4\pi\tau)^{\frac{n}{2}}}{\omega_n}.
\end{aligned}$$

This proves that

$$\int_{\overline{r}}^{\infty} \frac{r^{n+1}}{2\tau} \mathrm{e}^{-\frac{r^2}{4\tau}} \frac{\mathrm{V}(r)}{r^n} dr \leq (\mathrm{AVR}(g) + \varepsilon) (4\pi\tau)^{\frac{n}{2}}.$$

Therefore, we obtain from (2.255) that

$$h(\tau) \leq \frac{\mathrm{V}(\overline{\tau})}{(4\pi\tau)^{\frac{n}{2}}} + (\mathrm{AVR}(g) + \varepsilon).$$

By taking the limit $\tau \to \infty$, we obtain that

$$\lim_{\tau\to\infty} h(\tau) \leq \mathrm{AVR}(g) + \varepsilon.$$

Since this inequality holds for all $\varepsilon > 0$, we conclude that

$$\lim_{\tau\to\infty} h(\tau) \leq \mathrm{AVR}(g). \tag{2.256}$$

By (2.253) and (2.256), we obtain (2.250); that is,

$$\lim_{\tau\to\infty} h(\tau) = \mathrm{AVR}(g).$$

STEP 3: *Completion of the proof of Theorem* 2.60.

By Li and Wang [**LW20**], we have that

$$1 \geq \mathrm{e}^{\mu} := \int_M (4\pi)^{-\frac{n}{2}} \mathrm{e}^{-f} dv = h(1) \geq \lim_{\tau\to\infty} h(\tau). \tag{2.257}$$

By combining this with (2.250), we obtain the theorem.

□

Now assume that (M^n, g, f) is not the Gaussian shrinking gradient Ricci soliton. By Li and Wang [**LW20**, Theorem 3], there exists a positive constant $\delta_0(n)$ depending only on n such that $\mu \leq -\delta_0(n)$. Hence, we have the gap theorem:

$$\mathrm{AVR}(g) \leq \mathrm{e}^{\mu} \leq \mathrm{e}^{-\delta_0(n)} < 1. \tag{2.258}$$

Furthermore, suppose that we have the equality $\mathrm{AVR}(g) = \mathrm{e}^{\mu}$. In this case, by the monotonicity of $h(\tau)$, we have that $h'(\tau) \equiv 0$ for $\tau \geq 1$. By (2.248), this implies that $R \equiv 0$ on M^n. We conclude from Theorem I-2.14 that (M^n, g, f) is the Gaussian shrinking gradient Ricci soliton. This is a contradiction.

2.9. Open problems for shrinking solitons

OPEN PROBLEM 2.1 (Volume dichotomy). *For any* 4*-dimensional complete noncompact shrinking gradient Ricci soliton with bounded scalar curvature and* $\mathrm{AVR}(g) = 0$, *must we have* $\mathrm{Vol}(B(o,r)) \leq Cr^2$ *for some constant* C*? See Theorem* 2.33 *for progress on this problem. However, the hypothetical* 4*-dimensional Bamler shrinking solitons mentioned in §1.3.2.2.e present a possible counterexample to this inequality.*

OPEN PROBLEM 2.2. *Must any shrinking gradient Ricci soliton with positive Ricci curvature be compact?*

Note that, according to Theorem 4.47 in Volume 1, shrinking gradient Ricci solitons with non-negative sectional curvature and positive Ricci curvature are necessarily compact.

OPEN PROBLEM 2.3. *What is the best negative lower bound for the sectional curvatures of a* 4*-dimensional shrinking gradient Ricci soliton?*

At the time of the writing of this book, Theorem 2.38 *provides the best lower bound known.*

Open Problem 2.4 (Asymptotic geometry trichotomy). *Let* (M^4, g, f) *be a 4-dimensional shrinking gradient Ricci soliton with bounded scalar curvature and let* E *be an end of* M^4*. Then* E *satisfies one of the following:*

(a) E *is asymptotically conical, or*

(b) E *is* C^∞*-asymptotic to* $\mathbb{R} \times (\mathbb{S}^3/\Gamma)$*, or*

(c) *For any sequence* $x_i \in E$ *going to infinity along an integral curve of* ∇f*, the sequence* (M^4, g, x_i) *converges in the smooth pointed Cheeger–Gromov sense to* $\mathbb{R}^2 \times \mathbb{S}^2$ *or its* $\mathbb{Z}_2$ *quotient.*

In view of Theorem 2.30*, the question can be rephrased as the dichotomy that either there exists* $C > 0$ *such that* $R \geq \frac{1}{C}$ *on* E*, or* $R \to 0$ *at infinity on* E*.*

Open Problem 2.5. *Must a shrinking gradient Ricci soliton have finite topological type? Specifically, is the underlying manifold diffeomorphic to the interior of a compact manifold with boundary?*

This result is true for shrinkers with sub-quadratic curvature growth, see Fang, Man, and Zhang [**FMZ08**].

Open Problem 2.6 (Bamler). *Do there exist non-rotationally symmetric 4-dimensional* $\mathrm{O}(3) \times \mathbb{Z}_2$*-symmetric shrinking gradient Ricci solitons with unbounded curvature?*

Lai [**Lai24**] *proved the existence of non-rotationally symmetric* 4*-dimensional* $\mathrm{O}(3) \times \mathbb{Z}_2$*-symmetric steady gradient Ricci solitons.*

Open Problem 2.7. *Consider the closed* 4*-dimensional orbifold* $\mathcal{O}^4$ *with a single orbifold point with isotropy group* Q_8 *obtained as in Example* 1.65*. That is,* $\mathcal{O}^4$ *is diffeomorphic to* B^4/Q_8 *attached to the normal disk bundle of the Veronese embedding of* $\mathbb{R}P^2$ *into* S^4*. Does* $\mathcal{O}^4$ *admit an* $\mathrm{SO}(3)$*-invariant shrinking gradient Ricci soliton structure?*

CHAPTER 3

Counting Ends of Ricci Solitons

Perhaps the most fundamental invariant of the topology at infinity of a complete noncompact Riemannian manifold is the number of topological ends of the manifold. In this chapter, we will explore what can be said about this invariant in the case of complete noncompact Ricci solitons.

Just as the theory of harmonic functions has proven to be an effective tool to study the number and the nature of the ends of general Riemannian manifolds, the analogous theory of f-harmonic functions — particularly the properties of f-Green's functions — will provide insights about the ends of a gradient Ricci soliton (M^n, g, f).

3.1. Function theory on weighted manifolds

In this section, we present some general results about function theory on weighted manifolds. These results provide basic tools for our study of the ends of Ricci solitons.

DEFINITION 3.1. A **weighted manifold** $\left(M^n, g, \mathrm{e}^{-f}dv\right)$ is a complete noncompact Riemannian manifold (M^n, g) together with a smooth measure

$$dv_f := \mathrm{e}^{-f}dv,$$

where dv denotes the Riemannian volume measure with respect to g.

Later, we will specialize to the case of a gradient Ricci soliton, where f is the potential function.

3.1.1. The Green's function and (non)parabolicity.

We now introduce the key concepts of Green's function and nonparabolicity. Let

$$\Delta_f = \Delta - \nabla f \cdot \nabla$$

be the f-Laplacian defined by (1.38).

DEFINITION 3.2. Let $\left(M^n, g, \mathrm{e}^{-f}dv\right)$ be a weighted manifold. A function $G_f(x, y)$ satisfying

$$\Delta_f G_f(x, \cdot) = -\delta(x, \cdot) \quad \text{on } M^n \tag{3.1}$$

for each $x \in M^n$ is called a **Green's function** of $\left(M^n, g, \mathrm{e}^{-f}dv\right)$.

The equation (3.1) means that for any continuous function φ with compact support in M^n, the function

$$\rho(x) := \int_{M^n} G_f(x, y)\, \varphi(y)\, \mathrm{e}^{-f(y)}dv(y) \tag{3.2}$$

solves the Poisson equation

$$\Delta_f \rho = -\varphi \quad \text{on } M^n. \tag{3.3}$$

On a compact Riemannian manifold (Ω, g) with boundary $\partial\Omega$, a Green's function G_f^Ω satisfying the Dirichlet boundary condition $G_f^\Omega = 0$ on $\partial\Omega$ always exists and it is unique. Moreover, G_f^Ω is positive and symmetric:

$$G_f^\Omega(x, y) = G_f^\Omega(y, x) > 0; \tag{3.4}$$

see Aubin [**Aub82**, Chapter 4, §2]. The Green's function can be obtained from the Dirichlet heat kernel H_f^Ω by (see Chapter 5 in Grigor'yan [**Gri06**])

$$G_f^\Omega(x, y) = \int_0^\infty H_f^\Omega(x, y, t)\, dt. \tag{3.5}$$

However, a positive symmetric Green's function may not always exist on a noncompact weighted manifold. See Li [**Li12**, Theorem 17.3] for a necessary and sufficient condition for the existence of a positive symmetric Green's function of the Laplacian on a complete noncompact manifold.

DEFINITION 3.3. A weighted manifold $\left(M^n, g, \mathrm{e}^{-f} dv\right)$ that admits a positive Green's function G_f is called ***f*-nonparabolic**.

A weighted manifold $\left(M^n, g, \mathrm{e}^{-f} dv\right)$ that does not admit any positive Green's function G_f is called ***f*-parabolic**.

If a positive Green's function does exist, then it can be computed from the heat kernel by the following formula:

$$G_f(x, y) = \int_0^\infty H_f(x, y, t)\, dt.$$

Moreover, in this case, a **minimal symmetric Green's function** can be constructed by an exhaustion procedure and defining

$$G_f(x, y) = \lim_{i \to \infty} G_f^{\Omega_i}(x, y), \tag{3.6}$$

where $G_f^{\Omega_i}$ is the Green's function of Ω_i satisfying the Dirichlet boundary condition, and $\{\Omega_i\}$ is an increasing sequence of smooth compact subsets of M^n such that $\bigcup_i \Omega_i = M^n$. This minimal symmetric Green's function is unique by the maximum principle.

Whenever $\left(M^n, g, \mathrm{e}^{-f} dv\right)$ is f-nonparabolic, the Green's function will refer to the minimal one constructed by (3.6).

The Green's function of a weighted manifold is the same as the Green's function of a suitably conformally changed Riemannian manifold. In particular, in any dimension $n \geq 3$, let

$$\overline{g} := \mathrm{e}^{-\frac{2}{n-2} f} g$$

and let $\overline{\nabla}$ and $\overline{\Delta}$ denote the gradient and Laplacian with respect to this conformal metric, respectively. Then we have that

$$\begin{aligned}\int_{M^n} (\overline{\Delta} u) w d\overline{v} &= -\int_{M^n} \overline{g}\left(\overline{\nabla} u, \overline{\nabla} w\right) d\overline{v} \\ &= -\int_{M^n} \langle \nabla u, \nabla w \rangle \, \mathrm{e}^{-f} dv \\ &= \int_{M^n} (\Delta_f u) w \mathrm{e}^{-f} dv\end{aligned}$$

for any $u, w \in C_0^\infty(M^n)$. Consequently, $\Delta_f G_f = -\delta$ is equivalent to $\overline{\Delta}\,\overline{G} = -\delta$. Hence, G_f is the minimal Green's function $\overline{G}$ of the Riemannian manifold $(M^n, \overline{g})$; i.e.,

$$G_f = \overline{G}.$$

In particular, for any $n \geq 3$,

$$G_f(p,x) \sim \mathrm{e}^{f(p)} \frac{c(n)}{d^{n-2}(p,x)} \quad \text{as } x \to p, \tag{3.7}$$

where d is the distance on (M^n, g) and $c(n) := \frac{1}{(n-2)\,\mathrm{Vol}(\mathbb{S}^{n-1})}$.

Our goal for this section is to present a useful criterion for whether a weighted manifold is f-nonparabolic, i.e., admits a positive Green's function. While any weighted complete manifold admits a Green's function (see Chapter 17 in Li [**Li12**]), the existence of a positive Green's function presents some significant advantages for estimating it via gradient estimates and the Harnack property, which we recall in the next subsection.

3.1.2. Gradient estimates for positive f-harmonic functions.

Recall that

$$\mathrm{Ric}_f = \mathrm{Ric} + \nabla^2 f$$

denotes the Bakry–Emery curvature tensor. The following gradient estimate for positive f-harmonic functions requires a Bakry–Emery curvature lower bound, as well as a bound on ∇f.

THEOREM 3.4. *Let $\left(M^n, g, \mathrm{e}^{-f} dv\right)$ be a weighted manifold of dimension n with the Bakry–Emery curvature bounded below by $\mathrm{Ric}_f \geq -(n-1) K g$, for some constant $K \geq 0$. Assume that $x \in M^n$, $r > 0$, and there exists a constant $a > 0$ such that*

$$\sup_{B(x,2r)} |\nabla f| \leq a.$$

Then there exists a constant $C = C(n,a,K)$ such that for any $u > 0$ with $\Delta_f u = 0$ in $B(x, 2r)$, we have

$$\sup_{B(x,r)} |\nabla \ln u| \leq C\left(1 + \frac{1}{r}\right). \tag{3.8}$$

In particular,

$$u(y) \leq u(z)\, \mathrm{e}^{C(1+r)} \tag{3.9}$$

for all $y, z \in B(x,r)$.

PROOF. The argument is an adaptation of Cheng and Yau's gradient estimate [**CY75**]. Consider a smooth cutoff function $\phi = \phi(t)$ such that $\phi(t) = 1$ for $t \in [0, r]$ and $\phi(t) = 0$ for $t \geq 2r$. We may assume that

$$-\frac{C}{r} \leq \phi' \leq 0 \quad \text{and} \quad |\phi''| \leq \frac{C}{r^2}$$

for some universal constant $C > 0$. We consider $\phi(y) := \phi(d(x, y))$, which is a cutoff function on M^n, satisfying $\phi = 1$ on $B(x, r)$ and $\phi = 0$ on $M^n \setminus B(x, 2r)$.

The Laplacian comparison theorem (see Wei and Wylie [**WW09**]) implies that

$$\begin{aligned}\Delta_f d &\leq (n-1)\sqrt{K}\coth\left(\sqrt{K}d\right) + a \\ &\leq \frac{n-1}{d}\left(1 + \sqrt{K}d\right) + a.\end{aligned}$$

It therefore follows that

$$|\nabla\phi| \leq \frac{C}{r} \quad \text{and} \quad \Delta_f \phi \geq -\frac{(n-1)C}{r^2} - \frac{C}{r}\left((n-1)\sqrt{K} + a\right), \tag{3.10}$$

where the last inequality is understood in the support sense (a.k.a. barrier sense; cf. Definition 2.17 in Volume 1), and $C > 0$ is a universal constant.

Let $h := \ln u$ and let us consider the function $W : B(x, 2r) \to \mathbb{R}$ defined by

$$W := \phi^2 |\nabla h|^2 .$$

Let x_0 be a maximum point of W. At x_0 we have, by the maximum principle, that

$$\begin{aligned}\nabla W(x_0) &= 0, \\ \Delta_f W(x_0) &\leq 0.\end{aligned} \tag{3.11}$$

By the f-Bochner formula (1.44) and since $\Delta_f h = -|\nabla h|^2$, we get

$$\begin{aligned}\frac{1}{2}\Delta_f |\nabla h|^2 &= |\nabla^2 h|^2 + \langle \nabla h, \nabla(\Delta_f h)\rangle + \mathrm{Ric}_f(\nabla h, \nabla h) \\ &\geq |\nabla^2 h|^2 - \left\langle \nabla h, \nabla |\nabla h|^2 \right\rangle - (n-1) K |\nabla h|^2 .\end{aligned}$$

Moreover, by the Cauchy–Schwarz inequality, we have

$$\begin{aligned}|\nabla^2 h|^2 &\geq \frac{1}{n}(\Delta h)^2 \\ &= \frac{1}{n}\left(|\nabla h|^2 - \langle \nabla f, \nabla h\rangle\right)^2 .\end{aligned}$$

Therefore, we obtain that

$$\begin{aligned}\frac{1}{2}\Delta_f |\nabla h|^2 \geq{} & \frac{1}{n}\left(|\nabla h|^2 - \langle \nabla f, \nabla h\rangle\right)^2 \\ & - \left\langle \nabla h, \nabla |\nabla h|^2 \right\rangle - (n-1) K |\nabla h|^2 .\end{aligned}$$

As $W = \phi^2 |\nabla h|^2$, it follows that

$$\begin{aligned}\frac{1}{2}\phi^2\Delta_f W \geq& \frac{1}{n}\phi^4\left(|\nabla h|^2 - \langle\nabla f, \nabla h\rangle\right)^2 - \phi^4\left\langle \nabla h, \nabla\left(\phi^{-2}W\right)\right\rangle \\ &- (n-1)K\phi^2 W + \frac{1}{2}\left(\Delta_f\phi^2\right)W + \phi^2\left\langle\nabla\phi^2, \nabla\left(\phi^{-2}W\right)\right\rangle \\ \geq& \frac{1}{n}\phi^4\left(|\nabla h|^2 - \langle\nabla f, \nabla h\rangle\right)^2 - (n-1)KW + 2\phi\langle\nabla\phi, \nabla h\rangle W \\ &+ \frac{1}{2}\left(\Delta_f\phi^2\right)W - 4|\nabla\phi|^2 W + \left\langle\nabla\phi^2, \nabla W\right\rangle - \phi^2\langle\nabla h, \nabla W\rangle.\end{aligned}$$

By (3.10), we know that

$$\frac{1}{2}\left(\Delta_f\phi^2\right)W - 4|\nabla\phi|^2 W \geq -\frac{C}{r}\left(\frac{1}{r} + \sqrt{K} + a\right)W,$$

where $C > 0$ depends only on the dimension n. Moreover, we have

$$2\phi\langle\nabla\phi, \nabla h\rangle W \geq -\frac{C}{r}W^{\frac{3}{2}}.$$

Using (3.11), it follows that at x_0 we have that

$$\begin{aligned}0 \geq& \frac{1}{n}\phi^4\left(|\nabla h|^2 - \langle\nabla f, \nabla h\rangle\right)^2 - (n-1)KW \\ &- \frac{C}{r}\left(\frac{1}{r} + \sqrt{K} + a\right)W - \frac{C}{r}W^{\frac{3}{2}}.\end{aligned}$$

Since

$$\frac{C\sqrt{K}}{r}W \leq KW + \frac{C}{r^2}W,$$

we infer from the above inequality that at x_0,

$$W - \phi^2\langle\nabla f, \nabla h\rangle \leq \left(n\sqrt{K} + \frac{C}{r} + C\sqrt{\frac{a}{r}}\right)\sqrt{W} + \frac{C}{\sqrt{r}}W^{\frac{3}{4}},$$

where $C = C(n)$. As $|\nabla f| \leq a$, we get

$$\phi^2\langle\nabla f, \nabla h\rangle \leq a\sqrt{W},$$

and therefore

$$\sqrt{W}(x_0) \leq a + n\sqrt{K} + \frac{C}{r} + C\sqrt{\frac{a}{r}} + \frac{C}{\sqrt{r}}W^{\frac{1}{4}}(x_0),$$

where $C = C(n) > 0$. Solving this inequality in $W(x_0)$, it follows that

$$\sqrt{W}(x_0) \leq a + n\sqrt{K} + C\left(\frac{1}{\sqrt{r}} + \frac{1}{r}\right),$$

for a constant $C = C(n, a, K) > 0$ depending on n, a and K. As $\phi = 1$ on $B(x, r)$, we get that

$$\sup_{B(x,r)}|\nabla \ln u| \leq a + n\sqrt{K} + C\left(\frac{1}{\sqrt{r}} + \frac{1}{r}\right) \tag{3.12}$$

for $C = C(n, a, K)$. This proves the gradient estimate (3.8).

Finally, (3.9) follows by integrating (3.8) along minimizing geodesics starting from x. □

Some variants of this result can be obtained by replacing the bound on $|\nabla f|$ with a bound on f; see Wang and one of the authors [**MW11, MW12**].

The following sharp form of the above gradient estimate will be important below in the study of ends of steady gradient Ricci solitons.

COROLLARY 3.5 (Sharp gradient estimate). *Let $\left(M^n, g, \mathrm{e}^{-f} dv\right)$ be a weighted manifold of dimension n with $\mathrm{Ric}_f \geq 0$. Assume that there exists a constant $a > 0$ such that*

$$\sup_{M^n} |\nabla f| \leq a.$$

Then, for any $u > 0$ with $\Delta_f u = 0$ on M^n, we have

$$\sup_{M^n} |\nabla \ln u| \leq a.$$

PROOF. We apply (3.12) for $K = 0$ and get

$$\sup_{B(x,r)} |\nabla \ln u| \leq a + C(n,a)\left(\frac{1}{\sqrt{r}} + \frac{1}{r}\right)$$

for any $r > 0$. Taking $r \to \infty$, we conclude that $|\nabla \ln u| \leq a$ on M^n. □

To verify that this estimate is sharp, consider $M^n = \mathbb{R} \times N^{n-1}$, where N^{n-1} has nonnegative Ricci curvature. For $f(t,x) = at$, for all $t \in \mathbb{R}$ and $x \in N^{n-1}$, we have $\mathrm{Ric}_f \geq 0$ and $|\nabla f| = a$. The function $u(t,x) = \mathrm{e}^{at}$ satisfies $\Delta_f u = 0$ and $|\nabla \ln u| = a$.

Let $D \subset M^n$ be a compact domain with smooth boundary $\Sigma = \partial D$. For $p \in M^n$ fixed, let $t_0 > 0$ be large enough so that $D \subset B(p, t_0)$. The following result is a gradient estimate up to the boundary Σ for a special class of f-harmonic functions.

PROPOSITION 3.6. *Let $\left(M^n, g, \mathrm{e}^{-f} dv\right)$ be a weighted manifold of dimension n with the Bakry–Emery curvature bounded below by $\mathrm{Ric}_f \geq -(n-1)Kg$ and potential f satisfying*

$$\sup_{M^n} |\nabla f| \leq a.$$

For $T > 2t_0$ arbitrary, consider the f-harmonic function u defined by

$$\begin{aligned} \Delta_f u &= 0 \quad \text{on } B(p,T) \setminus D, \\ u &= 0 \quad \text{on } \partial B(p,T), \\ u &= 1 \quad \text{on } \Sigma. \end{aligned}$$

Then, for any $t_0 < r < \frac{T}{2}$, there exists $C > 0$ independent of T such that

$$|\nabla \ln u| \leq C \quad \text{on } B(p,r) \setminus D. \tag{3.13}$$

Consequently,

$$u(y) \leq u(z)\, \mathrm{e}^{Cr}$$

for all $y, z \in B(p,r) \setminus D$.

PROOF. We first assume that the maximum of $|\nabla \ln u|^2$ is achieved on Σ. By the strong maximum principle, $|\nabla u| > 0$ on Σ. Therefore the unit outer normal to $\Sigma = \partial\left(M^n \setminus D\right)$ is

$$e_n := \frac{\nabla u}{|\nabla u|}.$$

Hence, it follows that

$$\left\langle \nabla \left|\nabla \ln u\right|^2, \nabla u \right\rangle \geq 0$$

at any maximum point $x_0 \in \Sigma$ of $\left|\nabla \ln u\right|^2$. We rewrite this inequality as

$$\left|\nabla \ln u\right|^4 \leq \frac{\nabla^2 u(\nabla u, \nabla u)}{u^3}. \tag{3.14}$$

The mean curvature H_Σ of Σ satisfies

$$H_\Sigma \left|\nabla u\right| = \Delta u - \nabla^2_{nn} u,$$

which proves that at x_0,

$$\begin{aligned}\frac{\nabla^2 u(\nabla u, \nabla u)}{u^3} &= \frac{\left|\nabla u\right|^2}{u^3} \nabla^2_{nn} u \\ &= \frac{\left|\nabla u\right|^2}{u^3} \left(\langle \nabla u, \nabla f\rangle - H_\Sigma \left|\nabla u\right|\right) \\ &\leq C \left|\nabla \ln u\right|^3,\end{aligned}$$

where C depends on $\sup_\Sigma \left|\nabla f\right|$ and on $\sup_\Sigma \left|H_\Sigma\right|$. According to (3.14), this proves that

$$\left|\nabla \ln u\right|(x_0) \leq C. \tag{3.15}$$

Now consider the function $W := \phi^2 \left|\nabla \ln u\right|^2$ as in Theorem 3.4, where $\phi = 1$ on $B(p, r)$ and $\phi = 0$ on $M^n \setminus B(p, 2r)$, and let x_0 be a maximum point of W in $B(p, 2r) \setminus D$.

If $x_0 \in \Sigma$, then (3.15) shows that $W(x_0) \leq C$. If $x_0 \notin \Sigma$, then (3.12) proves that $W(x_0) \leq C$. The constant C depends on n, K, a, and r; hence it is independent of T. In either case, we get that W is bounded uniformly in T, which proves the gradient estimate (3.13) up to the boundary. □

3.1.3. Characterizing f-nonparabolicity.

We are ready to prove the following result; our presentation follows Chapter 17 of Li [**Li12**].

THEOREM 3.7 (Characterization of f-nonparabolicity). *Let $\left(M^n, g, \mathrm{e}^{-f} dv\right)$ be a weighted manifold and let $D \subset M^n$ be a compact domain with smooth boundary $\Sigma = \partial D$. For $p \in D$ fixed and $r_i \to \infty$, consider the sequence of f-harmonic functions $\{u_i\}$ defined by*

$$\begin{aligned}\Delta_f u_i &= 0 \quad \text{on } B(p, r_i) \setminus D, \\ u_i &= 0 \quad \text{on } \partial B(p, r_i), \\ u_i &= 1 \quad \text{on } \Sigma.\end{aligned}$$

Then a subsequence of $\{u_i\}$ converges uniformly and smoothly on compact subsets of $M^n \setminus D$ to an f-harmonic function $u : M^n \setminus D \to (0, 1]$. This function satisfies $u = 1$ on Σ and has finite weighted Dirichlet energy:

$$\int_{M^n \setminus D} \left|\nabla u\right|^2 \mathrm{e}^{-f} dv < \infty.$$

The manifold is f-nonparabolic if and only if u is not identically 1 on $M^n \setminus D$. Moreover, if $(M^n, g, \mathrm{e}^{-f} dv)$ is f-nonparabolic, then there exists a constant $C > 0$ such that the Green's function satisfies

$$\frac{1}{C} G_f(p, x) \leq u(x) \leq C G_f(p, x) \quad \text{for all } x \in M^n \setminus D.$$

PROOF. By the maximum principle, $0 \leq u_i \leq 1$ on $B(p, r_i) \setminus D$. Proposition 3.6 implies that $|\nabla u_i| \leq C(r)$ on $B(p, r) \setminus D$ for any $r < \frac{r_i}{2}$. By the Arzelà–Ascoli theorem (see Rudin [**Rud76**, Theorem 7.17]), this implies that a subsequence of $\{u_i\}$ converges in $C^{0,1}$ on compact subsets of $B(p, r_i) \setminus D$ to a function $u > 0$ which is a weak solution of $\Delta_f u = 0$. By standard elliptic theory, u is in fact smooth and satisfies $\Delta_f u = 0$.

Moreover, integrating by parts, we have that

$$\int_{B(p,r_i)\setminus D} |\nabla u_i|^2 \mathrm{e}^{-f} dv = \int_\Sigma |\nabla u_i| \mathrm{e}^{-f} dv,$$

using that $u_i = 1$ on Σ and that the outer unit normal to Σ is $\frac{\nabla u_i}{|\nabla u_i|}$. Proposition 3.6, applied for a fixed r_0 large enough such that $D \subset B(p, r_0)$, implies that $|\nabla u_i| \leq C$ on Σ, where $C > 0$ is a constant independent of i.

We obtain that

$$\int_{B(p,r)\setminus D} |\nabla u_i|^2 \mathrm{e}^{-f} dv \leq C$$

for any $r < r_i$. Taking the limit as $i \to \infty$ proves that

$$\int_{B(p,r)\setminus D} |\nabla u|^2 \mathrm{e}^{-f} dv \leq C$$

for a constant $C > 0$ independent of r. As r is arbitrary, we conclude that u has finite weighted energy.

Let us now assume that the limit u is nonconstant, and prove that $(M^n, g, \mathrm{e}^{-f} dv)$ is f-nonparabolic. We first note that

$$\liminf_{x \to \infty} u(x) = 0. \tag{3.16}$$

Indeed, for a contradiction, assume that

$$\alpha := \liminf_{x \to \infty} u(x) > 0.$$

Then, by the maximum principle, $1 \geq u > \alpha$ on $M^n \setminus D$. The function

$$\eta := \frac{u - \alpha}{1 - \alpha}$$

then satisfies

$$\begin{aligned} \Delta_f \eta &= 0 \quad \text{on } M^n \setminus D, \\ \eta &= 1 \quad \text{on } \Sigma, \\ \liminf_{x \to \infty} \eta(x) &= 0. \end{aligned}$$

By the maximum principle, $u_i \leq \eta$ on $B(p, r_i) \setminus D$, which implies that $u \leq \eta$ on $M^n \setminus D$. This contradicts $\alpha > 0$. In conclusion, (3.16) is true. Hence, u satisfies

$$\begin{aligned} \Delta_f u &= 0 \quad \text{on } M^n \setminus D, \\ u &= 1 \quad \text{on } \Sigma, \\ \liminf_{x \to \infty} u(x) &= 0. \end{aligned}$$

For a compact exhaustion Ω_i of M^n, denote by G_f^i the Green's function of Ω_i that satisfies the Dirichlet boundary condition on $\partial \Omega_i$. Fix $r > 0$ and assume that i_0 is large enough so that

$$D \subset B(p, r) \subset \Omega_i \quad \text{for all } i > i_0.$$

For $q \in B(p, r)$ and for $\varepsilon > 0$ fixed, note that

$$(1 + \varepsilon) G_f^{i+1}(q, x) > G_f^i(q, x) = 0 \quad \text{for all } x \in \partial \Omega_i.$$

By the asymptotics (3.7) of the Green's function near the pole q, it follows that

$$(1 + \varepsilon) G_f^{i+1}(q, x) > G_f^i(q, x) \text{ as } x \to q.$$

Furthermore, both $G_f^{i+1}(q, \cdot)$ and $G_f^i(q, \cdot)$ are f-harmonic. Therefore, by the maximum principle, we get that

$$G_f^i(q, x) < (1 + \varepsilon) G_f^{i+1}(q, x)$$

for all $x \in \Omega_i$. Since ε is arbitrary, it follows that

$$G_f^i(q, x) < G_f^{i+1}(q, x) \tag{3.17}$$

for all $x \in \Omega_i$.

Let

$$A_i := \max_{x \in \partial B(p,r)} G_f^i(q, x). \tag{3.18}$$

We claim that $\{A_i\}$ must be uniformly bounded in i. Let us assume the contrary, that $\{A_i\}$ is unbounded. To obtain a contradiction, we will construct two barrier functions, one on $\Omega_i \setminus B(p, r)$ and the other on $B(p, r)$.

First, let

$$\theta := \frac{u}{\min_{\partial B(p,r)} u},$$

which satisfies

$$\begin{aligned} \Delta_f \theta &= 0 \quad \text{on } M^n \setminus B(p, r), \\ \theta &\geq 1 \quad \text{on } \partial B(p, r), \\ \liminf_{x \to \infty} \theta(x) &= 0. \end{aligned} \tag{3.19}$$

The function $A_i \theta$ is f-harmonic on $\Omega_i \setminus B(p, r)$, is positive on $\partial \Omega_i$, and satisfies $A_i \theta \geq G_f^i(q, \cdot)$ on $\partial B(p, r)$. We obtain by the maximum principle that

$$G_f^i(q, x) \leq A_i \theta(x) \quad \text{for all } x \in \Omega_i \setminus B(p, r).$$

Moreover, since $G_f^i(q, x) = 0$ for $x \in \partial \Omega_i$, the maximum principle and (3.18) imply that $G_f^i(q, x) \leq A_i$ for all $x \in \Omega_i \setminus B(p, r)$. In conclusion,

$$G_f^i(q, x) \leq A_i \min\{1, \theta(x)\} \quad \text{for all } x \in \Omega_i \setminus B(p, r). \tag{3.20}$$

Now denote by G_f^r the Green's function of $B(p,r)$. By (3.18), we have that

$$G_f^i(q,x) \le A_i \quad \text{for all } x \in \partial B(p,r).$$

Thus the maximum principle applied as in the proof of (3.17) implies that

$$G_f^i(q,x) \le G_f^r(q,x) + A_i \tag{3.21}$$

for all $x \in B(p,r) \setminus \{q\}$. Hence, by (3.20) and (3.21), we get

$$\begin{aligned} \frac{1}{A_i} G_f^i(q,x) &\le \frac{1}{A_i} G_f^r(q,x) + 1, \quad \text{for all } x \in B(p,r) \setminus \{q\}, \\ \frac{1}{A_i} G_f^i(q,x) &\le \min\{1, \theta(x)\}, \quad \text{for all } x \in \Omega_i \setminus B(p,r). \end{aligned}$$

Since $\{A_i\}$ is unbounded, we conclude that $\left\{\frac{1}{A_i} G_f^i(q,x)\right\}$ converges up to a subsequence to a bounded f-harmonic function on $M^n \setminus \{q\}$.

By the standard removable singularities theorem for f-harmonic functions (the removable singularities theorem for f-harmonic functions can be proved in the same way as for harmonic functions; see Bochner [**Boc56**]), this function can be extended to a bounded f-harmonic function H on M^n that satisfies

$$\begin{aligned} H &\le 1 \quad \text{on } M^n, \\ H &\le \theta \quad \text{on } M^n \setminus B(p,r), \\ \max_{\partial B(p,r)} H &= 1, \end{aligned}$$

where the last property follows from (3.18). By the strong maximum principle, $H = 1$ on M^n, therefore $\theta \ge 1$ on $M^n \setminus B(p,r)$.

This contradicts (3.19). The contradiction shows that indeed $\{A_i\}$ is bounded uniformly in i. By (3.20) and (3.21), we obtain uniform estimates for G_f^i. Theorem 3.4 and the Arzelà–Ascoli theorem imply that a positive Green's function exists on M^n. Hence, we have proved that if u is nonconstant, then (M^n, g) is f-nonparabolic.

Moreover, by (3.20), we find that

$$G_f(p,x) \le Cu(x) \quad \text{for all } x \in M^n \setminus D.$$

In particular, (3.16) implies that

$$\liminf_{x \to \infty} G_f(p,x) = 0. \tag{3.22}$$

The reverse inequality between G_f and u also holds, as can be seen by inverting their roles in the maximum principle. Indeed, since

$$v(x) := \frac{G_f(p,x)}{\inf_{y \in \Sigma} G_f(p,y)}$$

is harmonic on $M^n \setminus D$, v is positive on $\partial B(p,r_i)$, and $v \ge 1$ on Σ, we conclude by the maximum principle that

$$u_i(x) \le \frac{G_f(p,x)}{\inf_{y \in \Sigma} G(p,y)} \quad \text{for all } x \in B(p,r_i) \setminus D.$$

Taking the limit as $i \to \infty$, this proves the same inequality for u, hence

$$u(x) \le CG_f(p,x) \quad \text{for all } x \in M^n \setminus D. \tag{3.23}$$

Finally, note that (3.22) and (3.23) imply that if M^n is f-nonparabolic, then u cannot be constant. This concludes the proof of the theorem. □

DEFINITION 3.8. The f-harmonic function $u : M^n \setminus D \to (0,1]$ defined as the limit of the sequence $\{u_i\}$ in Theorem 3.7 is called the **minimal barrier function** on $M^n \setminus D$.

The minimality of u refers to the following property:

PROPOSITION 3.9. *Let $(M^n, g, \mathrm{e}^{-f}dv)$ be a weighted manifold and let u be the minimal barrier function on $M^n \setminus D$. Suppose that $\eta > 0$ is an f-harmonic function on $M^n \setminus D$ such that $\eta = 1$ on $\Sigma = \partial D$. Then*

$$\eta \geq u \quad \text{on } M^n \setminus D. \tag{3.24}$$

PROOF. Let $\{u_i\}$ be the sequence used to define u. Since η and u_i are f-harmonic on $M^n \setminus D$ and since

$$\begin{aligned} \eta - u_i &= 0 \quad \text{on } \Sigma, \\ \eta - u_i &> 0 \quad \text{on } \partial B(p, r_i), \end{aligned}$$

the maximum principle implies that $\eta \geq u_i$ on $B(p,r_i) \setminus D$. Taking the limit $i \to \infty$, we see that

$$\eta \geq u \quad \text{on } M^n \setminus D.$$

□

In the following result, we establish that if a function is bounded in absolute value from above by the Green's function, then its weighted Dirichlet energy is finite.

THEOREM 3.10. *Let $\left(M^n, g, \mathrm{e}^{-f}dv\right)$ be a weighted manifold and $D \subset M^n$ a compact domain. Assume that ζ is an f-harmonic function on $M^n \setminus D$ such that*

$$|\zeta|(x) \leq C G_f(p,x) \quad \text{for all } x \in M^n \setminus D,$$

where $C > 0$ is a positive constant and $p \in D$ is fixed. Then ζ has finite weighted Dirichlet energy:

$$\int_{M^n \setminus D} |\nabla \zeta|^2 \mathrm{e}^{-f} dv < \infty.$$

PROOF. Let u be the minimal barrier function on $M^n \setminus D$. According to Theorem 3.7, there exists a constant $C > 0$ such that

$$|\zeta| \leq Cu \quad \text{on } M^n \setminus D.$$

Let $\{B(p, r_i)\}$ be a compact exhaustion of M^n and let $\{\theta_i\}$ be the sequence of f-harmonic functions that satisfy the boundary conditions

$$\begin{aligned} \theta_i &= \zeta \quad \text{on } \Sigma = \partial D, \\ \theta_i &= 0 \quad \text{on } \partial B(p, r_i). \end{aligned} \tag{3.25}$$

A subsequence of $\{\theta_i\}$ converges uniformly on compact subsets of M^n to an f-harmonic function θ. Since $|\theta_i| \leq Cu$ on Σ and on $\partial B(p, r_i)$, the maximum principle implies that $|\theta_i| \leq Cu$ on $B(p, r_i) \setminus D$. Taking $i \to \infty$ gives us that

$$|\theta|(x) \leq Cu(x) \quad \text{for all } x \in M^n \setminus D.$$

CLAIM 3.11. *We have that $\theta = \zeta$ on $M^n \setminus D$.*

PROOF OF CLAIM 3.11. The function

$$\sigma := \theta - \zeta$$

is f-harmonic on $M^n \setminus D$ and satisfies

$$\sigma = 0 \quad \text{on } \Sigma$$
$$|\sigma| \le C_0 u \quad \text{on } M^n \setminus D,$$

for some constant $C_0 > 0$. Therefore, the function

$$\eta := u - \frac{1}{C_0}\sigma$$

is f-harmonic on $M^n \setminus D$ and satisfies

$$\eta = 1 \quad \text{on } \Sigma,$$
$$\eta \ge 0 \quad \text{on } M^n \setminus D.$$

By Proposition 3.9, we get that $\eta \ge u$, which means that $\sigma \le 0$ on $M^n \setminus D$.

Since $-\sigma$ is also f-harmonic, vanishes on Σ, and $|-\sigma| \le C_0 u$, we obtain from above that $-\sigma \le 0$ on $M^n \setminus D$. In conclusion, $\sigma = 0$. This proves that $\theta = \zeta$ as claimed. □

Now, by (3.25), we have

$$\int_{B(p,r_i)\setminus D} |\nabla \theta_i|^2 \, \mathrm{e}^{-f} dv = \int_\Sigma \theta_i \frac{\partial \theta_i}{\partial \nu} \mathrm{e}^{-f} da.$$

According to Claim 3.11, θ_i converges to ζ uniformly on compact subsets; therefore

$$\int_{B(p,r)\setminus D} |\nabla \zeta|^2 \, \mathrm{e}^{-f} dv \le \int_\Sigma \zeta |\nabla \zeta| \mathrm{e}^{-f} da$$

for any $r > 0$ large enough so that $D \subset B(p,r)$. This proves the result. □

3.1.4. Characterizing f-parabolicity.

We have the following useful characterization of f-parabolic manifolds.

COROLLARY 3.12. *A weighted manifold $(M^n, g, \mathrm{e}^{-f} dv)$ is f-parabolic if and only if any positive f-superharmonic function on M^n is constant.*

PROOF. Assume first that the manifold is f-parabolic and $\zeta > 0$ is f-superharmonic; that is,

$$\Delta_f \zeta \le 0 \quad \text{on } M^n.$$

By the maximum principle, ζ achieves its infimum value at infinity. By subtracting this value from ζ if necessary, we may assume that

$$\liminf_{x \to \infty} \zeta(x) = 0. \tag{3.26}$$

We assume for a contradiction that ζ is not constant; hence $\zeta > 0$ on M^n. Consider the sequence $\{u_i\}$ defined in Theorem 3.7. Since

$$\Delta_f \left(\frac{\zeta}{\inf_\Sigma \zeta} - u_i \right) \le 0 \quad \text{on } B(p, r_i) \setminus D,$$
$$\frac{\zeta}{\inf_\Sigma \zeta} - u_i \ge 0 \quad \text{on } \Sigma = \partial D,$$
$$\frac{\zeta}{\inf_\Sigma \zeta} - u_i > 0 \quad \text{on } \partial B(p, r_i),$$

we find from the maximum principle that

$$u_i \leq \frac{\zeta}{\inf_\Sigma \zeta} \quad \text{on } B(p, r_i) \setminus D.$$

Taking the limit as $i \to \infty$, we have that u satisfies the same estimate. By (3.26), it follows that u is not constant; hence M^n is f-nonparabolic. This contradicts our assumption that M^n is f-parabolic, and proves that ζ must be constant.

We now prove the converse. Assume that any f-superharmonic function on M^n is constant. Let u be the minimal barrier function defined in Theorem 3.7. Then $u = 1$ on $\Sigma = \partial D$ and

$$\Delta_f u = 0 \quad \text{on } M^n \setminus D.$$

Let v be the function which equals u on $M^n \setminus D$ and $v = 1$ on D. Then v is positive, is defined on M^n, and satisfies

$$\Delta_f v \leq 0.$$

Therefore, v must be constant by our assumption that any positive f-superharmonic function on M^n is constant. This proves that $u = 1$ on $M^n \setminus D$, which means that M^n is f-parabolic. □

We also have the following sufficient geometric condition for parabolicity due to Cheng and Yau [**CY75**]; see Grigor'yan [**Gri85**] and Lyons and Sullivan [**LS84**] for a sharp sufficient condition in terms of volume growth.

Denote by

$$\mathrm{Vol}_f\left(B(p,r)\right) = \int_{B(p,r)} \mathrm{e}^{-f} dv$$

the weighted volume of the geodesic ball $B(p,r)$.

THEOREM 3.13. *Let $(M^n, g, \mathrm{e}^{-f} dv)$ be a weighted manifold. Assume that*

$$\mathrm{Vol}_f\left(B(p,r)\right) \leq Cr^2 \quad \textit{for all } r \geq r_0.$$

Then the weighted manifold is f-parabolic.

PROOF. Let u be the minimal barrier function defined by Theorem 3.7, so that $0 < u \leq 1$ on $M^n \setminus D$ and $u = 1$ on $\Sigma = \partial D$, where $D \subset M^n$ is a fixed domain. Since u is f-harmonic, it follows that

$$\Delta_f \ln u = -|\nabla \ln u|^2.$$

Let ϕ be the cutoff function with support in $B(p, r^2)$ defined by

$$\phi(x) := \begin{cases} 1 & \text{on } B(p,r), \\ \frac{1}{\ln r}\left(2\ln r - \ln d(p,x)\right) & \text{on } B\left(p, r^2\right) \setminus B(p,r), \\ 0 & \text{on } M^n \setminus B(p, r^2). \end{cases}$$

We have that

$$\begin{aligned} \int_{M^n \setminus D} |\nabla \ln u|^2 \phi^2 \mathrm{e}^{-f} dv &= -\int_{M^n\setminus D} \left(\Delta_f \ln u\right) \phi^2 \mathrm{e}^{-f} dv \\ &= \int_{M^n \setminus D} \left\langle \nabla \ln u, \nabla \phi^2 \right\rangle \mathrm{e}^{-f} dv \\ &\quad - \int_\Sigma \partial_\nu \left(\ln u\right) \mathrm{e}^{-f} da, \end{aligned}$$

where $\nu := \frac{\nabla u}{|\nabla u|}$ is the normal to $\Sigma = \partial\left(M^n \setminus D\right)$. Since $\partial_\nu \ln u \geq 0$, it follows that

$$\begin{aligned}\int_{M^n\setminus D} |\nabla \ln u|^2 \phi^2 \mathrm{e}^{-f} dv &\leq 2\int_{M^n\setminus D} \phi |\nabla \ln u| |\nabla \phi| \mathrm{e}^{-f} dv \\ &\leq \frac{1}{2}\int_{M^n\setminus D} |\nabla \ln u|^2 \phi^2 \mathrm{e}^{-f} dv \\ &\quad + 2\int_{M^n\setminus D} |\nabla \phi|^2 \mathrm{e}^{-f} dv.\end{aligned}$$

Therefore, we have proved that

$$\int_{M^n\setminus D} |\nabla \ln u|^2 \phi^2 \mathrm{e}^{-f} dv \leq 4\int_{M^n\setminus D} |\nabla \phi|^2 \mathrm{e}^{-f} dv. \tag{3.27}$$

Using the definition of ϕ and the co-area formula, we see that

$$\begin{aligned}\int_{M^n\setminus D} |\nabla \phi|^2 \mathrm{e}^{-f} dv &= \frac{1}{\ln^2 r}\int_{B(p,r^2)\setminus B(p,r)} \frac{1}{d^2} \mathrm{e}^{-f} dv \\ &= \frac{1}{\ln^2 r}\int_r^{r^2} \frac{1}{t^2}\left(\int_{\partial B(p,t)} \mathrm{e}^{-f} da\right) dt.\end{aligned}$$

Integrating the last term by parts and applying the weighted volume estimate

$$\mathrm{Vol}_f\left(B(p,t)\right) \leq Ct^2 \quad \text{for all } t \geq r_0,$$

we obtain that

$$\begin{aligned}\int_r^{r^2} \frac{1}{t^2}\left(\int_{\partial B(p,t)} \mathrm{e}^{-f} da\right) dt &= \frac{1}{r^4}\int_{B(p,r^2)} \mathrm{e}^{-f} dv - \frac{1}{r^2}\int_{B(p,r)} \mathrm{e}^{-f} dv \\ &\quad + \int_r^{r^2} \frac{2}{t^3}\left(\int_{B(p,t)} \mathrm{e}^{-f} dv\right) dt \\ &\leq C \ln r.\end{aligned}$$

Plugging this into (3.27), we conclude that

$$\int_{B(p,r)\setminus D} |\nabla \ln u|^2 \mathrm{e}^{-f} dv \leq \frac{C}{\ln r}.$$

Taking $r \to \infty$ implies that $\nabla \ln u = 0$ on $M^n \setminus D$; hence u is constant. This proves that the weighted manifold is f-parabolic. □

3.1.5. f-Nonparabolic and f-parabolic ends.

Our goal is to localize on each end the corresponding concepts on complete manifolds.

Recall from §1.4.2.1 that for a manifold M^n, an **end** E with respect to a compact subset $\Omega \subset M^n$ is an unbounded connected component of $M^n \setminus \Omega$. We will always assume that ∂E is smooth.

We distinguish two types of ends. Motivated by Theorem 3.7, given an end E of M^n with smooth boundary ∂E, we consider the sequence of functions $\{u_i\}$

defined by

$$\begin{aligned} \Delta_f u_i &= 0 \quad \text{on } E(p, r_i), \\ u_i &= 0 \quad \text{on } \partial E(p, r_i), \\ u_i &= 1 \quad \text{on } \partial E, \end{aligned} \tag{3.28}$$

where $r_i \to \infty$ and where we denote

$$\begin{aligned} E(p,r) &:= E \cap B(p,r), \\ \partial E(p,r) &:= E \cap \partial B(p,r), \end{aligned}$$

for r sufficiently large so that $\partial E \subset B(p,r)$. The maximum principle implies that

$$0 \leq u_i \leq 1 \quad \text{on } E(p, r_i).$$

By Proposition 3.6, a subsequence of $\{u_i\}$ converges uniformly on compact subsets of E to an f-harmonic function u on E.

DEFINITION 3.14. The f-harmonic function $u : E \to (0,1]$ defined as the limit of $\{u_i\}$ in (3.28) is called the **minimal barrier function** on the end E.

Motivated by Theorem 3.7, we define when an end is f-nonparabolic.

DEFINITION 3.15. We say that an end E is **f-parabolic** if the minimal barrier function on E is constant, i.e., $u = 1$ on E. Otherwise, if the minimal barrier function is nonconstant, then we say that E is **f-nonparabolic**.

When E is f-nonparabolic, we can follow (3.16) to see that

$$\liminf_{x \to \infty,\, x \in E} u(x) = 0.$$

Regarding applications to Ricci solitons, we will be interested mostly in f-nonparabolic manifolds. By applying the maximum principle argument as in Theorem 3.7, we obtain the following result:

PROPOSITION 3.16. *Let $(M^n, g, \mathrm{e}^{-f} dv)$ be f-nonparabolic and let E be an end of M^n. Then the minimal barrier function u of E satisfies*

$$\frac{1}{C} G_f(p,x) \leq u(x) \leq C G_f(p,x) \quad \text{for all } x \in E,$$

where $C > 0$ is a constant.

In particular, along any f-nonparabolic end E_1, there exists a sequence $x_i \to \infty$ such that $G_f(p, x_i) \to 0$, and along any f-parabolic end E_2, the weighted Green's function has a positive lower bound.

Note also that by (3.22) and Proposition 3.16, an f-nonparabolic manifold admits at least one f-nonparabolic end.

An f-nonparabolic end admits a positive non-constant and bounded f-harmonic function. At a first glance, an f-parabolic end seems to lack similar information, because the limiting f-harmonic function is constant. However, the next result shows that this is not the case; cf. Li and Tam [**LT92**].

PROPOSITION 3.17 (Nontrivial f-harmonic function on an f-parabolic end). *Let E be an f-parabolic end of $\left(M^n, g, \mathrm{e}^{-f} dv\right)$. Then there exists a non-negative f-harmonic function $w : E \to [0, \infty)$ such that $w = 0$ on ∂E and*

$$\limsup_{x \to \infty} w(x) = \infty. \tag{3.29}$$

Moreover, $w+1$ has finite Dirichlet energy for its logarithm:

$$\int_E |\nabla \ln(w+1)|^2 \mathrm{e}^{-f} dv < \infty. \tag{3.30}$$

PROOF. For the sequence $\{u_i\}$ defined in (3.28), we have that $u_i \to 1$ uniformly on compact subsets. Fix some r_1 large enough so that $\partial E(p, r_1)$ is well-defined and let

$$C_i := \frac{1}{1 - \min_{\partial E(p,r_1)} u_i}. \tag{3.31}$$

Note that $u_i(x) < 1$ for $x \in \partial E(p, r_1)$ by the strong maximum principle, so C_i is well-defined. Since $u_i \to 1$ on $\partial E(p, r_1)$, it follows that $C_i \to \infty$.

Define the sequence $\{w_i\}$ of f-harmonic functions by

$$w_i := C_i(1 - u_i).$$

Clearly, w_i solves the Dirichlet problem

$$\begin{aligned} \Delta_f w_i &= 0 \quad \text{on } E(p, r_i), \\ w_i &= C_i \quad \text{on } \partial E(p, r_i), \\ w_i &= 0 \quad \text{on } \partial E, \end{aligned} \tag{3.32}$$

for each i. According to (3.31),

$$\max_{\partial E(p,r_1)} w_i = 1.$$

For any $r \leq \frac{r_i}{2}$, the Harnack estimate (3.9) in Theorem 3.4 implies a uniform C^0-bound for $\{w_i\}$ on $E(p,r) \setminus E(p, \frac{r_1}{2})$, which by (3.8) implies that

$$|\nabla w_i| \leq C(r) \quad \text{on } E(p,r) \setminus E(p, \tfrac{r_1}{2}).$$

We claim that there exists $C > 0$ independent of i such that

$$A_i := \sup_{E(p,r_1)} |\nabla w_i| \leq C. \tag{3.33}$$

Indeed, if this is not true, then $A_i \to \infty$ as $i \to \infty$. The functions

$$\theta_i := \frac{w_i}{A_i}$$

are bounded uniformly in i on $E(p,r)$ in the C^1-norm. Theorem 3.4 and the Arzelà–Ascoli theorem imply that a subsequence of $\{\theta_i\}$ converges to an f-harmonic function $\theta \geq 0$ on E. Moreover,

$$\theta = 0 \quad \text{on } \partial E(p, r_1) \quad \text{and} \quad \sup_{E(p,r_1)} |\nabla \theta| = 1.$$

This is a contradiction. In conclusion, (3.33) holds true.

Applying Theorem 3.4 and the Arzelà–Ascoli theorem, we infer that a subsequence of $\{w_i\}$ converges to a non-constant f-harmonic function $w \geq 0$ on E so that $w = 0$ on ∂E. Extending w to be zero on $M^n \setminus E$, we obtain an f-subharmonic function on M^n. Corollary 3.12, applied to $C - w$ for $C > 0$ arbitrary, implies that w is unbounded from above on E.

We now prove the integral estimate (3.30). By (3.32), we have that

$$\Delta_f \ln(w_i + 1) = -|\nabla \ln(w_i + 1)|^2.$$

Hence

$$\begin{aligned}\int_{E(p,r_i)} |\nabla \ln(w_i+1)|^2 \mathrm{e}^{-f} dv &= -\int_{E(p,r_i)} \Delta_f \ln(w_i+1)\mathrm{e}^{-f} dv \\ &= -\int_{\partial E(p,r_i)} \frac{\partial}{\partial \nu} \ln(w_i+1)\mathrm{e}^{-f} da \\ &\quad + \int_{\partial E} \frac{\partial}{\partial \nu} \ln(w_i+1)\mathrm{e}^{-f} da,\end{aligned}$$

where $\frac{\partial}{\partial \nu} = \frac{\nabla w_i}{|\nabla w_i|}$ is the unit normal to the level set of w_i. As

$$\int_{\partial E(p,r_i)} \frac{\partial}{\partial \nu} \ln(w_i+1)\mathrm{e}^{-f} da = \int_{\partial E(p,r_i)} \frac{|\nabla w_i|}{w_i+1} \mathrm{e}^{-f} da \geq 0$$

and $w_i = 0$ on ∂E, it follows that

$$\int_{E(p,r_i)} |\nabla \ln(w_i+1)|^2 \mathrm{e}^{-f} dv \leq \int_{\partial E} |\nabla w_i| \mathrm{e}^{-f} da.$$

For $r > 0$ arbitrary, we let i be large enough so that $r_i > r$ and obtain from above that

$$\int_{E(p,r)} |\nabla \ln(w_i+1)|^2 \mathrm{e}^{-f} dv \leq \int_{\partial E} |\nabla w_i| \mathrm{e}^{-f} da.$$

Taking $i \to \infty$, we conclude that

$$\int_{E(p,r)} |\nabla \ln(w+1)|^2 \mathrm{e}^{-f} dv \leq \int_{\partial E} |\nabla w| \mathrm{e}^{-f} da.$$

As r is arbitrary, this proves the result. □

Although it will not be needed in this book, in the special case where f is constant, a result stronger than Proposition 3.17 asserts that an end E is parabolic if and only if it admits an *Evans potential*, i.e., a harmonic function ζ on E such that $\zeta = 0$ on ∂E and $\lim_{x\to\infty} \zeta(x) = \infty$. See Nakai [**Nak62**] for surfaces and Hansen and Netuka [**HN13**] for arbitrary dimensions.

The following result corresponds to Corollary 3.12.

PROPOSITION 3.18. *An end E of a weighted manifold $(M^n, g, \mathrm{e}^{-f} dv)$ is f-parabolic if and only if any positive f-superharmonic function on E achieves its infimum on ∂E.*

PROOF. Assume first that E is an f-parabolic end and $\eta > 0$ is f-superharmonic; that is to say that

$$\Delta_f \eta \leq 0 \quad \text{on } E.$$

By the maximum principle, η achieves its infimum value either on ∂E or at infinity on E. Assume for a contradiction that η achieves its infimum value at infinity of E. By subtracting this value from η if necessary, we may assume that

$$\eta > 0 \quad \text{on } E \quad \text{and} \quad \liminf_{x\to\infty,\, x\in E} \eta(x) = 0. \tag{3.34}$$

Consider the sequence $\{u_i\}$ defined in (3.28). Applying the maximum principle to the f-superharmonic function $\frac{\eta}{\inf_{\partial E}\eta} - u_i$, which is nonnegative on ∂E and on $\partial E(p, r_i)$, it follows that

$$u_i \leq \frac{\eta}{\inf_{\partial E}\eta} \quad \text{on } E(p, r_i).$$

Taking the limit as $i \to \infty$, we have that u satisfies the same estimate; hence u is not constant in view of (3.34). This contradicts our assumption that E is f-parabolic. In conclusion, η achieves its infimum on ∂E.

We now prove the converse. Assume that any f-superharmonic function on E achieves its infimum on ∂E. Let u be the minimal barrier function on E, defined as the limit of $\{u_i\}$ in (3.28). Then $u = 1$ on ∂E and $\Delta_f u = 0$ on E. Therefore, u must be constant by our assumption that any f-superharmonic function on E achieves its infimum on ∂E. This proves that E is f-parabolic. □

3.1.6. Manifolds with positive spectrum.

The behavior of the minimal barrier functions is better understood on the class of manifolds satisfying a weighted Poincaré inequality.

DEFINITION 3.19. We say that a weighted manifold $\left(M^n, g, \mathrm{e}^{-f} dv\right)$ has **positive spectrum** if

$$\lambda_1\left(\Delta_f\right) := \inf_{\phi \in C_0^\infty(M^n)} \frac{\int_{M^n} |\nabla \phi|^2 \mathrm{e}^{-f} dv}{\int_{M^n} \phi^2 \mathrm{e}^{-f} dv} > 0. \tag{3.35}$$

Under the assumption above, we have the following weighted Poincaré inequality:

$$\int_{M^n} \phi^2 \mathrm{e}^{-f} dv \le \frac{1}{\lambda_1(\Delta_f)} \int_{M^n} |\nabla \phi|^2 \mathrm{e}^{-f} dv \tag{3.36}$$

for all $\phi \in C_0^\infty(M^n)$.

Following Li and Wang [**LW01**], we establish an L^2-decay estimate for the minimal barrier function in Definition 3.14.

THEOREM 3.20 (Weighted L^2-decay estimate for barrier functions). *If $\left(M^n, g, \mathrm{e}^{-f} dv\right)$ is a weighted manifold with positive spectrum, then:*

1. *$\left(M^n, g, \mathrm{e}^{-f} dv\right)$ is f-nonparabolic.*
2. *For any end E of M^n and for u the minimal barrier function on E, we have*

$$\int_{E \setminus E(p,r)} u^2 \mathrm{e}^{-f} dv + \int_{E \setminus E(p,r)} |\nabla u|^2 \mathrm{e}^{-f} dv \le C \mathrm{e}^{-2\sqrt{\lambda_1(\Delta_f)}\, r},$$

where $C > 0$ is a constant independent of r.

PROOF.

STEP 1: *$\left(M^n, g, \mathrm{e}^{-f} dv\right)$ is f-nonparabolic.*

For $D \subset M^n$ a fixed compact subset, consider the sequence $\{\zeta_i\}$ defined as in Theorem 3.7 by

$$\begin{aligned} \Delta_f \zeta_i &= 0 \quad \text{on } B(p, r_i) \setminus D, \\ \zeta_i &= 0 \quad \text{on } \partial B(p, r_i), \\ \zeta_i &= 1 \quad \text{on } \Sigma = \partial D, \end{aligned}$$

where $r_i \to \infty$. Extend ζ_i to M^n by setting $\zeta_i = 1$ on D and $\zeta_i = 0$ on $M^n \backslash B(p, r_i)$. Then $\zeta_i : M^n \to \mathbb{R}$ is a test function for (3.35), so we have

$$\begin{aligned}\lambda_1(\Delta_f) \int_{M^n} \zeta_i^2 \mathrm{e}^{-f} dv &\le \int_{M^n} |\nabla \zeta_i|^2 \mathrm{e}^{-f} dv \\ &= \int_{B(p,r_i)\backslash D} |\nabla \zeta_i|^2 \mathrm{e}^{-f} dv \\ &= \int_\Sigma \frac{\partial \zeta_i}{\partial \nu} \mathrm{e}^{-f} da,\end{aligned}$$

where $\frac{\partial}{\partial \nu} = \frac{\nabla \zeta_i}{|\nabla \zeta_i|}$ is the unit normal to $\Sigma = \partial(M^n \setminus D)$.

Now let ζ be the limit of ζ_i, which exists by Theorem 3.7. Taking the limit as $i \to \infty$, we conclude that

$$\lambda_1(\Delta_f) \int_{M^n} \zeta^2 \mathrm{e}^{-f} dv \le \int_\Sigma |\nabla \zeta| \mathrm{e}^{-f} da.$$

In particular, ζ is not constant, and so $\left(M^n, g, \mathrm{e}^{-f} dv\right)$ is f-nonparabolic by Theorem 3.7.

STEP 2: *Weighted L^2-decay for u.*

Let E be an end and let u be the limit of the sequence $\{u_i\}$ in (3.28). Let $r_0 > 0$ be large enough so that $E(p, r_0) = E \cap B(p, r_0)$ is well-defined. For $t > r_0$ arbitrary let $r(x) := d(p, x)$ and define (similarly to Hua and Lu [**HL23**])

$$h(x) := \begin{cases} \sqrt{\lambda_1(\Delta_f)}\, r(x) & \text{for } 0 \le r(x) \le t, \\ \sqrt{\lambda_1(\Delta_f)}\, t & \text{for } r(x) > t. \end{cases} \tag{3.37}$$

We let ψ be a cutoff function so that $\psi = 1$ on $E \setminus E(p, r_0)$ and $\psi = 0$ on $M^n \setminus E$. Then ψu_i has compact support in M^n and hence we have

$$\begin{aligned}\lambda_1(\Delta_f) \int_{M^n} \psi^2 u_i^2 \mathrm{e}^{2h} \mathrm{e}^{-f} dv &\le \int_{M^n} \left|\nabla\left(\psi u_i \mathrm{e}^h\right)\right|^2 \mathrm{e}^{-f} dv \\ &= C_i + \int_{M^n} \psi^2 \left|\nabla\left(u_i \mathrm{e}^h\right)\right|^2 \mathrm{e}^{-f} dv,\end{aligned} \tag{3.38}$$

where

$$C_i := \int_{E(p,r_0)} |\nabla \psi|^2 u_i^2 \mathrm{e}^{2h} \mathrm{e}^{-f} dv + \frac{1}{2} \int_{E(p,r_0)} \left\langle \nabla \psi^2, \nabla\left(u_i^2 \mathrm{e}^{2h}\right)\right\rangle \mathrm{e}^{-f} dv.$$

As u_i converges uniformly on compact subsets of E to u, it follows that $C_i \le C$, for a constant $C > 0$ independent of i. We expand the right-hand side of (3.38) as

$$\begin{aligned}\int_{M^n} \psi^2 \left|\nabla\left(u_i \mathrm{e}^h\right)\right|^2 \mathrm{e}^{-f} dv = &\int_{M^n} \psi^2 |\nabla u_i|^2 \mathrm{e}^{2h} \mathrm{e}^{-f} dv \\ &+ \int_{M^n} \psi^2 \left|\nabla \mathrm{e}^h\right|^2 u_i^2 \mathrm{e}^{-f} dv \\ &+ \frac{1}{2} \int_{M^n} \psi^2 \left\langle \nabla u_i^2, \nabla \mathrm{e}^{2h}\right\rangle \mathrm{e}^{-f} dv.\end{aligned}$$

Integrating by parts, we obtain

$$\begin{aligned}\frac{1}{2}\int_{M^n}\psi^2\left\langle\nabla u_i^2,\nabla \mathrm{e}^{2h}\right\rangle \mathrm{e}^{-f}dv &= -\frac{1}{2}\int_{M^n}\psi^2\left(\Delta_f u_i^2\right)\mathrm{e}^{2h}\mathrm{e}^{-f}dv\\ &\quad -\frac{1}{2}\int_{M^n}\left\langle\nabla u_i^2,\nabla\psi^2\right\rangle \mathrm{e}^{2h}\mathrm{e}^{-f}dv\\ &\leq -\int_{M^n}\psi^2\left|\nabla u_i\right|^2\mathrm{e}^{2h}\mathrm{e}^{-f}dv + C,\end{aligned}$$

where $C > 0$ is independent of i. For the last line, we used that u_i is f-harmonic, so $\Delta_f u_i^2 = 2|\nabla u_i|^2$, and that

$$-\frac{1}{2}\int_{M^n}\left\langle\nabla u_i^2,\nabla\psi^2\right\rangle \mathrm{e}^{2h}\mathrm{e}^{-f}dv = -\frac{1}{2}\int_{E(p,r_0)}\left\langle\nabla u_i^2,\nabla\psi^2\right\rangle \mathrm{e}^{2h}\mathrm{e}^{-f}dv \leq C.$$

Hence, it follows that

$$\lambda_1\left(\Delta_f\right)\int_{M^n}\psi^2 u_i^2\mathrm{e}^{2h}\mathrm{e}^{-f}dv \leq \int_{M^n}\left|\nabla h\right|^2 u_i^2\mathrm{e}^{2h}\mathrm{e}^{-f}dv + C.$$

According to (3.37), $|\nabla h| = \sqrt{\lambda_1(\Delta_f)}$ on $E(p,t)$ and $|\nabla h| = 0$ on $E\setminus E(p,t)$. As $\psi = 1$ on $E\setminus E(p,r_0)$, it follows that

$$\lambda_1\left(\Delta_f\right)\int_{E(p,r_i)\setminus E(p,r_0)} u_i^2\mathrm{e}^{2h}\mathrm{e}^{-f}dv \leq \lambda_1\left(\Delta_f\right)\int_{E(p,t)} u_i^2\mathrm{e}^{2h}\mathrm{e}^{-f}dv + C.$$

In conclusion, we have proved that

$$\lambda_1\left(\Delta_f\right)\int_{E(p,r_i)\setminus E(p,t)} u_i^2\mathrm{e}^{2\sqrt{\lambda_1(\Delta_f)}\,t}\mathrm{e}^{-f}dv \leq C,$$

which in particular implies that

$$\int_{E(p,t+1)\setminus E(p,t)} u_i^2\mathrm{e}^{-f}dv \leq C\mathrm{e}^{-2\sqrt{\lambda_1(\Delta_f)}\,t}$$

for all $t < r_i - 1$. As $C > 0$ is independent of i, taking $i\to\infty$, we conclude that

$$\int_{E(p,t+1)\setminus E(p,t)} u^2\mathrm{e}^{-f}dv \leq C\mathrm{e}^{-2\sqrt{\lambda_1(\Delta_f)}\,t}, \tag{3.39}$$

where $C > 0$ is independent of t. This proves that

$$\int_{E\setminus E(p,r)} u^2\mathrm{e}^{-f}dv \leq C\mathrm{e}^{-2\sqrt{\lambda_1(\Delta_f)}\,r}.$$

STEP 3: *Weighted energy decay for u.*

To establish the corresponding energy decay estimate, we now let

$$\varphi(x) := \begin{cases} 1 & \text{for } t-1\leq r(x)\leq t,\\ t+1-r(x) & \text{for } t< r(x)\leq t+1,\\ r(x)-(t-2) & \text{for } t-2\leq r(x)< t-1,\\ 0 & \text{otherwise.}\end{cases}$$

We use integration by parts and the equation $\Delta_f u = 0$ to obtain that

$$\begin{aligned}\int_E |\nabla u|^2 \mathrm{e}^{2\sqrt{\lambda_1(\Delta_f)}r}\varphi^2 \mathrm{e}^{-f} dv &= -\int_E u \left\langle \nabla u, \nabla\left(\mathrm{e}^{2\sqrt{\lambda_1(\Delta_f)}r}\varphi^2\right)\right\rangle \mathrm{e}^{-f} dv \\ &\leq 2\sqrt{\lambda_1(\Delta_f)}\int_E u|\nabla u|\mathrm{e}^{2\sqrt{\lambda_1(\Delta_f)}r}\varphi^2\mathrm{e}^{-f}dv \\ &\quad + 2\int_E u|\nabla u|\mathrm{e}^{2\sqrt{\lambda_1(\Delta_f)}r}\varphi|\nabla\varphi|\mathrm{e}^{-f}dv \\ &\leq \frac{1}{2}\int_E |\nabla u|^2 \mathrm{e}^{2\sqrt{\lambda_1(\Delta_f)}r}\varphi^2\mathrm{e}^{-f}dv \\ &\quad + 4\lambda_1(\Delta_f)\int_E u^2\mathrm{e}^{2\sqrt{\lambda_1(\Delta_f)}r}\varphi^2\mathrm{e}^{-f}dv \\ &\quad + 4\int_E u^2\mathrm{e}^{2\sqrt{\lambda_1(\Delta_f)}r}|\nabla\varphi|^2\mathrm{e}^{-f}dv.\end{aligned}$$

By (3.39), it follows that

$$\int_{E(p,t)\setminus E(p,t-1)} |\nabla u|^2 \mathrm{e}^{2\sqrt{\lambda_1(\Delta_f)}r}\mathrm{e}^{-f}dv \leq C.$$

This proves that

$$\int_{E\setminus E(p,r)} |\nabla u|^2 \mathrm{e}^{-f}dv \leq C\mathrm{e}^{-2\sqrt{\lambda_1(\Delta_f)}r},$$

which proves the theorem.

□

The **weighted volume** is defined by

$$\mathrm{Vol}_f(S) := \int_S \mathrm{e}^{-f}dv$$

for any measurable subset S of M^n. We obtain the following useful consequence regarding the weighted volume of ends, see Li and Wang [**LW01**]:

THEOREM 3.21 (Weighted volume estimates). *Assume that the weighted manifold $(M^n, g, \mathrm{e}^{-f}dv)$ has positive spectrum and let E be an end of M^n.*

(1) *If E is f-nonparabolic, then it has exponential weighted volume growth:*

$$\mathrm{Vol}_f(E(p,r)) \geq C\mathrm{e}^{2\sqrt{\lambda_1(\Delta_f)}r},$$

for a constant $C > 0$ independent of r.

(2) *If E is f-parabolic, then it has exponential weighted volume decay at infinity:*

$$\mathrm{Vol}_f(E \setminus E(p,r)) \leq C\mathrm{e}^{-2\sqrt{\lambda_1(\Delta_f)}r},$$

for a constant $C > 0$ independent of r.

In particular, E is f-parabolic if and only if it has finite weighted volume.

PROOF. Assume first that E is f-parabolic. By Theorem 3.20, we get, as $u = 1$ on E, that

$$\int_{E\setminus E(p,r)} \mathrm{e}^{-f}dv \leq C\mathrm{e}^{-2\sqrt{\lambda_1(\Delta_f)}r}.$$

This proves that E has exponential weighted volume decay at infinity as claimed in the theorem.

We now prove that if E is f-nonparabolic, then it has exponential weighted volume growth. On one hand, by Theorem 3.20 we know that

$$\int_{E\setminus E(p,r)} |\nabla u|^2 \, e^{-f} dv \le C e^{-2\sqrt{\lambda_1(\Delta_f)}r}. \tag{3.40}$$

On the other, as u is f-harmonic,

$$\begin{aligned}\int_{\partial E(p,t)} \frac{\partial u}{\partial r} e^{-f} da &= -\int_{\partial E} \frac{\partial u}{\partial \nu} e^{-f} da + \int_{E(p,t)} (\Delta_f u)\, e^{-f} dv \\ &= -\int_{\partial E} \frac{\partial u}{\partial \nu} e^{-f} da,\end{aligned}$$

where $\frac{\partial}{\partial \nu} = \frac{\nabla u}{|\nabla u|}$ is the outer unit normal to ∂E. Moreover, by the Hopf lemma (see Gilbarg and Trudinger [**GT01**]), we know that

$$\frac{\partial u}{\partial \nu} = |\nabla u| > 0 \quad \text{on } \partial E.$$

Therefore, denoting

$$C := \int_{\partial E} |\nabla u| e^{-f} da > 0,$$

it follows that

$$\begin{aligned} C &= -\int_{\partial E(p,t)} \frac{\partial u}{\partial r} e^{-f} da \\ &\le \int_{\partial E(p,t)} |\nabla u| \, e^{-f} da \\ &\le \left(\int_{\partial E(p,t)} |\nabla u|^2 \, e^{-f} da \right)^{\frac{1}{2}} \operatorname{Area}_f \left(\partial E\left(p,t\right)\right)^{\frac{1}{2}}. \end{aligned}$$

Hence, there exists $C > 0$ so that

$$\frac{1}{\operatorname{Area}_f \left(\partial E\left(p,t\right)\right)} \le C \int_{\partial E(p,t)} |\nabla u|^2 e^{-f} da.$$

The co-area formula and (3.40) now imply that

$$\begin{aligned}\int_{r-1}^{r} \frac{dt}{\operatorname{Area}_f \left(\partial E\left(p,t\right)\right)} &\le C \int_{r-1}^{r} \left(\int_{\partial E(p,t)} |\nabla u|^2 \, e^{-f} da \right) dt \\ &= C \int_{E(p,r)\setminus E(p,r-1)} |\nabla u|^2 \, e^{-f} dv \\ &\le C e^{-2\sqrt{\lambda_1(\Delta_f)}r}.\end{aligned}$$

By the Cauchy–Schwarz inequality, we have

$$\begin{aligned} 1 &\le \left(\int_{r-1}^{r} \operatorname{Area}_f \left(\partial E\left(p,t\right)\right) dt \right) \left(\int_{r-1}^{r} \frac{dt}{\operatorname{Area}_f \left(\partial E\left(p,t\right)\right)} \right) \\ &\le C e^{-2\sqrt{\lambda_1(\Delta_f)}r} \int_{r-1}^{r} \operatorname{Area}_f \left(\partial E\left(p,t\right)\right) dt \\ &\le C e^{-2\sqrt{\lambda_1(\Delta_f)}r} \operatorname{Vol}_f \left(E\left(p,r\right)\right). \end{aligned}$$

In conclusion, we have

$$\mathrm{Vol}_f\left(E(p,r)\right) \geq C\mathrm{e}^{2\sqrt{\lambda_1(\Delta_f)}r}$$

for any f-nonparabolic end E. This proves the theorem. □

It is possible to obtain an L^1-estimate for the minimal barrier function of an end by applying the Hölder inequality and Theorem 3.20; however, such an estimate depends additionally on the weighted volume growth of the manifold. The following argument, inspired by Li and Wang [**LW06**], avoids this additional complication, and can also be adapted to estimate the f-harmonic functions defined on f-parabolic ends by Proposition 3.17. Recall that the Bakry–Emery curvature tensor is defined by

$$\mathrm{Ric}_f = \mathrm{Ric} + \nabla^2 f.$$

THEOREM 3.22 (Weighted L^1 estimates). *Assume that a weighted manifold* $(M^n, g, \mathrm{e}^{-f}dv)$ *has positive spectrum, its Bakry–Emery curvature is bounded from below, and*

$$\sup_{M^n} |\nabla f| < \infty.$$

(1) *Let u be the minimal barrier function on an f-nonparabolic end E (see Definition* 3.14*). Then there exists $C > 0$ so that*

$$\int_{E(p,r)} u\mathrm{e}^{-f}dv \leq Cr \quad \text{for all } r > r_0,$$

where r_0 is large enough so that $E(p,r_0) = E \cap B(p,r_0)$ is well-defined.

(2) *Let w be the non-negative f-harmonic function defined by Proposition* 3.17 *on some f-parabolic end E. Then there exists $C > 0$ so that*

$$\int_{E(p,r)} w\mathrm{e}^{-f}dv \leq Cr \quad \text{for all } r > r_0,$$

where r_0 is large enough so that $E(p,r_0) = E \cap B(p,r_0)$ is well-defined.

PROOF. We first prove the statement concerning the minimal barrier function of an f-nonparabolic end E.

STEP 1: *Setting up the test function in the Poincaré inequality.*

For $\alpha < 1$ fixed, define a function χ that is constant on each level set of u by

$$\chi(x) := \begin{cases} 1 & \text{on } \{\alpha \leq u \leq 1\}, \\ \ln(u(x)) - \ln\left(\mathrm{e}^{-1}\alpha\right) & \text{on } \{\mathrm{e}^{-1}\alpha < u < \alpha\}, \\ 0 & \text{on } \{u \leq \mathrm{e}^{-1}\alpha\}. \end{cases}$$

Since u may not converge to zero uniformly at infinity, χ is not a cutoff function on E. For this reason, for any $t > r_0 + 1$, we let

$$\psi(x) := \begin{cases} d(p,x) - r_0 & \text{on } E\left(p, r_0+1\right) \setminus E\left(p, r_0\right), \\ 1 & \text{on } E\left(p,t\right) \setminus E\left(p, r_0+1\right), \\ t+1-d(p,x) & \text{on } E\left(p,t+1\right) \setminus E\left(p,t\right), \\ 0 & \text{otherwise.} \end{cases}$$

Consequently, $\phi := \chi\psi$ is now a function with compact support on E. By the weighted Poincaré inequality (3.36), it follows that

$$\begin{aligned}\lambda_1(\Delta_f)\int_{M^n}\phi^2 u e^{-f}dv &\le \int_{M^n}\left|\nabla\left(\phi u^{\frac{1}{2}}\right)\right|^2 e^{-f}dv\\&= \frac{1}{4}\int_{M^n}\phi^2|\nabla u|^2u^{-1}e^{-f}dv + \int_{M^n}u|\nabla\phi|^2e^{-f}dv\\&\quad + \frac{1}{2}\int_{M^n}\langle\nabla u,\nabla\phi^2\rangle e^{-f}dv\\&= \frac{1}{4}\int_{M^n}\phi^2|\nabla u|^2u^{-1}e^{-f}dv + \int_{M^n}u|\nabla\phi|^2e^{-f}dv,\end{aligned}\tag{3.41}$$

where in the last line we integrated by parts and used that $\Delta_f u = 0$.

STEP 2: *The co-area formula.*

To compute the integrals on the right-hand side of (3.41), we use the co-area formula

$$\begin{aligned}\int_{\{b_0<u<b_1\}}|\nabla u|^2u^{-1}e^{-f}dv &= \int_{b_0}^{b_1}\frac{1}{s}\left(\int_{\{u=s\}}|\nabla u|e^{-f}da\right)ds\\&= C\ln\left(\frac{b_1}{b_0}\right)\end{aligned}\tag{3.42}$$

for any $0<b_0<b_1\le 1$, where

$$C := \int_{\partial E}|\nabla u|e^{-f}da > 0.\tag{3.43}$$

The fact that $\int_{\{u=s\}}|\nabla u|e^{-f}da$ is independent of s follows from $\Delta_f u = 0$ by the argument in Li and Wang [**LW06**], which we now present.

Note that the level set $\{u=s\}$ may be noncompact in general; hence, define η to be the cutoff function

$$\eta(x) := \begin{cases}1 & \text{on } E(p,t),\\ t+1-d(p,x) & \text{on } E(p,t+1)\setminus E(p,t),\\ 0 & \text{on } E\setminus E(p,t+1).\end{cases}$$

For any $s<1$, we have

$$\begin{aligned}&\left|\int_{\partial E}|\nabla u|e^{-f}da - \int_{\{u=s\}}|\nabla u|\eta e^{-f}da\right|\\&\quad = \left|\int_{\{s<u<1\}}(\Delta_f u)\,\eta e^{-f}dv - \int_{\{s<u<1\}}\langle\nabla u,\nabla\eta\rangle e^{-f}dv\right|\\&\quad \le \int_{\{s<u<1\}\cap(E(p,t+1)\setminus E(p,t))}|\nabla u|e^{-f}dv.\end{aligned}$$

The gradient estimate in Theorem 3.4 implies that

$$|\nabla u|\le Cu \quad \text{on } E(p,t+1)\setminus E(p,t),$$

for a constant $C>0$ depending on the Bakry–Emery curvature lower bound and the upper bound for $|\nabla f|$. Therefore, by the decay estimate in Theorem 3.20 it follows that

$$\begin{aligned}\int_{\{s<u<1\}\cap(E(p,t+1)\setminus E(p,t))} |\nabla u|\mathrm{e}^{-f}\,dv &\leq C\int_{\{s<u<1\}\cap(E(p,t+1)\setminus E(p,t))} u\mathrm{e}^{-f}\,dv\\ &\leq \frac{C}{s}\int_{\{s<u<1\}\cap(E(p,t+1)\setminus E(p,t))} u^2\mathrm{e}^{-f}\,dv\\ &\leq \frac{C}{s}\int_{E(p,t+1)\setminus E(p,t)} u^2\mathrm{e}^{-f}\,dv\\ &\leq \frac{C}{s}\mathrm{e}^{-2\sqrt{\lambda_1(\Delta_f)}\,t}\end{aligned}$$

for a constant $C>0$. Therefore, this proves that

$$\left|\int_{\partial E}|\nabla u|\mathrm{e}^{-f}\,da-\int_{\{u=s\}}|\nabla u|\eta\mathrm{e}^{-f}\,da\right|\leq\frac{C}{s}\mathrm{e}^{-2\sqrt{\lambda_1(\Delta_f)}\,t}.$$

Taking $t\to\infty$ shows that the integral

$$\int_{\{u=s\}}|\nabla u|\mathrm{e}^{-f}\,da=\int_{\partial E}|\nabla u|\mathrm{e}^{-f}\,da=C$$

is independent of s as claimed.

STEP 3: *Estimating the terms in* (3.41).

The co-area formula (3.42) implies that

$$\begin{aligned}\int_{M^n}\phi^2\,|\nabla u|^2\,u^{-1}\mathrm{e}^{-f}\,dv &\leq \int_{\{\mathrm{e}^{-1}\alpha<u<1\}}|\nabla u|^2\,u^{-1}\mathrm{e}^{-f}\,dv\\ &= C\left(1+\ln(\alpha^{-1})\right).\end{aligned}\tag{3.44}$$

To estimate the second term of the right-hand side of (3.41), note that

$$\int_{M^n}u\,|\nabla\phi|^2\,\mathrm{e}^{-f}\,dv\leq 2\int_{M^n}u\,|\nabla\psi|^2\,\chi^2\mathrm{e}^{-f}\,dv+2\int_{M^n}u\,|\nabla\chi|^2\,\psi^2\mathrm{e}^{-f}\,dv.$$

Since $u>\mathrm{e}^{-1}\alpha$ on the support of χ, it follows that

$$\begin{aligned}\int_{M^n}u\,|\nabla\psi|^2\,\chi^2\mathrm{e}^{-f}\,dv &\leq \int_{E(p,r_0+1)\setminus E(p,r_0)}u\mathrm{e}^{-f}\,dv\\ &\quad+\int_{\{\mathrm{e}^{-1}\alpha<u<1\}\cap(E(p,t+1)\setminus E(p,t))}u\mathrm{e}^{-f}\,dv\\ &\leq C+\frac{\mathrm{e}}{\alpha}\int_{E(p,t+1)\setminus E(p,t)}u^2\mathrm{e}^{-f}\,dv\\ &\leq C+\frac{C}{\alpha}\mathrm{e}^{-2\sqrt{\lambda_1(\Delta_f)}\,t},\end{aligned}\tag{3.45}$$

where in the last line we have used Theorem 3.20. Furthermore, (3.42) yields

$$\begin{aligned}\int_{M^n}u\,|\nabla\chi|^2\,\psi^2\mathrm{e}^{-f}\,dv &\leq \int_{\{\mathrm{e}^{-1}\alpha<u<\alpha\}}|\nabla u|^2\,u^{-1}\mathrm{e}^{-f}\,dv\\ &\leq C.\end{aligned}$$

Therefore, we obtain that

$$\int_{M^n}u\,|\nabla\phi|^2\,\mathrm{e}^{-f}\,dv\leq\frac{C}{\alpha}\mathrm{e}^{-2\sqrt{\lambda_1(\Delta_f)}\,t}+C.$$

Together with (3.41) and (3.44), this implies that

$$\lambda_1(\Delta_f)\int_{M^n}\phi^2 u e^{-f}dv \le \frac{C}{\alpha}e^{-2\sqrt{\lambda_1(\Delta_f)}t} + C\left(1+\ln(\alpha^{-1})\right).$$

In particular, we obtain that

$$\lambda_1(\Delta_f)\int_{\{\alpha<u<1\}\cap E(p,t)} u e^{-f}dv \le \frac{C}{\alpha}e^{-2\sqrt{\lambda_1(\Delta_f)}t} + C\left(1+\ln(\alpha^{-1})\right).$$

This holds for any $t > r_0+1$. Letting $t\to\infty$, we conclude that

$$\int_{\{\alpha<u<1\}} u e^{-f}dv \le C\left(1+\ln(\alpha^{-1})\right) \tag{3.46}$$

for any $\alpha\in(0,1)$.

STEP 4: *Concluding the proof of the L^1 estimate.*

For $r > r_0$, define

$$\alpha := \inf_{E(p,r)} u.$$

According to Theorem 3.4, there exists $C>0$ such that $u(x)\ge e^{-Cd(p,x)}$. Therefore, α satisfies

$$\alpha^{-1}\le e^{Cr}.$$

By (3.46), this implies that

$$\begin{aligned}\int_{E(p,r)} u e^{-f}dv &\le \int_{\{\alpha<u<1\}} u e^{-f}dv\\ &\le C\left(1+\ln(\alpha^{-1})\right)\\ &\le Cr.\end{aligned}$$

This concludes the proof of the L^1-estimate for the minimal barrier function on an f-nonparabolic end.

The argument above can be readily adapted to the situation where w is the f-harmonic function defined by an f-parabolic end E by Proposition 3.17. By applying Theorem 3.21, we obtain the following bound corresponding to (3.45):

$$\begin{aligned}\int_{\{0<w<\beta\}\cap(E(p,t+1)\setminus E(p,t))} w e^{-f}dv &\le \beta\int_{E(p,t+1)\setminus E(p,t)} e^{-f}dv\\ &\le C\beta e^{-2\sqrt{\lambda_1(\Delta_f)}t}\end{aligned}$$

for any $\beta>0$. The rest of the argument is the same and proves the estimate

$$\int_{\{0<w<\beta\}} w e^{-f}dv \le C(1+\ln\beta) \tag{3.47}$$

corresponding to (3.46). Since Theorem 3.4 implies that $w(x)\le e^{Cd(p,x)}$, it follows that

$$E(p,r)\subset\{0<w<e^{Cr}\}.$$

Applying (3.47) for $\beta := e^{Cr}$ we conclude that

$$\int_{E(p,r)} w e^{-f}dv \le Cr.$$

This proves the theorem.

∎

We remark that similar methods were applied by Sung, Wang, and one of the authors [**MSW19**] to establish sharp Green's function estimates of the form

$$\int_{B(p,r)} G_f(x,y)\,\mathrm{e}^{-f(y)}dv(y) \leq C\,(1+r)\,,$$

for any weighted complete manifold $(M^n, g, \mathrm{e}^{-f}dv)$ with positive spectrum, $\mathrm{Ric}_f \geq -(n-1)\,Kg$, and $|\nabla f| \leq a$, with C depending only on n, K, a, and $\lambda_1(\Delta_f)$. Such estimates can prove to be particularly useful in solving the Poisson equation $\Delta_f u = \varphi$ under some appropriate decay conditions at infinity for the source φ.

3.1.7. f-Harmonic functions and ends.

Using f-harmonic functions, we can detect the topology at infinity of a manifold, that is, the number of ends. This is done by patching together local f-harmonic functions defined on each end. We will construct f-harmonic functions defined on M^n that on each end are asymptotically close to the local f-harmonic function defined by the respective end. This theory was developed by Li and Tam in [**LT92**]; here, we follow the exposition in Sung, Tam, and Wang [**STW00**].

Throughout this section, u_0 will denote the minimal nonconstant barrier function on an f-nonparabolic end, see Definition 3.14, and w_0 will denote the f-harmonic function constructed in Proposition 3.17 on an f-parabolic end. Theorem 3.23 below will enable us to extend u_0 and w_0 to global f-harmonic functions u and w on M^n, see Corollary 3.25.

We will focus on the case where M^n is f-nonparabolic, which is what we need for applications to gradient Ricci solitons. Recall that G_f denotes the symmetric Green's function of Δ_f on M^n. We have the following result:

THEOREM 3.23 (Global extensions of f-harmonic functions). *Suppose that $(M^n, g, \mathrm{e}^{-f}dv)$ is a weighted manifold that is f-nonparabolic. Assume that η is an f-harmonic function, defined on $M^n \backslash B(p, r_0)$, that is smooth up to $\partial B(p, r_0)$. Then there exists an f-harmonic function ζ defined on M^n and a constant $C > 0$ so that*

$$|\zeta - \eta|\,(x) \leq CG_f(p,x)\,.$$

Moreover, $\zeta - \eta$ has finite weighted Dirichlet energy and

$$\liminf_{x\to\infty} \eta(x) \leq \zeta \leq \limsup_{x\to\infty} \eta(x)\,.$$

PROOF. Let ϕ be a smooth non-negative function such that $\phi = 0$ on $B(p, r_0)$ and $\phi = 1$ on $M^n \setminus B(p, 2r_0)$. The function $\phi\eta$ is defined everywhere on M^n and is f-harmonic on $M^n \setminus B(p, 2r_0)$. We aim to perturb it to an f-harmonic function defined on M^n. To this end, let h be defined by

$$h := \Delta_f(\phi\eta)\,,$$

which has support in $B(p, 2r_0)$. By (3.3), a solution of the Poisson equation

$$\Delta_f \rho = h$$

is given by

$$\rho(x) := -\int_{M^n} G_f(x,y)\,h(y)\,\mathrm{e}^{-f(y)}dv(y). \tag{3.48}$$

Therefore, the function

$$\zeta := \phi\eta - \rho \tag{3.49}$$

satisfies

$$\Delta_f \zeta = h - \Delta_f \rho = 0 \quad \text{on } M^n.$$

We now check that

$$|\zeta - \eta|(x) \leq C G_f(p, x).$$

For $x \in M^n \setminus B(p, 3r_0)$, we have that $\phi(x)\eta(x) = \eta(x)$ and therefore $|\zeta - \eta|(x) = |\rho|(x)$ by (3.49). Using (3.48) and the fact that h has support in $B(p, 2r_0)$, we obtain that for any $x \in M^n \setminus B(p, 3r_0)$

$$\begin{aligned} |\zeta - \eta|(x) &= \left| \int_M G_f(x, y)\, h(y) \mathrm{e}^{-f(y)} dv(y) \right| \\ &= \left| \int_{B(p,2r_0)} G_f(x, y)\, \Delta_f(\phi\eta)(y)\, \mathrm{e}^{-f(y)} dv(y) \right| \\ &\leq \left(\sup_{y \in B(p,2r_0)} G_f(x, y) \right) \int_{B(p,2r_0)} |\Delta_f(\phi\eta)|(y)\, \mathrm{e}^{-f(y)} dv(y) \\ &\leq C \sup_{y \in B(p,2r_0)} G_f(x, y). \end{aligned} \tag{3.50}$$

Finally, since $G_f(x, \cdot)$ is f-harmonic on $B(p, 3r_0)$ for any $x \in M^n \setminus B(p, 3r_0)$, we can apply the gradient estimate from Theorem 3.4 to get that

$$\sup_{y \in B(p,2r_0)} G_f(x, y) \leq C G_f(x, p),$$

where $C > 0$ depends on the Bakry-Emery curvature lower bound and the $|\nabla f|$ upper bound on $B(p, 3r_0)$. Since G_f is symmetric, by using (3.50), we conclude that

$$|\zeta - \eta|(x) \leq C G_f(p, x) \quad \text{for all } x \in M^n \setminus B(p, 3r_0). \tag{3.51}$$

According to Theorem 3.10, $\zeta - \eta$ has finite weighted Dirichlet energy, i.e.,

$$\int_{M^n \setminus D} |\nabla(\zeta - \eta)|^2 \mathrm{e}^{-f} dv < \infty.$$

We now prove that

$$\liminf_{x \to \infty} \eta(x) < \zeta < \limsup_{x \to \infty} \eta(x). \tag{3.52}$$

Let $\{B(p, r_i)\}$ be a compact exhaustion of M^n and let G_f^i be the Green's function of $B(p, r_i)$ that satisfies the Dirichlet boundary condition. The functions

$$\zeta_i(x) := \phi(x)\eta(x) + \int_{M^n} G_f^i(x, y)\, \Delta_f(\phi\eta)(y)\, \mathrm{e}^{-f(y)} dv(y)$$

are f-harmonic on $B(p, r_i)$ and satisfy

$$\zeta_i = \eta \quad \text{on } \partial B(p, r_i).$$

By the maximum principle,

$$\inf_{\partial B(p,r_i)} \eta \leq \zeta_i \leq \sup_{\partial B(p,r_i)} \eta.$$

According to (3.48) and (3.49),

$$\zeta(x) = \phi(x)\,\eta(x) + \int_{M^n} G_f(x,y)\,\Delta_f(\phi\eta)(y)\,\mathrm{e}^{-f(y)}dv(y).$$

Consequently, $\zeta_i \to \zeta$ uniformly on compact subsets. We conclude that (3.52) holds. This proves the theorem. □

Theorem 3.23 is very useful in our context. Having more than one end allows us to construct f-harmonic functions on each end with prescribed asymptotic behavior, which we then glue together using Theorem 3.23.

THEOREM 3.24 (Counting topological ends with f-harmonic functions). *Suppose that $(M^n, g, \mathrm{e}^{-f}dv)$ is a weighted manifold that is f-nonparabolic. Assume that M^n has more than one end. Then there exist spaces of f-harmonic functions V and W such that*

(1) $\dim V$ *is equal to the number of f-nonparabolic ends of M^n, and each function in V is bounded and has finite weighted Dirichlet energy.*

(2) $\dim W$ *is equal to the number of f-parabolic ends of M^n, and each function in W is one-side bounded on each end of M^n.*

PROOF. Since M^n is f-nonparabolic, Proposition 3.16 implies that M^n admits at least one f-nonparabolic end. If this is the unique f-nonparabolic end of M^n, then we can define V as the constant function $\zeta = 1$.

Suppose now that M^n has more than one f-nonparabolic end. Then there exists an f-nonparabolic end E of M^n such that $M^n \setminus E$ is also f-nonparabolic. Define the function

$$\eta(x) := \begin{cases} 1 & \text{on } E, \\ 0 & \text{on } M^n \setminus E. \end{cases}$$

This function is f-harmonic on $M^n \setminus B(p, r_0)$ for some $r_0 > 0$ large enough. Consequently, Theorem 3.23 implies that there exists an f-harmonic function ζ on M^n satisfying

$$|\zeta - \eta|(x) \le CG_f(p,x) \quad \text{for all } x \in M^n, \tag{3.53}$$

such that ζ has finite weighted energy and $0 < \zeta < 1$. Since both E and $M^n \setminus E$ are f-nonparabolic, Proposition 3.16 and (3.53) imply that $\zeta(x_i) \to 1$ for some sequence $x_i \to \infty$ on E and $\zeta(y_i) \to 0$ for some sequence $y_i \to \infty$ on $M^n \setminus E$.

If $E_1, \ldots, E_k$ are f-nonparabolic ends of M^n, then we can repeat the process to obtain corresponding f-harmonic functions $\zeta_1, \ldots, \zeta_k$ satisfying (3.52). By Proposition 3.16 and (3.53), we know that ζ_i converges to 1 along a sequence in E_i, and for each $j \ne i$ converges to zero along a sequence in E_j. Hence, the functions $\{\zeta_i\}$ are linearly independent. This proves the first part of the theorem.

We now assume that E is an f-parabolic end, and hence $M^n \setminus E$ is f-nonparabolic. Denote with w_0 the f-harmonic function defined on E by Proposition 3.17. Let

$$\eta(x) := \begin{cases} w_0(x) & \text{on } E, \\ 0 & \text{on } M^n \setminus E. \end{cases}$$

By Theorem 3.23, there exists an f-harmonic function ξ on M^n such that

$$|\xi - \eta|(x) \le CG_f(p,x).$$

In particular, this means that

$$\limsup_{x\to\infty,\ x\in E} \xi(x) = \infty,$$

and ξ is bounded on $M^n \setminus E$.

If $E_1, \ldots, E_k$ are f-parabolic ends of M^n, then the construction above yields f-harmonic functions $\xi_1, \ldots, \xi_k$. Each of these functions is unbounded on E_k and bounded on $M^n \setminus E_k$. Hence, they are linearly independent. This proves the theorem. □

The following consequence of Theorem 3.24 is particularly useful in studying if a weighted manifold is connected at infinity (i.e., $M^n \setminus \Omega$ has one unbounded component for any compact subset $\Omega \subset M^n$).

COROLLARY 3.25. *Let $\left(M^n, g, \mathrm{e}^{-f} dv\right)$ be a weighted manifold that is f-nonparabolic. Assume that E is an f-nonparabolic end of M^n such that $M^n \setminus E$ is also an end of M^n.*

(1) *If $M^n \setminus E$ is f-nonparabolic, then there exists an f-harmonic function $0 < u < 1$ on M^n such that $\int_{M^n} |\nabla u|^2 \mathrm{e}^{-f} dv < \infty$ and*

$$\liminf_{x\in E,\ x\to\infty} u(x) = 0,$$
$$\limsup_{x\in M^n\setminus E,\ x\to\infty} u(x) = 1.$$

(2) *If $M^n \setminus E$ is f-parabolic, then there exists an f-harmonic function $w > 0$ on M^n such that*

$$\liminf_{x\in E,\ x\to\infty} w(x) = 0,$$
$$\limsup_{x\in M^n\setminus E,\ x\to\infty} w(x) = \infty.$$

Note that in the above corollary the roles of E and $M^n \setminus E$ are reversed compared to Theorem 3.24; this adjustment was made for convenience in exposition.

It is useful to record that for manifolds with positive spectrum we have additional information concerning the harmonic functions constructed in Corollary 3.25.

PROPOSITION 3.26 (Integral estimates for the f-harmonic functions defined by ends). *Assume that a weighted manifold $(M^n, g, \mathrm{e}^{-f} dv)$ has positive spectrum, its Bakry–Emery curvature is bounded from below, and*

$$\sup_{M^n} |\nabla f| < \infty.$$

Assume that E is an f-nonparabolic end of M^n such that $M^n \setminus E$ is also an end of M^n.

(1) *If $M^n \setminus E$ is f-nonparabolic, then the f-harmonic function u in Corollary 3.25 satisfies*

$$\int_{B(p,r)} |\nabla u| \mathrm{e}^{-f} dv \leq Cr \quad \text{for all } r > 1.$$

(2) *If $M^n \setminus E$ is f-parabolic, then the f-harmonic function w in Corollary* 3.25 *satisfies*

$$\int_{B(p,r)} (w + |\nabla w|)\, \mathrm{e}^{-f} dv \leq Cr \quad \textit{for all } r > 1.$$

PROOF. We prove the second statement first. Hence, let us assume that E is an f-nonparabolic end and $M^n \setminus E$ is f-parabolic. Let w be the positive f-harmonic function from Corollary 3.25, satisfying

$$\liminf_{x\in E,\ x\to\infty} w(x) = 0 \quad \text{and} \quad \limsup_{x\in M^n\setminus E,\ x\to\infty} w(x) = \infty.$$

According to Theorem 3.23,

$$|w - \eta|(x) \leq C G_f(p, x),$$

where

$$\eta(x) := \begin{cases} w_0\,(x) & \text{on } M^n \setminus E, \\ 0 & \text{on } E, \end{cases}$$

and w_0 is the f-harmonic function defined on $M^n \setminus E$ by Proposition 3.17.

Let u_0 be the minimal barrier function on E, see Definition 3.14. Proposition 3.16 implies that $G_f(p, x) \leq C u_0(x)$ for all $x \in E$; therefore

$$w \leq C u_0 \quad \text{on } E.$$

According to Theorem 3.22,

$$\int_{E(p,r)} u_0 \mathrm{e}^{-f} dv \leq Cr.$$

Therefore, it follows that

$$\int_{B(p,r)\cap E} w\mathrm{e}^{-f} dv \leq Cr \quad \text{for all } r > 1.$$

The corresponding integral of w on $B(p, r) \cap (M^n \setminus E)$ can be estimated similarly, using that

$$|w - w_0|(x) \leq C \quad \text{on } M^n \setminus E,$$

together with Theorems 3.21 and 3.22. This proves that

$$\int_{B(p,r)} w\mathrm{e}^{-f} dv \leq Cr \quad \text{for all } r > 1. \tag{3.54}$$

The gradient estimate Theorem 3.4 and (3.54) imply that

$$\int_{B(p,r)} |\nabla w|\mathrm{e}^{-f} dv \leq Cr$$

for all $r > 1$. This proves the result for w, i.e., one f-nonparabolic and one f-parabolic end.

Certainly, the same argument applies for the function u in Corollary 3.25 when both E and $M^n \setminus E$ are f-nonparabolic, and proves that

$$\int_{B(p,r)\cap E} u\mathrm{e}^{-f} dv \leq Cr \quad \text{and} \quad \int_{B(p,r)\cap(M^n\setminus E)} (1 - u)\, \mathrm{e}^{-f} dv \leq Cr.$$

Theorem 3.4 then implies that

$$\int_{B(p,r)} |\nabla u| \mathrm{e}^{-f} dv \leq Cr \quad \text{for all } r > 1.$$

□

3.2. Ends of steady gradient Ricci solitons

In this section we prove that any complete noncompact steady gradient Ricci soliton either has exactly one topological end or splits as the product of $\mathbb{R}$ and a compact Ricci-flat manifold (see Theorem 3.29 below). This solves the problem of counting ends for steady gradient Ricci solitons.

We start with a weighted volume estimate, which holds for a more general class of weighted manifolds. Recall that $\lambda_1(M)$ denotes the best constant in the Poincaré inequality; see Definition 3.19.

LEMMA 3.27. *Let $\left(M^n, g, \mathrm{e}^{-f} dv\right)$ be a weighted manifold with nonnegative Bakry–Emery curvature $\mathrm{Ric}_f \geq 0$. Assume that*

$$\sup_{M^n} |\nabla f| \leq a,$$

where $a \geq 0$. Then, for any $r \geq 1$, we have

$$\mathrm{Vol}_f \left(B\left(p, r\right)\right) \leq r^n \mathrm{e}^{ar} \mathrm{Vol}_f \left(B\left(p, 1\right)\right).$$

Moreover,

$$\lambda_1 \left(\Delta_f\right) \leq \frac{1}{4} a^2.$$

PROOF. Let γ be a minimizing geodesic from p to x such that $\gamma(0) = p$ and $\gamma(r) = x$. By the f-Laplacian comparison theorem, see Wei and Wylie [**WW09**] or (4.31) in Volume 1, we have

$$\begin{aligned} \Delta_f r &\leq \frac{n-1}{r} - \frac{2}{r^2} \int_0^r t f'(t)\, dt \\ &\leq \frac{n-1}{r} + a, \end{aligned} \tag{3.55}$$

where $f(t) := f(\gamma(t))$. In the last line, we used that

$$\begin{aligned} |f'(t)| &= \left| \left\langle \nabla f(\gamma(t)), \frac{d\gamma}{dt}(t) \right\rangle \right| \\ &\leq |\nabla f|(\gamma(t)) \\ &\leq a. \end{aligned}$$

Integrating (3.55) in r yields

$$\ln\left(\frac{J_f(p, r, \xi)}{J_f(p, t, \xi)}\right) \leq (n-1) \ln \frac{r}{t} + a(r - t) \tag{3.56}$$

for any $t < r$, where

$$J_f(p, r, \xi) = \mathrm{e}^{-f} J(p, r, \xi)$$

denotes the **weighted volume form** in geodesic coordinates (r, ξ) based at p (a.k.a. the ***f*-Jacobian**; cf. (4.34) in Volume 1). Integrating (3.56) over ξ implies the following weighted area estimate for geodesic spheres:

$$t^{n-1}\mathrm{Area}_f(\partial B(p,r)) \leq r^{n-1}\mathrm{e}^{ar}\mathrm{Area}_f(\partial B(p,t)),$$

for all $t \leq 1 \leq r$. Integrating in $0 < t < 1$ and in $r > 1$ proves the volume estimate.

We now prove the spectral estimate. If $\lambda_1(\Delta_f) = 0$, then there is nothing to prove. Therefore, we may assume that $\lambda_1(\Delta_f) > 0$. Theorem 3.20 implies that $(M^n, g, \mathrm{e}^{-f}dv)$ is f-nonparabolic. By Theorem 3.21, we have that

$$r^n\mathrm{e}^{ar}\mathrm{Vol}_f\left(B\left(p,1\right)\right) \geq \mathrm{Vol}_f\left(B\left(p,r\right)\right) \geq C\mathrm{e}^{2\sqrt{\lambda_1(\Delta_f)}\,r}.$$

Since $r \geq 1$ is arbitrary, it follows that

$$\lambda_1(\Delta_f) \leq \frac{a^2}{4}.$$

The lemma is proved. □

It is known (see (1.37)) that

$$|\nabla f|^2 + R = 1,$$

where R is the scalar curvature of M^n. Hence, a steady gradient Ricci soliton satisfies the following properties

$$\begin{aligned} |\nabla f|^2 + R &= 1, \\ \Delta f + R &= 0, \\ R &\geq 0. \end{aligned} \tag{3.57}$$

By Lemma 3.27 with $a = 1$, we have that

$$\lambda_1(\Delta_f) \leq \frac{1}{4}.$$

We will prove that this inequality is in fact an equality.

PROPOSITION 3.28. *Let (M^n, g, f) be a complete noncompact gradient steady Ricci soliton normalized as in* (3.57). *Then*

$$\lambda_1\left(\Delta_f\right) = \frac{1}{4}.$$

PROOF. According to (3.57), we have that

$$\Delta_f(f) = \Delta f - |\nabla f|^2 = -1.$$

For any function ϕ with compact support in M^n, it follows that

$$\begin{aligned} \int_{M^n} \phi^2\mathrm{e}^{-f}dv &= -\int_{M^n} \phi^2\Delta_f(f)\mathrm{e}^{-f}dv \\ &= \int_{M^n} \langle\nabla\phi^2, \nabla f\rangle\mathrm{e}^{-f}dv \\ &\leq 2\int_{M^n} \phi|\nabla\phi||\nabla f|\mathrm{e}^{-f}dv \\ &\leq \frac{1}{2}\int_{M^n} \phi^2\mathrm{e}^{-f}dv + 2\int_{M^n} |\nabla f|^2|\nabla\phi|^2\mathrm{e}^{-f}dv. \end{aligned}$$

This proves that

$$\frac{1}{4}\int_{M^n} \phi^2\mathrm{e}^{-f}dv \leq \int_{M^n} |\nabla f|^2|\nabla\phi|^2\mathrm{e}^{-f}dv.$$

Since $|\nabla f| \leq 1$ on M^n, it follows that

$$\frac{1}{4}\int_{M^n} \phi^2 \mathrm{e}^{-f} dv \leq \int_{M^n} |\nabla \phi|^2 \mathrm{e}^{-f} dv$$

for any $\phi \in C_0^\infty(M^n)$. This proves that $\lambda_1(\Delta_f) \geq \frac{1}{4}$. By Lemma 3.27, we conclude that

$$\lambda_1(\Delta_f) = \frac{1}{4}.$$

This proves the result. □

We are now ready to prove the splitting theorem for steady gradient Ricci solitons with more than one end. This result is due to Wang and one of the authors [**MW11**, Theorem 4.2].

THEOREM 3.29 (Splitting theorem for steady solitons). *Let (M^n, g, f) be a complete noncompact steady gradient Ricci soliton. If M^n has more than one end, then (M^n, g) is isometric to $\mathbb{R} \times N^{n-1}$ for some compact Ricci-flat manifold N^{n-1}.*

PROOF. Assume that M^n has at least two ends. Proposition 3.28 and Theorem 3.20 imply that (M^n, g, f) is f-nonparabolic. Hence, at least one end of M^n is f-nonparabolic. We distinguish the following two cases:

CASE 1: *Two or more f-nonparabolic ends.*

If M^n has at least two f-nonparabolic ends, then there exists an f-nonparabolic end E of M^n such that $M^n \setminus E$ is also an f-nonparabolic end. Corollary 3.25 implies that M^n admits a nonconstant bounded f-harmonic function u with finite weighted energy $\int_{M^n} |\nabla u|^2 \mathrm{e}^{-f} dv < \infty$. According to the Bochner formula, we have that

$$\begin{aligned}\frac{1}{2}\Delta_f |\nabla u|^2 &= |\nabla^2 u|^2 + \langle \nabla(\Delta_f u), \nabla u\rangle + \mathrm{Ric}_f(\nabla u, \nabla u)\\ &= |\nabla^2 u|^2.\end{aligned}$$

Let ψ be the cutoff function defined by

$$\psi(x) := \begin{cases} 1 & \text{on } B(p,r), \\ \frac{2r - d(p,x)}{r} & \text{on } B(p,2r) \setminus B(p,r), \\ 0 & \text{on } M^n \setminus B(p,2r). \end{cases} \tag{3.58}$$

By the Bochner formula above, we get

$$\begin{aligned} 2\int_{M^n} |\nabla^2 u|^2 \psi^2 \mathrm{e}^{-f} dv &= \int_{M^n} \left(\Delta_f |\nabla u|^2\right) \psi^2 \mathrm{e}^{-f} dv \\ &= -\int_{M^n} \left\langle \nabla |\nabla u|^2, \nabla \psi^2 \right\rangle \mathrm{e}^{-f} dv \\ &\leq 4\int_{M^n} |\nabla |\nabla u|| |\nabla u| \psi |\nabla \psi| \mathrm{e}^{-f} dv \\ &\leq \int_{M^n} |\nabla^2 u|^2 \psi^2 \mathrm{e}^{-f} dv + 4\int_{M^n} |\nabla u|^2 |\nabla \psi|^2 \mathrm{e}^{-f} dv. \end{aligned}$$

Thus, we obtain that

$$\int_{M^n} |\nabla^2 u|^2 \psi^2 \mathrm{e}^{-f} dv \leq 4\int_{M^n} |\nabla u|^2 |\nabla \psi|^2 \mathrm{e}^{-f} dv. \tag{3.59}$$

Since $\int_{M^n} |\nabla u|^2 \mathrm{e}^{-f} dv < \infty$, the right-hand side of (3.59) tends to 0 after taking the limit $r \to \infty$. This implies that $\nabla^2 u = 0$. Consequently, $|\nabla u|$ must be a constant on M^n. The weighted volume

$$\mathrm{Vol}_f(M^n) = \int_{M^n} \mathrm{e}^{-f} dv = \infty,$$

as M^n is f-nonparabolic. However, since $\int_{M^n} |\nabla u|^2 \mathrm{e}^{-f} dv < \infty$, we conclude that $|\nabla u| = 0$; hence u is a constant. This is a contradiction.

CASE 2: *Only one f-nonparabolic end.*

If M^n has exactly one f-nonparabolic end, then there exists an f-nonparabolic end E of M^n such that $M^n \setminus E$ is an f-parabolic end of M^n. By Corollary 3.25, there exists a positive f-harmonic function w satisfying

$$\liminf_{x \in E,\, x \to \infty} w(x) = 0,$$
$$\limsup_{x \in M^n \setminus E,\, x \to \infty} w(x) = \infty.$$

According to Proposition 3.26, we have

$$\int_{B(p,r)} w \mathrm{e}^{-f} dv \leq Cr \tag{3.60}$$

for all $r > 1$, where $C > 0$ is independent of r. Let ψ be the cutoff function defined by (3.58). We have

$$\begin{aligned}
\lambda_1(\Delta_f) \int_{M^n} w\psi^2 \mathrm{e}^{-f} dv &\leq \int_{M^n} \left|\nabla \left(w^{\frac{1}{2}}\psi\right)\right|^2 \mathrm{e}^{-f} dv \\
&\leq \frac{1}{4} \int_{M^n} \frac{|\nabla w|^2}{w} \psi^2 \mathrm{e}^{-f} dv + \int_{M^n} w |\nabla \psi|^2 \mathrm{e}^{-f} dv \\
&\quad + \frac{1}{2} \int_{M^n} \langle \nabla w, \nabla \psi^2 \rangle \mathrm{e}^{-f} dv \\
&= \frac{1}{4} \int_{M^n} \frac{|\nabla w|^2}{w} \psi^2 \mathrm{e}^{-f} dv + \int_{M^n} w |\nabla \psi|^2 \mathrm{e}^{-f} dv,
\end{aligned}$$

where in the last line we have integrated by parts and used that $\Delta_f w = 0$.

Now, by Proposition 3.28, we know that

$$\lambda_1(\Delta_f) = \frac{1}{4}.$$

Moreover, since (M^n, g, f) is normalized such that $R + |\nabla f|^2 = 1$ and $R \geq 0$, it follows that $|\nabla f| \leq 1$ on M^n. According to Corollary 3.5, we have the gradient estimate

$$|\nabla \ln w| \leq 1 \quad \text{on } M^n.$$

This implies that

$$P := \lambda_1(\Delta_f) - \frac{1}{4} |\nabla \ln w|^2 \geq 0.$$

Therefore, we obtain that

$$\begin{aligned}\int_{M^n} Pw\psi^2 \mathrm{e}^{-f} dv &\leq \int_{M^n} w\,|\nabla\psi|^2\,\mathrm{e}^{-f} dv\\ &= \frac{C}{r^2}\int_{B(p,2r)\setminus B(p,r)} w\mathrm{e}^{-f} dv\\ &\leq \frac{C}{r},\end{aligned}$$

where the last line follows from (3.60). Letting $r \to \infty$ and using that $P \geq 0$, we obtain that $P = 0$ on M^n.

Following the proof of Corollary 3.5, we obtain that

$$|\nabla \ln w| = 1 \quad \text{and} \quad \nabla^2 \ln w = 0.$$

This immediately implies the splitting $(M^n, g) = \mathbb{R}\times N^{n-1}$. If N^{n-1} is noncompact, then M^n has exactly one end. Thus we may assume that N^{n-1} is compact. By Lemma 2.5 in Volume 1, N^{n-1} admits a steady gradient Ricci soliton structure. Since N^{n-1} is compact, it must be Ricci-flat. This completes the proof of the theorem.

□

Theorem 3.29 can be placed in the following context. According to Lemma 3.27 and Proposition 3.28, steady gradient Ricci solitons achieve maximal bottom spectrum for the weighted Laplacian among the class of all weighted manifolds $(M^n, g, \mathrm{e}^{-f}dv)$ with non-negative Bakry–Emery curvature $\mathrm{Ric}_f \geq 0$ and with potential of bounded gradient. Hence, Theorem 3.29 can be interpreted as a rigidity result for weighted manifolds achieving maximal bottom spectrum.

Moreover, the methods applied in the proof of Theorem 3.29 are robust enough to extend to non-smooth settings such as the steady soliton conifolds defined in [**CW17**]; see [**LW24b**].

REMARK 3.30. Historically, Cheng [**Che75**] proved a sharp upper bound of the bottom spectrum of the Laplacian on a complete manifold with Ricci curvature bounded below, and more recently Li and Wang studied the rigidity issue when the sharp upper bound is achieved [**LW01, LW02**]. The approach followed in proving Theorem 3.29 is in the same spirit as these works of Li and Wang.

3.3. Ends of Kähler shrinking Ricci solitons

Our goal in this section is to prove that complete noncompact Kähler shrinking gradient Ricci solitons have precisely one end (see Theorem 3.36 below). This result is due to Wang and one of the authors [**MW15c**]. Kähler shrinkers have an additional property that the vector field ∇f is real-holomorphic, which plays a crucial role here. First, we prove a key vanishing theorem for weighted harmonic functions, which holds in the general setting of weighted manifolds admitting real-holomorphic vector fields [**MW15b**].

3.3.1. A vanishing theorem for Kähler manifolds.

In this subsection we assume that (M^n, g) is a complete noncompact Kähler manifold of complex dimension $m = \frac{n}{2}$ and that f is a smooth function on M^n such that ∇f is real holomorphic; i.e., the $(1,0)$-part of ∇f is a holomorphic vector field. In terms of local complex coordinates $\{z^1, z^2, \ldots, z^m\}$, the fact that ∇f is real holomorphic means that the complex vector field

$$X := g^{i\bar{k}} \frac{\partial f}{\partial \bar{z}^k} \frac{\partial}{\partial z^i}$$

is holomorphic. In terms of a local unitary frame $\{v_k\}_{k=1,\ldots,m}$, this is equivalent to

$$f_{kl} := \nabla^2 f(v_k, v_l) = 0 \quad \text{for all } k, l \in \{1, 2, \ldots, m\}. \tag{3.61}$$

An alternative characterization is that $J(\nabla f)$ is a Killing vector field, where $J : TM \to TM$ is the complex structure on M^n.

Given a Kähler metric

$$ds^2 = g_{k\bar{j}} dz^k \otimes d\bar{z}^j,$$

the associated Riemannian metric is defined as follows:

$$\langle \cdot, \cdot \rangle := 4\mathrm{Re}\left(ds^2\right).$$

We will often work in normal complex coordinates $\{z^1, z^2, \ldots, z^m\}$; that is, $g_{i\bar{j}} = \delta_{i\bar{j}}$ and $\partial g_{i\bar{j}} = 0$ at a given point. Since

$$\frac{\partial}{\partial z^j} = \frac{1}{2}\left(\frac{\partial}{\partial x^j} - \sqrt{-1}\frac{\partial}{\partial x^{m+j}}\right) \quad \text{and} \quad dz^j = dx^j + \sqrt{-1}dx^{m+j},$$

we see that

$$\left\langle \frac{\partial}{\partial x^\alpha}, \frac{\partial}{\partial x^\beta} \right\rangle = 4\delta_{\alpha\beta} \quad \text{for all } \alpha, \beta = 1, \ldots, 2m.$$

Hence, with respect to normal complex coordinates, we have that

$$|\nabla u|^2 = \frac{1}{4}\sum_{\alpha=1}^{2m}\left(\frac{\partial u}{\partial x^\alpha}\right)^2 = \sum_{j=1}^{m}\left|\frac{\partial u}{\partial z^j}\right|^2.$$

Therefore, we can see that for the associated Riemannian metric, we have that

$$|\nabla u|^2 = g^{k\bar{j}} u_k u_{\bar{j}} \quad \text{and} \quad \Delta u = g^{k\bar{j}} u_{k\bar{j}}.$$

Any (p, q)-form $\omega \in A^{p,q}(M^n)$ will be written locally as

$$\omega = \frac{1}{p!q!}\omega_{I\bar{J}} dz^I \wedge d\bar{z}^J,$$

where $|I| = p$ and $|J| = q$. On the vector space $A^{p,q}(M^n)$, we use the metric to define a pointwise Hermitian product by

$$\langle \omega, \theta \rangle := \frac{1}{2^{p+q}}\frac{1}{p!q!} g^{I\bar{K}} g^{L\bar{J}} \omega_{I\bar{J}} \overline{\theta_{K\bar{L}}}.$$

The differential $d : \Omega^k(M^n) \to \Omega^{k+1}(M^n)$ acting on k-forms on M^n, given by

$$d\omega = dx^\alpha \wedge \nabla_{\frac{\partial}{\partial x^\alpha}} \omega,$$

is decomposed as

$$d = \partial + \bar{\partial},$$

where $\partial : A^{p,q}(M^n) \to A^{p+1,q}(M^n)$ and $\bar{\partial} : A^{p,q}(M^n) \to A^{p,q+1}(M^n)$ are given by

$$\partial\omega = dz^j \wedge \nabla_{\frac{\partial}{\partial z^j}} \omega \quad \text{and} \quad \bar{\partial}\omega = d\bar{z}^j \wedge \nabla_{\frac{\partial}{\partial \bar{z}^j}} \omega.$$

Henceforth, we denote $\partial_j := \frac{\partial}{\partial z^j}$ and $\bar{\partial}_j := \frac{\partial}{\partial \bar{z}^j}$.

The above operators d, ∂, and $\bar{\partial}$ have adjoints $d^* : \Omega^k(M^n) \to \Omega^{k-1}(M^n)$, $\partial^* : A^{p,q}(M^n) \to A^{p-1,q}(M^n)$, and $\bar{\partial}^* : A^{p,q}(M^n) \to A^{p,q-1}(M^n)$, respectively. We also have that

$$d^* = \partial^* + \bar{\partial}^*.$$

These operators are specified by the formulas

$$\begin{aligned} d^*\omega &= -g^{\alpha\beta}\, \iota_{\frac{\partial}{\partial x^\alpha}} \nabla_{\frac{\partial}{\partial x^\beta}} \omega, \\ \partial^*\omega &= -\frac{1}{2} g^{k\bar{j}}\, \iota_{\partial_k} \nabla_{\bar{\partial}_j} \omega, \\ \bar{\partial}^*\omega &= -\frac{1}{2} g^{k\bar{j}}\, \iota_{\bar{\partial}_j} \nabla_{\partial_k} \omega. \end{aligned}$$

Here $\iota_X \omega$ denotes the interior product of ω with respect to X, and $\alpha, \beta \in \{1, \ldots, 2m\}$ are used to denote real coordinate indices.

Define the Hodge Laplacian by

$$\Delta_d := -\left(dd^* + d^*d\right),$$

which is a self-adjoint and non-positive operator. If u is a smooth function, then $\Delta_d u = \Delta u = g^{k\bar{j}} u_{k\bar{j}}$, which is our convention for the Laplacian acting on functions.

One can also define two other operators

$$\Delta^{\partial} := -\left(\partial\partial^* + \partial^*\partial\right) \quad \text{and} \quad \Delta^{\bar{\partial}} := -\left(\bar{\partial}\bar{\partial}^* + \bar{\partial}^*\bar{\partial}\right),$$

which map $A^{p,q}(M^n)$ into itself. As (M^n, g) is Kähler,

$$\Delta_d = 2\Delta^{\partial} = 2\Delta^{\bar{\partial}}.$$

Certainly, Δ_d preserves the space $A^{p,q}(M^n)$.

The function $f \in C^\infty(M^n)$ gives us a weighted volume form

$$dv_f := \mathrm{e}^{-f} dv.$$

We then have the corresponding adjoint operators d_f^*, ∂_f^* and $\bar{\partial}_f^*$ defined with respect to the L^2-inner product associated to dv_f. These first-order differential operators satisfy the following formulas:

$$\begin{aligned} d_f^* &= d^* + \iota_{\nabla f}, \\ \partial_f^* &= \partial^* + \iota_{\nabla^{1,0} f}, \\ \bar{\partial}_f^* &= \bar{\partial}^* + \iota_{\nabla^{0,1} f}. \end{aligned} \tag{3.62}$$

We have denoted with

$$\nabla^{1,0} f := \frac{1}{2} g^{j\bar{k}} f_{\bar{k}} \partial_j \quad \text{and} \quad \nabla^{0,1} f := \frac{1}{2} g^{j\bar{k}} f_j \partial_{\bar{k}},$$

where $f_{\bar{k}} := \frac{\partial f}{\partial \bar{z}^k}$ and $f_j := \frac{\partial f}{\partial z^j}$. Indeed, we have that

$$\nabla f = \nabla^{1,0} f + \nabla^{0,1} f \quad \text{and} \quad d_f^* = \partial_f^* + \bar{\partial}_f^*.$$

From (3.62) and Cartan's formula $\mathcal{L}_{\nabla f} = \iota_{\nabla f} \circ d + d \circ \iota_{\nabla f}$, we deduce that the weighted Hodge Laplacian

$$\Delta_f := -\left(dd_f^* + d_f^* d\right)$$

satisfies

$$\Delta_f = \Delta_d - \mathcal{L}_{\nabla f},$$

where $\mathcal{L}$ denotes the Lie derivative. Furthermore, for

$$\Delta_f^{\partial} := -\left(\partial\partial_f^* + \partial_f^*\partial\right) \quad \text{and} \quad \Delta_f^{\bar\partial} := -\left(\bar\partial\bar\partial_f^* + \bar\partial_f^*\bar\partial\right),$$

we have

$$\begin{aligned}\Delta_f^{\partial} &= \Delta^{\partial} - \partial\iota_{\nabla^{1,0}f} - \iota_{\nabla^{1,0}f}\partial,\\ \Delta_f^{\bar\partial} &= \Delta^{\bar\partial} - \bar\partial\iota_{\nabla^{0,1}f} - \iota_{\nabla^{0,1}f}\bar\partial.\end{aligned}$$

We now state a necessary and sufficient condition on f for Δ_f to map (p,q)-forms to (p,q)-forms. Since this is clearly true for functions, from now on we assume that $0 < p+q < 2m$.

PROPOSITION 3.31. *The weighted Hodge Laplacian Δ_f preserves the space of (p,q)-forms $A^{p,q}(M^n)$ if and only if ∇f is real holomorphic. In this case,*

$$\Delta_f = \Delta_f^{\partial} + \Delta_f^{\bar\partial}.$$

PROOF. Note that

$$\begin{aligned}\mathcal{L}_{\nabla f} &= \left(d\iota_{\nabla^{1,0}f} + \iota_{\nabla^{1,0}f}d\right) + \left(d\iota_{\nabla^{0,1}f} + \iota_{\nabla^{0,1}f}d\right)\\ &= \left(\partial\iota_{\nabla^{1,0}f} + \iota_{\nabla^{1,0}f}\partial\right) + \left(\bar\partial\iota_{\nabla^{0,1}f} + \iota_{\nabla^{0,1}f}\bar\partial\right)\\ &\quad + \left(\bar\partial\iota_{\nabla^{1,0}f} + \iota_{\nabla^{1,0}f}\bar\partial\right) + \left(\partial\iota_{\nabla^{0,1}f} + \iota_{\nabla^{0,1}f}\partial\right)\\ &= \left(\Delta^{\partial} - \Delta_f^{\partial}\right) + \left(\Delta^{\bar\partial} - \Delta_f^{\bar\partial}\right) + S\\ &= \Delta_d - \Delta_f^{\partial} - \Delta_f^{\bar\partial} + S,\end{aligned}$$

where

$$S := \left(\bar\partial\iota_{\nabla^{1,0}f} + \iota_{\nabla^{1,0}f}\bar\partial\right) + \left(\partial\iota_{\nabla^{0,1}f} + \iota_{\nabla^{0,1}f}\partial\right).$$

Denote

$$S_1 := \bar\partial\iota_{\nabla^{1,0}f} + \iota_{\nabla^{1,0}f}\bar\partial \quad \text{and} \quad S_2 := \partial\iota_{\nabla^{0,1}f} + \iota_{\nabla^{0,1}f}\partial,$$

so that

$$S = S_1 + S_2.$$

Then

$$\Delta_f = \Delta_f^{\partial} + \Delta_f^{\bar\partial} - S.$$

Hence, we can finish the proof by computing $S(\omega)$ for $\omega \in A^{p,q}(M^n)$. We first note that in local coordinates,

$$\begin{aligned}S_1(\omega) &= \bar\partial\iota_{\nabla^{1,0}f}\omega + \iota_{\nabla^{1,0}f}\bar\partial\omega\\ &= \frac{1}{2}f_{\bar k\bar j}d\bar z^j \wedge \iota_{\partial_k}\omega + \frac{1}{2}f_{\bar k}\left(\bar\partial\iota_{\partial_k} + \iota_{\partial_k}\bar\partial\right)\omega\\ &= \frac{1}{2}f_{\bar k\bar j}d\bar z^j \wedge \iota_{\partial_k}\omega,\end{aligned}$$

where we have used the fact that $\bar\partial\iota_{\partial_k} + \iota_{\partial_k}\bar\partial = 0$. Here, we use the notation $f_{\bar k\bar j} := \frac{\partial^2 f}{\partial\bar z^k\partial\bar z^j}$. We compute S_2 in a similar fashion and obtain

$$S_2(\omega) = \frac{1}{2}f_{kj}dz^j \wedge \iota_{\partial_{\bar k}}\omega.$$

Hence, $S(\omega) = 0$ if and only if $S_1(\omega) = S_2(\omega) = 0$ for all $\omega \in A^{p,q}(M^n)$. This happens if and only if $f_{jk} = f_{\bar j\bar k} = 0$, which is the same as ∇f being real holomorphic. ∎

Let us denote by $B(x_0, r)$ the geodesic ball centered at a point x_0 and with radius $r > 0$.

The main result of this section is a vanishing theorem for f-harmonic functions with finite energy. For $f = 0$, this was established by Li in [**Li90**]. It was later generalized by Wang and one of the authors to non-constant f in [**MW15c, MW15b**].

THEOREM 3.32 (A vanishing theorem for f-harmonic functions). *Let (M^n, g) be a complete noncompact Kähler manifold and let $f \in C^\infty(M^n)$ with ∇f real holomorphic. Suppose that u is an f-harmonic function on M^n and that there exists a point $x_0 \in M^n$ and a constant $C > 0$ so that*

$$\int_{B(x_0,r)} |\nabla u|^2 \, \mathrm{e}^{-f} dv \le Cr^2$$

for all $r \ge r_0$. Then u is pluriharmonic; that is,

$$u_{\alpha\bar{\beta}} = 0,$$

with respect to any unitary frame. If additionally f is proper, then u is constant on M^n.

PROOF. Let u be an f-harmonic function. Then the 1-form

$$\omega := du$$

is also f-harmonic because

$$\begin{aligned} \Delta_f \omega &= -\left(dd_f^* + d_f^* d\right)(du) \\ &= -d\Delta_f u \\ &= 0. \end{aligned}$$

However, by splitting $\omega = \partial u + \bar{\partial} u$ into its $(1,0)$ and $(0,1)$ components, we see that $\Delta_f \partial u = 0$ because Δ_f preserves the $(1,0)$ and $(0,1)$-forms by Proposition 3.31. Consequently, the $(1,0)$-form

$$\theta := \partial u$$

satisfies

$$\Delta_f \theta = 0$$

and has growth rate

$$\int_{B(x_0,t)} |\theta|^2 \, \mathrm{e}^{-f} dv \le Ct^2 \quad \text{for all } t \ge r_0.$$

For $r > 1$ let ϕ be the cutoff function with support in $B\left(x_0, r^2\right)$ defined by

$$\phi(x) := \begin{cases} 1 & \text{on } B(x_0, r), \\ \frac{1}{\ln r}\left(2\ln r - \ln d(x_0, x)\right) & \text{on } B\left(x_0, r^2\right) \setminus B(x_0, r), \\ 0 & \text{on } M^n \setminus B(x_0, r^2). \end{cases}$$

Since $\Delta_f \theta = 0$, we see that

$$\begin{aligned} 0 &= \int_{M^n} \left\langle \left(dd_f^* + d_f^* d\right)\theta, \phi^2 \theta \right\rangle \mathrm{e}^{-f} dv \\ &= \int_{M^n} \left\langle d_f^*\theta, d_f^*\left(\phi^2\theta\right)\right\rangle \mathrm{e}^{-f} dv + \int_{M^n} \left\langle d\theta, d\left(\phi^2\theta\right)\right\rangle \mathrm{e}^{-f} dv \\ &\geq \int_{M^n} |d\theta|^2 \phi^2 \mathrm{e}^{-f} dv + \int_{M^n} \left|d_f^*\theta\right|^2 \phi^2 \mathrm{e}^{-f} dv \\ &\quad - 2\int_{M^n} \left|d_f^*\theta\right| |\theta| \, \phi \, |\nabla\phi| \, \mathrm{e}^{-f} dv - 2\int_{M^n} |d\theta| \, |\theta| \, \phi \, |\nabla\phi| \, \mathrm{e}^{-f} dv, \end{aligned} \tag{3.63}$$

where in the last line we have used the Cauchy–Schwarz inequality and that

$$\begin{aligned} d\left(\phi^2\theta\right) &= \phi^2 d\theta + d\phi^2 \wedge \theta, \\ d_f^*\left(\phi^2\theta\right) &= \phi^2 d_f^*\theta - \iota_{\nabla\phi^2}\theta. \end{aligned}$$

We have the inequalities

$$2\int_{M^n} \left|d_f^*\theta\right| |\theta| \, \phi \, |\nabla\phi| \, \mathrm{e}^{-f} dv \leq \frac{1}{2}\int_{M^n} |d_f^*\theta|^2 \phi^2 \mathrm{e}^{-f} dv + 2\int_{M^n} |\theta|^2 |\nabla\phi|^2 \mathrm{e}^{-f} dv$$

and

$$2\int_{M^n} |d\theta| \, |\theta| \, \phi \, |\nabla\phi| \, \mathrm{e}^{-f} dv \leq \frac{1}{2}\int_{M^n} |d\theta|^2 \phi^2 \mathrm{e}^{-f} dv + 2\int_{M^n} |\theta|^2 |\nabla\phi|^2 \mathrm{e}^{-f} dv.$$

It follows from (3.63) that

$$\int_{M^n} |d\theta|^2 \phi^2 \mathrm{e}^{-f} dv + \int_{M^n} \left|d_f^*\theta\right|^2 \phi^2 \mathrm{e}^{-f} dv \leq 8\int_{M^n} |\theta|^2 |\nabla\phi|^2 \mathrm{e}^{-f} dv. \tag{3.64}$$

According to the definition of the cutoff function ϕ and the co-area formula, we have that

$$\begin{aligned} \int_{M^n} |\theta|^2 |\nabla\phi|^2 \mathrm{e}^{-f} dv &\leq \frac{1}{\ln^2 r}\int_{B(x_0,r^2)\setminus B(x_0,r)} \frac{1}{d^2}|\theta|^2 \mathrm{e}^{-f} dv \\ &= \frac{1}{\ln^2 r}\int_r^{r^2} \frac{1}{t^2}\left(\int_{\partial B(x_0,t)} |\theta|^2 \mathrm{e}^{-f} da\right) dt. \end{aligned}$$

By integrating the last term by parts, we obtain that

$$\begin{aligned} \int_r^{r^2} \frac{1}{t^2}\left(\int_{\partial B(x_0,t)} |\theta|^2 \mathrm{e}^{-f} da\right) dt = &\frac{1}{r^4}\int_{B(x_0,r^2)} |\theta|^2 \mathrm{e}^{-f} dv \\ &- \frac{1}{r^2}\int_{B(x_0,r)} |\theta|^2 \mathrm{e}^{-f} dv \\ &+ \int_r^{r^2} \frac{2}{t^3}\left(\int_{B(x_0,t)} |\theta|^2 \mathrm{e}^{-f} dv\right) dt. \end{aligned}$$

Since

$$\int_{B(x_0,t)} |\theta|^2 \mathrm{e}^{-f} dv \leq Ct^2,$$

it follows that

$$\int_r^{r^2} \frac{1}{t^2}\left(\int_{\partial B(x_0,t)} |\theta|^2 \mathrm{e}^{-f} da\right) dt \leq C \ln r.$$

In conclusion, we have that

$$\int_{M^n} |\theta|^2 |\nabla\phi|^2 \mathrm{e}^{-f} dv \leq \frac{C}{\ln r}.$$

Plugging this into (3.64), we find that

$$\int_{B(x_0,r)} |d\theta|^2 \mathrm{e}^{-f} dv + \int_{B(x_0,r)} \left|d_f^*\theta\right|^2 \mathrm{e}^{-f} dv \leq \frac{C}{\ln r}.$$

Letting $r \to \infty$, we conclude that $d\theta = d_f^*\theta = 0$. Now that u is pluriharmonic follows immediately from

$$\overline{\partial}\partial u = d\partial u = d\theta = 0.$$

We now prove the second conclusion that u is constant if f is proper. Indeed, since u is pluriharmonic, we have in particular that $\Delta u = 0$. Together with $\Delta_f u = 0$, this implies that $\langle \nabla u, \nabla f\rangle = 0$. For $s < t$ arbitrary, the set $\{s \leq f \leq t\}$ is compact. Integrating by parts, it therefore follows that

$$\begin{aligned}\int_{\{s\leq f\leq t\}} |\nabla u|^2 \mathrm{e}^{-f} dv &= -\int_{\{s\leq f\leq t\}} u(\Delta_f u)\mathrm{e}^{-f} dv \\ &\quad + \int_{\{f=t\}} u\frac{\langle \nabla u, \nabla f\rangle}{|\nabla f|}\mathrm{e}^{-f} da \\ &\quad - \int_{\{f=s\}} u\frac{\langle \nabla u, \nabla f\rangle}{|\nabla f|}\mathrm{e}^{-f} da \\ &= 0.\end{aligned}$$

In conclusion, $\nabla u = 0$ on M^n, which means that u is constant. □

3.3.2. Ends of Kähler shrinking solitons.

To apply the results of the previous subsection, we will construct a weighted harmonic function assuming that the shrinking Ricci soliton has more than one end. Since $\int_{M^n} \mathrm{e}^{-f} dv < \infty$, Theorem 3.13 implies that $(M^n, g, \mathrm{e}^{-f} dv)$ is f-parabolic, so the framework of Theorem 3.24 cannot be applied to this weighted manifold.

For this reason we will change the sign of the weight and prefer to work with $(M^n, g, \mathrm{e}^{-F} dv)$, where $F := -f$. Our main result in this section is that all ends of a shrinking gradient Ricci soliton (M^n, g, f) are F-nonparabolic, a conclusion that does not require the Kähler condition. This motivates us to apply the results of the previous section for the weight function F.

Let us observe that

$$\begin{aligned}\Delta_F \mathrm{e}^{-\frac{1}{2}f} &= -\left(\frac{1}{4}|\nabla f|^2 + \frac{1}{2}\Delta f\right)\mathrm{e}^{-\frac{1}{2}f} \\ &= -\left(\frac{1}{4}f - \frac{3}{4}R + \frac{n}{4}\right)\mathrm{e}^{-\frac{1}{2}f}.\end{aligned}$$

Hence, in the special case where

$$\sup_{M^n} R < \infty,$$

we can conclude that

$$\Delta_F \mathrm{e}^{-\frac{1}{2}f} \leq 0 \quad \text{on } M^n \setminus D$$

for a large enough compact subset set $D \subset M^n$ such that $f > 3R$ on $M^n \setminus D$. Since the function $\mathrm{e}^{-\frac{1}{2}f}$ is F-superharmonic on $M^n \setminus D$ and achieves its infimum at infinity along each end of M^n, Proposition 3.18 implies that each end of M^n is F-nonparabolic.

Our goal is to prove that this conclusion is true for any shrinking Ricci soliton, without any assumptions on R.

PROPOSITION 3.33 (F-nonparabolicity of shrinking soliton ends). *Any end of a complete noncompact gradient shrinking Ricci soliton (M^n, g, f) is F-nonparabolic, where $F = -f$.*

PROOF. Let E be an end of M^n. Assume by contradiction that E is F-parabolic. By Proposition 3.17, there exists an F-harmonic function $w : E \to [0, \infty)$ such that $w = 0$ on ∂E, w is unbounded on E, and

$$\int_E |\nabla \ln(w+1)|^2 \mathrm{e}^{-F} dv < \infty.$$

In particular, for any $x \in E$ with $B(x,1) \subset E$, we have by (1.61) that

$$\int_{B(x,1)} |\nabla \ln(w+1)|^2 \, dv \leq C\mathrm{e}^{-\frac{1}{4}r^2(x)+cr(x)},$$

where $r(x) := d(o, x)$, for o a minimum point of f. Here, $c = c(n) > 0$ and $C > 0$ is independent of x. We may assume that $r(x) \geq r_0$ for all $x \in E$, where $r_0 > 0$ is large enough to be specified later. It follows that the function

$$h := \ln(w+1)$$

satisfies the following properties:

$$\limsup_{x \in E,\ x \to \infty} h(x) = \infty, \tag{3.65}$$

$$\Delta h = -\langle \nabla h, \nabla f \rangle - |\nabla h|^2,$$

and

$$\int_{B(x,1)} |\nabla h|^2 \, dv \leq C\mathrm{e}^{-\frac{1}{4}r^2(x)+cr(x)} \quad \text{for all } x \in E. \tag{3.66}$$

Our goal is to transform the L^2 gradient estimate (3.66) into a pointwise estimate. For that, we first establish a differential inequality for $|\nabla h|^2$.

CLAIM 3.34. *The function $\sigma := |\nabla h|^2$ satisfies the following inequality on E:*

$$\frac{1}{2n}\sigma^2 \leq \frac{3}{2} f\sigma + 2\langle \nabla \sigma, \nabla h \rangle + 4\nabla^2 f(\nabla h, \nabla h) + \Delta \sigma. \tag{3.67}$$

PROOF OF CLAIM 3.34. According to the Bochner formula, we have

$$\frac{1}{2}\Delta |\nabla h|^2 = \left|\nabla^2 h\right|^2 + \langle \nabla(\Delta h), \nabla h \rangle + \mathrm{Ric}(\nabla h, \nabla h).$$

By (3.65) we have that

$$\langle \nabla(\Delta h), \nabla h \rangle = -\nabla^2 h(\nabla f, \nabla h) - \nabla^2 f(\nabla h, \nabla h) - \left\langle \nabla |\nabla h|^2, \nabla h \right\rangle,$$

and by the shrinking soliton equation it follows that

$$\mathrm{Ric}(\nabla h, \nabla h) = \frac{1}{2}|\nabla h|^2 - \nabla^2 f(\nabla h, \nabla h).$$

Therefore, we have

$$\frac{1}{2}\Delta\left|\nabla h\right|^2 = \left|\nabla^2 h\right|^2 - \nabla^2 h(\nabla f, \nabla h) - \langle\nabla|\nabla h|^2, \nabla h\rangle + \frac{1}{2}\left|\nabla h\right|^2 - 2\nabla^2 f\left(\nabla h, \nabla h\right).$$

Using the inequality

$$|\nabla^2 h(\nabla f, \nabla h)| \leq \frac{1}{2}|\nabla^2 h|^2 + \frac{1}{2}|\nabla f|^2|\nabla h|^2,$$

it follows that

$$\frac{1}{2}\Delta\left|\nabla h\right|^2 \geq \frac{1}{2}\left|\nabla^2 h\right|^2 - \frac{1}{2}f\left|\nabla h\right|^2 - \left\langle\nabla\left|\nabla h\right|^2, \nabla h\right\rangle - 2\nabla^2 f\left(\nabla h, \nabla h\right).$$

By the Cauchy–Schwarz inequality and (3.65), we obtain

$$\begin{aligned}\left|\nabla^2 h\right|^2 &\geq \frac{1}{n}\left(\Delta h\right)^2 \\ &= \frac{1}{n}\left(\langle\nabla h, \nabla f\rangle + |\nabla h|^2\right)^2 \\ &\geq \frac{1}{n}\left(\frac{1}{2}|\nabla h|^4 - \langle\nabla h, \nabla f\rangle^2\right) \\ &\geq \frac{1}{2n}\left|\nabla h\right|^4 - \frac{1}{n}f\left|\nabla h\right|^2.\end{aligned}$$

In conclusion,

$$\frac{1}{2}\Delta\left|\nabla h\right|^2 \geq \frac{1}{4n}\left|\nabla h\right|^4 - \frac{n+1}{2n}f\left|\nabla h\right|^2 - \left\langle\nabla\left|\nabla h\right|^2, \nabla h\right\rangle - 2\nabla^2 f\left(\nabla h, \nabla h\right).$$

This proves Claim 3.34. □

Our goal is to apply the De Giorgi–Nash–Moser iteration scheme (see Chapter 19 in [**Li12**] or Chapter 4 in [**HL97**]) to σ satisfying (3.67). This process will additionally allow us to conveniently estimate the terms on the right side of (3.67).

CLAIM 3.35. *For any function ϕ with support in $B(x, 1) \subset E$ and for any $q \geq 20(n+1)$ we have*

$$(q-2)\int_{M^n}\left|\nabla\sigma\right|^2\sigma^{q-2}\phi^2 dv \leq 30q\int_{M^n} f\sigma^q\phi^2 dv + 14\int_{M^n}\sigma^q\left|\nabla\phi\right|^2 dv. \tag{3.68}$$

PROOF OF CLAIM 3.35. We multiply (3.67) by $\phi^2\sigma^{q-1}$ to obtain

$$\begin{aligned}\frac{1}{2n}\int_{M^n}\sigma^{q+1}\,\phi^2 dv \leq{}& \frac{3}{2}\int_{M^n} f\sigma^q\phi^2 dv + 2\int_{M^n}\langle\nabla\sigma, \nabla h\rangle\,\sigma^{q-1}\phi^2 dv \\ &+ 4\int_{M^n}\nabla^2 f\left(\nabla h, \nabla h\right)\sigma^{q-1}\phi^2 dv \\ &+ \int_{M^n}\sigma^{q-1}\left(\Delta\sigma\right)\phi^2 dv.\end{aligned} \tag{3.69}$$

We aim to estimate the last three terms on the right-hand side of (3.69). Note that according to (3.65),

$$\begin{aligned}2\int_{M^n}\langle\nabla\sigma,\nabla h\rangle\,\sigma^{q-1}\phi^2dv &= \frac{2}{q}\int_{M^n}\langle\nabla\sigma^q,\nabla h\rangle\,\phi^2dv\\ &= -\frac{2}{q}\int_{M^n}\sigma^q(\Delta h)\phi^2dv\\ &\quad -\frac{2}{q}\int_{M^n}\sigma^q\left\langle\nabla h,\nabla\phi^2\right\rangle dv\\ &= \frac{2}{q}\int_{M^n}\sigma^q\left(\langle\nabla h,\nabla f\rangle+\sigma\right)\phi^2dv\\ &\quad -\frac{2}{q}\int_{M^n}\sigma^q\left\langle\nabla h,\nabla\phi^2\right\rangle dv.\end{aligned}$$

Using the inequalities

$$\begin{aligned}2\sigma^q\langle\nabla h,\nabla f\rangle &\le 2|\nabla f|\sigma^{q+\frac{1}{2}}\\ &\le \sigma^{q+1}+f\sigma^q\end{aligned}$$

and

$$\begin{aligned}2\left|\sigma^q\left\langle\nabla h,\nabla\phi^2\right\rangle\right| &\le 4\sigma^{q+\frac{1}{2}}\phi|\nabla\phi|\\ &\le \sigma^{q+1}\phi^2+4|\nabla\phi|^2\sigma^q,\end{aligned}$$

it follows that

$$\begin{aligned}2\int_{M^n}\langle\nabla\sigma,\nabla h\rangle\,\sigma^{q-1}\phi^2dv &\le \frac{4}{q}\int_{M^n}\sigma^{q+1}\phi^2dv+\frac{4}{q}\int_{M^n}|\nabla\phi|^2\,\sigma^qdv\\ &\quad +\frac{1}{q}\int_{M^n}f\sigma^q\phi^2dv.\end{aligned}\tag{3.70}$$

We continue with estimating the last term on the right side of (3.69). Integrating by parts, we find that

$$\begin{aligned}\int_{M^n}\sigma^{q-1}(\Delta\sigma)\,\phi^2dv &= -(q-1)\int_{M^n}|\nabla\sigma|^2\,\sigma^{q-2}\phi^2dv\\ &\quad -\int_{M^n}\sigma^{q-1}\left\langle\nabla\sigma,\nabla\phi^2\right\rangle dv\\ &\le -(q-2)\int_{M^n}|\nabla\sigma|^2\,\sigma^{q-2}\phi^2dv\\ &\quad +\int_{M^n}\sigma^q\,|\nabla\phi|^2\,dv.\end{aligned}\tag{3.71}$$

Similarly, integration by parts gives

$$\begin{aligned}\int_{M^n}\nabla^2 f\,(\nabla h,\nabla h)\,\sigma^{q-1}\phi^2dv &= -\int_{M^n}\nabla^2 h\,(\nabla h,\nabla f)\,\sigma^{q-1}\phi^2dv\\ &\quad -\int_{M^n}(\Delta h)\,\langle\nabla f,\nabla h\rangle\,\sigma^{q-1}\phi^2dv\\ &\quad -\int_{M^n}\langle\nabla f,\nabla h\rangle\left\langle\nabla h,\nabla\sigma^{q-1}\right\rangle\phi^2dv\\ &\quad -\int_{M^n}\langle\nabla f,\nabla h\rangle\left\langle\nabla h,\nabla\phi^2\right\rangle\sigma^{q-1}dv.\end{aligned}\tag{3.72}$$

We now study each term on the right-hand side. First, note that

$$\begin{aligned}
-\int_{M^n} \nabla^2 h\left(\nabla h, \nabla f\right) \sigma^{q-1}\phi^2 dv &= -\frac{1}{2q}\int_{M^n} \left\langle \nabla\sigma^q, \nabla f\right\rangle \phi^2 dv \\
&= \frac{1}{2q}\int_{M^n} \sigma^q \left(\Delta f\right) \phi^2 dv \\
&+ \frac{1}{2q}\int_{M^n} \sigma^q \left\langle \nabla f, \nabla\phi^2\right\rangle dv.
\end{aligned}$$

We have the following estimates:

$$\begin{aligned}
\frac{1}{2q}\int_{M^n} \sigma^q \left\langle \nabla f, \nabla\phi^2\right\rangle dv &\leq \frac{1}{q}\int_{M^n} |\nabla f||\phi||\nabla\phi|\sigma^q dv \\
&\leq \frac{1}{2q}\int_{M^n} |\nabla f|^2\sigma^q\phi^2 dv + \frac{1}{2q}\int_{M^n} \sigma^q|\nabla\phi|^2 dv.
\end{aligned}$$

We may assume that $r(x) \geq r_0$ for all $x \in E$, where $r_0 > 0$ is large enough such that $\Delta f + |\nabla f|^2 \leq 2f$. Therefore, it results that

$$-\int_{M^n} \nabla^2 h\left(\nabla h, \nabla f\right) \sigma^{q-1}\phi^2 dv \leq \frac{1}{q}\int_{M^n} f\sigma^q\phi^2 dv + \frac{1}{2q}\int_{M^n} \sigma^q \left|\nabla\phi\right|^2 dv.$$

Next, (3.65) implies that

$$\begin{aligned}
-\int_{M^n} \left(\Delta h\right)\left\langle \nabla f, \nabla h\right\rangle \sigma^{p-1}\phi^2 dv &= \int_{M^n} \left\langle \nabla h, \nabla f\right\rangle^2 \sigma^{q-1}\phi^2 dv \\
&+ \int_{M^n} \left\langle \nabla h, \nabla f\right\rangle \sigma^q\phi^2 dv \\
&\leq \int_{M^n} \left(f\sigma^q + |\nabla f|\sigma^{q+\frac{1}{2}}\right)\phi^2 dv \\
&\leq \frac{1}{16n}\int_{M^n} \sigma^{q+1}\phi^2 dv + 5n\int_{M^n} f\sigma^q\phi^2 dv.
\end{aligned}$$

The third term on the right-hand side of (3.72) is estimated by

$$\begin{aligned}
-\int_{M^n} \left\langle \nabla f, \nabla h\right\rangle \left\langle \nabla h, \nabla\sigma^{q-1}\right\rangle \phi^2 dv &\leq (q-1)\int_{M^n} \left|\nabla f\right|\left|\nabla\sigma\right|\sigma^{q-1}\phi^2 dv \\
&\leq \frac{q-2}{8}\int_{M^n} \left|\nabla\sigma\right|^2 \sigma^{q-2}\phi^2 dv \\
&+ 3q\int_{M^n} f\sigma^q\phi^2 dv
\end{aligned}$$

for any $q \geq 3$. The last term in (3.72) is bounded by

$$\begin{aligned}
-\int_{M^n} \left\langle \nabla f, \nabla h\right\rangle \left\langle \nabla h, \nabla\phi^2\right\rangle \sigma^{q-1} dv &\leq 2\int_{M^n} \left|\nabla f\right|\sigma^q\phi\left|\nabla\phi\right| dv \\
&\leq \int_{M^n} f\sigma^q\phi^2 dv + \int_{M^n} \sigma^q \left|\nabla\phi\right|^2 dv.
\end{aligned}$$

Plugging all these estimates into (3.72), we conclude that

$$
\begin{aligned}
(3.73)\qquad 4\int_{M^n} \nabla^2 f\left(\nabla h, \nabla h\right)\sigma^{q-1}\phi^2 dv \leq\ & \frac{1}{4n}\int_{M^n} \sigma^{q+1}\phi^2 dv \\
&+ \frac{q-2}{2}\int_{M^n} \left|\nabla\sigma\right|^2 \sigma^{q-2}\phi^2 dv \\
&+ 5\int_{M^n} \sigma^q \left|\nabla\phi\right|^2 dv \\
&+ 13q\int_{M^n} f\sigma^q\phi^2 dv,
\end{aligned}
$$

for any $q \geq 20(n+1)$. Plugging (3.70), (3.71), and (3.73) into (3.69), it follows that

$$
(q-2)\int_{M^n} \left|\nabla\sigma\right|^2 \sigma^{q-2}\phi^2 dv \leq 30q\int_{M^n} f\sigma^q\phi^2 dv + 14\int_{M^n} \sigma^q \left|\nabla\phi\right|^2 dv
$$

for any $q \geq 20(n+1)$ and for any ϕ with support in $B\left(x, 1\right)$. This proves Claim 3.35. $\square$

We now combine the inequality (3.68) with the Sobolev inequality to set up the iteration process. The Sobolev inequality for shrinking gradient Ricci solitons asserts that there exist constants $\mu := \frac{n}{n-2} > 1$ and $C > 0$ such that

$$
\left(\int_{M^n} \psi^{2\mu} dv\right)^{\frac{1}{\mu}} \leq C\left(\int_{M^n} \left|\nabla\psi\right|^2 dv + \int_{M^n} R\psi^2 dv\right)
$$

for all $\psi \in C_0^\infty(M^n)$; see Li and Wang [**LW20**]. In particular, it follows that

$$
(3.74)\qquad \left(\int_{M^n} \psi^{2\mu} dv\right)^{\frac{1}{\mu}} \leq Cr^2(x)\left(\int_{M^n} \left|\nabla\psi\right|^2 dv + \int_{M^n} \psi^2 dv\right)
$$

for all $x \in E$ and for all $\psi \in C_0^\infty\left(B\left(x, 1\right)\right)$.

Note that a weaker Sobolev inequality, established in [**MW11**], involves the Sobolev constant depending exponentially on the distance function, i.e., of the form $Ce^{c(n)r(x)}$ rather than quadratically $Cr^2(x)$ as in (3.74). The argument presented below remains valid even for this weaker Sobolev inequality.

Letting $\psi := \sigma^{\frac{q}{2}}\phi$ in (3.74), we obtain

$$
\begin{aligned}
\left(\int_{M^n} \sigma^{q\mu}\phi^{2\mu} dv\right)^{\frac{1}{\mu}} &\leq Cr^2(x)\int_{M^n} \left|\nabla\left(\sigma^{\frac{q}{2}}\phi\right)\right|^2 dv + Cr^2(x)\int_{M^n} \sigma^q\phi^2 dv \\
&\leq Cr^2(x)q^2\int_{M^n} |\nabla\sigma|^2\sigma^{q-2}\phi^2 dv \\
&\quad + Cr^2(x)\int_{M^n} \left(\phi^2 + |\nabla\phi|^2\right)\sigma^q dv.
\end{aligned}
$$

Using (3.68), it follows that

$$
\left(\int_{M^n} \sigma^{q\mu}\phi^{2\mu} dv\right)^{\frac{1}{\mu}} \leq Cr^4(x)q^2\int_{M^n} \sigma^q\phi^2 dv + Cr^2(x)q\int_{M^n} |\nabla\phi|^2\sigma^q dv
$$

for any $x \in E$ and $\phi \in C_0^\infty(B(x, 1)$.

For any $0 < t < 1$, and any $\delta > 0$ such that $t + \delta < 1$, we let

$$\phi(z) := \begin{cases} 1 & \text{on } B(x,t), \\ \frac{1}{\delta}(t + \delta - d(x,z)) & \text{on } B(x,t+\delta) \setminus B(x,t), \\ 0 & \text{on } M^n \setminus B(x,t+\delta). \end{cases}$$

It follows that

$$\left(\int_{B(x,t)} \sigma^{q\mu} dv\right)^{\frac{1}{\mu}} \leq \frac{Cr^4(x)q^2}{\delta^2} \int_{B(x,t+\delta)} \sigma^q dv$$

for any $x \in E$ and $q \geq c(n) := 20(n+1)$.

By the standard De Giorgi–Nash–Moser iteration (see Chapter 19 in [**Li12**] or Chapter 4 in [**HL97**]), we obtain for any $0 \leq \theta, \rho < 1$ and any $q \geq c(n)$ that

$$\sup_{B(x,\theta\rho)} \sigma \leq C \left(\frac{r^2(x)}{(1-\theta)\rho}\right)^{\frac{2\mu}{\mu-1}\frac{1}{q}} \left(\int_{B(x,\rho)} \sigma^q dv\right)^{\frac{1}{q}},$$

where $C > 0$ is independent of $r(x)$. Fixing $q = c(n)$ it follows that

$$\sup_{B(x,\theta\rho)} \sigma \leq C \left(\frac{r^2(x)}{(1-\theta)\rho}\right)^{\frac{2\mu}{\mu-1}\frac{1}{q}} \left(\sup_{B(x,\rho)} \sigma\right)^{\frac{q-1}{q}} \left(\int_{B(x,\rho)} \sigma dv\right)^{\frac{1}{q}}.$$

Now, another standard iteration argument as in [**Li12**, **HL97**] implies that

$$\sigma(x) \leq Cr(x)^{\frac{4\mu}{\mu-1}} \int_{B(x,1)} \sigma\, dv \tag{3.75}$$

for any $x \in E$. Since $\sigma = |\nabla h|^2$, it follows by (3.66) that

$$\int_{B(x,1)} \sigma dv \leq \mathrm{e}^{-\frac{1}{4}r^2(x)+cr(x)}.$$

One concludes from (3.75) that

$$|\nabla h|(x) \leq C\mathrm{e}^{-\frac{1}{10}r^2(x)} \quad \text{for all } x \in E. \tag{3.76}$$

Integrating (3.76) along minimizing geodesics, and using that $h = \ln(w+1)$, we see that w is bounded on the end E. This contradicts (3.65). We have proven that all ends of (M^n, g, f) are F-nonparabolic. □

We are now ready to prove that Kähler shrinking gradient Ricci solitons are connected at infinity, which was established by Wang and one of the authors in [**MW15c**].

THEOREM 3.36 (Connectedness at infinity of Kähler shrinking solitons). *Any complete noncompact Kähler shrinking gradient Ricci soliton (M^n, g, f) has only one end.*

PROOF. We assume, for the sake of a contradiction, that (M^n, g) has more than one end. Proposition 3.33 implies that each end of M^n must be F-nonparabolic, where $F = -f$. Corollary 3.25 implies that there exists a non-constant bounded function $u : M^n \to \mathbb{R}$ such that

$$\Delta_F u = 0,$$
$$\int_{M^n} |\nabla u|^2 \, \mathrm{e}^{-F} dv < \infty.$$

Let $\{v_i\}$ be a local unitary frame on a Kähler shrinking gradient Ricci soliton. Since $\mathrm{Ric}(v_i, v_j) = g(v_i, v_j) = 0$ on any Kähler manifold, we see that a Kähler shrinking gradient Ricci soliton satisfies

$$\nabla^2 f(v_i, v_j) = 0 \quad \text{for all } 1 \leq i, j \leq \frac{n}{2}.$$

Consequently, ∇f is real holomorphic; cf. [**Cao96**]. Therefore, we also have that ∇F is a real holomorphic vector field. Moreover, F is proper by (1.61). By Theorem 3.32, applied with the weight F, we conclude that u is constant, which is a contradiction. This shows that M^n has only one end. □

Theorem 3.36 settles the problem of counting the number of ends for Kähler shrinking gradient Ricci solitons. Some partial results in the Riemannian setting were obtained in [**MSW21, MW22**]. For example, it is known that a shrinking gradient Ricci soliton (M^n, g, f) whose scalar curvature satisfies the bound

$$\sup_{M^n} R \leq \frac{n}{3} \tag{3.77}$$

has one end; see Wang and one of the authors [**MW22**]. The proof relies on Proposition 3.33 and on a vanishing theorem for F-harmonic functions with finite weighted energy. This vanishing theorem requires the assumption (3.77). Other than the cylinder $N^{n-1} \times \mathbb{R}$, with N^{n-1} compact and Ricci flat, whose scalar curvature is constant $R = \frac{n-1}{2}$, all current examples of shrinking gradient Ricci solitons have one end.

Let us point out that the methods presented in this chapter can also be applied to study the number of ends of expanding Ricci solitons. We refer to [**MW12, MW15c**] for such results.

3.4. Open problems on counting ends of solitons

OPEN PROBLEM 3.1. *Does there exist a shrinking gradient Ricci soliton with more than one end that does not split off an $\mathbb{R}$-factor? Such examples cannot be Kähler, see Theorem* 3.36.

OPEN PROBLEM 3.2. *Suppose that the answer to Problem* 3.1 *is affirmative; then:*

(a) *Can an asymptotically conical shrinking Ricci soliton have more than one end?*

(b) *If a shrinking Ricci soliton has two ends, each asymptotic to $\mathbb{S}^{n-1} \times \mathbb{R}$, is the soliton isometric to $\mathbb{S}^{n-1} \times \mathbb{R}$?*

(c) *Can a shrinking Ricci soliton have one end asymptotic to a cone and another end asymptotic to the round cylinder $\mathbb{S}^{n-1} \times \mathbb{R}$?*

OPEN PROBLEM 3.3. *If the answer to Problem 3.1 is affirmative, can one prove that any shrinking gradient Ricci soliton can have at most a finite number of ends? This is known to be true if the scalar curvature grows sub-quadratically in the distance function.*

Moreover, can the number of ends of a shrinker be bounded only in terms of dimension?

OPEN PROBLEM 3.4. *Do steady gradient Ricci solitons that contain a line necessarily isometrically split off an $\mathbb{R}$-factor?*

This result would imply Theorem 3.29 because any manifold with more than one end admits a line.

OPEN PROBLEM 3.5. *Do steady non-gradient Ricci solitons with more than one end split off an $\mathbb{R}$-factor?*

CHAPTER 4

Asymptotically Conical Shrinking Solitons

One of the themes of the previous two chapters is that the geometry and topology of complete noncompact gradient shrinking solitons tend to simplify near spatial infinity, particularly in dimensions four and below. As we have noted, complete gradient shrinkers are fully classified in dimensions two and three, and the only noncompact examples among these are the flat Gaussian solitons on $\mathbb{R}^2$ and $\mathbb{R}^3$ and certain $\mathbb{Z}_2$-quotients of the round cylinder $\mathbb{S}^2 \times \mathbb{R}$. In Theorem 2.30, we have proven that if the scalar curvature R along the end V of a 4-dimensional shrinker (M^4, g, f) is bounded above and below by positive constants, then either the end is smoothly asymptotic to a quotient of the round cylinder $\mathbb{S}^3 \times \mathbb{R}$ or, along any sequence of points x_i tending to infinity on an integral curve of ∇f, the sequence (M^4, g, x_i) converges to a quotient of $\mathbb{S}^2 \times \mathbb{R}^2$ in the Cheeger-Gromov sense. We will prove below in Theorem 4.11 that if instead $\lim_{x\to\infty} R(x) = 0$ on the end V, then (M^4, g) is smoothly asymptotic to a cone along V.

In this chapter, we will focus on the *asymptotically conical shrinking solitons* identified in this latter alternative. We will see that this class is internally rather rigid: there can be no more than one shrinker asymptotic to any regular cone (in any dimension), and, when such a shrinker exists, it is effectively uniquely determined by its asymptotic cone. One of our primary objectives will be to prove that any two ends of shrinking solitons (complete or not) which are asymptotic to the same cone must be *isometric* near infinity. The strength of the correspondence between the shrinker and its asymptotic cone identified by this result is peculiar to the shrinking case. By contrast, there are constructions of arbitrarily large families of nonisometric expanding Ricci solitons which are asymptotic to the same cone at infinity[1] and analogous statements of uniqueness fail in general for asymptotically conical minimal surfaces and asymptotically conical Kähler Ricci-flat manifolds.

The disparity between the flexibility of asymptotically conical shrinking and expanding solitons is perhaps most easily explained from a dynamical perspective. When reframed in terms of the self-similar solutions to the Ricci flow associated to the solitons, what is originally a problem of unique continuation at infinity for two apparently similar elliptic systems becomes two transparently different problems of uniqueness for parabolic systems in the shrinking and expanding cases.

To see this, we first need to consider these self-similar solutions in greater detail. While all solitons give rise to locally-defined self-similar solutions to the Ricci flow (see, e.g., Section 2.2 of [**Cho23**]), something special happens in the asymptotically conical case. On a sufficiently remote neighborhood of infinity on an asymptotically conical shrinking end, the self-similar solution $g(t) = -t\varphi_t^* g$

[1] Nevertheless, there are a variety of uniqueness statements for expanding Ricci solitons subject to additional restrictions. See, e.g., [**CD20**, **CDS24**] for a uniqueness result for Kähler expanders, and [**DS23**] for a general sufficient condition for uniqueness.

associated to the soliton metric g is well-defined for all $t \in [-1, 0)$ and converges locally smoothly to the conical metric as $t \nearrow 0$. Thus, near infinity, the self-similar solution $g(t)$ on $[-1, 0)$ extends to a *smooth* solution on all of $[-1, 0]$ which emanates from the end of the shrinking soliton at $t = -1$ and terminates in the end of the cone at $t = 0$. In other words, the ends of the cone and the shrinker can be realized as time-slices of a common smooth (incomplete) Ricci flow. The situation is similar for asymptotically conical expanding ends: provided the curvature and its covariant derivatives decay suitably rapidly, the self-similar solution $g(t) = t\varphi_t^* g$ associated with the expander extends near infinity to a smooth solution on $[0, 1]$ which emanates from an end of the cone at $t = 0$ and terminates in an end of the expander at $t = 1$.

Using these solutions, we can reformulate the questions of whether more than one shrinker or expander can be asymptotic to the same cone along an end into questions of uniqueness for solutions to the Ricci flow. However, these questions are now evidently different: in the shrinking case, the question is one of uniqueness in *backward* time, while, in the expanding case, it is one of uniqueness in *forward* time. On the domain in question – an open neighborhood of infinity on an asymptotically conical background – uniqueness in forward time and in backward time are fundamentally different propositions, and only the latter turns out to be valid in general.

The difference between these propositions can be understood by considering their analogs for solutions to the linear heat equation on the complement of a closed ball in $\mathbb{R}^n$. On one hand, there can be no uniqueness in forward time for such solutions, even if they are assumed to be bounded: simply consider any solution to the heat equation with non-trivial initial data supported in the ball. On the other hand, subject to suitable growth conditions at infinity, such solutions do enjoy backward uniqueness; this fact is not at all obvious, however (see Theorem 4.20 and §4.3.2.1 below). The greater rigidity of asymptotically conical shrinking solitons as compared to asymptotically conical expanding solitons is, in some sense, only a more elaborate expression of the fundamental asymmetry in the uniqueness of solutions to forward and backward parabolic equations on exterior domains.

We will use this dynamical perspective throughout this chapter to study the rigidity of shrinking solitons as a question of backward uniqueness for the Ricci flow. The key ingredient in the proof of our uniqueness result for asymptotically conical shrinkers, and the main technical result in this chapter, is a general backward uniqueness principle for sections of vector bundles over asymptotically conical shrinking ends satisfying certain systems of coupled parabolic and ordinary differential inequalities. We will prove this principle by the method of Carleman inequalities, using as our model the work of Escauriaza, Seregin, and Šverák [**ESŠ03a**] for linear parabolic equations.

Later in the chapter, we will apply this general backward uniqueness principle to further explore the relationship between an asymptotically conical shrinking soliton and its asymptotic cone. We will show, for example, that the isometry group of the cross-section of the cone embeds into that of the end of the shrinking soliton, and also that the shrinking soliton must be Kähler if the cone is Kähler. From the dynamical perspective, these results are just manifestations of the propagation of symmetry and restricted holonomy backwards in time along the Ricci flow.

4.1. Asymptotically conical shrinking ends

4.1.1. Asymptotically conical ends of manifolds.

The basic objects of interest in this chapter will be solitons with ends modeled on *regular* or *Euclidean cones* over closed Riemannian manifolds. We will use the same notation for these model spaces throughout the discussion which follows. Namely, (Σ, g_Σ) will denote a closed $(n-1)$-dimensional Riemannian manifold, and for $a \geq 0$, we will write

$$\mathcal{C}_a = \mathcal{C}_a^\Sigma := (a, \infty) \times \Sigma.$$

We will denote the metric cone over Σ by $\mathcal{C} = \mathcal{C}^\Sigma$, and write

$$\hat{g} := dr^2 + r^2 g_\Sigma$$

for the conical metric over the open half-cylinder $\mathcal{C}_0 = \mathcal{C}_0^\Sigma$. The symbol r will represent both the global coordinate on the first factor of $\mathcal{C}_0$ and the radial distance function $r : \mathcal{C}_0 \to \mathcal{C}_0$ given by $r(s, \sigma) = s$ on the cone from the (deleted) apex.

The conical metric $\hat{g}$ is characterized by the property that it scales by a factor of λ^2 when the interval $(0, \infty)$ is scaled by a factor of $\lambda > 0$. That is, if $\rho_\lambda : \mathcal{C}_0 \to \mathcal{C}_0$ is the map

$$\rho_\lambda(r, \sigma) := (\lambda r, \sigma)$$

which dilates the radial direction by a factor of $\lambda > 0$, then $\hat{g}$ transforms as

$$\rho_\lambda^* \hat{g} = \lambda^2 \hat{g} \tag{4.1}$$

under pullback by ρ_λ on $\mathcal{C}_0$. This characterization motivates the following definition:

DEFINITION 4.1 (Asymptotically Conical Ends). Let $a \geq 0$. A metric g on $\mathcal{C}_a$ is **C^k-asymptotic** to $\hat{g}$ if

$$\lambda^{-2} \rho_\lambda^* g \to \hat{g} \text{ in } C^k_{\mathrm{loc}}(\mathcal{C}_0, \hat{g}) \text{ as } \lambda \to \infty.$$

A Riemannian manifold (M^n, g) is **C^k- asymptotic to $\mathcal{C}^\Sigma$ along the end** $V \subset (M^n, g)$ if, for some $a \geq 0$, there is a diffeomorphism $F : \mathcal{C}_a \to V$ such that $F^* g$ is C^k-asymptotic to $\hat{g}$ on $\mathcal{C}_a$.

As we observe next, the assertion that a metric g on $\mathcal{C}_a$ is C^k-asymptotic to the conical metric $\hat{g}$ is equivalent to saying that the difference $g - \tilde{g}$ of the metrics and its covariant derivatives relative to $\hat{\nabla} = \nabla_{\hat{g}}$ of order $0 \leq m \leq k$ are $o(r^{-m})$ as $r \to \infty$.

LEMMA 4.2 (Characterization of C^k-conical asymptoticity). *Let $\hat{g} = dr^2 + r^2 g_\Sigma$ be a smooth conical metric on $\mathcal{C}_0$ and g a smooth metric on $\mathcal{C}_a$ for some $a > 0$. Then g is C^k-asymptotic to $\hat{g}$ on $\mathcal{C}_a$ if and only if, for all $0 \leq m \leq k$,*

$$\lim_{r \to \infty} \left(r^m \left| \widehat{\nabla}^m (g - \hat{g}) \right|_{\hat{g}} (r, \sigma) \right) = 0 \tag{4.2}$$

uniformly in σ as $r \to \infty$.

PROOF. Recall that if a metric g is dilated by a constant $c > 0$ and pulled back by a diffeomorphism ϕ, then the norm that g induces on a covariant l-tensor field T transforms according to the equation

$$|\phi^* T|_{c\phi^* g} = c^{-l/2} |T|_g \circ \phi.$$

Therefore, unwinding the notation in Definition 4.1, and using the identity (4.1) together with the invariance $\rho_\lambda^*\widehat{\nabla} = \widehat{\nabla}$ of the Levi-Civita connection under scaling, we see that

$$\begin{aligned}|\widehat{\nabla}^m(g-\hat{g})|_{\hat{g}} \circ \rho_\lambda &= |\widehat{\nabla}^m \rho_\lambda^*(g-\hat{g})|_{\rho_\lambda^*\hat{g}} = \lambda^{-(m+2)}|\widehat{\nabla}^m(\rho_\lambda^* g - \lambda^2\hat{g})|_{\hat{g}} \\ &= \lambda^{-m}|\widehat{\nabla}^m(\lambda^{-2}\rho_\lambda^* g - \hat{g})|_{\hat{g}}\end{aligned}$$

for all $\lambda \geq 1$. In particular, fixing $b > a$ and taking $\lambda = r/b$ for $r > b$, we see that

$$r^m|\widehat{\nabla}^m(g-\hat{g})|_{\hat{g}}(r,\sigma) = b^m|\widehat{\nabla}^m(\lambda^{-2}\rho_\lambda^* g - \hat{g})|_{\hat{g}}(b,\sigma)$$

for any $\sigma \in \Sigma$. It follows that a metric g on $\mathcal{C}_a$ is C^k-asymptotic to $\hat{g}$ if and only if, for all $0 \leq m \leq k$,

$$\lim_{r\to\infty}\left(r^m\left|\widehat{\nabla}^m(g-\hat{g})\right|_{\hat{g}}(r,\sigma)\right) = 0$$

uniformly in σ. □

4.1.1.1. *Uniqueness of asymptotic cones.* Lemma 4.2 shows that if a metric on $\mathcal{C}_a$ is C^0-asymptotic to a cone, the norm of the difference of the given metric and the conical metric tends to 0 uniformly as $r \to \infty$. We will need to know later (e.g., in the proof of Lemma 4.7) that a given metric on $\mathcal{C}_0$ can be C^0-asymptotic to at most one cone. First, we prove a lemma:

LEMMA 4.3. *Suppose g is a smooth metric on $\mathcal{C}_a = \mathcal{C}_a^\Sigma$ which is C^0-asymptotic to the cone $\hat{g} = dr^2 + r^2 g_\Sigma$. Then g and $\hat{g}$ are uniformly equivalent on $\overline{\mathcal{C}_b}$ for any $b > a$, and, for all $\varepsilon > 0$, there exists $b > a$ such that, for $(r,\sigma) \in \mathcal{C}_b$,*

$$(1-\varepsilon)|r-b| \leq \bar{r}_b(r,\sigma) \leq (1+\varepsilon)|r-b|, \tag{4.3}$$

where $\bar{r}_b(x) := \operatorname{dist}_g(x, \partial\mathcal{C}_b)$.

PROOF. By Lemma 4.2, our assumption implies that $\lim_{b\to\infty}\|g-\hat{g}\|_{C^0(\mathcal{C}_b,\hat{g})} = 0$. This implies that g and $\hat{g}$ are uniformly equivalent on any $\overline{\mathcal{C}_b} \subset \mathcal{C}_a$ and, further, that, for any given $\varepsilon > 0$, there is $b > a+1$ such that

$$(1-\varepsilon)^2\hat{g} \leq g \leq (1+\varepsilon)^2\hat{g} \tag{4.4}$$

on $\overline{\mathcal{C}_{b-1}}$. Let $x = (r,\sigma)$ be an arbitrary point of $\mathcal{C}_b$. Any curve γ in $\mathcal{C}_b$ connecting x to a point $y = (b,\sigma_0) \in \partial\mathcal{C}_b$ will satisfy that

$$(1-\varepsilon)L_{\hat{g}}[\gamma] \leq L_g[\gamma] \leq (1+\varepsilon)L_{\hat{g}}[\gamma].$$

Thus, as $d_{\hat{g}}(x, \partial\mathcal{C}_b) = r - b$, we have

$$(1-\varepsilon)|r-b| \leq \bar{r}_b(x) \leq (1+\varepsilon)|r-b|$$

as claimed. □

Now we prove the uniqueness statement. Essentially, the argument is that if a metric g is C^0-asymptotic to two cones along an end, then the cones are themselves C^0-asymptotic to each other, and that (given their rigid structure) this is only possible if the cones are isometric.

LEMMA 4.4 (Uniqueness of asymptotic cones). *Let (Σ, g_Σ) and $(\tilde{\Sigma}, g_{\tilde{\Sigma}})$ be closed $(n-1)$-dimensional manifolds and let $\hat{g} = dr^2 + r^2 g_\Sigma$ and $\tilde{g} = ds^2 + s^2 g_{\tilde{\Sigma}}$ be the associated conical metrics on $\mathcal{C}_0 = \mathcal{C}_0^\Sigma$ and $\tilde{\mathcal{C}}_0 = \mathcal{C}_0^{\tilde{\Sigma}}$. Let g be a smooth metric on*

$\mathcal{C}_a$ for some $a > 0$. Assume that:

(i) *The metric g is C^0-asymptotic to $\hat{g}$ on $\mathcal{C}_a$.*

(ii) *There is a $b > 0$ and a diffeomorphism $G : \tilde{\mathcal{C}}_b \to V$ onto an open subset $V \subset \mathcal{C}_a$ with compact complement in $\overline{\mathcal{C}_a}$ such that G^*g is C^0-asymptotic to $\tilde{g}$ on $\tilde{\mathcal{C}}_b$.*

Then there exists a smooth isometry $\varphi : (\mathcal{C}_0, \hat{g}) \to (\tilde{\mathcal{C}}_0, \tilde{g})$.

PROOF. Write $F = G^{-1} : V \to \tilde{\mathcal{C}}_b$ and fix (using the compactness of $\overline{\mathcal{C}_a} \setminus V$) an $a_0 > a$ such that $\overline{\mathcal{C}_{a_0}} \subset V$.

STEP 1: We claim that $\lim_{t\to\infty} \|F^*\tilde{g} - \hat{g}\|_{C^0(\mathcal{C}_t,\hat{g})} = 0$, that is (per Lemma 4.2), that the pull-back $F^*\tilde{g}$ of the cone $\tilde{g}$ to $\mathcal{C}_{a_0}$ is also C^0-asymptotic to $\hat{g}$.

By Lemma 4.2, our assumptions tell us that

$$\lim_{t\to\infty} \|g - \hat{g}\|_{C^0(\mathcal{C}_t,\hat{g})} = 0 = \lim_{t\to\infty} \|G^*g - \tilde{g}\|_{C^0(\tilde{\mathcal{C}}_t,\tilde{g})}. \tag{4.5}$$

By Lemma 4.3, the pairs g, $\hat{g}$ and G^*g, $\tilde{g}$ are therefore uniformly equivalent on any proper closed neighborhoods of infinity of $\mathcal{C}_a$ and $\tilde{\mathcal{C}}_b$, respectively. In particular, there is a constant N such that for any $t \geq a_0$,

$$\begin{aligned}\|F^*\tilde{g} - \hat{g}\|_{C^0(\mathcal{C}_t,\hat{g})} &\leq N\|F^*\tilde{g} - g\|_{C^0(\mathcal{C}_t,g)} + \|g - \hat{g}\|_{C^0(\mathcal{C}_t,\hat{g})} \\ &= N\|\tilde{g} - G^*g\|_{C^0(F(\mathcal{C}_t),G^*g)} + \|g - \hat{g}\|_{C^0(\mathcal{C}_t,\hat{g})}.\end{aligned}$$

Now, (4.5) says that the second term on the right tends to zero as $t \to \infty$, and it will also imply that the first term tends to zero provided we can show that, for all $u > 0$, there is $t > 0$ sufficiently large such that $F(\mathcal{C}_t) \subset \tilde{\mathcal{C}}_u$. To prove this, we argue by contradiction. If this claim is not true, there must exist some $u > b$ and a sequence $x_j = (r_j, \sigma_j) \in \mathcal{C}_{a_0}$ such that $r_j \to \infty$ but $F(x_j) \in \tilde{\mathcal{C}}_b \setminus \tilde{\mathcal{C}}_u$ for all j. However, by Lemma 4.3, the fact that $d_{\hat{g}}(x_j, \mathcal{C}_{a_0}) = r_j - a_0$ tends to infinity implies that

$$\lim_{j\to\infty} d_g(x_j, \partial\mathcal{C}_{a_0}) = \infty,$$

hence that

$$\lim_{j\to\infty} d_{G^*g}(F(x_j), F(\partial\mathcal{C}_{a_0})) = \infty,$$

and hence also that

$$\lim_{j\to\infty} d_{\tilde{g}}(F(x_j), F(\partial\mathcal{C}_{a_0})) = \infty.$$

This is a contradiction. Thus, $\lim_{t\to\infty} \|\tilde{g} - G^*g\|_{C^0(F(\mathcal{C}_t),G^*g)} = 0$, and then $\lim_{t\to\infty} \|F^*\tilde{g} - \hat{g}\|_{C^0(\mathcal{C}_t,\hat{g})} = 0$ as claimed.

Denote the dilation maps on $\mathcal{C}_0$ and $\tilde{\mathcal{C}}_0$ by ρ_λ and $\tilde{\rho}_\lambda$, respectively. According to Lemma 4.2, what we have shown says that $\lambda^{-2}\rho_\lambda^*F^*\tilde{g}$ converges to $\hat{g}$ in $C^0_{\text{loc}}(\mathcal{C}_0, \hat{g})$ as $\lambda \to \infty$. This, in turn, is equivalent to the assertion that $F_\lambda^*\tilde{g} \to \hat{g}$ in $C^0_{\text{loc}}(\mathcal{C}_0, \hat{g})$, where

$$F_\lambda := \tilde{\rho}_{\lambda^{-1}} \circ F \circ \rho_\lambda : \mathcal{C}_{\lambda^{-1}a_0} \to \tilde{\mathcal{C}}_{\lambda^{-1}b}.$$

We will write $F_\lambda(x) = (s_\lambda(x), \tilde{\sigma}_\lambda(x))$ for $x \in \mathcal{C}_{\lambda^{-1}a_0}$.

STEP 2: We claim that for any fixed $t > 0$, $s_\lambda(r, \cdot)/r \to 1$ uniformly as $\lambda \to \infty$ on any $\mathcal{C}_t$.

Applying Lemma 4.3, we see that for all $\varepsilon > 0$, there exists $t = t(\varepsilon) > a_0$ such that for all $x = (r, \sigma) \in \mathcal{C}_t$, the inequality

$$(1-\varepsilon)|r-t| \leq d_{F^*\tilde{g}}(x, \partial\mathcal{C}_t) = d_{\tilde{g}}(F(r,\sigma), F(\partial\mathcal{C}_t)) \leq (1+\varepsilon)\,|r-t|$$

holds. Since $\lambda^{-1} d_{\tilde{g}}(x, y) = d_{\tilde{g}}(\tilde{\rho}_{\lambda^{-1}}(x), \tilde{\rho}_{\lambda^{-1}}(y))$ for $\lambda > 0$, we then have

$$(1-\varepsilon)\left|r - \frac{t}{\lambda}\right| \leq d_{\tilde{g}}(F_\lambda(r,\sigma), (\tilde{\rho}_{\lambda^{-1}} \circ F)(\partial\mathcal{C}_t)) \leq (1+\varepsilon)\left|r - \frac{t}{\lambda}\right|,$$

for all $\lambda \geq 1$ and $(r, \sigma) \in \mathcal{C}_{\lambda^{-1}t}$.

Now, $F(\partial\mathcal{C}_t)$ is compact, so the distance from the vertex $\tilde{o}$ in the metric completion $(\tilde{\mathcal{C}}, d_{\tilde{\mathcal{C}}})$ of $(\tilde{\mathcal{C}}_0, \tilde{g})$ must satisfy

$$d_{\tilde{\mathcal{C}}}((\tilde{\rho}_{\lambda^{-1}} \circ F)(\partial\mathcal{C}_t), \tilde{o}) \leq C\lambda^{-1}$$

for some constant C independent of λ. It follows that

$$\frac{s_\lambda(r,\sigma)}{r} = \frac{d_{\tilde{\mathcal{C}}}(F_\lambda(r,\sigma), \tilde{o})}{r} \to 1$$

uniformly on $\mathcal{C}_t$ as $\lambda \to \infty$ for any fixed $t > 0$. In particular, there is $a_1 > a_0$ and $\lambda_0 > 0$ such that $F_\lambda(r, \sigma) \in \tilde{\mathcal{C}}_{r/2} \setminus \tilde{\mathcal{C}}_{2r}$ whenever $\lambda \geq \lambda_0$ and $r \geq a_1/\lambda$.

STEP 3: The cones $(\mathcal{C}_0, \hat{g})$ and $(\tilde{\mathcal{C}}_0, \tilde{g})$ are isometric.

With the control on F_λ we have just obtained and the local uniform convergence of $F_\lambda^* \tilde{g}$ to $\hat{g}$, we can find a sequence $\{\lambda_i\}_{i=1}^\infty$ such that $F_{\lambda_i}(\{r\} \times \Sigma) \subset \tilde{\mathcal{C}}_{r/2} \setminus \tilde{\mathcal{C}}_{2r}$ and

$$|d_{\hat{g}}(x_1, x_2) - d_{\tilde{g}}(F_{\lambda_i}(x_1), F_{\lambda_i}(x_2))| \leq \frac{N}{i} \tag{4.6}$$

for all $x_1, x_2 \in \mathcal{C}_{1/4i} \setminus \mathcal{C}_{4i}$ for some N depending only on the diameters of (Σ, g_Σ) and $(\tilde{\Sigma}, g_{\tilde{\Sigma}})$.

Let o denote the vertex in the metric completion $(\mathcal{C}, d_\mathcal{C})$ of $(\mathcal{C}_0, \hat{g})$. We define a sequence of maps $H_i : (B_\mathcal{C}(o, i), d_\mathcal{C}) \to (\tilde{\mathcal{C}}, d_{\tilde{\mathcal{C}}})$ by

$$H_i(x) := \begin{cases} F_{\lambda_i}(x) & x \in \mathcal{C}_{1/i} \setminus \mathcal{C}_i, \\ o & x \in \mathcal{C} \setminus \mathcal{C}_{1/i}. \end{cases}$$

Given (4.6) and the convergence of $r_{\lambda_i}(r, \sigma)$ to $\hat{r}$, it follows that the constant sequence $\{(\mathcal{C}, d_\mathcal{C}, o)\}_{i=1}^\infty$ of pointed metric spaces converges to $(\tilde{\mathcal{C}}, d_{\tilde{\mathcal{C}}}, \tilde{0})$ in the Gromov-Hausdorff sense. This implies that there exists an isometry $\varphi : (\mathcal{C}, d_\mathcal{C}, o) \to (\tilde{\mathcal{C}}, d_{\tilde{\mathcal{C}}}, \tilde{o})$ of pointed metric spaces (see, e.g, Theorem 8.1.7 in [**BBI01**]). But $(\mathcal{C}_0, \hat{g})$ and $(\tilde{\mathcal{C}}_0, \tilde{g})$ are smooth Riemannian manifolds, so, by the classical theorem of Myers and Steenrod [**MS39**], the restriction $\varphi|_{\mathcal{C}_0} : (\mathcal{C}_0, d_{\hat{g}}) \to (\tilde{\mathcal{C}}_0, d_{\tilde{g}})$ must in fact be a *smooth* isometry of the Riemannian manifolds $(\mathcal{C}_0, \hat{g})$ and $(\tilde{\mathcal{C}}_0, \tilde{g})$. This completes the proof.

□

4.1.2. Asymptotically conical shrinking ends.

Now we specialize to the case of asymptotically conical solitons.

DEFINITION 4.5. A gradient Ricci soliton (M^n, g, f, λ) is **C^k-asymptotic to $\mathcal{C}^\Sigma$ along the end** $V \subset M^n$ if (M^n, g) is C^k-asymptotic to $\mathcal{C}^\Sigma$ along V.

Note that the definition does not explicitly constrain the potential function f. We will return to this point shortly.

The asymptotically conical condition carries with it a number of added implications for shrinking solitons, many of which are most easily obtained by considering the self-similar solutions to the Ricci flow associated to the solitons. For example, as we will see, the standard local derivative-of-curvature estimates for the Ricci flow (see Shi [**Shi89**] and Hamilton [**Ham95b**]) imply that if a shrinking gradient Ricci soliton g is C^2-asymptotic to $\hat{g}$ on $\mathcal{C}_a$, it is actually **smoothly asymptotic** to $\hat{g}$, that is, C^k-asymptotic to $\hat{g}$ for all $k \geq 0$.

In fact, something even stronger is true: if the curvature tensor of a shrinker decays quadratically on an end, the shrinker must be smoothly asymptotic to *some* cone on that end (a cone which is unique up to isometry by Lemma 4.4). On the other hand, Wang and one of the authors [**MW17**] have shown that if the scalar curvature of a 4-dimensional shrinker tends to zero on an end, then the curvature tensor must decay quadratically. Thus any complete 4-dimensional shrinker whose scalar curvature tends to zero at infinity must be smoothly asymptotic to a cone. We will give a proof of this latter statement in Theorem 4.11 below. It relies on the following general proposition whose conclusions we will need elsewhere in our analysis.

It will be convenient to frame this and other of our statements in this chapter in terms of the **backward Ricci flow**

$$\partial_\tau g = 2\mathrm{Ric}(g). \tag{4.7}$$

One can convert any solution to the backward Ricci flow into a solution to the usual Ricci flow and vice versa via the transformation $\tau \leftrightarrow -t + \text{const}$.

PROPOSITION 4.6. *Let $(M^n, \bar{g}, \bar{f})$ be a (possibly incomplete) shrinking soliton and $E_0 \subset M^n$ a connected unbounded open set with smooth compact connected boundary $\partial E_0 \subset M^n$. Assume that:*

(i) *The function $\bar{r} := d_{\bar{g}}(\partial E_0, \cdot)$ is proper on $\overline{E_0}$.*
(ii) *E_0 is connected at infinity.*
(iii) *The curvature $\overline{\mathrm{Rm}}$ of $\bar{g}$ satisfies the quadratic decay bound*

$$|\overline{\mathrm{Rm}}|_{\bar{g}}(x) \leq \frac{K}{(\bar{r}(x)+1)^2} \tag{4.8}$$

on E_0 for some constant K.

Then:

(a) *There exists a connected open set $V \subset E_0$ with compact connected smooth boundary $\partial V \subset E_0$ satisfying the following properties: the set $\overline{E_0} \setminus V$ is compact, the integral curves of $\overline{\nabla} \bar{f}$ preserve V for $t > 0$, and every point of V sits on an integral curve of $\overline{\nabla} \bar{f}$ originating from ∂V.*

(b) *The maps* $\Phi_\tau : V \to V$ *defined by*

$$\partial_\tau \Phi_\tau = -\frac{1}{\tau}\overline{\nabla} \bar{f} \circ \Phi_\tau, \quad \Phi_1 = \mathrm{Id},$$

are well-defined injective local diffeomorphisms for $\tau \in (0,1]$.

(c) *The family of metrics* $g = g(\tau) = \tau \Phi_\tau^* \bar{g}$ *solves the backward Ricci flow* (4.7) *on* $V \times (0,1]$, *and satisfies the equations*

$$\mathrm{Ric}(g) + \nabla^2 f = \frac{g}{2\tau} \quad \text{and} \quad R + |\nabla f|^2 = \frac{f}{\tau}, \tag{4.9}$$

where $f = f(\tau) = \bar{f} \circ \Phi_\tau + c$ *for an appropriately chosen constant* c.

(d) *The curvature tensor of* $g(\tau)$ *satisfies the bounds*

$$\bar{r}^{m+2}(x)|\nabla^m \mathrm{Rm}|_{g(\tau)}(x,\tau) \leq K_m, \quad m \geq 0, \tag{4.10}$$

on $V \times (0,1]$ *for some constants* K_m.[2]

(e) *As* $\tau \searrow 0$, $g(\tau)$ *converges locally smoothly to a smooth metric* g_0 *and* $\tau f(\tau)$ *converges locally smoothly to a positive smooth proper function* q_0 *on* $\bar{V}$ *satisfying the identities*

$$\nabla^2_{g_0} q_0 = \frac{g_0}{2}, \quad |\nabla_{g_0} q_0|^2_{g_0} = q_0, \ \text{and}\ \mathrm{grad}_{\bar{g}} \bar{f} = \mathrm{grad}_{g_0} q_0.$$

(f) *The metric* g_0 *is conical in the sense that there exists* $c_0 > 0$, *a compact connected* $(n-1)$*-manifold* (Σ, g_Σ), *and a diffeomorphism* $F : \mathcal{C}_{c_0} \to W$ *between*

$$\mathcal{C}_{c_0} := (c_0, \infty) \times \Sigma \quad \text{and} \quad W := \{\, x \in V \mid q_0(x) > c_0^2/4 \,\},$$

such that

$$q_0(F(r,\sigma)) = \frac{r^2}{4}, \quad F^* g_0 = \hat{g} := dr^2 + r^2 g_\Sigma,$$

and

$$\mathrm{grad}_{F^* g_0} F^* q_0 = \frac{r}{2}\frac{\partial}{\partial r}.$$

(g) *The metric* $F^* \bar{g}$ *is smoothly asymptotic to* $\hat{g}$ *on* $\mathcal{C}_{c_0}$ *and*

$$\mathrm{grad}_{F^* \bar{g}} F^* \bar{f} = \frac{r}{2}\frac{\partial}{\partial r}.$$

4.1.2.1. *Proof of Proposition* 4.6. Since E_0 is connected, by adding a constant to $\bar{f}$, we may assume as usual that

$$\bar{R} + |\overline{\nabla} \bar{f}|^2 = \bar{f} \tag{4.11}$$

on E_0. For $a > 0$, define

$$E_a = \{\, x \in E_0 \mid \bar{r}(x) > a \,\}.$$

We will break up the proof into a sequence of smaller claims. In the argument that follows, we will use C to denote a sequence of positive constants which may change from line to line.

[2]We will revisit these bounds in Lemma 4.8 below.

CLAIM 1. *There is a constant C such that*

$$\frac{1}{4}(\bar{r}-C)^2 \leq \bar{f} \leq \frac{1}{4}(\bar{r}+C)^2 \tag{4.12}$$

and

$$\frac{\bar{r}}{2} - C \leq \langle \overline{\nabla} \bar{f}, \overline{\nabla} \bar{r} \rangle_{\bar{g}} \leq \frac{\bar{r}}{2} + C \tag{4.13}$$

on E_0. In particular, there is $a_0 \geq 4$ such that

$$\frac{\bar{r}^2}{8} \leq \bar{f} \leq \frac{\bar{r}^2}{2}, \quad \frac{\bar{r}}{4} \leq \langle \overline{\nabla} \bar{f}, \overline{\nabla} \bar{r} \rangle \leq \bar{r}, \quad \textit{and} \quad |\overline{\mathrm{Rm}}|_{\bar{g}} \leq C \bar{f}^{-1}, \tag{4.14}$$

on E_{a_0}.[3]

For (4.12)–(4.13), let $x \in E_0$ and let $\gamma : [0, \bar{r}(x)] \to \overline{E_0}$ be any unit speed geodesic in E_0 with $\gamma(0) \in \partial E_0$ which realizes the distance from ∂E_0 to x. Then

$$(\bar{f} \circ \gamma)''(s) = -\overline{\mathrm{Ric}}_{\gamma(s)}(\gamma'(s), \gamma'(s)) + \frac{1}{2}$$

along γ, so that, using the quadratic curvature decay assumption (4.8), we have

$$-\frac{C}{(s+1)^2} \leq (\bar{f} \circ \gamma)''(s) - \frac{1}{2} \leq \frac{C}{(s+1)^2}.$$

Integrating once along γ yields

$$-C \leq \langle \overline{\nabla} \bar{f}, \overline{\nabla} \bar{r} \rangle_{\bar{g}} - \frac{\bar{r}}{2} \leq C$$

for some constant C now also depending on $\sup_{\partial E_0} |\overline{\nabla} \bar{f}|$. This is (4.13). Integrating the previous inequality along γ twice, meanwhile, gives

$$\frac{\bar{r}^2}{4} - C(\bar{r}+1) \leq \bar{f} \leq \frac{\bar{r}^2}{4} + C\,(\bar{r}+1)\,,$$

for some C depending on $\sup_{\partial E_0}(|\bar{f}| + |\overline{\nabla} \bar{f}|)$. This implies (4.12). From (4.12) and (4.13), we can then achieve the bounds in (4.14) by taking a_0 sufficiently large.

CLAIM 2. *Let $D(b)$ and $S(b)$ denote the sets*

$$D(b) = \{\, x \in E_{a_0} \mid \bar{f}(x) > b \,\}, \quad S(b) = \partial D(b).$$

There is $b_0 > 1$ such that, for $b \geq b_0$, the sets $S(b)$ are nonempty smooth hypersurfaces in E_{a_0} diffeomorphic to a common compact connected manifold S. Also $D(b) \approx (b, \infty) \times S$, and

$$E_{\sqrt{8b}} \subset D(b) \subset E_{\sqrt{2b}} \tag{4.15}$$

for such b.

Take, e.g., $b_0 = 8 \times \sup\{\, \bar{f}(x) \mid x \in \overline{E_0} \setminus E_{a_0}\}$; by (4.14), this ensures in particular that $b_0 > 1$. The bounds in (4.14), which imply that $\bar{f}$ is proper and $\overline{\nabla} \bar{f} \neq 0$ on $\overline{E_{a_0}}$, also imply that $S(b)$ and $D(b)$ are nonempty and strictly contained in E_{a_0} for all $b \geq b_0$. Thus, for $b \geq b_0$, the level sets $S(b)$ of $\bar{f}$ are compact smooth hypersurfaces contained in E_{a_0}. Using the flow of the Morse vector field $\overline{\nabla} \bar{f} / |\overline{\nabla} \bar{f}|^2_{\bar{g}}$, we deduce that these $S(b)$ are all diffeomorphic to $S := S(b_0)$ and that there is a diffeomorphism $\Psi : (b_0, \infty) \times S \to D(b_0)$ with $\bar{f} \circ \Psi(b, \sigma) = b$.

[3]Compare with the estimates of Cao and Zhou [**CZ10**] in §1.6.4. Here, we do not assume that $(E_0, \bar{g}, \bar{f})$ is the restriction to E_0 of a complete shrinking soliton.

The inclusions in (4.15) follow immediately from the bounds (4.14) on f. The left inclusion implies in particular that $\overline{E_0} \setminus D(b)$ is compact for all $b \geq b_0$. Since E_0 is connected at infinity, $D(b_0)$ must be connected, and as $D(b_0) \approx (b_0, \infty) \times S$, S must be connected, too.

CLAIM 3. *Define $V := D(b_0)$. Then $V \subset E_0$ is open and connected with connected smooth boundary $\partial V = S(b_0)$, and $\overline{E_0} \setminus V$ is compact. Moreover, the integral curves $\sigma(s)$ of $\overline{\nabla} \bar{f}$ preserve $\overline{V}$ for $s > 0$ and every point of V lies on an integral curve of $\overline{\nabla} \bar{f}$ originating at a point of ∂V.*

We have already verified the first several assertions about V in Claim 2, and the fact that the integral curves of $\overline{\nabla} \bar{f}$ preserve $\overline{V}$ for positive time follows immediately from the definition of $\overline{V}$ as a sublevel set of $\bar{f}$. Thus, the flow φ_s of $\overline{\nabla} \bar{f}$ defined by

$$\partial_s \varphi_s = \overline{\nabla} \bar{f} \circ \varphi_s, \quad \varphi_0 = \mathrm{Id},$$

maps V diffeomorphically onto its image in V for all $s \in [0, \infty)$. (In fact, since ∂V is compactly contained in E_{a_0}, φ_s is actually defined on $\overline{V}$ for $s \in [-\delta, \infty)$ for some $\delta > 0$.)

To see that every point of V lies on an integral curve of $\overline{\nabla} \bar{f}$ starting at a point in ∂V, suppose $p \in V = D(b_0)$. Then $\bar{f}(p) > b_0$. If $\sigma(s) = \varphi_s(p)$, $s \in I$, is the maximally-defined integral curve of $\overline{\nabla} f$ in E_0 with $\sigma(0) = p$, then, in view of (4.14), we have

$$(f \circ \sigma)'(s) = |\overline{\nabla} f|^2_{\bar{g}}(\sigma(s)) \geq 1$$

for all $s \in I$ such that $\sigma(s) \in E_{a_0}$. Hence,

$$f(\sigma(s)) \leq f(p) + s$$

for all such s with $s < 0$. This implies that $f(\sigma(s_0)) = b_0$ for some largest $s_0 < 0$. Then $y := \sigma(s_0) \in \partial V = S(b_0)$ and $p = \varphi_{|s_0|}(y)$. This completes the proof of part (a) of the proposition.

CLAIM 4. *The self-similar solution $g(\tau)$ to the backward Ricci flow associated to $\bar{g}$ on V is well-defined for $\tau \in (0, 1]$.*

Define $s(\tau) = -\ln \tau$ for $\tau \in (0, 1]$ and let $\Phi_\tau = \varphi_{s(\tau)}$. Then

$$\partial_\tau \Phi = -\frac{1}{\tau} \overline{\nabla}_{\bar{g}} \bar{f} \circ \Phi_\tau, \quad \Phi_1 = \mathrm{Id},$$

and it follows from the previous claim that $\Phi_\tau : V \to V$ is well-defined and an injective local diffeomorphism for $\tau \in (0, 1]$. If

$$g = g(\tau) = \tau \Phi_\tau^* \bar{g}, \quad f = f(\tau) = \bar{f} \circ \Phi_\tau,$$

then (see, e.g., Proposition 1.7 of [**CCG$^+$07**])

$$\partial_\tau g = 2\mathrm{Ric}(g), \quad g(1) = \bar{g},$$

and

$$\mathrm{Ric}(g) + \nabla\nabla f = \frac{1}{2\tau}, \quad R + |\nabla f|^2 = \frac{f}{\tau},$$

on $V \times (0, 1]$, that is, $g(\tau)$ solves the backward Ricci flow and $(V, g(\tau), f(\tau), \tau^{-1})$ is a normalized gradient shrinking soliton structure for each $\tau \in (0, 1]$. This verifies parts (b) and (c) of the proposition.

CLAIM 5. *Increasing the value of* b_0 *in* $V = D(b_0)$, *if necessary, the solution* $g(\tau)$ *will satisfy the bounds*

$$\bar{r}^{m+2}(x)|\nabla^{(m)}\mathrm{Rm}|(x,\tau) \le K_m, \quad m \ge 0,$$

on $V \times (0,1]$ *for some constants* K_m.

First we verify that the curvature of the solution $g(\tau)$ is bounded on $V \times (0,1]$. Fix $x \in V$. Writing $\rho(\tau) = \bar{r}(\Phi_\tau(x))$, we see from (4.8) that

$$|\mathrm{Rm}|(x,\tau) = \frac{1}{\tau}\cdot|\overline{\mathrm{Rm}}|_{\bar{g}}(\Phi_\tau(x)) \le \frac{K}{\tau\rho^2(\tau)}. \tag{4.16}$$

On the other hand, we have the differential inequality

$$-\frac{1}{2\tau}(\rho(\tau)+C) \le \rho'(\tau) \le -\frac{1}{2\tau}(\rho(\tau)-C) \tag{4.17}$$

for $\tau \in (0,1]$ in view of (4.13). Thus, for $\tau \in (0,1]$,

$$\frac{1}{\sqrt{\tau}}(\bar{r}(x)-C)+C \le \rho(\tau) \le \frac{1}{\sqrt{\tau}}(\bar{r}(x)+C)-C,$$

so that

$$\bar{r}(x)+C\sqrt{\tau}-C \le \sqrt{\tau}\rho(\tau) \le \bar{r}(x)+C-C\sqrt{\tau}. \tag{4.18}$$

By Claim 1, we may increase b_0 if necessary so that $V = D(b_0) \subset E_{2C}$, where C is the constant on the left-hand side of the above inequality. This will ensure us that

$$\frac{1}{2}\bar{r}(x) \le \sqrt{\tau}\bar{r}(\Phi_\tau(x)) \le 2\bar{r}(x) \tag{4.19}$$

on $V \times (0,1]$. But then (4.16) gives

$$|\mathrm{Rm}|(x,\tau) \le \frac{K_0}{\bar{r}^2(x)},$$

for $K_0 = 4K$, so that the curvature of $g(\tau)$ is bounded and moreover decays quadratically with respect to $\bar{r}$ on $V \times (0,1]$.

Increasing b_0 slightly in $V = D(b_0)$, we may assume that our solutions $g(\tau)$ are actually well-defined on $V \times (0, 1+\delta]$ for some $\delta > 0$. Then Shi's local derivative estimates imply that (increasing b_0 slightly again) we will have the uniform estimates

$$|\nabla^m \mathrm{Rm}|(x,\tau) \le \tilde{K}_m$$

for each $m \ge 1$ on $V \times (0,1]$ for some constants $\tilde{K}_m$.

Using the self-similarity of the solution, we can parlay these bounds into derivative bounds for the curvature tensor of the original soliton. Suppose $x \in V$. Then it follows from Claim 2 that there is $y \in \partial V$ and a unique $\tau \in (0,1]$ such that $\Phi_\tau(y) = x$. Using (4.19), we can rescale to see that, for each $m \ge 1$,

$$\begin{aligned}|\overline{\nabla}^m\overline{\mathrm{Rm}}|_{\bar{g}}(x) &= |\nabla^m\overline{\mathrm{Rm}}|_{\bar{g}}(\Phi_\tau(y)) = \tau^{(m/2+1)}|\nabla^m\mathrm{Rm}|(y,\tau)\\ &\le \tau^{(m/2+1)}\tilde{K}_m \le C_m\tilde{K}_m\left(\frac{\bar{r}^{m+2}(y)}{\bar{r}^{m+2}(x)}\right)\\ &\le \frac{C_m\tilde{K}_m}{\bar{r}^{m+2}(x)}\end{aligned}$$

for some constant C_m depending on $\mathrm{dist}_{\bar{g}}(\partial V, \partial E_0)$. But $x \in V$ was arbitrary, so these bounds hold on all of V.

Finally, we can rescale one last time to see that the derivatives of the curvature tensor of the solution also decay of order $O(\bar{r}^{-(m+2)})$:

$$
\begin{aligned}
|\nabla^m \mathrm{Rm}|(x,\tau) = \tau^{-(m/2+1)}|\overline{\nabla}^m \overline{\mathrm{Rm}}|_{\bar{g}}(\Phi_\tau(x)) &\leq \frac{C_m \tilde{K}_m \tau^{-(m/2+1)}}{\bar{r}^{m+2}(\Phi_\tau(x))} \\
&\leq \frac{C_m \tilde{K}_m}{\bar{r}^{m+2}(x)}
\end{aligned}
$$

on $V \times (0,1]$. This proves part (d) of the proposition.

CLAIM 6. *The solution $g = g(\tau)$ converges locally smoothly to a smooth metric g_0 on V, and $\tau f(\tau) = \tau \bar{f} \circ \Phi_\tau$ converges locally smoothly to a smooth positive proper function q_0 on V which satisfies the equations*

$$
\nabla^2_{g_0} q_0 = \frac{g_0}{2}, \quad |\nabla_{g_0} q_0|^2_{g_0}, = q_0, \tag{4.20}
$$

on V.

With the pieces we already have in place, the local smooth convergence of $g(\tau)$ to a smooth limit g_0 follows from a standard sequence of arguments. First, one uses the uniform bound on $|\mathrm{Rm}|(x,\tau)$ on $V \times (0,1]$ to estimate the speed $\partial_\tau g = 2\mathrm{Ric}$ and argue that $g(\tau)$ converges to a continuous metric g_0 as $\tau \to 0$. Then, using the bounds on the norms $|\nabla^m \mathrm{Rm}|(x,\tau)$ of the higher derivatives of curvature, and working on fixed local charts, one shows that $g(\tau)$ converges locally uniformly in C^k to g_0 for every k, and therefore that g_0 is smooth. For details, see, e.g., §6.7 in the text [**CK04**] by Knopf and one of the authors.

Once we have obtained the local smooth convergence of $g(\tau)$ to g_0, we can use the identity (4.11) to see that

$$
\frac{\partial}{\partial \tau}(\tau f(x,\tau)) = -\tau R(x,\tau)
$$

at each $x \in V$ so that

$$
\bar{f}(x) - \tau f(x,\tau) = -\int_\tau^1 sR(x,s)\,ds. \tag{4.21}
$$

From this it follows that τf converges locally uniformly as $\tau \to 0$ to a continuous limit q_0. As in §6.7 of [**CK04**], the bounds on the covariant derivatives of the curvature tensor deduced in the last step can be converted into bounds on the derivatives of the curvature with respect to local coordinate systems. Then the local smooth convergence of τf to q_0 and the smoothness of q_0 can be obtained by successively differentiating (4.21).

Next, combining (4.14) and (4.19), we obtain that

$$
\frac{1}{32}\bar{r}^2(x) \leq \tau f(x,\tau) \leq 2\bar{r}^2(x)
$$

for all $(x,\tau) \in V \times (0,1]$. Sending $\tau \to 0$, it follows that q_0 satisfies the same bounds and is both positive and proper on V. In fact,

$$
\frac{1}{C} \cdot \bar{f}(x) \leq q_0(x) \leq C\bar{f}(x) \tag{4.22}
$$

for some $C > 0$.

Finally, using the local smooth convergence of $g = g(\tau)$ to g_0 and τf to q_0, we may take the limit as $\tau \to 0$ in the equations

$$\tau \mathrm{Ric}(g) + \nabla^2(\tau f) = \frac{g}{2}, \quad |\nabla(\tau f)|^2 = \tau f - \tau^2 R,$$

to obtain the identities (4.20). This proves all but the last assertion in part (e) of the proposition, which we address next.

CLAIM 7. *The vector field of the potential satisfies*

$$\tau \, \mathrm{grad}_{g(\tau)} f = \mathrm{grad}_{g_0} q_0$$

on V for each $\tau \in (0, 1]$.

The local smooth convergence of $g(\tau)$ to g_0 and τf to q_0 imply that $\mathrm{grad}_{g(\tau)}(\tau f)$ converges to $\mathrm{grad}_{g_0} q_0$ as $\tau \to 0$. Thus it suffices to show that $\mathrm{grad}_{g(\tau)}(\tau f)$ is independent of time. For this, we can simply compute that

$$\begin{aligned}\partial_\tau \nabla(\tau f) &= -2\tau \mathrm{Ric}(\nabla f) - \tau \nabla |\nabla f|^2 + \nabla f \\ &= -2\left(\mathrm{Ric} + \nabla^2 f - \frac{g}{2\tau}\right)(\tau \nabla f) = 0,\end{aligned}$$

using the soliton identities and the evolution equations for the Ricci flow.

CLAIM 8. *There is $c_0 > 0$, a compact connected $(n-1)$-manifold (Σ, g_Σ), and a diffeomorphism $F : \mathcal{C}_0 \to W$ from $\mathcal{C}_{c_0} = (c_0, \infty) \times \Sigma$ onto $W = \{ x \in V \mid q_0(x) > c_0^2/4 \}$ such that*

$$F^*(g_0) = \hat{g} := dr^2 + r^2 g_\Sigma \quad \textit{and} \quad \mathrm{grad}_{F^* g_0} F^* q_0 = \frac{r}{2} \partial_r$$

on $\mathcal{C}_{c_0}$.

Define $r(x) := 2\sqrt{q_0(x)}$ on V. By (4.20) and (4.22),

$$|\nabla_{g_0} r|^2_{g_0} = 1 \quad \text{and} \quad \alpha^{-1} \bar{f}(x) \le r^2(x) \le \alpha \bar{f}(x) \tag{4.23}$$

for some $\alpha > 1$. Since

$$\mathrm{grad}_{g_0}(r^2/4) = \mathrm{grad}_{g_0} q_0 = \mathrm{grad}_{\bar{g}} \bar{f},$$

it follows that

$$d\bar{f}(\nabla_{g_0} r) = \frac{2}{r} |\overline{\nabla} \bar{f}|^2_{\bar{g}} \ge Cr \tag{4.24}$$

on V for some $C > 0$, so that $\bar{f}$ is increasing at a uniform rate along the integral curves of $\nabla_{g_0} r$. Arguing as we did in Claim 3, we see that the integral curves of $\nabla_{g_0} r$ preserve V and every point of V lies on an integral curve of $\nabla_{g_0} r$ passing through ∂V.

Define the sets

$$\tilde{D}(c) = \{ x \in V \mid r(x) > c \}, \quad \tilde{S}(c) = \{ x \in V \mid r(x) = c \}.$$

For $c \ge c_0 > 0$ sufficiently large, the sets $\tilde{D}(c)$ and $\tilde{S}(c)$ are contained in V and $\tilde{S}(c)$ is diffeomorphic to some common compact manifold $\Sigma \approx \Sigma(c_0)$. For such c,

$$D(c^2 \alpha) \subset \tilde{D}(c) \subset D(c^2/\alpha).$$

Then $\overline{V} \setminus \tilde{D}(c_0)$ is compact, and therefore so too is $\overline{E_0} \setminus \tilde{D}(c_0)$. As before, $\tilde{D}(c_0)$ is connected.

Define $\mathcal{C}_{c_0} := (c_0, \infty) \times \Sigma$ and $W := \tilde{D}(c_0)$. The flow of $\nabla_{g_0} r$ produces a diffeomorphism

$$F : \mathcal{C}_{c_0} \to W$$

satisfying

$$F(\{s\} \times \Sigma) = \tilde{S}(s) \quad \text{and} \quad F_*(\partial_r) = \nabla_{g_0} r.$$

Since $W = \tilde{D}(c_0)$ is connected, Σ is connected. This verifies the first assertions in part (f) of the proposition.

CLAIM 9. *The metric $F^* g_0$ is conical.*

It is standard to deduce from the identities in part (f) that $\hat{g} = F^* g_0$ is a cone. First, we note that, at any point of $\mathcal{C}_0$,

$$\hat{g}(\partial_r, \partial_r) = (F^* g_0)(\partial_r, \partial_r) = g_0(F_* \partial_r, F_* \partial_r) = |\nabla_{g_0} r|^2_{g_0} = 1.$$

Next, we note that if $v \in T_{(s,\sigma)} \mathcal{C}_{c_0}$ is tangent to $\{s\} \times \Sigma$, then

$$0 = v(r \circ F) = dr(F_* v) = g_0(\nabla_{g_0} r, F_* v) = g_0(F_* \partial_r, F_* v) = \hat{g}(\partial_r, v).$$

For the components of $F^* g_0$ tangent to the slices $\{\text{const}\} \times \Sigma$, fix any $c_1 > c_0$ and define

$$g_\Sigma := c_1^{-2} \iota^* (F^* g_0),$$

where $\iota : \Sigma \to \{c_1\} \times \Sigma$ is the map $\iota(\sigma) = (c_1, \sigma)$. Let $U \subset \Sigma$ be a coordinate neighborhood about $p \in \Sigma$ and let v, w be the natural extensions of local coordinate vector fields on U to vector fields on $(c_0, \infty) \times U$ tangent to the slices $\{\text{const}\} \times \Sigma$. Define $\tilde{v} = F_* v$ and $\tilde{w} = F_* w$. Then $[\partial_r, v] = 0 = [\nabla_{g_0} r, \tilde{v}]$, with analogous identities for w and $\tilde{w}$.

From the equation

$$g_0 = r \nabla^2_{g_0} r + dr \otimes dr,$$

which follows from (4.20), we see that

$$\mathcal{L}_{\nabla_{g_0} r}(g_0(\tilde{v}, \tilde{w})) = \mathcal{L}_{\nabla_{g_0} r}(g_0)(\tilde{v}, \tilde{w}) = 2\nabla^2_{g_0} r(\tilde{v}, \tilde{w}) = \frac{2}{r} g_0(\tilde{v}, \tilde{w}).$$

That is,

$$\partial_r((F^* g_0)_{(r,\sigma)}(v, w)) = \frac{2}{r}(F^* g_0)_{(r,\sigma)}(v, w)$$

for all $\sigma \in U$. This implies that

$$(F^* g_0)_{(r,\sigma)}(v, w) = \frac{r^2}{c_1^2}(F^* g_0)_{(c_1,\sigma)}(v, w) = r^2 (g_\Sigma)_\sigma(v, w)$$

on $(c_0, \infty) \times U$. But $U \subset \Sigma$ was arbitrary, so, taken together, the cases above show that

$$F^* g_0 = \hat{g} = dr^2 + r^2 g_\Sigma$$

as claimed. This completes the proof of part (f) of the proposition. (See §4.3.2 of [**Pet16**] for a more general version of this characterization.)

CLAIM 10. *$F^* \bar{g}$ is smoothly asymptotic to $\hat{g}$ on $\mathcal{C}_{c_0}$ and* $\operatorname{grad}_{F^* \bar{g}} F^* \bar{f} = (r/2)\partial_r$.

We have already seen that $\operatorname{grad}_{\bar{g}} \bar{f} = \operatorname{grad}_{g_0} q_0$ in Claim 7. Combining this with Claim 8 yields

$$\operatorname{grad}_{F^* \bar{g}} F^* \bar{f} = (F^{-1})_* \operatorname{grad}_{\bar{g}} \bar{f} = (F^{-1})_* \operatorname{grad}_{g_0} q_0 = (r/2)\partial_r.$$

Thus, the family of maps $G_\tau : \mathcal{C}_{c_0} \to \mathcal{C}_{c_0}$ defined by

$$G_\tau(r,\sigma) = (r/\sqrt{\tau}, \sigma)$$

for $\tau \in (0,1]$ satisfy

$$\partial_\tau G_\tau(r,\sigma) = -\frac{r}{2\tau^{3/2}} \frac{\partial}{\partial r}\bigg|_{G_\tau(r,\sigma)} = -\frac{1}{\tau}\left((F^{-1})_* \operatorname{grad}_{\bar{g}} \bar{f}\right)\big|_{G_\tau(r,\sigma)}$$

and that $G_1 = \mathrm{Id}$. Hence $F \circ G_\tau = \Phi_\tau \circ F$, where Φ_τ is as defined in Claim 4, and

$$F^* g(\tau) = \tau G_\tau^*(F^*\bar{g}).$$

Since $\rho_\lambda : \mathcal{C}_0 \to \mathcal{C}_0$ satisfies

$$\rho_\lambda(r,\sigma) = (\lambda r, \sigma) = G_{\lambda^{-2}}(r,\sigma)$$

for $\lambda \geq 1$, we see that

$$\lambda^{-2}\rho_\lambda^* F^*\bar{g} = \lambda^{-2} G_{\lambda^{-2}}^* F^*\bar{g} = F^* g(\lambda^{-2}).$$

But we know that $g(\tau)$ converges locally smoothly on W to g_0 as $\tau \to 0$ (a fact valid with respect to the local C^k-norms defined by any metric, including $\hat{g}$ – see §6.7 of [**CK04**]), so it follows from the above identity that $\lambda^{-2}\rho_\lambda^* F^*\bar{g}$ converges locally smoothly on $\mathcal{C}_c$ to $F^* g_0 = \hat{g}$ as $\lambda \to \infty$. That is, $F^*\bar{g}$ is smoothly asymptotic to $\hat{g}$ on $\mathcal{C}_c$. This completes the proof of Proposition 4.6. □

4.1.2.2. *Smooth conical asymptoticity for Ricci shrinkers.*

Any shrinker which is C^2-asymptotic to some cone on an end E will (in particular) have quadratic curvature decay. Thus by the above proposition, it will be smoothly asymptotic to some (a priori, different) cone along that E. However, Lemma 4.4 implies that these two cones must be isometric, so our shrinker must have been *smoothly* asymptotic to the original cone all along. We summarize these observations in the following lemma.

LEMMA 4.7 (From C^2-asymptoticity to smooth asymptoticity). *Suppose that the shrinker (M^n, g, f) is C^2-asymptotic to the cone $(\mathcal{C}^\Sigma, \hat{g})$ along the end $V \subset M^n$. Then g is smoothly asymptotic to $\mathcal{C}^\Sigma$ on V.*

PROOF. It suffices to assume that $V = \mathcal{C}_a$ for some $a > 0$. By Lemma 4.2, we know that

$$\lim_{s\to\infty} s^m \|\widehat{\nabla}^m h\|_{C^0(\mathcal{C}_s,\hat{g})} = 0$$

for $m = 0,1,2$, where $h := g - \hat{g}$. Since the difference of the $(4,0)$-curvature tensors Rm and $\widehat{\mathrm{Rm}}$ satisfy the schematic equation

$$\mathrm{Rm} - \widehat{\mathrm{Rm}} = \widehat{\nabla}^2 h + g^{-2} * \widehat{\nabla} h * \widehat{\nabla} h, \tag{4.25}$$

and $\widehat{\mathrm{Rm}}$ has quadratic curvature decay, it follows that, fixing $b > a$, there is a constant K such that $r^2 |\mathrm{Rm}|_{\hat{g}}(r,\sigma) \leq K$ on $\mathcal{C}_b$.

This says that Rm has quadratic curvature decay relative to the $\hat{g}$-radial distance. However, by Lemma 4.3, the metrics $\hat{g}$ and g are uniformly equivalent and the radial distance r is comparable to the distance $\bar{r} = d_g(\cdot, \partial\mathcal{C}_b)$ induced by g. Therefore, $\bar{r}^2(r,\sigma)|\mathrm{Rm}|_g(r,\sigma) \leq K$ for some constant K as well. It follows from Proposition 4.6 that there is an open set $W \subset \mathcal{C}_b$ such that $\overline{\mathcal{C}_b} \setminus W$ is compact and a cone $(\tilde{\mathcal{C}}_0^{\tilde{\Sigma}}, \tilde{g})$ such that g is smoothly asymptotic to $\tilde{\mathcal{C}}^{\tilde{\Sigma}}$ along W. However, by Lemma 4.4, $(\mathcal{C}_0, \hat{g})$ and $(\tilde{\mathcal{C}}_0, \tilde{g})$ must be isometric. Thus g is smoothly asymptotic to $\hat{g}$ on $\mathcal{C}_a$ as claimed. □

4.1.2.3. *A normalizing gauge and improved asymptotics.*

According to Definition 4.5, a shrinking soliton is asymptotic to a cone along an end if the underlying Riemannian manifold is asymptotic to a cone along that end. The definition does not impose any explicit constraints on the potential function. The proof of Proposition 4.6 demonstrates that interpreting the asymptotically conical condition in terms of the solution to the Ricci flow associated to the shrinker on the end naturally introduces a "gauge" on the end of the soliton in which the gradient of the potential and the generator of the dilation diffeomorphisms $\rho_{e^{s/2}}$ coincide. The following lemma shows that there are some added advantages to working within this gauge, including the ability to easily obtain (and express) improved asymptotics for the metric, curvature, and potential.

LEMMA 4.8 (Normalization Lemma). *Suppose that the shrinker (M^n, g, f) is smoothly asymptotic to $\mathcal{C}^{\Sigma}$ along the end V. Then there exists $a > 0$, an end $W \subset V$ of M^n, and a diffeomorphism $F : \mathcal{C}_a \to W$ such that $\tilde{g} := F^*g$ is smoothly asymptotic to $\hat{g}$ and $\tilde{f} = F^*f$ satisfies*

$$\operatorname{grad}_{\tilde{g}} \tilde{f} = \frac{r}{2}\frac{\partial}{\partial r} \tag{4.26}$$

on $\mathcal{C}_a$. Moreover, for all $m \geq 0$, we have the improved decay estimates

$$|\nabla^m(\tilde{g} - \hat{g})| \leq L_m r^{-(m+2)}, \quad \left|\nabla^m\left(\widetilde{\mathrm{Rm}} - \widehat{\mathrm{Rm}}\right)\right| \leq L_m r^{-(m+4)}, \tag{4.27}$$

and

$$\left|f(r,\sigma) - \frac{r^2}{4}\right| \leq L r^{-2} \tag{4.28}$$

on $\mathcal{C}_a$ for some positive constants L_m and L.

REMARK 4.9. The estimates in (4.27) can be sharpened further by feeding them back into the evolution equation for $\tilde{g}(\tau)$. Indeed, one can show that

$$\left|\hat{\nabla}^m\left(\tilde{g} - \hat{g} - 2\widehat{\mathrm{Ric}}\right)\right| = O(r^{-(m+4)}).$$

For further details, and an alternative framing of the reductions in Proposition 4.6 and Lemmas 4.7 and 4.8, see Theorem 3.8 in [**CDS24**].

PROOF. The first two statements follow from Proposition 4.6 and Lemma 4.7. For the decay estimates and asymptotics, we return to the setting of the latter part of the proof of Proposition 4.6. We may assume that we have passed already to the smooth solution $\tilde{g}(\tau) = \tau G_\tau^* \tilde{g}_1$ to the backward Ricci flow on $\mathcal{C}_a \times (0, 1]$ for some $a > 0$ which satisfies $\tilde{g}_1 = \tilde{g}(1)$ and converges smoothly to $\hat{g}$ on $\mathcal{C}_a$ as $\tau \to 0$. Here, as before, $G_\tau(r, \sigma) = (r/\sqrt{\tau}, \sigma)$.

From the fact that $\tilde{g}_1$ is smoothly asymptotic to $\hat{g}$, we know that, for all $k \geq 0$,

$$\lim_{b\to\infty} \sup_{x \in \mathcal{C}_b} \left(r^k(x)|\hat{\nabla}^k h|_{\hat{g}}(x, 1)\right) = 0 \tag{4.29}$$

where $h(x, \tau) = \tilde{g}(x, \tau) - \hat{g}(x)$. We will use the fact that the Ricci flow $\tilde{g}(\tau)$ converges to $\hat{g}$ to parlay this into more rapid decay. Since the norms $|\cdot|_g$ and $|\cdot|_{\hat{g}}$ are equivalent on $\mathcal{C}_c$, we will only use a subscript to distinguish between them when the distinction is important.

Just as in Lemma 4.7, we have a bound of the form $|\widetilde{\mathrm{Rm}}|(x,1) \leq C/r^2(x)$. Differentiating the schematic equation

$$\widetilde{\mathrm{Rm}} - \widehat{\mathrm{Rm}} = \widehat{\nabla}^2 h + \tilde{g}^{-2} * \widehat{\nabla} h * \widehat{\nabla} h$$

with respect to the connection $\widehat{\nabla}$ and using (4.29) and that $|\widehat{\nabla}^m \widehat{\mathrm{Rm}}| \leq Cr^{-(m+2)}$, we conclude initially that

$$|\widehat{\nabla}^{(m)} \widetilde{\mathrm{Rm}}|(x,1) \leq \frac{C}{r^{m+2}(x)}. \tag{4.30}$$

But, on $\mathcal{C}_a \times (0,1]$, we have that

$$G_\tau(x) = \rho_{1/\sqrt{\tau}}(x), \text{ and } r(G_\tau(x)) = \frac{r(x)}{\sqrt{\tau}}, \tag{4.31}$$

so

$$|\widetilde{\mathrm{Rm}}|(x,\tau) = \tau^{-1}|\widetilde{\mathrm{Rm}}|(G_\tau(x),1) \leq C\tau^{-1}r^{-2}(G_\tau(x)) = Cr^{-2}(x),$$

and, more generally, from (4.30), that

$$|\widehat{\nabla}^m \widetilde{\mathrm{Rm}}|(x,\tau) = \tau^{-(m/2+1)}|\widehat{\nabla} \widetilde{\mathrm{Rm}}|(G_\tau(x),1) \leq Cr^{-(m+2)}(x),$$

on all of $\mathcal{C}_a \times (0,1]$. Using these bounds, we can then integrate at any $x \in \mathcal{C}_a$ to obtain

$$|h|_{\hat{g}}(x) = |\tilde{g}_1 - \hat{g}|_{\hat{g}}(x) \leq 2\int_0^1 |\widetilde{\mathrm{Ric}}|_{\hat{g}}(x,s) \leq \frac{C}{r^2(x)},$$

which gives the improved C^0-estimate in (4.27).

For the C^1-estimate, we start with the identity

$$\widetilde{\nabla}\widetilde{\mathrm{Ric}} = \widehat{\nabla}\widetilde{\mathrm{Ric}} + \tilde{g}^{-1} * \widehat{\nabla} h * \widetilde{\mathrm{Ric}}. \tag{4.32}$$

Then, at any $x \in \mathcal{C}_a$,

$$\partial_\tau |\widehat{\nabla} h|^2_{\hat{g}} = 4\langle \widehat{\nabla}\widetilde{\mathrm{Ric}}, \widehat{\nabla} h\rangle \leq 2|\widehat{\nabla}\widetilde{\mathrm{Ric}}|^2 + 2|\widehat{\nabla} h|^2 \leq Cr^{-6}(x) + 2|\widehat{\nabla} h|^2,$$

so that

$$\partial_\tau (e^{-2\tau}|\widehat{\nabla} h|^2(x,\tau)) \leq Ce^{-2\tau}r^{-6}(x).$$

Integrating this inequality over $\tau \in [0,1]$ and using that $h(x,0) = 0$ implies that $|\widehat{\nabla} h|^2(x,1) \leq C/r^{-6}(x)$.

The claimed higher derivative bounds on h and $\widetilde{\mathrm{Rm}} - \widehat{\mathrm{Rm}}$ can then be obtained inductively from these estimates. Indeed, the bound on $\widetilde{\mathrm{Rm}} - \widehat{\mathrm{Rm}}$ follows immediately from the improved C^2-estimates on h and the expression (4.32) on the difference of the curvature tensors.

For the bounds on $\tilde{f}$, note that we can return to (4.9) and integrate to see that

$$\left|\tau \tilde{f}(x,\tau) - \frac{r^2(x)}{4}\right| \leq \left|\int_0^\tau s\tilde{R}(x,s)\,ds\right| \leq \frac{C\tau^2}{r^2(x)}.$$

It follows that $\tilde{f}(\tau)$ satisfies the bounds

$$\frac{r^2(x)}{4\tau} - \frac{C\tau}{r^2} \leq \tilde{f}(x,\tau) \leq \frac{r^2(x)}{4\tau} + \frac{C\tau}{r^2(x)}$$

on $\mathcal{C}_a \times (0,1]$, and (4.28) follows by setting $\tau = 1$. □

The upshot of Lemmas 4.7 and 4.8 is that there is no loss of generality in assuming that a shrinking soliton on $\mathcal{C}_a$ which is at least C^2-asymptotic to some cone $\hat{g}$ is actually smoothly asymptotic to that cone, nor in assuming that the gradient of the potential function coincides with the generator of the dilation diffeomorphisms $\rho_{e^{s/2}}$. We will give this normalization a name.

DEFINITION 4.10. A shrinking soliton structure $(\mathcal{C}_a, g, f)$ is **dynamically asymptotic to** $\hat{g}$ if g is smoothly asymptotic to $\hat{g}$ on $\mathcal{C}_a$ and satisfies the estimates (4.27) and (4.28), and the identity $\operatorname{grad}_g f = \frac{r}{2}\frac{\partial}{\partial r}$.

The name reflects that (in the gauge implicitly fixed by the definition) the self-similar solution to the Ricci flow associated to the soliton metric g extends to a smooth solution on $\mathcal{C}_a \times [-1, 0]$ which interpolates between the soliton g at $t = -1$ and the cone $\hat{g}$ at $t = 0$. The normalizations implied by this definition allow us to state the uniqueness of asymptotically conical shrinking ends in a precise form (see Theorem 4.17 below).

4.1.3. 4-Dimensional shrinkers with scalar curvature decay.

Before we turn to the uniqueness of asymptotically conical shrinkers, we observe that the estimates proven in Chapter 2 bounding the norms of the covariant derivatives of the curvature tensor of a 4-dimensional shrinker in terms of the scalar curvature imply a succinct characterization of 4-dimensional asymptotically conical shrinkers.

THEOREM 4.11 (Scalar curvature decay implies asymptotically conical). *Let (M^4, g, f) be a complete 4-dimensional shrinking gradient Ricci soliton with bounded scalar curvature. Assume that $\lim_{s\to\infty} R(\gamma(s)) \to 0$ along any unbounded integral curve $\gamma(s)$ of ∇f. (In particular, this hypothesis holds if $\lim_{x\to\infty} R(x) = 0$.) Then (M^4, g, f) is smoothly asymptotic to a cone $\mathcal{C}^\Sigma$ along every end V of M^4.*

PROOF. As in Chapter 2, let

$$\Sigma(t) = \{x \in M^4 \mid f(x) = t\} \quad \text{and} \quad D(t) = \{x \in M^4 \mid f(x) \leq t\}$$

denote the level sets and sublevel sets of f, respectively. According to the quadratic lower bound (1.61) for the potential (cf. (I-4.7) and Theorem I-4.3), f is proper, so each level set $\Sigma(t)$ is compact. Since R is bounded, this lower bound also implies (via the identity $R + |\nabla f|^2 = f$), that there is $t_0 > 0$ such that $|\nabla f| > 1$ on the set $M^4 \setminus D(t_0)$, In particular, all integral curves of ∇f starting in $\overline{M^4 \setminus D(t_0)}$ remain in this set and are unbounded.

Let ϕ_t denote the family of diffeomorphisms on $\overline{M^4 \setminus D(t_0)}$ defined as in (2.53), that is, the solution to the ODE

$$\begin{aligned} \frac{\partial}{\partial t}\phi_t &= \frac{\nabla f}{|\nabla f|^2} \circ \phi_t, \\ \phi_{t_0} &= \mathrm{Id}\,. \end{aligned}$$

Then the level sets $\Sigma(t)$ are diffeomorphic for $t \geq t_0$ and $\phi_t(\Sigma(t_0)) = \Sigma(t)$.

Let $x_0 \in \Sigma(t_0)$ be arbitrary. By our choice of t_0, the integral curves of both ∇f and $\nabla f / |\nabla f|^2$ emanating from x_0 will be unbounded, and our assumption implies that R will tend to zero along either of them. In particular, there exists $x_1 = \phi_{t_1}(x_0) \in \Sigma(t_1)$ such that

$$R(x_1) \leq \frac{1}{4C_0},$$

where C_0 is the constant in (2.55). On the other hand, there exists a ball $B(x_1, \varepsilon_1)$ such that

$$R(x) \leq \frac{1}{2C_0}$$

for all $x \in B(x_1, \varepsilon_1) \cap \Sigma(t_1)$. As $\Sigma(t_0)$ is diffeomorphic to $\Sigma(t_1)$ via the diffeomorphism ϕ_t defined in (2.53), it follows that there exists $\varepsilon > 0$ such that

$$\phi_{t_1}\left(B(x_0, \varepsilon) \cap \Sigma(t_0)\right) \subset B(x_1, \varepsilon_1) \cap \Sigma(t_1).$$

Therefore,

$$R(\phi_{t_1}(z)) \leq \frac{1}{2C_0}$$

for all $z \in B(x_0, \varepsilon) \cap \Sigma(t_0)$. By (2.56), we conclude that for every $x_0 \in \Sigma(t_0)$ there exists $\varepsilon(x_0) > 0$ and $t_1(x_0) > t_0$ such that

$$R(\phi_t(z)) \leq \frac{1}{C_0}\frac{t_1}{t} \tag{4.33}$$

for any $z \in B(x_0, \varepsilon(x_0)) \cap \Sigma(t_0)$ and any $t > t_1(x_0)$.

As $\Sigma(t_0)$ is compact, we may cover it with finitely many sets

$$\left\{B(x_i, \varepsilon(x_i)) \cap \Sigma(t_0)\right\}_{i=1}^{m}.$$

Let

$$T = \max\{t_1(x_1), \ldots, t_1(x_m)\}.$$

According to (4.33), we have that

$$R(\phi_t(z)) \leq \frac{1}{C_0}\frac{T}{t}$$

for all $z \in \Sigma(t_0)$ and all $t > T$. In particular, there exists a constant $C > 0$ such that

$$R \leq \frac{C}{f}.$$

By Theorem 2.16, we obtain that

$$\sup_{M^4}\left(1 + r^{k+2}(x)\right)\left|\nabla^k \mathrm{Rm}\right|(x) \leq C_k.$$

The result now follows from Proposition 4.6. □

It is unknown whether the weaker hypothesis

$$\liminf_{x\to\infty} R(x) = 0$$

implies that a 4-dimensional shrinker with bounded scalar curvature must be asymptotically conical. However, we have seen in Theorem 2.18 that in this case the asymptotic volume ratio must at least be positive.

In [**MW17**], Wang and one of the authors have proven that, in all dimensions, the conclusion of Theorem 4.11 holds provided that $|\mathrm{Ric}|(x) \to 0$ as $x \to \infty$. In particular, every complete shrinker on which the curvature tensor decays to zero is asymptotically conical.

4.2. Rigidity of asymptotically conical shrinking solitons

The next several sections will be devoted to the rigidity principle for asymptotically conical shrinking solitons discussed in the introduction to this chapter. Let us state it now and observe some of its immediate consequences. We will give two versions of this principle. The first version is the original statement from [**KW15**] by Wang and one of the authors.

THEOREM 4.12 (Rigidity of asymptotically conical shrinkers). *Suppose that the shrinking gradient Ricci solitons* (M^n, g, f) *and* $(\tilde{M}^n, \tilde{g}, \tilde{f})$ *are* C^2*-asymptotic to the cone* $\mathcal{C}^\Sigma$ *along the ends* $V \subset M^n$ *and* $\tilde{V} \subset \tilde{M}^n$*, respectively. Then there exist ends* $W \subset V$ *and* $\tilde{W} \subset \tilde{V}$ *of* M^n *and* $\tilde{M}^n$ *and a diffeomorphism* $\varphi : W \to \tilde{W}$ *such that* $\varphi^*\tilde{g} = g$.

This is an analog of a result of Wang for asymptotically conical self-shrinking solutions to the mean curvature flow [**Wan14**]. Corresponding results for other curvature flows have been proven by Guo [**Guo18**], Khan [**Kha23**], and Haskins, Khan, and Payne [**HKP25**]. We will shortly derive Theorem 4.12 from a more specialized statement (see Theorem 4.17 below).

One feature of Theorem 4.12 is that it is completely local to the ends of the manifolds: it makes no assumptions on the manifolds (M^n, g) and $(\tilde{M}^n, \tilde{g})$ off of V and $\tilde{V}$, nor imposes any conditions on their inner spatial boundaries ∂V and $\partial\tilde{V}$. In particular, the manifolds (M^n, g) and $(\tilde{M}^n, \tilde{g})$ need be neither complete nor connected at infinity. However, Ricci solitons are real-analytic (see Ivey [**Ive96**] or §B.3 of Volume 1), and so, when the underlying manifolds are complete, it follows from Theorem 4.12 and the usual analytic continuation argument for isometries (see, e.g., Myers [**Mye35**] or Kobayashi and Nomizu [**KN96**]) that their universal covers must be isometric.

COROLLARY 4.13 (Uniqueness of universal covers). *Suppose that the two shrinking gradient Ricci solitons* (M_i^n, g_i, f_i)*,* $i = 1, 2$*, are complete and* C^2*-asymptotic to* $\mathcal{C}^\Sigma$ *along the ends* $V_i \subset M_i^n$*. Then the universal covers* $(\tilde{M}_1^n, \tilde{g}_1)$ *and* $(\tilde{M}_2^n, \tilde{g}_2)$ *are isometric.*

Still, this leaves the question: if the manifolds (M_1^n, g_1) and (M_2^n, g_2) are complete, when may we conclude that they are globally isometric? By Corollary 4.13, this is the same as asking when the isometry between the universal covers $\tilde{M}_1^n$ and $\tilde{M}_2^n$ descends to the bases M_1^n and M_2^n. A slight extension of the isometry continuation argument in [**KN96**] shows that the isometry will descend from the covers provided at least that the ends V_i are topologically "large" in the sense that the homomorphisms $\pi_1(V_i, x_0) \to \pi_1(M_i^n, x_0)$ induced by the inclusions $V_i \hookrightarrow M_i^n$ are surjective. (This is equivalent to the condition that the preimages $p_i^{-1}(V_i) \subset \tilde{M}_i^n$ of the ends V_i under the covering maps $p_i : \tilde{M}_i^n \to M_i^n$ are connected.) For a precise statement and proof of this fact, see Lemma 4.47 in Section 4.9 below.

For now, we observe that this topological condition is met at least when the shrinkers and their universal covers are known to be connected at infinity.

COROLLARY 4.14. *Assume that the shrinking solitons* (M_i^n, g_i, f_i)*,* $i = 1, 2$*, are complete and are* C^2*-asymptotic to* $\mathcal{C}^\Sigma$ *along the ends* $V_i \subset M_i^n$*. If the manifolds* (M_i^n, g_i) *and their universal covers* $(\tilde{M}_i^n, \tilde{g}_i)$ *are connected at infinity, then* (M_1^n, g_1) *and* (M_2^n, g_2) *are isometric.*

PROOF. By Theorem 4.12, we have an isometry

$$\varphi : (W_1, g_1|_{W_1}) \to (W_2, g_2|_{W_2})$$

between the ends $W_i \subset V_i$. Since (being Ricci solitons) the manifolds (M_i^n, g_i) are real-analytic, the statement will follow from Lemma 4.47 once we establish the following claim.

CLAIM. The preimage $p_i^{-1}(W_i)$ of W_i under the universal covering map $p_i : \tilde{M}_i^n \to M_i^n$ is connected.

To see this, note that since (M_i^n, g_i) and $(\tilde{M}_i^n, \tilde{g}_i)$ are connected at infinity, we may write $W_i = M_i^n \backslash D_i$, where D_i is compact. Since the fundamental group of M_i^n is finite (see Wylie [**Wyl08**]), Theorem 1.78), the covering map p_i is proper, and so $\tilde{D}_i = p_i^{-1}(D_i)$ is compact. Consider the connected components of $\tilde{W}_i = \tilde{M}_i^n \setminus \tilde{D}_i$. On one hand, any two of them are isometric. On the other hand, exactly one of them is unbounded. Thus $\tilde{W}_i = p_i^{-1}(W_i)$ is connected, and this is equivalent to the fact that, for any $x_0 \in W_i$, the homomorphism $\pi_1(W_i, x_0) \to \pi_1(M_i^n, x_0)$ induced by the inclusion $W_i \hookrightarrow M_i^n$ is surjective. □

REMARK 4.15. Presently, all known complete asymptotically conical shrinking solitons are connected at infinity; see Problem 3.1.

Recall that we have seen in Theorem 3.36 that a complete noncompact Kähler shrinker is connected at infinity. Thus the shrinking soliton structure induced by such a soliton on its universal cover must also be connected at infinity. Corollary 4.14 then implies the following definitive statement in the Kähler case.

COROLLARY 4.16. *If the Kähler shrinking solitons (M_i^n, g_i, f_i), $i = 1, 2$, are complete and are C^2-asymptotic to $\mathcal{C}^\Sigma$ along the ends $V_i \subset M_i^n$, then the underlying manifolds (M_1^n, g_1) and (M_2^n, g_2) are isometric.*

Alternatively, by a result of Esparza [**Esp25**], any complete Kähler shrinker is simply connected. Thus, one can deduce Corollary 4.16 directly from Corollary 4.13.

4.2.1. A restatement of the rigidity principle.

The rigidity principle expressed by Theorem 4.12 implies that an asymptotically conical shrinking soliton is effectively determined by its asymptotic cone. Currently, there are few restrictions known on the kinds of compact manifolds (Σ, g_Σ) which can arise as the cross-sections of the asymptotic cones of complete shrinking solitons. We will take up this question further in Section 4.9 below. However, for some of these applications, we will need to understand more precisely how the ends W and $\tilde{W}$ in the conclusion of the statement of Theorem 4.12 depend on the original ends V and $\tilde{V}$. Using the normalization guaranteed by Lemma 4.8, we can make this relationship completely explicit.

THEOREM 4.17. *Suppose the shrinking gradient Ricci solitons $(\mathcal{C}_a, g, f)$ and $(\mathcal{C}_a, \tilde{g}, \tilde{f})$ are dynamically asymptotic to $\hat{g}$. Then $g \equiv \tilde{g}$ on $\mathcal{C}_a$.*

Most of the rest of this chapter will be devoted to the proof of this theorem (or, rather its parabolic restatement, Theorem 4.19). Let us observe now that Theorem 4.17 implies Theorem 4.12.

Proof of Theorem 4.12, assuming Theorem 4.17. Suppose that the two shrinkers (M^n, g, f) and $(\tilde{M}^n, \tilde{g}, \tilde{f})$ are C^2-asymptotic to the cone $\mathcal{C}^{\Sigma}$ along the ends $V \subset M^n$ and $\tilde{V} \subset \tilde{M}^n$, respectively. According to the Normalization Lemma 4.8, there is an $a > 0$ and diffeomorphisms $F : \mathcal{C}_a \to W$ and $\tilde{F} : \mathcal{C}_a \to \tilde{W}$ onto ends $W \subset V$ and $\tilde{W} \subset \tilde{V}$ such that the soliton structures $(\mathcal{C}_a, F^*g, F^*f)$ and $(\mathcal{C}_a, \tilde{F}^*\tilde{g}, \tilde{F}^*\tilde{f})$ are dynamically asymptotic to $\hat{g}$. By Theorem 4.17, $F^*g = \tilde{F}^*\tilde{g}$ on $\mathcal{C}_a$, so the claim follows with $\varphi = \tilde{F} \circ F^{-1}$. □

4.3. The Ricci flow on an asymptotically conical shrinking end

We will prove Theorem 4.17 via Theorem 4.19 below by transforming the statement into a *parabolic* problem of backward uniqueness for the Ricci flow. We first observe that the self-similar solutions $g(t)$ and $\tilde{g}(t)$ associated to the soliton structures $(\mathcal{C}_a, g, f)$ and $(\mathcal{C}_a, \tilde{g}, \tilde{f})$ in the statement are well-defined on $\mathcal{C}_a$ for all $t \in [-1, 0)$ and converge locally smoothly to the cone $\hat{g}$ as $t \nearrow 0$. Thus, $g(t)$ and $\tilde{g}(t)$ will extend smoothly to solutions to the Ricci flow on $\mathcal{C}_a \times [-1, 0]$ which share the terminal condition $g(0) = \hat{g} = \tilde{g}(0)$.

4.3.1. The self-similar solution associated to the shrinker.

For the discussion that follows, let us assume that we are given a closed $(n-1)$-dimensional manifold (Σ, g_Σ) and a shrinking soliton structure $(\mathcal{C}_a, \bar{g}, \bar{f})$ on $\mathcal{C}_a = \mathcal{C}_a^{\Sigma} = (a, \infty) \times \Sigma$ for some $a > 0$ which is dynamically asymptotic to $\hat{g} = dr^2 + r^2 g_\Sigma$. We want to record some of the implications that the normalizations in Definition 4.10 have for the solution to the Ricci flow associated to this structure.

Given that we are interested in a problem of *backward* uniqueness, it will be convenient to introduce the backward time parameter $\tau = -t$ and work with the solution $g(\tau)$ to the *backward Ricci flow* (4.7) associated to $(\mathcal{C}_a, \bar{g}, \bar{f})$ as we did in the proof of Proposition 4.6. Let us recall the construction of that solution here.

Rewriting the usual correspondence between a shrinker and its associated self-similar solution (see, e.g., Proposition 1.7 of [**CCG**$^+$**07**]) in terms of τ leads to the ordinary differential initial value problem

$$\frac{\partial \Phi}{\partial \tau} = -\frac{1}{\tau}\overline{\nabla} \bar{f} \circ \Phi, \quad \Phi_1 = \mathrm{Id}. \tag{4.34}$$

Solving (4.34) for a general shrinking soliton structure on $\mathcal{C}_a$ produces a family $\Phi_\tau : \mathcal{C}_a \to \mathcal{C}_a$ of local diffeomorphisms which are defined on a neighborhood of each point for τ near 1. The family of metrics defined by $g(\tau) = \tau \Phi_\tau^* \bar{g}$ will then satisfy $g(1) = \bar{g}$ and (4.7) at least locally in space and time. Here, we can say much more. Since $(\mathcal{C}_a, \bar{g}, \bar{f})$ is dynamically asymptotic to $\hat{g}$, we have that $\overline{\nabla} \bar{f} = \frac{r}{2}\frac{\partial}{\partial r}$ on $\mathcal{C}_a$, and so the maps Φ_τ are actually given explicitly by the formula

$$\Phi_\tau(r, \sigma) = (r/\sqrt{\tau}, \sigma) = \rho_{1/\sqrt{\tau}}(r, \sigma).$$

In particular, both Φ_τ and $g(\tau)$ are well-defined on all of $\mathcal{C}_a$ for all $\tau \in (0, 1]$. Since

$$g(\lambda^{-2}) = \lambda^{-2}\rho_\lambda^* \bar{g}, \quad \lambda \geq 1,$$

it follows from Definition 4.1 that

$$g(\tau) \to \hat{g} \text{ locally smoothly as } \tau \to 0.$$

Thus, $g(\tau)$ extends to a smooth solution to (4.7) on all of $\mathcal{C}_a \times [0, 1]$ with $g(0) = \hat{g}$. (Of course, $g(\tau)$ is only self-similar on $\mathcal{C}_a \times (0, 1]$.)

We summarize the properties of this solution we will need below. (Refer to Proposition 4.6 and Lemma 4.8 for further details.) Here, $|\cdot| = |\cdot|_{g(\tau)}$ and $\nabla = \nabla_{g(\tau)}$.

PROPOSITION 4.18. *Suppose $(\mathcal{C}_a, \bar{g}, \bar{f})$ is dynamically asymptotic to $\hat{g}$. Let $\Phi : \mathcal{C}_a \times (0,1] \to \mathcal{C}_a$ be the map $\Phi(r,\sigma,\tau) = \Phi_\tau(r,\sigma) = (r/\sqrt{\tau}, \sigma)$. Then the family of metrics $g = g(\tau)$ defined by*

$$g(\tau) = \begin{cases} \tau \Phi_\tau^* \bar{g} & \tau \in (0,1], \\ \hat{g} & \tau = 0, \end{cases}$$

solves (4.7) *on $\mathcal{C}_a \times [0,1]$ with $g(0) = \hat{g}$ and $g(1) = \bar{g}$, and satisfies the estimates*

$$|\nabla^m \mathrm{Rm}|(r,\sigma,\tau) \leq \frac{K_m}{r^{m+2}+1}, \quad |\nabla^m (g - \hat{g})| \leq \frac{L_m \tau}{r^{m+2}+1}, \tag{4.35}$$

for each $m \geq 0$ for some constants K_m, $L_m > 0$. Also, if $f = f(\tau) = \bar{f} \circ \Phi_\tau$, then

$$r^2 - \frac{N\tau^2}{r^4} \leq 4\tau f \leq r^2 + \frac{N\tau^2}{r^4} \tag{4.36}$$

for some constant $N > 0$ on $\mathcal{C}_a \times [0,1]$ and $\tau f \to r^2/4$ locally smoothly on $\mathcal{C}_a$ as $\tau \to 0$. Finally, g and f together satisfy the identities

$$\mathrm{Ric}(g) + \nabla^2 f = \frac{g}{2\tau}, \quad |\nabla f|^2 + R = \frac{f}{\tau}, \quad \nabla f = \frac{r}{2\tau}\frac{\partial}{\partial r}, \tag{4.37}$$

on $\mathcal{C}_a \times (0,1]$.

4.3.2. Backward uniqueness for the Ricci flow on a conical end. Using Proposition 4.18, we can recast Theorem 4.17 as a problem of backward uniqueness.

THEOREM 4.19. *Suppose that $(\mathcal{C}_a, g_1, f_1)$ and $(\mathcal{C}_a, \tilde{g}_1, \tilde{f}_1)$ are dynamically asymptotic to $\hat{g}$. Let $g = g(\tau)$ and $\tilde{g} = g(\tau)$ be the associated solutions to* (4.7) *on $\mathcal{C}_a \times [0,1]$ described in Proposition* 4.18 *with $g(1) = g_1$, $\tilde{g}(1) = \tilde{g}_1$, and $g(0) = \tilde{g}(0) = \hat{g}$. Then $g(\tau) \equiv \tilde{g}(\tau)$ on $\mathcal{C}_a \times [0,1]$. In particular, $g_1 \equiv \tilde{g}_1$ on $\mathcal{C}_a$.*

A backward uniqueness principle for general complete Ricci flows of uniformly bounded curvature was proven by one of the authors in [**Kot10**]; its proof was later simplified somewhat in [**Kot16b**]. This principle is inapplicable here, however, since the solutions $(\mathcal{C}_a, g(\tau))$ and $(\mathcal{C}_a, \tilde{g}(\tau))$ are incomplete for every τ, and are unconstrained on the interior spatial boundary $\partial\mathcal{C}_a$. We need a backward uniqueness principle specialized to this particular kind of exterior domain. Our model will be the following result of Escauriaza, Seregin, and Šverák [**ESŠ03a**] for solutions to linear backward parabolic inequalities in Euclidean space.

THEOREM 4.20 (Escauriaza, Seregin, and Šverák). *For $R > 0$, define $D_R = \mathbb{R}^n \setminus \overline{B(0,R)}$. Suppose that the function $u : D_R \times [0,T] \to \mathbb{R}$ satisfies*

$$|\partial_\tau u + \Delta u| \leq C(|u| + |\nabla u|) \quad \text{and} \quad |u(x,\tau)| \leq Ce^{C|x|^2}$$

on $D_R \times [0,T]$ for some constant $C > 0$. If $u(x,0) \equiv 0$ on D_R, then $u(x,\tau) \equiv 0$ on $D_R \times [0,T]$.

The important feature of their result is that it makes no restriction on the behavior of u on the inner spatial boundary of the domain. In [**ESŠ03a**], the authors applied this result to resolve an open question regarding the regularity of solutions to the Navier–Stokes equations in three dimensions. One can ask whether a similar statement of backward uniqueness holds for linear parabolic inequalities on exterior domains of the form $\mathbb{R}^n \setminus S$ for more general regions S. The question turns out to be a subtle one, depending in general on the geometry of ∂S. See, e.g., Li and Šverák [**LŠ12**], and the related result in [**ESŠ03b**] for S the closed half-space $\mathbb{R}^n_+$.

4.3.2.1. *Uniqueness and nonuniqueness in the expanding case.*

As we noted in the introduction to this chapter, any solution to the heat equation with initial data equal to some bump function supported in the ball $B(0, R)$ will be positive everywhere on $\mathbb{R}^n \setminus \overline{B(0, R)}$ for all $t > 0$, so the analog of Theorem 4.20 for *forward* parabolic inequalities is trivially false. Thus Theorem 4.20 reflects a fundamental asymmetry in the flexibility of solutions to the heat equation on exterior domains in forward and backward time. By contrast, on the whole space $\mathbb{R}^n$, uniqueness in forward and backward time are valid for solutions satisfying the same, essentially sharp, growth condition of the form $|u(x,t)| \leq Ce^{C|x|^2}$.

Given the analogy between the heat equation and the Ricci flow, it is therefore not surprising that there should be a corresponding asymmetry in the flexibility of solutions to forward and backward Ricci flows on asymptotically conical ends, and also in the flexibility of asymptotically conical expanders and asymptotically conical shrinkers. The recent work of Angenent and Knopf [**AK22**] contains, among other things, a rigorous construction of an infinite family of cones over $\Sigma = \mathbb{S}^p \times \mathbb{S}^q$, for p, $q \geq 2$ satisfying $4 \leq p + q \leq 8$, each of which occur as the asymptotic cone of multiple distinct complete asymptotically conical expanding solitons. The expanders asymptotic to a given cone occur in families of arbitrarily large size, and may even have different topologies (namely, $\mathbb{R}^{p+1} \times \mathbb{S}^q$ or $\mathbb{S}^p \times \mathbb{R}^{q+1}$).

The authors in [**AK22**] also prove the existence of complete shrinking solitons on $\mathbb{R}^{p+1} \times \mathbb{S}^q$ asymptotic to the same cones, each of which (according to Theorem 4.12) are the unique such example. Fixing one such cone and gluing the spacetime of the self-similar solution of the associated shrinker to the spacetimes of any of the distinct expanding self-similar solutions asymptotic to that cone along the conical time-slice at $t = 0$, produces families of immortal Ricci flows which resolve the same isolated conical singularity at $t = 0$ in different ways. These examples show that, at least in dimensions $n \geq 5$, a Ricci flow spacetime may fail to be unique after the initial singularity. See Problem 4.5 at the end of this chapter.

As a special case, Angenent and Knopf prove that there are infinitely many complete expanders asymptotic to the Ricci-flat cone

$$g_{\mathrm{RF}} = dr^2 + \left(\frac{p-1}{n-1}\right) r^2 g_{\mathbb{S}^p} + \left(\frac{q-1}{n-1}\right) r^2 g_{\mathbb{S}^q}.$$

By contrast, it follows from Theorem 4.12 that there is exactly one shrinking soliton asymptotic to g_{RF}, namely g_{RF} itself, which, with the potential $f(r) = \frac{r^2}{4}$, yields an (incomplete) shrinking soliton structure on the cone minus the vertex.

There are a variety of uniqueness results for asymptotically conical expanding solitons under additional constraints on the asymptotic behavior of the expander; see, e.g., the work of Deruelle [**Der16**], [**Der17b**], Deruelle and Schulze [**DS23**],

Conlon and Deruelle [**CD20**], and Conlon, Deruelle, and Sun [**CDS24**]. Asymptotically conical expanding solitons will be discussed in greater detail in a sequel to this volume.

4.3.3. Addressing the weak-parabolicity of the flow.

We would like to approach Theorem 4.19 as though it were a problem of backward uniqueness for a strictly parabolic system as in Theorem 4.20, and adapt the Carleman estimates for linear parabolic inequalities on Euclidean space from the proof of that result to our asymptotically conical setting. This is not entirely straightforward, however, as the Ricci flow is a nonlinear weakly parabolic system, and the difference of two solutions to the equation does not naturally satisfy a strictly parabolic equation.

4.3.3.1. *The inapplicability of DeTurck's trick in backward time.*

The lack of strict parabolicity of the Ricci flow is a consequence of the diffeomorphism invariance of the equation (see, e.g., [**Ham82**] or §3.2 of [**CK04**]) and is also an obstacle one must overcome when attempting to prove the uniqueness of solutions to the Ricci flow forward in time. In the forward-time case, the standard approach is to reduce the problem to one of the uniqueness of solutions for a related strictly parabolic equation via a method known as the **DeTurck Trick** (see DeTurck [**DeT03**] and Hamilton [**Ham95b**]). It is not possible to directly apply this approach to the backward-time case, however. To explain the basic obstacle, let us briefly review how the method is used in the usual proof of the forward uniqueness of solutions.

Suppose for simplicity that M^n is closed and that $g_1(t)$ and $g_2(t)$ are solutions to the Ricci flow on $M^n \times [0,T]$ with $g_1(0) = g_2(0)$. Let $\bar{g}$ be any fixed metric on M^n. Solving the *harmonic map heat flow*

$$\partial_t \varphi_i = \Delta_{g(t),\bar{g}} \varphi_i, \quad \varphi_i|_{t=0} = \mathrm{Id}, \tag{4.38}$$

yields smooth families of diffeomorphisms $(\varphi_i)_t : M^n \to M^n$, $i = 1, 2$, defined for $t \in [0,T]$. Here, $\Delta_{g,\bar{g}}$ is the *map Laplacian* or *tension field*, which acts on a map $F : (M^n, g) \to (N^m, \bar{g})$ by $\Delta_{g,\bar{g}} F = \mathrm{tr}_g(\nabla df)$.[4] The pullback metrics $h_i(t) := ((\varphi_i)_t^{-1})^* g_i(t)$, $i = 1, 2$, then, are solutions to the **Ricci-DeTurck flow**

$$\partial_t h_i = -2\mathrm{Ric}(h_i) + \mathcal{L}_{W[h_i,\bar{g}]} h_i, \tag{4.39}$$

where

$$W^r = (h_i)^{pq} \left((\Gamma_{h_i})_{pq}^r - (\Gamma_{\bar{g}})_{pq}^r \right)$$

for $i = 1, 2$, and satisfy $h_1(0) = g_1(0) = g_2(0) = h_2(0)$. This is a strictly parabolic system. Indeed, (4.39) has the form

$$\partial_t h_i = (h_i)^{pq} \overline{\nabla}_p \overline{\nabla}_q h_i + \dots,$$

where the ellipsis denotes lower-order terms. Since M^n is compact, one then concludes from standard parabolic theory that $h_1(t) = h_2(t)$ on $M^n \times [0,T]$.

[4] In local coordinates (x^i) and (y^α) on M^n and N^m, this is

$$(\Delta_{g,\bar{g}} F)^\gamma = g^{ij} \left(\frac{\partial^2 F^\gamma}{\partial x^i \partial x^j} - (\Gamma_g)_{ij}^k \frac{\partial F^\gamma}{\partial x^k} + \left((\Gamma_{\bar{g}})_{\alpha\beta}^\gamma \circ F \right) \frac{\partial F^\alpha}{\partial x^i} \frac{\partial F^\beta}{\partial x^j} \right).$$

Now we write $h(t) = h_1(t) = h_2(t)$ and observe that we can also express the diffeomorphisms φ_i as solutions to an ordinary differential initial value problem depending only on $h(t)$ and $\bar{g}$. Namely,

$$\partial_t \varphi_i = -W[h(t), \bar{g}] \circ \varphi_i, \quad \varphi_i|_{t=0} = \mathrm{Id},$$

for $i = 1, 2$. From the theory of ordinary differential equations, φ_1 and φ_2 must be identical, so

$$g_1(t) = (\varphi_1)_t^* h(t) = (\varphi_2)_t^* h(t) = g_2(t)$$

on $M^n \times [0, T]$ as desired. This proves the uniqueness of compact solutions to the Ricci flow. (In his original paper [**Ham82**], Hamilton proved both the short-time existence and uniqueness of compact solutions via a modification of the Nash-Moser inverse function theorem.)

While the above prescription can be applied on noncompact manifolds, the short-time existence of well-controlled solutions to (4.38) and the uniqueness of solutions to (4.39) are no longer standard and require their own separate justification, subject to restrictions on the behavior of the solutions at infinity. Chen and Zhu [**CZ12**] have used this prescription to prove the uniqueness of complete solutions to the Ricci flow with uniformly bounded curvature.[5]

To see where this method fails for uniqueness in backward time, suppose again that M^n is closed and let $g_1(t)$ and $g_2(t)$ be solutions to the Ricci flow on $M^n \times [0, T]$ with $g_1(T) = g_2(T)$. Following the template above, we should seek to convert $g_1(t)$ and $g_2(t)$ into solutions $h_1(t)$ and $h_2(t)$ to the Ricci-DeTurck flow (4.39) with $h_1(T) = h_2(T)$. This leads us to try to solve the corresponding problem

$$\frac{\partial \varphi_i}{\partial t} = \Delta_{g_i(t), \bar{g}} \varphi_i, \quad \varphi_i|_{t=T} = \mathrm{Id}, \tag{4.40}$$

for diffeomorphisms φ_i, $i = 1, 2$, relative to some fixed metric $\bar{g}$. However, this parabolic *terminal-value* problem is ill-posed, and does not admit solutions in general. Thus the applicability of DeTurck's trick to problems of backward uniqueness for geometric evolution equations is obstructed by the failure of short-time *existence* for backward parabolic problems. For such problems another approach is needed.

4.3.4. Backward uniqueness for a PDE-ODE System.

The backward uniqueness of complete solutions to the Ricci flow of bounded curvature was proven by one of the authors in [**Kot10**]. In that paper, the weak-parabolicity of the equation was circumvented by embedding the problem of backward uniqueness for solutions to the Ricci flow into one for a larger, prolonged "PDE-ODE" system of mixed differential inequalities which is sufficiently close to being parabolic that it can be treated with a modification of the parabolic theory. The use of PDE-ODE systems in this connection has precedence in the work of Alexakis [**Ale09**] on the problem of unique continuation for hyperbolic systems arising in the study of the vacuum Einstein equations, and the work of Wong and Yu [**WY12**] on elliptic systems arising in the study of coupled Einstein-scalar equations.

We will implement the method of [**Kot10**] in our proof of Theorem 4.19 with some minor changes. We review the basic idea now.

[5]One of the authors has given an alternative approach to these results which avoids the passage through the Ricci-DeTurck flow [**Kot16a, Kot17**].

4.3.4.1. *The strategy.*

In a nutshell, the idea is to center the analysis on the difference of the curvature tensors Rm and $\widetilde{\mathrm{Rm}}$ of the solutions $g = g(\tau)$ and $\tilde{g} = \tilde{g}(\tau)$ rather than the difference of the solutions themselves. This has the advantage that the curvature tensor Rm of a solution to (4.7) and all of its covariant derivatives satisfy *strictly* (backward) parabolic equations of the form

$$(\partial_\tau + \Delta)\nabla^k \mathrm{Rm} = P(g^{-1}, \mathrm{Rm}, \ldots, \nabla^k \mathrm{Rm}),$$

where $\nabla = \nabla_{g(\tau)}$ and $\Delta = \Delta_{g(\tau)}$ are the Levi-Civita connection and the Laplacian of the metric $g(\tau)$ and P is a polynomial in its entries. Provided that we could show that the curvature tensors Rm and $\widetilde{\mathrm{Rm}}$ of the solutions are equal for all τ, we could then deduce that $g(\tau) = \tilde{g}(\tau)$ from the evolution equation (4.7).

There is an obstacle to be overcome first, however. While the curvature tensors Rm and $\widetilde{\mathrm{Rm}}$ and their covariant derivatives satisfy heat-type equations, they do so relative to the Laplacians $\Delta = \Delta_{g(\tau)}$ and $\widetilde{\Delta} = \Delta_{\tilde{g}(\tau)}$ associated to their own solutions. This is problematic since, locally, the Laplacian acting on a tensor of rank at least one depends not only on the metric but also on its first and second derivatives, and so the difference of the two Laplacians will contribute terms up to second order in the difference of the metrics. Thus, for example (see Lemma 4.21 below),

$$\begin{aligned}\Delta \mathrm{Rm} - \widetilde{\Delta}\widetilde{\mathrm{Rm}} = \Delta(\mathrm{Rm} - \widetilde{\mathrm{Rm}}) &+ P_1\left(g^{-1}, \tilde{g}^{-1}, g - \tilde{g}\right) * \nabla^2 \widetilde{\mathrm{Rm}} \\ &+ P_2\left(g^{-1}, \tilde{g}^{-1}, \nabla(g - \tilde{g})\right) * \nabla \widetilde{\mathrm{Rm}} \\ &+ P_3\left(g^{-1}, \tilde{g}^{-1}, \nabla(g - \tilde{g}), \nabla^2(g - \tilde{g})\right) * \widetilde{\mathrm{Rm}},\end{aligned}$$

where the P_i are polynomial expressions. In particular, the evolution equation for the difference $\mathrm{Rm} - \widetilde{\mathrm{Rm}}$ of the curvature tensors involves the derivatives of the difference $g - \tilde{g}$ of the solutions up to second order. But our reason for considering the difference of the curvature tensors in the first place was to avoid having to treat the difference of the solutions directly, so it appears at first that we have only made our problem more complicated.

However, we can rescue this strategy if we relax our expectations for the system we eventually hope to obtain. The key observation (which can be confirmed with a few back-of-the-envelope calculations) is that the tensors $g - \tilde{g}$, $\nabla(g - \tilde{g})$, and $\nabla^2(g - \tilde{g})$ can be controlled pointwise, via their evolution equations, in terms of their own norms and those of the differences $\mathrm{Rm} - \widetilde{\mathrm{Rm}}$, $\nabla \mathrm{Rm} - \widetilde{\nabla}\widetilde{\mathrm{Rm}}$, and $\nabla^2 \mathrm{Rm} - \widetilde{\nabla}^2 \widetilde{\mathrm{Rm}}$. These latter three tensors, on the other hand, satisfy a closed system of heat-type equations modulo terms involving $g - \tilde{g}$, $\nabla(g - \tilde{g})$ and $\nabla^2(g - \tilde{g})$.

This suggests that, rather than seeking to frame the problem as one of backward uniqueness for a system of parabolic inequalities, we instead ought to set our sights on framing it as one for a two-tiered "PDE-ODE" system, where the principal components (the differences $\nabla^k \mathrm{Rm} - \widetilde{\nabla}^k \widetilde{\mathrm{Rm}}$) satisfy a closed system of heat-type equations modulo a collection of secondary components (the differences $\nabla^k(g - \tilde{g})$) which can be controlled pointwise – in an ordinary differential sense – by the other components in the system. In this case, when the solutions have bounded curvature, these six tensors together satisfy a system of parabolic and ordinary differential inequalities which resembles an actual parabolic system closely enough that one might reasonably hope to adapt parabolic techniques to prove a backward uniqueness statement for its solutions.

Logically, the vanishing of the six tensors described above is redundant for the purposes of proving Theorem 4.19, and, while some prolongation is necessary to obtain a closed system, we don't really need to include them all. For example, the difference $\mathrm{Rm} - \widetilde{\mathrm{Rm}}$ can be controlled by the difference of the metrics and its derivatives up to order two, and (as it will turn out) we can control $\nabla^2\mathrm{Rm} - \widetilde{\nabla}^2\widetilde{\mathrm{Rm}}$ in terms of $\nabla(\nabla\mathrm{Rm} - \widetilde{\nabla}\widetilde{\mathrm{Rm}})$ and other elements of the system via L^2-absorbing inequalities. Thus it will ultimately suffice to include only the tensor $\nabla\mathrm{Rm} - \widetilde{\nabla}\widetilde{\mathrm{Rm}}$ and the differences $g - \tilde{g}$, $\nabla(g - \tilde{g})$, and $\nabla^2(g - \tilde{g})$ of the metrics up to second order.[6]

4.3.4.2. *The PDE-ODE system and Uhlenbeck's trick.*

We will now give a precise definition of the system we arrived at in our heuristic discussion in the previous section. As before, $g(\tau)$ and $\tilde{g}(\tau)$ will denote the solutions to (4.7) on $\mathcal{C}_a \times [0,1]$ from Theorem 4.19. We will choose the solution $g(\tau)$ to serve as a family of reference metrics, and write $|\cdot| = |\cdot|_{g(\tau)}$, $\nabla = \nabla_{g(\tau)}$, and $\Delta = \Delta_{g(\tau)}$, for the norms, connections, and Laplace operators induced by $g(\tau)$.

The first step is to set up the bundles which the elements of the system will inhabit. While we will regard these elements in our system as families of sections of tensor bundles over $\mathcal{C}_a$ depending smoothly on the parameter $\tau \in [0,1]$, there is some computational advantage to be gained by being slightly more formal about what we mean by a *family* of tensors. We will in effect be using a form of **Uhlenbeck's trick** (see, e.g., §6.2 of [**CK04**]).

Let $\pi : \mathcal{C}_a \times [0,1] \to \mathcal{C}_a$ denote the projection $\pi(x,t) = x$ onto the first factor, and consider the pullback bundles $\pi^*(T^\nu(T^*\mathcal{C}_a))$ associated to the bundle $T^\nu(T^*\mathcal{C}_a)$ of covariant ν-tensors over $\mathcal{C}_a$. The elements of the bundles $\pi^*(T\mathcal{C}_a)$ and $\pi^*(T^\nu(T^*\mathcal{C}_a))$ are sometimes called **spatial vectors** and **spatial tensors**. The smooth sections of these bundles over $\mathcal{C}_a \times [0,1]$ correspond precisely to the smooth families of vector fields and tensor fields over $\mathcal{C}_a$

Now, the family of Riemannian metrics $g(\tau)$ induces a metric on the bundle $\pi^*(T^\nu(T^*\mathcal{C}_a))$ of spatial tensors. The family of Levi-Civita connections $\nabla = \nabla_{g(\tau)}$ can be extended to a *metric* connection on $\pi^*(T\mathcal{C}_a)$, and hence on all of the bundles $\pi^*(T^\nu(T^*\mathcal{C}_a))$, with the definition

$$\nabla_{\frac{\partial}{\partial\tau}} V = \partial_\tau V - \mathrm{Ric}(V)$$

for sections V of $\pi^*(T\mathcal{C}_a)$. Going forward, we will write

$$D_\tau = \nabla_{\frac{\partial}{\partial\tau}},$$

and reserve ∇ for differentiation with respect to directions tangent to $\mathcal{C}_a$.

Equivalently, we can regard D_τ as an operator which acts on smooth families of sections of $T^\nu(T^*\mathcal{C}_a)$ by

$$D_\tau V_{\alpha_1\cdots\alpha_\nu} = \partial_\tau V_{\alpha_1\cdots\alpha_\nu} - R^\beta_{\alpha_1} V_{\beta\alpha_2\cdots\alpha_\nu} - \cdots - R^\beta_{\alpha_k} V_{\alpha_1\cdots\alpha_{\nu-1}\beta}.$$

[6] It also isn't necessary to include $\nabla(g - \tilde{g})$, as its weighted L^2-norm can be controlled by those of the other quantities in the system via interpolation. We find it simpler just to include it in the system, however.

It thus coincides with the "total" τ-derivative taken relative to evolving $g(\tau)$-orthonormal frames. In particular, $D_\tau g = 0$. This operator is characterized by the property that, for any sections V, W of $\pi^*(T^\nu(T^*M))$ one has

$$\frac{\partial}{\partial \tau} \langle V, W\rangle = \langle D_\tau V, W\rangle + \langle V, D_\tau W\rangle$$

in terms of the time-varying inner product $\langle\cdot,\cdot\rangle = \langle\cdot,\cdot\rangle_{g(\tau)}$. For an alternative interpretion of this operator, see Appendix F of [**CCG$^+$08**].

Returning to the purpose at hand, we define the bundles

$$\mathcal{X} := \pi^*\left(T^5(T^*\mathcal{C}_a)\right),$$
$$\mathcal{Y} := \pi^*\left(T^2(T^*\mathcal{C}_a)\right) \oplus \pi^*\left(T^3(T^*\mathcal{C}_a)\right) \oplus \pi^*\left(T^4(T^*\mathcal{C}_a)\right),$$

over $\mathcal{C}_a \times [0,1]$, and the sections

$$\mathbf{X} = T, \quad \mathbf{Y} = (h, \nabla h, \nabla^2 h), \tag{4.41}$$

of $\mathcal{X}$ and $\mathcal{Y}$, where

$$T = \nabla \mathrm{Rm} - \widetilde{\nabla \mathrm{Rm}} \quad \text{and} \quad h = g - \tilde{g}.$$

4.3.5. The evolution equations.

Above, we argued loosely that $\mathbf{X}$ and $\mathbf{Y}$ ought to satisfy a closed system of differential inequalities. We confirm this now by computing the evolution equations for the components of $\mathbf{X}$ and $\mathbf{Y}$.

4.3.5.1. *Notational conventions.*

The evolution equations for the components of $\mathbf{X}$ and $\mathbf{Y}$ are straightforward, if somewhat tedious, to derive. To simplify the computations, we will use the standard "asterisk" notation to denote contractions of tensor products whose particular internal structure is unimportant for our purposes.

Since we have two metrics g and $\tilde{g}$ lurking in the background, we will stipulate that, in this section, $A*B$ will exclusively denote linear combinations of contractions of terms of the form $A^*\otimes B^*$ with the metric g, where A^* and B^* are obtained from A and B by potentially raising and lowering indices with the metric g and where the coefficients of the terms are constants depending at most on the dimension. In particular, we will not conceal any factor of $\tilde{g}$ or $\tilde{g}^{-1}$ with the asterisk notation. We will also use the shorthands

$$A^k = \underbrace{A * A * \cdots * A}_{k}$$

and

$$(A_1 + A_2 + \cdots + A_k) * B = A_1 * B + A_2 * B + \cdots + A_k * B$$

to reduce some of the clutter in the expressions.

Note that, for all families of tensors A and B, we have

$$|A * B| \leq C|A||B|$$

for some constant $C = C(n)$. Here the norms $|\cdot| = |\cdot|_g$ are those induced by g.

Additionally, under our convention, we have

$$\nabla(A * B) = \nabla A * B + A * \nabla B$$

and

$$D_\tau(A * B) = D_\tau A * B + A * D_\tau B,$$

where $\nabla = \nabla_g$ and $D_\tau = D_{\tau,g}$ are the operators determined by g.

4.3.5.2. *The computations.*

We begin by recording some expressions for the differences of basic quantities related to g and $\tilde{g}$. Again, according to our convention, the asterisk notation may conceal factors of g and g^{-1}, but *not* factors of $\tilde{g}$ and $\tilde{g}^{-1}$.

LEMMA 4.21. *Let $h = g - \tilde{g}$. Then*

$$g^{ij} - \tilde{g}^{ij} = -\tilde{g}^{ia} g^{jb} h_{ab} = \tilde{g}^{-1} * h, \tag{4.42}$$

$$\nabla_k \tilde{g}^{ij} = \tilde{g}^{ia} \tilde{g}^{jb} \nabla_k h_{ab} = \tilde{g}^{-2} * \nabla h, \tag{4.43}$$

$$\widetilde{\mathrm{Rm}} - \mathrm{Rm} = \nabla^2 h + \tilde{g}^{-1} * (\nabla h)^2 + \mathrm{Rm} * h, \tag{4.44}$$

where Rm *and* $\widetilde{\mathrm{Rm}}$ *denote the* $(4,0)$ *curvature tensors of* g *and* $\tilde{g}$*, respectively. In addition, we have the schematic identities*

$$\widetilde{\nabla} V - \nabla V = \tilde{g}^{-1} * V * \nabla h, \tag{4.45}$$

$$\begin{aligned}\widetilde{\Delta} V - \Delta V = \tilde{g}^{-2} * V * \nabla^2 h + \left\{ \tilde{g}^{-3} * V * \nabla h + \tilde{g}^{-2} * \nabla V \right\} * \nabla h \\ + \tilde{g}^{-1} * \nabla^2 V * h,\end{aligned} \tag{4.46}$$

for any tensor V of rank at least 1, and the commutator relation

$$[D_\tau, \nabla] V = \nabla \mathrm{Ric} * V + \mathrm{Ric} * \nabla V. \tag{4.47}$$

We leave the proof of Lemma 4.21 as an exercise. Next, we recall the evolution equations for the covariant derivatives of a solution under the Ricci flow (see, e.g., §7.2 of [**CK04**]).

LEMMA 4.22. *Let $g(\tau)$ be any smooth solution to the backward Ricci flow* (4.7) *on $M^n \times [0, T]$. Then, for all $m \geq 0$, the $(m+4)$-tensor $\nabla^m \mathrm{Rm}$ evolves according to the schematic equation*

$$(\partial_\tau + \Delta) \nabla^m \mathrm{Rm} = \sum_{l=0}^{m} \left(g^{-2} * \nabla^l \mathrm{Rm} * \nabla^{m-l} \mathrm{Rm} \right) \tag{4.48}$$

on $M^n \times [0, T]$, where here the asterisk represents ***does not*** *conceal any factors of g or g^{-1}.*

Using equations (4.42)–(4.47) together with (4.48), we obtain the following expressions for the components of **X** and **Y**.

PROPOSITION 4.23. *The tensors $h = g - \tilde{g}$ and $T = \nabla\mathrm{Rm} - \widetilde{\nabla}\widetilde{\mathrm{Rm}}$ satisfy the schematic equations*

$$D_\tau h = \nabla^2 h + \tilde{g}^{-1} * (\nabla h)^2 + \{\tilde{g}^{-1} * \widetilde{\mathrm{Rm}} + \mathrm{Rm}\} * h, \tag{4.49}$$

$$\begin{aligned} D_\tau \nabla h = T &+ \{\tilde{g}^{-2} * \widetilde{\mathrm{Rm}} * h + \tilde{g}^{-1} * \widetilde{\mathrm{Rm}} + \mathrm{Rm}\} * \nabla h \\ &+ \{\tilde{g}^{-1} * T + \tilde{g}^{-1} * \nabla\mathrm{Rm} + \nabla\mathrm{Rm}\} * h, \end{aligned} \tag{4.50}$$

$$\begin{aligned} D_\tau \nabla^2 h = \nabla T &+ \left\{\tilde{g}^{-2} * \widetilde{\mathrm{Rm}} + \tilde{g}^{-1} * \widetilde{\mathrm{Rm}} + \mathrm{Rm}\right\} * \nabla^2 h \\ &+ \left\{\tilde{g}^{-1} * T + \tilde{g}^{-2} * h * T + \tilde{g}^{-2} * \widetilde{\mathrm{Rm}} * \nabla h + \tilde{g}^{-3} * \widetilde{\mathrm{Rm}} * h * \nabla h \right. \\ &\qquad \left. + \tilde{g}^{-2} * \nabla\mathrm{Rm} * h + \tilde{g}^{-1} * \nabla\mathrm{Rm} + \nabla\mathrm{Rm}\right\} * \nabla h \\ &+ \left\{\tilde{g}^{-1} * \nabla T + \tilde{g}^{-1} * \nabla^2\mathrm{Rm} + \nabla^2\mathrm{Rm}\right\} * h, \end{aligned} \tag{4.51}$$

and

$$\begin{aligned} (D_\tau + \Delta)T = (\mathrm{Rm} &+ \widetilde{\mathrm{Rm}}) * T \\ &+ \left\{\tilde{g}^{-2} * \widetilde{\nabla}\widetilde{\mathrm{Rm}} + \tilde{g}^{-2} * \widetilde{\nabla}\widetilde{\mathrm{Rm}} * h + \nabla\mathrm{Rm}\right\} * \nabla^2 h \\ &+ \left\{\tilde{g}^{-3} * \widetilde{\nabla}\widetilde{\mathrm{Rm}} * h * \nabla h + \tilde{g}^{-3} * \widetilde{\nabla}\widetilde{\mathrm{Rm}} * \nabla h + \tilde{g}^{-1} * \nabla\mathrm{Rm} * \nabla h \right. \\ &\qquad \left. + \tilde{g}^{-2} * \widetilde{\nabla}^2\widetilde{\mathrm{Rm}} * h + \tilde{g}^{-2} * \widetilde{\nabla}^2\widetilde{\mathrm{Rm}}\right\} * \nabla h \\ &+ \left\{\nabla\mathrm{Rm} * \widetilde{\mathrm{Rm}} + (\tilde{g}^{-1} + \tilde{g}^{-2}) * \widetilde{\nabla}\widetilde{\mathrm{Rm}} * \widetilde{\mathrm{Rm}} + \tilde{g}^{-1} * \widetilde{\nabla}^3\widetilde{\mathrm{Rm}}\right\} * h. \end{aligned} \tag{4.52}$$

PROOF. Let us temporarily write $S = \mathrm{Rm} - \widetilde{\mathrm{Rm}}$. For (4.49), we compute directly that

$$\begin{aligned} D_\tau h_{ij} &= 2(R_{ij} - \tilde{R}_{ij}) + R_i^p h_{pj} + R_j^p h_{ip} \\ &= 2g^{ab}(R_{iabj} - \tilde{R}_{iabj}) + 2(g^{ab} - \tilde{g}^{ab})\tilde{R}_{iabj} + R_i^p h_{pj} + R_j^p h_{ip} \\ &= S + (\tilde{g}^{-1} * \widetilde{\mathrm{Rm}} + \mathrm{Rm}) * h. \end{aligned} \tag{4.53}$$

Replacing $S = \mathrm{Rm} - \widetilde{\mathrm{Rm}}$ with the expression in (4.44), we obtain (4.49).

For (4.50), we use (4.47) with (4.53) to compute that

$$\begin{aligned} D_\tau \nabla h &= [D_\tau, \nabla]h + \nabla D_\tau h \\ &= \nabla\mathrm{Rm} * h + \mathrm{Rm} * \nabla h + \nabla S + \nabla(\tilde{g}^{-1} * \widetilde{\mathrm{Rm}} * h + \mathrm{Rm} * h). \end{aligned}$$

Now, on the one hand, we have

$$\nabla\widetilde{\mathrm{Rm}} = \widetilde{\nabla}\widetilde{\mathrm{Rm}} + (\nabla - \widetilde{\nabla})\widetilde{\mathrm{Rm}} = -T + \nabla\mathrm{Rm} + \tilde{g}^{-1} * \widetilde{\mathrm{Rm}} * \nabla h, \tag{4.54}$$

so that

$$\nabla S = \nabla\mathrm{Rm} - \nabla\widetilde{\mathrm{Rm}} = T + \tilde{g}^{-1} * \widetilde{\mathrm{Rm}} * \nabla h.$$

On the other hand, again using the expression (4.54) for $\nabla\widetilde{\mathrm{Rm}}$, we have

$$\begin{aligned}\nabla(\tilde{g}^{-1} * \widetilde{\mathrm{Rm}} * h + \mathrm{Rm} * h) &= (\tilde{g}^{-2} * \widetilde{\mathrm{Rm}} * h + \tilde{g}^{-1} * \widetilde{\mathrm{Rm}} + \mathrm{Rm}) * \nabla h \\ &\quad + (\tilde{g}^{-1} * T + \tilde{g}^{-1} * \nabla\mathrm{Rm} + \nabla\mathrm{Rm}) * h.\end{aligned}$$

Combining these equations yields (4.50). Similarly, writing

$$D_\tau \nabla^2 h = \nabla(D_\tau \nabla h) + \mathrm{Rm} * \nabla^2 h + \nabla\mathrm{Rm} * \nabla h,$$

and using (4.50) with (4.54), we arrive at (4.51).

For (4.52), we first note that (4.48) says

$$\left(\partial_\tau + \tilde{\Delta}\right)\widetilde{\tilde{\nabla}\mathrm{Rm}} = \tilde{g}^{-2} * \widetilde{\tilde{\nabla}\mathrm{Rm}} * \widetilde{\mathrm{Rm}}$$

in the case $m = 1$, with an analogous equation for $\nabla\mathrm{Rm}$. Then, since, schematically, we have

$$\tilde{g}^{-2} - g^{-2} = \tilde{g}^{-2} * h + \tilde{g}^{-1} * h,$$

it follows that

$$\begin{aligned}(\partial_\tau + \Delta)T &= (\tilde{\Delta} - \Delta)\widetilde{\tilde{\nabla}\mathrm{Rm}} + (\tilde{g}^{-1} + \tilde{g}^{-2}) * \widetilde{\tilde{\nabla}\mathrm{Rm}} * \widetilde{\mathrm{Rm}} * h \\ &\quad + \nabla\mathrm{Rm} * S + \widetilde{\mathrm{Rm}} * T.\end{aligned}$$

By (4.46), the first term on the right is

$$\begin{aligned}(\tilde{\Delta} - \Delta)\widetilde{\tilde{\nabla}\mathrm{Rm}} &= \tilde{g}^{-2} * \widetilde{\tilde{\nabla}\mathrm{Rm}} * \nabla^2 h + \tilde{g}^{-1} * \nabla^2\widetilde{\tilde{\nabla}\mathrm{Rm}} * h \\ &\quad + \left\{\tilde{g}^{-2} * \nabla\widetilde{\tilde{\nabla}\mathrm{Rm}} + \tilde{g}^{-3} * \widetilde{\tilde{\nabla}\mathrm{Rm}} * \nabla h\right\} * \nabla h.\end{aligned}$$

Using (4.45), we compute that

$$\nabla\widetilde{\tilde{\nabla}\mathrm{Rm}} = \widetilde{\tilde{\nabla}^2\mathrm{Rm}} + \tilde{g}^{-1} * \widetilde{\tilde{\nabla}\mathrm{Rm}} * \nabla h$$

and

$$\begin{aligned}\nabla^2\widetilde{\tilde{\nabla}\mathrm{Rm}} &= \widetilde{\tilde{\nabla}^3\mathrm{Rm}} + \tilde{g}^{-1} * \widetilde{\tilde{\nabla}\mathrm{Rm}} * \nabla^2 h \\ &\quad + \left\{\tilde{g}^{-2} * \widetilde{\tilde{\nabla}\mathrm{Rm}} * \nabla h + \tilde{g}^{-1} * \widetilde{\tilde{\nabla}^2\mathrm{Rm}}\right\} * \nabla h,\end{aligned}$$

so that

$$\begin{aligned}(\tilde{\Delta} - \Delta)\widetilde{\tilde{\nabla}\mathrm{Rm}} &= \left\{\tilde{g}^{-2} * \widetilde{\tilde{\nabla}\mathrm{Rm}} + \tilde{g}^{-2} * \widetilde{\tilde{\nabla}\mathrm{Rm}} * h\right\} * \nabla^2 h \\ &+ \left\{\tilde{g}^{-3} * \widetilde{\tilde{\nabla}\mathrm{Rm}} * h * \nabla h + \tilde{g}^{-3} * \widetilde{\tilde{\nabla}\mathrm{Rm}} * \nabla h + \tilde{g}^{-2} * \widetilde{\tilde{\nabla}^2\mathrm{Rm}} * h \right. \\ &\quad \left. + \tilde{g}^{-2} * \widetilde{\tilde{\nabla}^2\mathrm{Rm}}\right\} * \nabla h + \tilde{g}^{-1} * \widetilde{\tilde{\nabla}^3\mathrm{Rm}} * h.\end{aligned}$$

Also, by (4.44),

$$\nabla\mathrm{Rm} * S = \nabla\mathrm{Rm} * \nabla^2 h + \tilde{g}^{-1} * \nabla\mathrm{Rm} * (\nabla h)^2 + \nabla\mathrm{Rm} * \mathrm{Rm} * h.$$

So, putting things together, we have

$$
\begin{aligned}
(\partial_\tau + \Delta)T = \widetilde{\mathrm{Rm}} * T + \Big\{\tilde{g}^{-2} * \tilde{\nabla}\widetilde{\mathrm{Rm}} + \tilde{g}^{-2} * \tilde{\nabla}\widetilde{\mathrm{Rm}} * h + \nabla \mathrm{Rm}\Big\} * \nabla^2 h \\
+ \Big\{\tilde{g}^{-3} * \tilde{\nabla}\widetilde{\mathrm{Rm}} * h * \nabla h + \tilde{g}^{-3} * \tilde{\nabla}\widetilde{\mathrm{Rm}} * \nabla h + \tilde{g}^{-1} * \nabla \mathrm{Rm} * \nabla h \\
+ \tilde{g}^{-2} * \tilde{\nabla}^2\widetilde{\mathrm{Rm}} * h + \tilde{g}^{-2} * \tilde{\nabla}^2\widetilde{\mathrm{Rm}}\Big\} * \nabla h \\
+ \Big\{\nabla \mathrm{Rm} * \mathrm{Rm} + (\tilde{g}^{-1} + \tilde{g}^{-2}) * \tilde{\nabla}\widetilde{\mathrm{Rm}} * \widetilde{\mathrm{Rm}} + \tilde{g}^{-1} * \tilde{\nabla}^3\widetilde{\mathrm{Rm}}\Big\} * h,
\end{aligned}
$$

which, since

$$D_\tau T = \partial_\tau T + \mathrm{Rm} * T,$$

gives (4.52). □

The important feature of the equations (4.49)–(4.52) is that each term on the right contains at least one factor of h, ∇h, $\nabla^2 h$, T, or ∇T. Under the assumptions of Theorem 4.19, the other factors in these terms will, at minimum, be bounded, and some (for example, those involving the curvature tensors of the solutions and their derivatives) will decay quadratically in space.

PROPOSITION 4.24. *Let $g(\tau)$, $\tilde{g}(\tau)$, $\mathbf{X} = \mathbf{X}(\tau)$, and $\mathbf{Y} = \mathbf{Y}(\tau)$ be defined as above. Then there is a constant $N > 0$ such that $\mathbf{X}$ and $\mathbf{Y}$ satisfy*

$$|\mathbf{X}| + |\nabla \mathbf{X}| + |\mathbf{Y}| \leq \frac{N\tau}{r^2 + 1}, \tag{4.55}$$

and

$$
\begin{aligned}
|D_\tau \mathbf{X} + \Delta \mathbf{X}| &\leq \frac{N}{r^2+1}(|\mathbf{X}| + |\mathbf{Y}|), \\
|D_\tau \mathbf{Y}| &\leq N(|\mathbf{X}| + |\nabla \mathbf{X}| + |\mathbf{Y}|),
\end{aligned} \tag{4.56}
$$

on $\mathcal{C}_a \times [0,1]$. Here $|\cdot| = |\cdot|_{g(\tau)}$, $\nabla = \nabla_{g(\tau)}$, and $\Delta = \Delta_{g(\tau)}$.

PROOF. According to Proposition 4.18, there is some $N > 0$ such that

$$|\nabla^k(g - \hat{g})| + |\tilde{\nabla}^k(\tilde{g} - \hat{g})|_{\tilde{g}} \leq \frac{N\tau}{r^{k+2}+1}, \quad |\nabla^k \mathrm{Rm}| + |\tilde{\nabla}^k\widetilde{\mathrm{Rm}}|_{\tilde{g}} \leq \frac{N}{r^{k+2}+1},$$

on $\mathcal{C}_a \times [0,1]$ for all $k = 0,1,2,3$. The bounds on the curvature tensors imply that the metrics $g(\tau)$ and $\tilde{g}(\tau)$ are uniformly equivalent on $\mathcal{C}_a$ for all τ (see, e.g., Lemma 6.49 of [**CK04**]); in particular, they are uniformly equivalent to the conical metric $\hat{g} = g(0) = \tilde{g}(0)$. With this, it follows from the estimates on the differences of the metrics that

$$|\widehat{\nabla}^k(g - \hat{g})|_{\hat{g}} + |\widehat{\nabla}^k(\tilde{g} - \hat{g})|_{\hat{g}} \leq \frac{N\tau}{r^{k+2}+1}$$

for some N and $k = 0,1,2,3$. For example, in the case $k = 1$, we have

$$|\widehat{\nabla}(g - \hat{g})| \leq |\nabla(g - \hat{g})|\left(1 + C|\hat{g}^{-1}|\,|g - \hat{g}|\right) \leq \frac{N\tau}{r^3+1},$$

using (4.45). So,

$$|\widehat{\nabla}^k h| \leq |\widehat{\nabla}^k(g - \hat{g})| + |\widehat{\nabla}^k(\tilde{g} - \hat{g})| \leq \frac{N\tau}{r^{k+2}+1},$$

and hence

$$|\nabla^k h| \leq \frac{N\tau}{r^{k+2}+1},$$

for $k = 0, 1, 2, 3$. In particular,

$$|\mathbf{Y}| \leq \frac{N\tau}{r^2+1} \tag{4.57}$$

on $\mathcal{C}_a \times [0,1]$ for some N.

On the other hand, using the evolution equations (4.48) for $\nabla\mathrm{Rm}$ and $\widetilde{\nabla}\widetilde{\mathrm{Rm}}$, and the above derivative estimates, we see that $\mathbf{X} = T = \nabla\mathrm{Rm} - \widetilde{\nabla}\widetilde{\mathrm{Rm}}$ satisfies

$$|\partial_\tau \mathbf{X}| \leq \frac{N\tau}{r^5+1},$$

and thus

$$|\mathbf{X}| \leq \frac{N\tau}{r^5+1}. \tag{4.58}$$

Similarly, for $\nabla\mathbf{X}$, we first apply (4.48) to see that

$$\partial_\tau \widetilde{\nabla}^2\widetilde{\mathrm{Rm}} = -\widetilde{\Delta}\widetilde{\nabla}^2\widetilde{\mathrm{Rm}} + \tilde{g}^{-2} * \widetilde{\nabla}^2\widetilde{\mathrm{Rm}} * \widetilde{\mathrm{Rm}} + \tilde{g}^{-2} * \widetilde{\nabla}\widetilde{\mathrm{Rm}} * \widetilde{\nabla}\widetilde{\mathrm{Rm}}$$

with an analogous expression for $\partial_\tau\nabla^2\mathrm{Rm}$. It follows that $|\nabla^2\mathrm{Rm} - \widetilde{\nabla}^2\widetilde{\mathrm{Rm}}|$ is bounded above by $N\tau/(r^6+1)$ for some constant N. Then, using (4.45) to estimate

$$|\nabla\mathbf{X}| \leq |\nabla^2\mathrm{Rm} - \widetilde{\nabla}^2\widetilde{\mathrm{Rm}}| + C|g^{-1}||\nabla h||\widetilde{\nabla}\widetilde{\mathrm{Rm}}|,$$

we obtain that

$$|\nabla\mathbf{X}| \leq \frac{N\tau}{r^6+1} \tag{4.59}$$

for some N. Combining (4.57), (4.58), and (4.59) then gives (4.55).

Finally, to verify (4.56), we use the bounds just derived on the components of $\mathbf{X}$ and $\mathbf{Y}$ with the evolution equations (4.49)–(4.52) to see that

$$\begin{aligned}
|D_\tau h| &\leq N(|h| + |\nabla h| + |\nabla^2 h|),\\
|D_\tau \nabla h| &\leq N(|h| + |\nabla h| + |T|),\\
|D_\tau \nabla^2 h| &\leq N(|h| + |\nabla h| + |\nabla^2 h| + |\nabla T|),
\end{aligned}$$

and

$$|(D_\tau + \Delta)T| \leq \frac{N\tau}{r^2+1}(|h| + |\nabla h| + |\nabla^2 h| + |T|)$$

on $\mathcal{C}_a \times [0,1]$ for some constant N. □

4.4. A general backward uniqueness principle

The key ingredient in the proof of Theorem 4.19 is a general backward uniqueness principle for solutions to systems of differential inequalities of the form (4.56) over an asymptotically conical self-similarly shrinking background. To simplify the statement of this principle and its proof, we will fix some notation and put in place some standing assumptions for the next several sections.

4.4.1. Setup and notation.

For the discussion that follows, let us fix a cone $(\mathcal{C}_0, \hat{g})$ and a shrinking soliton $(\mathcal{C}_{\mathcal{R}}, g_1, f_1)$ which is dynamically asymptotic to $\hat{g}$ for some $\mathcal{R} \geq 1$. We will write $g(\tau)$ for the solution to the backward Ricci flow (4.7) on

$$\mathcal{C}_{\mathcal{R},1} := \mathcal{C}_{\mathcal{R}} \times [0, 1]$$

associated to this soliton which satisfies $g(0) = \hat{g}$ and $g(1) = g_1$.

The properties of this solution we will need are summarized in Proposition 4.18. Our analysis will only involve finitely many of the derivatives of $g(\tau)$, so (in view of (4.27)) let us fix a uniform $K > 0$ such that

$$|\nabla^m \mathrm{Rm}| \leq \frac{K}{r^{m+2} + 1}, \tag{4.60}$$

for $0 \leq m \leq 6$, and

$$|\nabla^m (g - \hat{g})| \leq \frac{K\tau}{r^{m+2} + 1}, \tag{4.61}$$

for $0 \leq m \leq 4$ on $\mathcal{C}_{\mathcal{R},1}$.

Finally, let $\mathcal{X}$, $\mathcal{Y}$ be vector bundles of the form $\bigoplus \pi^*(T^{\nu_i}(T^*\mathcal{C}_{\mathcal{R}}))$ over $\mathcal{C}_{\mathcal{R},1}$, where $\pi : \mathcal{C}_{\mathcal{R},1} \to \mathcal{C}_{\mathcal{R}}$ is the projection. We will assume that the bundles $\mathcal{X}$ and $\mathcal{Y}$ are equipped with the metrics and connections induced by $g = g(\tau)$, and write $|\cdot| = |\cdot|_{g(\tau)}$, $\nabla = \nabla_{g(\tau)}$, $\Delta = \Delta_{g(\tau)}$, etc.

THEOREM 4.25. *Suppose* $\mathbf{X}$ *and* $\mathbf{Y}$ *are smooth sections of* $\mathcal{X}$ *and* $\mathcal{Y}$ *satisfying*

$$|\mathbf{X}| + |\nabla \mathbf{X}| + |\mathbf{Y}| \leq L \tag{4.62}$$

and

$$\begin{aligned} |D_\tau \mathbf{X} + \Delta \mathbf{X}| &\leq \varepsilon\big(|\mathbf{X}| + |\nabla \mathbf{X}| + |\mathbf{Y}|\big), \\ |D_\tau \mathbf{Y}| &\leq L\big(|\mathbf{X}| + |\nabla \mathbf{X}| + |\mathbf{Y}|\big), \end{aligned} \tag{4.63}$$

on $\mathcal{C}_{\mathcal{R},1}$ *for some constant* $L > 0$ *and some function* $\varepsilon = \varepsilon(r) > 0$ *with* $\lim_{r\to\infty} \varepsilon(r) = 0$. *Then there are* $\mathcal{S} = \mathcal{S}(n, K, L) \geq \mathcal{R}$ *and* $T = T(n) \in (0, 1]$ *such that if* $\mathbf{X}$ *and* $\mathbf{Y}$ *vanish identically on* $\mathcal{C}_{\mathcal{S}} \times \{0\}$, *they vanish identically on* $\mathcal{C}_{\mathcal{S},T}$.

4.4.2. Proof of Theorem 4.19.

We will prove Theorem 4.25 over the next several sections. First, however, we show how one may deduce Theorem 4.19 from it. We will need the following result:

THEOREM 4.26. *Suppose that* $g(t)$ *is a smooth solution to the Ricci flow on* $M^n \times [0, T]$. *Then there exists a unique real-analytic structure* $\mathcal{A}$ *relative to which* $(M^n, g(t))$ *is analytic for all* $t \in (0, T]$. *This structure contains the atlas of* $g(t_0)$-*normal coordinate charts for any* $t_0 \in (0, T]$.

The real-analyticity of the time-slices $(M^n, g(t))$, $t > 0$, of a solution to Ricci flow was proven by Bando [**Ban87**] in the compact case. Bando's proof extends, essentially verbatim, to the case of complete solutions of bounded curvature. Later, Bando's Bernstein-type estimates were localized by one of the authors [**Kot13**], proving the instantaneous real-analyticity of arbitrary smooth solutions. The additional content we need from Theorem 4.26 is the assurance that the Riemannian manifolds $(M^n, g(t))$ are analytic with respect to a fixed (time-independent) atlas for $t > 0$. See, e.g, Theorem 3.1 of [**KW24**] by Wang and one of the authors for a proof.

In passing, we note that Theorem 4.26 implies that the asymptotic cone – and therefore the cross-section – of an asymptotically conical shrinking soliton must be real-analytic.

COROLLARY 4.27. *If a shrinking gradient Ricci soliton* (M^n, g, f) *is* C^2*-asymptotic to a cone* $\mathcal{C}^\Sigma$ *along an end* $V \subset M^n$*, then the cross-section* (Σ, g_Σ) *is real-analytic. Moreover, both* g *and the conical metric are real-analytic with respect to a common atlas on* V.

Currently, there are very few restrictions known on the manifolds (Σ, g_Σ) which may occur as the cross-section of the asymptotic cone $\mathcal{C}^\Sigma$ of an asymptotically conical shrinking soliton.

PROOF OF THEOREM 4.19, ASSUMING THEOREM 4.25. Given the solutions $g = g(\tau)$ and $\tilde{g} = \tilde{g}(\tau)$ to (4.7) on $\mathcal{C}_{a,1} = \mathcal{C}_a \times [0,1]$ with $g(1) = g_1$, $\tilde{g}(1) = \tilde{g}_1$, and $g(0) = \tilde{g}(0) = \hat{g}$, we define $h = g - \tilde{g}$ and form the sections

$$\mathbf{X} = \nabla \mathrm{Rm} - \widetilde{\nabla \mathrm{Rm}}, \quad \mathbf{Y} = (h, \nabla h, \nabla^2 h),$$

of the bundles $\mathcal{X}$ and $\mathcal{Y}$ as above. Then

$$\mathbf{X}(\cdot, 0) \equiv 0, \quad \mathbf{Y}(\cdot, 0) \equiv 0,$$

and, since Proposition 4.24 ensures that $\mathbf{X}$ and $\mathbf{Y}$ meet the other hypotheses of Theorem 4.25, it follows from this theorem that there are $b \geq a$ and $T = T(n) \in (0,1]$ such that $\mathbf{X} \equiv 0$ and $\mathbf{Y} \equiv 0$ on $\mathcal{C}_{b,T}$. In particular, $h = g - \tilde{g} \equiv 0$ on $\mathcal{C}_b$ for all $\tau \in [0,T]$.

Now, by Theorem 4.26, $g(\tau)$ and $\tilde{g}(\tau)$ are real-analytic with respect to some common (time-independent) real-analytic structure on $\mathcal{C}_a$ for each $\tau \in [0,1]$. Therefore $h(\tau) = g(\tau) - \tilde{g}(\tau)$ is real-analytic on $\mathcal{C}_a$ for each $\tau \in [0,1]$. As $h(\tau)$ vanishes on $\mathcal{C}_b$ for each $\tau \in [0,T]$, it must vanish on $\mathcal{C}_a \times \{\tau\}$ for each $\tau \in [0,T]$ and therefore on all of $\mathcal{C}_{a,T}$.

But, for $0 < \tau \leq 1$, we have $g(\tau) = \tau \Phi_\tau^* g_1$ and $\tilde{g}(\tau) = \tau \Phi_\tau^* \tilde{g}_1$ on $\mathcal{C}_a$, where $\Phi_\tau : \mathcal{C}_a \to \mathcal{C}_a$ is the map $\Phi_\tau(r, \sigma) = (r/\sqrt{\tau}, \sigma)$. So

$$h(T) = T\Phi_T^* g_1 - T\Phi_T^* \tilde{g}_1 = T\Phi_T^* h(1) \tag{4.64}$$

on $\mathcal{C}_a$. Since $h(T) \equiv 0$ on $\mathcal{C}_b$, it follows from (4.64) that $h(1) \equiv 0$ on $\mathcal{C}_{b/\sqrt{T}}$. But, invoking again the real-analyticity of $h(1)$, we have $g_1 - \tilde{g}_1 = h(1) \equiv 0$ on $\mathcal{C}_a$, and therefore

$$g(\tau) - \tilde{g}(\tau) = h(\tau) = \tau \Phi_\tau^* h(1) \equiv 0$$

on $\mathcal{C}_a \times [T, 1]$ also. □

4.5. Decay estimates for asymptotically conical Ricci flows

The proof of Theorem 4.25 will involve a series of L^2-estimates weighted by functions of the conical distance function r. In order to derive these estimates, we will need bounds on the derivatives $\nabla^k r$ of r with respect to the time-dependent connection $\nabla = \nabla_{g(\tau)}$. We will continue to work in the setting of §4.4.1, with $g = g(\tau)$ representing the solution to (4.7) on $\mathcal{C}_{\mathcal{R},1} = \mathcal{C}_\mathcal{R} \times [0,1]$ described by Proposition 4.18 which satisfies (4.60)–(4.61).

The estimate (4.61) implies in particular that the norms $|\cdot|_{g(\tau)}$, $\tau \in [0,1]$, induced by the solution (including $|\cdot|_{\hat{g}} = |\cdot|_{g(0)}$) are uniformly equivalent on $\mathcal{C}_\mathcal{R}$.

Moreover, by combining (4.60) and (4.61) and iterating (4.45), we have bounds on the mixed derivatives of the curvature and the difference of metrics of the form

$$|\widehat{\nabla}^k \nabla^{m-k} \mathrm{Rm}| \leq \frac{CK}{r^{m+2}+1}, \tag{4.65}$$

for $0 \leq k \leq m \leq 6$, and

$$|\widehat{\nabla}^k \nabla^{m-k} (g - \hat{g})| \leq \frac{CK\tau}{r^{m+2}+1}, \tag{4.66}$$

for $0 \leq k \leq m \leq 4$ on $\mathcal{C}_{\mathcal{R},1}$.

In the estimates below, we will use C to denote a sequence of positive constants depending at most on the dimension n, and use N to denote a sequence of constants which potentially also depend on the parameter K from (4.60) and (4.61).

4.5.1. Properties of the radial distance on the cone.

We begin by recalling some geometric properties of conical metrics.

LEMMA 4.28. *On $(\mathcal{C}_0, \hat{g})$ we have the equations*

$$|\widehat{\nabla} r|^2_{\hat{g}} = 1, \quad \widehat{\nabla}^2 r^2 = 2\hat{g}, \quad \widehat{\mathrm{Rm}}(\widehat{\nabla} r, \cdot, \cdot, \cdot) = 0, \tag{4.67}$$

and, for all $m \geq 0$, the bounds

$$|\widehat{\nabla}^m r|_{\hat{g}} \leq \frac{N_m}{r^{m-1}}, \quad |\widehat{\nabla}^m \widehat{\mathrm{Rm}}|_{\hat{g}} \leq \frac{N_m}{r^{m+2}}, \tag{4.68}$$

for some constants N_m depending on (Σ, g_Σ). We also have the identities

$$\begin{aligned} \widehat{\nabla}_i \hat{R}_{jk} \widehat{\nabla}_i r &= 2\widehat{\nabla}_j \hat{R}_{ik} \widehat{\nabla}_i r = -\frac{2}{r} \hat{R}_{jk}, \\ \hat{\Delta} \hat{R}_{ij} \widehat{\nabla}_i r \widehat{\nabla}_j r &= \frac{2\hat{R}}{r^2}, \end{aligned} \tag{4.69}$$

where the repeated indices here denote contraction with the metric $\hat{g}$.

PROOF. The identities in (4.67) and the bounds in (4.68) are consequences of the standard formulas for warped products (see, e.g., §J of Chapter 9 of [**Bes08**]). They imply the identities in (4.69) as we now see. First, from $\hat{R}_{ji\ell k} \widehat{\nabla}_k r = 0$, we obtain that $\hat{R}_{jk} \widehat{\nabla}_k r = 0$, and, with

$$r \widehat{\nabla}^2_{ik} r = \hat{g}_{ik} - \widehat{\nabla}_i r \widehat{\nabla}_k r,$$

that

$$\widehat{\nabla}_i \hat{R}_{jk} \widehat{\nabla}_k r = -\hat{R}_{jk} \widehat{\nabla}^2_{ik} r = -\frac{1}{r} \hat{R}_{jk} (\hat{g}_{ik} - \widehat{\nabla}_i r \widehat{\nabla}_k r) = -\frac{\hat{R}_{ij}}{r}.$$

Similarly, since $\hat{R}_{ijk\ell} \widehat{\nabla}_\ell r = 0$ implies that

$$\widehat{\nabla}_m \hat{R}_{ijk\ell} \widehat{\nabla}_\ell r = -\hat{R}_{ijk\ell} \widehat{\nabla}^2_{m\ell} r,$$

we find that

$$\begin{aligned} \widehat{\nabla}_i \hat{R}_{jk} \widehat{\nabla}_i r &= \widehat{\nabla}_j \hat{R}_{ik} \widehat{\nabla}_i r + (\widehat{\nabla}_i \hat{R}_{jk} - \widehat{\nabla}_j \hat{R}_{ik}) \widehat{\nabla}_i r = -\frac{\hat{R}_{jk}}{r} + \widehat{\nabla}_p \hat{R}_{pkji} \widehat{\nabla}_i r \\ &= -\frac{\hat{R}_{jk}}{r} - \hat{R}_{pkji} \widehat{\nabla}_p \widehat{\nabla}_i r = -\frac{\hat{R}_{jk}}{r} - \frac{1}{r} \hat{R}_{pkji} (\hat{g}_{pi} - \widehat{\nabla}_p r \widehat{\nabla}_i r) \\ &= -2\frac{\hat{R}_{jk}}{r}. \end{aligned}$$

Then also

$$\begin{aligned}\hat{\Delta}\hat{R}_{ij}\widehat{\nabla}_i r\widehat{\nabla}_j r &= \widehat{\nabla}_p(\widehat{\nabla}_p\hat{R}_{ij}\widehat{\nabla}_i r\widehat{\nabla}_j r) - 2\widehat{\nabla}_p\hat{R}_{ij}\widehat{\nabla}^2_{ip}r\widehat{\nabla}_j r\\ &= \frac{2}{r^2}\hat{R}_{ip}\left(\hat{g}_{ip} - \widehat{\nabla}_i r\widehat{\nabla}_p r\right) = \frac{2\hat{R}}{r^2},\end{aligned}$$

as claimed. □

4.5.2. Estimates on the radial distance with respect to $g(\tau)$. Now we estimate the derivatives of the radial distance function with respect to the evolving metric $g(\tau)$. Since $g(\tau)$ closely approximates the cone where r is very large and τ is very close to zero, r ought to approximately satisfy the identities in (4.67) relative to the norm and connection induced by $g(\tau)$ in these regions. We will need to control the error of these approximations. To this end, we define the quantities

$$\eta_1 := |\nabla r|^2 - 1, \quad \eta_2 := \nabla^2\left(\frac{r^2}{2}\right) - g, \tag{4.70}$$

$$\eta_3 := \Delta r - \frac{n-1}{r} = \frac{\mathrm{tr}_g(\eta_2) - \eta_1}{r}, \tag{4.71}$$

and

$$(\eta_4)_{ijk} := R_{ijk}^{\ell}\nabla_\ell r. \tag{4.72}$$

Note that $\eta_i \equiv 0$ for $1 \leq i \leq 4$ on $\mathcal{C}_{\mathcal{R}} \times \{0\}$.

PROPOSITION 4.29. *There are constants N depending on n and K such that*

$$|\nabla^m\eta_1| \leq \frac{N\tau^2}{r^{m+4}+1}, \quad |\partial_\tau\eta_1| \leq \frac{N\tau}{r^4+1}, \tag{4.73}$$

$$|\nabla^m\eta_2| \leq \frac{N\tau}{r^{m+2}+1}, \quad |\partial_\tau\eta_2| \leq \frac{N}{r^2+1}, \tag{4.74}$$

$$|\nabla^m\eta_3| \leq \frac{N\tau}{r^{m+3}+1}, \quad |\partial_\tau\eta_3| \leq \frac{N}{r^3+1}, \tag{4.75}$$

on $\mathcal{C}_{\mathcal{R},1}$ for $m = 0, 1, 2$. Additionally, we have the estimates

$$|\eta_4| \leq \frac{N\tau}{r^4+1} \tag{4.76}$$

and

$$|\nabla^m r| \leq \frac{N}{r^{m-1}+1}, \tag{4.77}$$

on $\mathcal{C}_{\mathcal{R},1}$ for $1 \leq m \leq 4$, for N with the same dependencies as above.

Of course, there are analogous estimates for larger m; the range here (which is sufficient for our purposes later) is restricted by the uniform choice of the constant K in (4.60)–(4.61) made at the start of the section.

PROOF. For (4.73), we claim that we can use Taylor's theorem to write

$$\eta_1(x,\tau) = 2\frac{\tau^2}{r^2}\hat{R}(x) + \frac{1}{2}\int_0^\tau Q_0(x,s)^{ij}\widehat{\nabla}_i r\widehat{\nabla}_j r(\tau - s)^2\,ds, \tag{4.78}$$

where the tensor Q_0 satisfies the schematic equation

$$\begin{aligned}Q_0 = g^{-5} * \nabla^4\mathrm{Rm} + g^{-6} * \nabla^2\mathrm{Rm} * \mathrm{Rm} + g^{-6} * (\nabla\mathrm{Rm})^2\\ + g^{-7} * (\mathrm{Rm})^3.\end{aligned} \tag{4.79}$$

(Here, and in the rest of this proof, we use $A * B$ to represent linear combinations of *metric independent* contractions of $A \otimes B$.) To verify this claim, first recall that, under the backward Ricci flow, the Ricci curvature tensor satisfies the equation

$$\partial_\tau R_{ij} = -\Delta R_{ij} - 2g^{kp}g^{lq}R_{ipqj}R_{kl} + 2g^{pq}R_{ip}R_{jq}, \tag{4.80}$$

the $(4,0)$-Riemann curvature tensor satisfies the schematic equation

$$\partial_\tau \mathrm{Rm} = g^{-1} * \nabla^2 \mathrm{Rm} + g^{-2} * (\mathrm{Rm})^2, \tag{4.81}$$

and the inverse of the metric satisfies

$$\partial_\tau g^{ij} = -2g^{ia}g^{jb}R_{ab} = g^{-3} * \mathrm{Rm}. \tag{4.82}$$

Moreover, for any tensor V of rank at least one, we have the commutation equations

$$\begin{aligned} [\partial_\tau, \nabla]V &= g^{-2} * \nabla \mathrm{Rm} * V, \\ [\partial_\tau, \nabla^2]V &= g^{-2} * \nabla \mathrm{Rm} * \nabla V + g^{-2} * \nabla^2 \mathrm{Rm} * V, \end{aligned} \tag{4.83}$$

where, as above, Rm is the $(4,0)$-curvature tensor (see, e.g, Lemma 13.26 of [**CCG⁺08**]).

To see that (4.78) holds, note that at $\tau = 0$ we have the identities

$$\eta_1 = |\widehat{\nabla} r|^2_{\hat{g}} - 1 = 0, \quad \partial_\tau \eta_1 = -2\hat{R}^{ij}\widehat{\nabla}_i r \widehat{\nabla}_j r = 0, \tag{4.84}$$

and

$$\begin{aligned} \partial^2_{\tau\tau}\eta_1 &= -2(\partial_\tau R^{ij})\nabla_i r \nabla_j r \\ &= 2(\hat{\Delta}\hat{R}^{ij} + 2\hat{R}^{ipqj}\hat{R}_{pq} + 2\hat{R}^i_p \hat{R}^{pj})\widehat{\nabla}_i r \widehat{\nabla}_j r \\ &= \frac{4\hat{R}}{r^2}, \end{aligned} \tag{4.85}$$

in view of (4.67), (4.69), (4.80), and (4.82). Also, for any τ, we have

$$\partial^3_{\tau\tau\tau}\eta_1 = -2(\partial^2_{\tau\tau}R^{ij})\widehat{\nabla}_i r \widehat{\nabla}_j r := Q_0^{ij}\widehat{\nabla}_i r \widehat{\nabla}_j r. \tag{4.86}$$

Using (4.81), (4.82), and (4.83), we see that

$$\begin{aligned} Q_0 &= \partial^2_{\tau\tau}(g^{-3} * \mathrm{Rm}) = \partial_\tau \left(g^{-4} * \nabla^2 \mathrm{Rm} + g^{-5} * (\mathrm{Rm})^2\right) \\ &= g^{-6} * \nabla^2 \mathrm{Rm} * \mathrm{Rm} + g^{-4} * \nabla^2(\partial_\tau \mathrm{Rm}) + g^{-4} * [\partial_\tau, \nabla^2]\mathrm{Rm} + g^{-7} * (\mathrm{Rm})^3 \\ &\quad + g^{-5} * (\partial_\tau \mathrm{Rm}) * \mathrm{Rm} \\ &= g^{-5} * \nabla^4 \mathrm{Rm} + g^{-6} * \nabla^2 \mathrm{Rm} * \mathrm{Rm} + g^{-6} * (\nabla \mathrm{Rm})^2 + g^{-7} * (\mathrm{Rm})^3, \end{aligned}$$

which is (4.79). Equation (4.78) then follows from (4.84), (4.85), (4.86), and Taylor's formula.

Note that, differentiating (4.79) with respect to the (τ-independent) connection $\widehat{\nabla}$ and using (4.65) and (4.66), we have

$$|\widehat{\nabla}^k Q_0| \leq \frac{N}{r^{k+6}+1} \tag{4.87}$$

on $\mathcal{C}_{\mathcal{R},1}$ for $0 \leq k \leq 2$. Thus, differentiating (4.78) and using (4.87), we obtain bounds of the form

$$|\widehat{\nabla}^k \eta_1|_{\hat{g}} \leq N\left(\frac{\tau^2}{r^{k+4}+1} + \frac{\tau^3}{r^{k+6}+1}\right) \leq \frac{N\tau^2}{r^{k+4}+1}$$

on $\mathcal{C}_{\mathcal{R},1}$ for $0 \leq k \leq 2$. Then, since

$$|\nabla \eta_1| \leq N|\widehat{\nabla}\eta_1|_{\hat{g}} \quad \text{and} \quad |\nabla^2 \eta_1| \leq N\big(|\widehat{\nabla} h|_{\hat{g}}|\nabla \eta_1|_{\hat{g}} + |\widehat{\nabla}^2 \eta_1|_{\hat{g}}\big),$$

the estimates on the spatial derivatives in (4.73) follow. For the estimate on the temporal derivative in (4.73), we use (4.78) to see that

$$\partial_\tau \eta_1(x,\tau) = \frac{4\tau\hat{R}}{r^2} + \int_0^\tau Q(x,s)^{ij}\widehat{\nabla}_i r \widehat{\nabla}_j r(\tau - s)\,ds$$

and then apply (4.69) and (4.87).

We argue similarly for (4.74), computing first that

$$\begin{aligned}(\partial_\tau \eta_2)_{ij} &= -r\partial_\tau \Gamma_{ij}^k \nabla_k r - 2R_{ij} \\ &= r(\nabla^k R_{ij} - \nabla_i R_j^k - \nabla_j R_i^k)\nabla_k r - 2R_{ij}.\end{aligned} \tag{4.88}$$

In particular, the first two identities in (4.69) imply that

$$(\partial_\tau \eta_2)_{ij} = -2\hat{R}_{ij}$$

at $\tau = 0$. Since, schematically,

$$\partial_\tau \eta_2 = r(g^{-2} * \nabla \mathrm{Rm} * \nabla r) + g^{-1} * \mathrm{Rm},$$

using (4.81), (4.82), and (4.83), we find that

$$\partial^2_{\tau\tau}\eta_2 = rQ_1(\nabla r) + Q_2,$$

where

$$Q_1 = g^{-3} * \nabla^3 \mathrm{Rm} + g^{-4} * \nabla \mathrm{Rm} * \mathrm{Rm}$$

and

$$Q_2 = g^{-2} * \nabla^2 \mathrm{Rm} + g^{-3} * (\mathrm{Rm})^2.$$

With (4.69), we then have

$$\eta_2(x,\tau) = -2\tau\widehat{\mathrm{Ric}}(x) + \int_0^\tau \big(rQ_1(x,s)(\nabla r) + Q_2(x,s)\big)(\tau - s)\,ds.$$

Differentiating this equation in space with respect to $\widehat{\nabla}$ and using (4.65), (4.66), and (4.68), we thus obtain

$$|\widehat{\nabla}^k \eta_2|_{\hat{g}} \leq N\left(\frac{\tau}{r^{k+2}+1} + \frac{\tau^2}{r^{k+4}+1}\right) \leq \frac{N\tau}{r^{k+2}+1}$$

for $0 \leq k \leq 2$. The estimates on $|\eta_2|$, $|\nabla \eta_2|$, and $|\nabla^2 \eta_2|$ then follow as above via (4.66). For the estimate on the temporal derivative in (4.74), one may apply (4.60) directly to (4.88).

Next, differentiating the identity

$$\nabla^2 r = \frac{g - \nabla r \otimes \nabla r + \eta_2}{r},$$

we obtain the cases $2 \leq m \leq 4$ of (4.77) from (4.74) and induction (the case $m = 1$ follows from (4.73)). Then using (4.73) and (4.74) together with the identity

$$\eta_3 = \frac{\mathrm{tr}_g\, \eta_2 - \eta_1}{r},$$

we obtain the estimates in (4.75).

Finally, for (4.76), note that

$$\nabla_i(\eta_2)_{jk} - \nabla_j(\eta_2)_{ik} = -rR_{ijk}^l\nabla_\ell r = -r(\eta_4)_{ijk},$$

so, by (4.74), we have

$$|\eta_4| \le \frac{2}{r}|\nabla\eta_2| \le \frac{N\tau}{r^4+1}$$

as claimed. □

4.6. Carleman estimates to prove backward uniqueness

Now we are ready to prove the estimates needed for Theorem 4.25. We will continue to assume that $g(\tau)$ is the solution to (4.7) on $\mathcal{C}_{\mathcal{R},1} = \mathcal{C}_{\mathcal{R}} \times [0,1]$ associated to the shrinking soliton dynamically asymptotic to $\hat{g}$. In particular, we assume that $g(\tau)$ satisfies the estimates (4.60) and (4.61). Below, $\mathcal{Z}$ will denote a pulled-back bundle of the form $\pi^*(T^\nu(T^*\mathcal{C}_{\mathcal{R}}))$ over $\mathcal{C}_{\mathcal{R},1}$, where $\pi : \mathcal{C}_{\mathcal{R},1} \to \mathcal{C}_{\mathcal{R}}$ is the projection.

4.6.1. The strategy.

We will derive two pairs of **Carleman-type estimates** for sections of $\mathcal{Z}$ with respect to the background metric $g(\tau)$. The idea behind these estimates is as follows. For given smooth functions ϕ, $F : \mathcal{C}_{\mathcal{R},1} \to \mathbb{R}$, consider the operators

$$\mathcal{A} := D_\tau - \nabla_{\nabla\phi} + \frac{F}{2}\,\mathrm{Id} \quad \text{and} \quad \mathcal{S} := \Delta + \nabla_{\nabla\phi} - \frac{F}{2}\,\mathrm{Id}, \tag{4.89}$$

acting on sections Z of $\mathcal{Z}$. Then

$$D_\tau + \Delta = \mathcal{A} + \mathcal{S},$$

and, for an appropriate choice of F (depending on ϕ and $\partial_\tau dv$), the operators $\mathcal{A}$ and $\mathcal{S}$ will be respectively, anti-symmetric and self-adjoint on $L^2(G dv\, d\tau)$, where $G = e^\phi$. These properties imply that, for compactly supported sections Z which vanish at $\tau = 0$ and $\tau = 1$ we have

$$\begin{aligned}\|(D_\tau + \Delta)Z\|^2 &= \|(\mathcal{A}+\mathcal{S})Z\|^2 \\ &= \|\mathcal{A}Z\|^2 + \|\mathcal{S}Z\|^2 + 2\langle\mathcal{A}Z, \mathcal{S}Z\rangle \\ &\ge \langle[\mathcal{S},\mathcal{A}]Z, Z\rangle,\end{aligned} \tag{4.90}$$

where $\langle\cdot,\cdot\rangle$ and $\|\cdot\|$ represent the (space-time) weighted $L^2(G dv\, d\tau)$ inner product and norm for sections of $\mathcal{Z}$ over $\mathcal{C}_{\mathcal{R},1}$, and $[\mathcal{S},\mathcal{A}]$ denotes the commutator $\mathcal{S}\mathcal{A}-\mathcal{A}\mathcal{S}$ of the operators $\mathcal{A}$ and $\mathcal{S}$. Roughly speaking, the object is then to choose the function $\phi = \log G$ (and, accordingly, F) in such a way that the commutator term dominates the H^1-norm of Z up to some other controllable error/boundary terms

$$\|Z\|^2 + \|\nabla Z\|^2 \lesssim \langle[\mathcal{S},\mathcal{A}]Z, Z\rangle,$$

and so that one ultimately obtains an estimate of the form

$$\|Z\|^2 + \|\nabla Z\|^2 \lesssim \varepsilon\,\|(D_\tau+\Delta)Z\|^2. \tag{4.91}$$

up to boundary terms, where ε is a "small" constant.

In the situation (near to our own) in which $Z = \eta W$ for η a cutoff function and W is some section satisfying a *parabolic* inequality of the form

$$|(D_\tau+\Delta)W| \le N(|W| + |\nabla W|),$$

one also has matching inequalities of the form

$$\|(D_\tau+\Delta)Z\|^2 \lesssim C\left(\|Z\|^2 + \|\nabla Z\|^2\right) \tag{4.92}$$

up to error/boundary terms. Then one can leverage the small constant ε to play the inequality in (4.91) off of that in (4.92) to bound the L^2-norm of W on a smaller region $\Omega \subset \mathcal{C}_{\mathcal{R},1}$ by various terms which one hopes to control within the context of the problem using the structure of the weight. Our proof of Theorem 4.25 will require some adjustment to handle the system described by (4.63).

These types of estimates have their origins in the work [**Car39**] of T. Carleman on the strong unique continuation of elliptic PDE in two dimensions. Their use in unique continuation and backward uniqueness problems for parabolic equations dates at least to the work of S. Mizohata [**Miz58**], and is now ubiquitous (for a small sample, see, e.g., [**Esc00**], [**EF03**], [**EFV06**], [**EKPV06**], [**LP61**], [**SS87**], [**Sog90**], [**Tat97**], and the references therein).

For the proof of Theorem 4.25, we will use modified versions of the Carleman estimates proven in [**KW15**] by Wang and one of the authors, which are in turn modeled on the estimates of Escauriaza, Seregin, and Šverák [**ESŠ03a**] for the exterior backward uniqueness problem in $\mathbb{R}^n$ discussed earlier. Whereas the estimates proven in [**KW15**] use weight functions defined in terms of the soliton potential of the background metric $g(\tau)$, the estimates in our presentation here will use weights defined in terms of the radial distance function. This simplifies the arguments in some places. We will closely follow the template in [**Kot24b**], written by one of the authors, in which similar, somewhat more general, estimates are derived.

4.6.2. A general divergence identity.

In our applications, we will work with sections do not necessarily vanish on the terminal time-slices and with operators $\mathcal{S}$ and $\mathcal{A}$ which are not quite symmetric and antisymmetric. Thus, we will start with the following general divergence identity which encodes the integrations-by-parts used to obtain the estimate by the commutator in (4.90).

LEMMA 4.30. *Let Z be a section of $\mathcal{Z}$, and F, G be smooth functions on $\mathcal{C}_{\mathcal{R},1}$ with $G > 0$. Then we have the identity*

$$
\begin{aligned}
&\nabla_i\Big\{2\langle D_\tau Z, \nabla_i Z\rangle G + |\nabla Z|^2 \nabla_i G - 2\langle \nabla_{\nabla G} Z, \nabla_i Z\rangle + \langle \nabla_i Z, Z\rangle FG \\
&\qquad + \frac{1}{2}|Z|^2(F\nabla_i G - G\nabla_i F)\Big\}\, dv \\
&\qquad - \frac{\partial}{\partial \tau}\left\{\left(|\nabla Z|^2 G + \frac{1}{2}|Z|^2 FG\right) dv\right\} \\
&= \Big\{2\big\langle (D_\tau + \Delta)Z, \mathcal{A}Z\big\rangle - 2|\mathcal{A}Z|^2 - 2\nabla^2_{ij}\phi\,\langle \nabla_i Z, \nabla_j Z\rangle \\
&\qquad + \left(F - G^{-1}(\partial_\tau G - \Delta G) - R\right)\left(|\nabla Z|^2 + \frac{F}{2}|Z|^2\right) G \\
&\qquad - \frac{1}{2}(\partial_\tau F + \Delta F)|Z|^2 + E_\phi(Z, \nabla Z)\Big\} G\, dv,
\end{aligned} \tag{4.93}
$$

where $\phi = \ln G$ and

$$
E_\phi(Z, \nabla Z) = 2\langle [\nabla_i, D_\tau], \nabla_i Z\rangle + 2\nabla_j\phi\,\langle [\nabla_i, \nabla_j]Z, \nabla_i Z\rangle. \tag{4.94}
$$

Here and throughout, we write $dv = dv_{g(\tau)}$. This is a vector bundle version of the identity in Lemma 1 of [**ESŠ03a**]; it can be verified with a straightforward if somewhat lengthy computation. The key ingredient is the following **Rellich–Nečas** type identity:

$$\begin{aligned}\Delta G|\nabla Z|^2 = 2\langle \Delta Z, \nabla_{\nabla G} Z\rangle - 2\nabla_i G\langle [\nabla_i, \nabla_j]Z, \nabla_j Z\rangle + 2\nabla^2_{ij} G\langle \nabla_i Z, \nabla_j Z\rangle \\ + \nabla_i\left(\nabla_i G|\nabla Z|^2 - 2\langle \nabla_{\nabla G} Z, \nabla_i Z\rangle\right).\end{aligned}$$

Regarding the term E_ϕ in (4.94), note that standard commutation identities for the Ricci flow[7] imply that

$$[\nabla_i, D_\tau]Z = (\nabla_p R_{iq} - \nabla_q R_{ip})\Lambda_{pq} Z + R_{ip}\nabla_p Z, \tag{4.95}$$

and

$$\nabla_j\phi[\nabla_i, \nabla_j]Z = -\nabla_j \phi R_{ijpq}\Lambda_{pq} Z, \tag{4.96}$$

where Λ_{pq} is the operator which acts on covariant ν-tensors by

$$\Lambda_{pq}(Z_\alpha) = \delta_{p\alpha_1} Z_{q\alpha_2\cdots\alpha_\nu} + \delta_{p\alpha_2} Z_{\alpha_1 q\alpha_3\cdots\alpha_\nu} + \cdots + \delta_{p\alpha_\nu} Z_{\alpha_1\alpha_2\cdots q}$$

Hence we have the estimate

$$|E_\phi(Z, \nabla Z)| \leq C(|\nabla \mathrm{Ric}| + |\nabla\phi||\mathrm{Rm}|)|\nabla Z||Z| + C|\mathrm{Ric}||\nabla Z|^2 \tag{4.97}$$

for some $C = C(n)$. Later we will need a slightly more careful estimate on the second term in E_ϕ. In our applications, the function $\phi = \ln G$ will have the form $\phi(x,\tau) = \Phi(r(x), \tau)$ for some $\Phi : \mathbb{R}^2 \to \mathbb{R}$, and in this case,

$$\begin{aligned}|E_\phi(Z, \nabla Z)| \leq C\left(|\nabla \mathrm{Ric}| + |\partial_r \Phi||\mathrm{Rm}(\nabla r, \cdot)|\right)|\nabla Z||Z| \\ + C|\mathrm{Ric}||\nabla Z|^2.\end{aligned} \tag{4.98}$$

4.6.3. Integral inequalities for the first Carleman estimate.

The identity (4.93) is valid for any smooth F and G with $G > 0$. Assuming G to be some fixed positive (but otherwise yet unspecified) function, let us now choose F in terms of G so that the quantity in the penultimate line of (4.93) vanishes. Then, fixing any compactly supported section Z which vanishes at $\tau = 0$ and integrating the identity over space and time, we obtain the following consequence:

PROPOSITION 4.31. *Let $0 < T \leq 1$ and let G be any smooth positive function on $\mathcal{C}_{\mathcal{R},T}$. For any compactly supported section Z of $\mathcal{Z}$ which vanishes identically for $\tau = 0$, we have the inequality*

$$\begin{aligned}\frac{1}{2}\int_{\mathcal{C}_{\mathcal{R},T}} |D_\tau Z + \Delta Z|^2 G\, dv\, d\tau + \int_{\mathcal{C}_{\mathcal{R}}\times\{T\}} \left(|\nabla Z|^2 + \frac{F}{2}|Z|^2\right) G\, dv \\ \geq \int_{\mathcal{C}_{\mathcal{R},T}} (Q_1(\nabla Z, \nabla Z) + Q_2(Z, Z) - E_\phi(Z, \nabla Z))G\, dv\, d\tau,\end{aligned} \tag{4.99}$$

where $\phi = \ln G$,

$$F = G^{-1}(\partial_\tau G - \Delta G + RG), \tag{4.100}$$

$$Q_1(\nabla Z, \nabla Z) = 2\nabla^2_{ij}\phi\langle \nabla_i Z, \nabla_j Z\rangle, \tag{4.101}$$

$$Q_2(Z, Z) = \frac{1}{2}(\partial_\tau F + \Delta F)|Z|^2, \tag{4.102}$$

and E_ϕ is as defined in (4.94).

[7]See, e.g., Appendix F of [**CCG$^+$08**]. The frame bundle formulation in that reference produces an equivalent operator $D_t = -D_\tau$ acting on smooth families of tensors.

PROOF. Assume that the section Z has compact support and vanishes at $\tau = 0$. Then, integrating both sides of (4.93) over $\mathcal{C}_{\mathcal{R},T}$ we are left only with the term

$$-\int_{\mathcal{C}_{\mathcal{R}}\times\{T\}} \left(|\nabla Z|^2 + \frac{F}{2}|Z|^2\right) G\, dv$$

on the left-hand side, whereas, on the right-hand side, we have

$$2\int_{\mathcal{C}_{\mathcal{R},T}} \left(\langle D_\tau Z + \Delta Z, \mathcal{A}Z\rangle - |\mathcal{A}Z|^2\right)\, dv\, d\tau$$
$$- \int_{\mathcal{C}_{\mathcal{R},T}} \left(Q_1(\nabla Z, \nabla Z) + Q_2(Z, Z) + E_\phi(Z, \nabla Z)\right)\, dv\, d\tau.$$

Combining the two yields (4.99), as

$$2\int_{\mathcal{C}_{\mathcal{R},T}} \left(\langle D_\tau Z + \Delta Z, \mathcal{A}Z\rangle - |\mathcal{A}|^2\right)\, dv\, d\tau \leq \frac{1}{2}\int_{\mathcal{C}_{\mathcal{R},T}} |D_\tau Z + \Delta Z|^2\, dv\, d\tau.$$

□

Later, we will also make use of the following elementary inequality.

PROPOSITION 4.32. *Let $0 < T \leq 1$ and let G be any smooth positive function on $\mathcal{C}_{\mathcal{R},T}$. Write $\phi = \ln G$. Then there exists a constant N depending on n and K such that*

$$(4.103)\qquad -\int_{\mathcal{C}_{\mathcal{R},T}} (N + \partial_\tau \phi)\, |Z|^2 G\, dv\, d\tau \leq \int_{\mathcal{C}_{\mathcal{R},T}} |D_\tau Z|^2\, G\, dv\, d\tau$$

for any compactly supported section Z of $\mathcal{Z}$ which vanishes identically for $\tau = 0$.

PROOF. We start with the identity

$$(4.104)\qquad \frac{\partial}{\partial\tau}\left(|Z|^2 G\, dv\right) = \left(2\,\langle D_\tau Z, Z\rangle\, G + |Z|^2 \partial_\tau G + R|Z|^2 G\right) dv.$$

The inequality (4.103) then follows upon integrating (4.104) over $\mathcal{C}_{\mathcal{R},T}$ and applying the Cauchy–Schwarz inequality together with (4.60). (Note that the integrand in the boundary term on $\mathcal{C}_{\mathcal{R}} \times \{T\}$ is nonnegative.) □

4.6.4. The weight function.

The weight G we will use in our first Carleman inequality is analogous to the corresponding choice in [**ESŠ03a**]. For fixed $\alpha > 0$, $0 < \delta < 1$, and $0 < \tau_0 \leq 1$, we define

$$(4.105)\qquad G_1 := G_{1;\alpha,\delta,\tau_0} := \exp\left(\alpha(\tau_0 - \tau)r^{2-\delta} + r^2\right)$$

and

$$(4.106)\qquad \phi_1 := \phi_{1;\alpha,\delta,\tau_0} := \ln G_{1;\alpha,\delta,\tau_0}.$$

PROPOSITION 4.33. *The function* $\phi_1 = \ln G_1$ *satisfies*

$$\partial_\tau \phi_1 = -\alpha r^{2-\delta}, \tag{4.107}$$

$$\nabla \phi_1 = \big(\alpha(2-\delta)(\tau_0-\tau)r^{1-\delta} + 2r\big)\nabla r, \tag{4.108}$$

$$\begin{aligned} |\nabla\phi_1|^2 = \big(&4r^2 + 4\alpha(2-\delta)(\tau_0-\tau)r^{2-\delta} \\ &+ \alpha^2(2-\delta)^2(\tau_0-\tau)^2 r^{2-2\delta}\big)(1+\eta_1), \end{aligned} \tag{4.109}$$

$$\begin{aligned} \nabla^2\phi_1 &= 2(g+\eta_2) \\ &\quad + \frac{\alpha(2-\delta)(\tau_0-\tau)}{r^\delta}\left(g - \delta\nabla r\otimes\nabla r + \eta_2\right), \end{aligned} \tag{4.110}$$

$$\begin{aligned} \Delta\phi_1 &= 2(n+\eta_1+r\eta_3) \\ &\quad + \frac{\alpha(2-\delta)(\tau_0-\tau)}{r^\delta}\left(n-\delta+(1-\delta)\eta_1 + r\eta_3\right), \end{aligned} \tag{4.111}$$

where the quantities η_i *are as defined in* (4.70)–(4.71). *In particular, there is* $\mathcal{S} = \mathcal{S}(\delta, n, K) \geq \mathcal{R}$, *independent of* $\alpha > 0$, *such that*

$$\nabla^2\phi_1 \geq g \tag{4.112}$$

on $\mathcal{C}_{\mathcal{S},T}$.

PROOF. The identities (4.107) and (4.108) follow directly from the definition of ϕ_1, and (4.109) follows immediately from (4.108). For (4.110), we compute from (4.108) that

$$\begin{aligned} \nabla^2\phi_1 &= \alpha(2-\delta)(\tau_0-\tau)r^{-\delta}\left((1-\delta)\nabla r\otimes\nabla r + r\nabla^2 r\right) + \nabla^2 r^2 \\ &= \alpha(2-\delta)(\tau_0-\tau)r^{-\delta}\left(g+\eta_2-\delta\nabla r\otimes\nabla r\right) + 2(g+\eta_2), \end{aligned} \tag{4.113}$$

whereas for (4.111), we take the trace of (4.113) and compute that

$$\begin{aligned} \Delta\phi_1 &= \alpha(2-\delta)(\tau_0-\tau)r^{-\delta}\left((1-\delta)|\nabla r|^2 + r\Delta r\right) + 2(r\Delta r + |\nabla r|^2) \\ &= \alpha(2-\delta)(\tau_0-\tau)r^{-\delta}\left(n-\delta+(1-\delta)\eta_1+r\eta_3\right) + 2(n+\eta_1+r\eta_3). \end{aligned}$$

For the lower bound in (4.112), we combine (4.73) and (4.74) to choose $\mathcal{S}$ sufficiently large to ensure that $|\eta_1|$ and $|\eta_2| < (1-\delta)/2$ on $\mathcal{C}_{\mathcal{S},T}$. Then the first term on the right of (4.110) will be greater than or equal to g, while the second term will be nonnegative-definite. Indeed, if $V = \nabla r/|\nabla r|$, then

$$(g - \delta\nabla r\otimes\nabla r - \eta_2)(V,V) \geq \frac{1}{2}(1+\delta) - \delta(1+|\eta_1|) > \delta,$$

while, if V is perpendicular to ∇r, then

$$(g - \delta\nabla r\otimes\nabla r - \eta_2)(V,V) = (g-\eta_2)(V,V) \geq \frac{1}{2}(1+\delta)|V|^2 \geq 0,$$

and the bound follows. □

4.6.5. Estimating the quadratic terms.

Our first Carleman estimate will follow from a lower bound on the quadratic terms Q_1, Q_2, and E_{ϕ_1} in (4.99). From the lower bound on the Hessian of ϕ_1 in (4.112), we may choose $\mathcal{S}$ sufficiently large, depending on δ and K so that

$$Q_1(\nabla Z, \nabla Z) = 2\nabla^2_{ij}\phi_1\,\langle\nabla_i Z, \nabla_j Z\rangle \geq 2|\nabla Z|^2 \tag{4.114}$$

on $\mathcal{C}_{\mathcal{S},T}$. To estimate Q_2, we will need bounds on the function

$$F_1 := G_1^{-1}(\partial_\tau - \Delta)G_1 + R \tag{4.115}$$

and its derivatives.

PROPOSITION 4.34. *There exist constants $\mathcal{S} \geq \mathcal{R}$ and $N > 0$ depending on n and K such that, for all $\alpha > 0$, $0 < \delta < 1$, and $0 < \tau_0 \leq 1$, we have*

$$0 \geq F_1 \geq -Nr^2\left(1 + \alpha r^{-\delta} + \alpha^2(\tau_0 - \tau)^2 r^{-2\delta}\right) \tag{4.116}$$

and

$$\partial_\tau F_1 + \Delta F_1 \geq 3\alpha r^{2-\delta} + \alpha^2(\tau_0 - \tau)r^{2-2\delta} \tag{4.117}$$

on $\mathcal{C}_{\mathcal{S},\tau_0}$.

PROOF. For (4.116), we start with the identity

$$F_1 = \partial_\tau\phi_1 - \Delta\phi_1 - |\nabla\phi_1|^2 + R.$$

The third term on the right-hand side is nonpositive and, by (4.107), so too is the first. From (4.111) (or from (4.112)), we have $\Delta\phi_1 \geq n$ on $\mathcal{C}_{\mathcal{S},\tau_0}$ for $\mathcal{S}$ large enough. Since also $|R| \leq N/(r^2+1)$ on $\mathcal{C}_{\mathcal{R},\tau_0}$, we see that

$$F_1 \leq -n + \frac{N}{r^2+1} \leq -\frac{n}{2}$$

on $\mathcal{C}_{\mathcal{S},\tau_0}$ provided $\mathcal{S}$ is taken larger still. This gives the upper bound in (4.116).

For the lower bound, we group terms according to their powers of α, and use (4.107), (4.108), and (4.111) to write

$$F_1 = H_0 + \alpha H_1 + \alpha^2 H_2,$$

where

$$\begin{aligned}
H_0 =& -2n - 4r^2 - 2(1+2r^2)\eta_1 - 2r\eta_3 + R,\\
H_1 =& -(1 + 4(2-\delta)(\tau_0 - \tau))r^{2-\delta}\\
& - r^{-\delta}(2-\delta)(\tau_0-\tau)(n - \delta + (1-\delta)\eta_1 + r\eta_3),\\
H_2 =& -(2-\delta)^2(\tau_0-\tau)^2 r^{2-2\delta}(1+\eta_1).
\end{aligned}$$

According to the bounds (4.60), (4.73), and (4.75), we have that

$$H_0 \geq -C(1+r^2), \quad H_1 \geq -Cr^{2-\delta}, \quad \text{and} \quad H_2 \geq -4(\tau_0-\tau)^2 r^{2-2\delta},$$

on $\mathcal{C}_{\mathcal{S},\tau_0}$ for some $C = C(n)$, provided that $\mathcal{S} = \mathcal{S}(n, K)$ is taken large enough. The lower bound in (4.116) follows.

Similarly, for the bound (4.117), we write

$$\partial_\tau F_1 + \Delta F_1 = B_0 + \alpha B_1 + \alpha^2 B_2,$$

where

$$B_i = \partial_\tau H_i + \Delta H_i$$

for $i = 0, 1, 2$, and estimate each of the B_i separately. First, we compute that, for any $\beta > 0$,

$$\begin{aligned}
\Delta r^\beta &= \beta r^{\beta-1}\Delta r + \beta(\beta-1)r^{\beta-2}|\nabla r|^2\\
&= \beta r^{\beta-2}(n+\beta-2) + \beta r^{\beta-2}\left((\beta-1)\eta_1 + r\eta_3\right)\\
&= \beta r^{\beta-2}(n+\beta-2) + O(r^{\beta-4})
\end{aligned} \tag{4.118}$$

in view of (4.73) and (4.75). Here we use the big-O notation $\mathrm{O}(r^a)$ to denote a term Q for which $|Q(r,\sigma,\tau)| \leq N r^a$ on $\mathcal{C}_{\mathcal{S},\tau_0}$ as $r \to \infty$. In particular, $\Delta r^2 = 2n + \mathrm{O}(r^{-2})$ and $\Delta r = \mathrm{O}(r^{-1})$. Since the quantities $|\partial_\tau \eta_1|$, $|\partial_\tau \eta_3|$, $|\Delta \eta_1|$, $|\Delta \eta_3|$, $|\nabla \eta_1|$, and $|\nabla \eta_3|$ are at least $\mathrm{O}(r^{-3})$ (recall that the η_i are defined in (4.70) and (4.71)), and

$$(\partial_\tau + \Delta) R = -2|\mathrm{Ric}|^2 = \mathrm{O}(r^{-4}),$$

applying the backward heat operator $\partial_\tau + \Delta$ to the expression above for H_0, we see that all terms in B_0 decay at least at the rate $\mathrm{O}(r^{-2})$, with the exception of

$$-4\Delta r^2 = -8n + \mathrm{O}(r^{-2}).$$

Hence

$$B_0 \geq -16n$$

on $\mathcal{C}_{\mathcal{S},\tau_0}$ for $\mathcal{S}$ sufficiently large.

Computing similarly, we see that

$$B_1 \geq 4(2-\delta) r^{2-\delta} - C r^{-\delta},$$

and

$$B_2 \geq 2(2-\delta)^2 (\tau_0 - \tau) r^{2-2\delta} - C(\tau_0 - \tau) r^{-2\delta},$$

for some $C = C(n)$ and for all $\mathcal{S}$ sufficiently large. Thus, since $\alpha \geq 1$ and $0 < \delta < 1$, we can choose $\mathcal{S}$ large enough to ensure that

$$\begin{aligned}(\partial_\tau + \Delta) F_1 &= B_0 + \alpha B_1 + \alpha^2 B_2 \\ &\geq 3\alpha(2-\delta) r^{2-\delta} + \alpha^2 (2-\delta)^2 (\tau_0 - \tau) r^{2-2\delta}\end{aligned}$$

on $\mathcal{C}_{\mathcal{S},\tau_0}$. □

4.6.6. The Carleman estimates.

Now we use the bounds on Q_1 and Q_2 to derive a Carleman estimate from Proposition 4.31. Here and below, we will use $\|\cdot\|_\Omega$ to denote the space-time (respectively, spatial) L^2-norm induced by the time-dependent family of measures $dv = dv_{g(\tau)}$ on $\Omega \subset \mathcal{C}_{\mathcal{R},1}$ (respectively, $\Omega \subset \mathcal{C}_{\mathcal{R}} \times \{\tau\}$ for some fixed τ).

PROPOSITION 4.35. *Let $0 < \delta < 1$. Then there exists $\mathcal{S} \geq \mathcal{R}$ depending on n and K, such that, for all $\alpha \geq 1$ and $0 < \tau_0 \leq 1$, and all compactly supported smooth sections Z of $\mathcal{Z}$ which vanish on $\mathcal{C}_{\mathcal{S}} \times \{0\}$, we have the estimates*

$$\begin{aligned}&\alpha \|Z G_1^{\frac12}\|^2_{\mathcal{C}_{\mathcal{S},\tau_0}} + \|\nabla Z G_1^{\frac12}\|^2_{\mathcal{C}_{\mathcal{S},\tau_0}} \\ &\quad \leq \frac{1}{2} \|(D_\tau + \Delta) Z G_1^{\frac12}\|^2_{\mathcal{C}_{\mathcal{S},\tau_0}} + \|\nabla Z G_1^{\frac12}\|^2_{\mathcal{C}_{\mathcal{S}} \times \{\tau_0\}}\end{aligned} \tag{4.119}$$

and

$$\alpha \|Z G_1^{\frac12}\|^2_{\mathcal{C}_{\mathcal{S},\tau_0}} \leq \|D_\tau Z G_1^{\frac12}\|^2_{\mathcal{C}_{\mathcal{S},\tau_0}}, \tag{4.120}$$

where $G_1 = G_{1;\alpha,\tau_0}$.

PROOF. Recall the inequality in (4.99). By (4.114) and Proposition 4.34, for $\mathcal{S}$ sufficiently large, we have

$$Q_1(\nabla Z, \nabla Z) + Q_2(Z, Z) \geq 2|\nabla Z|^2 + \frac{1}{2}\left(3\alpha + \alpha^2(\tau_0 - \tau) r^{2-2\delta}\right)|Z|^2.$$

We just need to bound the term involving the commutator quantity E_{ϕ_1} from below. For this, recall that, by (4.97), we have

$$|E_{\phi_1}(Z,\nabla Z)| \leq C\left(|\nabla \mathrm{Ric}| + (1+|\nabla\phi_1|)|\mathrm{Rm}|\right)|\nabla Z||Z| + C|\mathrm{Ric}||\nabla Z|^2$$

for some $C = C(n)$. Since

$$|\nabla\phi_1| = \big(\alpha(2-\delta)(\tau_0-\tau)r^{1-\delta} + 2r\big)|\nabla r| \leq Nr\big(1+\alpha(\tau_0-\tau)r^{-\delta}\big),$$

it follows from (4.60) that

$$\begin{aligned} |E_{\phi_1}(Z,\nabla Z)| &\leq \frac{N}{r}\left(1+\frac{\alpha(\tau_0-\tau)}{r^{\delta}}\right)|Z||\nabla Z| + \frac{N}{r^2}|\nabla Z|^2 \\ &\leq \frac{\alpha^2(\tau_0-\tau)}{2}r^{2-2\delta}|Z|^2 + |\nabla Z|^2 \end{aligned}$$

on $\mathcal{C}_{\mathcal{S},\tau_0}$ for $\mathcal{S}$ sufficiently large. Thus, taking these bounds together, we have

$$Q_1(\nabla Z,\nabla Z) + Q_2(Z,Z) - E_{\phi_1}(Z,\nabla Z) \geq |\nabla Z|^2 + \alpha|Z|^2$$

on $\mathcal{C}_{\mathcal{S},\tau_0}$, for large enough $\mathcal{S}$. Equation (4.119) then follows from (4.99), using that $F_1 \leq 0$.

For (4.120), note that by (4.107), we have

$$\partial_\tau\phi_1 = -\alpha r^{2-\delta},$$

so that the left-hand side of (4.103) is greater than $\alpha\|ZG^{1/2}\|^2_{\mathcal{C}_{\mathcal{S},\tau_0}}$ provided that $\mathcal{S}$ is taken large enough. □

4.7. Carleman estimates to imply rapid decay

The weight G_1 in Proposition 4.35 grows at an exponential-quadratic rate. In order to apply this estimate in the proof of Theorem 4.25, we must prove that bounded sections which vanish at $\tau = 0$ decay rapidly enough to overcome this growth. We will establish this decay with the help of another Carleman estimate modeled on inequality (1.4) in [**ESŠ03a**]. Their inequality, which involves the Euclidean heat kernel and the family of temporal weights

$$\sigma_a(\tau) = (\tau+a)\mathrm{e}^{-(\tau+a)/3}$$

says that there is some constant $C = C(n)$ such that

$$\begin{aligned} &\sqrt{\alpha}\left\|\sigma_a^{-\alpha-\frac{1}{2}}\mathrm{e}^{-\frac{|x-y|^2}{8(\tau+a)}}u\right\|_{L^2(\mathbb{R}^n\times(0,1))} + \left\|\sigma_a^{-\alpha}\mathrm{e}^{-\frac{|x-y|^2}{8(\tau+a)}}\nabla u\right\|_{L^2(\mathbb{R}^n\times(0,1))} \\ &\qquad \leq C\left\|\sigma_a^{-\alpha}\mathrm{e}^{-\frac{|x-y|^2}{8(\tau+a)}}(\partial_\tau+\Delta)u\right\|_{L^2(\mathbb{R}^n\times(0,1))} \end{aligned} \tag{4.121}$$

for all $\alpha \geq 0$, $y \in \mathbb{R}^n$, $a \in (0,1)$, and $u \in C_c^\infty(\mathbb{R}^n\times[0,1))$ satisfying $u(\cdot,0)\equiv 0$.

The proof of this inequality makes use of the translational invariance of the Euclidean metric and heat kernel. In order to generalize this to an asymptotically conical background, we first look to find an alternative intermediate inequality for the Euclidean case which does not make use of these properties. It turns out that, with some minor modifications to the argument of [**ESŠ03a**], one can establish the following alternative "radial" version of their statement: there are constants $\alpha_0 = \alpha_0(n) \geq 1$ and $C = C(n)$ such that for all $\rho \geq 1$ and $u \in C_c^\infty(\{|x| > \rho/16\} \times [0,1))$ which vanish at $\tau = 0$, one has

$$\begin{aligned}\sqrt{\alpha}\left\|\sigma_a^{-\alpha-\frac{1}{2}}\mathrm{e}^{-\frac{(|x|-\rho)^2}{8(\tau+a)}}u\right\|_{L^2(\mathbb{R}^n\times(0,1))} &+ \left\|\sigma_a^{-\alpha}\mathrm{e}^{-\frac{(|x|-\rho)^2}{8(\tau+a)}}\nabla u\right\|_{L^2(\mathbb{R}^n\times(0,1))} \\ &\leq C\left\|\sigma_a^{-\alpha}\mathrm{e}^{-\frac{(|x|-\rho)^2}{8(\tau+a)}}(\partial_\tau+\Delta)u\right\|_{L^2(\mathbb{R}^n\times(0,1))}\end{aligned} \tag{4.122}$$

for all $\alpha \geq \alpha_0$ and $0 < a < 1$. Below, we will generalize this inequality for sections of tensor bundles over an asymptotically conical shrinking background.

4.7.1. Another divergence identity.

The strategy to prove our desired Carleman estimate is largely the same as that behind the proof of the earlier estimate in Proposition 4.31. Here, however, the situation is complicated by the lack of log-convexity of the weight $G = G_2$ (which, as in (4.121)–(4.122), will have as its principal part an approximate heat kernel): the reader will recall that in the proof of Proposition 4.31, we made important use of the fact that

$$Q_1(\nabla Z, \nabla Z) = 2\nabla_{ij}^2 \log G_1 \langle \nabla_i Z, \nabla_j Z\rangle \gtrsim |\nabla Z|^2$$

As in [**ESŠ03a**], we will introduce some additional temporal factors to increase the convexity of the weight. The effect of these factors is encoded in the following perturbation of the divergence identity (4.93). It can be obtained directly from (4.93) by substituting H for F and replacing G with $\theta\sigma_a^{-\alpha}G$, where θ and σ are smooth positive functions of τ.

LEMMA 4.36. *Let G and H be smooth positive functions on $\mathcal{C}_{\mathcal{R},1}$ and let θ, σ be smooth positive functions on $[0,1]$. Write $\phi = \ln G$ as before. Then, for all $Z \in C^\infty(\mathcal{Z} \times [0,T])$ and $\alpha > 0$, we have the identity*

$$\begin{aligned}&\theta\sigma^{-\alpha}\nabla_i\Big\{2\langle D_\tau Z, \nabla_i Z\rangle G + |\nabla Z|^2\nabla_i G - 2\langle \nabla_{\nabla G} Z, \nabla_i Z\rangle \\ &\quad + \langle \nabla_i Z, Z\rangle GH + \frac{1}{2}|Z|^2(H\nabla_i G - G\nabla_i H)\Big\}\,dv \\ &\quad - \frac{\partial}{\partial\tau}\left\{\left(|\nabla Z|^2 + \frac{1}{2}|Z|^2 H\right)\theta\sigma^{-\alpha}G\,dv\right\} \\ &= \Big\{2\langle(D_\tau+\Delta)Z, \mathcal{A}Z\rangle - 2|\mathcal{A}Z|^2 - \left(2\nabla_{ij}^2\phi + \frac{\dot\theta}{\theta}g_{ij}\right)\langle\nabla_i Z, \nabla_j Z\rangle \\ &\quad + \left(H - G^{-1}(\partial_\tau G - \Delta G) - R + \alpha\frac{\dot\sigma}{\sigma}\right)\left(|\nabla Z|^2 + \frac{H}{2}|Z|^2\right) \\ &\quad - \frac{1}{2}\left(\partial_\tau H + \Delta H + \frac{\dot\theta}{\theta}H\right)|Z|^2 - E_\phi(Z, \nabla Z)\Big\}\theta\sigma^{-\alpha}G\,dv.\end{aligned} \tag{4.123}$$

Here, E_ϕ is defined as in (4.94).

Let us make a few choices to simplify the expression above. As in [**ESŠ03a**], if we take $\theta = \sigma/\dot\sigma$ for σ some strictly increasing (but otherwise as yet unspecified) positive smooth function, we will have

$$\frac{\dot\theta}{\theta}\cdot\frac{\dot\sigma}{\sigma} = -\ddot{\widehat{\ln\sigma}}, \tag{4.124}$$

where the dots indicate derivatives taken relative to the parameter τ. Taking G to be a given (but also as yet unspecified) positive smooth function, and choosing H in the form $H = F - \alpha\dot{\sigma}/\sigma$ where (as before)

$$F = G^{-1}\left(\partial_\tau G - \Delta G\right) + R,$$

the penultimate line in (4.123) will vanish and the quantity in the coefficient of $|Z|^2$ in the first term on the last line will satisfy

$$\begin{aligned} \partial_\tau H + \Delta H + \frac{\dot{\theta}}{\theta}H &= \partial_\tau H + \Delta F + \frac{\dot{\theta}}{\theta}\left(F - \alpha\frac{\dot{\sigma}}{\sigma}\right) - \alpha\widehat{\ln\sigma}^{\cdot\cdot} \\ &= \partial_\tau F + \Delta F + \frac{\dot{\theta}}{\theta}F. \end{aligned}$$

4.7.2. Two integral identities.

With θ and H chosen as in the preceding paragraph, we may apply (4.123) to compactly supported sections of $\mathcal{Z}$, integrate over $\mathcal{C}_{\mathcal{R},T}$, and use the Cauchy–Schwarz inequality as in Proposition 4.31 to obtain the following inequality:

PROPOSITION 4.37. *Let σ be a smooth positive increasing function on $[0,1]$ and let G be a smooth positive function on $\mathcal{C}_{\mathcal{R},1}$. Then, for any $\alpha > 0$, and any smooth section Z of $\mathcal{Z}$ which vanishes for $\tau = 0$ and has compact support in $\mathcal{C}_{\mathcal{R}} \times [0,1)$, we have the inequality*

$$\begin{aligned} &\int_{\mathcal{C}_{\mathcal{R},1}} \frac{\sigma^{1-\alpha}}{\dot{\sigma}}\left|D_\tau Z + \Delta Z\right|^2 G\,dv\,d\tau \\ &\geq \int_{\mathcal{C}_{\mathcal{R},1}} \frac{\sigma^{1-\alpha}}{\dot{\sigma}}(Q_3(\nabla Z, \nabla Z) + Q_4(Z,Z) - E_\phi(Z, \nabla Z))G\,dv\,d\tau, \end{aligned} \tag{4.125}$$

where $\theta = \sigma/\dot{\sigma}$, $F = G^{-1}(\partial_\tau G - \Delta G) + R$,

$$Q_3(\nabla Z, \nabla Z) = \left(2\nabla^2_{ij}\phi + \frac{\dot{\theta}}{\theta}g_{ij}\right)\langle\nabla_i Z, \nabla_j Z\rangle, \tag{4.126}$$

$$Q_4(Z,Z) = \frac{1}{2}\left(\partial_\tau F + \Delta F + \frac{\dot{\theta}}{\theta}F\right)|Z|^2, \tag{4.127}$$

and E_ϕ is as defined in (4.94) for $\phi = \log G$.

We will use (4.125) in combination with the following absorbing-type inequality.

PROPOSITION 4.38. *For σ, G, and Z satisfying the same hypotheses as Proposition 4.37, we have the inequality*

$$\begin{aligned} &\int_{\mathcal{C}_{\mathcal{R},1}} \sigma^{-\alpha}\left(\frac{\alpha\dot{\sigma}}{2\sigma} - F\right)|Z|^2 G\,dv\,d\tau \\ &\leq 2\int_{\mathcal{C}_{\mathcal{R},1}} \sigma^{-\alpha}\left(|\nabla Z|^2 + \frac{\sigma}{\alpha\dot{\sigma}}\left|D_\tau Z + \Delta Z\right|^2\right)G\,dv\,d\tau \end{aligned} \tag{4.128}$$

for all $\alpha > 0$.

PROOF. This follows from integrating the identity

$$\begin{aligned} &\frac{\partial}{\partial\tau}\left\{\sigma^{-\alpha}|Z|^2 G\,dv\right\} + \sigma^{-\alpha}\nabla_i\left(\nabla_i|Z|^2 G - |Z|^2\nabla_i G\right)\,dv \\ &= \sigma^{-\alpha}\left\{\left(\partial_\tau + \Delta\right)|Z|^2 + \left(F - \alpha\frac{\dot{\sigma}}{\sigma}\right)\right\}G\,dv \end{aligned} \tag{4.129}$$

over $\mathcal{C}_{\mathcal{R},1}$ and applying the estimate

$$\begin{aligned}(\partial_\tau + \Delta)\,|Z|^2 &= 2|\nabla Z|^2 + 2\,\langle (D_\tau + \Delta)Z, Z\rangle \\ &\leq 2|\nabla Z|^2 + \frac{\alpha\dot\sigma}{2\sigma}|Z|^2 + \frac{2\sigma}{\alpha\dot\sigma}|D_\tau Z + \Delta Z|^2\end{aligned}$$

to the first term on the right-hand side. □

4.7.3. The weight function for the second inequality.

Now we define σ and G and specialize these general identities and inequalities to our case of interest. As in [**ESŠ03a**], we choose for σ the functions

$$\sigma_a(\tau) = (\tau + a)\mathrm{e}^{-\frac{\tau+a}{3}} \tag{4.130}$$

for $0 < a \leq 1$. For G we choose the functions

$$G_2(x,\tau) := G_{2;a,\rho}(x,\tau) := (\tau + a)^{-n/2}\exp\left(-\frac{(r(x)-\rho)^2}{4(\tau+a)}\right) \tag{4.131}$$

for fixed $0 < a \leq 1$ and $\rho \geq 0$.

The function σ_a is positive, increasing, and approximately linear for small τ in the sense that

$$\frac{\tau + a}{3} \leq \sigma_a(\tau) \leq \tau + a, \quad \frac{1}{10} \leq \dot\sigma_a(\tau) \leq 1, \tag{4.132}$$

for a, $\tau \in [0,1]$. Also – recalling (4.124) – we have the identity

$$\frac{\dot\theta}{\theta} = -\frac{\sigma_a}{\dot\sigma_a}\cdot\widehat{\ln\sigma_a}^{\,\cdot\cdot} = \frac{1}{\tau+a}\left(\frac{1}{1-(\tau+a)/3}\right), \tag{4.133}$$

where $\theta = \sigma_a/\dot\sigma_a$.

When $\rho = 0$, the function $G_2 = G_{2;a,\rho}$ is an exact solution to the heat equation with respect to the conical metric $\hat g = g(0)$. When $\rho > 0$, G_2 will approximately solve the heat equation with respect to $g(\tau)$ on $\mathcal{C}_{\mathcal{R},1}$. We will estimate the error in this approximation next.

4.7.4. Estimating the weight function.

Define the functions

$$\phi_2 := \phi_{2;a,\rho} := \ln G_{2;a,\rho} = -\frac{(r-\rho)^2}{4(\tau+a)} - \frac{n}{2}\ln(\tau + a)$$

and

$$F_2 := G_2^{-1}(\partial_\tau G_2 - \Delta G_2) + R.$$

PROPOSITION 4.39. *Let* $0 < a < 1$ *and* $\rho > 0$. *The function* $\phi_2 = \phi_{2;a,\rho}$ *satisfies the identities*

$$\partial_\tau \phi_2 = \frac{(r-\rho)^2}{4(\tau+a)^2} - \frac{n}{2(\tau+a)}, \tag{4.134}$$

$$\nabla \phi_2 = -\left(\frac{r-\rho}{2(\tau+a)}\right)\nabla r, \tag{4.135}$$

$$|\nabla \phi_2|^2 = \frac{(r-\rho)^2}{4(\tau+a)^2}(1+\eta_1), \tag{4.136}$$

$$\nabla^2 \phi_2 = -\frac{1}{2(\tau+a)}\left\{\left(1-\frac{\rho}{r}\right)g + \frac{\rho}{r}\nabla r \otimes \nabla r + \left(1-\frac{\rho}{r}\right)\eta_2\right\}, \tag{4.137}$$

$$\Delta \phi_2 = -\frac{1}{2(\tau+a)}\left\{n - \frac{(n-1)\rho}{r} + \eta_1 + (r-\rho)\eta_3\right\}, \tag{4.138}$$

where η_i, $i = 1,2,3$, *are as defined in* (4.70)–(4.71). *In particular, there is a constant* N *depending on* n *and* K, *(but independent of* ρ *and* a*) such that*

$$\nabla^2 \phi_2 \geq -\left(\frac{1}{2(\tau+a)} + \frac{N}{r^2+1}\right)g \tag{4.139}$$

on $\mathcal{C}_{\mathcal{R},1} \cap \mathcal{C}_{\rho/16,1}$.

PROOF. The identities (4.134)–(4.138) follow by direct computation. For inequality (4.139), we argue as for (4.112). First, consider a unit vector V orthogonal to ∇r. According to (4.137) and (4.74), we have

$$\begin{aligned}\nabla^2 \phi_2(V,V) &= -\frac{1}{2(\tau+a)}\left(\left(1-\frac{\rho}{r}\right) + \left(1-\frac{\rho}{r}\right)\eta_2(V,V)\right)\\ &\geq -\frac{1}{2(\tau+a)} - \frac{N}{r^2+1}\end{aligned}$$

for $r > \rho/16$. On the other hand, if $V = \nabla r/|\nabla r|$, then again

$$\begin{aligned}\nabla^2 \phi_2(V,V) &= -\frac{1}{2(\tau+a)}\left\{1 + \frac{\rho}{r}\eta_1 + \left(1-\frac{\rho}{r}\right)\eta_2(V,V)\right\}\\ &\geq -\frac{1}{2(\tau+a)} - \frac{N}{r^2+1}\end{aligned}$$

by (4.73) and (4.74). Combining the two cases yields (4.139). □

PROPOSITION 4.40. *Let* $0 < a < 1$ *and* $\rho > 0$. *There is a constant* N *depending on* n *and* K *such that* $F_2 = F_{2;a,\rho}$ *satisfies*

$$\left|F_2 + \frac{(n-1)\rho}{2r(\tau+a)}\right| \leq \frac{N}{r^2+1} \tag{4.140}$$

and

$$\left|(\partial_\tau + \Delta)F_2 - \frac{(n-1)\rho}{2r(\tau+a)^2}\right| \leq \frac{N}{(\tau+a)(r^2+1)} \tag{4.141}$$

on the set $\mathcal{C}_{\mathcal{R},1} \cap \mathcal{C}_{\rho/16,1}$.

PROOF. From Proposition 4.39, we see that $F_2 = \partial_\tau \phi_2 - \Delta\phi_2 - |\nabla\phi_2|^2 + R$ satisfies

$$F_2 = -\frac{(n-1)\rho}{2(\tau+a)r} + \left(\frac{1}{2(\tau+a)} - \frac{(r-\rho)^2}{4(\tau+a)^2}\right)\eta_1 + \left(\frac{r-\rho}{2(\tau+a)}\right)\eta_3 + R. \tag{4.142}$$

We then obtain (4.140) directly from the decay bounds in (4.60), (4.73), and (4.75) on R, η_1, and η_3, respectively.

For (4.141), we argue as for (4.117). First of all, from (4.142), we have

$$F_2 = A_0 + A_1\eta_1 + A_2\eta_3 + R, \tag{4.143}$$

where

$$A_0 = -\frac{(n-1)\rho}{2(\tau+a)r}, \quad A_1 = \frac{1}{2(\tau+a)} - \frac{(r-\rho)^2}{4(\tau+a)^2}, \quad \text{and} \quad A_2 = \frac{r-\rho}{2(\tau+a)}.$$

Now, for any β, recall that from (4.118) we have

$$\Delta r^\beta = \beta r^{\beta-2}(n+\beta-2+(\beta-1)\eta_1 + r\eta_3),$$

so that (noting the decay in τ) we have

$$\Delta r^\beta = \beta(n+\beta-2)r^{\beta-2} + \mathrm{O}\left(\tfrac{\tau}{r^{4-\beta}}\right), \tag{4.144}$$

where, here, we write $Q = \mathrm{O}(\tilde{Q})$ if $|Q| \le N|\tilde{Q}|$ as $r \to \infty$ and τ, $a \to 0$ for some N depending on n and K (and independent of a and ρ). Thus,

$$\begin{aligned}(\partial_\tau + \Delta)A_0 &= \frac{(n-1)\rho}{2(\tau+a)r}\left(\frac{1}{\tau+a} + \frac{n-3}{r^2} + \mathrm{O}\left(\tfrac{\tau}{r^4}\right)\right) \\ &= \frac{(n-1)\rho}{2r(\tau+a)^2} + \mathrm{O}\left(\frac{1}{(\tau+a)r^2}\right)\end{aligned} \tag{4.145}$$

for $r > \rho/16$.

Similarly,

$$\Delta(r-\rho)^2 = 2\left(n - \frac{(n-1)\rho}{r}\right) + \mathrm{O}\left(\tfrac{\tau}{r^2}\right) + \rho\,\mathrm{O}\left(\tfrac{\tau}{r^3}\right) = \mathrm{O}(1),$$

so that

$$(\partial_\tau + \Delta)A_1 = \mathrm{O}\left(\frac{r^2}{(\tau+a)^3}\right), \quad |\nabla A_1| = \mathrm{O}\left(\frac{r}{(\tau+a)^2}\right),$$

for $r > \rho/16$. Since, by (4.73),

$$\eta_1 = \mathrm{O}\left(\frac{\tau^2}{r^4}\right), \quad |\nabla\eta_1| = \mathrm{O}\left(\frac{\tau^2}{r^5}\right), \quad (\partial_\tau + \Delta)\eta_1 = \mathrm{O}\left(\tfrac{\tau}{r^4}\right),$$

it follows that

$$\begin{aligned}(\partial_\tau + \Delta)(A_1\eta_1) &= \eta_1(\partial_\tau + \Delta)A_1 + A_1(\partial_\tau + \Delta)\eta_1 + 2\langle\nabla A_1, \nabla\eta_1\rangle \\ &= \mathrm{O}\left(\frac{1}{(\tau+a)r^2}\right).\end{aligned} \tag{4.146}$$

Finally, in the same way, we find

$$(\partial_\tau + \Delta)A_2 = \mathrm{O}\left(\frac{r}{(\tau+a)^2}\right), \quad |\nabla A_2| = \mathrm{O}\left(\frac{1}{\tau+a}\right),$$

which, since

$$\eta_3 = \mathrm{O}\left(\frac{\tau}{r^3}\right), \quad |\nabla \eta_3| = \mathrm{O}\left(\frac{\tau}{r^4}\right), \quad (\partial_\tau + \Delta)\eta_3 = \mathrm{O}\left(\frac{1}{r^3}\right),$$

implies that

$$(\partial_\tau + \Delta)(A_2\eta_3) = \mathrm{O}\left(\frac{1}{(\tau+a)r^2}\right). \tag{4.147}$$

Combining (4.145) with (4.146) and (4.147), and using the fact that

$$(\partial_\tau + \Delta)R = \mathrm{O}\left(\frac{1}{r^4}\right),$$

we see that

$$(\partial_\tau + \Delta)F_2 = \frac{(n-1)\rho}{2(\tau+a)^2 r} + \mathrm{O}\left(\frac{1}{(\tau+a)r^2}\right).$$

Equation (4.141) then follows. □

4.7.5. Carleman estimates to establish rapid decay.

With the estimates from Proposition 4.40 in hand, we return now to Proposition 4.37 to prove the crucial decay estimate we need for the proof of Theorem 4.25. As before, $\mathcal{Z}$ denotes a bundle of spatial tensors over $\mathcal{C}_{\mathcal{R},1}$ of the form $\pi^*(T^\nu(T^*\mathcal{C}_{\mathcal{R}}))$.

PROPOSITION 4.41. *There are constants $C = C(n) > 0$, and $\mathcal{S} = \mathcal{S}(n, K) \geq \mathcal{R}$ such that, for any section Z of $\mathcal{Z}$ which vanishes on $\mathcal{C}_{\mathcal{S}} \times \{0\}$ and is compactly supported in $(\mathcal{C}_{\mathcal{S}} \cap \mathcal{C}_{\rho/16}) \times [0,1)$, we have the estimate*

$$\begin{aligned}\sqrt{k}\,\|\sigma_a^{-k-1/2} Z G_2^{1/2}\|_{\mathcal{C}_{\mathcal{S},1}} &+ \|\sigma_a^{-k}\nabla Z G_2^{1/2}\|_{\mathcal{C}_{\mathcal{S},1}} \\ &\leq C\|\sigma_a^{-k}(D_\tau Z + \Delta Z)G_2^{1/2}\|_{\mathcal{C}_{\mathcal{S},1}}\end{aligned} \tag{4.148}$$

for all $0 < a \leq 1$, $k \geq 1$, and $\rho \geq 1$, where σ_a and $G_2 = G_{2;a,\rho}$ are as defined in (4.130) and (4.131).

PROOF. Fix $0 < a < 1$ and $\rho \geq 1$. We will specify $\mathcal{S}$ over the course of the proof, assuming to begin with that $\mathcal{S} \geq \mathcal{R}$. As before, we will use C to denote a series of positive constants depending at most on the dimension n, and use N to indicate a series of constants depending potentially also on the parameter K from (4.60) and (4.61).

First, using (4.139) and the identity (4.133) for $\theta = \sigma_a/\dot{\sigma}_a$, we see that the tensor Q_3 from the inequality (4.125) satisfies

$$\begin{aligned}Q_3(\nabla Z, \nabla Z) &= \left(2\nabla^2_{ij}\phi + \frac{\dot{\theta}}{\theta}g_{ij}\right)\langle \nabla_i Z, \nabla_j Z\rangle \\ &\geq \left(\frac{1}{\tau+a}\left(\frac{1}{1-(\tau+a)/3} - 1\right) - \frac{N}{r^2+1}\right)|\nabla Z|^2 \\ &= \left(\frac{1}{3} - \frac{N}{r^2+1}\right)|\nabla Z|^2 \\ &\geq \frac{7}{24}|\nabla Z|^2\end{aligned} \tag{4.149}$$

on $\mathcal{C}_{\mathcal{S},1}$, provided $\mathcal{S}$ is chosen large enough. Also, using (4.140) and (4.141), we may estimate the tensor Q_4 by

$$\begin{aligned} Q_4(Z,Z) &= \frac{1}{2}\left(\partial_\tau F_2 + \Delta F_2 + \frac{\dot\theta}{\theta}F_2\right)|Z|^2 \\ &\geq \frac{1}{2}\left(\frac{(n-1)\rho}{2r(\tau+a)^2} - \frac{N}{(\tau+a)(r^2+1)}\right)|Z|^2 \\ &\quad - \frac{1}{2}\left(\frac{1}{1-(\tau+a)/3}\right)\left(\frac{(n-1)\rho}{2r(\tau+a)^2} + \frac{N}{(\tau+a)(r^2+1)}\right)|Z|^2 \\ &\geq \frac{1}{\tau+a}\left(\frac{(n-1)\rho}{12r} - \frac{N}{r^2+1}\right)|Z|^2, \end{aligned}$$

so that there is some $C > 0$ such that

$$Q_4(Z,Z) \geq \frac{C}{(\tau+a)r}|Z|^2 \tag{4.150}$$

on $\mathcal{C}_{\mathcal{S},1}$ for all $\rho \geq 1$, $0 < a \leq 1$, and $\mathcal{S}$ sufficiently large.

It remains to estimate the commutator term E_{ϕ_2}. We will need a somewhat finer estimate than was required for Proposition 4.35 to absorb the contribution from $\nabla\phi_2$. Note that, by (4.94) and (4.98), we have

$$\begin{aligned} |E_{\phi_2}(Z,\nabla Z)| &\leq C\left(|\nabla \mathrm{Ric}| + \left(\frac{|r-\rho|}{\tau+a}\right)|\eta_4|\right)|\nabla Z||Z| \\ &\quad + C|\mathrm{Ric}||\nabla Z|^2, \end{aligned} \tag{4.151}$$

where η_4 is as defined in (4.72). But in (4.76) we have already seen that

$$|\eta_4| \leq \frac{N\tau}{r^4+1},$$

so

$$\begin{aligned} |E_\phi(Z,\nabla Z)| &\leq \frac{N}{r^3+1}|Z||\nabla Z| + \frac{N}{r^2+1}|\nabla Z|^2 \\ &\leq \frac{N}{r^3+1}|Z|^2 + \frac{1}{24}|\nabla Z|^2 \end{aligned} \tag{4.152}$$

for $r > 16\rho$. Putting (4.149), (4.150), and (4.152) together, we obtain that

$$Q_3(Z,Z) + Q_4(\nabla Z,\nabla Z) - E_{\phi_2}(Z,\nabla Z) \geq \frac{1}{4}|\nabla Z|^2 \tag{4.153}$$

on $\mathcal{C}_{\mathcal{S},1}$ if $\mathcal{S}$ is taken large enough, and hence, via (4.125), that

$$\frac{1}{4}\int_{\mathcal{C}_{\mathcal{S},1}} \frac{\sigma_a^{1-\alpha}}{\dot\sigma_a}|\nabla Z|^2 G_2\,dv\,d\tau \leq \int_{\mathcal{C}_{\mathcal{S},1}} \frac{\sigma_a^{1-\alpha}}{\dot\sigma_a}\left|D_\tau Z + \Delta Z\right|^2 G_2\,dv\,d\tau. \tag{4.154}$$

Using (4.132) to bound $\dot\sigma$ from above and below, we may rewrite (4.154) with $k = (\alpha-1)/2$ as

$$\|\sigma_a^{-k}\nabla Z G_2^{1/2}\|^2_{\mathcal{C}_{\mathcal{S},1}} \leq 40\,\|\sigma_a^{-k}(\partial_\tau Z + \Delta Z)G_2^{1/2}\|^2_{\mathcal{C}_{\mathcal{S},1}}, \tag{4.155}$$

valid for all $k > 0$. On the other hand, applying the absorbing inequality in Proposition 4.38 with the choice $\alpha = 2k$, we have

$$\begin{aligned}&\int_{\mathcal{C}_{\mathcal{S},1}} \sigma^{-2k}\left(\frac{k\dot{\sigma}_a}{\sigma_a} - F_2\right)|Z|^2 G_2\, dv\, d\tau \\ &\quad\le \int_{\mathcal{C}_{\mathcal{S},1}} \sigma^{-2k}\left(2|\nabla Z|^2 + \frac{\sigma_a}{k\dot{\sigma}_a}\left|D_\tau Z + \Delta Z\right|^2\right) G_2\, dv\, d\tau.\end{aligned} \tag{4.156}$$

By (4.140), we have that $F_2 \le N/(r^2+1)$, and hence that $F_2 \le (1/20)\sigma_a^{-1}$ on $\mathcal{C}_{\mathcal{S},1}$ if $\mathcal{S}$ is large enough. Since $\dot{\sigma}_a \ge 1/10$ on $[0,1]$, it follows that there is some universal constant C_0 such that

$$\begin{aligned}&k\|\sigma_a^{-k-1/2} Z G_2^{1/2}\|^2_{\mathcal{C}_{\mathcal{S},1}} \\ &\quad\le C_0\|\sigma_a^{-k}\nabla Z G_2^{1/2}\|^2_{\mathcal{C}_{\mathcal{S},1}} + C_0\|\sigma_a^{-k}(D_\tau Z + \Delta Z)G_2^{1/2}\|^2_{\mathcal{C}_{\mathcal{S},1}},\end{aligned} \tag{4.157}$$

for all $k \ge 1$.

Finally, multiplying both sides of (4.155) by $C_0 + 1$ and adding it to (4.157), we obtain that

$$\begin{aligned}&k\|\sigma_a^{-k-1/2} Z G_2^{1/2}\|^2_{\mathcal{C}_{\mathcal{S},1}} + \|\sigma_a^{-k}\nabla Z G_2^{1/2}\|^2_{\mathcal{C}_{\mathcal{S},1}} \\ &\quad\le C\|\sigma_a^{-k}(D_\tau Z + \Delta Z)G_2^{1/2}\|^2_{\mathcal{C}_{\mathcal{S},1}}\end{aligned} \tag{4.158}$$

for some universal constant C. The inequality is valid for all $0 < a \le 1$, $k \ge 1$, and $\rho \ge 1$ provided that $\mathcal{S} = \mathcal{S}(n, K)$ is taken sufficiently large. The estimate (4.148) follows. □

4.7.6. The inequality for the ODE component.

Next we prove a matching inequality with the weights σ_a and G_2 to bound the norm of $D_\tau Z$. This is more complicated to prove than (4.120), as the weight function is no longer pointwise monotonic in τ.

PROPOSITION 4.42. *Assume that $0 < T \le 1/4$ and $\rho \ge 1$. Then there are constants $\beta > 1$ and $c > 0$ depending at most on n, and a constant N depending potentially also on K, such that, for any section Z of $\mathcal{Z}$ which vanishes at $\tau = 0$ and has compact support in $\mathcal{C}_{\mathcal{R},T}$, the inequality*

$$\begin{aligned}&\|\sigma_a^{-k-\frac{1}{2}} Z G_2^{\frac{1}{2}}\|_{\mathcal{C}_{\mathcal{R},T}} \\ &\quad\le N k^{-1}\|\sigma_a^{-k} D_\tau Z G_2^{\frac{1}{2}}\|_{\mathcal{C}_{\mathcal{R},T}} + N(\beta a)^{-(k+c)}\rho^{\frac{n-1}{2}}\|Z\|_\infty\end{aligned} \tag{4.159}$$

holds for all $k \ge 1$ and all $0 < a \le 1/8$. Here, $\|Z\|_\infty = \sup_{\mathcal{C}_{R,T}}|Z|$.

PROOF. Fix $0 < a \le 1/8$ and $0 < T \le 1/4$. We start with the identity

$$\begin{aligned}&\sigma_a^{-2k}\frac{\partial}{\partial\tau}|Z|^2 G_2 - \frac{\partial}{\partial\tau}\left(\sigma_a^{-2k}|Z|^2 G_2\right) \\ &\quad= \sigma_a^{-2k}\left(\frac{2k + n/2}{(\tau + a)} - \frac{2k}{3} - \frac{(r-\rho)^2}{4(\tau+a)^2}\right)|Z|^2 G_2,\end{aligned} \tag{4.160}$$

valid for any k, $\rho > 0$. Since

$$\frac{\partial}{\partial\tau}|Z|^2 = 2\langle D_\tau Z, Z\rangle \le \frac{3}{k}|D_\tau Z|^2 + \frac{k}{3}|Z|^2,$$

we have

$$\begin{aligned}&\frac{3}{k}\sigma_a^{-2k}\left|D_\tau Z\right|^2 G_2 - \frac{\partial}{\partial \tau}\left(\sigma_a^{-2k}|Z|^2 G_2\right)\\ &\qquad \geq \sigma_a^{-2k}\left(\frac{k}{(\tau+a)} - \frac{(r-\rho)^2}{4(\tau+a)^2}\right)|Z|^2 G_2\end{aligned} \tag{4.161}$$

on $\mathcal{C}_{\mathcal{R},T}$, and consequently that

$$\begin{aligned}&\frac{3}{k}\int_{\mathcal{C}_{\mathcal{R},T}} \sigma_a^{-2k}|D_\tau Z|^2 G_2\, d\hat{\mu}\, d\tau\\ &\qquad \geq \int_{\mathcal{C}_{\mathcal{R},T}} \frac{\sigma_a^{-2k}}{\tau+a}\left(k - \frac{(r-\rho)^2}{4(\tau+a)}\right)|Z|^2 G_2\, d\hat{v}\, d\tau,\end{aligned} \tag{4.162}$$

where $d\hat{v}$ is the volume form for the (time-independent) conical metric $\hat{g} = g(0)$.

Define the set

$$\Omega = \Omega_{a,k,\rho} := \left\{\, (x,\tau) \,\middle|\, (r(x)-\rho)^2 > 2k(\tau+a) \,\right\}. \tag{4.163}$$

On $\Omega \cap \operatorname{supp} Z$, we have

$$\begin{aligned}&\frac{(r-\rho)^2}{\tau+a}\exp\left(-\frac{1}{4}\cdot\frac{(r-\rho)^2}{\tau+a}\right) \leq C\cdot\exp\left(-\frac{7}{32}\cdot\frac{(r-\rho)^2}{\tau+a}\right)\\ &= C\cdot\exp\left(-\frac{1}{16}\cdot\frac{(r-\rho)^2}{\tau+a} - \frac{5}{32}\cdot\frac{(r-\rho)^2}{\tau+a}\right) \leq C\cdot\exp\left(\frac{-(r-\rho)^2}{6} - \frac{5k}{16}\right).\end{aligned}$$

Here we have used that $s\mathrm{e}^{-s/32} \leq C$ for $s \geq 0$ and that

$$\exp\left(-\frac{(r-\rho)^2}{16(\tau+a)}\right) \leq \exp\left(-\frac{(r-\rho)^2}{6}\right),$$

as $\tau + a \leq T + a < 3/8$. Hence, using the definition of G_2, we see that

$$\begin{aligned}\frac{(r-\rho)^2}{(\tau+a)^2}G_2 &= \frac{(r-\rho)^2}{(\tau+a)^{\frac{n}{2}+2}}\exp\left(-\frac{1}{4}\cdot\frac{(r-\rho)^2}{\tau+a}\right)\\ &\leq C\mathrm{e}^{-\frac{5k}{16}}a^{-(\frac{n}{2}+1)}\exp\left(-\frac{(r-\rho)^2}{6}\right)\end{aligned} \tag{4.164}$$

on Ω. Also, as

$$\sigma_a(\tau) = (\tau+a)\mathrm{e}^{-\frac{\tau+a}{3}} \geq a\mathrm{e}^{-\frac{1}{8}},$$

we have

$$\sigma_a(\tau)^{-2k} \leq a^{-2k}\mathrm{e}^{\frac{k}{4}}. \tag{4.165}$$

Combining (4.164) and (4.165), we obtain the following upper bound on the second term on the right side of (4.162):

$$\begin{aligned}&\frac{1}{4}\int_\Omega \sigma_a^{-2k}\frac{(r-\rho)^2}{(\tau+a)^2}|Z|^2 G_2\, d\hat{v}\, d\tau\\ &\qquad \leq C\mathrm{e}^{-\frac{k}{16}}a^{-(2k+\frac{n}{2}+1)}\|Z\|_\infty^2\int_{\mathcal{C}_{\mathcal{R},T}} \mathrm{e}^{-\frac{(r-\rho)^2}{6}}\, d\hat{v}\, d\tau\\ &\qquad \leq C\mathrm{e}^{-\frac{k}{16}}a^{-(2k+\frac{n}{2}+1)}\|Z\|_\infty^2 \operatorname{Vol}_{g_\Sigma}\Sigma\cdot\int_0^\infty \mathrm{e}^{-\frac{(s-\rho)^2}{6}}s^{n-1}\, ds.\end{aligned} \tag{4.166}$$

Estimating (for $\rho \geq 1$) that

$$\int_0^\infty e^{\frac{-(s-\rho)^2}{6}} s^{n-1}\, ds = \int_{-\rho}^\infty e^{\frac{-s^2}{6}} (s+\rho)^{n-1}\, ds \leq \rho^{n-1} \int_{-\infty}^\infty e^{\frac{-s^2}{6}} |s+1|^{n-1}\, ds$$
$$\leq C\rho^{n-1},$$

we conclude from (4.166) that

$$\frac{1}{4}\int_\Omega \sigma_a^{-2k} \frac{(r-\rho)^2}{(\tau+a)^2} |Z|^2 G_2\, d\hat{v}\, d\tau \leq N(ae^{\frac{1}{32}})^{-(2k+\frac{n}{2}+1)} \rho^{n-1} \|Z\|_\infty^2 \tag{4.167}$$

for some N depending on $\hat{g}$.

On the other hand, on the complement Ω^c of Ω, we have

$$\frac{(r-\rho)^2}{4(\tau+a)^2} \leq \frac{k}{2(\tau+a)},$$

so

$$\frac{1}{4}\int_{\Omega^c} \sigma^{-2k} \frac{(r-\rho)^2}{(\tau+a)^2} |Z|^2 G_2\, d\hat{v}\, d\tau \leq \frac{k}{2}\int_{\mathcal{C}_{\mathcal{R},T}} \frac{\sigma^{-2k}}{\tau+a} |Z|^2 G_2\, d\hat{v}\, d\tau. \tag{4.168}$$

Thus, taking (4.167) and (4.168) together and returning to (4.162), we have

$$\begin{aligned}
&\frac{3}{k}\int_{\mathcal{C}_{\mathcal{R},T}} |D_\tau Z|^2 \sigma_a^{-2k} G\, d\hat{v}\, d\tau \\
&\quad\geq \frac{k}{2}\int_{\mathcal{C}_{\mathcal{R},T}} \frac{\sigma_a^{-2k}}{\tau+a}|Z|^2 G_2\, d\hat{v}\, d\tau - N(ae^{\frac{1}{32}})^{-(2k+\frac{n}{2}+1)}\rho^{n-1}\|Z\|_\infty^2 \\
&\quad\geq \frac{k}{6}\int_{\mathcal{C}_{\mathcal{R},T}} \sigma_a^{-2k-1}|Z|^2 G_2\, d\hat{v}\, d\tau - N(ae^{\frac{1}{32}})^{-(2k+\frac{n}{2}+1)}\rho^{n-1}\|Z\|_\infty^2.
\end{aligned}$$

Using the uniform equivalence of the metrics $g(\tau)$ and $\hat{g}$, we thus obtain the bound

$$\begin{aligned}
&\|\sigma_a^{-k-\frac{1}{2}} Z G_2^{\frac{1}{2}}\|_{\mathcal{C}_{\mathcal{R},T}} \\
&\qquad\leq Nk^{-2}\|\sigma_a^{-k} D_\tau Z G_2^{\frac{1}{2}}\|_{\mathcal{C}_{\mathcal{R},T}}^2 + N(ae^{\frac{1}{32}})^{-(2k+\frac{n}{2}+1)}\rho^{n-1}\|Z\|_\infty^2,
\end{aligned}$$

in terms of the space-time L^2-norms $\|\cdot\|$ induced by $g(\tau)$. Equation (4.159) then follows with $\beta := e^{1/32}$ and $c := n/4 + 1/2$. □

4.8. Proof of the backward uniqueness principle

Now we assemble the proof of Theorem 4.25 from the Carleman estimates in Propositions 4.35, 4.41, and 4.42. We will continue to assume that $g(\tau)$ is the solution to the backward Ricci flow (4.7) on $\mathcal{C}_{\mathcal{R},1}$ associated to a shrinking soliton $(\mathcal{C}_\mathcal{R}, g_1, f_1)$ which is dynamically asymptotic to $\hat{g}$, and that $g(\tau)$ satisfies the bounds (4.60) and (4.61) on $\mathcal{C}_{\mathcal{R},1}$ for a uniform $K > 0$.

The proof has two parts. In the first part, we show that sections $\mathbf{X}$ and $\mathbf{Y}$ which satisfy the assumptions of the theorem decay at least at an exponential-quadratic rate as $r \to \infty$. In the second, we establish the vanishing of $\mathbf{X}$ and $\mathbf{Y}$.

4.8.1. Establishing the rapid decay of the sections X and Y.

PROPOSITION 4.43. *There are constants $0 < A \le 1$ and $B > 0$ depending on n, and constants $N > 0$ and $\mathcal{S} \ge \mathcal{R}$ depending in addition on K and L, such that, if $\mathbf{X}$ and $\mathbf{Y}$ satisfy the assumptions of Theorem 4.25 on $\mathcal{C}_{\mathcal{R},1}$, then*

$$\| |\mathbf{X}| + |\nabla \mathbf{X}| + |\mathbf{Y}| \|_{\mathcal{D}_\rho(\rho\sqrt{a})\times[0,a]} \le N e^{-\frac{B\rho^2}{a}} \tag{4.169}$$

for all $\rho > 16\mathcal{S}$ and $0 < a \le A$. Here $\mathcal{D}_x(t)$ denotes the annular region

$$\mathcal{D}_u(t) = \{ (r,\sigma) \in \mathcal{C}_{\mathcal{R}} \mid |r - u| < t \}.$$

PROOF. We will specify the constants A, B, N, and $\mathcal{S}$ over the course of the proof. Let us assume initially that we at least have $0 < A \le 1$ and that $\mathcal{S}$ is large enough to meet the assumptions of Propositions 4.35, 4.41, and 4.42, and to satisfy $\mathcal{S} \ge 2\mathcal{R}$. As in previous arguments, we will use C to denote a sequence of constants depending only on n, and use N to indicate a sequence of constants which may depend in addition on the parameters K and L.

Fix an arbitrary $\rho > 16\mathcal{S}$. We will need two cutoff functions, one temporal and one spatial. For the temporal cutoff, we select $\varphi \in C^\infty(\mathbb{R})$ satisfying

$$\varphi(\tau) = \begin{cases} 1 & \tau \le 1/6, \\ 0 & \tau \ge 1/5, \end{cases} \qquad -C \le \varphi'(\tau) \le 0.$$

For the spatial cutoff, we select for each $\xi > 4\rho$ a function $\psi_{\rho,\xi}$ in the form

$$\psi_{\rho,\xi}(x) = \Psi_{\rho,\xi}(r(x)),$$

where $\Psi_{\rho,\xi} \in C^\infty(\mathbb{R},[0,1])$ satisfies

$$\Psi_{\rho,\xi}(r) = \begin{cases} 1 & \rho/3 \le r \le 2\xi, \\ 0 & r < \rho/6 \text{ or } r > 3\xi, \end{cases} \qquad \rho|\Psi'_{\rho,\xi}| + \rho^2|\Psi''_{\rho,\xi}| \le N.$$

(For the right-hand inequality, we have used that $\xi > 4\rho$ and $\rho > 16\mathcal{S}$.) In view of Proposition 4.29, we have the bounds

$$\begin{aligned} |\nabla\psi_{\rho,\xi}| &\le |\Psi'_{\rho,\xi}|(1+\eta_1)^{1/2} \le N\rho^{-1}, \\ |\Delta\psi_{\rho,\xi}| &\le |\Psi'_{\rho,\xi}||(n-1)r^{-1} + \eta_3| + |\Psi''_{\rho,\xi}|(1+|\eta_1|) \le N\rho^{-2}, \end{aligned} \tag{4.170}$$

on the derivatives of $\psi_{\rho,\xi}$.

The sections $\mathbf{X}_{\rho,\xi}$ and $\mathbf{Y}_{\rho,\xi}$ defined by

$$\mathbf{X}_{\rho,\xi} = \varphi\psi_{\rho,\xi}\mathbf{X}, \quad \mathbf{Y}_{\rho,\xi} = \varphi\psi_{\rho,\xi}\mathbf{Y},$$

will then have compact support in the annular cylinder $\mathcal{A}(\rho/6, 3\xi) \times [0,1/4)$, where $\mathcal{A}(r_1,r_2)$ denotes the set

$$\mathcal{A}(r_1,r_2) = \{ (r,\sigma) \in \mathcal{C}_{\mathcal{R}} \mid r_1 < r < r_2 \},$$

and will vanish identically for $\tau = 0$. Applying Proposition 4.41 to the components of $\mathbf{X}_{\rho,\xi}$, we obtain

$$\begin{aligned} &\sqrt{k}\|\sigma_a^{-k-\frac{1}{2}}\mathbf{X}_{\rho,\xi}G^{\frac{1}{2}}\|_{\mathcal{A}(\frac{\rho}{6},3\xi)\times[0,\frac{1}{5}]} + \|\sigma_a^{-k}\nabla\mathbf{X}_{\rho,\xi}G^{\frac{1}{2}}\|_{\mathcal{A}(\frac{\rho}{6},3\xi)\times[0,\frac{1}{5}]} \\ &\le C\|\sigma_a^{-k}(D_\tau + \Delta)\mathbf{X}_{\rho,\xi}G^{\frac{1}{2}}\|_{\mathcal{A}(\frac{\rho}{6},3\xi)\times[0,\frac{1}{5}]}, \end{aligned} \tag{4.171}$$

for all $0 < a \le A$ and $\rho > 16\mathcal{S}$. Here and below, we use $\|\cdot\|$ to denote either the space-time L^2-norm or the spatial L^2-norm induced by the family of metrics $g(\tau)$.

Using (4.63) and (4.170), we have

$$\begin{aligned}|D_\tau \mathbf{X}_{\xi,\rho} + \Delta \mathbf{X}_{\rho,\xi}| &\leq \varepsilon(|\mathbf{X}_{\rho,\xi}| + |\nabla \mathbf{X}_{\rho,\xi}| + |\mathbf{Y}_{\rho,\xi}|) + \psi_{\rho,\xi}|\varphi'||\mathbf{X}| \\ &\quad + \varphi\left(|\Delta\psi_{\rho,\xi}||\mathbf{X}| + (2+\varepsilon)|\nabla\psi_{\rho,\xi}||\nabla \mathbf{X}|\right) \\ &\leq \varepsilon(|\mathbf{X}_{\rho,\xi}| + |\nabla \mathbf{X}_{\rho,\xi}| + |\mathbf{Y}_{\rho,\xi}|) + C|\mathbf{X}|\chi_{\operatorname{supp}\varphi'} \\ &\quad + N\rho^{-1}\left(|\mathbf{X}| + |\nabla \mathbf{X}|\right)\chi_{\operatorname{supp}\nabla\psi_{\rho,\xi}}.\end{aligned}$$

Integrating this inequality over space and time, and combining it with (4.171), we see that if $\mathcal{S}$ is taken large enough, then

$$\begin{aligned}\sqrt{k}\|\sigma_a^{-k-\frac{1}{2}}\mathbf{X}_{\rho,\xi}G^{\frac{1}{2}}_2\|_{\mathcal{A}(\frac{\rho}{6},3\xi)\times[0,\frac{1}{5}]} &+ \|\sigma_a^{-k}\nabla\mathbf{X}_{\rho,\xi}G^{\frac{1}{2}}_2\|_{\mathcal{A}(\frac{\rho}{6},3\xi)\times[0,\frac{1}{5}]} \\ &\leq \frac{1}{2}\|\sigma_a^{-k}\mathbf{Y}_{\rho,\xi}G^{\frac{1}{2}}_2\|_{\mathcal{A}(\frac{\rho}{6},3\xi)\times[0,\frac{1}{5}]} \\ &\quad + N\|\sigma_a^{-k}\mathbf{X}G^{\frac{1}{2}}_2\|\|_{\mathcal{A}(\frac{\rho}{6},3\xi)\times[\frac{1}{6},\frac{1}{5}]} \\ &\quad + N\|\sigma_a^{-k}(|\mathbf{X}|+|\nabla\mathbf{X}|)G^{\frac{1}{2}}_2\|_{\mathcal{A}(\frac{\rho}{6},\frac{\rho}{3})\times[0,\frac{1}{5}]} \\ &\quad + N\|\sigma_a^{-k}(|\mathbf{X}|+|\nabla\mathbf{X}|)G^{\frac{1}{2}}_2\|_{\mathcal{A}(2\xi,3\xi)\times[0,\frac{1}{5}]},\end{aligned} \tag{4.172}$$

since $\varepsilon(r) \to 0$ as $r \to \infty$ by assumption, and $\rho > 16\mathcal{S}$. Here, we have also used that $\sigma_a(\tau) \leq 1$.

On the other hand, assuming now that $0 < A \leq 1/8$ and applying Proposition 4.42 to $\mathbf{Y}_{\rho,\xi}$, we obtain

$$\begin{aligned}\|\sigma_a^{-k-\frac{1}{2}}\mathbf{Y}_{\rho,\xi}G^{\frac{1}{2}}_2\|_{\mathcal{A}(\frac{\rho}{6},3\xi)\times[0,\frac{1}{5}]} &\leq Nk^{-1}\|\sigma_a^{-k}D_\tau\mathbf{Y}_{\rho,\xi}G^{\frac{1}{2}}_2\|_{\mathcal{A}(\frac{\rho}{6},3\xi)\times[0,\frac{1}{5}]} \\ &\quad + N(\beta a)^{-(k+c)}\rho^{\frac{n-1}{2}}\|\mathbf{Y}\|_\infty,\end{aligned} \tag{4.173}$$

for some $\beta > 1$ and $c > 0$ and all $k > 0$ and $0 < a < A$. Since

$$\begin{aligned}|D_\tau\mathbf{Y}_{\rho,\xi}| &\leq L\left(|\mathbf{X}_{\rho,\xi}| + |\nabla\mathbf{X}_{\rho,\xi}| + |\mathbf{Y}_{\rho,\xi}|\right) + L\varphi|\nabla\psi_{\rho,\xi}||\mathbf{X}| + \psi_{\rho,\xi}|\varphi'||\mathbf{Y}| \\ &\leq N\left(|\mathbf{X}_{\rho,\xi}| + |\nabla\mathbf{X}_{\rho,\xi}| + |\mathbf{Y}_{\rho,\xi}|\right) + N\rho^{-1}|\mathbf{X}|\chi_{\operatorname{supp}\nabla\psi_{\rho,\xi}} \\ &\quad + C\psi_{\rho,\xi}|\mathbf{Y}|\chi_{\operatorname{supp}\varphi'},\end{aligned}$$

returning to (4.173), we see that, for suitably large k_0 depending on n, K, and L, we will have

$$\begin{aligned}&\|\sigma_a^{-k-\frac{1}{2}}\mathbf{Y}_{\rho,\xi}G^{\frac{1}{2}}_2\|_{\mathcal{A}(\frac{\rho}{6},3\xi)\times[0,\frac{1}{5}]} \\ &\leq \frac{1}{2}\left(\|\sigma_a^{-k}\mathbf{X}_{\rho,\xi}G^{\frac{1}{2}}_2\|_{\mathcal{A}(\frac{\rho}{6},3\xi)\times[0,\frac{1}{5}]} + \|\sigma_a^{-k}\nabla\mathbf{X}_{\rho,\xi}G^{\frac{1}{2}}_2\|_{\mathcal{A}(\frac{\rho}{6},3\xi)\times[0,\frac{1}{5}]}\right) \\ &\quad + N\|\sigma_a^{-k}\mathbf{X}G^{\frac{1}{2}}_2\|_{\mathcal{A}(\frac{\rho}{6},\frac{\rho}{3})\times[0,\frac{1}{5}]} + N\|\sigma_a^{-k}\mathbf{X}G^{\frac{1}{2}}_2\|_{\mathcal{A}(2\xi,3\xi)\times[0,\frac{1}{5}]} \\ &\quad + N\|\sigma_a^{-k}\mathbf{Y}G^{\frac{1}{2}}_2\|\|_{\mathcal{A}(\frac{\rho}{6},3\xi)\times[\frac{1}{6},\frac{1}{5}]} + N(\beta a)^{-(k+c)}\rho^{\frac{n-1}{2}}\|\mathbf{Y}\|_\infty,\end{aligned} \tag{4.174}$$

for all $k \geq k_0$ and $0 < a < A$.

Now add (4.174) to (4.172) and use (4.174) again with the fact that $\sigma_a(\tau) \leq 1$ to estimate the first term on the right of (4.172). After absorbing the terms on the right involving L^2-norms of $\mathbf{X}_{\rho,\xi}$, $\nabla\mathbf{X}_{\rho,\xi}$, and $\mathbf{Y}_{\rho,\xi}$ over $\mathcal{A}(\rho/6, 3\xi) \times [0, \frac{1}{5}]$ into their counterparts on the left, we arrive at the inequality

$$\begin{aligned}
&\|\sigma_a^{-k-\frac{1}{2}}(|\mathbf{X}_{\rho,\xi}| + |\mathbf{Y}_{\rho,\xi}|)G_2^{\frac{1}{2}}\|_{\mathcal{A}(\frac{\rho}{6},3\xi)\times[0,\frac{1}{5}]} + \|\sigma_a^{-k}\nabla\mathbf{X}_{\rho,\xi}G_2^{\frac{1}{2}}\|_{\mathcal{A}(\frac{\rho}{6},3\xi)\times[0,\frac{1}{5}]}\\
&\quad\le N\|\sigma_a^{-k}(|\mathbf{X}| + |\mathbf{Y}|)G_2^{\frac{1}{2}}\|_{\mathcal{A}(\frac{\rho}{6},3\xi)\times[\frac{1}{6},\frac{1}{5}]}\\
&\qquad + N\|\sigma_a^{-k}(|\mathbf{X}| + |\nabla\mathbf{X}| + |\mathbf{Y}|)G_2^{\frac{1}{2}}\|_{\mathcal{A}(\frac{\rho}{6},\frac{\rho}{3})\times[0,\frac{1}{5}]}\\
&\qquad + N\|\sigma_a^{-k}(|\mathbf{X}| + |\nabla\mathbf{X}| + |\mathbf{Y}|)G_2^{\frac{1}{2}}\|_{\mathcal{A}(2\xi,3\xi)\times[0,\frac{1}{5}]} + N\rho^{\frac{n-1}{2}}(\beta a)^{-(k+c)},
\end{aligned}$$

which is valid for all $0 < a < A$, $k \ge k_0$, and ρ and ξ satisfying $\xi \ge 4\rho$ and $\rho \ge 16\mathcal{S}$.

Since the metrics $g(\tau)$ are uniformly equivalent to the conical metric $\hat{g}$, we have

$$\mathrm{Vol}_{g(\tau)}(\mathcal{A}(r,s)) \le Ns^n$$

for all $\mathcal{R} < r < s$. Thus, given (4.55) and the exponential-quadratic decay of G_2, if we send $\xi \to \infty$, the limits of all terms in the above inequality will exist and be finite, and the penultimate term on the right will vanish. Shrinking the domain of the left side of the resulting inequality, and using (4.132) to compare $\sigma_a(\tau)$ with $\tau + a$, we obtain

$$\begin{aligned}
&\|(\tau + a)^{-k}(|\mathbf{X}| + |\nabla\mathbf{X}| + |\mathbf{Y}|)G_2^{\frac{1}{2}}\|_{\mathcal{C}_{\frac{\rho}{3},\frac{1}{6}}}\\
&\quad\le C^kN\|(\tau + a)^{-k}(|\mathbf{X}| + |\mathbf{Y}|)G_2^{\frac{1}{2}}\|_{\mathcal{C}_{\frac{\rho}{6}}\times[\frac{1}{6},\frac{1}{5}]}\\
&\qquad + C^kN\|(\tau + a)^{-k}(|\mathbf{X}| + |\nabla\mathbf{X}| + |\mathbf{Y}|)G_2^{\frac{1}{2}}\|_{\mathcal{A}(\frac{\rho}{6},\frac{\rho}{3})\times[0,\frac{1}{5}]}\\
&\qquad + N(\beta a)^{-(k+c)}\rho^{\frac{n-1}{2}}\\
&\quad:= \mathrm{I} + \mathrm{II} + \mathrm{III}.
\end{aligned} \tag{4.175}$$

Consider the term I. Since $0 < a < A \le 1/8$, we have $1/6 < \tau + a < 1/3$ on the domain of the integral and

$$G_2(x,\tau) \le C\mathrm{e}^{-\frac{3(r-\rho)^2}{4}}$$

on the domain $\mathcal{C}_{\rho/6} \times [1/6, 1/5]$ of the integral. Thus, we can estimate

$$\begin{aligned}
\mathrm{I}^2 &= C^kN\|(\tau + a)^{-k}(|\mathbf{X}| + |\mathbf{Y}|)G_2^{\frac{1}{2}}\|^2_{\mathcal{C}_{\frac{\rho}{6}}\times[\frac{1}{6},\frac{1}{5}]}\\
&\le C^kN\int_{\frac{\rho}{6}}^{\infty} \mathrm{e}^{-\frac{3(s-\rho)^2}{4}} s^{n-1}\,ds \le C^kN\rho^{n-1},
\end{aligned} \tag{4.176}$$

using (4.62).

As for the term II, note that, on one hand, we have

$$G_2(x,\tau) \le (\tau + a)^{-\frac{n}{2}}\mathrm{e}^{-\frac{\rho^2}{9(\tau+a)}}$$

on the domain of integration. On the other hand, for any integer $m > 0$, we have

$$\max_{s>0} s^{-m}\mathrm{e}^{-\frac{\rho^2}{18s}} = \rho^{-2m}(18m)^m\mathrm{e}^{-m} \le \rho^{-2m}C^m m!$$

by Stirling's formula. Taking $m = k + n/4$, it follows then that

$$(\tau + a)^{-k}\sqrt{G_2(x,\tau)} \le C^k\rho^{-2k-\frac{n}{2}}k!$$

on that domain. Hence

$$\begin{aligned} \mathrm{II} &= C^k N \|(\tau+a)^{-k}(|\mathbf{X}|+|\nabla\mathbf{X}|+|\mathbf{Y}|)G_2^{\frac{1}{2}}\|_{\mathcal{A}(\frac{\rho}{6},\frac{\rho}{3})\times[0,\frac{1}{5}]} \\ &\leq C^k N \rho^{-2k-\frac{n}{2}} k! \left(\int_0^{\frac{1}{5}} \int_{\mathcal{A}(\frac{\rho}{6},\frac{\rho}{3})} dv_{g(\tau)}\, d\tau\right)^{\frac{1}{2}} \\ &\leq C^k N \rho^{-2k} k!. \end{aligned} \tag{4.177}$$

Combining (4.175) with (4.176) and (4.177), we obtain that

$$\|(\tau+a)^{-k}(|\mathbf{X}|+|\nabla\mathbf{X}|+|\mathbf{Y}|)G_2^{\frac{1}{2}}\|_{\mathcal{C}_{\frac{\rho}{3},\frac{1}{6}}} \leq N\rho^{\frac{n}{2}}\left(C^k(1+\rho^{-2k}k!)+(\beta a)^{-(k+c)}\right)$$

for all $k \geq k_0$; that is,

$$\begin{aligned} &\|(\tau+a)^{-(l+k_0)}(|\mathbf{X}|+|\nabla\mathbf{X}|+|\mathbf{Y}|)G_2^{\frac{1}{2}}\|_{\mathcal{C}_{\frac{\rho}{3},\frac{1}{6}}} \\ &\quad \leq N\rho^{\frac{n}{2}}\left(C_1^{\ell}\left(1+\rho^{-2\ell}(\ell+k_0)!\right)+(\beta a)^{-(\ell+k_0+c)}\right), \end{aligned} \tag{4.178}$$

for all $\ell \geq 0$, where C_1 is some universal constant.

Now, there is another universal constant C_2 such that

$$(\ell+k_0)! \leq C^{k_0+\ell}(k_0!)(\ell!) \leq NC_2^{\ell}(\ell!)$$

for all $\ell \geq 0$. Thus, if we multiply both sides of (4.178) by $\rho^{2\ell}/(C^{\ell}\ell!)$ for some $C \geq 2\cdot\max\{C_1, C_2\}$ (and also discard the factors $(\tau+a)^{-(k_0+n/4)} \geq 1$ on the left), we obtain that

$$\begin{aligned} &\left\|\frac{\rho^{2\ell}}{(C(\tau+a))^{\ell}\ell!}(|\mathbf{X}|+|\nabla\mathbf{X}|+|\mathbf{Y}|)\mathrm{e}^{-\frac{(r-\rho)^2}{8(\tau+a)}}\right\|_{\mathcal{C}_{\frac{\rho}{3},\frac{1}{6}}} \\ &\quad \leq N\rho^{\frac{n}{2}}\left(\frac{\rho^{2\ell}}{2^{\ell}\ell!}+\frac{1}{2^{\ell}}+a^{-(k_0+c)}\frac{\rho^{2\ell}}{(C\beta a)^{\ell}\ell!}\right) \end{aligned}$$

for all $\ell \geq 0$. Summing over ℓ then implies that

$$\begin{aligned} &\left\|(|\mathbf{X}|+|\nabla\mathbf{X}|+|\mathbf{Y}|)\mathrm{e}^{\frac{\rho^2}{C(\tau+a)}-\frac{(r-\rho)^2}{8(\tau+a)}}\right\|_{\mathcal{C}_{\frac{\rho}{3},\frac{1}{6}}} \\ &\quad \leq N\rho^{\frac{n}{2}}\left(1+\mathrm{e}^{\frac{\rho^2}{2}}+a^{-(k_0+c)}\mathrm{e}^{\frac{\rho^2}{C\beta a}}\right). \end{aligned} \tag{4.179}$$

Recall that the constant β from Proposition 4.42 satisfies $\beta > 1$. Define $\delta > 0$ by $\beta = 1+2\delta$, and multiply both sides of (4.179) by $\mathrm{e}^{-\frac{\rho^2}{C(1+\delta)a}}$ to see that

$$\begin{aligned} &\left\|(|\mathbf{X}|+|\nabla\mathbf{X}|+|\mathbf{Y}|)\mathrm{e}^{-\frac{(r-\rho)^2}{8(\tau+a)}}\right\|_{\mathcal{C}_{\frac{\rho}{3}}\times[0,\delta a]} \\ &\quad \leq N\rho^{\frac{n}{2}}\left(\mathrm{e}^{\frac{-\rho^2}{C(1+\delta)a}}(1+\mathrm{e}^{\frac{\rho^2}{2}})+a^{-(k_0+c)}\mathrm{e}^{\frac{-\delta\rho^2}{C(1+2\delta)(1+\delta)a}}\right). \end{aligned} \tag{4.180}$$

Now (given the dependencies of k_0), we can estimate

$$\rho^{\frac{n}{2}}a^{-(k_0+c)}\mathrm{e}^{\frac{-\delta\rho^2}{C(1+2\delta)(1+\delta)a}} \leq N\mathrm{e}^{\frac{-\delta\rho^2}{2C(1+2\delta)(1+\delta)a}}$$

for N depending on n, K, and L, but not on a or ρ. Also, if $A = A(n) \leq 1/8$ is taken small enough, we can bound

$$\rho^{\frac{n}{2}}\mathrm{e}^{\frac{-\rho^2}{C(1+\delta)a}}(1+\mathrm{e}^{\frac{\rho^2}{2}}) \leq N\mathrm{e}^{\frac{-\rho^2}{2C(1+\delta)a}}$$

independently of $\rho \geq 1$ for all $0 < a < A$. Thus we obtain some $C_3 = C_3(n) > 0$ such that

$$\left\| (|\mathbf{X}| + |\nabla \mathbf{X}| + |\mathbf{Y}|) e^{-\frac{(r-\rho)^2}{8(\tau+a)}} \right\|_{\mathcal{C}_{\frac{\rho}{3}} \times [0, \delta a]} \leq N e^{-\frac{\rho^2}{C_3 a}} \tag{4.181}$$

for all $0 < a < A$ and $\rho > 16\mathcal{S}$.

On the other hand, $e^{-\frac{(r-\rho)^2}{8(\tau+a)}} \geq e^{-\frac{\delta\rho^2}{8}}$ on $\mathcal{D}_\rho(\rho\sqrt{\delta a}) \times [0, \delta a]$, so, shrinking the domain of integration in the term on the left of (4.181) and multiplying both sides by $e^{\frac{\delta\rho^2}{8}}$, we obtain the bound

$$\| |\mathbf{X}| + |\nabla \mathbf{X}| + |\mathbf{Y}| \|_{\mathcal{D}_\rho(\rho\sqrt{\delta a}) \times [0, \delta a]} \leq N e^{\frac{-\rho^2}{2C_3 a}} \tag{4.182}$$

for all $a \in (0, A]$ provided we decrease A if necessary to satisfy $A < 4/(\delta C_3)$. Finally, replacing a with δa and A with δA, the claim follows with $B = \delta/2C_3$. □

4.8.2. The sections X and Y vanish identically.

With the above decay estimates in hand, we can now apply the Carleman estimates in Proposition 4.35 to complete the proof of Theorem 4.25.

PROOF OF THEOREM 4.25. Let $\mathbf{X}$ and $\mathbf{Y}$ be as in the statement of Theorem 4.25, and $g(\tau)$ a solution to (4.7) as described in §4.4.1. In particular, we assume that both (4.60) and (4.61) hold for some fixed K. We will show that there is $\mathcal{S} \geq \mathcal{R}$ depending on n, K, and L and $T = T(n) \in (0, 1]$ such that $\mathbf{X} \equiv 0$ and $\mathbf{Y} \equiv 0$ on $\mathcal{C}_{\mathcal{S},T}$. Below, we will again use C to denote a series of constants depending only on n and use N to denote constants which may depend in addition on the parameters K and L.

Let A_0 and $\mathcal{S}_0$ denote the values of A and $\mathcal{S}$ specified in Proposition 4.43, and let us here initially assume that $\mathcal{S} \geq 1 + 16\mathcal{S}_0$. It follows from (4.169) that there is $A = A(n) \in (0, A_0]$ such that

$$\| |\mathbf{X}| + |\nabla \mathbf{X}| + |\mathbf{Y}| \|_{\mathcal{D}_s(\sqrt{A}) \times [0, A]} \leq N e^{-4s^2} \tag{4.183}$$

for all $s \geq \mathcal{S}$. Expressing $\mathcal{C}_\mathcal{S}$ as the union of countably many annular regions $\mathcal{D}_{s_i}(\sqrt{A})$, applying (4.183) to each region, and summing the result, we obtain that

$$\| (|\mathbf{X}| + |\nabla \mathbf{X}| + |\mathbf{Y}|) e^{2r^2} \|_{\mathcal{C}_{\mathcal{S},A}} \leq N_0 \tag{4.184}$$

for some $N_0 = N_0(n, K, L)$.

Now, (4.184) implies that there is a sequence $\tau_m \in [A/2, A]$ with $\tau_m \to A$ as $m \to \infty$ such that[8]

$$\int_{\mathcal{C}_\mathcal{S} \times \{\tau_m\}} (|\mathbf{X}| + |\nabla \mathbf{X}| + |\mathbf{Y}|)^2 \, e^{2r^2} \, dv := N'_m < \infty. \tag{4.185}$$

Fix such a sequence $\{\tau_m\}_{m=1}^\infty$. We will show below that $\mathbf{X}$ and $\mathbf{Y}$ vanish on $\mathcal{C}_\mathcal{S} \times [0, \tau_m/2]$ for some $\mathcal{S}$ which is independent of m. Theorem 4.25 will follow with $T_0 = A/2$ after sending $m \to \infty$, and hence $\tau_m \to A$.

Now fix m and define

$$G_1(x, \tau) = G_{1;\alpha,\tau_m}(x, \tau) = \exp\left(\alpha(\tau_m - \tau) r^{2-\delta}(x) + r^2(x)\right)$$

on $\mathcal{C}_{\mathcal{R},1}$ for $\alpha \geq 0$. Note that we have $G_1(x, \tau_m) = e^{r^2}$ on $\mathcal{C}_\mathcal{R}$, and, for all $\alpha > 0$ and $M > 1$, we have $G_1(x, \tau) \leq e^{Mr^2}$ on $\mathcal{C}_{\mathcal{R}',1}$ for $\mathcal{R}' = \mathcal{R}'(\alpha, M)$ sufficiently large.

---

[8] Indeed, we may choose $A(1 - 1/m) < \tau_m < A$ such that $N'_m \leq 2mN_0^2/A$.

Recalling (4.63), we may choose $\mathcal{S}$ so large (depending on L) such that

$$\begin{aligned} |D_\tau \mathbf{X} + \Delta \mathbf{X}| &\leq \frac{1}{100}(|\mathbf{X}| + |\nabla \mathbf{X}| + |\mathbf{Y}|), \\ |D_\tau \mathbf{Y}| &\leq N(|\mathbf{X}| + |\nabla \mathbf{X}| + |\mathbf{Y}|), \end{aligned} \tag{4.186}$$

on $\mathcal{C}_{\mathcal{S},1}$. Let $\rho > \mathcal{S}$ and $\xi > 4\rho$. As in the proof of Proposition 4.43, we can construct a cutoff function $\psi_{\rho,\xi} : \mathcal{C}_{\mathcal{R}} \to [0,1]$ with

$$\psi_{\rho,\xi}(x) = \begin{cases} 1 & 2\rho < r(x) < \xi, \\ 0 & r(x) < \rho \text{ or } r(x) > 2\xi, \end{cases}$$

and

$$|\nabla \psi_{\rho,\xi}| + |\Delta \psi_{\rho,\xi}| \leq N$$

on $\mathcal{C}_{\mathcal{R},1}$. With these choices, the sections

$$\mathbf{X}_{\rho,\xi} := \psi_{\rho,\xi}\mathbf{X}, \quad \mathbf{Y}_{\rho,\xi} := \psi_{\rho,\xi}\mathbf{Y},$$

will be compactly supported in $\mathcal{C}_{\mathcal{S},\tau_m}$ and vanish at $\tau = 0$.

Applying (4.119) to $\mathbf{X}_{\rho,\xi}$, it follows that

$$\begin{aligned} \alpha\|\mathbf{X}_{\rho,\xi}G_1^{\frac{1}{2}}\|^2_{\mathcal{C}_{\mathcal{S},\tau_m}} &+ \|\nabla \mathbf{X}_{\rho,\xi}G_1^{\frac{1}{2}}\|^2_{\mathcal{C}_{\mathcal{S},\tau_m}} \\ &\leq \|(D_\tau + \Delta)\mathbf{X}_{\rho,\xi}G_1^{\frac{1}{2}}\|^2_{\mathcal{C}_{\mathcal{S},\tau_m}} + 2\|\nabla \mathbf{X}_{\rho,\xi}G_1^{\frac{1}{2}}\|^2_{\mathcal{C}_{\mathcal{S}}\times\{\tau_m\}}, \end{aligned}$$

and, applying (4.120) to $\mathbf{Y}_{\rho,\xi}$, that

$$\|\mathbf{Y}_{\rho,\xi}G_1^{\frac{1}{2}}\|^2_{\mathcal{C}_{\mathcal{S},\tau_m}} \leq \alpha^{-1}\|D_\tau \mathbf{Y}_{\rho,\xi}G_1^{\frac{1}{2}}\|^2_{\mathcal{C}_{\mathcal{S},\tau_m}}.$$

Summing, it follows that

$$\begin{aligned} &\left\|\left(\sqrt{\alpha}|\mathbf{X}_{\rho,\xi}| + |\nabla \mathbf{X}_{\rho,\xi}| + |\mathbf{Y}_{\rho,\xi}|\right) G_1^{\frac{1}{2}}\right\|_{\mathcal{C}_{\mathcal{S},\tau_m}} \\ &\qquad \leq \|(D_\tau + \Delta)\mathbf{X}_{\rho,\xi}G_1^{\frac{1}{2}}\|_{\mathcal{C}_{\mathcal{S},\tau_m}} + \frac{1}{\sqrt{\alpha}}\|D_\tau \mathbf{Y}_{\rho,\xi}G_1^{\frac{1}{2}}\|_{\mathcal{C}_{\mathcal{S},\tau_m}} \\ &\qquad\quad + 2\|\nabla \mathbf{X}_{\rho,\xi}G_1^{\frac{1}{2}}\|_{\mathcal{C}_{\mathcal{S}}\times\{\tau_m\}} \end{aligned} \tag{4.187}$$

for all $\alpha \geq 1$.

Now, using (4.186), we have the estimates

$$\begin{aligned} |(D_\tau + \Delta)\, \mathbf{X}_{\rho,\xi}| &\leq \frac{1}{100}\left(|\mathbf{X}_{\rho,\xi}| + |\nabla \mathbf{X}_{\rho,\xi}| + |\mathbf{Y}_{\rho,\xi}|\right) + 2|\nabla \psi_{\rho,\xi}||\nabla \mathbf{X}| \\ &\quad + \left(|\Delta \psi_{\rho,\xi}| + \frac{1}{100}|\nabla \psi_{\rho,\xi}|\right)|\mathbf{X}|, \\ |\nabla \mathbf{X}_{\rho,\xi}| &\leq |\nabla \psi_{\rho,\xi}||\mathbf{X}| + \psi_{\rho,\xi}|\nabla \mathbf{X}|, \end{aligned}$$

and

$$|D_\tau \mathbf{Y}_{\rho,\xi}| \leq N(|\mathbf{X}_{\rho,\xi}| + |\nabla \mathbf{X}_{\rho,\xi}| + |\mathbf{Y}_{\rho,\xi}|) + N|\nabla \psi_{\rho,\xi}||\mathbf{X}|$$

on $\mathcal{C}_{\mathcal{S},\tau_m}$. Thus, the terms on the right of (4.187) can be estimated by

$$\begin{aligned} \|(D_\tau + \Delta)\mathbf{X}_{\rho,\xi}G_1^{\frac{1}{2}}\|_{\mathcal{C}_{\mathcal{S},\tau_m}} &\leq \frac{1}{100}\|(|\mathbf{X}_{\rho,\xi}| + |\nabla \mathbf{X}_{\rho,\xi}| + |\mathbf{Y}_{\rho,\xi}|)G_1^{\frac{1}{2}}\|_{\mathcal{C}_{\mathcal{S},\tau_m}} \\ &\quad + N\|(|\mathbf{X}| + |\nabla \mathbf{X}|)G_1^{\frac{1}{2}}\|_{\mathcal{A}(\rho,2\rho)\times[0,\tau_m]} \\ &\quad + N\|(|\mathbf{X}| + |\nabla \mathbf{X}|)G_1^{\frac{1}{2}}\|_{\mathcal{A}(\xi,2\xi)\times[0,\tau_m]}, \end{aligned}$$

$$\|\nabla \mathbf{X}_{\rho,\xi} G_1^{\frac{1}{2}}\|_{\mathcal{C}_{\mathcal{S}}\times\{\tau_m\}} \leq N\|\mathbf{X}e^{\frac{r^2}{2}}\|_{\mathcal{A}(\rho,2\rho)\times\{\tau_m\}} + N\|\mathbf{X}e^{\frac{r^2}{2}}\|_{\mathcal{A}(\xi,2\xi)\times\{\tau_m\}} + N\|\nabla\mathbf{X}e^{\frac{r^2}{2}}\|_{\mathcal{A}(\rho,2\xi)\times\{\tau_m\}},$$

and

$$\begin{aligned}\frac{1}{\sqrt{\alpha}}\|D_\tau \mathbf{Y}_{\rho,\xi} G_1^{\frac{1}{2}}\|_{\mathcal{C}_{\mathcal{S},\tau_m}} \leq{}& \frac{1}{100}\|(|\mathbf{X}_{\rho,\xi}| + |\nabla\mathbf{X}_{\rho,\xi}| + |\mathbf{Y}_{\rho,\xi}|)G_1^{\frac{1}{2}}\|_{\mathcal{C}_{\mathcal{S},\tau_m}} \\ &+ N\|(|\mathbf{X}| + |\nabla\mathbf{X}|)G_1^{\frac{1}{2}}\|_{\mathcal{A}(\rho,2\rho)\times[0,\tau_m]} \\ &+ N\|(|\mathbf{X}| + |\nabla\mathbf{X}|)G_1^{\frac{1}{2}}\|_{\mathcal{A}(\xi,2\xi)\times[0,\tau_m]}\end{aligned}$$

for all $\alpha \geq \alpha_0 = \alpha_0(n, K)$ sufficiently large. So, returning to (4.187), and absorbing the terms involving $\mathbf{X}_{\rho,\xi}$, $\nabla\mathbf{X}_{\rho,\xi}$, and $\mathbf{Y}_{\rho,\xi}$ into the left-hand side of the inequality, we obtain that

$$\begin{aligned}&\|(|\mathbf{X}| + |\mathbf{Y}|)G_1^{\frac{1}{2}}\|_{\mathcal{A}(2\rho,\xi)\times[0,\tau_m]} \\ &\quad\leq \|(|\mathbf{X}_{\rho,\xi}| + |\nabla\mathbf{X}_{\rho,\xi}| + |\mathbf{Y}_{\rho,\xi}|)G_1^{\frac{1}{2}}\|_{\mathcal{C}_{\mathcal{S},\tau_m}} \\ &\quad\leq N\|(|\mathbf{X}| + |\nabla\mathbf{X}|)G_1^{\frac{1}{2}}\|_{\mathcal{A}(\rho,2\rho)\times[0,\tau_m]} \\ &\qquad + N\|(|\mathbf{X}| + |\nabla\mathbf{X}|)G_1^{\frac{1}{2}}\|_{\mathcal{A}(\xi,2\xi)\times[0,\tau_m]} \\ &\qquad + N\|\mathbf{X}e^{\frac{r^2}{2}}\|_{\mathcal{A}(\rho,2\rho)\times\{\tau_m\}} + N\|\mathbf{X}e^{\frac{r^2}{2}}\|_{\mathcal{A}(\xi,2\xi)\times\{\tau_m\}} \\ &\qquad + N\|\nabla\mathbf{X}e^{\frac{r^2}{2}}\|_{\mathcal{A}(\rho,2\xi)\times\{\tau_m\}}\end{aligned} \tag{4.188}$$

for all $\alpha \geq \alpha_0$. Given the bounds (4.184) and (4.185), all of the integrals in (4.188) tend to finite limits as $\xi \to \infty$, and the limits of the second and fourth terms on the right vanish. Thus, sending $\xi \to \infty$ and shrinking the domain on the left side, we obtain that

$$\begin{aligned}\|(|\mathbf{X}| + |\mathbf{Y}|)G_1^{\frac{1}{2}}\|_{\mathcal{C}_{4\rho,\frac{\tau_m}{2}}} &\leq N\|(|\mathbf{X}| + |\nabla\mathbf{X}|)G_1^{\frac{1}{2}}\|_{\mathcal{A}(\rho,2\rho)\times[0,\tau_m]} \\ &\quad + N\|(|\mathbf{X}| + |\nabla\mathbf{X}|e^{\frac{r^2}{2}}\|_{\mathcal{C}_\rho\times\{\tau_m\}} \\ &\leq N\|(|\mathbf{X}| + |\nabla\mathbf{X}|)G_1^{\frac{1}{2}}\|_{\mathcal{A}(\rho,2\rho)\times[0,\tau_m]} + N\cdot N'_m.\end{aligned} \tag{4.189}$$

Now, on $\mathcal{A}(\rho, 2\rho) \times [0, \tau_m]$, we have that

$$G_1(x,\tau) \leq \exp\left(4\rho^2(1 + \alpha\tau_m\rho^{-\delta})\right)$$

while, on $\mathcal{C}_{4\rho,\tau_m/2}$, that

$$G_1(x,\tau) \geq \exp\left(8\rho^2(1 + \alpha\tau_m\rho^{-\delta})\right).$$

So, (4.189) implies that

$$\|(|\mathbf{X}| + |\mathbf{Y}|)\|_{\mathcal{C}_{4\rho,\frac{\tau_m}{2}}} \leq N(1 + N'_m)e^{-4\rho^2\left(1+\alpha\tau_m\rho^{-\delta}\right)} \tag{4.190}$$

for all $\alpha > \alpha_0$, $\rho > \mathcal{S}$, and $m > 0$. Sending first $\alpha \to \infty$ and then $\rho \to \mathcal{S}$ in (4.190), we obtain that $\mathbf{X} \equiv 0$ and $\mathbf{Y} \equiv 0$ on $\mathcal{C}_{4\mathcal{S},\tau_m/2}$ for all $m > 0$. Sending at last $m \to \infty$, we conclude that $\mathbf{X} \equiv 0$ and $\mathbf{Y} \equiv 0$ on $\mathcal{C}_{4\mathcal{S},A/2}$. □

4.8.3. Comments on the proof.

The reader may have noticed that the proof of Theorem 4.25 makes little use of the specific properties of the background metric $g(\tau)$. In fact, neither the self-similarity of $g(\tau)$ for $\tau \in (0,1]$ nor even that $g(\tau)$ solves the backward Ricci flow are essential to the proof; the statement extends to sections $\mathbf{X}$ and $\mathbf{Y}$ of bundles over ends of manifolds with evolving metrics that are are asymptotically conical in a more general sense.

A generalization of Theorem 4.25 proven in [**Kot24b**] by one of the authors shows that it is enough to assume that the family of metrics $g(\tau)$ emanates smoothly from the cone at $\tau = 0$ and that the first few derivatives of both the speed $\partial_\tau g(\tau)$ the curvature tensor $\mathrm{Rm}(g(\tau))$ satisfy uniform *quadratic* decay bounds in r. These assumptions imply weaker versions of the bounds on the error quantities η_i in (4.73)–(4.76), but the proof given above can be adapted to accommodate them. Another variation on this theorem can be found in the work of Haskins, Khan, and Payne [**HKP25**] on asymptotically conical shrinking solitons to the Laplacian flow of closed G_2-structures.

At the same time, according to Theorem 4.59 below, the only solutions to the Ricci flow on $[0,T)$ which have quadratic curvature decay on an end and converge smoothly to a cone as $t \to T$ are those which arise from asymptotically conical shrinking solitons. Thus, there is no essential loss in generality in working with a self-similar background metric.

4.9. Geometric properties inherited from the asymptotic cone

The rigidity principle expressed in Theorems 4.12 and 4.17 effectively reduces the problem of classifying complete asymptotically conical shrinkers to the problem of classifying the possible asymptotic cones of such shrinkers, and therefore to the problem of classifying the manifolds (Σ, g_Σ) which may arise as the cross-sections of those cones.[9] There are few general restrictions on such manifolds currently known. We have already seen in Corollary 4.27 that the cross-sections must at least be real-analytic. Another, which follows from the work of Chen [**Che09**] and of the work of Lu, Yang, and one of the authors [**CLY11**], is that the scalar curvature of the cross-section must be strictly positive.

THEOREM 4.44 (Cross-sections have positive scalar curvature). *Let (Σ, g_Σ) be a closed $(n-1)$-manifold and suppose that (M^n, g, f) is a complete shrinker which is C^2-asymptotic to the cone $\mathcal{C}^\Sigma$ along some end $V \subset M^n$. Then the scalar curvature of g_Σ satisfies*

$$R_\Sigma \geq (n-1)(n-2) \tag{4.191}$$

and the inequality is strict unless (M^n, g, f) is the Gaussian soliton.

PROOF. Shrinking the end V if necessary, it follows from the main result of [**CLY11**] (cf. Theorem I-4.31) that the scalar curvature R of g satisfies $R \geq c_0/f$ on V for some $c_0 \geq 0$. Moreover, $c_0 > 0$ unless (M^n, g) is flat and (therefore) the Gaussian shrinker on $\mathbb{R}^n$. Appealing to Definition 4.5 (and suppressing the diffeomorphism to assume that g is defined on $V = \mathcal{C}_a = (a, \infty) \times \Sigma$), we see that

[9]This is of particular interest in dimension four; see Problem 4.1 at the end of this section.

the scalar curvature $\hat{R}$ of $\hat{g}$ satisfies

$$\hat{R}(r,\sigma) = \lim_{\lambda\to\infty} \lambda^2 R(\lambda r, \sigma) \geq \lim_{\lambda\to\infty} \lambda^2 \left(\frac{c_0 c_1}{(\lambda r)^2}\right) = \frac{c_0 c_1}{r^2}$$

where we have used (1.59) to estimate $f^{-1} \geq c_1 r^{-2}$ on $\mathcal{C}_a$ for some $c_1 = c_1(n,a) > 0$. By standard expressions for warped product metrics (see, e.g., §J of Chapter 9 of Besse [**Bes08**]), the scalar curvature $\hat{R}$ of the cone is related to the scalar curvature R_Σ of g_Σ by

$$\hat{R}(r,\sigma) = \frac{1}{r^2}\left(R_\Sigma(\sigma) - (n-1)(n-2)\right).$$

Hence, $R_\Sigma \geq c_0 c_1 + (n-1)(n-2)$ and the result follows. □

4.9.1. Inheritance of structure from propagation of structure.

In order to understand what restrictions there are on the possible asymptotic cones of asymptotically conical shrinking solitons, one needs to understand the extent to which the geometric properties of the cone are realized in the geometry of the body of the soliton. Thus, a natural first question is to ask *what geometric features does a shrinker inherit from its asymptotic cone?*

We should clarify that here we are interested in the inheritance of *rigid* geometric features. For example, when one assumes that the asymptotic cone of some shrinker is rotationally symmetric, one already knows that the shrinker is *approximately* rotationally symmetric at infinity in some sense, just from the definition of what it means to be asymptotically conical. What we would like to know is when the soliton must be *exactly* rotationally symmetric under such circumstances.

It is useful to reformulate this question from a dynamical perspective. If g is a shrinker on $\mathcal{C}_a$ which is dynamically asymptotic to the cone $\hat{g}$ and $g(\tau)$ is the smooth backward Ricci flow on $\mathcal{C}_{a,1} = \mathcal{C}_a \times [0,1]$ which interpolates between the cone $g(0) = \hat{g}$ and the shrinker $g(1) = g$, then to ask which geometric features are shared by cone and shrinker is to ask which features of of the $\tau = 0$ time-slice are necessarily present in the $\tau = 1$ time-slice. This leads to a kind of heuristic principle:

> *The properties of the cone which are inherited by the shrinker should correspond to those which propagate backward in time under the Ricci flow.*

Of course, whether or not some geometric feature of a solution is preserved under the flow is a question of uniqueness in some guise and as such is crucially dependent on the domain of the flow and the nature of the solution near the boundary of the domain. According to [**Kot10**] and [**CK20, Kot14**] by Cook and one of the authors, both symmetry and restricted holonomy will propagate backward in time along a solution $(M, g(t))$ to the Ricci flow if it is complete and of bounded curvature. Using the general backward uniqueness principle in Theorem 4.25, we will see below that the same is true for asymptotically conical Ricci flows on the domain $\mathcal{C}_a$.

We note that the reverse question of what structure the *cone inherits from the shrinker* is almost trivial from a dynamical point of view: any rigid geometric property possessed by $g = g(1)$ will be possessed by $g(\tau) = \tau\Phi_\tau^* g$ for $\tau \in (0,1]$ by self-similarity, and will be therefore be transferred to the cone $\hat{g} = g(0)$ by continuity.

4.9.2. Inheritance of symmetry.

Let us first consider the case of symmetry, beginning with two simple observations about cones:

(1) Any isometry of a metric cone over a smooth manifold (Σ, g_Σ) must preserve the vertex of the cone unless the cone is flat.

(2) The group of vertex-preserving isometries of the cone is isomorphic to the group of isometries of the cross-section $(\Sigma, g_{\Sigma}))$.

Using the version of the rigidity principle in Theorem 4.17 and the fact that the shrinking soliton metric and the conical metric are real-analytic with respect to some common atlas (see Theorem 4.26), we obtain the following precise correspondence between the isometry group of the cross-section and the symmetries of the shrinker near infinity.

THEOREM 4.45 (Inheritance of symmetry). *Let (M^n, g, f) be a shrinking gradient Ricci soliton which is C^2-asymptotic to $\mathcal{C}^\Sigma$ along the end $V \subset M^n$. Then*

$$\mathrm{Isom}(\Sigma, g_\Sigma) \cong \mathrm{Isom}(W, g|_W)$$

for some $a \geq 0$ and end $W \subset V$. If, in addition, (M^n, g) is complete and the map $\pi_1(V, x_0) \to \pi_1(M^n, x_0)$ induced by inclusion is surjective for some $x_0 \in V$, then $\mathrm{Isom}(\Sigma, g_\Sigma)$ *embeds into* $\mathrm{Isom}(M^n, g)$.

We first need a lemma which says that isometries of the end of a nontrivial asymptotically conical shrinker must preserve the gradient of the potential f.

LEMMA 4.46. *Let $g(\tau)$ be the solution to (4.7) on $\mathcal{C}_a \times [0,1]$ associated to the soliton $(M^n, f(1), g(1))$ which is dynamically asymptotic to $\hat{g} = g(0)$ on $\mathcal{C}_a$. Suppose that $\phi : \mathcal{C}_a \to \mathcal{C}_a$ is a local isometry of $g = g(\tau)$ for some $\tau \in (0,1]$. Then either $\phi^*(\nabla f) = \nabla f$ or g is flat.*

PROOF. Write $f = f(\tau)$ and $h = \phi^* f - f$. Since $g = g(\tau)$ satisfies the τ-scaled soliton equation (4.37) with either $\phi^* f$ or f as the potential, it follows that $\nabla^2 h = 0$ on $\mathcal{C}_a$. Thus ∇h is parallel and $|\nabla h| = c_0$ for some constant $c_0 \geq 0$. If $c_0 = 0$, then $\phi^* \nabla f = \nabla f$. So suppose that $c_0 > 0$. We will show that $\mathrm{Rm} \equiv 0$.

First, covariantly differentiating $\nabla^2 h = 0$ and commuting derivatives yields that $\mathrm{Rm}(\nabla h, \cdot, \cdot, \cdot) = 0$ and $\nabla \mathrm{Rm}(\nabla h, \cdot, \cdot, \cdot, \cdot) = 0$, and hence also that $\nabla_{\nabla h} \mathrm{Rm} = 0$. In particular, $|\mathrm{Rm}|^2$ is constant on the integral curves of ∇h. Now, using the quadratic growth of f, we can find $a_0 > 2a$ such that $\inf_{\mathcal{C}_{a_0}} f > \sup_{\partial \mathcal{C}_{2a}} f$. Fix $p \in \mathcal{C}_{a_0}$ and let $c_1 = \langle \nabla f, \nabla h \rangle(p)$ and $c_2 = f(p)$. If $c_1 \geq 0$, define $X = \nabla h$. Otherwise, define $X = -\nabla h$. Let σ denote the maximally defined integral curve of X in $\mathcal{C}_a$ with $\sigma(0) = p$. Using (4.37) we have

$$\begin{aligned} \frac{d^2}{ds^2} f(\sigma(s)) &= \left(\nabla^2 f(\nabla h, \nabla h) + \nabla^2 h(\nabla f, \nabla h)\right)(\sigma(s)) \\ &= -\mathrm{Ric}(\nabla h, \nabla h)(\sigma(s)) + \frac{|\nabla h|^2(\sigma(s))}{2\tau} = \frac{c_0^2}{2\tau}, \end{aligned}$$

and therefore $f(\sigma(s)) = c_0^2 s^2/(4\tau) + |c_1| s + c_2$, for all s such that $\sigma(s)$ is defined and belongs to $\mathcal{C}_{a_0}$. In particular, $f(\sigma(s)) \geq c_2 > \sup_{\partial \mathcal{C}_{2a}} f$ for $s \geq 0$, so σ is defined and remains in $\mathcal{C}_{2a}$ for $s \in [0, \infty)$. Since $f(\sigma(s)) \to \infty$ as $s \to \infty$, we must have $r(\sigma(s)) \to \infty$ as $s \to \infty$ as well. Consequently, $|\mathrm{Rm}|^2(p) = \lim_{s\to\infty} |\mathrm{Rm}|^2(\sigma(s)) = 0$, since, on one hand, $|\mathrm{Rm}|^2$ is constant along σ and, on the other, $|\mathrm{Rm}| \leq C/r^2$ on $\mathcal{C}_a$. This implies that $\mathrm{Rm} \equiv 0$ on $\mathcal{C}_{a_0}$, and, by analyticity, that $\mathrm{Rm} \equiv 0$ on $\mathcal{C}_a$. $\square$

PROOF OF THEOREM 4.45. The first statement follows almost immediately from the version of the rigidity principle in Theorem 4.17. Indeed, by the Normalization Lemma 4.8, there are $W \subset V$ and $F : \mathcal{C}_a \to W$ such that the shrinking soliton $(\mathcal{C}_a, \tilde{g}, \tilde{f})$ is dynamically asymptotic to $\hat{g}$ on $\mathcal{C}_a$, where $\tilde{g} := F^*g$ and $\tilde{f} := F^*f$. Let $\gamma \in \mathrm{Isom}(\Sigma, g_\Sigma)$. Then $\Gamma = \mathrm{Id} \times \gamma$ is an isometry of $(\mathcal{C}_a, \hat{g})$. But, since Γ leaves r and $\frac{\partial}{\partial r}$ invariant, it follows that $(\mathcal{C}_a, \Gamma^*\tilde{g}, \Gamma^*\tilde{f})$ is also dynamically asymptotic to $\hat{g} = \Gamma^*\hat{g}$ on $\mathcal{C}_a$. By Theorem 4.17 we then have

$$(F \circ \Gamma)^* g = \Gamma^* \tilde{g} = \tilde{g} = F^* g$$

on $\mathcal{C}_a$. Thus, $\gamma \mapsto F \circ (\mathrm{Id} \times \gamma) \circ F^{-1}$ is well-defined and an injective homomorphism of $\mathrm{Isom}(\Sigma, g_\Sigma)$ into $\mathrm{Isom}(W, g)$.

Conversely, suppose φ is an isometry of (W, g). Then $\Gamma := F^{-1} \circ \varphi \circ F$ is an isometry of $(\mathcal{C}_a, \tilde{g})$; we claim $\Gamma = \mathrm{Id} \times \gamma$ for some $\gamma \in \mathrm{Isom}(\Sigma, g_\Sigma)$. For this, note first that if $\tilde{g}$ is flat, then so too is $\hat{g}$ and the associated solution $\tilde{g}(\tau)$ to the backward Ricci flow which interpolates between $\tilde{g} = \tilde{g}(1)$ and $\hat{g} = \tilde{g}(0)$ is static. Then Γ belongs to $\mathrm{Isom}(\mathcal{C}_a, \hat{g})$ and is therefore of the form $\Gamma = \mathrm{Id} \times \gamma$ for some $\gamma \in \mathrm{Isom}(\Sigma, g_\Sigma)$.

So let us assume that $\tilde{g}$ is not flat, in which case we have $\Gamma^* \tilde{\nabla} \tilde{f} = \tilde{\nabla} \tilde{f}$ by Lemma 4.46. The family of diffeomorphisms $\tilde{\Phi}_\tau$ generated by $-\tau^{-1} \tilde{\nabla} \tilde{f}$ will therefore be invariant under Γ, and by self-similarity we will have that $\Gamma^* \tilde{g}(\tau) = \tilde{g}(\tau)$ on $\mathcal{C}_a$ for all $\tau \in (0, 1]$. But then

$$\Gamma^* \hat{g} = \Gamma^* (\lim_{\tau \to 0} \tilde{g}(\tau)) = \lim_{\tau \to 0} \tilde{g}(\tau) = \hat{g}$$

on $\mathcal{C}_a$ by continuity, so $\Gamma \in \mathrm{Isom}(\mathcal{C}_a, \hat{g})$ and therefore $\Gamma = \mathrm{Id} \times \gamma$ for some $\gamma \in \mathrm{Isom}(\Sigma, g_\Sigma)$. It follows that the correspondence $\gamma \mapsto F \circ (\mathrm{Id} \times \gamma) \circ F^{-1}$ is an isomorphism from $\mathrm{Isom}(\Sigma, g_\Sigma)$ to $\mathrm{Isom}(W, g|_W)$.

To prove the second statement of the theorem, we just need to show that if the fundamental group of the end surjects on to the fundamental group of the manifold, then any isometry of the end extends to an isometry of the manifold. This is the content of Lemma 4.47 below. □

The following (see [**KW24**]) is just a slight generalization of the usual analytic continuation argument for the extension of isometries, which has its origins in the classical work of Myers [**Mye35**].

LEMMA 4.47 (Isometry Extension Lemma). *Suppose that (M^n, g) and (N^n, h) are complete connected real-analytic Riemannian manifolds and $\varphi : U \to V$ is an isometry between connected open sets $U \subset M^n$ and $V \subset N^n$. Let $x_0 \in U$.*

(1) *If the homomorphism $\iota_* : \pi_1(U, x_0) \to \pi_1(M^n, x_0)$ induced by the inclusion $\iota : U \hookrightarrow M^n$ is surjective, then φ extends to a surjective local isometry $\Phi : M^n \to N^n$.*
(2) *If, in addition, $\pi_1(V, \varphi(x_0))$ surjects onto $\pi_1(N^n, \varphi(x_0))$, then φ extends to an isometry $\Phi : (M^n, g) \to (N^n, h)$.*

PROOF. Let $\varphi : U \to V$ be an isometry and $x_0 \in U$. Assume first that the homomorphism $\iota_* : \pi_1(U, x_0) \to \pi_1(M^n, x_0)$ induced by inclusion $\iota : U \hookrightarrow M^n$ is surjective. Let $y_0 = \varphi(x_0)$ and write $\tilde{g} = p_M^* g$ and $\tilde{h} = p_N^* h$, where $p_M : \tilde{M}^n \to M^n$ and $p_N : \tilde{N}^n \to N^n$ are the universal covering maps.

Let U_0 and V_0 be evenly-covered neighborhoods of x_0 and y_0 and fix arbitrary points $\tilde{x}_0 \in p_M^{-1}(x_0)$ and $\tilde{y}_0 \in p_N^{-1}(y_0)$. Let $\tilde{U}_0$ and $\tilde{V}_0$ be connected neighborhoods of $\tilde{x}_0$ and $\tilde{y}_0$ mapped diffeomorphically onto U_0 and V_0 by p_M and p_N and denote by $\sigma_0 : V_0 \to \tilde{V}_0$ the inverse of $p_N|_{\tilde{V}_0}$. Then $\tilde{\varphi} = \sigma_0 \circ \varphi \circ p_M|_{\tilde{U}_0}$ is an isometry between $(\tilde{U}_0, \tilde{g}|_{\tilde{U}_0})$ and $(\tilde{V}_0, \tilde{h}|_{\tilde{V}_0})$. Since $(\tilde{M}^n, \tilde{g})$ and $(\tilde{N}^n, \tilde{h})$ are also real-analytic, it follows from Myers's theorem [**Mye35**] that $\tilde{\varphi}$ extends to an isometry $\tilde{\Phi} : (\tilde{M}^n, \tilde{g}) \to (\tilde{N}^n, \tilde{h})$.

We claim that $\tilde{U} := p_M^{-1}(U)$ is connected. For this, it is enough to verify that any two points in the fiber $p_M^{-1}(x_0)$ are connected by a path contained in $\tilde{U}$. So suppose $\tilde{x}_1 \in p_M^{-1}(x_0)$, and let $\tilde{\sigma}$ be any path in $\tilde{M}^n$ connecting $\tilde{x}_0$ and $\tilde{x}_1$. Then $\sigma = p_M \circ \tilde{\sigma}$ is a loop at x_0. By assumption, there is a loop η in U based at x_0 such that $\iota_*[\eta] = [\sigma]$. If $\tilde{\eta}$ is the lift of η to $\tilde{M}^n$ starting at $\tilde{x}_0$, then $\tilde{\eta}$ is contained in $\tilde{U} = p_M^{-1}(U)$ and must terminate at $\tilde{x}_1$ since $[\sigma] = [\eta]$.

Now we show that

$$p_N \circ \tilde{\Phi}|_{\tilde{U}} = \varphi \circ p_M|_{\tilde{U}}. \tag{4.192}$$

Fix any $\tilde{x} \in \tilde{U}$ and let $x = p_M(\tilde{x})$. Since $\tilde{U}$ is connected, we can find a continuous path $\tilde{\gamma} : [0,1] \to \tilde{U}$ with $\tilde{\gamma}(0) = \tilde{x}_0$ and $\tilde{\gamma}(1) = \tilde{x}$. Then $\gamma = p_M \circ \tilde{\gamma}$ is a continuous path in U with $\gamma(0) = x_0$ and $\gamma(1) = x$. For each $t \in [0,1]$, choose an evenly-covered neighborhood $U_t \subset U$ of $x_t = \gamma(t)$ and let $\tau_t : U_t \to \tilde{U}_t$ be a local section of p_M which maps U_t diffeomorphically onto a connected neighborhood $\tilde{U}_t \subset \tilde{U}$ of $\tilde{x}_t = \tilde{\gamma}(t)$. (For $t = 0$ we take U_0 to be the neighborhood defined previously with the same notation.) Note that the sets $V_t = p_N(\tilde{\Phi}(\tilde{U}_t))$ are open and connected as $\tilde{\Phi}$ and p_N are continuous open maps.

Now we have two continuations of the isometry $\varphi|_{U_0} : U_0 \to V_0$ along γ: one of them being the collection $\{\psi_t\}_{t\in[0,1]}$ defined by the isometries $\psi_t = p_N \circ \tilde{\Phi} \circ \tau_t : U_t \to V_t$, and the other being the collection $\{\varphi_t\}_{t\in[0,1]}$ defined by the restrictions $\varphi_t = \varphi|_{U_t} : U_t \to \varphi(U_t)$ of φ to each U_t. These continuations must coincide near $x = \gamma(1)$. In particular,

$$(\varphi \circ p_M)(\tilde{x}) = \varphi(x) = \varphi_1(x) = \psi_1(x) = p_N(\tilde{\Phi}(\tau_1(x))) = (p_N \circ \tilde{\Phi})(\tilde{x}),$$

so that the identity (4.192) holds.

To complete the proof of the first claim, suppose $\alpha \in \mathrm{Aut}_{p_M}(\tilde{M}^n)$ is any covering automorphism of $\tilde{M}^n$. By (4.192), we have that

$$p_N \circ \tilde{\Phi} \circ \alpha|_{\tilde{U}} = \varphi \circ p_M \circ \alpha|_{\tilde{U}} = \varphi \circ p_M|_{\tilde{U}} = p_N \circ \tilde{\Phi}|_{\tilde{U}};$$

that is, $p_N \circ \tilde{\Phi} \circ \alpha$ and $p_N \circ \tilde{\Phi}$ agree on $\tilde{U}$. But both maps are local isometries of $(\tilde{M}^n, \tilde{g})$ into (N^n, h), so they must be equal everywhere. Since the element $\alpha \in \mathrm{Aut}_{p_M}(\tilde{M}^n)$ was arbitrary, the map $p_N \circ \tilde{\Phi}$ descends to a local isometry $\Phi : (M^n, g) \to (N^n, h)$. We must also have $\Phi|_U = \varphi$ since U is connected and Φ and φ agree on $U_0 \subset U$ by construction. Finally, since Φ is a local isometry and (M^n, g) is complete, $\Phi(M^n)$ is open and closed in N^n. Therefore Φ is also surjective, completing the proof of the first claim.

For the second claim, assume in addition that the homomorphism induced by the inclusion $V \hookrightarrow N^n$ maps $\pi_1(V, \varphi(x_0))$ onto $\pi_1(N^n, \varphi(x_0))$. Then we may apply the first claim twice to extend the maps φ and φ^{-1} to surjective local isometries $\Phi : (M^n, g) \to (N^n, h)$ and $\Psi : (N^n, h) \to (M^n, g)$. Since $\Psi \circ \Phi$ and $\Phi \circ \Psi$ are local isometries and agree with the identity maps on U and V, they must agree with the identity maps on all of M^n and N^n, respectively. In particular, Φ is a diffeomorphism and hence an isometry of (M^n, g) onto (N^n, h). $\square$

As an application, we observe that Theorem 4.45 implies (with the help of other classification results) that there are no complete rotationally symmetric asymptotically conical shrinking solitons other than the Gaussian soliton.

COROLLARY 4.48 (Gaussian shrinker characterization). *A complete connected shrinking gradient Ricci soliton* (M^n, g, f) *is* C^2*-asymptotic to a rotationally symmetric cone* $((0,\infty)\times\mathbb{S}^{n-1}, dr^2 + \alpha r^2 g_{\mathbb{S}^{n-1}})$ *along some end* $V \subset M^n$ *if and only if* $M^n = \mathbb{R}^n$ *and* g *is flat.*

PROOF. In view of the classification of complete gradient shrinking solitons in dimensions two and three (see, e.g., Theorem 3.12 and Corollary 5.8 of [**Cho23**]), it suffices to consider the case that $n \geq 4$.[10] According to Theorem 4.45, if (M^n, g, f) is C^2-asymptotic to a rotationally symmetric cone along an end $V \subset M^n$, it is rotationally symmetric itself on some end $W \subset V$. In particular, g is locally conformally flat on W. As $n \geq 4$, the Weyl tensor vanishes on W and therefore vanishes everywhere by real-analyticity. But then it follows from the classification in [**Zha09a**] (see also [**NW08**] and [**CC13**]) that g itself must be flat. □

Nevertheless, Wang and one of the authors have shown [**KW15**] that for each $\alpha > 0$, there is a (unique, necessarily incomplete) rotationally symmetric shrinking soliton asymptotic to the cone $\hat{g}_\alpha = dr^2 + \alpha r^2 g_{\mathbb{S}^{n-1}}$ on $\mathcal{C}_a^{\mathbb{S}^{n-1}} = (a,\infty)\times\mathbb{S}^{n-1}$ for some $a = a(\alpha)$.

4.9.3. Inheritance of Kählerity.

Next, let us consider the case of holonomy. From Hamilton [**Ham86**] and the work of Cook and one of the authors [**CK20, Kot14**], it is known that if $(M^n, g(t))$ is a complete solution to the Ricci flow for $t \in [0,T]$ of uniformly bounded curvature, then the holonomy group of $g(\tau)$ is preserved up to isomorphism: $\mathrm{Hol}(g(0)) \cong \mathrm{Hol}(g(T))$. In particular, if $(M^n, g(T))$ is Kähler, then so is $(M^n, g(0))$; that is, Kählerity propagates backward in time along the Ricci flow. By the guiding heuristic of this section, we should expect, then, that if the asymptotic cone of an asymptotically conical shrinker is Kähler, the shrinker must itself be Kähler.

However (for reasons we will discuss in §4.9.3.1 below), the backward uniqueness principle for *solutions* to the Ricci flow in Theorem 4.19 cannot in general be directly applied to the question of propagation of Kählerity. Instead, the argument in [**Kot14**] makes use of an encoding of the question of into one of backward uniqueness an extended PDE-ODE system which measures the potential failure of the holonomy of the solution to be preserved. Using this encoding in conjunction with Theorem 4.25, we can obtain the equivalent statement for the Ricci flow on an asymptotically conical end. This has the following consequence for asymptotically conical shrinking solitons [**Kot18**].

THEOREM 4.49. *Assume that* $(\mathcal{C}_0, \hat{g})$ *is Kähler with complex structure* $\hat{J}$ *and suppose that the shrinking gradient Ricci soliton* (M^n, g, f) *is* C^2*-asymptotic to* $\mathcal{C}^\Sigma$ *along* $V \subset M^n$*. Then there is an end* $W \subset V$ *and a complex structure* J *defined on* W *relative to which* $g|_W$ *is Kähler and* (W, J) *is biholomorphic to* $(\mathcal{C}_a, \hat{J})$ *for some* $a > 0$*. If* (M^n, g) *is complete, the Kähler structure on* W *extends to all of* M^n*.*

[10]In fact, the classification shows that the Gaussian soliton is the *only* complete asymptotically conical shrinking soliton in those dimensions.

Combined with Corollary 4.16, this result implies that there is at most one shrinking soliton asymptotic to a Kähler cone. Below, we will give the idea of how Theorem 4.49 is proven via Theorem 4.25. The key is to understand first how one might measure the failure of a solution to the Ricci flow to remain Kähler.

4.9.3.1. *Backward propagation of Kählerity.*

Let us assume that we are given a Kähler structure (M^n, g_0, J_0) on a (possibly incomplete) manifold (M^n, g_0) and a smooth family of metrics $g = g(\tau)$ satisfying the backward Ricci flow (4.7) on $M^n \times [0, T]$ with $g(0) = g_0$. We do not assume here that $g(\tau)$ is self-similar. We want to understand under what conditions the solution $g(\tau)$ will be Kähler for $\tau > 0$.

Note that the problem of the propagation of the Kähler structure along (4.7) cannot be simply reduced to the problem of the backward uniqueness of *solutions* to (4.7). All of the standard methods available to construct a Kähler solution to the (forward) Ricci flow from a given Kähler initial metric involve the solution of some auxiliary parabolic equation. But the corresponding *terminal-value* problems for such equations are all ill-posed, and so there is no way in general to construct from a given Kähler metric g_0 a Kähler solution $\tilde{g}(\tau)$ to the backward Ricci flow for $\tau \in [0, \varepsilon]$ that we might use to compare against the given solution $g(\tau)$. Nevertheless, the propagation of Kählerity can be recast as a problem of backward uniqueness for an appropriate PDE-ODE system involving the curvature operator.

The first step is to extend J_0 to a family of endomorphisms $J = J(\tau) : TM^n \to TM^n$ on $[0, T]$ such that the algebraic relationship between $J(\tau)$ and $g(\tau)$ is preserved. Later, we will prove that $J(\tau) \equiv J_0$ for all τ – that is, that the complex structure is independent of time – but it is convenient to allow it to evolve a priori nontrivially along with the metric. We define $J(\tau)$ as the solution to the linear system

$$\begin{cases} \frac{\partial}{\partial \tau} J_b^a = R_b^c J_c^a - R_c^a J_b^c & \text{on } M^n \times (0, T], \\ J(x, 0) = J_0(x) & \text{on } M^n \end{cases} \tag{4.193}$$

of ODE in each fiber $\mathrm{End}_x(TM^n)$ of $\mathrm{End}(TM^n)$.

If we regard the family $J = J(\tau)$ as a section of the pulled-back bundle $\pi^*(\mathrm{End}(TM^n))$ over $M^n \times [0, T]$ (the map $\pi : M^n \times [0, T] \to M^n$ being the projection as usual), then (4.193) may be restated concisely as

$$D_\tau J = 0, \quad J(0) = J_0, \tag{4.194}$$

where D_τ is as defined in §4.3.4.2 above. With this definition, we will have

$$J^2 = -\mathrm{Id} \quad \text{and} \quad g(\cdot, \cdot) = g(J\cdot, J\cdot)$$

on $M^n \times [0, T]$, that is, $J = J(\tau)$ will remain an almost complex structure and $g = g(\tau)$ will remain Hermitian with respect to J along the flow.[11]

4.9.3.2. *A time-dependent splitting of* $\wedge^2(T^*M^n)$.

Now, using $J = J(\tau)$, we define several associated families of endomorphisms of $\wedge^2(T^*M^n)$. Let us write

$$\theta \wedge \sigma = \theta \otimes \sigma - \sigma \otimes \theta$$

[11]This is again, effectively, an implementation of *Uhlenbeck's trick* (see, e.g., §6.2 of [**CK04**]). For an implementation of this framework which is closer to the original version of the trick as described in [**Ham86**], see [**CK20**] by Cook and one of the authors.

for θ, $\sigma \in T^*M^n$, and use the metric

$$\langle \theta \wedge \sigma, \phi \wedge \psi \rangle = \det \begin{pmatrix} \langle \theta, \phi \rangle & \langle \theta, \psi \rangle \\ \langle \sigma, \phi \rangle & \langle \sigma, \psi \rangle \end{pmatrix} \tag{4.195}$$

induced on $\wedge^2(T^*M^n)$ by $g(\tau)$. Note that $\langle \cdot, \cdot \rangle$ is 1/2 times the usual metric on $T^2(T^*M)$.

Define $\tilde{\mathrm{P}}$, $\hat{\mathrm{P}} \in \mathrm{End}(TM^n)$ by

$$\tilde{\mathrm{P}} := \frac{1}{2}(\mathrm{Id} + \mathrm{G}), \quad \hat{\mathrm{P}} := \frac{1}{2}(\mathrm{Id} - \mathrm{G}),$$

where $\mathrm{G} \in \mathrm{End}(\wedge^2(T^*M^n))$ is given by

$$(\mathrm{G}\eta)(X, Y) := \eta(JX, JY)$$

for $\eta \in \wedge^2(T^*M^n)$ and X, Y in TM^n. Then, $\tilde{\mathrm{P}}$ and $\hat{\mathrm{P}}$ are complementary orthogonal projections and we have the orthogonal direct sum decomposition $\wedge^2(T^*M^n) = \mathcal{H} \oplus \mathcal{K}$, where

$$\mathcal{H} := \mathcal{H}(\tau) = \mathrm{im}\, \tilde{\mathrm{P}}(\tau), \quad \mathcal{K} := \mathcal{K}(\tau) = \mathrm{im}\, \hat{\mathrm{P}}(\tau).$$

Moreover, $\mathcal{H}$ is closed under the bracket

$$[\omega, \eta]_{ij} := \omega_{ik}\eta_{kj} - \omega_{jk}\eta_{ki}$$

induced by $g(\tau)$.

REMARK 4.50. Though we will not use this framing for our discussion, we note that with respect to the complexification

$$\wedge^2_{\mathbb{C}}(T^*M^n) := \wedge^2(T^*M^n) \otimes_{\mathbb{R}} \mathbb{C}$$

of the bundle of 2-forms, we have

$$\mathrm{P}^{(2,0)} = \frac{1}{2}\left(\hat{\mathrm{P}} - \sqrt{-1}\mathrm{F}\right), \quad \mathrm{P}^{(0,2)} = \frac{1}{2}\left(\hat{\mathrm{P}} + \sqrt{-1}\mathrm{F}\right), \quad \mathrm{P}^{(1,1)} = \tilde{\mathrm{P}},$$

where $\mathrm{F} \in \mathrm{End}(\wedge^2(T^*M^n))$ is given by

$$(\mathrm{F}\eta)(X, Y) = \frac{1}{2}\left(\eta(X, JY) + \eta(JX, Y)\right) \tag{4.196}$$

and $\mathrm{P}^{(2,0)}$, $\mathrm{P}^{(0,2)}$, and $\mathrm{P}^{(1,1)}$ are the projections of $\wedge^2_{\mathbb{C}}(T^*M^n)$ onto $\wedge^{2,0}(T^*M^n)$, $\wedge^{0,2}(T^*M^n)$, and $\wedge^{1,1}(T^*M^n)$, respectively.

4.9.3.3. *The decomposition of the curvature operator.*

If $g(\tau)$ is Kähler relative to $J = J(\tau)$, then the Kähler identities imply that the curvature operator has a nontrivial kernel. In fact, it will annihilate the subspace $\mathcal{K} = \mathcal{K}(\tau)$. To see this, let us use

$$\mathrm{R} : \wedge^2(T^*M^n) \to \wedge^2(T^*M^n)$$

to denote the curvature operator of $g(\tau)$. Since our convention for the $(4,0)$-curvature tensor is such that $R(e_i, e_j, e_j, e_i) > 0$ on the sphere, we will define

$$\mathrm{R}(\eta)_{ij} := -R_{ijab}\eta_{ab}, \quad \eta \in \wedge^2(T^*M^n). \tag{4.197}$$

Note that, relative to an orthonormal frame $\{e_i\}_{i=1}^n$, we have

$$\mathrm{R}_{ijk\ell} = \mathrm{R}_{k\ell ij} = \left\langle \mathrm{R}(e_i^* \wedge e_j^*), e_k^* \wedge e_\ell^* \right\rangle = -2R_{abij} \left\langle e_a^* \wedge e_b^*, e_k^* \wedge e_\ell^* \right\rangle = -2R_{ijk\ell}.$$

Similarly, we will use S for the operator $\nabla\mathrm{R} \in T^*M \otimes \mathrm{End}(\wedge^2(T^*M^n))$ defined by

$$\mathrm{S}(X, \eta)_{ij} := -2\nabla_a R_{ijbc} X_a \eta_{bc}$$

and write

$$\mathrm{S}_{mijk\ell} = \langle \mathrm{S}(e_m, e_i^* \wedge e_j^*), e_k^* \wedge e_\ell^* \rangle = -2\nabla_m R_{ijk\ell}.$$

Now define $\tilde{\mathrm{R}}$, $\hat{\mathrm{R}} : \wedge^2(T^*M^n) \to \wedge^2(T^*M^n)$ by

$$\tilde{\mathrm{R}} := \mathrm{R} \circ \tilde{\mathrm{P}}, \quad \hat{\mathrm{R}} := \mathrm{R} \circ \hat{\mathrm{P}},$$

and $\tilde{\mathrm{S}}$, $\hat{\mathrm{S}} \in T^*M^n \otimes \mathrm{End}(\wedge^2(T^*M^n))$ by

$$\tilde{\mathrm{S}} := \nabla\mathrm{R} \circ \tilde{\mathrm{P}}, \quad \hat{\mathrm{S}} := \nabla\mathrm{R} \circ \hat{\mathrm{P}},$$

that is,

$$\tilde{\mathrm{S}}(X, \eta) = (\nabla_X \mathrm{R}) \circ \tilde{\mathrm{P}}(\eta), \quad \hat{\mathrm{S}}(X, \eta) = (\nabla_X \mathrm{R}) \circ \hat{\mathrm{P}}(\eta),$$

for $X \in TM^n$ and $\eta \in \wedge^2(T^*M^n)$.

The following identities can be obtained by direct computation; see [**Kot18**].

LEMMA 4.51. *At $\tau = 0$, we have*

$$\nabla\tilde{\mathrm{P}} = \nabla\hat{\mathrm{P}} = 0, \quad \hat{\mathrm{R}} = 0, \quad \tilde{\mathrm{R}} = \mathrm{R}, \quad \hat{\mathrm{S}} = 0, \quad \textit{and} \quad \tilde{\mathrm{S}} = \nabla\mathrm{R}.$$

These relations can be understood in terms of the holonomy of the solution. Indeed, the holonomy Lie algebra of $g(0)$ defines a parallel subalgebra $\mathfrak{hol}(g(0))$ of $\wedge^2(T^*M^n)$ and

$$\operatorname{im} \mathrm{R}(p, 0) \subset \mathfrak{hol}_p(g(0)) \subset \mathcal{H}_p(0), \quad \ker \mathrm{R}(p, 0) \subset \mathcal{H}_p^\perp(0) = \mathcal{K}_p(0).$$

at all $p \in M^n$. The last two identities then follow from the previous ones.

4.9.3.4. *The PDE-ODE system.*

The tensors $\mathrm{A} = \nabla\hat{\mathrm{P}}$ and $\hat{\mathrm{R}} = \mathrm{R} \circ \hat{\mathrm{P}}$ measure the failure of the metric g to remain Kähler for $\tau > 0$. According to the computations in [**Kot14**], if we include the tensors $\mathrm{P} = \nabla^2\hat{\mathrm{P}}$ and $\hat{\mathrm{S}}$, the aggregate system will satisfy a closed PDE-ODE system of differential inequalities:

PROPOSITION 4.52. *The sections* A, B, $\hat{\mathrm{R}}$, $\hat{\mathrm{S}}$ *satisfy*

$$|D_\tau \mathrm{A}| \leq C|\mathrm{R}||\mathrm{A}| + C|\hat{\mathrm{S}}|, \tag{4.198}$$

$$|D_\tau \mathrm{B}| \leq C|\nabla\mathrm{R}||\mathrm{A}| + C|\mathrm{R}| \left(|\mathrm{B}| + |\nabla\hat{\mathrm{S}}|\right), \tag{4.199}$$

$$\left|(D_\tau + \Delta)\,\hat{\mathrm{R}}\right| \leq C|\nabla\mathrm{R}||\mathrm{A}| + C|\mathrm{R}| \left(|\mathrm{B}| + |\hat{\mathrm{R}}|\right), \tag{4.200}$$

$$\left|(D_\tau + \Delta)\,\hat{\mathrm{S}}\right| \leq C|\nabla^2\mathrm{R}||\mathrm{A}| + C|\nabla\mathrm{R}||\hat{\mathrm{R}}| + C|\mathrm{R}| \left(|\mathrm{B}| + |\hat{\mathrm{R}}| + |\hat{\mathrm{S}}|\right) \tag{4.201}$$

for some $C = C(n)$ on $M^n \times [0, T]$.

The computations require a more careful analysis of the reaction terms in the evolution equations for R and ∇R than what was needed to obtain Proposition 4.24. The key feature is that each term on the right is proportional to at least one of the norms of the principal quantities A, B, $\hat{\mathrm{R}}$, $\hat{\mathrm{S}}$, or the derivatives of the latter two. We refer the reader to [**Kot14, Kot18**] for the details.

Let us here at least observe that if the elements of the above system vanish identically along the flow, then $(M^n, g(\tau))$ is Kähler relative to the *fixed* complex structure J_0.

LEMMA 4.53. *Let g, J, $\hat{\mathrm{R}}$, and $\hat{\mathrm{P}}$ be as above. If $\hat{\mathrm{R}} \equiv 0$ and $\nabla\hat{\mathrm{P}} \equiv 0$ on $M^n \times [0, T]$, and $\nabla J \equiv 0$ on $M^n \times \{0\}$, then $\nabla J \equiv 0$ and $\frac{\partial}{\partial\tau} J \equiv 0$ on $M^n \times [0, T]$.*

PROOF. Let F be defined as in (4.196). We note first that our assumptions imply that $\mathrm{S}\circ\mathrm{F}=0$. Indeed, a direct computation shows that

$$\mathrm{F}\hat{\mathrm{P}} = \hat{\mathrm{P}}\mathrm{F} = \mathrm{F}$$

on all of $M^n \times [0,T]$, so, as $\nabla\hat{\mathrm{P}} \equiv 0$,

$$\mathrm{S}\circ\mathrm{F} = \mathrm{S}\circ(\hat{\mathrm{P}}\circ\mathrm{F}) = \hat{\mathrm{S}}\circ\mathrm{F} = (\nabla\mathrm{R}\circ\hat{\mathrm{P}})\circ\mathrm{F} = \nabla\hat{\mathrm{R}}\circ\mathrm{F} = 0.$$

Using this, we see that $\nabla J \equiv 0$ for all τ by the following pointwise calculation: Fix $p \in M^n$ and let $W_{ib}^a = \nabla_i J_a^b$. Since $D_\tau J = 0$, we have

$$\begin{aligned}
\partial_\tau W_{ib}^a &= \nabla_i\left(R_c^a J_b^c - R_b^c J_c^a\right) - \partial_\tau \Gamma_{ib}^c J_c^a + \partial_\tau \Gamma_{ic}^a J_b^c\\
&= J_c^a g^{cd}\left(\nabla_d R_{bi} - \nabla_b R_{di}\right) + J_b^c g^{ad}\left(\nabla_c R_{di} - \nabla_d R_{ci}\right) - R_c^a W_{ib}^c + R_b^c W_{ic}^a\\
&= -g^{ac} g^{\ell m} \nabla_\ell R_{mibd} J_c^d + g^{ad} g^{\ell m} \nabla_\ell R_{midc} J_b^c - R_c^a W_{ib}^c + R_b^c W_{ic}^a\\
&= g^{ad} g^{\ell m}\left(\nabla_\ell R_{micb} J_d^c + \nabla_\ell R_{midc} J_b^c\right) - R_c^a W_{ib}^c + R_b^c W_{ic}^a\\
&= \frac{1}{2} g^{ad} g^{\ell m} (\mathrm{S}\circ\mathrm{F})_{\ell mibd} - R_c^a W_{ib}^c + R_b^c W_{ic}^a\\
&= -R_c^a W_{ib}^c + R_b^c W_{ic}^a.
\end{aligned}$$

Since $W(p,0)=0$, it follows that $(\nabla J)(p,\tau) = W(p,\tau) = 0$.

The Kähler identities then imply that $\mathrm{Ric}\circ J = J\circ\mathrm{Ric}$ as endomorphisms of TM for each τ. So

$$\partial_\tau J = J\circ\mathrm{Ric} - \mathrm{Ric}\circ J = 0,$$

and $J(\cdot,\tau) = J(\cdot,0)$ for all $\tau\in[0,T]$.. □

For generalizations of this result, see [**CK20, Kot14**].

4.9.3.5. *Shrinkers asymptotic to Kähler cones.*

So far, everything we have done in the previous four subsections is valid for any solution $g(\tau)$ to the backward Ricci flow which is Kähler at $\tau = 0$. We now specialize to the case of a solution associated to an asymptotically conical shrinking soliton and use (4.60) to bound the coefficients of the elements A, B, $\hat{\mathrm{R}}$, and $\hat{\mathrm{S}}$ in (4.198)–(4.201) through their evolution equations. This yields the following analog of Proposition 4.24, and clears the way for an application of Theorem 4.25.

PROPOSITION 4.54. *Assume $(\mathcal{C}_0,\hat{g})$ is Kähler and the shrinking gradient Ricci soliton $(\mathcal{C}_a, g_1, f_1)$ is dynamically asymptotic to $\hat{g}$ on $\mathcal{C}_a$. If $g(\tau)$ is the associated solution to (4.7) on $\mathcal{C}_{a,1}$ with $g(0)=\hat{g}$ and $g(1)=g_1$, and* A, B, $\hat{\mathrm{R}}$, *and* $\hat{\mathrm{S}}$ *are as defined above in terms of $g(\tau)$, then the sections*

$$\mathbf{X} = (\hat{\mathrm{R}},\hat{\mathrm{S}}), \quad \mathbf{Y} = (\mathrm{A},\mathrm{B}),$$

over $\mathcal{C}_{a,1}$ are uniformly bounded, vanish identically at $\tau=0$, and satisfy

$$|(D_\tau+\Delta)\mathbf{X}| \le \frac{N}{r^2+1}(|\mathbf{X}|+|\mathbf{Y}|),$$

$$|D_\tau \mathbf{Y}| \le N(|\mathbf{X}|+|\nabla\mathbf{X}|) + \frac{N}{r^2+1}|\mathbf{Y}|,$$

for some constant N.

The proof of Theorem 4.49 is now just a matter of assembly. First, we apply Theorem 4.25 to the pair $(\mathbf{X},\mathbf{Y})$ described above to prove that a shrinker asymptotic to a Kähler cone on an end must be Kähler near spatial infinity. Then, via a continuation principle akin to Lemma 4.47, we argue that if the shrinker is complete, the Kähler structure must extend to the entire manifold.

PROOF OF THEOREM 4.49. We assume that $(\mathcal{C}_0, \hat{g})$ is Kähler with complex structure $\hat{J}$, and that (M^n, g, f) is a shrinking soliton which is C^2-asymptotic to $\mathcal{C}^\Sigma$ along the end $V \subset M^n$. According to Lemma 4.8, there is an $a > 0$ and a diffeomorphism $F : \mathcal{C}_a \to W$ onto some end $W \subset V$ such that $(\mathcal{C}_a, \tilde{g}_1, \tilde{f}_1)$ is dynamically asymptotic to $\hat{g}$, where $\tilde{g}_1 = F^*g$ and $\tilde{f}_1 = F^*f$.

Let $\Phi_\tau : \mathcal{C}_a \to \mathcal{C}_a$ be the map $\Phi_\tau(r, \sigma) = (r/\sqrt{\tau}, \sigma)$ for $\tau \in (0, 1]$. By Proposition 4.18, the solution $\tilde{g}(\tau) = \tau\Phi_\tau^*\tilde{g}_1$ to (4.7) extends to a solution on $[0, 1]$ with $\tilde{g}(0) = \hat{g}$ and $\tilde{g}(1) = \tilde{g}_1$. Forming the tensors A, B, $\hat{\mathrm{R}}$, and $\hat{\mathrm{S}}$ associated to $\tilde{g}(\tau)$, it follows from Proposition 4.54 and Theorem 4.25 that

$$\mathbf{X} = (\hat{\mathrm{R}}, \hat{\mathrm{S}}), \quad \mathbf{Y} = (\mathrm{A}, \mathrm{B}),$$

vanish identically on $\mathcal{C}_{b,T}$ for some $b \geq a$ and $0 < T \leq 1$. In particular, $\nabla\hat{\mathrm{P}} \equiv 0$ and $\hat{\mathrm{R}} \equiv 0$ on $\mathcal{C}_{b,T}$, so Lemma 4.53 implies that $\tilde{g}(\tau)$ is Kähler with complex structure $\hat{J}$ on $\mathcal{C}_b$ for each $0 \leq \tau \leq T$.

We claim that $\tilde{g}_1$ is Kähler with respect to $\hat{J}$ on all of $\mathcal{C}_a$. To see this, note that, by Theorem 4.26, the metrics $\tilde{g}(\tau)$ for $\tau \in [0, 1]$ (including $\hat{g} = \tilde{g}(0)$) are real-analytic with respect to a common atlas on $\mathcal{C}_a$. Thus, as $\hat{J}$ is real-analytic, the vanishing of the tensors

$$Q := \tilde{g}(\cdot, \cdot) - \tilde{g}(\hat{J}\cdot, \hat{J}\cdot), \quad W := \nabla\hat{J},$$

on $\mathcal{C}_b \times \{\tau\}$ for each $\tau \in [0, T]$ implies that they vanish on all of $\mathcal{C}_{a,T}$.

On the other hand, the complex structure $\hat{J}$ is invariant under the flow of $\frac{\partial}{\partial r}$, and therefore under pull-back by Φ_τ. Since $Q(\tau) = \tau\Phi_\tau^*Q(1)$ and $W(\tau) = \Phi_\tau^*W(1)$, it follows that $Q(1)$ and $W(1)$ vanish on $\mathcal{C}_{a/\sqrt{T}}$, and hence (by analyticity, as above) on all of $\mathcal{C}_a$. Therefore $\tilde{g}_1$ is Kähler with respect to $\hat{J}$ on $\mathcal{C}_a$. Then $\bar{g} = (F^{-1})^*\tilde{g}_1$ is Kähler on W with respect to $J_W = (F^{-1})^*\hat{J}$ and $F : (\mathcal{C}_a, \hat{J}) \to (W, J_W)$ is a biholomorphism. This completes the proof of the first assertion in Theorem 4.49.

For the second assertion, in which (M^n, g) is now additionally assumed to be complete, we may apply what we have just proven to obtain that g is Kähler on some end $W \subset M^n$ with respect to some complex structure J_W. By Corollary 4.16, for any $x_0 \in W$, the homomorphism of $\pi_1(W, x_0)$ into $\pi_1(M^n, x_0)$ induced by inclusion is a surjection. It thus follows from Lemma 4.55 below that J_W extends to a complex structure compatible with g on all of M^n. □

LEMMA 4.55. *Suppose that (M^n, g) is a complete connected real-analytic manifold and E is a connected nonempty open subset of M^n. Let $p \in E$. If the homomorphism $\pi_1(E, p) \to \pi_1(M^n, p)$ induced by inclusion is surjective, then any parallel almost-complex structure J_E on E extends in a unique way to a parallel almost-complex structure J on M^n.*

PROOF. Fix $p \in E$ and let $\bar{J} = J_E(p)$. Any extension J of J_E to M^n is completely determined by parallel translation of $\bar{J}$ along paths and is therefore unique: if $q \in M^n$, then

$$J(q) = P_\gamma \bar{J} P_\gamma^{-1} \tag{4.202}$$

where $P_\gamma : T_pM^n \to T_qM^n$ is parallel transport along any piecewise smooth path γ from p to q.

Now let $q \in M^n$, choose any smooth path γ from p to q and define $J(q) : T_qM \to T_qM$ by (4.202). We claim that $J(q)$ is independent of the path γ. From this it will follow that J is a smooth parallel complex structure on M^n. Since J_E must coincide with parallel transport of $\bar{J}$ along paths in E, we will have $J|_E = J_E$.

To see that J is well-defined, let σ be any other path from p to q, and denote its reverse parametrization by $\bar{\sigma}$. Under our assumption on $\pi_1(E, p)$, the loop $\gamma \cdot \bar{\sigma}$ is homotopic to a loop α based at p which lies entirely inside E. Then

$$P_\gamma = P_\sigma P_{\bar{\sigma}} P_\gamma P_{\bar{\alpha}} P_\alpha = P_\sigma P_\beta P_\alpha,$$

where $\beta = \bar{\alpha} \cdot \gamma \cdot \bar{\sigma}$ is a null-homotopic loop based at p.

Since (M^n, g) is real-analytic, the reduced holonomy $\mathrm{Hol}_p^0(M^n, g)$ at p is determined by the local holonomy (see, e.g., §10 of Chapter II of [**KN96**]). In particular, $\mathrm{Hol}_p^0(M^n, g)$ leaves $\bar{J}$ invariant. The same is true of $\mathrm{Hol}_p(E, g)$ by assumption. Therefore,

$$P_\beta \bar{J} P_\beta^{-1} = \bar{J} = P_\alpha^{-1} \bar{J} P_\alpha,$$

and $P_\gamma \bar{J} P_\gamma^{-1} = P_\sigma \bar{J} P_\sigma^{-1}$ as desired. □

REMARK 4.56. Alternatively, to prove the second assertion in Theorem 4.49, one can appeal to the result of [**Esp25**]: since M^n is simply connected, one has $\mathrm{Hol}_p(M^n, g) \cong \mathrm{Hol}_p^0(M^n, g)$. As noted in the proof of the preceding lemma, the real-analyticity of (M^n, g) implies that $\mathrm{Hol}_p^0(M^n, g)$ is determined by the local holonomy about any p, and therefore the Kählerity of (M^n, g) on an open subset implies its global Kählerity.

4.9.4. Further applications.

The main results we have established so far for asymptotically shrinking solitons have natural "dynamical" interpretations in terms of the Ricci flow. Theorems 4.12 and 4.45 are manifestations of the propagation of symmetry backwards along the Ricci flow, and Theorem 4.49 is a manifestation of the propagation of restricted holonomy. These interpretations hinge on the fact that the self-similar solution associated to a shrinker on an asymptotically conical end extends to a smooth solution near infinity which originates at the shrinker and terminates in the end of the cone. Now we take a step back and ask a more basic question: *What can we conclude about a Ricci flow just from the knowledge that it terminates in a cone?*

This leads to at least two different questions of propagation of structure. On one hand, a cone (minus the vertex) is, in particular, a warped product. Thus, we might wish to ask when a warped product structure is guaranteed to propagate backwards in time under the Ricci flow. On the other hand, a cone is not only a warped product, but a very special kind of one, characterized by the manner in which it scales under pull-back by the radial dilation map. As it seems doubtful that a well-behaved solution to a parabolic equation could simply and *spontaneously* acquire a property as rigid as this scaling invariance in finite time, it is natural to ask what trace of this scaling property must be present in the solution at previous times. We consider both of these questions below.

4.9.4.1. *Backward propagation of warped product structures.*

Warped products (and more general submersive structures) are typically *not* preserved by the Ricci flow either forwards or backwards in time. In fact, a short computation shows that if a family of metrics on a product $B \times F$ has the form

$$g(b, x, t) = g_1(b, t) + h(b, t)^2 g_2(x, t)$$

and solves the Ricci flow for $t \in (0, a)$, then either $h = h(t)$ and g splits as a product of solutions on B and F, respectively, or the fiber metric $g_2(x, t)$ is Einstein for all $t \in (0, a)$.

Thus, for a nontrivial warped product structure to be preserved under the Ricci flow, *it is necessary that the fiber metric be Einstein.* In fact, with some mild restrictions on the solution in the noncompact case, the condition that the fibers be Einstein is also sufficient for the warped product structure to propagate forwards in time. In typical applications, this is justified in two steps. First one uses the warped-product ansatz to solve a coupled system for the warping function and the family of metrics on the base and combines the two to produce a warped product solution to the flow with the given initial data. Then one invokes a uniqueness theorem (e.g., [**Ham82**, **CZ12**]) to conclude it is the only solution.

The condition that the fibers of the warped product be Einstein also turns out to be sufficient to prove that warped product structures propagate backward in time under the flow (with the same caveats in the noncompact case). However, one must overcome the same complications with the gauge which plague the other backward uniqueness problems above. In particular, the above two step procedure fails in the backward time case, as one cannot solve the (backward parabolic) equations for the warping function and the family of metrics on the base that one needs to construct a warped product solution.

In [**Kot24a**], one of the authors has given an argument using O'Neill's submersion invariants ([**O'N66**], [**Bes08**]) to encode the problem as one of backward uniqueness for solutions to a PDE-ODE system akin to those in (4.186). We will not state the general theorem here, but instead just note an application to solitons which follows from applying Theorem 4.25 to the particular PDE-ODE system along the backward Ricci flow associated to an asymptotically conical shrinker.

THEOREM 4.57. *Assume that (Σ, g_Σ) is a product $\Sigma = \Sigma_1 \times \cdots \times \Sigma_k$ of compact Einstein manifolds (Σ_i, g_{Σ_i}). If the shrinking gradient Ricci soliton (M^n, g, f) is C^2-asymptotic to C^Σ along the end $V \subset M^n$, then there is an $a > 0$ and an end $W \subset V$ for which $(W, g|_W)$ is isometric to a multiply warped product on $(a, \infty) \times \Sigma_1 \times \cdots \times \Sigma_k$ of the form*

$$g = dr^2 + h_1^2(r) g_{\Sigma_1} + \cdots + h_k^2(r) g_{\Sigma_k}.$$

For the proof, see [**Kot24a**] by one of the authors. Theorem 4.57 can be used to obtain some non-existence results.

COROLLARY 4.58. *Suppose (Σ, g_Σ) is a compact Einstein manifold with positive scalar curvature. If there is a complete shrinking gradient Ricci soliton (M^n, g, f) which is C^2-asymptotic to $\mathcal{C}^\Sigma$ along an end, then (Σ, g_Σ) is isometric to the sphere of radius $\sqrt{2(n-1)}$, and (M^n, g, f) is the Gaussian soliton on $\mathbb{R}^n$.*

Indeed, the classification of rotationally symmetric shrinking solitons in [**Kot08**] is based on a classification of solutions to a system of ODEs which is identical to the system for warped product shrinking soliton metrics $dr^2 + h^2(r) g_\Sigma$ with (appropriately scaled) Einstein fibers of positive scalar curvature. That analysis shows that if the shrinking soliton is complete and noncompact, then either the solution splits as $\mathbb{R} \times \Sigma$ or must close up smoothly as $r \to 0$. The former is impossible if the shrinking soliton is asymptotically conical and nonflat, and the latter is possible only if Σ is the round sphere of appropriate radius.

4.9.4.2. *Ricci flows which terminate in cones.*

Now we investigate more generally what restrictions there are for a Ricci flow that terminates in the end of a cone near infinity. As we have observed, a conical metric is characterized by the scaling property (4.1). Since this property involves only the actions of the diffeomorphism group and the positive real numbers on the space of metrics – both symmetries of the Ricci flow equation – it is natural to ask whether a solution which flows into a cone must possess some vestige of this scaling property at earlier times.

The answer to this question is yes, and the influence of this scaling property is essentially definitive: if a solution terminates in a cone on some end (and does so in a reasonably well-behaved way), it must be *self-similar* at previous times. The following is a local version of the result, proven in [**Kot24b**] by one of the authors. Note that the statement does not presuppose the existence of a shrinking soliton asymptotic to the cone.

THEOREM 4.59. *Let $(\mathcal{C}_0, \hat{g})$ be a cone with a smooth compact cross-section (Σ, g_Σ), and $g(\tau)$ a smooth solution to the backward Ricci flow on $\mathcal{C}_a \times [0,1]$ which satisfies $g(0) = \hat{g}$ on $\mathcal{C}_a$ and*

$$|\mathrm{Rm}|(x,t) \leq \frac{N}{r^2(x)+1} \quad \text{on} \quad \mathcal{C}_a \times [0,1]$$

for some constant $N > 0$. Then there is $f \in C^\infty(\mathcal{C}_a)$ such that $(\mathcal{C}_a, g(1), f)$ is a shrinking gradient Ricci soliton which is dynamically asymptotic to $\hat{g}$.

The underlying problem is again one of backward uniqueness. For suppose $g(\tau)$ is a solution to the backward Ricci flow (4.7) which initiates in the cone $\hat{g}$ at $\tau = 0$. Let $\rho_\lambda : \mathcal{C}_0 \to \mathcal{C}_0$ denote the dilation map $\rho(r,\sigma) = (\lambda r, \sigma)$ as before. Then, for each $\lambda \geq 1$, the metrics

$$g_\lambda(\tau) = \lambda^{-2}\rho_\lambda^* g(\lambda^2\tau) \tag{4.203}$$

define solutions to the Ricci flow on $\mathcal{C}_a \times [0, \lambda^{-2}]$ for which

$$g_\lambda(0) = \lambda^{-2}\rho_\lambda^*\hat{g} = \hat{g}$$

for all λ. That is, the solutions g_λ all emanate from the same cone $\hat{g}$! Provided one can show (by some analog of Theorem 4.19) that the solutions g_λ coincide on $\mathcal{C}_a \times [0, \lambda^{-2}]$ for all $\lambda \geq 1$, it will follow from (4.203) that

$$g(\tau) = \tau\Phi_\tau^* g(1)$$

on $\mathcal{C}_a \times [0,1]$, where $\Phi_\tau(r,\sigma) = (r/\sqrt{\tau}, \sigma) = \rho_{1/\sqrt{\tau}}(r,\sigma)$ as before. This says that $g(\tau)$ is a Ricci soliton flowing along $X = \frac{r}{2\tau}\frac{\partial}{\partial r}$. Then it is just a matter of showing that X is the gradient of some function with respect to g, a problem which itself can be formulated as another problem of backward uniqueness.

For these backward uniqueness results, we need a modification of Theorem 4.25 in which the background metric $g(\tau)$ is assumed neither to be self-similar for $\tau > 0$ nor to arise from a shrinking soliton which is dynamically asymptotic to the cone. As we noted following the proof of Theorem 4.25, these assumptions provided us with the sharp decay rates (4.60) and (4.61) which simplified some estimates, but were not really essential to the argument.

Note that since Theorem 4.59 is local to the end $\mathcal{C}_a$, we can apply it to solutions to (1.3) which flow, in a controlled way, into a cone on some end as $t \to T$, but which may become singular elsewhere on the manifold at the terminal time T. This leads to a global version of Theorem 4.59, which we state in terms of the *forward* Ricci flow.

THEOREM 4.60. *Let $(\mathcal{C}_0, \hat{g})$ be a regular cone and suppose that $g(t)$ is a smooth complete solution to the Ricci flow on $M^n \times [0,T)$. Assume that there is an end $V \subset M^n$ and a diffeomorphism $F : V \to \mathcal{C}_a$ for some $a > 0$ such that $g(t)$ has quadratic curvature decay on $V \times [0,T)$ and converges locally smoothly to $F^*\hat{g}$ on V as $t \to T$. Then there exists $f_0 \in C^\infty(M^n)$ such that $g_0 = g(0)$ satisfies*

$$\operatorname{Ric}(g_0) + \nabla^2_{g_0} f_0 = \frac{1}{2T} g_0.$$

Moreover, if Φ_t is the family of diffeomorphisms generated by the vector fields $(T-t)^{-1} \operatorname{grad}_{g_0} f_0$, then, as $t \to T$, $f_t = f_0 \circ \Phi_t$ converges locally smoothly on V to f_T, where

$$f_T \circ F^{-1}(r, \sigma) = \frac{r^2}{4}.$$

The proof relies in part on a general extension principle for soliton structures, akin to the classical extension principle for Killing vector fields due to Nomizu [**Nom60**].

THEOREM 4.61. *Suppose (M^n, g) is a connected, simply-connected real-analytic manifold and $U \subset M^n$ is a nonempty connected open set. If X is a smooth vector field on U for which (U, g, X, λ) is a Ricci soliton, then X extends to a unique vector field on M^n such that (M^n, g, X, λ) is a Ricci soliton. If (U, g, X, λ) is a gradient soliton, then so is (M^n, g, X, λ).*

For the proofs of these theorems, see the paper of [**Kot24b**] by one of the authors.

4.10. Open problems for asymptotically conical shrinkers

In principle, Theorem 4.12 reduces the classification of complete 4-dimensional asymptotically conical shrinking solitons to the classification of the possibilities for the asymptotic cones of such spaces. This raises the following open-ended question.

OPEN PROBLEM 4.1. *Assume that the shrinker (M^4, g, f) is complete and asymptotic to the cone $\mathcal{C}^\Sigma$ along some end. What can be said about the geometry and topology of the link (Σ, g_Σ)?*

Regarding the geometry, one might ask (optimistically) whether (Σ, g_Σ) must be homogeneous, or, at least, have constant scalar curvature. By Theorem 4.45, if (Σ, g_Σ) is homogeneous, then the shrinker is of cohomogeneity one.

OPEN PROBLEM 4.2. *Determine the complete 4-dimensional asymptotically conical shrinkers of cohomogeneity one.*

As for the topology of (Σ, g_Σ), since the scalar curvature of g_Σ is positive, it follows from Perelman's resolution of the Geometrization Conjecture [**Per02, Per03a, Per03b**] that the link can only be a connected sum of $\mathbb{S}^3$'s and $\mathbb{S}^2 \times \mathbb{S}^1$'s. It is unknown whether the $\mathbb{S}^2 \times \mathbb{S}^1$ factors may actually occur. Of course, by Hamilton's theorem [**Ham82**], if the link had not just positive scalar curvature but positive *Ricci* curvature, it would be a spherical space form. Note that by Wylie's theorem [**Wyl08**], complete shrinking solitons have finite fundamental group; however, the fundamental group of the distant level sets of the potential function (which are diffeomorphic to the link) need not necessarily inject into the fundamental group of the soliton.

In a different direction, Stolarski [**Sto22**] has shown that every complete asymptotically conical gradient Ricci soliton arises as a finite-time singularity model of a compact Ricci flow. For a potential formulation of 4-dimensional Ricci flow with surgery, it would be useful to know whether, given any complete 4-dimensional asymptotically conical shrinking soliton, there is a corresponding complete expanding soliton with the same asymptotic cone and whether the expander is unique.

OPEN PROBLEM 4.3. *Suppose* (M^4_-, g_-, f_-) *is a complete shrinker which is asymptotic to the cone* $\mathcal{C}^\Sigma$ *along an end. Does there also exist a complete expander* (M^4_+, g_+, f_+) *which is asymptotic to* $\mathcal{C}^\Sigma$ *along an end?*

Examples of Angenent and Knopf [**AK22**] show that the uniqueness may fail in dimensions $n \geq 5$. For important recent progress on this problem, see Bamler and Chen [**BC23**].

One might also ask for a partial converse of Stolarski's theorem.

OPEN PROBLEM 4.4. *Suppose* $(M^4, g(t))$ *is a smooth ancient solution which arises as a finite-time singularity model for a closed* 4*-dimensional Ricci flow. Is* $(M^4, g(0))$ *a gradient shrinking soliton?*

OPEN PROBLEM 4.5. *Does there exist a singular solution of the Ricci flow on a closed manifold that forms a singularity model asymptotic to a cone for which there exist distinct expanding solitons asymptotic to this cone?*

CHAPTER 5

Curvature and Volume Estimates for Steady Solitons

Thus far, the bulk of our discussion has focused on shrinking solitons. In this chapter, we turn our attention to steady solitons and establish curvature bounds and volume growth estimates similar to those obtained in Chapter 2 for the shrinking case.

In the first part, we show that the growth of the volume of a metric ball in a complete steady gradient Ricci soliton grows at least linearly and at most exponentially in the radius. These bounds follow from pointwise estimates on the potential function and the f-mean curvature. Additional volume estimates for steady gradient Ricci soliton singularity models using Bamler's theory will be discussed in Chapter 6.

In the second part of the chapter, we prove several bounds on the growth and decay of the curvature of complete steady gradient Ricci solitons. First, we show that the scalar curvature of a complete steady gradient Ricci soliton can decay at most exponentially; this is the optimal rate for steadies, as can be seen by considering the 2-dimensional cigar soliton. Next, using the same methods we used to prove the quadratic bound on shrinking 4-dimensional singularity models, we show that the curvature of a steady 4-dimensional singularity model must be bounded. Finally, we show that, just as for shrinkers, the curvature tensor of a complete nontrivial (i.e., non Ricci-flat) 4-dimensional steady gradient Ricci soliton satisfies a bound of the form $|\mathrm{Rm}| \leq CR$.

5.1. Estimates for the volumes of steady solitons

In this section, we discuss what is known about upper and lower estimates for the volume growth of steady solitons.

5.1.1. Upper estimates for the areas and volumes of geodesic spheres and balls.

We have the following upper estimates for the volumes and areas of geodesic balls and their boundaries, respectively.

PROPOSITION 5.1 (Area and volume estimates for steadies). *For any steady gradient Ricci soliton (M^n, g, f) and point $p \in M^n$, we have*

$$\mathrm{Area}(\partial B(p,r)) \leq C_1 \mathrm{e}^{C_2\sqrt{r}} \quad \textit{and} \quad \mathrm{Vol}(B(p,r)) \leq C_1 \mathrm{e}^{C_2\sqrt{r}} \tag{5.1}$$

for all $r > 0$, where $C_1, C_2 > 0$ depend only on the dimension n.

PROOF. Let $\{x^i\}$ be geodesic spherical coordinates centered at p, so that $x^n(x) = r(x) := d(x,p)$; see (I-4.15). Denote

$$g_{ij} := g\Big(\frac{\partial}{\partial x^i}, \frac{\partial}{\partial x^j}\Big) \qquad \text{for } 1 \leq i \leq n. \tag{5.2}$$

Let V be a unit tangent vector at p. Let $\gamma := \gamma_V : [0,\infty) \to M^n$ be the unit-speed geodesic emanating from p with $\gamma'(0) = V$. Let $g_{ij}(r) := g_{ij}(\gamma(r))$. The **Jacobian** of the exponential map in spherical coordinates for g is

$$\mathrm{J}(r) := \mathrm{J}_V(r) := \sqrt{\det(g_{ij}(r))}. \tag{5.3}$$

Let $f(r) := f(\gamma(r))$ and let $H(r)$ be the mean curvature at $\gamma(r)$ of the geodesic sphere centered at p. By the Riccati inequality (I-4.27) we have

$$\frac{dH}{dr} + \frac{H^2}{n-1} \leq -\mathrm{Ric}\left(\frac{\partial}{\partial r}, \frac{\partial}{\partial r}\right) = f'', \tag{5.4}$$

where the last equality follows from the steady gradient Ricci soliton equation. This implies (I-4.31); that is,

$$H(r) \leq \frac{n-1}{r} + f'(r) - \frac{2}{r^2}\int_0^r sf'(s)ds.$$

Since $|\nabla f| \leq 1$, we obtain that

$$H(r) \leq \frac{n-1}{r} + 2 \quad \text{for all } r > 0. \tag{5.5}$$

For $r = 1$, we get that $H(1) \leq n+1$. Moreover, since $H(r) = \frac{d}{dr}\ln \mathrm{J}(r)$ and $\lim_{r\to 0}\frac{\mathrm{J}(r)}{r^{n-1}} = 1$, (5.5) implies that $\mathrm{J}(1) \leq \mathrm{e}^2$.

On the other hand, integrating (5.4) from 1 to t, we obtain

$$H(t) + \frac{1}{n-1}\int_1^t H^2(r)dr \leq H(1) + f'(t) - f'(1) \leq n+3 \tag{5.6}$$

for all $t \geq 1$. Recall that $H(r) = \frac{d}{dr}\ln \mathrm{J}(r)$. Let

$$K(t) := \int_1^t H(r)dr = \ln\frac{\mathrm{J}(t)}{\mathrm{J}(1)}.$$

By the Cauchy–Schwarz inequality,

$$K^2(t) = \left(\int_1^t H(r)dr\right)^2 \leq (t-1)\int_1^r H^2(r)dr.$$

Therefore, (5.6) implies that

$$K'(t) + \frac{K^2(t)}{(n-1)\,t} \leq C_0 \quad \text{for all } t \geq 1,$$

where $C_0 := n+3$. Define the comparison difference quantity

$$L(t) := \sqrt{C_0(n-1)t} - K(t).$$

Then

$$\begin{aligned} L'(t) &\geq -K'(t) \\ &\geq \frac{K^2(t)}{(n-1)\,t} - C_0 \\ &\geq \left(\frac{K(t)}{(n-1)\,t} + \frac{\sqrt{C_0}}{\sqrt{(n-1)\,t}} \right) L(t) \end{aligned}$$

for all $t \geq 1$. Since $L(1) = \sqrt{C_0(n-1)} > 0$, we conclude that $L(t) > 0$ for all $t \geq 1$, and hence

$$\mathrm{J}(t) < \mathrm{e}^{\sqrt{C_0(n-1)t}}\,\mathrm{J}(1) \qquad \text{for all } t \geq 1. \tag{5.7}$$

By integrating this inequality over spheres and using that $\mathrm{J}(1) \leq \mathrm{e}^2$, we obtain the area estimate. The volume estimate then follows by integrating the area estimate with respect to the radii of the spheres. □

5.1.2. Lower volume bounds for steady solitons.

We have the following lower bound for the volume growth of steady gradient Ricci solitons due to Šešum and one of the authors, [**MŠ13**, Theorem 1.7]. This bound is qualitatively sharp by the examples of products of $\mathbb{R}$ with closed Ricci-flat manifolds, which have linear volume growth. In addition, the cigar soliton shows that this result is also sharp for the class of steady gradient Ricci solitons with nontrivial potential functions.

PROPOSITION 5.2. *Any steady gradient Ricci soliton has at least linear volume growth; i.e., for each $p \in M^n$, there exists a constant $c > 0$ such that*

$$\mathrm{Vol}(B(p,r)) \geq cr \qquad \textit{for all } r \geq 1.$$

PROOF. If $\mathrm{Ric} = 0$, then this follows from the linear volume growth result of Yau [**Yau76**] for $\mathrm{Ric} \geq 0$, which is proved using the relative volume comparison theorem. Thus, we may assume that $R > 0$. Let $p \in M^n$. Integrating the equation $R + \Delta f = 0$, we have

$$\int_{B(p,r)} R\,dv = -\int_{B(p,r)} \Delta f\,dv = -\int_{\partial B(p,r)} \langle \nabla f, \nu \rangle\,da \leq \mathrm{Area}\left(\partial B(p,r)\right)$$

since $|\nabla f| \leq 1$, where ν is the unit outward normal to $\partial B(p,r)$. Hence, for $r \geq 1$,

$$\mathrm{Area}(\partial B(p,r)) \geq \int_{B(p,1)} R\,dv =: c > 0.$$

Thus, $\mathrm{Vol}(B(p,r)) \geq c(r-1)$ for all $r > 1$. □

We have the following exponentially growing lower bound for the f-volume of geodesic balls.

PROPOSITION 5.3. *For any nontrivial complete noncompact steady gradient Ricci soliton (M^n, g, f), there exists $c > 0$ such that*

$$\mathrm{Vol}_f(B(p,r)) \geq c\mathrm{e}^r. \tag{5.8}$$

In particular,

$$\mathrm{Vol}_f(M^n) = \int_{M^n} \mathrm{e}^{-f}\,dv = \infty. \tag{5.9}$$

PROOF. Using $R + |\nabla f|^2 = 1$ from (1.37) and $R + \Delta f = 0$ from (1.34), we compute that

$$\Delta(\mathrm{e}^{-f}) = \mathrm{e}^{-f}(-\Delta f + |\nabla f|^2) = \mathrm{e}^{-f}.$$

Hence,

$$\int_{B(p,r)} \mathrm{e}^{-f} dv = \int_{B(p,r)} \Delta(\mathrm{e}^{-f})\, dv = \int_{\partial B(p,r)} \langle \nabla(\mathrm{e}^{-f}), \nu \rangle\, da \leq \int_{\partial B(p,r)} \mathrm{e}^{-f} da,$$

where we used that $|\nabla f| \leq 1$ and where ν is the unit outward normal vector field to $\partial B(p,r)$. This implies that the function $W(r) := \int_{B(p,r)} \mathrm{e}^{-f} dv$ satisfies the ordinary differential inequality

$$W'(r) \geq W(r).$$

Integrating this yields for $r \geq 1$ that

$$\int_{\partial B(p,r)} \mathrm{e}^{-f} da \geq \int_{B(p,r)} \mathrm{e}^{-f} dv \geq c\mathrm{e}^{r}, \tag{5.10}$$

where $c := \frac{1}{\mathrm{e}} \int_{B(p,1)} \mathrm{e}^{-f} dv > 0$. This proves (5.8), and consequently (5.9). □

REMARK 5.4. An alternative proof of Proposition 5.3 is to observe that Proposition 3.28 implies that $\lambda_1(\Delta_f) = \frac{1}{4}$, and by Theorem 3.21, it follows that

$$\operatorname{Vol}_f(B(p,r)) \geq c\mathrm{e}^{2\sqrt{\lambda_1(\Delta_f)}\, r} = c\mathrm{e}^{r}.$$

5.2. Estimates for the potential functions of steady solitons

Cao and Zhou's estimate (1.61) gives qualitatively sharp lower and upper bounds for the potential function on a shrinking soliton. For steady gradient Ricci solitons, which admit more complicated geometries at infinity, there are (at least as yet) no correspondingly definitive bounds for the potential function. In this section, we will discuss some of what is known about the potential function in the steady case.

Let (M^n, g, f) be a steady gradient Ricci soliton. Assume that g is not Ricci-flat and fix $p \in M^n$. By (1.36) and (1.37), we have $0 < R \leq 1$ and $|\nabla f| < 1$. Thus

$$-d(x,p) + f(p) < f(x) < d(x,p) + f(p) \quad \text{for } x \in M^n \setminus \{p\}, \tag{5.11}$$

with equality for $x = p$. To obtain further estimates for the potential function, we relate it to the area and volume estimates.

5.2.1. Estimate for the infimum of the potential function on geodesic spheres.

By the area and volume estimates above, we can obtain an upper bound for the infimum on geodesic spheres of the potential function.

THEOREM 5.5. *If (M^n, g, f) is a nontrivial complete noncompact steady gradient Ricci soliton and $p \in M^n$, then there exists a positive constant C such that*

$$-r - C \leq \inf_{\partial B(p,r)} f \leq -r + C\sqrt{r} \quad \textit{for all } r \geq 1. \tag{5.12}$$

PROOF. By (5.10) and Proposition 5.1, we have for $r \geq 1$ that

$$ce^r \leq \int_{\partial B(p,r)} e^{-f} da \leq \text{Area}(\partial B(p,r)) \sup_{\partial B(p,r)} e^{-f} \leq Ce^{C\sqrt{r}} \sup_{\partial B(p,r)} e^{-f}.$$

Hence

$$\inf_{\partial B(p,r)} f \leq -r + C\sqrt{r}.$$

Since $|\nabla f| \leq 1$, we get that

$$-r - C \leq \inf_{\partial B(p,r)} f,$$

and the theorem follows. □

5.2.2. Estimate for the potential function in the case of positive Ricci curvature.

In the case where a steady gradient Ricci soliton (M^n, g, f) has positive Ricci curvature, the potential function is strictly concave and consequently satisfies a pointwise upper estimate.

PROPOSITION 5.6. *Suppose that (M^n, g, f) is a complete noncompact steady gradient Ricci soliton with positive Ricci curvature and such that R attains its maximum at a point $p \in M^n$. Then M^n is diffeomorphic to $\mathbb{R}^n$, f is strictly concave, $f_{\max}$ is attained exactly at p, and there exists a positive constant C such that*

$$-d(x,p) + f(p) \leq f(x) \leq -C^{-1}d(x,p) + C. \tag{5.13}$$

PROOF. The lower bound in (5.13) is immediate from $|\nabla f| \leq 1$.

As for the upper bound, we first observe that since $\nabla^2 f = -\text{Ric} < 0$ is negative-definite, f is strictly concave. By our hypothesis, there exists a critical point p of R. Because the Ricci tensor is non-degenerate (it is positive), by $\nabla R = 2\text{Ric}(\nabla f)$ from (1.46), we have that p is a critical point of f. This implies that p is the unique critical point of f and p is a nondegenerate critical point of f (with index n).

Let $\gamma : [0, L] \to M^n$ be a unit speed geodesic emanating from p. We have

$$\frac{d^2}{ds^2} f(\gamma(s)) = \nabla^2 f(\gamma'(s), \gamma'(s)) = -\text{Ric}(\gamma'(s), \gamma'(s)) < 0. \tag{5.14}$$

Since $\frac{d}{ds} f(\gamma(s))\big|_{s=0} = 0$, we have $-\frac{d}{ds} f(\gamma(s)) > 0$ for $s > 0$. Let δ be the infimum of $-\frac{d}{ds} f(\gamma(s))\big|_{s=1}$ over all unit speed geodesics γ emanating from p. Clearly $\delta > 0$. Let $x \in M^n \setminus B(p, 1)$ and let $\gamma : [0, d(x,p)] \to M^n$ be a minimal geodesic from p to x. By (5.14), we have

$$-\frac{d}{ds} f(\gamma(s)) \geq \delta \quad \text{for } s \geq 1.$$

We conclude that f is a proper function and that

$$-f(\gamma(s)) \geq -f(\gamma(1)) + \delta\,(s-1)\,.$$

Hence, $f_{\max}$ is attained exactly at p and (5.13) follows by taking $s = d(x,p)$. Finally, since the only critical point of f is p, at which f attains its unique maximum, and since f is a proper function, we have that each of the level sets $\{f = c\}$, where $c < f_{\max}$, is diffeomorphic to $\mathbb{S}^{n-1}$. It therefore follows that M^n is diffeomorphic to $\mathbb{R}^n$. □

5.3. Scalar curvature decay estimate for steady solitons

We know from (1.54) that the scalar curvature of a complete steady soliton is always non-negative and is strictly positive unless the soliton is Ricci-flat. A natural question, then, is whether the scalar curvature of a nontrivial steady can decay to zero arbitrarily rapidly. To set our expectations, let us first look at the 2-dimensional setting.

Let (M^2, g, f) be a complete steady gradient Ricci soliton with positive scalar curvature on a surface. Since $2\mathrm{Ric}(\nabla f) = \nabla R$ and $\mathrm{Ric} = \frac{1}{2}Rg$, we obtain that $\nabla f = \nabla \ln R$. Thus, by adding a constant to f if necessary, we have that

$$R = \mathrm{e}^f \tag{5.15}$$

(cf. (I-3.42)). Since $f(x) \geq f(o) - d(x,o)$, this implies that R has a exponential lower bound. In fact, by Theorem I-3.11, which is a result of Hamilton, (M^2, g, f) must be the cigar soliton.

More generally, in all dimensions we have an estimate for the scalar curvature decay rate under the assumption $f(x) \to -\infty$ as $x \to \infty$ on the potential function, due to Lu, Yang, and one of the authors [**CLY11**]. The improved version below follows from [**MSW19**].

THEOREM 5.7. *There exists a constant $c > 0$ depending only on n such that if (M^n, g, f) is a steady gradient Ricci soliton with $f \leq A$ for some $A \in \mathbb{R}$, then*

$$R(x) \geq c\mathrm{e}^{f(x)-A} \quad \textit{for all } x \in M^n. \tag{5.16}$$

PROOF. Without loss of generality, we may assume that $f \leq 0$. We compute that

$$\Delta_f(\mathrm{e}^f) = (\Delta_f f + |\nabla f|^2)\mathrm{e}^f = -R\,\mathrm{e}^f < 0.$$

Let $a \in \left(0, \frac{1}{2}\right)$. By (1.42) with $\lambda = 0$, we obtain

$$\Delta_f\left(R - a\mathrm{e}^f\right) \leq -\frac{2}{n}R^2 + aR\,\mathrm{e}^f \leq \frac{na^2}{8}\mathrm{e}^{2f}.$$

Next we compute that

$$\Delta_f(\mathrm{e}^{2f}) = 2\mathrm{e}^{2f}\left(1 - 2R\right).$$

Hence

$$G := R - a\mathrm{e}^f - \frac{1-2a}{4}\mathrm{e}^{2f} \tag{5.17}$$

satisfies the elliptic inequality:

$$\Delta_f G \leq \left(\frac{na^2}{8} - \frac{1-2a}{2} + (1-2a)\,R\right)\mathrm{e}^{2f}. \tag{5.18}$$

Let $r_0 \geq 1$ and let $\phi : [0,\infty) \to [0,1]$ be a C^∞ cutoff function with $\phi(r) = 1$ for $r \leq r_0$, $\phi(r) = 0$ for $r \geq 2r_0$, $-Cr_0^{-1} \leq \phi'(s) \leq 0$, and $|\phi''(s)| \leq Cr_0^{-2}$. Fix $p \in M^n$ and denote $r(x) = d(x,p)$. Define

$$H := \phi^2 G = \phi^2\left(R - a\mathrm{e}^f - \frac{1-2a}{4}\mathrm{e}^{2f}\right),$$

where $\phi(x) := \phi(r(x))$ and where G is as in (5.17). Recall, by the f-Laplacian comparison theorem (see (I-4.37)), that $\Delta_f r \leq \frac{n-1}{r} + 1$; in particular,

$$\Delta_f r \leq n \quad \text{in } B(p, 2r_0) \setminus B(p, r_0) \subset M^n \setminus B(p, 1).$$

Hence

$$\Delta_f \phi = \phi' \Delta_f r + \phi'' |\nabla r|^2 \geq -C r_0^{-1}. \tag{5.19}$$

We compute, at points where $\phi > 0$, that

$$\Delta_f H = \phi^2 \Delta_f G + 4\phi^{-1} \langle \nabla \phi, \nabla H \rangle - 6|\nabla \phi|^2 G + 2\phi(\Delta_f \phi) G,$$

so that, by (5.18),

$$\begin{aligned} \Delta_f H - 4\phi^{-1} \langle \nabla \phi, \nabla H \rangle \leq \phi^2 \left(\frac{na^2}{8} - \frac{1-2a}{2} + (1-2a)\, R \right) \mathrm{e}^{2f} \\ - 6|\nabla \phi|^2 G + 2\phi(\Delta_f \phi) G. \end{aligned} \tag{5.20}$$

Suppose that $H < 0$ somewhere. Then H achieves its negative minimum at some point $x_0 \in B(p, 2r_0)$. Note that by (5.17),

$$G > -a\mathrm{e}^f - \frac{1-2a}{4}\mathrm{e}^{2f} \geq -\frac{2a+1}{4},$$

where for the second inequality we used that $f \leq 0$. Therefore, we have that $0 > G(x_0) > -1$. By (5.19), it follows that

$$\phi(\Delta_f \phi) G \leq C r_0^{-1} \quad \text{at} \quad x_0.$$

Moreover,

$$|\nabla \phi|^2 G \geq -C r_0^{-2} \geq -C r_0^{-1}.$$

Applying the maximum principle to (5.20) therefore yields

$$0 \leq \phi^2 \left(\frac{na^2}{8} - \frac{1-2a}{2} + (1-2a)\, R \right) \mathrm{e}^{2f} + C r_0^{-1} \quad \text{at} \quad x_0. \tag{5.21}$$

Now, $H(x_0) < 0$ and $f \leq 0$ implies that

$$R(x_0) < a\mathrm{e}^f + \frac{1-2a}{4}\mathrm{e}^{2f} < \frac{2a+1}{4}.$$

Thus (5.21) yields

$$0 \leq \phi^2(x_0) \left(\frac{na^2}{8} - \left(\frac{1}{2} - a \right)^2 \right) \mathrm{e}^{2f(x_0)} + C r_0^{-1}.$$

Taking $a > 0$ sufficiently small depending only on n (e.g., $a = \frac{1}{\sqrt{2n+2}}$), we obtain

$$\phi^2(x_0) \mathrm{e}^{2f(x_0)} \leq C r_0^{-1} \tag{5.22}$$

for some constant $C > 0$.

CLAIM 5.8. *There exists $C_0 > 0$ such that*

$$H(x_0) \geq -C_0 r_0^{-1/4}. \tag{5.23}$$

PROOF OF CLAIM 5.8. We study the following two cases.

CASE 1: $\mathrm{e}^{2f(x_0)} \leq r_0^{-1/2}$. Then, by the definition of H, we have that

$$\begin{aligned} H(x_0) &\geq -a\mathrm{e}^f - \frac{1-2a}{4}\mathrm{e}^{2f} \\ &\geq -a r_0^{-1/4} - \frac{1-2a}{4} r_0^{-1/2} \\ &\geq -C_0 r_0^{-1/4}. \end{aligned}$$

CASE 2: $e^{2f(x_0)} > r_0^{-1/2}$. Then (5.22) implies that

$$\phi^2(x_0) \leq C r_0^{-1/2},$$

so that $G > -\frac{2a+1}{4}$ yields

$$H(x_0) \geq -\frac{2a+1}{4}\phi^2(x_0) \geq -C r_0^{-1/2} \geq -C_0 r_0^{-1/4}.$$

This proves (5.23).

□

Since $\phi = 1$ in $B(p, r_0)$, we obtain from (5.23) that

$$\inf_{B(p,r_0)} \left(R - a e^f - \frac{1-2a}{4} e^{2f} \right) \geq -C r_0^{-1/4}.$$

Finally, taking $r_0 \to \infty$, we conclude that $R - a e^f - \frac{1-2a}{4} e^{2f} \geq 0$ on all of M^n. This completes the proof of the theorem. □

The following is an immediate consequence of Theorem 5.7.

COROLLARY 5.9. *If (M^n, g, f) is a complete noncompact steady gradient Ricci soliton with $f \leq 0$ and $\lim_{x\to\infty} R(x) = 0$, then*

$$\lim_{x\to\infty} f(x) = -\infty.$$

In particular, f is a proper function.

5.4. 4-Dimensional steady soliton singularity models have bounded curvature

In this section we discuss the proof of a result by Freedman, Shin, Zhang, and one of the authors [**CFSZ20**], which says that 4-dimensional steady gradient Ricci soliton singularity models must have bounded curvature.

THEOREM 5.10. *Any 4-dimensional steady gradient Ricci soliton singularity model must have bounded curvature tensor.*

The proof of this theorem, which we now present, follows the same method as to prove the quadratic curvature bound for 4-dimensional shrinking soliton singularity models in Theorem 2.41.

PROOF OF THEOREM 5.10. Suppose that the theorem is false. Then there would exist a 4-dimensional steady gradient Ricci soliton singularity model (M^4, g, f) and a sequence of points $\{y_i\}$ in M^4 with $|\mathrm{Rm}|(y_i) \to \infty$. By Corollary 2.44 with

$$A_i = \frac{1}{3}|\mathrm{Rm}|^{\frac{1}{2}}(y_i) \to \infty,$$

there would exist a sequence of points $\{x_i\}$ such that for each $i \geq 1$ we have that $x_i \in B\big(y_i, \frac{2}{3}\big)$,

$$Q_i := |\mathrm{Rm}|(x_i) \geq |\mathrm{Rm}|(y_i) \geq 100,$$

and

$$|\mathrm{Rm}|(x) \leq 4Q_i \quad \text{in } B\big(x_i, A_i Q_i^{-\frac{1}{2}}\big). \tag{5.24}$$

Since (M^4, g) is a singularity model, by Perelman's no local collapsing theorem, we have that

$$\text{Vol}\left(B\big(x_i, Q_i^{-\frac{1}{2}}\big)\right) \geq \kappa Q_i^{-2} \tag{5.25}$$

for some $\kappa > 0$ independent of i.

To see that the local derivative estimates hold, we consider the canonical form $(M^4, g(t), f(t))$ associated to (M^4, g, f). We have that $g(t) = \varphi_t^* g$ is a solution to the Ricci flow and $f(t) = f \circ \varphi_t$, where $\frac{\partial}{\partial t}\varphi_t(x) = (\nabla f)(\varphi_t(x))$. Recall that $|\nabla f|^2 \leq R + |\nabla f|^2 = 1$. For any $x \in M^4$ and $\bar{t} \in [-Q_i^{-1}, 0]$,

$$d(\varphi_{\bar{t}}(x), x) \leq \int_{\bar{t}}^0 \left|\frac{\partial}{\partial t}\varphi_t(x)\right| dt \leq |\bar{t}| \leq Q_i^{-1} \leq Q_i^{-\frac{1}{2}}. \tag{5.26}$$

Thus, if $x \in B\big(x_i, (A_i - 1)Q_i^{-\frac{1}{2}}\big)$ and $t \in [-Q_i^{-1}, 0]$, then

$$d(\varphi_t(x), x_i) \leq d(x, x_i) + d(\varphi_t(x), x) \leq (A_i - 1)Q_i^{-\frac{1}{2}} + Q_i^{-\frac{1}{2}} = A_i Q_i^{-\frac{1}{2}};$$

that is, $\varphi_t(x) \in B\big(x_i, A_i Q_i^{-\frac{1}{2}}\big)$. This implies that for such x and t,

$$|\text{Rm}|(x, t) = |\text{Rm}|(\varphi_t(x), 0) \leq 4Q_i.$$

Now, by Shi's local derivative of curvature estimates (Theorem 1.66), we have

$$|\nabla^m \text{Rm}| \leq C_m Q_i^{1+\frac{m}{2}} \quad \text{in } B\big(x_i, \tfrac{1}{2}A_i Q_i^{-\frac{1}{2}}\big) \times \big[-\tfrac{1}{2}Q_i^{-1}, 0\big].$$

By this and the noncollapsing estimate (5.25), we may apply Hamilton's Cheeger–Gromov compactness theorem to obtain that there exists a subsequence such the rescaled pointed solutions to the Ricci flow $(M^4, g_i(t), x_i)$, $t \in \big[-\frac{1}{2}, 0\big]$, where $g_i(t) := Q_i g(Q_i^{-1}t)$, converge in the pointed C^∞ Cheeger–Gromov sense to a complete solution

$$(M_\infty^4, g_\infty(t), x_\infty), \quad t \in \big[-\tfrac{1}{2}, 0\big],$$

to the Ricci flow satisfying $|\text{Rm}_{g_\infty(t)}| \leq 4$ and $|\text{Rm}_{g_\infty(0)}|(x_\infty) = 1$.

Since $g(t)$ satisfies $0 \leq R \leq 1$, we have that $0 \leq R_{g_i(t)} \leq Q_i^{-1}$. Since $Q_i \to \infty$, we see that $R_{g_\infty(t)} \equiv 0$. Since $g_\infty(t)$ is a solution to the Ricci flow, this implies that $\text{Ric}_{g_\infty(t)} \equiv 0$. Since $g_\infty(t)$ is also a singularity model, by Perelman's no local collapsing theorem, there exists a positive real number κ such that

$$\text{Vol}\left(B_r^{g_\infty(t)}(x)\right) \geq \kappa r^4 \tag{5.27}$$

for all $x \in M_\infty^4$ and $r > 0$. By a result of Cheeger and Naber [**CN15**, Corollary 8.85], this implies that $(M_\infty^4, g_\infty(t))$ is an asymptotically locally Euclidean space.

We can now follow the remainder of the proof of Theorem 2.41 to obtain a contradiction. This is due to the consequent existence of an unbounded number of disjoint copies of the ALE space in the original closed 4-manifold $\overline{M}^4$, for which the steady gradient Ricci soliton serves as a singularity model; this is ruled out by Theorem 2.46, and so completes the proof of Theorem 5.10. □

5.4.1. Limits at infinity of steady solitons with bounded curvature.

In view of Theorem 5.10 and Perelman's no local collapsing theorem, the following result applies to 4-dimensional steady gradient Ricci soliton singularity models as a special case. This result is similar in spirit to Proposition 2.57 for shrinking gradient Ricci solitons.

PROPOSITION 5.11 (Limits based at points with uniformly positive curvature norm). *Let (M^n, g, f) be a complete noncompact steady gradient Ricci soliton that has bounded sectional curvature and is noncollapsed. If $x_i \to \infty$ satisfies $|\mathrm{Rm}|(x_i) \geq c$ for some positive constant c, then the (unrescaled) sequence (M^n, g, x_i) subconverges in the C^∞-pointed Cheeger–Gromov sense to a complete noncompact pointed Riemannian manifold $(M^n_\infty, g_\infty, x_\infty)$ that has bounded sectional curvature and is noncollapsed and nonflat. Moreover, there exists a function f_∞ on M^n_∞ such that $(M^n_\infty, g_\infty, f_\infty)$ is also a steady gradient Ricci soliton.*

Furthermore, if $n = 4$ and if (M^4, g, f) is also a singularity model, then the limit (M^4_∞, g_∞) is not a non-flat Ricci-flat 4-manifold.

PROOF. The first part of the proposition regarding the existence and the properties of the limit $(M^n_\infty, g_\infty, x_\infty)$ follows from the C^∞-pointed Cheeger–Gromov compactness theorem and Shi's local derivatives of curvature estimates, where the nonflatness of g_∞ follows from $|\mathrm{Rm}_\infty|(x_\infty) \geq c > 0$.

To see the second part, let $f_i := f - f(x_i)$. We have the gradient bound $|\nabla f| \leq 1$. Furthermore, since the curvature of g is bounded, we have that $|\nabla^k f| = |\nabla^{k-2}\mathrm{Ric}| \leq C_k$ for all $k \geq 2$. Therefore, the sequence $\{f_i\}$ subconverges to some smooth function f_∞ on M^n_∞ with respect to the Cheeger–Gromov convergence above. Thus, $\mathrm{Ric}_{g_\infty} + \nabla^2_{g_\infty} f_\infty = 0$.

Finally, suppose that $n = 4$ and that (M^4, g, f) is a singularity model. If (M^4_∞, g_∞) were Ricci-flat, then by noncollapsing and Cheeger and Naber's result ([**CN15**, Corollary 8.85]), g_∞ would be asymptotically locally Euclidean. By the ALE exclusion principle (Theorem 2.46), we would obtain a contradiction. □

PROPOSITION 5.12 (Limits based at points with decaying scalar curvature). *Let (M^4, g, f) be a steady gradient Ricci soliton singularity model. If $x_i \to \infty$ satisfies $R(x_i) \to 0$, then (M^4, g, x_i) subconverges in the C^∞-pointed Cheeger–Gromov sense to $(\mathbb{R}^4, g_{\mathbb{E}}, 0^4)$.*

PROOF. By the C^∞-pointed Cheeger–Gromov compactness theorem and Shi's local derivatives of curvature estimates, (M^4, g, x_i) subconverges to some complete noncompact pointed Riemannian manifold $(M^4_\infty, g_\infty, x_\infty)$. Since $R(x_i) \to 0$, we have that $R_{g_\infty}(x_\infty) = 0$. In consideration of the associated canonical form of the steady gradient Ricci soliton, which is a Ricci flow, we may appeal to the strong maximum principle to obtain that $R_{g_\infty} \equiv 0$ and hence $\mathrm{Ric}_{g_\infty} \equiv 0$. Since g_∞ is noncollapsed, by the same result of Cheeger and Naber cited above, we have that g_∞ is asymptotically locally Euclidean. We may now apply the ALE exclusion principle to conclude that g_∞ is the Euclidean metric on $\mathbb{R}^4$. □

5.5. Curvature estimates for 4-dimensional steady solitons

In this section we prove the following result of Chan, Ma, and Zhang [**CMZ21**].

THEOREM 5.13. *Let (M^4, g, f) be a complete noncompact steady gradient Ricci soliton with bounded sectional curvature. If g is not Ricci-flat, then there exists a constant $C > 0$ such that*

$$|\mathrm{Rm}| \leq CR \quad \text{on } M^4. \tag{5.28}$$

PROOF. In the following, C will denote a positive universal constant that may change from line to line. Analogous to the derivation of (2.13), by using Proposition

2.3, we have that

$$\begin{aligned}\Delta_f |\mathrm{Ric}|^2 &\geq 2\,|\nabla \mathrm{Ric}|^2 - 4\,|\mathrm{Rm}|\,|\mathrm{Ric}|^2 \\ &\geq 2\,|\nabla \mathrm{Ric}|^2 - C\,|\mathrm{Ric}|^3 - \frac{C}{|\nabla f|}\,|\nabla \mathrm{Ric}|\,|\mathrm{Ric}|^2 \\ &\geq |\nabla \mathrm{Ric}|^2 - C\,|\mathrm{Ric}|^3 - \frac{C}{|\nabla f|^2}\,|\mathrm{Ric}|^4,\end{aligned} \tag{5.29}$$

where the last line follows from

$$\frac{C}{|\nabla f|}\,|\nabla \mathrm{Ric}|\,|\mathrm{Ric}|^2 \leq |\nabla \mathrm{Ric}|^2 + \frac{C}{|\nabla f|^2}|\mathrm{Ric}|^4.$$

Recall from (2.16) that we have

$$\Delta_f\,|\mathrm{Rm}| \geq -8\,|\mathrm{Rm}|^2.$$

By Proposition 2.3, this yields

$$\Delta_f\,|\mathrm{Rm}| \geq -C|\mathrm{Ric}|^2 - \frac{C}{|\nabla f|^2}|\nabla \mathrm{Ric}|^2. \tag{5.30}$$

From (5.29) and (5.30), we obtain that

$$\begin{aligned}\Delta_f\left(|\mathrm{Rm}| + \alpha|\mathrm{Ric}|^2\right) \geq{}& \left(\alpha - \frac{C}{|\nabla f|^2}\right)|\nabla \mathrm{Ric}|^2 - C\alpha\,|\mathrm{Ric}|^3 \\ &- \frac{C\alpha}{|\nabla f|^2}\,|\mathrm{Ric}|^4 - C|\mathrm{Ric}|^2\end{aligned} \tag{5.31}$$

for any $\alpha > 1$. Since M^4 has bounded Ricci curvature,

$$B := \sup_{M^4} |\mathrm{Ric}| < \infty.$$

Let

$$\Omega := \left\{x \in M^4 : |\nabla f|(x) \geq \frac{1}{2}\right\}.$$

Then, (5.31) implies the following inequality on Ω:

$$\Delta_f\left(|\mathrm{Rm}| + \alpha|\mathrm{Ric}|^2\right) \geq (\alpha - C)\,|\nabla \mathrm{Ric}|^2 - C\alpha\left(B^2+1\right)|\mathrm{Ric}|^2 \tag{5.32}$$

for some universal constant $C > 0$. In (5.32), we choose $\alpha = C + 1$ and conclude that

$$\Delta_f\left(|\mathrm{Rm}| + \alpha|\mathrm{Ric}|^2\right) \geq |\nabla \mathrm{Ric}|^2 - C\left(B^2+1\right)|\mathrm{Ric}|^2 \quad \text{on } \Omega.$$

According to $\Delta_f R = -2|\mathrm{Ric}|^2$, we then have that

$$\Delta_f\left(|\mathrm{Rm}| + \alpha|\mathrm{Ric}|^2 - \beta R\right) \geq |\nabla \mathrm{Ric}|^2 + \left(2\beta - C\left(B^2+1\right)\right)|\mathrm{Ric}|^2 \quad \text{on } \Omega.$$

For $\beta \geq C\left(B^2+1\right)$, it follows that

$$\Delta_f\left(|\mathrm{Rm}| + \alpha|\mathrm{Ric}|^2 - \beta R\right) \geq |\nabla \mathrm{Ric}|^2 + |\mathrm{Ric}|^2 \quad \text{on } \Omega. \tag{5.33}$$

Moreover, we further assume that β is chosen so that

$$\frac{1}{2}\beta \geq \sup_{M^4} |\mathrm{Rm}| + \alpha B^2 + 1, \tag{5.34}$$

where the right-hand side is finite by the assumption that M^4 has bounded sectional curvature.

By Proposition 2.3, there exists a universal constant $C > 0$ such that

$$|\nabla \mathrm{Ric}|^2 + |\mathrm{Ric}|^2 \geq \frac{1}{C}|\mathrm{Rm}|^2 \quad \text{on } \Omega.$$

Hence, we conclude that the function

$$\Phi := |\mathrm{Rm}| + \alpha|\mathrm{Ric}|^2 - \beta R \tag{5.35}$$

satisfies

$$\Delta_f \Phi \geq \frac{1}{C}\left(|\mathrm{Rm}|^2 + |\mathrm{Ric}|^2\right) \quad \text{on } \Omega. \tag{5.36}$$

Since (M^4, g) is complete, has Ricci curvature bounded from below, and Φ is bounded, by the Omori–Yau maximum principle (see Omori [**Omo67**] and Yau [**Yau75**]), there exists a sequence of points $\{x_i\}_{i=1}^{\infty}$ such that

$$\lim_{i\to\infty} \Phi(x_i) = \sup_{M^4} \Phi, \quad \lim_{i\to\infty} |\nabla\Phi(x_i)| = 0, \text{ and } \limsup_{i\to\infty} \Delta\Phi(x_i) \leq 0.$$

In particular, since $|\nabla f| \leq 1$, there exist $\{x_i\}_{i=1}^{\infty}$ such that

$$\Phi(x_i) \geq \sup_{M^4} \Phi - \frac{1}{i} \quad \text{and} \quad \Delta_f\Phi(x_i) \leq \Delta\Phi(x_i) + |\nabla\Phi|(x_i) \leq \frac{1}{i}.$$

It is worth mentioning that the above weak maximum principle for Δ_f holds for any gradient Ricci soliton, not just steady, and without any additional assumptions on curvature, see Pigola, Rimoldi, and Setti [**PRS11**].

CASE 1: $|\nabla f|(x_{i_j}) \geq \frac{1}{2}$ *for some subsequence* $\{x_{i_j}\}$. Then, since $x_{i_j} \in \Omega$ for the subsequence, by (5.36) we have

$$|\mathrm{Rm}|^2(x_{i_j}) + |\mathrm{Ric}|^2(x_{i_j}) \leq C\Delta_f\Phi(x_{i_j}) \leq \frac{C}{i_j}.$$

Thus,

$$\sup_{M^4} \Phi = \lim_{j\to\infty} \Phi(x_{i_j}) = 0.$$

CASE 2: $|\nabla f|(x_i) \leq \frac{1}{2}$ *for all i sufficiently large.* Then $R(x_i) \geq \frac{1}{2}$. From (5.34), we deduce that

$$\begin{aligned}
\Phi(x_i) &\leq |\mathrm{Rm}|(x_i) + \alpha|\mathrm{Ric}|^2(x_i) - \frac{1}{2}\beta \\
&\leq \sup_{M^4} |\mathrm{Rm}| + \alpha B^2 - \frac{1}{2}\beta \\
&\leq -1.
\end{aligned}$$

This proves that

$$\sup_{M^4} \Phi \leq \Phi(x_i) + \frac{1}{i} \leq -1 + \frac{1}{i} \leq 0.$$

Thus, we conclude in either case that $\Phi \leq 0$ on M^4. This completes the proof of the proposition. □

We remark that, by Theorem 5.10, the hypothesis of Theorem 5.13 holds if (M^4, g, f) is a steady gradient Ricci soliton singularity model.

COROLLARY 5.14. *Let (M^4, g, f) be a complete non-Ricci flat gradient steady Ricci soliton singularity model. Then there exists a positive constant C such that*

$$|\mathrm{Rm}| \leq CR \quad \textit{on } M^4.$$

Cao and Cui [**CC20**] proved that if (M^4, g, f) is a non-Ricci flat steady gradient Ricci soliton with scalar curvature convergent to zero at infinity, then the sectional curvatures are bounded. Under this hypothesis, Chan [**Cha19**] then proved that $|\mathrm{Rm}| \leq CR$. Here, we use Theorem 5.13 to obtain this conclusion.

THEOREM 5.15. *Let (M^4, g, f) be a complete noncompact steady gradient Ricci soliton with*

$$\lim_{x\to\infty} R(x) = 0.$$

If g is not Ricci-flat, then M^4 has bounded sectional curvature. Hence, we have that

$$|\mathrm{Rm}| \leq CR \quad on \;\; M^4. \tag{5.37}$$

PROOF. We adapt the proof of Proposition 2.5 to our setting here. By

$$\lim_{x\to\infty} R(x) = 0,$$

it follows that there exists a compact set $D \subset M^4$ so that

$$|\nabla f| \geq \frac{1}{2} \quad \text{on} \;\; M^4 \setminus D. \tag{5.38}$$

By (5.29), we obtain that

$$\begin{aligned}\Delta_f |\mathrm{Ric}|^2 &\geq 2|\nabla \mathrm{Ric}|^2 - 4|\mathrm{Rm}|\,|\mathrm{Ric}|^2 \\ &\geq 2|\nabla \mathrm{Ric}|^2 - C|\mathrm{Ric}|^3 - C|\nabla \mathrm{Ric}|\,|\mathrm{Ric}|^2\end{aligned} \tag{5.39}$$

on $M^4 \setminus D$, where $C > 0$ is a universal constant. We have

$$\begin{aligned}\Delta_f\left(|\mathrm{Ric}|^2 R^{-\frac{1}{2}}\right) &= R^{-\frac{1}{2}}\Delta_f\left(|\mathrm{Ric}|^2\right) + |\mathrm{Ric}|^2\Delta_f\left(R^{-\frac{1}{2}}\right) \\ &\quad + 2\left\langle \nabla R^{-\frac{1}{2}}, \nabla|\mathrm{Ric}|^2\right\rangle.\end{aligned}$$

Since $\Delta_f R = -2|\mathrm{Ric}|^2$, we obtain that

$$\begin{aligned}\Delta_f\left(R^{-\frac{1}{2}}\right) &= -\frac{1}{2}(\Delta_f R)R^{-\frac{3}{2}} + \frac{3}{4}|\nabla R|^2 R^{-\frac{5}{2}} \\ &= |\mathrm{Ric}|^2 R^{-\frac{3}{2}} + \frac{3}{4}|\nabla R|^2 R^{-\frac{5}{2}}.\end{aligned}$$

Furthermore, the following holds:

$$\begin{aligned}2\left\langle \nabla R^{-\frac{1}{2}}, \nabla|\mathrm{Ric}|^2\right\rangle &\geq -2|\nabla \mathrm{Ric}|\,|\nabla R|\,R^{-\frac{3}{2}}|\mathrm{Ric}| \\ &\geq -\frac{3}{4}|\nabla R|^2 R^{-\frac{5}{2}}|\mathrm{Ric}|^2 - \frac{4}{3}|\nabla \mathrm{Ric}|^2 R^{-\frac{1}{2}}.\end{aligned}$$

Combining these estimates with (5.39), it follows that

$$\begin{aligned}\Delta_f\left(|\mathrm{Ric}|^2 R^{-\frac{1}{2}}\right) &\geq \frac{2}{3}|\nabla \mathrm{Ric}|^2 R^{-\frac{1}{2}} + |\mathrm{Ric}|^4 R^{-\frac{3}{2}} \\ &\quad - C|\mathrm{Ric}|^3 R^{-\frac{1}{2}} - C|\nabla \mathrm{Ric}|\,|\mathrm{Ric}|^2 R^{-\frac{1}{2}}.\end{aligned}$$

Applying the inequality

$$C|\nabla \mathrm{Ric}|\,|\mathrm{Ric}|^2 R^{-\frac{1}{2}} \leq \frac{2}{3}|\nabla \mathrm{Ric}|^2 R^{-\frac{1}{2}} + C|\mathrm{Ric}|^4 R^{-\frac{1}{2}},$$

we conclude that

$$\Delta_f\left(|\mathrm{Ric}|^2 R^{-\frac{1}{2}}\right) \geq (1 - CR)|\mathrm{Ric}|^4 R^{-\frac{3}{2}} - C|\mathrm{Ric}|^3 R^{-\frac{1}{2}}$$

holds on $M^4 \setminus D$, where $C > 0$ is a universal constant. As

$$\lim_{x\to\infty} R(x) = 0,$$

we may assume that the compact set $D \subset M^4$ is large enough so that $1 - CR \geq \frac{1}{2}$. Together with

$$C|\mathrm{Ric}|^3 R^{-\frac{1}{2}} \leq \frac{1}{4}|\mathrm{Ric}|^4 R^{-\frac{3}{2}} + CR^{\frac{1}{2}},$$

it follows that

$$\Delta_f \left(|\mathrm{Ric}|^2 R^{-\frac{1}{2}}\right) \geq \frac{1}{4}|\mathrm{Ric}|^4 R^{-\frac{3}{2}} - CR^{\frac{1}{2}}.$$

Since $0 < R \leq 1$ on M^4, we conclude that the function

$$u = |\mathrm{Ric}|^2 R^{-\frac{1}{2}}$$

satisfies

$$\Delta_f u \geq \frac{1}{4}u^2 - C \quad \text{on } M^4 \setminus D. \tag{5.40}$$

Lemma 2.4 proves that u is bounded on M^4. Thus, M^4 has bounded Ricci curvature.

Following the proof of (5.31) and noting that $|\nabla f| \geq \frac{1}{2}$ on $M^4 \setminus D$, we infer that

$$\Delta_f \left(|\mathrm{Rm}| + \alpha|\mathrm{Ric}|^2\right) \geq (\alpha - C)\, |\nabla \mathrm{Ric}|^2 - C\alpha\, |\mathrm{Ric}|^3 - C\alpha\, |\mathrm{Ric}|^4 - C|\mathrm{Ric}|^2.$$

We apply this for $\alpha = C+1$, and use that M^4 has bounded Ricci curvature. Then, there exists a constant $C_0 > 0$ such that

$$\Delta_f \left(|\mathrm{Rm}| + \alpha|\mathrm{Ric}|^2\right) \geq |\nabla \mathrm{Ric}|^2 + |\mathrm{Ric}|^2 - C_0.$$

By Proposition 2.3, there exists a universal constant $C > 0$ such that

$$|\nabla \mathrm{Ric}|^2 + |\mathrm{Ric}|^2 \geq \frac{1}{C}|\mathrm{Rm}|^2 \quad \text{on } \Omega.$$

Hence, the function

$$u := |\mathrm{Rm}| + \alpha|\mathrm{Ric}|^2$$

satisfies

$$\Delta_f u \geq \frac{1}{C}u^2 - C_1 \quad \text{on } M^4 \setminus D.$$

Lemma 2.4 implies that u is bounded on M^4, and so M^4 has bounded Riemann curvature. The conclusion that $|\mathrm{Rm}| \leq CR$ now follows from Theorem 5.13. This proves the result. □

Theorem 5.15 may be viewed as an analogue of Theorem 2.12 in the setting of steady Ricci solitons.

We say that a steady gradient Ricci soliton (M^4, g, f) has proper potential if

$$\lim_{x\to\infty} f(x) = -\infty,$$

and that (M^4, g, f) has linear scalar curvature decay if

$$R \leq \frac{C}{r+1},$$

where r is the distance function to a fixed point in M^4 and $C > 0$ is a constant. The following result, obtained by Chan, Ma, and Zhang in [**CMZ21**], is a Hamilton–Ivey-type pinching estimate analogous to Theorem 2.38 for shrinkers.

THEOREM 5.16. *Assume that* (M^4, g, f) *is a complete and non-Ricci flat gradient steady Ricci soliton with proper potential and linear scalar curvature decay. Then there exists a constant* $C > 0$ *such that*

$$-\frac{C}{(r+1)^{\frac{3}{2}}}\,\mathrm{id}_{\wedge^2(T^*M)} \leq \mathrm{Rm} \leq \frac{C}{r+1}\,\mathrm{id}_{\wedge^2(T^*M)}.$$

5.6. Asymptotically cylindrical steady solitons with positive sectional curvature

In this section, survey a pair of results regarding steady solitons satisfying positivity of curvature conditions.

DEFINITION 5.17 (Asymptotically quotient cylindrical). A complete noncompact steady gradient Ricci soliton (M^n, g, f) is **asymptotically quotient cylindrical** if for any sequence of points $p_i \to \infty$ there is a subsequence of

$$(M^n, g_i(t), p_i), \quad \text{where } g_i(t) = R(p_i)\, g\big(tR(p_i)^{-1}\big),$$

converging to a shrinking round (quotient) cylinder solution $(\mathbb{S}^{n-1}/\Gamma) \times \mathbb{R}$ to the Ricci flow (where Γ may depend on the sequence). Here, $g(t)$ is the associated canonical form. If $\Gamma = 1$, then we say that (M^n, g, f) is **asymptotically cylindrical**.

Brendle [**Bre14**] proved the following result:

THEOREM 5.18. *If* (M^n, g, f), $n \geq 4$, *is an asymptotically cylindrical steady gradient Ricci soliton with positive sectional curvature, then* (M^n, g, f) *is the rotationally symmetric Bryant soliton.*

Deng and Zhu [**DZ20b**], which in part used Brendle's result, proved the following.

THEOREM 5.19. *Let* (M^4, g, f) *be a 4-dimensional complete noncompact* κ-*noncollapsed steady gradient Ricci soliton with nonnegative sectional curvature. If* g *has linear curvature decay, i.e.,*

$$R(x) \leq \frac{C}{d(x,o)+1} \quad \textit{for all } x \in M^4, \tag{5.41}$$

then (M^4, g, f) *must be rotationally symmetric.*

For the study of 4-dimensional steady gradient Ricci solitons under the assumption that they "dimension reduce", see Deng, Ma, and one of the authors [**CDM22**].

5.7. Open problems for curvature and volume estimates of steady solitons

OPEN PROBLEM 5.1. *Let* $(M^4_\infty, g_\infty, f_\infty)$ *be a steady gradient Ricci soliton limit as in Proposition* 5.11; *that is, it is the limit of a steady gradient Ricci soliton based at points tending to infinity. Must* $(M^4_\infty, g_\infty, f_\infty)$ *split off an* $\mathbb{R}$-*factor?*

A weaker question is: Can one show that the tangent flow at infinity (see Definition 6.67 *below) of* $(M^4_\infty, g_\infty, f_\infty)$ *cannot be* $\mathbb{S}^3/\Gamma \times \mathbb{R}$? *For example, can one show that* $(M^4_\infty, g_\infty, f_\infty)$ *cannot be the* 4-*dimensional Bryant soliton?*

OPEN PROBLEM 5.2. *In dimension* 4, *must any steady gradient Ricci soliton* (M^4, g, f) *satisfy* $|\mathrm{Rm}| \leq CR$*? Corollary* 5.14 *says that this is true if* (M^4, g) *is a singularity model, whereas Theorem* 5.15 *says that this is true if* (M^4, g) *instead satisfies* $R \to 0$.

OPEN PROBLEM 5.3. *In dimensions* 5 *and higher, must steady gradient Ricci soliton singularity models have bounded sectional curvature?*

OPEN PROBLEM 5.4. *In dimension* 4, *must a steady gradient Ricci soliton singularity model with positive asymptotic volume ratio be a Ricci-flat ALE* 4-*manifold?*

OPEN PROBLEM 5.5. *What are the volume growths of Lai's* 4-*dimensional flying wing steady gradient Ricci solitons?*

CHAPTER 6

Conjugate Heat Kernel Methods for Steady Solitons

In this chapter, we continue our study of the curvature and volume of gradient Ricci solitons under the general hypotheses of being singularity models, sometimes in dimension 4. The fundamental new tools we introduce to our study of solitons are found in Bamler's higher-dimensional Ricci flow theory [**Bam20a**, **Bam23**, **Bam20b**]; a cursory summary of this is given in §6.1 and §6.2. This theory is based in part on the conjugate heat kernel and introduces new a priori estimates, synthetic compactness methods, and novel structure and regularity theories. The structure and regularity theories draw inspiration from and extend to Ricci flow in certain directions the theory of Cheeger, Colding, and Naber for Riemannian manifolds with Ricci curvature bounds, all the while being adapted in a manner intrinsic to Ricci flow. See Bamler, Jiang, Naber, and one of the authors [**BCJN**] for a detailed treatment in book form of spaces with Ricci curvature bounds.

In short, Bamler's Ricci flow compactness, structure, and regularity theories partially solve the microscope problem discussed in Chapter 1, with particularly effective results in dimension 4. This, in turn, has a significant impact on our understanding of 4-dimensional shrinking and steady gradient Ricci soliton singularity models. We will describe some of this impact in this chapter.

Related to our study of steady gradient Ricci solitons is Bamler's notion of "tangent flow at infinity"; this is reviewed in §6.3. Roughly speaking, the tangent flow at infinity of a noncompact ancient solution of the Ricci flow reflects what may be considered as its principal asymptotic behavior. It is related to Perelman's notion of an asymptotic soliton. For 4-dimensional *steady* gradient Ricci soliton singularity models on orbifolds with isolated singularities, we classify in §6.4 their tangent flows at infinity. We also show in §6.5 that if the steady gradient Ricci soliton singularity models are not Ricci-flat ALE 4-orbifolds with isolated singularities, then their scalar curvatures have at most quadratic decay.

For steady gradient Ricci soliton singularity models in any dimension, we discuss in §6.6 the results that their volume growth rate is qualitatively bounded from below by the volume growth rate of the Bryant steady soliton and is bounded from above by the volume growth rate of Euclidean space, the latter of which is a special case of a result of Bamler.

6.1. Bamler's Ricci flow monotonicity theory

Originally, Perelman developed a space-time geometry for Ricci flow based on the reduced distance function [**Per02**, §7ff]. Bamler's theory provides a new perspective based on the conjugate heat kernel and probabilistic ideas.

In this section and the next, we provide a cursory survey of those aspects of Bamler's theory relevant to our study of steady gradient Ricci solitons. Generally, we will use his results as black boxes, only occasionally delving into the details when necessary to apply his theory. The reader is strongly encouraged to read Bamler's original papers [**Bam20a**, **Bam23**, **Bam20b**] for the theory of higher-dimensional Ricci flow singularity formation.

Beautiful applications are given in Bamler [**Bam21a**] (cf. §1.4.2.1 in this volume) and Bamler, Cifarelli, Conlon, and Deruelle [**BCCD24**] (cf. Theorem 1.25 in this volume). Further applications, due to Bamler, Chan, Deng, Ma, Zhang, and one of the authors, are discussed in this chapter.

In this chapter, $(\mathcal{O}^n, (g_t)_{t\in I})$, unless otherwise indicated, will be a Ricci flow on a closed orbifold with isolated singularities. Generally, the results we discuss also hold in the noncompact case for singularity models, which is the desired class of objects of study for singularity analysis.

Recall that in the classical viewpoint toward singularity analysis (see §1.1.4), different sequences of space-time points $\{(x_i, t_i)\}_i$, *provided* their associated rescalings converge, yield different perceptible glimpses into the singularity formation of the Ricci flow; see Hamilton [**Ham95b**, §16ff] and Perelman [**Per02**, §10ff]. A choice of sequence $\{(x_i, t_i)\}_i$ is called a **point selection**. In the Hamilton–Perelman theory, the aim is to obtain convergence for suitable point selections and then to use this to characterize singularity formation sufficiently well in order to successfully perform surgery for singular Ricci flows while keeping track of the topological changes. This viewpoint is fundamental to Ricci flow and Bamler's theory presents a critical modification.

Conjugate heat kernels, introduced by Perelman [**Per02**, §9,§10] in the study of Ricci flow, occupy a central position in Bamler's theoretical framework. Namely, given any point selection $\{(x_i, t_i)\}$, the associated conjugate heat kernels are used to essentially define a new notion of "Ricci flow world lines" backwards in time from each (x_i, t_i) and this facilitates the compactness theorems that hold in complete generality. We now turn to a more detailed exploration of these conjugate heat kernels, and their role in our analysis of Ricci flows that go singular, as well as of gradient Ricci soliton singularity models.

6.1.1. Heat equation and conjugate heat kernels.

The conjugate heat kernel, a fundamental analytic-geometric invariant of a Ricci flow, will be central to our ensuing discussions.

6.1.1.1. *Gradient estimate for the heat equation coupled to the Ricci flow.* We begin with the heat equation. Let $(\mathcal{O}^n, g_t)$ be a Ricci flow on a closed orbifold with isolated singularities. Let

$$\square := \partial_t - \Delta_{g_t} \tag{6.1}$$

be the **heat operator** associated to the Ricci flow g_t. Let $u \in C^3(\mathcal{O}^n \times I)$ be a solution to the heat equation

$$\square u = 0, \tag{6.2}$$

which is effectively coupled to the Ricci flow. We can compute that (see, for example Exercise 1.40 in [**CLN06**])

$$\square |\nabla u|^2 = -2|\nabla^2 u|^2 \leq 0. \tag{6.3}$$

A key point in deriving (6.3) is that the coupling with the Ricci flow introduces a $\mathrm{Ric}(\nabla u, \nabla u)$ term which cancels a like term obtained by commuting Δ and ∇ acting on u.[1] Thus (see, for example, Exercise 1.38 in [**CLN06**]),

$$\square|\nabla u| = -\frac{|\nabla^2 u|^2}{|\nabla u|} + \frac{|\nabla|\nabla u||^2}{|\nabla u|} \leq 0 \tag{6.4}$$

wherever $|\nabla u| \neq 0$, and where the last inequality is a consequence of Kato's inequality.

By applying the weak parabolic maximum principle to (6.3), we obtain the following (cf. [**Bam20a**, Lemma 2.5]):

PROPOSITION 6.1 (Gradient monotonicity). *For a solution u to the heat equation coupled to a Ricci flow $(\mathcal{O}^n, g_t)$ on a closed orbifold with isolated singularities, if $|\nabla u| \leq C$ at $t = t_0$, then $|\nabla u| \leq C$ for $t \geq t_0$.*

This estimate is sharp if we allow noncompact orbifolds, as is seen by taking u to be a static linear function on Euclidean space, which has constant gradient. We remark that the general philosophy in geometric analysis, with Ricci flow serving as a prime example, holds that sharp estimates are typically of high utility.

6.1.1.2. *Definition of conjugate heat kernel.*

Let

$$\square^* := -\partial_t - \Delta_{g_t} + R_{g_t} \tag{6.5}$$

be the **conjugate heat operator**; see [**Per02**, §9]. Since $(\mathcal{O}^n, g_t)$ satisfies the Ricci flow, for any C^2 functions ϕ, ψ on $\mathcal{O}^n \times I$ with compact support, where I is an open interval, by integration by parts we have that

$$\iint_{\mathcal{O}^n \times I} \square\phi\, \psi\, dv_t\, dt = \iint_{\mathcal{O}^n \times I} \phi\, \square^* \psi\, dv_t\, dt, \tag{6.6}$$

where dv_t denotes the Riemannian measure with respect to g_t. Thus, $\square^*$ is the L^2-adjoint of $\square$ in space-time. We remark that in the derivation of (6.6), one uses the fact that under the Ricci flow, $\partial_t dv_t = -R_{g_t} dv_t$.

We also observe that if $\square^* u = 0$, then for a Ricci flow background g_t, we have the conservation law:

$$\frac{d}{dt}\int_{\mathcal{O}^n} u(x,t)\, dv_t(x) = 0, \tag{6.7}$$

using the divergence theorem and the equation $\partial_t dv_t = -R\, dv_t$, as used in the previous paragraph.

DEFINITION 6.2 (Conjugate heat kernel). Let $K(x,t;y,s)$ denote the unique **minimal positive fundamental solution to the heat equation** coupled to g_t. We also call K the **(conjugate) heat kernel**. It is of fundamental importance to Ricci flow. By the definition of the fundamental solution,

$$\square_{x,t} K(x,t;y,s) = 0, \quad \lim_{t \to s_+} K(x,t;y,s) = \delta_y(x), \tag{6.8}$$

$$\square^*_{y,s} K(x,t;y,s) = 0, \quad \lim_{s \to t_-} K(x,t;y,s) = \delta_x(y), \tag{6.9}$$

[1]This offers computational advantage compared to working on a static Riemannian manifold. Correspondingly, in the static case, often non-negative Ricci curvature is assumed, whereas in the Ricci flow case, often no curvature sign condition is assumed.

for $x, y \in \mathcal{O}^n$, $s < t$, and where δ_y is the Dirac delta function centered at y. By K being **minimal**, we mean that for any positive fundamental solution H to the heat equation, we have that $K \leq H$.

Note that (6.8) implies (6.9), and vice versa. Indeed, if (6.8) holds, then for $s < r < t$ and $x, y \in \mathcal{O}^n$, we have by integrating by parts that

$$\begin{aligned} 0 &= \int_{\mathcal{O}^n} K(x,t;z,r) \square_{z,r} K(z,r;y,s)\, dv_r(z) \\ &= \int_{\mathcal{O}^n} \square^*_{z,r} K(x,t;z,r) K(z,r;y,s)\, dv_r(z). \end{aligned}$$

By letting $r \to s_+$, we obtain $\square^*_{y,s} K(x,t;y,s) = 0$, so that (6.9) holds. One proves the converse similarly. Observe that by (6.7) and (6.9), we have

$$\int_{\mathcal{O}^n} K(x,t;y,s)\, dv_s(y) \equiv 1. \tag{6.10}$$

Fix t, let $h_t : \mathcal{O}^n \to \mathbb{R}$, and let $h(y,s)$, where $y \in \mathcal{O}^n$ and $s \leq t$, be the solution to the conjugate heat equation with $h(y,t) = h_t(y)$ for $y \in \mathcal{O}^n$ as its initial data. Then we may represent the solution by

$$h(y,s) = \int_{\mathcal{O}^n} h_t(x) K(x,t;y,s)\, dv_t(x). \tag{6.11}$$

In this context, Bamler's philosophy suggests a point (x_0, t_0) in the space-time $\mathcal{O}^n \times I$ of a Ricci flow should be considered not merely as a point but as the Dirac delta function δ_{x_0} in the time-slice $\mathcal{O}^n \times t_0$. In this light, a point assumes the character of a probability measure. The "past" of the point should be considered as the conjugate heat kernel based at (x_0, t_0). In particular, geometrically the past is not simply the standard "world line" (x_0, t), where $t < t_0$. This viewpoint is related to Bamler's compactness theorem, which we will recall in §6.2.9 below, as well as his notion of tangent flow at infinity (see Definition 6.67 below).

REMARK 6.3 (Conjugate heat kernel of pullback and scaled flows). Let $(\mathcal{O}^n, g(t))$ be a Ricci flow, $\phi : \mathcal{O}^n \to \mathcal{O}^n$ a diffeomorphism, and $c \in \mathbb{R}_+$. Then $g^\phi(t) := \phi^* g(t)$ and $g_c(t) := c g(c^{-1}t)$ are Ricci flows.

(1) The conjugate heat kernel K^ϕ for $g^\phi(t)$ is given by

$$K^\phi(x,t;y,s) = K\big(\phi(x), t; \phi(y), s\big). \tag{6.12}$$

(2) The conjugate heat kernel K_c for $g_c(t)$ is given by

$$K_c(x,t;y,s) = c^{-n/2} K(x, c^{-1}t; y, c^{-1}s). \tag{6.13}$$

6.1.1.3. *Existence and uniqueness of conjugate heat kernels.*

When the underlying space is a smooth manifold, the following result is given in [**CCG$^+$10**, Theorem 24.40] (the compact case is standard and also discussed in [**CCG$^+$10**, Chapter 24]). See Chiang [**Chi90**] for the case of closed Riemannian orbifolds (with a static metric).

PROPOSITION 6.4 (Existence and uniqueness of conjugate heat kernels). *Without any assumption on g_t except smoothness, the existence and uniqueness of the (conjugate) heat kernel $K(x,t;y,s)$ holds when $\mathcal{O}^n$ is noncompact.*

The idea of the proof is to take the limit of Dirichlet heat kernels of an exhaustion of $\mathcal{O}^n$ by smooth compact domains. As usual, estimates are used to prove convergence. By the maximum principle, any Dirichlet heat kernel based at a point is less than any positive fundamental solution based at the same point. This implies the minimality and uniqueness.

DEFINITION 6.5 (Conjugate heat kernel measure [**Bam20a**, Definition 2.4]). The measure $\nu_{x_0,t_0}(t) := \nu_{x_0,t_0;t}$ defined by the heat kernel

$$d\nu_{x_0,t_0;t} := K(x_0,t_0;\cdot,t)\,dv_t \tag{6.14}$$

is the **conjugate heat kernel measure based at** (x_0,t_0).

DEFINITION 6.6 (Conjugate heat kernel potential). Let $\tau(t) := t_0 - t > 0$, and define $f_t := f(\cdot,t)$ by

$$d\nu_{x_0,t_0;t} =: (4\pi\tau(t))^{-\frac{n}{2}}\mathrm{e}^{-f_t}dv_t\,. \tag{6.15}$$

We call f_t the **potential of the conjugate heat kernel** based at (x_0,t_0).

For ancient solutions, we can think of f_t as a proxy for the potential function of a shrinking gradient Ricci soliton. Both of these functions may be considered as proxies for the distance squared function.

REMARK 6.7 (Conjugate heat kernel measure of pullback and scaled flows). By Remark 6.3, we have that if $\phi : \mathcal{O}^n \to \mathcal{O}^n$ a diffeomorphism and $c \in \mathbb{R}_+$, then the conjugate heat kernel measure $d\nu_c^\phi$ of the solution

$$g_c^\phi(t) := c\phi^* g(c^{-1}t) \tag{6.16}$$

is given by

$$(d\nu_c^\phi)_{x_0,t_0;t} = (\phi^{-1})_*\big(d\nu_{\phi(x_0),c^{-1}t_0;c^{-1}t}\big). \tag{6.17}$$

6.1.1.4. *Perelman's reduced distance function.*

Perelman introduced a space-time geometry for Ricci flows, which is closely related to the study of conjugate heat kernels. We now briefly discuss this geometry, known as **reduced geometry**, and encourage the reader to consult Perelman's original paper [**Per02**].

Let $(\mathcal{O}^n, g(t))$, $t \in (-\infty, 0]$, be an ancient Ricci flow on a smooth orbifold with isolated singularities. Given $\overline{\tau} > 0$ and $t_0 \leq 0$, let $\gamma : [0,\overline{\tau}] \to \mathcal{O}^n$ be a piecewise smooth path. Perelman's **$\mathcal{L}$-length** of γ is defined by (see [**Per02**, §7])

$$\mathcal{L}(\gamma) = \int_0^{\overline{\tau}} \sqrt{\tau}\big(R(\gamma(\tau), t_0 - \tau) + |\gamma'(\tau)|^2_{g(t_0-\tau)}\big)d\tau. \tag{6.18}$$

Note that the point $\gamma(\tau)$ is considered as being at the time $t_0 - \tau$. That is, we consider $\tau \mapsto (\gamma(\tau), t_0 - \tau)$ as the space-time path associated to γ.

Let $x_0 \in \mathcal{O}^n$. Perelman's **reduced distance** function ℓ based at (x_0,t_0) is defined by

$$\ell(x,\overline{\tau}) := \ell(x_0,t_0;x,t_0-\overline{\tau}) := \frac{1}{2\sqrt{\overline{\tau}}}\inf_\gamma \mathcal{L}(\gamma), \tag{6.19}$$

where the infimum is taken over all $\gamma : [0,\overline{\tau}] \to \mathcal{O}^n$ satisfying $\gamma(0) = x_0$ and $\gamma(\overline{\tau}) = x$. We think of this as being a notion of space-time distance between the points (x_0,t_0) and $(x,t_0-\overline{\tau})$.

EXAMPLE 6.8 (Reduced distance of Euclidean space). Consider Euclidean space $(\mathbb{R}^n, g_{\mathbb{E}})$ as a static solution to the Ricci flow. Given $\mathbf{x}_0 \in \mathbb{R}^n$, the Euclidean reduced distance is given by

$$\ell(\mathbf{x}, \tau) = \frac{|\mathbf{x} - \mathbf{x}_0|^2}{4\tau};$$

see e.g. [**CCG$^+$07**, Exercise 7.12].

By Perelman [**Per02**, §7.1], for any $\tau > 0$, there exists a point $q_\tau \in \mathcal{O}^n$ such that

$$\ell(q_\tau, \tau) \leq \frac{n}{2}. \tag{6.20}$$

Any such point q_τ is called an **ℓ-center** at time $t_0 - \tau$; we call τ the **scale** of q_τ. Instead of being captured by a single worldline $\{(x_0, t) : t < t_0\}$, the "past" of (x_0, t_0) is represented by a range of possible points $(q_\tau, t_0 - \tau)$, emphasizing that the past is not precisely localized to a single point in space.

6.1.1.5. *Perelman's differential Harnack estimate.*

In this subsection we discuss a remarkable differential Harnack-type estimate by Perelman for conjugate heat kernels, which is closely related to his space-time geometry.

We first derive the evolution equation for the potential $f := f_t$ of the conjugate heat kernel, which is a backward heat-type equation. Since $K(x_0, t_0; \cdot, t) = (4\pi\tau(t))^{-\frac{n}{2}} \mathrm{e}^{-f_t}$ and by (6.9), we have

$$\begin{aligned} 0 &= \square^* \left((4\pi\tau(t))^{-\frac{n}{2}} \mathrm{e}^{-f(x,t)} \right) \\ &= (4\pi\tau(t))^{-\frac{n}{2}} \mathrm{e}^{-f(x,t)} \left(-\frac{n}{2\tau} + \partial_t f + \Delta f - |\nabla f|^2 + R \right), \end{aligned}$$

where $\square^* = -\partial_t - \Delta_{g_t} + R_{g_t}$. Thus, the potential of the conjugate heat kernel satisfies (cf. [**Per02**, (3.3)])

$$\partial_t f = -\Delta f + |\nabla f|^2 - R + \frac{n}{2\tau}. \tag{6.21}$$

REMARK 6.9. In comparison, suppose that $(\mathcal{O}^n, g_t, f_t, \frac{1}{1-t})$ is the canonical form of a shrinking gradient Ricci soliton. Then, by (1.74), we have that (6.21) holds for $\tau(t) = 1 - t$, and with f_t being the potential of the shrinking soliton instead of the potential of the conjugate heat kernel.

Define Perelman's differential Harnack quantity by

$$w := \tau\big(2\Delta f - |\nabla f|^2 + R\big) + f - n. \tag{6.22}$$

By [**Per02**, §9], we have **Perelman's differential Harnack estimate** for the conjugate heat kernel:[2]

$$w \leq 0. \tag{6.23}$$

This estimate, presented above in its differential form, is fundamental for understanding Ricci flow. Remarkably, there is no curvature sign assumption on the Ricci flow solution, making the estimate completely general in this sense.

REMARK 6.10 (Model case for Perelman's differential Harnack estimate). In the case of the canonical form of a shrinker, we have by (1.73) and (1.75) that $w \equiv 0$.

[2] For an exposition, see e.g. [**CCG$^+$08**, Chapter 16].

The quantity

$$v := w(4\pi\tau)^{-\frac{n}{2}}\mathrm{e}^{-f} \tag{6.24}$$

satisfies the conjugate heat-type equation (see [**Per02**, §9]):

$$\Box^* v = -2\tau \left|\mathrm{Ric} + \nabla^2 f - \frac{1}{2\tau} g\right|^2 (4\pi\tau)^{-\frac{n}{2}}\mathrm{e}^{-f}. \tag{6.25}$$

This formula is one of the essential ingredients in Perelman's proof of (6.23). The idea is that $v \leq 0$ at $\tau = 0$ and hence by the maximum principle, $v \leq 0$ for all $\tau > 0$. However, the rigorous proof is not quite that simple since the conjugate heat kernel $(4\pi\tau)^{-\frac{n}{2}}\mathrm{e}^{-f}$ approaches a Dirac delta function as $\tau \to 0$; see [**CCG**$^+$**08**, Chapter 16] for details.

As a consequence of (6.23), one can show that (see [**Per02**, Corollary 9.5])

$$f(x,t) \leq \ell(x_0, t_0; x, t). \tag{6.26}$$

In particular, this inequality is obtained by integrating the differential Harnack estimate (6.23) and using the asymptotics of the conjugate heat kernel as $\tau \to 0$ (which tends to a δ function).

REMARK 6.11. By [**Per02**, (7.13)], the reduced distance ℓ satisfies

$$\frac{\partial \ell}{\partial \tau} - \Delta\ell + |\nabla\ell|^2 - R + \frac{n}{2\tau} \geq 0.$$

Thus,

$$\Box^*\left((4\pi\tau)^{-\frac{n}{2}}\mathrm{e}^{-\ell}\right) \leq 0 \tag{6.27}$$

in the weak sense. Since $\Box^*\left((4\pi\tau)^{-\frac{n}{2}}\mathrm{e}^{-f}\right) = 0$, and since f and ℓ have the same asymptotic behavior as $\tau \to 0$, this gives another, heuristic, proof that $f \leq \ell$.

Inequality (6.26) is equivalent to

$$K(x_0, t_0; x, t) \geq (4\pi(t_0 - t))^{-\frac{n}{2}}\mathrm{e}^{-\ell(x_0,t_0;x,t)}. \tag{6.28}$$

This inequality, which is an integrated form of Perelman's estimate, reduces estimating conjugate heat kernels from below to estimating the reduced distance from above.

Observe that for at ℓ-center q_τ, by (6.28) we have the conjugate heat kernel lower bound:

$$K\big(x_0, t_0; q_\tau, t_0 - \tau\big) \geq (4\pi\mathrm{e})^{-\frac{n}{2}}\tau^{-\frac{n}{2}}. \tag{6.29}$$

EXAMPLE 6.12 (Euclidean conjugate heat kernel potential). Let $(\mathbf{x}_0, t_0) \in \mathbb{R}^n \times (-\infty, \infty)$. In the case of the Euclidean conjugate heat kernel $K_{\mathbb{E}}$, the potential is

$$f(\mathbf{x}, t) = \frac{|\mathbf{x} - \mathbf{x}_0|^2}{4(t_0 - t)}. \tag{6.30}$$

The **Euclidean heat kernel measure** is $\nu_{\mathbf{x}_0,t_0}$, where $d\nu_{\mathbf{x}_0,t_0} = u\, dv_{g_{\mathbb{E}}}$, $g_{\mathbb{E}}$ denotes the Euclidean metric, and

$$K_{\mathbb{E}}(\mathbf{x}_0, t_0; \mathbf{x}, t) := (4\pi(t_0 - t))^{-\frac{n}{2}}\mathrm{e}^{-f(\mathbf{x},t)}. \tag{6.31}$$

This corresponds to the Gaussian shrinking soliton (cf. §2.3.1 in Volume 1).

For the purpose of subsequent calculations, observe that we may rewrite the backward heat-type equation (6.21) as a forward heat-type equation:

$$\begin{aligned}\tau\Box f &= \tau\left(-2\Delta f + |\nabla f|^2 - R + \frac{n}{2\tau}\right)\\ &= f - \frac{n}{2} - w;\end{aligned} \tag{6.32}$$

cf. (1.76) for the potential function of a shrinking gradient Ricci soliton. Of course, this is not an "honest" forward heat-type equation since w contains the second-order term $2\tau\Delta f$. However, by (6.23), the potential f of the conjugate heat kernel satisfies the following useful forward heat-type *inequality*:

$$\tau\Box f \geq f - \frac{n}{2}. \tag{6.33}$$

6.1.2. Entropy monotonicity.

We aim to study the fundamental invariants of the Ricci flow associated with conjugate heat kernels, and more generally, with probability measures. One such invariant, entropy, is a measure of *uncertainty* and was first studied in thermodynamics. The higher the entropy the greater the uncertainty. Entropy is also a measure of *information*; the higher the entropy the greater the information. For background on entropy, see [**Khi57**]. For its applications to parabolic and elliptic partial differential equations, see Nash [**Nas58**]. Entropy is a particularly fundamental invariant for Ricci flow.

6.1.2.1. *Nash entropy of a probability measure.*

Let $(\mathcal{O}^n, g)$ be a Riemannian orbifold with isolated singularities. Let

$$d\nu =: (4\pi\tau)^{-\frac{n}{2}} e^{-f} dv_g$$

be a **probability measure** on $\mathcal{O}^n$, where, dv_g denotes the Riemannian measure of g, so that

$$\int_{\mathcal{O}^n} d\nu = 1.$$

DEFINITION 6.13. The **Nash** (a.k.a. **Boltzmann**) **entropy** of a probability measure $d\nu$ (equivalently, of f) at scale $\tau > 0$ is defined by (see [**Per02**, §5])

$$\begin{aligned}\mathcal{N}[g, f, \tau] &:= \int_{\mathcal{O}^n} f d\nu - \frac{n}{2}\\ &= \int_{\mathcal{O}^n} (4\pi\tau)^{-\frac{n}{2}} f e^{-f} dv_g - \frac{n}{2}.\end{aligned} \tag{6.34}$$

Setting $u = e^{-f}$, we observe that the integrand in (6.34) can be rewritten as $fe^{-f} = -u\ln u$, which is a typical form of an entropy integrand.

It is easy to see that $\mathcal{N}$ satisfies the following scaling property: For any positive constant c, we have that

$$\mathcal{N}[cg, f, c\tau] = \mathcal{N}[g, f, \tau]. \tag{6.35}$$

We will assume the orbifold $\mathcal{O}^n$ is closed, or in the noncompact case conditions sufficient to ensure $\mathcal{N}$ is well-defined, such as being a singularity model. Entropies ($\mathcal{N}$ and Perelman's $\mathcal{W}$ defined in §6.1.2.3 below) are fundamental functionals and invariants in Ricci flow.

EXAMPLE 6.14 (Euclidean heat kernel Nash entropy). Consider the special case where $(\mathcal{O}^n, g)$ is Euclidean space $(\mathbb{R}^n, g_{\mathbb{E}})$. Let $\mathbf{x}_0 \in \mathbb{R}^n$ and let

$$f_\tau(\mathbf{x}) := \frac{|\mathbf{x} - \mathbf{x}_0|^2}{4\tau}. \tag{6.36}$$

Then $d\nu := (4\pi\tau)^{-\frac{n}{2}} \mathrm{e}^{-f_\tau} dv_{\mathbb{E}}$ satisfies

$$\int_{\mathbb{R}^n} d\nu = (4\pi\tau)^{-\frac{n}{2}} \int_{\mathbb{R}^n} \mathrm{e}^{-f_\tau} dv_{\mathbb{E}} = 1, \tag{6.37}$$

where $dv_{\mathbb{E}}$ denotes the Riemannian volume element of the Euclidean metric $g_{\mathbb{E}}$. Thus, ν is a probability measure, and we also have

$$\mathcal{N}[g_{\mathbb{E}}, f_\tau, \tau] \equiv 0. \tag{6.38}$$

6.1.2.2. *Pointed Nash entropy of conjugate heat kernels.*

Foundational to Bamler's theory and the Ricci flow is the pointed Nash entropy, which we now discuss.

Let $(\mathcal{O}^n, (g_t)_{t\in I})$ be a Ricci flow on a closed orbifold with isolated singularities. As in Definition 6.5, let $\nu_{x_0,t_0;t}$, where

$$d\nu_{x_0,t_0;t} = (4\pi\tau(t))^{-\frac{n}{2}} \mathrm{e}^{-f_t} dv_t = K(x_0, t_0; \cdot, t)\, dv_t, \tag{6.39}$$

be the conjugate heat kernel measure based at a point $(x_0, t_0) \in \mathcal{O}^n \times I$, where $\tau(t) = t_0 - t$. By definition (6.15), f_t is the potential of the conjugate heat kernel based at (x_0, t_0).

For the solution g_t, its **pointed Nash entropy** based at (x_0, t_0) with scale $\tau > 0$ is defined via (6.34) by

$$\begin{aligned} \mathcal{N}_{x_0,t_0}(\tau) &:= \mathcal{N}[g_{t_0-\tau}, f_{t_0-\tau}, \tau] \\ &= \int_{\mathcal{O}^n} f_{t_0-\tau}\, d\nu_{x_0,t_0;t_0-\tau} - \frac{n}{2} \\ &= -\int_{\mathcal{O}^n} K \ln K(x_0, t_0; y, t_0 - \tau)\, dv_{t_0-\tau}(y) - \frac{n}{2}\ln(4\pi\tau) - \frac{n}{2}, \end{aligned} \tag{6.40}$$

where $K := K(x_0, t_0; \cdot, t_0 - \tau)$ and where $dv_{t_0-\tau}$ is the Riemannian volume element of $g_{t_0-\tau}$. That is, the pointed Nash entropy is the Nash entropy of a conjugate heat kernel measure.

In the following, we observe that the pointed Nash entropy at scale 0 (i.e., as $\tau \to 0$) detects the cardinality of the isotropy group of a point in our orbifold. Let $\Gamma := \Gamma_{x_0}$ denote the isotropy group of $x_0 \in \mathcal{O}^n$; see (A.4) below. For $\tau = 0$, we define

$$\mathcal{N}_{x_0,t_0}(0) := -\ln|\Gamma_{x_0}|.$$

In particular, if $\Gamma_{x_0} = 1$ (that is, $\mathcal{O}^n$ is a smooth manifold near x_0), then $\mathcal{N}_{x_0,t_0}(0) = 0$. Now, as $\tau \to 0_+$, the geometry localizes at x_0, and the behavior of the heat kernel approaches that of the Γ-quotient of Euclidean space. Consequently, we obtain the limiting behavior:

$$\lim_{\tau\to 0_+} \mathcal{N}_{x_0,t_0}(\tau) = -\ln|\Gamma_{x_0}|, \tag{6.41}$$

which follows from the asymptotics of the conjugate heat kernel. For the case of a smooth manifold, see §3 in Chapter 24 of [**CCG**$^+$**10**]. In the case of an orbifold point, consider the heat kernel of $\mathbb{R}^n/\Gamma$. First, observe that for a finite subgroup Γ of $O(n)$,

$$\int_{\mathbb{R}^n/\Gamma}(4\pi\tau)^{-\frac{n}{2}}\mathrm{e}^{-\frac{|\mathbf{x}|^2}{4\tau}}d[\mathbf{x}]=\frac{1}{|\Gamma|}. \tag{6.42}$$

Thus the heat kernel $K_{\mathbb{R}^n/\Gamma}$ centered at the origin $[0^n]$ of $\mathbb{R}^n/\Gamma$ is given by

$$K_{\mathbb{R}^n/\Gamma}([0^n],t_0;[\mathbf{x}],t_0-\tau)=|\Gamma|(4\pi\tau)^{-\frac{n}{2}}\mathrm{e}^{-\frac{|\mathbf{x}|^2}{4\tau}}. \tag{6.43}$$

Next, we compute:

$$\int_{\mathbb{R}^n/\Gamma}\frac{|\mathbf{x}|^2}{4\tau}(4\pi\tau)^{-\frac{n}{2}}\mathrm{e}^{-\frac{|\mathbf{x}|^2}{4\tau}}d[\mathbf{x}]=\pi^{-\frac{n}{2}}\int_{\mathbb{R}^n/\Gamma}|\mathbf{y}|^2\mathrm{e}^{-|\mathbf{y}|^2}d[\mathbf{y}]=\frac{n}{2|\Gamma|}. \tag{6.44}$$

Using this, we compute that

$$\mathcal{N}^{\mathbb{R}^n/\Gamma}_{x_0,t_0}(\tau)\equiv-\ln|\Gamma|, \tag{6.45}$$

which generalizes (6.38). Thus, in the limit as $\tau\to0$, the Euclidean calculation dominates due to the localization near the orbifold point, and the resulting Nash entropy approaches $-\ln|\Gamma|$. By definition and (6.41), the function $\tau\mapsto\mathcal{N}_{x_0,t_0}(\tau)$ is continuous at $\tau=0$.

Viewing the potential of the heat kernel as a proxy for the distance squared function, we can consider the pointed Nash entropy as essentially the **second moment** of the heat kernel.

6.1.2.3. *Perelman's $\mathcal{W}$-entropy and its monotonicity.*

A seemingly more complicated notion of entropy, compared to the Nash entropy, is Perelman's $\mathcal{W}$-entropy, which we discuss in this subsection. It is intimately related to both the Nash entropy and Perelman's differential Harnack estimate.

Given a Riemannian orbifold, Perelman's **$\mathcal{W}$-entropy** of $(\mathcal{O}^n,g,f,\tau)$, where f is a function on $\mathcal{O}^n$ and $\tau\in\mathbb{R}_+$, is defined by

$$\mathcal{W}(g,f,\tau)=\int_{\mathcal{O}^n}\left(\tau(R_g+|\nabla f|_g^2)+f-n\right)(4\pi\tau)^{-\frac{n}{2}}\mathrm{e}^{-f}dv_g. \tag{6.46}$$

Here, we assume that $\int_{\mathcal{O}^n}(4\pi\tau)^{-\frac{n}{2}}\mathrm{e}^{-f}dv_g=1$. For any positive constant c, we have the scaling property that

$$\mathcal{W}(cg,f,c\tau)=\mathcal{W}(g,f,\tau). \tag{6.47}$$

If $f\equiv0$, then $\mathcal{W}$ is a version of the total scalar curvature, a.k.a. Einstein–Hilbert functional. On the other hand, the entropy aspect of $\mathcal{W}$ comes from the term $f\mathrm{e}^{-f}$ in the sense that this term is equal to $-\mathrm{e}^{-f}\ln(\mathrm{e}^{-f})$.

Perelman's **entropy monotonicity formula** (see [**Per02**, (3.4)]) says that under the coupled Ricci flow equations

$$\partial_t g_t=-2\mathrm{Ric}_{g_t}, \tag{6.48a}$$

$$\partial_t f_t=-R_{g_t}-\Delta_{g_t}f_t+|\nabla f|_{g_t}^2+\frac{n}{2\tau(t)}, \tag{6.48b}$$

$$\partial_t\tau(t)=-1, \tag{6.48c}$$

where $\tau(t)>0$, we have

(6.49)

$$\frac{d}{dt}\mathcal{W}(g_t,f_t,\tau(t))=2\tau\int_{\mathcal{O}^n}\left|\mathrm{Ric}_{g_t}+\nabla^2f_t-\frac{1}{2\tau(t)}g_t\right|^2_{g_t}(4\pi\tau(t))^{-\frac{n}{2}}\mathrm{e}^{-f}dv_t\geq0.$$

This monotonicity formula follows by integrating (6.25), which is the evolution equation used to prove Perelman's differential Harnack estimate. Observe that the right-hand side of (6.49) vanishes if and only if $(\mathcal{O}^n, g_t, f_t, \frac{1}{\tau(t)})$ is a shrinking soliton.

Moreover, since $u := (4\pi\tau(t))^{-\frac{n}{2}} e^{-f_t}$ satisfies $\square^* u = 0$, we have the conservation law

$$\frac{d}{dt}\int_{\mathcal{O}^n} (4\pi\tau(t))^{-\frac{n}{2}} e^{-f_t} dv_t = 0. \tag{6.50}$$

In particular, if $(\mathcal{O}^n, g(t))$ is a Ricci flow on a closed orbifold with isolated singularities, and if $d\nu_{x_0,t_0;t} = (4\pi(t_0 - t))^{-\frac{n}{2}} e^{-f_t} dv_t$ is a conjugate heat kernel measure, then by (6.49) we have the pointed $\mathcal{W}$-entropy monotonicity

$$\frac{d}{dt}\mathcal{W}(g_t, f_t, t_0 - t) \geq 0. \tag{6.51}$$

We also have that

$$\lim_{t\to(t_0)_-} \mathcal{W}(g_t, f_t, t_0 - t) = -\ln|\Gamma_{x_0}|. \tag{6.52}$$

The last equality is derived similarly to (6.41). Namely, in the case of $\mathbb{R}^n/\Gamma$, we have $f(\mathbf{x}) = \frac{|\mathbf{x}|^2}{4\tau} - \ln|\Gamma|$. Using this, we compute that

$$\mathcal{W}^{\mathbb{R}^n/\Gamma}(g_t, f_t, t_0 - t) = -\ln|\Gamma|. \tag{6.53}$$

From this, we can derive (6.52). Now, by (6.51) and (6.52), we have the pointed $\mathcal{W}$-entropy monotonicity upper bound

$$\mathcal{W}(g_t, f_t, t_0 - t) \leq -\ln|\Gamma_{x_0}| \tag{6.54}$$

for all $t < t_0$.

6.1.2.4. *Perelman's μ-invariant and its monotonicity.*

The **μ-invariant** of a Riemannian orbifold $(\mathcal{O}^n, g)$ at scale $\tau > 0$ is defined by

$$\mu(g,\tau) := \inf_f \mathcal{W}(g, f, \tau), \tag{6.55}$$

where the infimum of $\mathcal{W}$ is taken over all f satisfying the constraint

$$\int_{\mathcal{O}^n} (4\pi\tau)^{-\frac{n}{2}} e^{-f} dv_g = 1.$$

By taking this infimum, we have converted the entropy into an invariant of only the metric g and a scale $\tau > 0$. From (6.47), we have that for any $c > 0$,

$$\mu(cg, c\tau) = \mu(g,\tau). \tag{6.56}$$

Using (6.49) and (6.50), it is easy to show that under the Ricci flow coupled to the equation $\partial_t \tau(t) = -1$, we have the μ-monotonicity property:

$$\mu(g_{t_2}, \tau(t_2)) \geq \mu(g_{t_1}, \tau(t_1)) \quad \text{for all } t_2 \geq t_1. \tag{6.57}$$

In other words, for any scale $\tau > 0$, we have that

$$\mu(g_{t_2}, \tau) \geq \mu(g_{t_1}, \tau + t_2 - t_1) \quad \text{for all } t_2 \geq t_1. \tag{6.58}$$

6.1.2.5. *The relationship between the pointed Nash and pointed Perelman entropies.*

A straightforward calculation yields that the pointed Nash entropy $\mathcal{N}_{x_0,t_0}(\tau)$ defined by (6.40) satisfies ([**Bam20a**, (5.4)])

$$\frac{d}{d\tau}(\tau\mathcal{N}_{x_0,t_0}(\tau)) = \mathcal{W}(g_{t_0-\tau}, f_{t_0-\tau}, \tau) \leq -\ln|\Gamma_{x_0}|, \tag{6.59}$$

where the non-positivity of $\mathcal{W}$ follows from (6.54). Since $\lim_{\tau\to 0_+} \mathcal{N}_{x_0,t_0}(\tau) = 0$, we obtain from integrating (6.59) that

$$\begin{aligned} \mathcal{N}_{x_0,t_0}(\overline{\tau}) &= \frac{1}{\overline{\tau}}\int_0^{\overline{\tau}} \mathcal{W}(g_{t_0-\tau}, f_{t_0-\tau}, \tau)\, d\tau \\ &\geq \frac{1}{\overline{\tau}}\int_0^{\overline{\tau}} \mu(g_{t_0-\tau}, \tau) d\tau \\ &\geq \mu(g_{t_0-\overline{\tau}}, \overline{\tau}). \end{aligned} \tag{6.60}$$

In fact, by (6.51), we have that

$$\mathcal{N}_{x_0,t_0}(\overline{\tau}) \geq \mathcal{W}(g_{t_0-\overline{\tau}}, f_{t_0-\overline{\tau}}, \overline{\tau}). \tag{6.61}$$

By the inequality in (6.59) and the equality in (6.60), we also have the following upper estimate for the pointed Nash entropy:

$$\mathcal{N}_{x_0,t_0}(\tau) \leq -\ln|\Gamma_{x_0}| \leq 0. \tag{6.62}$$

By differentiating (6.59), we obtain

$$\begin{aligned} &\frac{d^2}{d\tau^2}(\tau\mathcal{N}_{x_0,t_0}(\tau)) \\ &\quad = \frac{d}{d\tau}\mathcal{W}(g_{t_0-\tau}, f_{t_0-\tau}, \tau) \\ &\quad = 2\tau\int_{\mathcal{O}^n} \left|\operatorname{Ric}_{g_{t_0-\tau}} + \nabla^2 f_{t_0-\tau} - \frac{1}{2\tau} g_{t_0-\tau}\right|^2 (4\pi\tau)^{-\frac{n}{2}} e^{-f_{t_0-\tau}} dv_{g_{t_0-\tau}} \\ &\quad \geq 0. \end{aligned} \tag{6.63}$$

Using this, one can prove that the pointed Nash entropy is monotone (see [**Bam20a**, (5.6)]):

$$\frac{d}{d\tau}\mathcal{N}_{x_0,t_0}(\tau) \leq 0. \tag{6.64}$$

Note that since $\lim_{\tau\to 0_+} \mathcal{N}_{x_0,t_0}(\tau) = -\ln|\Gamma_{x_0}|$, this gives another proof of (6.62).

6.1.2.6. *The asymptotic pointed Nash entropy of ancient solutions.*

Let $(\mathcal{O}^n, g(t))$, $t \in (-\infty, 0]$, be a complete ancient solution to the Ricci flow with bounded curvature on each compact time interval. Since the pointed Nash entropy is monotone, we can extract an invariant of an ancient solution by considering its asymptotic limit.

Definition 6.15. The **asymptotic pointed Nash entropy** is defined by

$$\mu_\infty := \lim_{\tau\to\infty} \mathcal{N}_{x_0,t_0}(\tau) = \inf_{\tau\in(0,\infty)} \mathcal{N}_{x_0,t_0}(\tau). \tag{6.65}$$

By Ma and Zhang [**MZ21**, Proposition 4.6], the asymptotic pointed Nash entropy μ_∞ is independent of the choice of $x_0 \in \mathcal{O}^n$ and $t_0 \in (-\infty, 0]$. In particular, the asymptotic pointed Nash entropy can be considered as a global invariant of the ancient solution, reflecting the geometry at infinity.

6.1.2.7. *Poincaré, energy, and second moment inequalities.*

In Riemannian geometry, the Poincaré and Sobolev inequalities are fundamental tools for applying analytical methods to the study of manifolds.

For the Ricci flow, we have the following fundamental Poincaré inequality. Let

$$d\nu_{x_0,t_0}(\tau) := d\nu_{x_0,t_0;t_0-\tau}.$$

The **Hein–Naber L^2-Poincaré inequality** with respect to the conjugate heat kernel measure is as follows (see Hein and Naber [**HN14**, Theorem 1.10] and [**Bam20a**, Theorem 11.1] for a generalization).

THEOREM 6.16 (Hein–Naber L^2-Poincaré inequality). *If $[t_0 - \tau, t_0] \subset I$, then for any $h \in C^1(\mathcal{O}^n)$,*

$$\int_{\mathcal{O}^n} h^2 d\nu_{x_0,t_0}(\tau) \leq 2\tau \int_{\mathcal{O}^n} |\nabla h|^2 d\nu_{x_0,t_0}(\tau) + \left(\int_{\mathcal{O}^n} h\, d\nu_{x_0,t_0}(\tau) \right)^2. \tag{6.66}$$

Given a closed Riemannian orbifold $(\mathcal{O}^n, g)$ with isolated singularities and a function ϕ, **Perelman's $\mathcal{F}$-energy** is defined by

$$\mathcal{F}[g, \phi] := \int_{\mathcal{O}^n} (|\nabla \phi|^2 + R) \mathrm{e}^{-\phi} dv = \int_{\mathcal{O}^n} (\Delta \phi + R) \mathrm{e}^{-\phi} dv. \tag{6.67}$$

The **λ-functional** is defined by

$$\lambda(g) := \inf_{\phi} \mathcal{F}[g, \phi], \quad \text{where } \int_M \mathrm{e}^{-\phi} dv = 1. \tag{6.68}$$

We have the bounds

$$R_{\min} \leq \lambda(g) \leq R_{\text{avg}}, \tag{6.69}$$

where R_{avg} is the average scalar curvature; see, for example, §3.1 of Chapter 5 in [**CCG$^+$07**].

Observe that we have the following relation between the $\mathcal{W}$-entropy, $\mathcal{F}$-energy, and Nash entropy:

$$\mathcal{W}(g, f, \tau) = \tau \mathcal{F}[g, \phi] + \mathcal{N}[g, f, \tau] - \frac{n}{2}. \tag{6.70}$$

By (6.69), we have that

$$\mathcal{W}(g, f, \tau) \geq \tau R_{\min} + \mathcal{N}[g, f, \tau] - \frac{n}{2}. \tag{6.71}$$

Now define f by $K =: (4\pi\tau)^{-\frac{n}{2}} \mathrm{e}^{-f}$, so that f is the potential for the conjugate heat kernel.

THEOREM 6.17 (Bamler's estimates [**Bam20a**, Proposition 5.13]). *The following holds true.*

(1) (Heat kernel measure $\mathcal{F}$-energy estimate)

$$\mathcal{F}[g, \phi] = \int_{\mathcal{O}^n} (|\nabla f|^2 + R)\, d\nu_{x_0,t_0}(\tau) \leq \frac{n}{2\tau}, \tag{6.72}$$

where $\phi = f + \frac{n}{2} \ln(4\pi\tau)$.

(2) (Second moment of $\mathcal{N}$-entropy bound)

$$\int_{\mathcal{O}^n} \left(f - \mathcal{N}_{x_0,t_0}(\tau) - \frac{n}{2}\right)^2 d\nu_{x_0,t_0}(\tau) \leq n - 2\tau R_{\min,t_0-\tau}, \tag{6.73}$$

where $R_{\min,t_0-\tau} := \inf_{x\in\mathcal{O}^n} R(x, t_0 - \tau)$.

6.1.3. Hein and Naber's logarithmic Sobolev inequality.

For the Ricci flow, we have the following logarithmic Sobolev inequuality, which is related to Nash entropy.

PROPOSITION 6.18 (Hein and Naber's logarithmic Sobolev inequality [**HN14**]). *If* $[t_0 - \tau, t_0] \subset I$, *then for any positive function* $h \in C^1(\mathcal{O}^n)$,

$$\int_{\mathcal{O}^n} h \ln h \, d\nu_{x_0,t_0}(\tau) \leq \int_{\mathcal{O}^n} h \, d\nu_{x_0,t_0}(\tau) \ln\left(\int_{\mathcal{O}^n} h \, d\nu_{x_0,t_0}(\tau)\right) + \tau \int_{\mathcal{O}^n} \frac{|\nabla h|^2}{h} d\nu_{x_0,t_0}(\tau). \tag{6.74}$$

The sharpness of the Poincaré and Sobolev inequalities above illustrate the naturality of considering the conjugate heat kernel measures $d\nu_{x_0,t_0}(\tau)$.

6.1.4. Variance and Wasserstein distance for probability measures.

The Ricci flow coupled to any two solutions of the conjugate heat equation, as flows of metric measure spaces, has fundamental monotonicity properties, which we now discuss.

6.1.4.1. *Definitions of variance and Wasserstein distance.*

In Riemannian geometry, the distance between points is a fundamental invariant. For Ricci flow, Perelman discovered a natural distance for the Ricci flow space-time of a solution, called the reduced distance. With applications to Ricci flow in mind, Bamler defined two natural notions of distance between probability measures.

Firstly, Bamler defined a notion of variance for probability measures on Riemannian manifolds, which easily extends to orbifolds with isolated singularities; see [**Bam20a**, Definition 3.1].

DEFINITION 6.19 (Variance of probability measures). Let ν_1, ν_2 be two probability measures on a Riemannian orbifold $(\mathcal{O}^n, g)$ with isolated singularities. Their **variance** is:

$$\mathrm{Var}(\nu_1, \nu_2) := \iint_{\mathcal{O}^n\times\mathcal{O}^n} d_g^2(x, y) d\nu_1(x) d\nu_2(y). \tag{6.75}$$

This definition clearly applies more generally to metric spaces. We also define the variance of a probability measure (with itself) by

$$\mathrm{Var}(\nu_1) := \mathrm{Var}(\nu_1, \nu_1). \tag{6.76}$$

So the variance of ν_1, ν_2 is defined as the *average squared distance* between two points in $\mathcal{O}^n$ with respect to the product measure $\nu_1 \times \nu_2$ on $\mathcal{O}^n \times \mathcal{O}^n$. In other words, it is the expected square distance between a point drawn randomly according to the probability distribution ν_1 and a point drawn randomly according to the probability distribution ν_2. The variance measures how spread out the two distributions are relative to each other on the Riemannian orbifold. In particular, $\mathrm{Var}(\nu_1)$ measures how spread out the probability distribution ν_1 is.

EXAMPLE 6.20. The variance of Dirac delta measures (which are point distributions) is the squared distance between their centers:

$$\operatorname{Var}(\delta_x, \delta_y) = d^2(x,y), \quad \text{i.e.,} \quad \sqrt{\operatorname{Var}(\delta_x, \delta_y)} = d(x,y). \tag{6.77}$$

Thus, we may consider the square root of the variance as a generalization of the Riemannian distance function.

Let ν be a probability measure. Observe that for all $x \in \mathcal{O}^n$,

$$\operatorname{Var}(\delta_x, \nu) = \iint_{\mathcal{O}^n \times \mathcal{O}^n} d^2(x_1, x_2)\, d\delta_x(x_1)\, d\nu(x_2) = \int_{\mathcal{O}^n} d^2(x, x_2)\, d\nu(x_2) \tag{6.78}$$

is the average squared distance to the point x. Hence,

$$\int_{\mathcal{O}^n} \operatorname{Var}(\delta_x, \nu)\, d\nu(x) = \iint_{\mathcal{O}^n \times \mathcal{O}^n} d^2(x, x_2)\, d\nu(x_2)\, d\nu(x) = \operatorname{Var}(\nu). \tag{6.79}$$

In other words, the average, with respect to $d\nu$, variance between Dirac delta measures centered at points in $\mathcal{O}^n$ and the probability distribution ν is equal to the variance of ν.

Secondly, Bamler considered the following notion of distance with a view toward applications to Ricci flow. Let ν_1, ν_2 be two probability measures on a complete Riemannian orbifold $(\mathcal{O}^n, g)$ with isolated singularities.

DEFINITION 6.21 (1-Wasserstein distance between probability measures). The **1-Wasserstein distance** between ν_1, ν_2 is defined by

$$d^g_{W_1}(\nu_1, \nu_2) := \sup_{f \in 1\text{-Lip}} \left(\int_{\mathcal{O}^n} f\, d\nu_1 - \int_{\mathcal{O}^n} f\, d\nu_2 \right), \tag{6.80}$$

where the supremum is taken over all bounded 1-Lipschitz functions $f : \mathcal{O}^n \to \mathbb{R}$.

This definition of W_1 is the same as taking the supremum over bounded C^∞ functions f with $|\nabla f| \leq 1$ on $\mathcal{O}^n$; see [**Bam20a**, §2.4] for more information.

Note, for example, that $d^g_{W_1}(\nu_1, \nu_2) = d^g_{W_1}(\nu_2, \nu_1)$ since we can replace f by $-f$ in the definition of 1-Wasserstein distance. It is also easy to see that the triangle inequality holds for $d^g_{W_1}$:

$$d^g_{W_1}(\nu_1, \nu_3) \leq d^g_{W_1}(\nu_1, \nu_2) + d^g_{W_1}(\nu_2, \nu_3), \tag{6.81}$$

for probability measures ν_1, ν_2, ν_3. In particular, if f satisfies $|\nabla f| \leq A$ on $\mathcal{O}^n$, then

$$\left| \int_{\mathcal{O}^n} f\, d\nu_1 - \int_{\mathcal{O}^n} f\, d\nu_2 \right| \leq A\, d^g_{W_1}(\nu_1, \nu_2). \tag{6.82}$$

EXAMPLE 6.22. Given points $x_1, x_2 \in \mathcal{O}^n$,

$$d^g_{W_1}(\delta_x, \delta_y) = d(x,y). \tag{6.83}$$

An idea for a proof of this is that there exists a 1-Lipschitz function f with $f(y) = 0$ and $f(x) = d(x,y)$, and we cannot make their difference larger.

So we may think of both the variance and the 1-Wasserstein distance as generalizing the notion of the Riemannian distance between points. In particular, we are interested in their behavior for conjugate heat equation measures.

Given a complete separable metric space (X, d), let $\mathcal{P}(X)$ denote the space of Borel probability measures on X.

REMARK 6.23. Let (X_i, d_i), $\nu_i \in \mathcal{P}(X_i)$. A **coupling** between ν_1 and ν_2 is a probability measure $q \in \mathcal{P}(X_1 \times X_2)$ with **marginals** ν_1 and ν_2, respectively. That is, let $\pi_i : X_1 \times X_2 \to X_i$, $i = 1, 2$, be the i-th projection map. By definition,

$$\nu_i(A) = q(\pi_i^{-1}(A)). \tag{6.84}$$

In other words,

$$\nu_1(A) = q(A \times X_2), \quad \nu_2(B) = q(X_1 \times B). \tag{6.85}$$

For example, the product measure $\nu_1 \times \nu_2$ is a coupling between ν_1 and ν_2, representing the case where the two measures are independent. In general, each coupling represents a different possible probabilistic relationship between ν_1 and ν_2.

For $p > 0$, the p-**Wasserstein distance** is defined by

$$d^g_{W_p}(\nu_1, \nu_2) = \inf_{\text{coup } q} \left(\iint_{\mathcal{O}^n \times \mathcal{O}^n} d_g(x_1, x_2)^p dq(x_1, x_2) \right)^{1/p},$$

where the infimum is taken over couplings q between ν_1 and ν_2. When $p = 1$, this definition agrees with the definition above of $d^g_{W_1}$ by the Kantorovich–Rubinstein theorem (see e.g. Villani [**Vil03**, Theorem 1.14]).

In particular,

$$d^g_{W_p}(\delta_{x_1}, \delta_{x_2}) = \left(\iint_{\mathcal{O}^n \times \mathcal{O}^n} d_g(y_1, y_2)^p d\delta_{(x_1,x_2)}(y_1, y_2) \right)^{1/p} = d_g(x_1, x_2).$$

The reason for why the p-Wasserstein distance, $p \geq 1$, between two Dirac delta functions δ_{x_1} and δ_{x_2} is equal to $d_g(x_1, x_2)$ is because the only coupling is $\delta_{(x_1,x_2)}$.

The 1-Wasserstein distance and the variance square root are in a sense comparable. Lemma 3.2 in [**Bam20a**] states the following:

LEMMA 6.24 (1-Wasserstein distance and variance comparison). *For any three probability measures ν_1, ν_2, ν_3 on a Riemannian orbifold $(\mathcal{O}^n, g)$ with isolated singularities, we have:*

(1) (Triangle inequality for the variance square root)

$$\sqrt{\mathrm{Var}}(\nu_1, \nu_3) \leq \sqrt{\mathrm{Var}}(\nu_1, \nu_2) + \sqrt{\mathrm{Var}}(\nu_2, \nu_3); \tag{6.86}$$

(2) (1-Wasserstein distance and variance square root comparison)

$$d^g_{W_1}(\nu_1, \nu_2) \leq \sqrt{\mathrm{Var}}(\nu_1, \nu_2) \leq d^g_{W_1}(\nu_1, \nu_2) + \sqrt{\mathrm{Var}}(\nu_1) + \sqrt{\mathrm{Var}}(\nu_2). \tag{6.87}$$

6.1.4.2. *Monotonicity of* 1*-Wasserstein distance.*

Monotonicity formulas for Ricci flow are of central important. They are used to control the qualitative behavior of solutions and in particular to understand the geometry of singularity formation. In this subsection we discuss Bamler's 1-Wasserstein distance monotonicity formula in the more general category of super Ricci flows.

DEFINITION 6.25 (Super Ricci flow). Let g_t be a 1-parameter family of metrics on a smooth orbifold $\mathcal{O}^n$ with isolated singularities. We say that g_t is a **super Ricci flow** if

$$\partial_t g_t \geq -2\mathrm{Ric}_t. \tag{6.88}$$

In view of definition (6.80) of d_{W_1}, the following sharp result hinges on the gradient estimate for the heat equation of Proposition 6.1; see [**Bam20a**, Lemma 2.7].

THEOREM 6.26 (1-Wasserstein distance monotonicity of conjugate heat solutions). *Suppose that $\nu_{1,t}, \nu_{2,t}$ are two smooth conjugate heat equation probability measures for a super Ricci flow $(\mathcal{O}^n, g(t))$ on a closed orbifold with isolated singularities; that is,*

$$d\nu_{i,t} = v_i(t)\,dv_t, \quad \text{where } \square^* v_i(t) = 0, \quad v_i(t) \geq 0, \quad \int_{\mathcal{O}^n} d\nu_{i,t} = 1 \tag{6.89}$$

for $i = 1, 2$. Then the W_1-distance function

$$t \mapsto d_{W_1}^{g_t}(\nu_{1,t}, \nu_{2,t}) \tag{6.90}$$

is non-decreasing (forward in time).

REMARK 6.27. Consider points p and q on an Einstein metric solution to the Ricci flow. The distance between p and q decreases, remains constant, or increases if the scalar curvature is positive, zero, or negative, respectively. Moreover, monotonicity does not generally hold. In contrast, monotonicity does hold for the 1-Wasserstein distance between conjugate heat equation probability measures, as shown in the theorem above.

Specializing to conjugate heat kernel measures, we obtain the following:

COROLLARY 6.28 (1-Wasserstein distance montonicity of conjugate heat kernels). *For any points $x_1, x_2 \in \mathcal{O}^n$ and time $t \leq t_0$,*

$$d_{W_1}^{g_t}(\nu_{x_1,t_0;t}, \nu_{x_2,t_0;t}) \leq d_{t_0}(x_1, x_2). \tag{6.91}$$

PROOF. Indeed, we have that the function

$$t \mapsto d_{W_1}^{g_t}(\nu_{x_1,t_0;t}, \nu_{x_2,t_0;t}) \tag{6.92}$$

is non-decreasing and limits to $d_{t_0}(x_1, x_2)$ as $t \to t_0$. □

If we think of $\nu_{x_i,t_0;t}$ as representing the past of (x_i, t_0) at time t, for $i = 1, 2$, then the corollary may be interpreted as saying the distance between x_1 and x_2 is a non-decreasing function of time t for $t \leq t_0$ under the Ricci flow. That is, as a consequence, we have distance control backwards in time.

6.1.5. H_n-centers.

Under Ricci flow, the "past" of a point at a given time is not simply the same point at earlier times. The past consists of the probability distributions associated to the heat kernel based at the space-time point. But, what if we want to represent the past by a *single point* at each past time? Although there is no best or unique point for such a representation, we can define a notion of "good enough" past points, called H_n-centers.

6.1.5.1. *Distance squared and the H_n-constant.*
Define the dimensional constant

$$H_n := \frac{(n-1)\pi^2}{2} + 4. \tag{6.93}$$

This strange constant naturally arises as follows: Theorem 3.5 in [**Bam20a**] implies the following 2-point distance function heat-type inequality.

THEOREM 6.29 (Distance-squared function heat-type inequality). *For any Ricci flow on $\mathcal{O}^n \times \mathcal{O}^n \times I$,*

$$(\partial_t - \Delta_x - \Delta_y) d_t^2(x, y) \geq -H_n \tag{6.94}$$

in both the barrier and viscosity senses. Equivalently,

$$\square \left(d_t^2(x, y) + H_n t\right) \geq 0, \tag{6.95}$$

*where $\square$ is the heat operator of $(\mathcal{O}^n \times \mathcal{O}^n, g_t \oplus g_t)$. This is the H_n***-scaled distance non-decreasing property** *of Ricci flow.*

We can think of this as the 2-point Laplacian comparison for the Ricci flow, or more aptly, the 2-point heat operator comparison theorem for the Ricci flow.

6.1.5.2. *Variance monotonicity formula.*

Using that the variance is defined in terms of the distance squared function, we obtain from Theorem 6.29 the following variance monotonicity formula, which is Corollary 3.7 in [**Bam20a**]:

THEOREM 6.30 (Variance of conjugate heat measures monotonicity). *Let $\nu_{1,t}$, $\nu_{2,t}$ be two conjugate heat equation measures for a Ricci flow. That is, $d\nu_{i,t} = v_i(t)\, dv_t$, where $\square^* v_i(t) = 0$ for $i = 1, 2$. Then the function*

$$t \mapsto \mathrm{Var}_t(\nu_{1,t}, \nu_{2,t}) + H_n t \tag{6.96}$$

is non-decreasing. In particular, $t \mapsto \mathrm{Var}_t(\nu_{1,t}) + H_n t$ is non-decreasing.

Applying the theorem to conjugate heat kernel measures yields the following (Corollary 3.8 in [**Bam20a**]):

COROLLARY 6.31 (Variance estimate for conjugate heat kernels). *For any $x_1, x_2 \in \mathcal{O}^n$, $t_0 \in I$, and $t \leq t_0$,*

$$\mathrm{Var}_t(\nu_{x_1,t_0;t}, \nu_{x_2,t_0;t}) \leq d_{t_0}^2(x_1, x_2) + H_n(t_0 - t). \tag{6.97}$$

In particular,

$$\mathrm{Var}_t(\nu_{x_1,t_0;t}) \leq H_n(t_0 - t). \tag{6.98}$$

We may think of $\mathrm{Var}_t(\nu_{x_1,t_0;t})$ as measuring the "concentration" of $\nu_{x_1,t_0;t}$; that is, the larger (smaller) $\mathrm{Var}_t(\nu_{x_1,t_0;t})$ is, the less (more) concentrated $\nu_{x_1,t_0;t}$ is. With this interpretation, the theorem and the corollary indicate that the *concentration* of conjugate heat kernels is, up to a linear function, non-decreasing as we move backwards in time.

6.1.5.3. *Definition of H_n-center.*

Heuristically, we can think of an H_n-center as a point where a conjugate heat kernel is concentrated. The following is [**Bam20a**, Definition 3.10]:

DEFINITION 6.32 (H_n-centers of a space-time point). A space-time point in a Ricci flow $(z, s) \in \mathcal{O}^n \times I$ is an H_n**-center** of $(x, t) \in \mathcal{O}^n \times I$ if $s \leq t$ and

$$\mathrm{Var}_s\left(\delta_z, \nu_{x,t;s}\right) \leq H_n(t - s), \tag{6.99}$$

where Var_s is the variance with respect to g_s. That is, the mean distance squared of points in $\mathcal{O}^n$ to z, with respect to the probability measure $\nu_{x,t;s}$, is at most $H_n(t - s)$.

By (6.87), we have the following estimate for the Wasserstein distance between a conjugate heat kernel measure and any associated H_n-center (as a Dirac delta measure):

$$d^{g_s}_{W_1}(\delta_z, \nu_{x,t;s}) \leq \sqrt{H_n(t-s)}. \tag{6.100}$$

We can think of this as a concentration result in a Wasserstein sense.

6.1.5.4. *Conjugate heat kernel concentration estimate.*
We call a ball centered at an H_n-center an H_n**-centered ball**. The following, which is Proposition 3.13 in [**Bam20a**], is an integral quantitative estimate for how much conjugate heat kernel measures are concentrated on their H_n-centered balls.

LEMMA 6.33 (Concentration of conjugate heat kernel on H_n-centered balls). *If (z, s) is an H_n-center of (x, t), then for $A > 0$ the conjugate heat kernel measure of an H_n-centered ball satisfies*

$$\nu_{x,t;s}\left(B\left(z, s, \sqrt{AH_n(t-s)}\right)\right) \geq 1 - \frac{1}{A}. \tag{6.101}$$

In other words,

$$\nu_{x,t;s}\left(\mathcal{O}^n \setminus B\left(z, s, \sqrt{AH_n(t-s)}\right)\right) \leq \frac{1}{A}. \tag{6.102}$$

6.1.5.5. *Existence and proximity estimate for H_n-centers.*
The following result, which is an immediate consequence of Corollary 6.31, says that any two H_n-centers are proximate to each other in a scaled sense; see [**Bam20a**, Proposition 3.12].

PROPOSITION 6.34 (Existence and proximity of H_n-centers). *Suppose $(\mathcal{O}^n, g(t))$ is a Ricci flow on a closed orbifold with isolated singularities, $(x, t) \in \mathcal{O}^n \times I$, and $s \leq t$. Then:*

(1) (Existence) *There exists $z_s \in \mathcal{O}^n$ such that (z_s, s) is an H_n-center of (x, t).*
(2) (Proximity) *For any two such H_n-centers $(z_{s,1}, s)$ and $(z_{s,2}, s)$,*

$$d_s(z_{s,1}, z_{s,2}) \leq 2\sqrt{H_n(t-s)}. \tag{6.103}$$

PROOF.
(1) By (6.79) and the variance monotonicity estimate (6.97), we have

$$\int_{\mathcal{O}^n} \mathrm{Var}_s\left(\delta_z, \nu_{x,t;s}\right) d\nu_{x,t;s}(z) = \mathrm{Var}_s(\nu_{x,t;s}) \leq H_n(t-s).$$

The existence of an H_n-center z_s now follows from $\int_{\mathcal{O}^n} d\nu_{x,t}(z, s) = 1$, and the fact that the infimum of a function is less than or equal to its average.

(2) Suppose that (z_1, s) and (z_2, s) are H_n-centers of (x, t). By using the triangle inequality (6.86) for $\sqrt{\mathrm{Var}}$, we have

$$\begin{aligned} d_s(z_1, z_2) &= \sqrt{\mathrm{Var}_s(\delta_{z_1}, \delta_{z_2})} \\ &\leq \sqrt{\mathrm{Var}_s(\delta_{z_1}, \nu_{x,t;s})} + \sqrt{\mathrm{Var}_s(\nu_{x,t;s}, \delta_{z_2})} \\ &\leq 2\sqrt{H_n(t-s)}. \end{aligned}$$

□

Using the heuristic concept that, from the point of view of time t, the parabolic scale at time s is $\sqrt{t-s}$, we have that the "normalized distance" between H_n-centers is bounded.

6.1.5.6. *Estimates for the pointed Nash entropy as a function of the base point.* Consider the pointed Nash entropy as a function of its base point. That is, consider the function

$$\mathcal{N}_s^* : (x,t) \mapsto \mathcal{N}_{x,t}(t-s)$$

for $x \in \mathcal{O}^n$ and $t > s$. A significant advantage of the Nash entropy over Perelman's $\mathcal{W}$-entropy is that $\mathcal{N}_s^*$ satisfies surprisingly nice properties. In particular, we have the following gradient and heat operator estimates for the pointed Nash entropy as a function of its base point [**Bam20a**, Theorem 5.9]:

THEOREM 6.35 (Gradient and heat operator estimates for $\mathcal{N}_s^*$). *Let $(\mathcal{O}^n, g(t))$, $t \in I$, be a Ricci flow. Let $s \in I$ and suppose that $R(\cdot, s) \geq R_{\min}$. Then*

$$|\nabla \mathcal{N}_s^*|(x,t) \leq \left(\frac{n}{2(t-s)} - R_{\min}\right)^{\frac{1}{2}}, \tag{6.104}$$

$$-\frac{n}{2(t-s)} \leq \square \mathcal{N}_s^*(x,t) \leq 0. \tag{6.105}$$

One computes that (see [**Bam20a**, §5.3])

$$\begin{aligned} \square \mathcal{N}_s^*(x,t)(\tau) &= -\int_M \square_{x,t}\big(K \ln K(x,t;y,s)\big)\, dg_s(y) - \frac{n}{2(t-s)} \\ &= \int_M |\nabla_x \ln K|^2(x,t;y,s)\, d\nu_{x,t;s}(y) - \frac{n}{2(t-s)}. \end{aligned} \tag{6.106}$$

The right-most inequality in (6.105) now follows from [**Bam20a**, Proposition 4.2].

We further observe, by a straightforward calculation, that

$$\square^2 \mathcal{N}_s^*(x,t) = \frac{n}{2(t-s)^2} - 2\int_M |\nabla_x^2 \ln K|^2(x,t;y,s)\, d\nu_{x,t;s}(y). \tag{6.107}$$

Although the result of this calculation is unexpectedly simple, we are not aware of any application of (6.107).

6.1.5.7. *Gaussian upper bounds for the conjugate heat kernel.*

Given the philosophy that the conjugate heat kernel represents the past of its base point, obtaining upper and lower bounds for it is crucial for understanding the Ricci flow.

Suppose that $(\mathcal{O}^n, g(t))$, $t \in [\bar{t}, 0]$, where $\bar{t} < 0$, is a solution to the Ricci flow on a closed orbifold with isolated singularities. The following estimate appears in Bamler [**Bam20a**, (7.3) in Theorem 7.2]:

THEOREM 6.36 (Gaussian upper bound for the conjugate heat kernel). *The following pointwise upper bound for K holds:*[3]

$$K(x_0,0;x,t) \leq C|t|^{-\frac{n}{2}} \exp\left(-\mathcal{N}_{x_0,0}(|t|) - \frac{d_t(x,z_t)^2}{9|t|}\right), \tag{6.108}$$

where (z_t, t) is an H_n-center of $(x_0, 0)$ and where C depends only on n and any lower bound for $|t| \min_{\mathcal{O}^n \times [t,0]} R$. In particular, if the scalar curvature R is non-negative, then C depends only on n.

The upper bound for the conjugate heat kernel corresponds to a lower bound for the space-time distance-type function f defined by $K(x_0,0;x,t) = (4\pi|t|)^{-\frac{n}{2}} e^{-f(x,t)}$.

---

[3]Notice the sharpness of the power $-\frac{n}{2}$ of $|t|$ in front of the exponential on the right-hand side of (6.108) and that the only significant dependence is on the pointed Nash entropy.

6.1.5.8. *Noncollapsing using the pointed Nash entropy.*

Analogous to Perelman's relation between the $\mathcal{W}$ entropy and no local collapsing, we have the following result; see Bamler [**Bam20a**, Theorem 6.1]. Effectively, this result indicates that to prove no local collapsing, one can replace Perelman's $\mathcal{W}$ entropy by the Nash entropy $\mathcal{N}$, and moreover, it is sufficient to use conjugate heat kernels as test functions for $\mathcal{N}$.

Recall that $\mathcal{N}$ reflects the main "entropy" part of $\mathcal{W}$ and also constitutes the main part to obtain the volume ratio in the no local collapsing argument. We have that $\mathcal{N} \geq \mathcal{W} \geq \mu$, and this result is a significant improvement, one which appears crucial to Bamler's theory. In particular, an important advantage of Bamler's use of the pointed Nash entropy $\mathcal{N}$ instead of $\mathcal{W}$ is that $\mathcal{N}$ has bounded oscillation on certain parabolic neighborhoods, so one only needs to assume a bound on $\mathcal{N}$ at a single point, which is analogous to only assuming a volume lower bound at a single point in Cheeger–Colding–Naber theory.

THEOREM 6.37 (Noncollapsing in terms of the pointed Nash entropy). *Let $(\mathcal{O}^n, (g_t)_{t\in I})$ be a Ricci flow on a closed orbifold with isolated singularities. If $[t-2r^2, t] \subset I$ and if we have the local scalar curvature upper bound $R \leq r^{-2}$ in $B(x,t,r)$, then*

$$\mathrm{Vol}_{g_t} B(x,t,r) \geq c(n)\, \mathrm{e}^{\mathcal{N}_{x,t}(r^2)} r^n, \tag{6.109}$$

where $c(n)$ is a positive constant depending only on n.

Suppose that $I = [0,T)$ and $t \in [0,T)$. Regarding the constant on the right-hand side of (6.109), we remark that by the μ monotonicity formula (6.58) and by (6.60), we have that

$$\mathcal{N}_{x,t}(r^2) \geq \mu(g_{t-r^2}, r^2) \geq \mu(g_0, t) \geq \inf_{\tau\in[0,T)} \mu(g_0,\tau) =: -Y_0. \tag{6.110}$$

We also have the following no local collapsing result, which crucially does not assume a local scalar curvature upper bound, for balls whose centers are H_n-centers (Theorem 6.17 is used in the proof of this result); see [**Bam20a**, Theorem 6.2].

THEOREM 6.38 (Noncollapsing at H_n-centers). *Let $(\mathcal{O}^n, (g_t)_{t\in I})$ be a Ricci flow on a closed orbifold with isolated singularities. Assume that $[t-r^2, t] \subset I$ and that $(z_{t-r^2}, t-r^2)$ is an H_n-center of $(x,t) \in \mathcal{O}^n \times I$. If $R(\cdot, t-r^2) \geq R_{\min,t-r^2}$ (a real number), then we have the H_n-centered ball volume estimate*

$$\mathrm{Vol}_{g_{t-r^2}} B\big(z_{t-r^2}, t-r^2, \sqrt{2H_n}\,r\big) \geq c\big(n, r^2 R_{\min,t-r^2}\big)\, \mathrm{e}^{\mathcal{N}_{x,t}(r^2)} r^n, \tag{6.111}$$

where $c\big(n, r^2R_{\min,t-r^2}\big) := c(n) \exp\big(-2(n-2r^2R_{\min,t-r^2})^{1/2}\big)$.

In particular, on a complete ancient solution to the Ricci flow, we have that[4]

$$\mathrm{Vol}_{g_{t-r^2}} B\big(z_{t-r^2}, t-r^2, \sqrt{2H_n}\,r\big) \geq c(n)\, \mathrm{e}^{\mathcal{N}_{x,t}(r^2)} r^n \geq c(n)\, \mathrm{e}^{\mu_\infty} r^n, \tag{6.112}$$

where μ_∞ is the asymptotic pointed Nash entropy defined by (6.65).

6.1.5.9. *Non-inflating estimate.*

We have the following upper bound for the volume of a ball centered at an arbitrary point in a Ricci flow space-time; see [**Bam20a**, Theorem 8.1]. This result complements the lower bound provided by noncollapsing.

[4]Recall that any complete ancient solution to the Ricci flow satisfies $R \geq 0$.

THEOREM 6.39 (Non-inflating estimate via the pointed Nash entropy). *Let $(\mathcal{O}^n, (g_t)_{t\in I})$ be a Ricci flow on a closed orbifold with isolated singularities. Assume that t and $r > 0$ are such that $[t - r^2, t] \subset I$ and let $A \in [1, \infty)$. If, on $\mathcal{O}^n \times [t - r^2, t]$, the scalar curvature lower bound $R \geq R_{\min,[t-r^2,t]}$ holds, where $R_{\min,[t-r^2,t]}$ is a real number, then for all $x \in \mathcal{O}^n$,*

$$\begin{aligned} \text{Vol}_{g_t} B(x, t, Ar) &\leq C(r^2 R_{\min,[t-r^2,t]})\, e^{\mathcal{N}_{x,t}(r^2)+C(n)A^2} r^n \\ &\leq C(r^2 R_{\min,[t-r^2,t]})\, e^{C(n)A^2} r^n, \end{aligned} \tag{6.113}$$

where $C(n) > 0$ is a constant depending only on n.

Earlier, pioneering work on non-inflating estimates, was done by Zhang [**Zha12**]. Later in this chapter, we will see applications of the non-inflating estimate to singularity models, particularly to steady soliton singularity models; see Proposition 6.65 and Theorem 6.90 below.

6.1.6. Scaled heat equation gradient estimates.

We now return to the study of the heat equation coupled to the Ricci flow. We consider how solutions to the (forward) heat equation coupled to the super Ricci flow smooth out via a sharp gradient estimate.

Our discussion begins with the 1-dimensional heat equation. The **error function** $\Phi : \mathbb{R} \to (0, 1)$ is uniquely defined by the following properties:

$$\Phi'(x) = (4\pi)^{-1/2} e^{-x^2/4}, \qquad \lim_{x\to-\infty} \Phi(x) = 0, \qquad \lim_{x\to\infty} \Phi(x) = 1. \tag{6.114}$$

Note that Φ' is the standard Gaussian function. We have that

$$\Phi_t(x) := \Phi(t^{-1/2}x), \tag{6.115}$$

$t > 0$, is a steady self-similar solution to the 1-dimensional heat equation

$$\partial_t \Phi_t = \partial_x^2 \Phi_t \tag{6.116}$$

with initial data equal to the Heaviside function and $\Phi_1 = \Phi$. In particular, $\Phi_t(x)$ represents how the heat equation on the real line smooths out a sudden jump discontinuity.

For any $t > 0$ denote by $\Phi_t^{-1} : (0, 1) \to \mathbb{R}$ the inverse function of $\Phi_t : \mathbb{R} \to (0, 1)$. We also observe that the spatial derivative of Φ_t,

$$\partial_x \Phi_t(x) = \partial_x \Phi(t^{-1/2}x) = t^{-1/2}\Phi'(t^{-1/2}x) = (4\pi t)^{-1/2} e^{-x^2/4t}, \tag{6.117}$$

is equal to the 1-dimensional heat kernel.

We have the following monotonicity formula (see Bamler [**Bam20a**, Theorem 4.1]), which is one of the bases for Bamler's *metric flow* definition which we give in the next section.

PROPOSITION 6.40 (Scaled gradient estimate for the heat equation). *Let $u : \mathcal{O}^n \times [t_0, t_1] \to (0, 1)$ be a smooth solution to the heat equation coupled to the super Ricci flow on a closed orbifold with isolated singularities, and let $T \geq 0$. If u satisfies $|\nabla(\Phi_T^{-1} \circ u_{t_0})| \leq 1$, then*

$$|\nabla(\Phi_{T+t-t_0}^{-1} \circ u_t)| \leq 1 \quad \text{for } t \in (t_0, t_1]. \tag{6.118}$$

In particular, when $T = 0$ we mean that without any hypotheses on u_{t_0} we have

$$|\nabla(\Phi_{t-t_0}^{-1} \circ u_t)| \leq 1 \quad \text{for } t \in (t_0, t_1]. \tag{6.119}$$

EXAMPLE 6.41 (The real line). Consider $\mathcal{O}^n = \mathbb{R}^1$, with the static Euclidean metric ds^2. Let $T = t_0$ and $u_t = \Phi_t$. Then

$$\Phi_{T+t-t_0}^{-1} \circ u_t = \Phi_t^{-1} \circ \Phi_t = \mathrm{id}_{\mathbb{R}},$$

which satisfies $|\nabla(\Phi_{T+t-t_0}^{-1} \circ u_t)| \equiv 1$.

REMARK 6.42. Proposition 6.40 is a monotonicity formula for a measure of the oscillation of a solution to the heat equation coupled to the super Ricci flow (with values in $(0,1)$) relative to the solution of the 1-dimensional heat equation starting at the Heaviside function. We further note that we are interested in functions on $\mathcal{O}^n$ with values in the open interval $(0,1)$ in view of considering probability measures of subsets of $\mathcal{O}^n$. Effectively, we can assume the values are in the closed interval $[0,1]$. In particular, we will apply this to heat equation solutions starting at a characteristic function.

PROOF OF PROPOSITION 6.40. Without loss of generality, we may assume that $T = t_0 > 0$. Let $u_t := \Phi_t \circ h_t$, so that $h_t = \Phi_t^{-1} \circ u_t$. Then

$$\frac{1}{2}\partial_t |\nabla h|^2 \leq \frac{1}{2}\triangle |\nabla h|^2 - \frac{1}{2t}\nabla h \cdot h \nabla |\nabla h|^2 + \frac{1}{2t}(1 - |\nabla h|^2)|\nabla h|^2. \tag{6.120}$$

By the maximum principle, the bound $|\nabla h|^2 \leq 1$ remains preserved, which finishes the proof; see [**Bam20a**, Theorem 4.1] for details. □

6.2. Bamler's synthetic metric flow and compactness theorems

In Riemannian geometry, to understand limit spaces, one needs to consider categories of spaces more general than smooth Riemannian manifolds. In the Cheeger–Colding–Naber theory of Ricci limit spaces, metric length spaces form a natural category, within which their theory systematically explores the fine structure of these spaces. In this section, we provide an overview for Bamler's category for studying the limit spaces (singularity models) of Ricci flow.

In [**Bam23**], Bamler introduced the concept of synthetic (abstract) 'metric flows', which can be viewed as flow analogues of metric measure spaces. These flows feature measures that satisfy a reproduction formula analogous to the one satisfied by the conjugate heat kernel measures along (super) Ricci flows. In particular, metric flows generalize (super) Ricci flows, equipped with their conjugate heat kernels, and form a natural and suitable class to take limits.

6.2.1. Definition of metric flow.

Extending the notion of (super) Ricci flow, a central class of flows for metric spaces is given by the following; see [**Bam23**, Definition 3.2].

DEFINITION 6.43 (Bamler's metric flow). Let $I \subset \mathbb{R}$ be a subset (called the **time set**),[5] let $\mathcal{X}$ be a set (called the **space-time**), and let $\mathbf{t} : \mathcal{X} \to I$ be a function (called the **time function**). A quadruple

$$\mathcal{X} := \big(\mathcal{X}, \mathbf{t}, (d_t)_{t\in I}, (\nu_{x;s})_{x\in\mathcal{X}, s\in I, s\leq \mathbf{t}(x)}\big), \tag{6.121}$$

[5] The subset I is not necessarily an interval.

where $x \in \mathcal{X}$, is called a **metric flow (over** I**)** if all of the following conditions are satisfied:

(1) $(\mathcal{X}_t, d_t)$ is a complete separable metric space for all $t \in I$, where
$$\mathcal{X}_t := \mathbf{t}^{-1}(t),$$
and $\mathcal{X}_t$ is called a **time slice**.

(2) For all space-time points $x \in \mathcal{X}$ and for all times $s \in I$, where $s \leq \mathbf{t}(x)$,[6] we have a probability measure $\nu_{x;s}$ on $\mathcal{X}_s$, called the **conjugate heat kernel measure based at** x, such that the following conditions hold:

(a) $\nu_{x;\mathbf{t}(x)} = \delta_x$, where δ_x is the Dirac delta measure of $\mathcal{X}_{\mathbf{t}(x)}$ at x.

(b) (Reproduction formula) For all $t_1, t_2, t_3 \in I$ in the time set with $t_1 \leq t_2 \leq t_3$, $x_3 \in \mathcal{X}_{t_3}$, and for any Borel set $\mathcal{B} \subset \mathcal{X}_{t_1}$, the conjugate heat kernel measure satisfies the **reproduction formula** (a.k.a. **semi-group property**)
$$\nu_{x_3;t_1}(\mathcal{B}) = \int_{\mathcal{X}_{t_2}} \nu_{y;t_1}(\mathcal{B})\, d\nu_{x_3;t_2}(y).$$

(c) (Lipschitz monotonicity) Let $t_1, t_2 \in I$, with $t_1 < t_2$, and let $u_{t_1} : \mathcal{X}_{t_1} \to [0,1]$ be a measurable function such that
$$\Phi^{-1} \circ u_{t_1} : \mathcal{X}_{t_1} \to \mathbb{R}$$
is a $T^{-1/2}$-Lipschitz function with respect to d_{t_1} for some $T \geq 0$, where $\Phi : \mathbb{R} \to (0,1)$ is the error function defined by (6.114).[7] Then the function $u_{t_2} : \mathcal{X}_{t_2} \to \mathbb{R}$, defined by
$$u_{t_2}(x) := \int_{\mathcal{X}_{t_1}} u_{t_1}(y)\, d\nu_{x;t_1}(y) \quad \text{for } x \in \mathcal{X}_{t_2},$$
is either constant or has the property that the function $\Phi^{-1} \circ u_{t_2} : \mathcal{X}_{t_2} \to \mathbb{R}$ is $(t_2 - t_1 + T)^{-1/2}$-Lipschitz with respect to d_{t_2}.[8] We call the function $(x,t) \mapsto u_t(x)$, for $t > t_1$, the **solution to the heat equation** or **heat flow** with initial data u_{t_1}.

Condition (2)(b) is a generalization of the standard reproduction formula for the heat kernels of (super) Ricci flows and condition (2)(c) is a generalization of the scaled gradient estimate in Proposition 6.40. Moreover, the gradient bound 2(c) actually characterizes super Ricci flows.

6.2.2. Basic properties of metric flows.

As a matter of course, certain basic properties of metric flows will be used. The following is [**Bam23**, Definition 3.10].

DEFINITION 6.44 (Restricting the time set of a metric flow). If $\mathcal{X}$ is a metric flow over a time set $I \subset \mathbb{R}$ and if I' is a subset of I, then the **restriction of** $\mathcal{X}$ **to** I' is defined by

$$\mathcal{X}|_{I'} := \big(\mathcal{X}_{I'}, \mathbf{t}|_{\mathcal{X}_{I'}}, (d_t)_{t \in I'}, (\nu_{x;s})_{x \in \mathcal{X}_{I'},\, s \in I',\, s \leq \mathbf{t}(x)}\big), \tag{6.122}$$

where $\mathcal{X}_{I'} := \mathbf{t}^{-1}(I')$.

[6]The inequality $s \leq \mathbf{t}(x)$ means that s is in the past or present of x.

[7]We define any function to be $0^{-1/2}$-Lipschitz.

[8]A function $f : (X,d) \to \mathbb{R}$ defined on a metric space is L**-Lipschitz** provided that $|f(x) - f(y)| \leq L\, d(x,y)$ for all $x, y \in X$.

It is straightforward to see that the restriction of a metric flow over a time set I to a smaller time set I' remains a metric flow.

The class of metric flows is invariant under space-time rescaling as well as time shift. Namely, we have the following (see [**Bam23**, Lemma 3.9]).

LEMMA 6.45 (Parabolic rescaling and time shift of a metric flow). *Let*

$$\big(\mathcal{X}, \mathbf{t}, (d_t)_{t\in I}, (\nu_{x;s})_{x\in\mathcal{X}, s\in I, s\leq \mathbf{t}(x)}\big),$$

where $\mathbf{t} : \mathcal{X} \to I$, *be a metric flow over* I.

(1) (Time shift) *Given any* $t_0 \in \mathbb{R}$, *define the shifted time set* $I_0 := I - t_0$ *and shifted time function* $\mathbf{t} - t_0 : \mathcal{X} \to I_0$ *by* $(\mathbf{t} - t_0)(x) := \mathbf{t}(x) - t_0$. *Then*

$$\mathcal{X} - t_0 := \big(\mathcal{X}, \mathbf{t} - t_0, (d_{t+t_0})_{t\in I_0}, (\nu_{x;s+t_0})_{x\in\mathcal{X}, s\in I_0, s\leq(\mathbf{t}-t_0)(x)}\big)$$

is a metric flow over I_0.

(2) (Parabolic rescaling) *For any positive real number* λ, *the quadruple*

$$\mathcal{X}^\lambda := \big(\mathcal{X}, \widetilde{\mathbf{t}}, (\widetilde{d}_t)_{t\in\lambda^2 I}, (\widetilde{\nu}_{x;s})_{x\in\mathcal{X},\, s\in\lambda^2 I,\, s\leq\widetilde{\mathbf{t}}(x)}\big), \tag{6.123}$$

called the λ***-rescaling of*** $\mathcal{X}$, *is also a metric flow. Here, the rescaled time function* $\widetilde{\mathbf{t}} : \mathcal{X} \to \lambda^2 I$ *is defined by* $\widetilde{\mathbf{t}}(x) := \lambda^2 \mathbf{t}(x)$, *the rescaled distance function is* $\widetilde{d}_t := \lambda d_{\lambda^{-2}t}$, *and the rescaled heat kernel measure is* $\widetilde{\nu}_{x;s} := \nu_{x;\lambda^{-2}s}$.

We can combine the time shift and parabolic rescaling, as is done in singularity analysis, to obtain the metric flow:

$$\begin{aligned}&(\mathcal{X} - t_0)^\lambda = \\ &\big(\mathcal{X}, \lambda^2(\mathbf{t} - t_0), (\lambda d_{\lambda^{-2}t+t_0})_{t\in\lambda^2(I-t_0)}, (\nu_{x;\lambda^{-2}s+t_0})_{x\in\mathcal{X},\, s\in\lambda^2(I-t_0),\, s\leq\lambda^2(\mathbf{t}-t_0)(x)}\big)\end{aligned} \tag{6.124}$$

over $\lambda^2(I - t_0)$. Note that $\big((\mathcal{X} - t_0)^\lambda\big)_t = \mathcal{X}_{\lambda^{-2}t+t_0}$.

REMARK 6.46. The metrics of $\mathcal{X}$ are also denoted by Bamler as $d_{\mathbf{t}}$, so that $d_{\mathbf{t}}(x, y) = d_t(x, y)$ for $x, y \in \mathcal{X}_t$. With this notation, the metrics of $(\mathcal{X} - t_0)^\lambda$ are denoted by $d_{\lambda^2(\mathbf{t}-t_0)}$.

6.2.3. Smooth (super) Ricci flows as metric flows.

By [**Bam23**, §3.7], any smooth super Ricci flow $(\mathcal{O}^n, g_t)$, $t \in I$, where $\mathcal{O}^n$ is an orbifold with isolated singularities and I is an interval, induces a metric flow by defining $\mathcal{X} = \mathcal{O}^n \times I$, $\mathbf{t}(x,t) = t$, d_t to be the Riemannian distance of g_t, and $(\nu_{(x,t);s})_{s\in I,\, s\leq t}$ to be the conjugate heat kernel measure based at (x, t) for each $(x, t) \in \mathcal{O}^n \times I$. We call $\mathcal{X}$ the metric flow **corresponding to** the Ricci flow $(\mathcal{O}^n, g_t)$.

The significance of the time shift in Lemma 6.45(1) is that if $t_0 \in I$, then t_0 is translated to time 0. In general, if $x \in \mathcal{X}_t$, then $x \in (\mathcal{X} - t_0)_{t-t_0}$. For the parabolic rescaling in Lemma 6.45(2), if $I = (-T, 0)$, then $\lambda^2 I = (-\lambda^2 T, 0)$. If the metric flow $\mathcal{X}$ is induced by a smooth Ricci flow $(\mathcal{O}^n, g(t) := g_t)$, then the rescaled metric flow $(\mathcal{X} - t_0)^\lambda$ is induced by the rescaled Ricci flow $g^\lambda(t) := \lambda^2 g(\lambda^{-2}t + t_0)$.

6.2.4. H-concentrated metric flows.

Motivated by the variance estimate for conjugate heat kernels (Corollary 6.31), we have the following:

DEFINITION 6.47 (H-concentrated metric flow). We say that a metric flow $\mathcal{X}$ is **H-concentrated**, where $H \in \mathbb{R}$, if for all $s, t \in I$ with $s \leq t$ and $x_1, x_2 \in \mathcal{X}_t$, we have that

$$\mathrm{Var}(\nu_{x_1;s}, \nu_{x_2;s}) \leq d_t^2(x_1, x_2) + H(t-s). \tag{6.125}$$

We remark that if $\mathcal{X}$ is H-concentrated, then $(\mathcal{X} - t_0)^\lambda$ is H-concentrated for all $t_0 \in \mathbb{R}$ and $\lambda > 0$ (see [**Bam23**, Remark 3.33]). By Corollary 6.31, for any n-dimensional smooth Ricci flow $(\mathcal{O}^n, g_t)$ on a closed orbifold with isolated singularities, its corresponding metric flow $\mathcal{X}$ is H_n-concentrated, where $H_n = \frac{(n-1)\pi^2}{2} + 4$ is defined by (6.93). The same statement is true for Ricci flow singularity models on orbifolds with isolated singularities.

6.2.5. Metric shrinking solitons.

The analogue of a solution to the conjugate heat equation coupled to the Ricci flow is the following (see [**Bam23**, Definition 3.13]):

DEFINITION 6.48 (Conjugate heat flow). Let $\mathcal{X}$ be a metric flow on a time set I. A family of measures μ_t, $t \in I$, where each μ_t is a probability measure on $\mathcal{X}_t$, is a **conjugate heat flow of measures** if for all $s, t \in I$ with $s \leq t$, we have

$$\mu_s = \int_{\mathcal{X}_t} \nu_{x;s}\, d\mu_t(x), \tag{6.126}$$

where $x \in \mathcal{X}_t$ and $\{\nu_{x;s}\}_{s \leq t}$ is the conjugate heat kernel measure of $\mathcal{X}$ based at x. Given $s \leq t$, by integrating the family of measures $\{\nu_{x;s}\}_{x \in \mathcal{X}_t}$ over the measure space $(\mathcal{X}_t, \mu_t)$, we mean that if A is a measurable subset of $\mathcal{X}_s$ with respect to the Borel σ-algebra $\mathcal{B}$ defined by $(\mathcal{X}_s, d_s)$, then for $s \leq t$,[9]

$$\mu_s(A) = \int_{\mathcal{X}_t} \nu_{x;s}(A)\, d\mu_t(x). \tag{6.127}$$

The essential property of shrinking gradient Ricci solitons are that they are *shrinking self-similar* solutions to the Ricci flow. In the category of metric flows, we have the following notion which generalizes the notion of shrinking gradient Ricci soliton (see [**Bam23**, Definition 3.42]).

DEFINITION 6.49 (Metric shrinking soliton). Let $I \subset \mathbb{R}$ be an interval with $\sup I = 0$ and $0 \notin I$. A pair $(\mathcal{X}, (\mu_t)_{t \in I})$ of a metric flow

$$\mathcal{X} := \big(\mathcal{X}, \mathfrak{t}, (d_t)_{t \in I}, (\nu_{x;s})_{x \in \mathcal{X}, s \in I, s \leq \mathfrak{t}(x)}\big),$$

and a conjugate heat flow of measures $(\mu_t)_{t \in I}$ on $\mathcal{X}$ is called a **metric shrinking soliton** if the following conditions hold:

There is a metric measure space (X, d, μ), a family of measures $(\nu'_{x;u})_{x \in X;\, u \in \mathbb{R}_{\geq 0}}$ on X, and a family of maps $\phi_t : \mathcal{X}_t \to X$, $t \in I$, such that:

(1) For all $t \in I$, $\phi_t : (\mathcal{X}_t, d_t, \mu_t) \to (X, \sqrt{|t|}\, d, \mu)$ is a **metric measure space isometry**; i.e., for all $x, y \in \mathcal{X}_t$,

$$d_t(x, y) = \sqrt{|t|}\, d(\phi_t(x), \phi_t(y)),$$

[9]Note that the $s = t$ case of (6.127) just says that $\mu_t(A) = \int_{\mathcal{X}_t} \delta_x(A)\, d\mu_t(x)$.

and $\mu = (\phi_t)_*(\mu_t)$, meaning that for all μ-measurable subsets $A \subset X$, the preimage $\phi_t^{-1}(A)$ is μ_t-measurable, and we have

$$\mu(A) = \mu_t(\phi_t^{-1}(A)).$$

(2) For all $t \in I$, $x \in \mathcal{X}_t$, and $s \in I$ with $s \leq t$, we have

$$(\phi_s)_*(\nu_{x;s}) = \nu'_{\phi_t(x);\,\ln(s/t)}. \tag{6.128}$$

$(\mu_t)_{t\in I}$ is called the **potential flow** of the metric shrinking soliton.

The following example illustrates how metric shrinking solitons generalize shrinking gradient solitons in the smooth case; see [**Bam23**, Example 3.44]. Thus, it motivates the abstract Definition 6.49 of a metric shrinking soliton.

EXAMPLE 6.50 (Shrinking gradient Ricci soliton is a metric shrinking soliton). Let $(\mathcal{O}^n, g, f)$ be a shrinking gradient Ricci soliton on an orbifold with isolated singularities. We will show that it is a metric shrinking soliton as defined by Definition 6.49. Let d_g be the Riemannian distance function and let

$$d\mu = (4\pi)^{-\frac{n}{2}} \mathrm{e}^{-f} dv_g,$$

where f is normalized so that $\int_{\mathcal{O}^n} (4\pi)^{-\frac{n}{2}} \mathrm{e}^{-f} dv_g = 1$. Then $(\mathcal{O}^n, d_g, \mu)$ is a metric measure space, where μ is the measure associated to the weighted volume element $d\mu$.

As given by Proposition 1.73, let $(\mathcal{O}^n, \widetilde{g}_t, \widetilde{f}_t)$, $t \in (-\infty, 1)$, be the associated canonical form, which is a solution to the Ricci flow, with diffeomorphisms $\widetilde{\varphi}_t : \mathcal{O}^n \to \mathcal{O}^n$ satisfying $\widetilde{g}_t = (1-t)\widetilde{\varphi}_t^* g$, $\widetilde{f}_t = f \circ \widetilde{\varphi}_t$, $\widetilde{\varphi}_0 = \mathrm{id}$, and

$$\frac{\partial}{\partial t}\widetilde{\varphi}_t(x) = \frac{1}{1-t}(\nabla_g f)(\widetilde{\varphi}_t(x)). \tag{6.129}$$

For $t \in (-\infty, 0) =: I$ (we translate the time interval of the solution by -1), define

$$\begin{aligned}
\phi_t &:= \widetilde{\varphi}_{t+1}, \\
g_t &:= \widetilde{g}_{t+1} = |t|\,\phi_t^* g, \\
f_t &:= \widetilde{f}_{t+1} = f \circ \phi_t, \\
d\mu_t &:= (4\pi|t|)^{-\frac{n}{2}} \mathrm{e}^{-f_t} dv_t = (\phi_t)^{-1}(d\mu),
\end{aligned}$$

where dv_t is the Riemannian measure of g_t. Hence, we see that

$$\phi_t : (\mathcal{O}^n, d_{g_t}, \mu_t) \to \left(\mathcal{O}^n, \sqrt{|t|}\, d_g, \mu\right)$$

is a metric measure space isometry for each $t \in (-\infty, 0)$. We also note that $\phi_{-1} = \mathrm{id}$, $g_{-1} = g$, $f_{-1} = f$, and $d\mu_{-1} = d\mu$.

Using (6.129), we compute that

$$\frac{\partial}{\partial u}\phi_{-\mathrm{e}^{-u}} = (\nabla_g f) \circ \phi_{-\mathrm{e}^{-u}}.$$

Thus, $\{\phi_{-\mathrm{e}^{-u}}\}_{u\in\mathbb{R}}$ is a 1-parameter *group* of diffeomorphisms of $\mathcal{O}^n$; that is,

$$\phi_{-\mathrm{e}^{-u_1-u_2}} = \phi_{-\mathrm{e}^{-u_1}} \circ \phi_{-\mathrm{e}^{-u_2}} \quad \text{for all } u_1, u_2 \in \mathbb{R}. \tag{6.130}$$

In contrast, $\{\phi_t\}_{t<0}$ is only a 1-parameter *family* of diffeomorphisms. We rewrite (6.130) as

$$\phi_{ct} = \phi_{-c} \circ \phi_t = \phi_t \circ \phi_{-c} \quad \text{for all } c > 0,\ t < 0. \tag{6.131}$$

In particular,

$$\phi_{-c^{-1}} = (\phi_{-c})^{-1}. \tag{6.132}$$

By (6.131), we have

$$\begin{aligned} g_{ct} &= c|t|\phi_{ct}^*(g) \\ &= c|t|\phi_{-c}^*(\phi_t^*(g)) \\ &= c\phi_{-c}^*(g_t). \end{aligned} \tag{6.133}$$

Thus,

$$g(t) = g_t = c\phi_{-c}^*(g_{c^{-1}t}) = g_c^{\phi_{-c}}(t), \tag{6.134}$$

where the last equality follows from definition (6.16). From this and (6.17), we deduce that

$$d\nu_{x_0,t_0;t} = (d\nu_c^{\phi_{-c}})_{x_0,t_0;t} = (\phi_{-c}^{-1})_*\big(d\nu_{\phi_{-c}(x_0),c^{-1}t_0;c^{-1}t}\big). \tag{6.135}$$

Hence, by applying (6.132), we obtain

$$(\phi_t)_*\big(d\nu_{x_0,t_0;t}\big) = (\phi_{c^{-1}t})_*\big(d\nu_{\phi^{-1}_{-c^{-1}}(x_0),c^{-1}t_0;c^{-1}t}\big). \tag{6.136}$$

Therefore,

$$(\phi_t)_*\big(d\nu_{\phi_{t_0}^{-1}(x_0),t_0;t}\big) = (\phi_{c^{-1}t})_*\big(d\nu_{\phi^{-1}_{c^{-1}t_0}(x_0),c^{-1}t_0;c^{-1}t}\big). \tag{6.137}$$

Let $\mathcal{X}$ be the metric flow induced by the Ricci flow $(\mathcal{O}^n, g_t)$, $t \in I = (-\infty, 0)$. By (6.137), the family of measures $(\nu'_{x;u})_{x\in\mathcal{O}^n,\, u\in\mathbb{R}_{\geq 0}}$ on $\mathcal{O}^n$ defined by

$$\nu'_{x;\,\ln(s/t)} := (\phi_s)_*\big(\nu_{(\phi_t^{-1}(x),t);s}\big)$$

for all $x \in \mathcal{O}^n$ and $s \leq t < 0$ is well-defined; i.e., it depends only on x and s/t. We conclude that the pair $\big(\mathcal{X}, (\mu_t)_{t\in I}\big)$ is a metric shrinking soliton.

6.2.6. Metric flow pairs.

A natural analogue, for metric flows, of the notion of a *pointed* solution to the Ricci flow, is the following; see [**Bam23**, Definition 5.1].

DEFINITION 6.51 (Pointed metric flow). Let $I \subset \mathbb{R}$ be an interval and let $I' \subset I$. We say that $(\mathcal{X}, (\mu_t)_{t\in I'})$ is a **metric flow pair** if:

(1) I' has full measure in I, i.e., $m(I \setminus I') = 0$, where m is the Lebesgue measure on $\mathbb{R}$;
(2) $\mathcal{X}$ is a metric flow over I';
(3) $(\mu_t)_{t\in I'}$ is a conjugate heat flow on $\mathcal{X}$ with $\operatorname{supp}(\mu_t) = \mathcal{X}_t$ for all $t \in I'$.

For each $J \subset I'$, we say that $(\mathcal{X}, (\mu_t)_{t\in I'})$ is **fully defined over** J.

A notable special case is when $I' = I$.

For example, if $(\mathcal{O}^n, g_t)$, $t \in [t_1, t_0]$, is a smooth Ricci flow and if $x_0 \in \mathcal{O}^n$, then the metric flow $\mathcal{X} = \mathcal{O}^n \times [t_1, t_0]$ with the conjugate heat kernel measure $(\nu_{(x_0,t_0);s})_{s\in[t_1,t_0]}$, $s \leq t_0$, based at (x_0, t_0) is a metric flow pair. This corresponds to the pointed Ricci flow $\big(\mathcal{O}^n, (g_t)_{t\in[t_1,t_0]}, (x_0, t_0)\big)$. In this case, we also call the metric flow pair a **pointed metric flow**. However, the notion of metric flow pair is more general since, even in this setting, the conjugate heat *flow* $(\mu_t)_{t\in I'}$ does not need to be a conjugate heat *kernel* (based at some point).

The following is [**Bam23**, Definition 5.2].

DEFINITION 6.52. $\mathbb{F}_I^J$ denotes the **space of metric flow pairs** over I that are fully defined over J, where two metric flow pairs are defined to be equivalent if there exists an almost always isometry between the two metric flow pairs that is fully defined over J.

In the spirit of the device of common isometric embeddings used in the definition of Gromov–Hausdorff distance, we have the following; see [**Bam23**, Definition 5.4].

DEFINITION 6.53 (Correspondence of metric flows). Let $\mathcal{X}^i$ be a (finite or infinite) sequence or a family of metric flows over subsets $I'^{,i} \subset \mathbb{R}$, $i \in \mathcal{I}$. A **correspondence** (between the metric flows) **over a subset** $I'' \subset \mathbb{R}$ consists of a pair

$$\mathfrak{C} := \big((Z_t, d_t^Z)_{t \in I''}, (\varphi_t^i)_{t \in I''^{,i},\, i \in \mathcal{I}}\big), \tag{6.138}$$

where:

(1) (Z_t, d_t^Z) is a metric space for each $t \in I''$;
(2) $I''^{,i} \subset I'^{,i} \cap I''$ for each $i \in \mathcal{I}$;
(3) $\varphi_t^i : (\mathcal{X}_t^i, d_t^i) \to (Z_t, d_t^Z)$ is an isometric embedding for each $t \in I''^{,i}$ and $i \in \mathcal{I}$.

If a subset $J \subset \mathbb{R}$ has the property that $J \subset I''^{,i}$ for all $i \in \mathcal{I}$, then $\mathfrak{C}$ is said to be **fully defined over** J.

6.2.7. $\mathbb{F}$-distance between metric flow pairs.

The fundamental notion of distance between pointed metric spaces is the pointed Gromov–Hausdorff distance. The following is a fundamental notion of distance between metric flow pairs; see [**Bam23**, Definition 5.5].

DEFINITION 6.54 (The $\mathbb{F}$-Distance within a correspondence between metric flow pairs). Suppose that $\mathfrak{X}^i := \big(\mathcal{X}^i, (\mu_t^i)_{t \in I'^{,i}}\big)$, for $i = 1, 2$, are metric flow pairs over subsets $I^i \subset \mathbb{R}$ which are fully defined over a subset $J \subset I^1 \cap I^2$. Let

$$\mathfrak{C} := \big((Z_t, d_t^Z)_{t \in I''}, (\varphi_t^i)_{t \in I''^{,i},\, i=1,2}\big) \tag{6.139}$$

be a correspondence between $\big(\mathcal{X}^i, (\mu_t^i)_{t \in I'^{,i}}\big)$, $i = 1, 2$, over a subset I'' which are fully defined over J. The **$\mathbb{F}$-distance between** $\big(\mathcal{X}^i, (\mu_t^i)_{t \in I'^{,i}}\big)$, $i = 1, 2$, **within** $\mathfrak{C}$ and **uniform over** J, denoted by

$$d_{\mathbb{F}}^{\mathfrak{C}, J}\big((\mathcal{X}^1, (\mu_t^1)_{t \in I'^{,1}}), (\mathcal{X}^2, (\mu_t^2)_{t \in I'^{,2}})\big), \tag{6.140}$$

is defined as the infimum of all positive real numbers r with the property that there exists a measurable subset $E \subset I''$ with $J \subset I'' \setminus E \subset I''^{,1} \cap I''^{,2}$ and there exists a family of couplings $(q_t)_{t \in I'' \setminus E}$ between μ_t^1 and μ_t^2 such that:

(1) $m(E) \leq r^2$.
(2) For all $s, t \in I'' \setminus E$ with $s \leq t$, we have that

$$\int_{\mathcal{X}_t^1 \times \mathcal{X}_t^2} d_{W_1}^{Z_s}\big((\varphi_s^1)_* \nu_{x^1;s}^1, (\varphi_s^2)_* \nu_{x^2;s}^1\big) dq_t(x^1, x^2) \leq r. \tag{6.141}$$

The following is [**Bam23**, Definition 5.7].

DEFINITION 6.55 (The $\mathbb{F}$-Distance between metric flow pairs). The **$\mathbb{F}$-distance** $d_{\mathbb{F}}^J$ between metric flow pairs $\mathcal{X}_1, \mathcal{X}_2$ over a common interval I is defined to be the infimum over all correspondences $\mathfrak{C}$ between $\mathcal{X}_1, \mathcal{X}_2$ over I fully defined over J of the $\mathbb{F}$-distance $d_{\mathbb{F}}^{\mathfrak{C},J}$ between $\mathcal{X}_1$ and $\mathcal{X}_2$ within $\mathfrak{C}$ and uniform over J.

The following is [**Bam23**, Theorems 5.11 and 5.18].

THEOREM 6.56. *$(\mathbb{F}_I^J, d_{\mathbb{F}}^J)$ is a complete metric space, where in the definition of metric space we allow the distance to be infinite.*

We say that a sequence of metric flow pairs $\{\mathcal{X}_i\}_{i\in\mathbb{N}}$ over a common interval I and fully defined over J **$\mathbb{F}$-converges** to a metric flow pair $\mathcal{X}_\infty$ over I and fully defined over J if $d_{\mathbb{F}}^J(\mathcal{X}_i, \mathcal{X}_\infty) \to 0$. We say that a sequence **$\mathbb{F}$-subconverges** if it has a subsequence that $\mathbb{F}$-converges.

6.2.8. $\mathbb{F}$-convergence of metric flow pairs within a correspondence.

Let $\{(\mathcal{X}^i, (\mu_t^i)_{t\in I',i})\}_{i\in\mathbb{N}}$ be a sequence of metric flow pairs over intervals I^i, and let $(\mathcal{X}^\infty, (\mu_t^\infty)_{t\in I',\infty})$ be a metric flow pair over an interval I^∞. Suppose that we have a correspondence

$$\mathfrak{C} = \left((Z_t, d_t^Z)_{t\in I''}, (\varphi_t^i)_{t\in I'',i,\, i\in\mathbb{N}\cup\{\infty\}}\right) \tag{6.142}$$

between the metric flow $\mathcal{X}^i$, $i \in \mathbb{N}\cup\{\infty\}$, over a subset I'' of $\mathbb{R}$. We assume that $I^\infty \setminus I''$ has measure zero. Let J be a subset of $\mathbb{R}$ and assume that both $\{(\mathcal{X}^i, (\mu_t^i)_{t\in I',i})\}_{i\in\mathbb{N}}$ and the correspondence $\mathfrak{C}$ are fully defined over J. The following is [**Bam23**, Definition 6.1].

DEFINITION 6.57 ($\mathbb{F}$-convergence within a correspondence). We say that a sequence of metric flow pairs $\{(\mathcal{X}^i, (\mu_t^i)_{t\in I',i})\}_{i\in\mathbb{N}}$ **$\mathbb{F}$-converges** to a metric flow pair $(\mathcal{X}^\infty, (\mu_t^\infty)_{t\in I',\infty})$ **within a correspondence $\mathfrak{C}$, and the convergence is uniform over J,** if

$$d_{\mathbb{F}}^{\mathfrak{C},J}\left((\mathcal{X}^i, (\mu_t^i)_{t\in I',i}), (\mathcal{X}^\infty, (\mu_t^\infty)_{t\in I',\infty})\right) \to 0 \tag{6.143}$$

as $i \to \infty$.

6.2.9. $\mathbb{F}$-Compactness for metric flows.

As a special case of his more general result, and based on a synthetic compactness theorem (see [**Bam23**, Theorem 7.4]), Bamler obtained the following $\mathbb{F}$-compactness theorem for super Ricci flows [**Bam23**, Theorem 1.2 (cf. Corollary 7.5)]. Here, we present his result for the specific case of Ricci flows.

THEOREM 6.58 ($\mathbb{F}$-Compactness theorem for pointed Ricci flows). *Suppose that $(\mathcal{O}_i^n, (g_{i,t})_{t\in(-T,0]}, (x_i, 0))$ is a sequence of pointed smooth Ricci flows on closed orbifolds with isolated singularities.*[10] *Then:*

(1) ($\mathbb{F}$-Precompactness) *The corresponding sequence of pointed metric flows over $(-T,0]$—$\mathcal{X}^i$, with conjugate heat kernels $(\nu_{(x_i,0);t})_{t\in(-T,0)}$ —$\mathbb{F}$-subconverges to some metric flow pair $(\mathcal{X}^\infty, (\mu_t^\infty)_{t\in(-T,0)})$.*

(2) (Agreement with Cheeger–Gromov convergence) *If this pointed sequence converges in the pointed C^∞ Cheeger–Gromov sense, then the $\mathbb{F}$-convergence is to the metric flow associated to the C^∞ limit.*

[10] When $\mathcal{O}_i^n$ is noncompact, assume that $T = \infty$ and that $(\mathcal{O}_i^n, (g_{i,t})_{t\in(-\infty,0]})$ is a singularity model in the sense of Definition 6.62 below.

To expand on the last paragraph of the theorem's statement, Bamler's $\mathbb{F}$-compactness theorem is a full generalization of the Cheeger–Gromov–Hamilton compactness theorem for Ricci flow in the following sense:

Let $\big(M_i^n, (g_{i,t})_{t\in(-T,0]}, (x_i,0)\big)$ be a sequence of pointed complete Ricci flows on smooth manifolds. Suppose that we have the following pointed C^∞ Cheeger–Gromov convergence:

$$\big(M_i^n, (g_{i,t})_{t\in(-T,0]}, (x_i,0)\big) \overset{d_{\mathrm{pCG}}}{\longrightarrow} \big(M_\infty^n, (g_{\infty,t})_{t\in(-T,0]}, (x_\infty,0)\big) \tag{6.144}$$

to a pointed complete Ricci flow on a smooth manifold. Then, the $\mathbb{F}$-convergence in Theorem 6.58 agrees with the pointed C^∞ Cheeger–Gromov convergence in (6.144). Namely, the corresponding sequence of pointed metric flows $\big(\mathcal{X}^i, (\nu_{(x_i,0);t})_{t\in(-T,0)}\big)$ $\mathbb{F}$-converges to the pointed metric flow

$$\big(\mathcal{X}^\infty, (\nu_{(x_\infty,0);t})_{t\in(-T,0)}\big)$$

associated to $\big(M_\infty^n, (g_{\infty,t})_{t\in(-T,0]}, (x_\infty,0)\big)$.

6.2.10. Examples of $\mathbb{F}$-convergence.

In this subsection we discuss some examples of $\mathbb{F}$-convergence in the context of singularity formation.

EXAMPLE 6.59 (Bryant steady soliton [**Bam23**, Example 1.3]). Denote by $(\Sigma^n, g_t^{\mathrm{Bry}})$, $t\in\mathbb{R}$, the Bryant steady soliton, which is an eternal solution to the Ricci flow. Let $o\in\Sigma^n$ denote the center of its rotational symmetry. Let $g_t^\lambda := \lambda^2 g^{\mathrm{Bry}}_{\lambda^{-2}t}$ denote the parabolic rescaling of g_t^{Bry} by $\lambda\in\mathbb{R}_+$. Then, we have the following $\mathbb{F}$-*convergence* of blow-downs of the Bryant steady soliton, considered as ancient solutions of the Ricci flow:

$$\big(\Sigma^n, (g_t^\lambda)_{t\in\mathbb{R}_-}, (\nu_{(o,0);t})_{t\in\mathbb{R}_-}\big) \overset{d_{\mathbb{F}}}{\longrightarrow} \big(\mathbb{S}^{n-1}\times\mathbb{R}, (\overline{g}_t)_{t\in\mathbb{R}_-}, (\mu_t^\infty)_{t\in\mathbb{R}_-}\big) \tag{6.145}$$

as $\lambda\to 0$, where

$$\overline{g}_t := 2(n-1)|t|\, g_{\mathbb{S}^{n-1}} + g_{\mathbb{R}}$$

is the shrinking cylinder solution and where $(\mu_t^\infty)_{t\in\mathbb{R}_-}$ is a conjugate heat flow on the cylinder limit. This limit is defined as a "tangent flow at infinity" in the next section; in this case, it is the same as Perelman's asymptotic soliton ([**Per03a**, §1]). In contrast, based at the same space-time point $(o,0)$, we have the pointed *Gromov–Hausdorff convergence* to a ray (which is the asymptotic cone of the Bryant steady soliton):

$$(\Sigma^n, g_0^\lambda, o) \overset{d_{\mathrm{pGH}}}{\longrightarrow} ([0,\infty),0)$$

as $\lambda\to 0$.

EXAMPLE 6.60 (Degenerate neckpinch). Let $(\mathbb{S}^n, g_t)$, $t\in[0,T)$, be a rotationally symmetric solution to the Ricci flow, and let $p\in\mathbb{S}^n$ be a fixed center of the rotational symmetries of g_t. Assume that the Bryant soliton is a pointed C^∞ Cheeger–Gromov limit of rescalings, as follows. For any sequence of times $t_i\to T$, let $\lambda_i := |\mathrm{Rm}|^{\frac12}(p,t_i)$. Define the rescaled solutions $g_{i,t} := \lambda_i^2 g_{t_i+\lambda_i^{-2}t}$. Assume we have the pointed C^∞ Cheeger–Gromov convergence:

$$\big(\mathbb{S}^n, (g_{i,t})_{t\in[-t_i\lambda_i^2,0]}, (p,0)\big) \overset{d_{\mathrm{pCG}}}{\longrightarrow} \big(\Sigma^n, (g_t^{\mathrm{Bry}})_{t\in(-\infty,0]}, (o,0)\big). \tag{6.146}$$

We say that this solution forms a **degenerate neckpinch** at p as $t \to T$; see Hamilton [**Ham95b**, §3], Gu and Zhu [**GZ08**], Angenent, Isenberg, and Knopf [**AIK15**], [**CCG$^+$15**, Chapter 36].

Furthermore, since the convergence in (6.146) is smooth, the corresponding sequence of pointed metric flows

$$\big(\mathcal{X}^i, (\nu^i_{(p,0);t})_{t\in[-t_i\lambda_i^2,0]}\big), \quad \text{where } \nu^i_{(p,0);t} := \nu_{(p,t_i);t_i+\lambda_i^{-2}t},$$

$\mathbb{F}$-converges to the pointed metric flow $\big(\mathcal{X}^\infty, (\nu_{(o,0);t})_{t\in(-\infty,0]}\big)$ associated to the *Bryant soliton* $\big(\Sigma^n, (g_t^{\mathrm{Bry}})_{t\in(-\infty,0]}, (o,0)\big)$. On the other hand, if take the base time to be the fixed singular time T instead of the sequence t_i, then we should obtain $\mathbb{F}$-convergence to a *shrinking cylinder*. Namely, let

$$\widetilde{g}_{i,t} := \lambda_i^2 g_{T+\lambda_i^{-2}t} \quad \text{and} \quad \widetilde{\nu}^i_{(p,0);t} := \nu_{(p,T);T+\lambda_i^{-2}t}$$

for $t \in [-T\lambda_i^2, 0)$; here, the conjugate heat kernel $\nu_{(p,T);s}$, $s < T$, based at the singular time T is defined as a limit of conjugate heat kernels based at times approaching T. Then the sequence of metric flows $\big(\widetilde{\mathcal{X}}^i, (\widetilde{\nu}^i_{(p,0);t})_{t\in[-T\lambda_i^2,0)}\big)$ should $\mathbb{F}$-converge to the cylinder $\mathbb{S}^{n-1}\times\mathbb{R}$. To the best of our knowledge, a formal proof of this convergence has not yet explicitly appeared in the literature.

A takeaway from these examples is that, in situations where we do not have Cheeger–Gromov convergence, the point selection via conjugate heat kernels can move the base point to where we have smooth convergence.

EXAMPLE 6.61 (The asymptotic cone of the FIK shrinker as a singularity model). Let $(\mathrm{FIK}, g_{\mathrm{FIK}}, f_{\mathrm{FIK}})$ denote the 4-dimensional FIK shrinking soliton. Let (M^4, g_t), $t \in [0,T)$, be a Ricci flow on a closed 4-manifold that goes singular at time T and forms an FIK asymptotically conical singularity model in the following sense. Let $p \in M^4$, $t_i \to T$, $\lambda_i := |\mathrm{Rm}|^{\frac{1}{2}}(p,t_i)$, and $g_{i,t} := \lambda_i^2 g_{t_i+\lambda_i^{-2}t}$. Suppose that

$$\big(M^4, g_{i,0}, p\big) \to (\mathrm{FIK}, g_{\mathrm{FIK}}, p_\infty),$$

converges in the pointed smooth Cheeger–Gromov sense. As in the previous example, define $\widetilde{g}_{i,t} := \lambda_i^2 g_{T+\lambda_i^{-2}t}$ and $\widetilde{\nu}^i_{(p,0);t} := \nu_{(p,T);T+\lambda_i^{-2}t}$. Then the sequence of metric flows $\big(\widetilde{\mathcal{X}}^i, (\widetilde{\nu}^i_{(p,0);t})_{t\in[-T\lambda_i^2,0)}\big)$ should $\mathbb{F}$-converge to the metric flow associated to the FIK shrinker. Furthermore, using a diagonal argument and by carefully choosing the rescalings, one can show that there exists a rescaled sequence of time slices of the Ricci flow on M^4 which converges in the Gromov–Hausdorff sense to the asymptotic cone of the FIK shrinker. It may be possible that in some situations, the rescaled sequence at the singular time $g_{i,\lambda_i^2(T-t_i)} = \lambda_i^2 g_T$ converges to the asymptotic cone of the FIK shrinker, but this has not been shown even in the rotationally symmetric case.

6.3. Existence and regularity of singularity models and tangent flows at infinity

In this section we discuss the definition, existence, and regularity of singularity models of solutions to the Ricci flow that go singular at a finite time and are defined on 4-dimensional closed orbifolds with isolated singularities. Furthermore, a primary vehicle to understand singularity models are their tangent flows at infinity.

6.3.1. Existence and regularity of Ricci flow singularity models.

6.3.1.1. *Definition of a metric flow singularity model.*

Bamler's theory provides a general definition of Ricci flow singularity models, which extends the classical Definition 1.8 and is crucial to understanding higher-dimensional singularity formation.

Suppose that $(\overline{\mathcal{O}}^4, \overline{g}(t))$, $t \in [0, T)$, where $0 < T < \infty$, is a singular Ricci flow on a 4-dimensional closed smooth orbifold with isolated singularities; see Definition 1.5. Let $\overline{\mathcal{X}}$ denote $(\overline{\mathcal{O}}^4, \overline{g}(t))$ as a metric flow (in the sense of Definition 6.43).[11]

DEFINITION 6.62 (Singularity models in the metric flow category). A metric flow pair $(\mathcal{X}, (\mu_t)_{t\in(-\infty,0]})$ is a **singularity model** of a Ricci flow on a 4-orbifold with isolated singularities $(\overline{\mathcal{O}}^4, \overline{g}(t))$, $t \in [0, T)$, if there exists a sequence of space-time points $\{(x_i, t_i)\}_{i=1}^\infty$ in $\overline{\mathcal{O}}^4 \times [0, T)$ with $t_i \to T$ and a sequence $\{\lambda_i\}_{i=1}^\infty$ with $\lambda_i \to \infty$ such that the time-shifted and rescaled pointed metric flows $\overline{\mathcal{X}}^{x_i,t_i;\lambda_i}$, as defined by (6.147), $\mathbb{F}$-converge to the metric flow pair $(\mathcal{X}, (\mu_t)_{t\in(-\infty,0]})$ in the sense of Definition 6.57. Moreover, we make the nontriviality assumption that the metric flow $\mathcal{X}$ is not the metric flow associated to static Euclidean space $\mathbb{R}^4$.[12]

Since Bamler's $\mathbb{F}$-compactness theorem generalizes the Cheeger–Gromov–Hamilton compactness theorem, this definition of singularity model is more general than the previous definitions of singularity model, such as the ones defined by Hamilton and Perelman using the curvature scale to rescale, while assuming or proving Cheeger–Gromov convergence. In particular, as a special case, we may take $\lambda_i^2 = |\mathrm{Rm}|(x_i, t_i)$ in Definition 6.62. If Cheeger–Gromov–Hamilton convergence holds for the associated rescaled sequence, then so does $\mathbb{F}$-convergence to the metric flow of the same limit.

6.3.1.2. *Existence of singularity models.*

We will consider a special case of Bamler's metric flow compactness theorem to establish the existence of singularity models.

Let $(\overline{\mathcal{O}}^4, \overline{g}(t))$, $t \in [0, T)$, where $T \in (0, \infty)$, be a smooth solution to the Ricci flow on a 4-dimensional closed orbifold with isolated singularities that goes singular at time T. Let $\{(x_i, t_i)\}_{i=1}^\infty$ be a sequence in $\overline{\mathcal{O}}^4 \times [0, T)$ with $t_i \to T$ and let $\lambda_i \to \infty$.[13] Let $\overline{\mathcal{X}}$ denote the metric flow over $[0, T)$ associated to the Ricci flow $(\overline{\mathcal{O}}^4, \overline{g}(t))$, $t \in [0, T)$. Consider the sequence of rescaled pointed metric flows:

$$\overline{\mathcal{X}}^{x_i,t_i;\lambda_i} := \left((\overline{\mathcal{X}} - t_i)^{\lambda_i}|_{[-\lambda_i^2 t_i, 0]}, \left(\overline{\nu}_{x_i,t_i;t_i+\lambda_i^{-2}t}\right)_{t\in[-\lambda_i^2 t_i, 0]}\right) \tag{6.147}$$

for $i \geq 1$. Observe that $\lambda_i^2 t_i \to \infty$.

By Theorem 6.58 (the compactness theorem for pointed Ricci flows), the sequence of pointed metric flows $\overline{\mathcal{X}}^{x_i,t_i;\lambda_i}$ in (6.147) $\mathbb{F}$-subconverges to some metric flow pair $(\mathcal{X}, (\mu_t)_{t\leq 0})$, where $\mathcal{X}$ is an ancient metric flow over $(-\infty, 0]$. In the next subsection, we discuss Bamler's regularity theory applied to this limit.

---

[11] The overlines in our notation are included to distinguish from the associated limits, which will be unadorned.

[12] However, one allows nontrivial flat orbifold quotients $\mathbb{R}^4/\Gamma$.

[13] For singularity analysis, one assumes that $|\mathrm{Rm}_{\overline{g}}|(x_i, t_i) \to \infty$.

6.3.1.3. *Regularity of singularity models.*

Let $\overline{g}_i(t) := \lambda_i^2\overline{g}(t_i + \lambda_i^{-2}t)$. We have by (6.60) that the pointed Nash entropies of the sequence in (6.147) satisfy

$$\begin{aligned}\overline{\mathcal{N}}_{x_i,0}(\tau) &:= \mathcal{N}\left[\overline{g}_i(-\tau), \overline{f}_i(-\tau), \tau\right] \\ &\geq \mu\left(\overline{g}_i(-\tau), \tau\right) \\ &= \mu\left(\lambda_i^2\overline{g}(-\lambda_i^{-2}\tau + t_i), \tau\right) \\ &= \mu\left(\overline{g}(-\lambda_i^{-2}\tau + t_i), \lambda_i^{-2}\tau\right)\end{aligned} \tag{6.148}$$

for $0 < \tau \leq \lambda_i^2 t_i$. Now, by the μ-monotonicity formula (6.58), we obtain

$$\overline{\mathcal{N}}_{x_i,0}(\tau) \geq \mu(\overline{g}(0), t_i) \geq \inf_{\tau' \in (0,T)} \mu(\overline{g}(0), \tau') =: -Y_0 > -\infty. \tag{6.149}$$

Bamler's regularity theorem ([**Bam20b**]), combined with his compactness theorem (Theorem 6.58), yields the following.

THEOREM 6.63. *Suppose that*

$$\left(\mathcal{O}_i^4, (g_{i,t})_{t\in(-T_i,0]}, (x_i, 0)\right), \quad \textit{where } T_i \to \infty,$$

is a sequence of smooth pointed Ricci flows on closed 4-dimensional orbifolds with isolated singularities. Assume that there exists a non-negative real number Y_0 *such that*

$$\mathcal{N}_{x_i,0}(\tau) \geq -Y_0 \quad \textit{for all } \tau \in (-T_i, 0) \textit{ and } i \geq 1. \tag{6.150}$$

Let $\mathcal{X}_i$ *denote the pointed metric flow over* $(-T_i, 0]$ *associated to the Ricci flow* $\left(\mathcal{O}_i^4, (g_{i,t})_{t\in(-T_i,0]}, (x_i, 0)\right)$.

(1) ($\mathbb{F}$-Precompactness) *Then the sequence of pointed metric flows* $\mathcal{X}_i$ $\mathbb{F}$*-subconverges to some ancient metric flow pair* $\left(\mathcal{X}^\infty, (\mu_t^\infty)_{t\leq 0}\right)$, *where* $\mathcal{X}^\infty$ *is a metric flow over* $(-\infty, 0]$.

(2) (Regularity of 4-dimensional metric shrinking soliton limits) *Suppose, in addition, that the metric flow* $\mathcal{X}^\infty$ *is a metric shrinking soliton. Then* $\mathcal{X}^\infty$ *is the metric flow associated to a pointed smooth complete ancient Ricci flow* $(\mathcal{O}^4, g(t), x_\infty)$, $t \in (-\infty, 0]$, *defined on a 4-dimensional smooth orbifold with isolated singularities.*[14]

REMARK 6.64. The philosophy in Bamler's papers [**Bam20a**], [**Bam23**], and [**Bam20b**] states that any facts about the regularity of limits of Ricci flows should also hold for singular Ricci flows in the sense of Kleiner and Lott [**KL17, KL20**].

Now, in the case where $\mathcal{X}^\infty$ is a metric shrinking soliton, if the limit orbifold $\mathcal{O}^4$ does not have singular points, that is, if $\mathcal{O}^4 = M^4$ is a smooth 4-manifold, then $(M^4, g(t))$ is the pointed C^∞ Cheeger–Gromov limit of rescalings of $\left(\overline{\mathcal{O}}^4, \overline{g}(t)\right)$; see [**Bam23**, §9.4] (this is true even if $\overline{\mathcal{O}}^4$ has singular points). Namely, $(M^4, g(t), x_\infty)$ is the pointed C^∞ Cheeger–Gromov limit of $\left(\overline{\mathcal{O}}^4, \lambda_i^2\overline{g}(\lambda_i^{-2}t + t_i), x_i\right)$. On the other hand, if the limit $\mathcal{O}^4$ has singular points, then we have Cheeger–Gromov convergence away from these singular points. We remark that the convergence may be smooth even at some of the orbifold points of the limit. In particular, the convergence is smooth wherever the curvature is bounded along the convergent sequence.

[14]This fact relies on a removable singularities theorem for Ricci solitons that has no analogue for a general (non-selfsimilar) Ricci flow.

6.3.1.4. *The non-inflating property of singularity models.*
Bamler's non-inflating estimate (Theorem 6.39) provides a Euclidean upper bound for the volume growth of singularity models, including those that are Ricci solitons. In particular, as a special case of Theorem 6.39, we have the following.

PROPOSITION 6.65 (Non-inflating estimate for Ricci flow singularity models). *If $(\mathcal{O}^n_\infty, (g^\infty_t)_{t\in(-\infty,0]})$ is a complete singularity model on an orbifold with isolated singularities, then the scalar curvature is non-negative*[15] *and*

$$\mathrm{Vol}_{g^\infty_t} B(x,t,r) \leq C(n) r^n \tag{6.151}$$

for all $x \in \mathcal{O}^n_\infty$, $t \in (-\infty, 0]$, and $r \in (0,\infty)$, where $B(x,t,r)$ denotes the open ball centered at x of radius r with respect to the metric g^∞_t.

6.3.1.5. *The conjugate heat kernel of singularity models.*
Let $(\mathcal{O}^n, g(t))$, $t \in (-\infty, 0]$, be a singularity model on an n-orbifold with isolated singularities. Note that if $n = 4$, then any metric flow singularity model must be on a 4-orbifold with isolated singularities.

Fix a base point $x_0 \in \mathcal{O}^n$ in the singularity model and consider the conjugate heat kernel based at $(x_0, 0)$, defined as

$$(x,t) \mapsto K(x_0, 0; x, t).$$

Let $f : \mathcal{O}^n \times (-\infty, 0) \to \mathbb{R}$ be the potential of the conjugate heat kernel defined by (6.15), so that

$$K(x_0, 0; x, t) = (4\pi|t|)^{-\frac{n}{2}} \mathrm{e}^{-f(x,t)}. \tag{6.152}$$

Let μ_∞ be the asymptotic pointed Nash entropy defined by (6.65). By Bamler's Gaussian upper bound (6.108) for the conjugate heat kernel,

$$K(x_0, 0; x, t) \leq C(n)|t|^{-\frac{n}{2}} \exp\left(-\mu_\infty - \frac{d_t(x, z_t)^2}{9|t|}\right), \tag{6.153}$$

where (z_t, t) is an H_n-center and where the constant $C(n)$ depends only on n and changes from line to line. Thus, the potential satisfies

$$f(x,t) \geq 1 - C(n) + \mu_\infty + \frac{d_t(x, z_t)^2}{9|t|}; \tag{6.154}$$

cf. (1.61) for the potential function of a shrinking gradient Ricci soliton.

6.3.1.6. *The pointed Nash entropy of singularity models.*
Recall from (6.65) that since $(\mathcal{O}^n, g(t))$ is a singularity model of the Ricci flow, we have that the asymptotic pointed Nash entropy satisfies

$$\inf_{\tau>0} \mathcal{N}_{x_0,t_0}(\tau) = \lim_{\tau\to\infty} \mathcal{N}_{x_0,t_0}(\tau) =: \mu_\infty > -\infty, \tag{6.155}$$

and μ_∞ is independent of the choice of x_0, t_0 (the result [**MZ21**, Proposition 4.6] is stated for the bounded curvature case, but holds also for singularity models by taking limits).

---

[15] By B.-L. Chen [**Che09**], any complete ancient solution has $R \geq 0$.

6.3.2. Tangent flows at infinity of singularity models.

Suppose that $(\mathcal{O}^n, g_t)$, $t \in (-\infty, 0]$, is a singularity model of the Ricci flow on a smooth orbifold with isolated singularities, so that by definition, the Riemannian metric g_t is smooth and complete for each $t \in (-\infty, 0]$.[16] Given a base point $(x_0, 0) \in \mathcal{O}^n \times (-\infty, 0]$, consider the metric flow pair consisting of the metric flow $\mathcal{X} = \mathcal{O}^n \times (-\infty, 0]$, $(d_t)_{t\leq 0}$, $(\nu_{x,t;s})_{x\in\mathcal{O}^n,\, s\leq t\leq 0}$, over $(-\infty, 0]$ and the conjugate heat kernel $(\nu_{x_0,0;s})_{x\in M^n,\, s\leq 0}$ based at $(x_0, 0)$.

By Theorem 6.58 (Bamler's compactness theorem), we have the following.

THEOREM 6.66 (Existence of tangent flows at infinity). *Let $(\mathcal{O}^n, g_t)$ be a singularity model of the Ricci flow on an orbifold, as described above. For any $x_0 \in \mathcal{O}^n$ and any sequence $\{\lambda_i\}$ of positive real numbers with $\lambda_i \to 0$, consider the sequence of rescaled metric flow pairs*

$$\left(\mathcal{X}^{\lambda_i^{1/2}}, (\nu_{x_0,0;\lambda_i^{-1}t})_{t\leq 0}\right)$$

based at $(x_0, 0)$, where $\mathcal{X}^{\lambda_i^{1/2}}$ represents the $\lambda_i^{1/2}$-rescaling of $\mathcal{X}$. Then this sequence $\mathbb{F}$-subconverges to some metric flow pair

$$\left(\mathcal{X}^\infty, (\mu_t^\infty)_{t\leq 0}\right),$$

where $\mathcal{X}^\infty$ is a metric flow over $(-\infty, 0]$.

The following is [**Bam23**, Definition 6.55].

DEFINITION 6.67 (Tangent flow at infinity). Any metric flow $\mathcal{X}^\infty$ associated to $\{\lambda_i\}$ as an $\mathbb{F}$-limit, after passing to a subsequence, with $\lambda_i \to 0$, and $(x_0, 0)$ as in Theorem 6.66, is called a **tangent flow at infinity** of the singularity model $(\mathcal{O}^n, g_t)$, $t \in (-\infty, 0]$.

So Theorem 6.66 says that for any singularity model, $x_0 \in \mathcal{O}^n$, and sequence $\lambda_i \to 0$ there exists an associated tangent flow at infinity. On the other hand, since a priori the $\mathbb{F}$-limit $\mathcal{X}^\infty$ may depend on the sequence $\{\lambda_i\}$ and the subsequence, tangent flows at infinity may not necessarily be unique.

In general, tangent flows at infinity have structures that generalize shrinking gradient Ricci solitons.

THEOREM 6.68 ([**Bam20b**, Theorem 2.40]). *Every tangent flow at infinity of a singularity model is a metric shrinking soliton.*

We have the following regularity and dichotomy theorem for 4-dimensional tangent flows at infinity ([**Bam20b**, Theorem 2.46]).

THEOREM 6.69 (4-Dimensional tangent flows at infinity).

(1) (Regularity) *In dimension 4, every tangent flow at infinity of a singularity model is isometric to the metric length space of a smooth complete 4-dimensional Riemannian orbifold $(\mathcal{O}^4_\infty, g_\infty)$ with isolated singularities.*
(2) (Dichotomy) *Additionally, $(\mathcal{O}^4_\infty, g_\infty)$ is either isometric to $(\mathbb{R}^4/\Gamma)$, where Γ is a finite subgroup of $O(4)$, or the scalar curvature of g_∞ is positive.*

---

[16] However, we do not assume that the curvatures of g_t are bounded for each $t \in (-\infty, 0]$; i.e., we do not assume bounded curvature on each time slice.

6.4. Tangent flows at infinity of 4-dimensional steady soliton singularity models

In this section we classify the tangent flows at infinity of 4-dimensional steady gradient Ricci solitons that are also singularity models.

DEFINITION 6.70. Let $(\mathcal{O}^4, g(t))$, $t \in (-\infty, \infty)$, be a complete noncompact steady gradient Ricci soliton on a 4-dimensional smooth orbifold with isolated singularities. We say that $(\mathcal{O}^4, g(t))$ is a **steady gradient Ricci soliton singularity model** if there exists $p \in \mathcal{O}^4$ such that $(\mathcal{O}^4, g(t), p)$, $t \in (-\infty, 0]$, has an associated pointed metric flow $\big(\mathcal{X}, (\mu_t)_{t\in(-\infty,0]}\big)$, which is the singularity model, in the metric-flow-pair sense of Definition 6.62, of some Ricci flow $\big(\overline{\mathcal{O}}^4, \overline{g}(t)\big)$, $t \in [0, T)$, that goes singular at time T and is defined on a closed 4-orbifold with isolated singularities.

EXAMPLE 6.71.

(1) By Gu and Zhu [**GZ08**], the Bryant steady soliton is a singularity model associated to a rotationally symmetric degenerate neckpinch.
(2) By Appleton [**App23**], the Eguchi–Hanson Ricci-flat ALE space is a singularity model of some 4-dimensional Ricci flow. Moreover, it is possible, but yet unproven, that the $\mathbb{Z}_2$-quotient of the 4-dimensional Bryant steady soliton, which is an orbifold with a single $\mathbb{Z}_2$ singular point, is a singularity model.

6.4.1. Statement of the classification of tangent flows at infinity of steady models.

In this section we prove the following result of Bamler, Deng, Ma, Zhang, and one of the authors [**BCD⁺22**, Propositions 3.1 and 3.2]. This result classifies the tangent flows at infinity of 4-dimensional steady gradient Ricci soliton singularity models.

THEOREM 6.72 (Tangent flows at infinity of 4-dimensional steady soliton singularity models). *Let $(\mathcal{O}^4, g(t), f(t))$, $t \in (-\infty, \infty)$, be a complete noncompact steady gradient Ricci soliton singularity model on a 4-dimensional smooth orbifold with isolated singularities.*

Then its tangent flow at infinity $(\mathcal{O}^4_\infty, g_\infty(t))$, $t \in (-\infty, 0]$, is unique (i.e., it is independent of x_0, of the sequence $\lambda_i \to 0$, as well as of the choice of the subsequence) and is isometric to either:

(1) (Orbifold Euclidean cone) $\mathbb{R}^4/\Gamma$ *where* $\Gamma \neq \{1\}$,
(2) (Shrinking 3-cylinder) $(\mathbb{S}^3/\Gamma) \times \mathbb{R}$,
(3) (Shrinking 2-cylinder) $\mathbb{S}^2 \times \mathbb{R}^2$ *or* $((\mathbb{S}^2 \times \mathbb{R})/\mathbb{Z}_2) \times \mathbb{R}$.

Furthermore, the tangent flow at infinity $(\mathcal{O}^4_\infty, g_\infty(t))$ is isometric to $\mathbb{R}^4/\Gamma$ if and only if the singularity model $(\mathcal{O}^4, g(t))$ is a static Ricci-flat ALE 4-orbifold with isolated singularities.

EXAMPLE 6.73.

(1) The quotient by Γ of the 4-dimensional Bryant steady soliton, which is an orbifold with a single isolated Γ singular point, has tangent flow at infinity $(\mathbb{S}^3/\Gamma) \times \mathbb{R}$.
(2) Appleton's cohomogeneity-one steady solitons (see §1.3.3.2.c) have tangent flow at infinity $(\mathbb{S}^3/\mathbb{Z}_k) \times \mathbb{R}$, where $k \geq 3$.

(3) The product of $\mathbb{R}$ and the 3-dimensional Bryant steady soliton has tangent flow at infinity $\mathbb{S}^2 \times \mathbb{R}^2$.

(4) Lai's 4-dimensional steady solitons (see §1.3.3.2.d) have tangent flow at infinity $\mathbb{S}^2 \times \mathbb{R}^2$.

6.4.2. Proof of the classification of tangent flows at infinity of steady models.

In the rest of this section we give a proof of Theorem 6.72. The main issues for the tangent flows at infinity of 4-dimensional steady soliton models are to prove existence, uniqueness, and a classification.

6.4.2.1. *Existence of tangent flows at infinity of steady soliton singularity models.*

Let $(\mathcal{O}^4, g(t), f(t))$, $t \in (-\infty, \infty)$, be a noncompact steady gradient Ricci soliton singularity model on a 4-dimensional smooth orbifold with isolated singularities. By Theorems 6.66, 6.68, and 6.69, we have the following:

Fix $x_0 \in \mathcal{O}^4$. The tangent flow at infinity $\mathcal{X}^\infty$ of $(\mathcal{O}^4, g(t), x_0)$ associated to any positive sequence $\lambda_i \to 0$, after passing to a subsequence, exists and $\mathcal{X}^\infty$ is a metric shrinking soliton in the sense of Definition 6.49. Moreover, $\mathcal{X}^\infty$ is isometric at each time $t \in (-\infty, 0]$ to the metric length space of a smooth complete solution to the Ricci flow $(\mathcal{O}^4_\infty, g_\infty(t))$ on a 4-dimensional orbifold with isolated singularities, where $(\mathcal{O}^4_\infty, g_\infty(t))$, with a potential function $f_\infty(t)$, is a smooth shrinking gradient Ricci soliton.

Furthermore, let $\mathcal{S}^\infty$ denote the set of singular points of $\mathcal{O}^4_\infty$ (which is a discrete set) and let $\mathcal{R}^\infty$ denote the set of regular points of $\mathcal{O}^4_\infty$ (which is the complement of $\mathcal{S}^\infty$ in $\mathcal{O}^4_\infty$). Since $\mathcal{S}^\infty$ is a discrete set, we may assume without loss of generality that $z_\infty \in \mathcal{R}^\infty$ and that the rescalings based at the H_4-centers $(\mathcal{O}^4, \lambda_i^2 g(-\lambda_i^{-2}), z_{-\lambda_i^{-2}})$ converge to $(\mathcal{R}^\infty, g_\infty(0), z_\infty)$ in the pointed C^∞ Cheeger–Gromov sense on compact subsets of $\mathcal{R}^\infty$ (one removes suitable subsets of $\mathcal{O}^4$, depending on i, to obtain this convergence); see [**Bam23**, §9.4] (cf. Chan, Ma, and Zhang [**CMZ24**, sentence after (4.1)] for why we may assume $z_\infty \in \mathcal{R}^\infty$).

By [**Bam23**, Theorem 6.24], in addition to the metric shrinking soliton $\mathcal{X}^\infty$ being an $\mathbb{F}$-limit of rescalings of the steady gradient Ricci soliton singularity model $(\mathcal{O}^4, g(t))$, we have that $\mathcal{X}^\infty$ (effectively, $(\mathcal{O}^4_\infty, g_\infty(t))$) is also an $\mathbb{F}$-limit of rescalings of the original solution $(\overline{\mathcal{O}}^4, \overline{g}(t))$ that goes singular at time T. Hence, we have the pointed C^∞ Cheeger–Gromov convergence of the space-time rescalings of subsets of the original solution $(\overline{\mathcal{O}}^4, \overline{g}(t))$ to the regular part $(\mathcal{R}^\infty, g_\infty(t))$.

6.4.2.2. *Independence of the tangent flow at infinity on the base point.* By Chan, Ma, and Zhang [**CMZ22**, Theorem 1.6], in all dimensions the tangent flow at infinity $\mathcal{X}^\infty$ is independent of the choice of the base point $(x_0, 0)$. Namely, if $x_0 \in \mathcal{O}^4$ and $\lambda_i \to 0$ are such that the metric flow pairs $\left(\mathcal{X}^{\lambda_i^{1/2}}, (\nu_{x_0,0;\lambda_i^{-1}t})_{t\leq 0}\right)$ $\mathbb{F}$-converge to some metric flow pair $\left(\mathcal{X}^\infty, (\mu_t^\infty)_{t\leq 0}\right)$, then for each $x_1 \in \mathcal{O}^4$, the sequence $\left(\mathcal{X}^{\lambda_i^{1/2}}, (\nu_{x_1,0;\lambda_i^{-1}t})_{t\leq 0}\right)$ also $\mathbb{F}$-converges to $\left(\mathcal{X}^\infty, (\mu_t^\infty)_{t\leq 0}\right)$. However, $\mathcal{X}^\infty$ may still potentially depend on the sequence λ_i.

6.4.2.3. *Dichotomy for tangent flows at infinity of steady soliton singularity models.* The following is [**BCD+22**, Proposition 3.1]:

PROPOSITION 6.74 (Dichotomy for tangent flows at infinity of steady models). *For any orientable 4-dimensional steady gradient Ricci soliton singularity model $(\mathcal{O}^4, g(t), f(t))$, $t \in (-\infty, \infty)$, on an orbifold with isolated singularities, any tangent flow at infinity, $\mathcal{X}^\infty$, either:*

(1) (Euclidean) *is isometric to the Gaussian shrinking soliton $\mathbb{R}^4/\Gamma$, where $\Gamma \neq \{1\}$ is a finite subgroup of* $\mathrm{O}(4)$, *or*

(2) (Splitting) $\mathcal{X}^\infty$ *splits off a line, in which case $(\mathcal{O}^4, g(t), f(t))$ is the product of $\mathbb{R}$ with a shrinking gradient Ricci soliton on a smooth 3-manifold.*

PROOF OF PROPOSITION 6.74. Let a metric flow pair $\left(\mathcal{X}^\infty, (\mu_t^\infty)_{t \leq 0}\right)$ be a tangent flow at infinity of an orientable 4-dimensional steady gradient Ricci soliton singularity model on an orbifold with isolated singularities. By definition, $\left(\mathcal{X}^\infty, (\mu_t^\infty)_{t \leq 0}\right)$ is the $\mathbb{F}$-limit of the metric flow pairs $\left(\mathcal{X}^{\lambda_i^{1/2}}, (\nu_{x_0,0;\lambda_i^{-1}t})_{t \leq 0}\right)$ associated to $(\mathcal{O}^4, \lambda_i^2 g(-\lambda_i^{-2}), z_i)$ for some sequence $\lambda_i \to 0$, where

$$z_i := z_{-\lambda_i^{-2}}$$

is the H_n-center of $(x_0, 0)$ at time $-\lambda_i^{-2}$.

Moreover, by the discussion above, $\mathcal{X}^\infty$ is the metric length space of a smooth complete shrinking gradient Ricci soliton $(\mathcal{O}^4_\infty, g_\infty(t), f_\infty(t))$ on an orbifold with isolated singularities.

We have that $(\mathcal{O}^4, \lambda_i^2 g(-\lambda_i^{-2}), z_i)$ converges to $(\mathcal{R}^\infty, g_\infty, z_\infty)$ in the pointed C^∞ Cheeger–Gromov sense on compact subsets of $\mathcal{R}^\infty$, where $g_\infty := g_\infty(-1)$, $z_\infty \in \mathcal{R}^\infty$, and $\mathcal{R}^\infty$ is the set of regular points of $\mathcal{O}^4_\infty$. Again, without loss of generality, we have $z_\infty \in \mathcal{R}^\infty$ by adjusting the H_4-centers in the sequence.

We divide the analysis of the $\mathbb{F}$-limit $\mathcal{X}^\infty$ into two cases:

(1) $|\nabla f|$ is bounded at the H_n-centers, and

(2) $|\nabla f|$ tends to infinity at the H_n-centers.

Let

$$g_i := \lambda_i^2 g(-\lambda_i^{-2}) \quad \text{and} \quad f_i := f(\,\cdot\,, -\lambda_i^{-2}) - f(z_i, -\lambda_i^{-2}), \tag{6.156}$$

where λ_i is above. Since $g(t)$ is a steady gradient Ricci soliton, we have that

$$\mathrm{Ric}_{g_i} + \nabla^2_{g_i} f_i = 0. \tag{6.157}$$

CASE 1: *There exists a subsequence for which $|df_i|_{g_i}(z_i)$ is uniformly bounded.*

Pass to this subsequence. In this case we will show that any associated tangent flow at infinity $\mathcal{X}^\infty$ is the Gaussian shrinking soliton $\mathbb{R}^4/\Gamma$, for some finite subgroup $\Gamma \neq \{1\}$ of $\mathrm{O}(4)$

Let $\{\phi_i\}$ be the pointed diffeomorphisms from compactly contained subsets of $(\mathcal{R}^\infty, z_\infty)$ into $(\mathcal{O}^4, z_i)$ used for the smooth Cheeger–Gromov convergence of subsets of $(\mathcal{O}^4, g_i, z_i)$ to $(\mathcal{R}^\infty, g_\infty, z_\infty)$. By the smooth Cheeger–Gromov convergence, for any point $y_\infty \in \mathcal{R}^\infty$, we have that $|\mathrm{Rm}_{\phi_i^* g_i}|$ is uniformly bounded in a uniform neighborhood of $y_\infty \in \mathcal{R}^\infty$. Thus, we have that $|\mathrm{Ric}_{\phi_i^* g_i}|$ is uniformly bounded in a uniform neighborhood of $y_\infty \in \mathcal{R}^\infty$. By the steady soliton equation, the same holds for $|\nabla^2_{\phi_i^* g_i}(f_i \circ \phi_i)|$.

Since $f_i(z_i) = 0$, we have $(f_i \circ \phi_i)(z_\infty) = 0$. Hence, $|\nabla_{\phi_i^* g_i}(f_i \circ \phi_i)|$ is uniformly bounded in a neighborhood of $z_\infty \in \mathcal{R}^\infty$. We then obtain that $|\nabla_{\phi_i^* g_i}(f_i \circ \phi_i)|$ is uniformly bounded on any compact subset of $\mathcal{R}^\infty$. Furthermore, by Shi's local derivative of curvature estimates, for each $k \geq 1$, $|(\nabla^k \mathrm{Rm})_{\phi_i^* g_i}|$ is uniformly bounded on any compact subset of $\mathcal{R}^\infty$.

This and the steady soliton equation imply that $|\nabla^\ell_{\phi_i^* g_i}(f_i \circ \phi_i)|$ is uniformly bounded on any compact subset of $\mathcal{R}^\infty$ for each $\ell \geq 0$. Therefore, the sequence $f_i \circ \phi_i$ subconverges to a smooth function $\overline{f}_\infty$ on $\mathcal{R}^\infty$, and we may pass to the limit in equation (6.157) and obtain the steady gradient Ricci soliton equation

$$\mathrm{Ric}_{g_\infty} + \nabla^2_{g_\infty} \overline{f}_\infty = 0 \tag{6.158}$$

on $\mathcal{R}^\infty$, where g_∞ is as above. Since g_∞ and $\overline{f}_\infty$ are smooth on all of $\mathcal{O}^4_\infty$, equation (6.158) holds on all of $\mathcal{O}^4_\infty$.

On the other hand, since $(\mathcal{O}^4_\infty, g_\infty, f_\infty)$ is a shrinking gradient Ricci soliton, where $f_\infty := f_\infty(-1)$, we have

$$\mathrm{Ric}_{g_\infty} + \nabla^2_{g_\infty} f_\infty = \frac{1}{2} g_\infty.$$

Let $h := \overline{f}_\infty - f_\infty$. Then, on $\mathcal{R}^\infty$ we have that

$$\mathcal{L}_{\nabla h} g_\infty = 2\nabla^2_{g_\infty} h = -g_\infty.$$

This implies, by adding a suitable constant to h, that

$$|\nabla_{g_\infty} h|^2 = -\frac{1}{2} h \geq 0.$$

Thus, the function $\rho := 2\sqrt{-h}$ satisfies the equations

$$|\nabla_{g_\infty} \rho| \equiv 1 \quad \text{and} \quad \nabla^2_{g_\infty} \rho(\nabla_{g_\infty} \rho, \cdot) = 0$$

on $\mathcal{R}^\infty$.

This implies that the tangent flow at infinity $(\mathcal{O}^4_\infty, g_\infty)$ is a flat cone; cf. the proof of Proposition 2.9 in Volume 1. On the other hand, since $\mathcal{O}^4_\infty$ is an orbifold with isolated singularities, we must have that $(\mathcal{O}^4_\infty, g_\infty)$ is isometric to $\mathbb{R}^4/\Gamma$, with the quotient Euclidean metric and where Γ is a finite subgroup of $\mathrm{O}(4)$.

We will defer the proof that $\Gamma \neq \{1\}$ to the next corollary.

CASE 2: *The sequence $|df_i|_{g_i}(z_i)$ is unbounded.*

In this case we will show that the tangent flow at infinity $\mathcal{X}^\infty$ isometrically splits off an $\mathbb{R}$ factor.

By hypothesis, there exists a subsequence such that

$$|df_i|_{g_i}(z_i) =: \alpha_i \to \infty.$$

Define

$$\widehat{f}_i := \alpha_i^{-1}\big(f(\cdot, -\lambda_i^{-2}) - f(z_i, -\lambda_i^{-2})\big).$$

Then we have that

$$\widehat{f}_i(z_i) = 0, \quad |d\widehat{f}_i|_{g_i}(z_i) = 1, \tag{6.159}$$

and, after pulling back by the diffeomorphisms ϕ_i,

$$|\nabla^2_{g_i} \widehat{f}_i|_{g_i} = \alpha_i^{-1} |\nabla^2_{g_i} f_i|_{g_i} = \alpha_i^{-1} |\mathrm{Ric}_{g_i}|_{g_i} \to 0 \quad \text{on compact subsets of } \mathcal{R}^\infty$$

by the smooth Cheeger–Gromov convergence on $\mathcal{R}^\infty$.

Analogous to Case 1, $|\nabla^{\ell}_{g_i}\widehat{f}_i|$ is uniformly bounded on compact subsets of $\mathcal{R}^\infty$ for all $\ell \geq 1$ (after pulling back by the diffeomorphisms ϕ_i). Therefore, $\widehat{f}_i \circ \phi_i$ subconverges to a smooth function $\widehat{f}_\infty$ on $\mathcal{R}^\infty$. Since $|\nabla^2_{g_i}\widehat{f}_i|_{g_i} \to 0$ uniformly on compact subsets of $\mathcal{R}^\infty$, we conclude that $\nabla^2_{g_\infty}\widehat{f}_\infty \equiv 0$ on $\mathcal{R}^\infty$. By (6.159), we also have that $|d\widehat{f}_\infty|_{g_\infty}(z_\infty) = 1$.

From this, we conclude that $\mathcal{R}^\infty$ locally splits off an $\mathbb{R}$ factor. Since $\mathcal{O}^4$ is an orbifold with isolated singularities, this implies that $\mathcal{O}^4 = \mathcal{R}^\infty$ is the global product of $\mathbb{R}$ and a smooth shrinking gradient Ricci soliton on a 3-manifold (if the 3-dimensional factor has singularities, then the singularities of $\mathcal{O}^4$ would be at least 1-dimensional, which is a contradiction).

□

By the classification of shrinking gradient Ricci solitons on smooth 3-manifolds, we obtain the following consequence of Proposition 6.74 (see [**BCD+22**, Proposition 3.1]):

COROLLARY 6.75 (Classification of tangent flows at infinity of steady models). *Let $(\mathcal{O}^4, g(t), f(t))$, $t \in (-\infty, \infty)$, be an orientable 4-dimensional steady gradient Ricci soliton singularity model, defined on an orbifold with isolated singularities. Then any tangent flow $\mathcal{X}^\infty$ at infinity of $(\mathcal{O}^4, g(t))$ is either:*

(1) (Gaussian) $\mathbb{R}^4/\Gamma$, $\Gamma \neq \{1\}$,
(2) (Quotient 3-cylinder) $(\mathbb{S}^3/\Gamma') \times \mathbb{R}$, *where* $\Gamma, \Gamma' < \mathrm{O}(4)$ *are finite subgroups acting freely on* $\mathbb{S}^3$,
(3) (2-Cylinder) $\mathbb{S}^2 \times \mathbb{R}^2$, *or*
(4) ($\mathbb{Z}_2$-Quotient 2-cylinder) $\big((\mathbb{S}^2 \times \mathbb{R})/\mathbb{Z}_2\big) \times \mathbb{R}$.[17]

Furthermore, if $\mathcal{X}^\infty$ is $\mathbb{R}^4/\Gamma$, then $(\mathcal{O}^4, g(t))$ is a Ricci-flat ALE space, as defined in Definition 2.42.

PROOF OF COROLLARY 6.75. By Proposition 6.74, we have two cases:

CASE 1: *$\mathcal{X}^\infty$ splits off a line.*

Then, $\mathcal{X}^\infty$ is isometric to the product of $\mathbb{R}$ and a 3-dimensional shrinking gradient Ricci soliton on a smooth 3-manifold N^3. Thus we may apply Corollary 1.75 to obtain that N^3 is either $\mathbb{S}^3/\Gamma$, $\mathbb{S}^2 \times \mathbb{R}$, or $(\mathbb{S}^2 \times \mathbb{R})/\mathbb{Z}_2$, since N^3 cannot be $\mathbb{R}^3$. Therefore, $\mathcal{X}^\infty = N^3 \times \mathbb{R}$ is $(\mathbb{S}^3/\Gamma) \times \mathbb{R}$, $\mathbb{S}^2 \times \mathbb{R}^2$, or $\big((\mathbb{S}^2 \times \mathbb{R})/\mathbb{Z}_2\big) \times \mathbb{R}$.

CASE 2: $\mathcal{X}^\infty$ is $\mathbb{R}^4/\Gamma$, *where* $\Gamma \neq \{1\}$.

Since $(\mathcal{O}^4, g(t), f(t))$ is a steady gradient Ricci soliton, there exist self-diffeomorphisms Φ_t of $\mathcal{O}^4$ such that $g(t) = \Phi_t^* g(0)$ and $f(t) = f(0) \circ \Phi_t$. Recall that $g_i = \lambda_i^2 g(-\lambda_i^{-2})$ and $f_i = f(\cdot, -\lambda_i^{-2}) - f(z_i, -\lambda_i^{-2})$, which converge to g_∞ and $\overline{f}_\infty$, respectively, satisfying the steady gradient Ricci soliton equation $\mathrm{Ric}_{g_\infty} + \nabla^2_{g_\infty}\overline{f}_\infty = 0$. We have

$$R_{g(0)}\big(\Phi_{-\lambda_i^{-2}}(z_i)\big) = R_{g(-\lambda_i^{-2})}(z_i) = \lambda_i^2 R_{g_i}(z_i) \longrightarrow 0 \cdot R_{g_\infty}(z_\infty) = 0.$$

Moreover,

$$|df(0)|^2_{g(0)}\big(\Phi_{-\lambda_i^{-2}}(z_i)\big) = \lambda_i^2 |df_i|^2_{g_i}(z_i) \longrightarrow 0 \cdot |d\overline{f}_\infty|^2_{g_\infty}(z_\infty) = 0.$$

[17]The action of $\mathbb{Z}_2$ on $\mathbb{S}^2 \times \mathbb{R}$ is generated by the isometry $(x, y) \mapsto (-x, -y)$ for $x \in \mathbb{S}^2$ and $y \in \mathbb{R}$.

Since $R_{g(0)} + |df(0)|^2_{g(0)} = C$ is a constant on $\mathcal{O}^4$, this implies that $C = 0$. Therefore $R_{g(0)} \equiv 0$ on $\mathcal{O}^4$. Since $g(t)$ is a solution to the Ricci flow, this, in turn, implies that $\mathrm{Ric}_{g(0)} \equiv 0$ on $\mathcal{O}^4$. On the other hand, by Bamler's no local collapsing theorem (Theorem 6.37), there exists a positive constant κ such that

$$\mathrm{Vol}_{g(0)} B_{g(0)}(x_0, r) \geq \kappa r^4 \quad \text{for all } r > 0.$$

By a result of Cheeger and Naber [**CN15**, Corollary 8.85], this implies that $(\mathcal{O}^4, g(0))$ is a Ricci-flat ALE space. In particular, Cheeger and Naber's result, which was originally stated in the category of smooth manifolds, applies essentially without modification in the category of 4-dimensional smooth orbifolds with isolated singularities.

This result holds because, in the orbifold category, we have non-negative Ricci curvature in the sense of RCD spaces. In the proof, one uses the stratification theory to rule out codimension 2 and 3 singularities. For codimension 2 singularities, the same blow up argument works as before. For codimension 3 singularities, a key point is that the cone over $\mathbb{R}P^2$ is not cobordantly trivial; again, the same arguments work as before. See Bamler, Jiang, Naber, and one of the authors [**BCJN**] for a detailed account.

Finally, the statement that $\Gamma \neq \{1\}$ follows from the equality case of the Bishop–Gromov volume comparison theorem applied to $(\mathcal{O}^4, g(0))$ (see, for example, [**CC24**, Theorem 11.32]). □

6.4.2.4. *Uniqueness of the tangent flows at infinity of steady soliton singularity models.*

PROPOSITION 6.76 (Uniqueness of tangent flows at infinity of steady models). *Let $(\mathcal{O}^4, g(t), f(t))$, $t \in (-\infty, \infty)$, be an orientable steady gradient Ricci soliton singularity model, defined on a 4-dimensional orbifold with isolated singularities. Then the tangent flow at infinity $\mathcal{X}^\infty$ is unique; i.e., it is independent of the choice of $\lambda_i \to 0$ for which the associated sequence of metric flow pairs $(\mathcal{X}^{\lambda_i^{1/2}}, (\nu_{x_0,0;\lambda_i^{-1}t})_{t\leq 0})$ converges.*

PROOF. By Corollary 6.75, any tangent flow at infinity $\mathcal{X}^\infty$ of $(\mathcal{O}^4, g(t))$ is either $\mathbb{R}^4/\Gamma$, $\Gamma \neq \{1\}$, $(\mathbb{S}^3/\Gamma') \times \mathbb{R}$, where $\Gamma, \Gamma' < \mathrm{O}(4)$ are finite subgroups acting freely on $\mathbb{S}^3$, $\mathbb{S}^2 \times \mathbb{R}^2$, or $\big((\mathbb{S}^2 \times \mathbb{R})/\mathbb{Z}_2\big) \times \mathbb{R}$.

Suppose that some tangent flow at infinity $\mathcal{X}^\infty$ is $\mathbb{R}^4/\Gamma$. Then, $(\mathcal{O}^4, g(t))$ is a Ricci-flat ALE space. This implies that any tangent flow at infinity is $\mathbb{R}^4/\Gamma$, proving the uniqueness of $\mathcal{X}^\infty$ in this case.

Associated to $(\mathcal{O}^4, g(t))$, let $\mathcal{T}$ denote the set of all tangent flows at infinity $(\mathcal{Y}^\infty, (\mu_t^\infty)_{t\leq 0})$, restricted to the compact time interval $I := [-2, -\frac{1}{2}]$. By the previous two paragraphs, we may assume that each $\mathcal{Y}^\infty \in \mathcal{T}$ splits off an $\mathbb{R}$ factor.

Let $J := \{-1\} \subset I$ and consider the $\mathbb{F}$-distance between elements of $\mathcal{T}$ (uniform over J)

$$d_{\mathbb{F}}^J\big((\mathcal{Y}^\infty, (\mu_t^\infty)_{t\in I}), (\mathcal{Y}'^\infty, (\mu_t'^\infty)_{t\in I})\big)$$

as defined in [**Bam23**, Definition 5.7]. By [**Bam23**, Theorem 7.4], the metric space $(\mathcal{T}, d_{\mathbb{F}}^J)$ is compact. Since $\mathcal{T}$ is contained in the collection of metric flows $(\mathbb{S}^3/\Gamma) \times \mathbb{R}$, $\mathbb{S}^2 \times \mathbb{R}^2$, $\big((\mathbb{S}^2 \times \mathbb{R})/\mathbb{Z}_2\big) \times \mathbb{R}$ over I, which is discrete with respect to $d_{\mathbb{F}}^J$, we have that the set $\mathcal{T}$ is finite. We want to show that $\mathcal{T}$ consists of a single element.

Let $k = |\mathcal{T}| < \infty$ and let

$$\mathcal{T} = \{(\mathcal{Y}_i^\infty, (\mu_{i,t}^\infty)_{t\in I}) : 1 \leq i \leq k\}.$$

Define

$$\varepsilon := \frac{1}{10} \min_{1\leq i\neq j\leq k} d_{\mathbb{F}}^J\big((\mathcal{Y}_i^\infty, (\mu_{i,t}^\infty)_{t\in I}), (\mathcal{Y}_j^\infty, (\mu_{j,t}^\infty)_{t\in I})\big)$$

to be $\frac{1}{10}$ times the minimum distance between distinct elements of $\mathcal{T}$.

Assume, for a contradiction, that $k = |\mathcal{T}| \geq 2$. Without loss of generality, we may then assume that the elements of $\mathcal{T}$ are ordered so that

$$\varepsilon = \frac{1}{10} d_{\mathbb{F}}^J\big((\mathcal{Y}_1^\infty, (\mu_{1,t}^\infty)_{t\in I}), (\mathcal{Y}_2^\infty, (\mu_{2,t}^\infty)_{t\in I})\big) > 0 \tag{6.160}$$

is the minimum $\mathbb{F}$-distance in $\mathcal{T}$. By the definition of $\mathcal{Y}_i^\infty$ being a tangent flow at infinity, for $i = 1, 2$, there exist sequences of scales $\{\lambda_{i,m}\}$ with $\lambda_{i,m} \to 0$ as $m \to \infty$ such that

$$\lim_{m\to\infty} d_{\mathbb{F}}^J\big((\mathcal{X}_I^{\lambda_{i,m}}, (\nu_{x_0;t}^{\lambda_{i,m}})), (\mathcal{Y}_i^\infty, (\mu_{i,t}^\infty)_{t\in I})\big) \to 0, \tag{6.161}$$

where $\mathcal{X}$ denotes the metric flow associated to $(\mathcal{O}^4, g(t), x_0)$, $t \leq 0$, and where $\mathcal{X}_I^{\lambda_{i,m}}$ denotes its scaling by $\lambda_{i,m}$ and restriction to I.

By discarding some of the scales $\lambda_{1,m}$, we may assume that $\lambda_{1,m} < \lambda_{2,m}$ for all $m \geq 1$. Furthermore, by (6.161) there exists an integer M such that for all $m \geq M$ we have for $i = 1, 2$ that

$$\lim_{m\to\infty} d_{\mathbb{F}}^J\big((\mathcal{X}_I^{\lambda_{i,m}}, (\nu_{x_0;t}^{\lambda_{i,m}})), (\mathcal{Y}_i^\infty, (\mu_{i,t}^\infty)_{t\in I})\big) < \varepsilon, \tag{6.162}$$

where $J = \{-1\} \subset I$. Thus, by (6.160) and by the triangle inequality, we obtain

$$d_{\mathbb{F}}^J\big((\mathcal{X}_I^{\lambda_{1,m}}, (\nu_{x_0;t}^{\lambda_{1,m}})), (\mathcal{X}_I^{\lambda_{2,m}}, (\nu_{x_0;t}^{\lambda_{2,m}}))\big) > 8\varepsilon. \tag{6.163}$$

Since, for each $m \geq M$, the function

$$\lambda \mapsto d_{\mathbb{F}}^J\big((\mathcal{X}_I^{\lambda_{1,m}}, (\nu_{x_0;t}^{\lambda_{1,m}})), (\mathcal{X}_I^{\lambda}, (\nu_{x_0;t}^{\lambda}))\big)$$

is continuous, there exists a parameter $\lambda_m \in [\lambda_{1,m}, \lambda_{2,m}]$ such that

$$d_{\mathbb{F}}^J\big((\mathcal{X}_I^{\lambda_{1,m}}, (\nu_{x_0;t}^{\lambda_{1,m}})), (\mathcal{X}_I^{\lambda_m}, (\nu_{x_0;t}^{\lambda_m}))\big) \in [2\varepsilon, 4\varepsilon]. \tag{6.164}$$

On the other hand, since we know that tangent flows at infinity exist, there exists a subsequence of the associated rescalings $(\mathcal{X}_I^{\lambda_m}, (\nu_{x_0;t}^{\lambda_m}))$ which converges to some element $(\mathcal{Y}_\ell^\infty, (\mu_{\ell,t}^\infty)_{t\in I})$ of $\mathcal{T}$.

By taking the limit as $m \to \infty$ in (6.164), we have that

$$d_{\mathbb{F}}^J\big((\mathcal{Y}_1^\infty, (\mu_{1,t}^\infty)_{t\in I}), (\mathcal{Y}_\ell^\infty, (\mu_{\ell,t}^\infty)_{t\in I})\big) \in [2\varepsilon, 4\varepsilon] \subset (0, 10\varepsilon). \tag{6.165}$$

This contradicts the definition of ε. The proposition is proved. □

6.4.3. 4-Dimensional steady singularity models whose tangent flows at infinity are 3-cylinders.

The following is [**BCD$^+$22**, Propositions 3.8 and 3.24].

THEOREM 6.77 (3-cylindrical steady soliton models). *Let $(\mathcal{O}^4, g(t), f(t))$, $t \in (-\infty, \infty)$, be a noncompact steady gradient Ricci soliton singularity model on a 4-dimensional smooth orbifold with isolated singularities. Suppose that the tangent* *flow at infinity* $(\mathcal{O}^4_\infty, g_\infty(t))$, $t \in (-\infty, 0]$, *is a 3-cylinder* $(\mathbb{S}^3/\Gamma) \times \mathbb{R}$. *Then:*

(1) (Γ-ε-Neck covering outside a compact set) *For any* $\varepsilon > 0$, *there exists a compact subset* $\mathcal{K}_\varepsilon$ *of* $\mathcal{O}^4$ *such that each point of* $\mathcal{O}^4 \setminus \mathcal{K}_\varepsilon$ *is the center of a* Γ*-quotient* ε*-neck, with respect to* $g := g(0)$.

(2) (Positive curvature with linear decay) *Given any* $p \in \mathcal{O}^4$, *there exist a positive constant* C *and a compact subset* $\mathcal{K}$ *of* $\mathcal{O}^4$ *such that* $(\mathcal{O}^4 \setminus \mathcal{K}, g)$ *has positive curvature operator and satisfies the estimates*

$$C^{-1} d(x,p)^{-1} \leq |\mathrm{Rm}|(x) \leq C d(x,p)^{-1} \quad \text{for all } x \in \mathcal{O}^4 \setminus \mathcal{K}. \tag{6.166}$$

6.4.3.1. *Γ-ε-Neck structure in the complement of a compact subset.*

In this subsection, we prove Theorem 6.77(1), which states that each point in the complement of some compact subset is the center of a Γ-quotient ε-neck (abbreviated as Γ-ε-neck); cf. [**BCD**$^+$**22**, Proposition 3.8].

By our hypothesis and the uniqueness result of Proposition 6.76, there exists a unique finite subgroup Γ of $O(4)$ such that any tangent flow at infinity of $(\mathcal{O}^4, g(t))$ is isometric to $(\mathbb{S}^3/\Gamma) \times \mathbb{R}$ with the product cylinder metric

$$g_\infty(0) = g_{\mathrm{cyl}} := 4 g_{\mathbb{S}^3/\Gamma} + g_{\mathbb{R}}.$$

Since this $\mathbb{F}$-limit is a smooth 4-manifold, the $\mathbb{F}$-convergence can be upgraded to smooth pointed Cheeger–Gromov convergence. Namely, we have the following:

Fix $x_0 \in \mathcal{O}^4$ and let $(z_\lambda, -1/\lambda)$ be an H_4-center of $(x_0, 0)$ at time $-1/\lambda$. Let $\varepsilon \in (0, 10^{-6})$. Since the unique tangent flow at infinity of $(\mathcal{O}^4, g(t))$ is $(\mathbb{S}^3/\Gamma) \times \mathbb{R}$, there exists

$$\overline{\lambda} = \overline{\lambda}(\varepsilon) > 0 \tag{6.167}$$

such that for all $\lambda \in \left(0, \overline{\lambda}\right]$, we have the following:

Define the parabolically rescaled solutions

$$g_\lambda(t) := \lambda g(t/\lambda). \tag{6.168}$$

Fix a base point $o_\infty \in (\mathbb{S}^3/\Gamma) \times \{0\} \subset (\mathbb{S}^3/\Gamma) \times \mathbb{R}$. Then there exists a smooth embedding

$$\Psi_\lambda : B^{\mathrm{cyl}}_{1/\varepsilon} := B^{\mathrm{cyl}}(o_\infty, 1/\varepsilon) \to \mathcal{O}^4$$

with

$$\Psi_\lambda(o_\infty) = z_\lambda \tag{6.169}$$

and

$$\left\| \Psi_\lambda^*(g_\lambda(-1)) - g_{\mathrm{cyl}} \right\|_{C^{\lceil 1/\varepsilon \rceil}(B^{\mathrm{cyl}}_{1/\varepsilon})} < \varepsilon. \tag{6.170}$$

Without loss of generality, we may assume that Ψ_λ is a diffeomorphism onto the ball $B_{g_\lambda(-1)}(z_\lambda, 1/\varepsilon) = \Psi_\lambda(B^{\mathrm{cyl}}_{1/\varepsilon})$. Hence, for all $\lambda \in \left(0, \overline{\lambda}\right]$, the H_4-center z_λ is the center of a Γ-ε-neck in $(\mathcal{O}^4, g_\lambda(-1))$.

Moreover, since $g(t)$ is a steady soliton, we can express the above in terms of $g(0)$. Namely, we have $g(t) = \Phi_t^*(g(0))$ for $t \in \mathbb{R}$, where $\Phi_t : \mathcal{O}^4 \to \mathcal{O}^4$ is the 1-parameter group of diffeomorphisms generated by $-\nabla_{g(0)} f(0)$. Thus, $g_\lambda(-1) = \lambda g(-1/\lambda) = \lambda \Phi^*_{-1/\lambda}(g(0))$. Hence

$$\Phi_{-1/\lambda} \circ \Psi_\lambda : B^{\mathrm{cyl}}_{1/\varepsilon} \to B_{g(0)}\left(w_\lambda, 1/(\sqrt{\lambda}\varepsilon)\right)$$

is a diffeomorphism and we may rewrite (6.170) as

$$\left\|\lambda(\Phi_{-1/\lambda}\circ\Psi_\lambda)^*(g(0)) - g_{\mathrm{cyl}}\right\|_{C^{\lceil 1/\varepsilon\rceil}(B^{\mathrm{cyl}}_{1/\varepsilon})} < \varepsilon. \tag{6.171}$$

That is, for all $\lambda \in \left(0, \overline{\lambda}\right]$, we have the Γ-ε-neck in $(\mathcal{O}^4, g(0))$:

$$\mathcal{N}_\lambda := (\Phi_{-1/\lambda}\circ\Psi_\lambda)(B^{\mathrm{cyl}}_{1/\varepsilon}) = B_{g(0)}\big(w_\lambda, 1/(\varepsilon\sqrt{\lambda})\big), \tag{6.172}$$

where $w_\lambda := (\Phi_{-1/\lambda}\circ\Psi_\lambda)(o) = \Phi_{-1/\lambda}(z_\lambda)$. We call $1/\lambda$ the **scale** of the neck $\mathcal{N}_\lambda$. Define

$$\mathcal{U}_\lambda := B_{g(0)}\left(w_\lambda, 10\sqrt{H_4}/\sqrt{\lambda}\right). \tag{6.173}$$

Since $\varepsilon < 10^{-6}$, we have $10\mathcal{U}_\lambda \subset \mathcal{N}_\lambda$.

Since, following an argument of Perelman, we can patch together Γ-ε-necks to form a Γ-ε-tube, Theorem 6.77(1) follows from the following result:

CLAIM 6.78 (Neck covering outside a compact set). *There exists a compact subset $\mathcal{K}_\varepsilon$ of $\mathcal{O}^4$ such that*

$$\mathcal{O}^4\setminus\mathcal{K}_\varepsilon \subset \bigcup_{\lambda\in(0,\overline{\lambda}]} 10\mathcal{U}_\lambda = \bigcup_{\lambda\in(0,\overline{\lambda}]} B_{g(0)}\left(w_\lambda, 100\sqrt{H_4}/\sqrt{\lambda}\right). \tag{6.174}$$

We now proceed to prove Claim 6.78. We can think of the following as an extension of Proposition 6.34.[18]

CLAIM 6.79 (Neck overlap continuity). *For any $\lambda \in \left(0, \overline{\lambda}\right)$, there exists $\delta(\lambda) > 0$ such that if $|\lambda' - \lambda| < \delta(\lambda)$ and $|\lambda'' - \lambda| < \delta(\lambda)$, then*

$$\mathcal{U}_{\lambda'}\cap\mathcal{U}_{\lambda''} \neq \emptyset. \tag{6.175}$$

PROOF OF CLAIM 6.79. Suppose, for a contradiction, that there exists $\lambda > 0$ and positive sequences $\lambda'_i, \lambda''_i \to \lambda$ such that

$$\mathcal{U}_{\lambda'_i}\cap\mathcal{U}_{\lambda''_i} = \emptyset. \tag{6.176}$$

Define

$$\mathcal{V}_\lambda := B_{g_\lambda(-1)}\left(z_\lambda, 9\sqrt{H_4}\right) = B_{g(-1/\lambda)}\left(z_\lambda, 9\sqrt{H_4}/\sqrt{\lambda}\right) = \frac{9}{10}\Phi_{1/\lambda}(\mathcal{U}_\lambda);$$

note that $\Phi_{1/\lambda} = \Phi^{-1}_{-1/\lambda}$. By applying $\Phi_{1/\lambda}$ to (6.176), it is easy to see that we have for i sufficiently large,

$$\frac{9}{10}\mathcal{V}_{\lambda'_i}\cap\frac{9}{10}\mathcal{V}_{\lambda''_i} \subset \Phi_{\delta'_i}(\mathcal{V}_{\lambda'_i})\cap\Phi_{\delta''_i}(\mathcal{V}_{\lambda''_i}) = \emptyset, \tag{6.177}$$

where $\delta'_i := 1/\lambda - 1/\lambda'_i \to 0$ and $\delta''_i := 1/\lambda - 1/\lambda''_i \to 0$.

On the other hand, by the concentration estimate of Lemma 6.33, with $A = 9^2 = 81$, the conjugate heat kernel measure of $(\mathcal{O}^4, g(t))$ satisfies

$$\begin{aligned}\nu_{x_0,0;-1/\lambda}\left(\frac{9}{10}\mathcal{V}_\lambda\right) &= \nu_{x_0,0;-1/\lambda}\left(B_{g(-1/\lambda)}\left(z_\lambda, 9\sqrt{H_4}/\sqrt{\lambda}\right)\right)\\ &\geq 1-\frac{1}{81} > \frac{9}{10}.\end{aligned} \tag{6.178}$$

[18]Note that the definition of $\mathcal{U}_\lambda$ depends on the choice z_λ of H_n-center at time $-1/\lambda$.

Now, from [**BCD**$^+$**22**, Lemma 3.3], we have that[19]

$$\nu_{x_0,0;-1/\lambda}\left(\frac{9}{10}\mathcal{V}_{\lambda_i'}\right) > \frac{1}{2} \quad \text{and} \quad \nu_{x_0,0;-1/\lambda}\left(\frac{9}{10}\mathcal{V}_{\lambda_i''}\right) > \frac{1}{2} \tag{6.179}$$

for all i sufficiently large. However, this contradicts $\frac{9}{10}\mathcal{V}_{\lambda_i'} \cap \frac{9}{10}\mathcal{V}_{\lambda_i''} = \emptyset$. Therefore, Claim 6.79 follows. □

Since the hypotheses of Theorem 6.77 imply that $\mathcal{O}^4$ does not split, Theorem 3.29 ensures that $\mathcal{O}^4$ is connected at infinity. Here, we remark that Theorem 3.29 extends to our setting of a steady soliton on a smooth 4-orbifold with isolated singularities and tangent flow at infinity $(\mathbb{S}^3/\Gamma) \times \mathbb{R}$; see [**BCD**$^+$**22**, Theorem A.1] for a proof of this.

Let $\lambda \in (0, \overline{\lambda})$. Thus, since $\mathcal{U}_\lambda$ is diffeomorphic to a cylinder, we have that $\mathcal{O}^4 \setminus \mathcal{U}_\lambda$ consists of two components, one bounded and one unbounded. Let $\mathcal{E}_\lambda$ denote the unbounded component, and let $\mathcal{K}_\lambda := \overline{\mathcal{O}^4 \setminus \mathcal{E}_\lambda}$, which is a compact set.

For each $\lambda \in (0, \overline{\lambda}]$, where $\overline{\lambda}$ is given by (6.167), there exists a largest real number $\delta(\lambda)$ such that

$$\frac{9}{10}\mathcal{U}_{\lambda'} \cap \frac{9}{10}\mathcal{U}_\lambda \neq \emptyset \quad \text{for all } \lambda' \in (\lambda - \delta(\lambda), \lambda] \tag{6.180}$$

(this is because of the interval $(\lambda - \delta(\lambda), \lambda]$ is open on the lower bound). By the neck overlap continuity established in Claim 6.79, we have $\delta(\lambda) > 0$. Define the positive decreasing sequence $\{\lambda_i\}_{i=1}^\infty$ inductively by $\lambda_1 := \overline{\lambda}$ and

$$\lambda_{i+1} := \lambda_i - \delta(\lambda_i) \quad \text{for all } i \geq 1. \tag{6.181}$$

We now show that

$$\lambda_i \to 0. \tag{6.182}$$

Suppose, for a contradiction, that $\lambda_i \to \underline{\lambda}$, where $\underline{\lambda} > 0$. Observe that $\delta(\lambda_i) \to 0$ since λ_i is a positive decreasing sequence. By Claim 6.79, for i sufficiently large, we have $\frac{1}{3}\mathcal{U}_{\lambda_i} \cap \frac{1}{3}\mathcal{U}_{\underline{\lambda}} \neq \emptyset$. We also have that there exists $\underline{\delta} > 0$ such that

$$\frac{1}{3}\mathcal{U}_\lambda \cap \frac{1}{3}\mathcal{U}_{\underline{\lambda}} \neq \emptyset \quad \text{for all } \lambda \in (\underline{\lambda} - \underline{\delta}, \underline{\lambda} + \underline{\delta}). \tag{6.183}$$

Thus, $\frac{9}{10}\mathcal{U}_\lambda \cap \frac{9}{10}\mathcal{U}_{\lambda_i} \neq \emptyset$ for all $\lambda \in (\underline{\lambda} - \underline{\delta}, \underline{\lambda} + \underline{\delta})$. This implies $\delta(\lambda_i) \geq \underline{\delta} > 0$ for i sufficiently large, which is a contradiction. This proves (6.182).

Furthermore, we have that $\frac{1}{3}\mathcal{U}_{\lambda_i} \cap \frac{1}{3}\mathcal{U}_{\lambda_{i+1}} = \emptyset$. From this, it follows that $\frac{1}{3}\mathcal{U}_{\lambda_{i+1}}$ is contained in the noncompact component of $\mathcal{O}^4 \setminus \frac{1}{3}\mathcal{U}_{\lambda_i}$, call it $\mathcal{F}_{\lambda_i}$. Indeed, if this statement is not true, then the smallest integer i_0 at least $i+1$ such that $\frac{1}{3}\mathcal{U}_{\lambda_{i_0}} \subset \mathcal{F}_{\lambda_i}$ satisfies $i_0 > i+1$. On the other hand, $\frac{1}{3}\mathcal{U}_{\lambda_{i_0-1}}$ is contained in the compact component of $\mathcal{O}^4 \setminus \frac{1}{3}\mathcal{U}_{\lambda_i}$. This implies that $\frac{1}{3}\mathcal{U}_{\lambda_i}$ separates $\frac{1}{3}\mathcal{U}_{\lambda_{i_0-1}}$ and $\frac{1}{3}\mathcal{U}_{\lambda_{i_0}}$, which is a contradiction. Thus, the corresponding infinite sequence of Γ-ε-necks $\{\mathcal{N}_{\lambda_i}\}$ are joined end-to-end. Since $\mathcal{O}^4$ has exactly one topological end, we conclude that the complement of the union $\bigcup_i \mathcal{N}_{\lambda_i}$ is a compact set. This completes the proof of Theorem 6.77(1).

[19] In [**BCD**$^+$**22**, Lemma 3.3], we can replace the assumption that $0 < t - s < \delta$ by $|t - s| < \delta$.

6.4.3.2. *Linearly decaying positive curvature operator outside a compact subset.*

In this subsection, we prove Theorem 6.77(2), which is a result of Deng and Zhu [**DZ20a**, Theorem 1.5] (cf. [**BCD**$^+$**22**, Proposition 3.24]).

6.4.3.2.a. *Precise estimates for potential function and scalar curvature.*

We begin with a weak estimate for the scalar curvature.

LEMMA 6.80 (Slower than quadratic scalar curvature decay). *The scalar curvature R of the steady soliton satisfies*

$$\lim_{x\to\infty} R(x)\, d_p(x)^2 = \infty, \tag{6.184}$$

where d_p denotes the distance to a fixed point p.

PROOF. Suppose, for a contradiction, that there exists a sequence $x_i \to \infty$ with quadratic scalar curvature decay: $R(x_i)\, d_g(x_i,p)^2 \leq C$ for some constant C. By Theorem 6.77(1), we have that

$$\left(B_g\left(x_i, \tfrac{A_i}{\sqrt{R(x_i)}}\right), R(x_i) g, x_i\right) \longrightarrow \left((S^3/\Gamma)\times\mathbb{R}, g_{\mathrm{cyl}}, o_\infty\right)$$

converges to the cylinder in the smooth pointed Cheeger–Gromov sense. Since $d_{R(x_i)g}(x_i,p)^2 \leq C$, by the CG-convergence we have that $\frac{R(x_i)}{R(p)} = 1 + o(1)$. This implies that $R(p)\, d_g(x_i,p)^2 \leq 2C$, which yields a contradiction since $R(p) > 0$. □

LEMMA 6.81 (Potential function is asymptotically the distance function). *The potential function f satisfies*

$$\lim_{x\to\infty} \frac{f(x)}{d_p(x)} = -1. \tag{6.185}$$

PROOF. Let $\varepsilon > 0$ be sufficiently small and let $\mathcal{K}_\varepsilon$ be the compact subset of $\mathcal{O}^4$ provided by Theorem 6.77(1). Without loss of generality, we may assume that $p \in \mathcal{K}_\varepsilon$. Choose r_0 sufficiently large such that $\mathcal{K}_\varepsilon \subset B(p, r_0)$. Let $r \geq r_0$.

Recall from Theorem 5.5 that the potential function satisfies the estimates

$$-r - C \leq \inf_{\partial B(p,r)} f \leq -r + C\sqrt{r}\,. \tag{6.186}$$

In particular, there exists $y_0 \in \partial B(p,r)$ such that

$$-r - C \leq f(y_0) \leq -r + C\sqrt{r}\,. \tag{6.187}$$

On the other hand, since $y_0 \notin \mathcal{K}_\varepsilon$ because $r \geq r_0$, we have that y_0 is the center of a Γ-ε-neck $\mathcal{N}$ in $\mathcal{O}^4$. Thus, there exists a diffeomorphism

$$\Psi : B^{\mathrm{cyl}}_{1/\varepsilon} \to \mathcal{N}$$

such that the rescaled pullback metric $R(y_0)\Psi^* g$ is ε-close in the $C^{\lceil 1/\varepsilon\rceil}$-norm to g_{cyl} on $B^{\mathrm{cyl}}_{1/\varepsilon}$, and

$$y_0 \in \Psi\big((S^3/\Gamma)\times\{0\}\big) =: \Sigma_0, \tag{6.188}$$

the "center sphere" of the neck $\mathcal{N}$, which is diffeomorphic to S^3/Γ.

The extrinsic diameter of Σ_0 satisfies

$$\mathrm{diam}_{\mathrm{extr}}(\Sigma_0) := \max_{x,y\in\Sigma_0} d(x,y) \leq \frac{C(n)}{\sqrt{R(y_0)}} \leq o\big(d(y_0,p)\big) = o(r), \tag{6.189}$$

where the last inequality is by Lemma 6.80.

CLAIM 6.82.

$$\partial B(p,r) \subset \Psi\big((S^3/\Gamma) \times [-L, L]\big) =: \mathcal{N}_L, \tag{6.190}$$

where $L := 10C(n)/\sqrt{R(y_0)}$.

PROOF OF CLAIM 6.82. Since $\mathcal{O}^4$ is connected at infinity by Theorem 3.29 and is asymptotically cylindrical, the geodesic sphere $\partial B(p,r)$ is connected. For a contradiction, suppose that there exists a point $y_1 \in \partial B(p,r) \setminus \mathcal{N}_L$. This implies that there exists a point

$$y_2 \in \partial B(p,r) \cap \pm\partial\mathcal{N}_L,$$

where $\pm\partial\mathcal{N}_L = \Psi\big((S^3/\Gamma) \times \{\pm L\}\big)$. By (6.188), we have that $y_0 = \Psi(z_0, 0)$, where $z_0 \in S^3/\Gamma$. Let

$$y_3 := \Psi(z_0, \pm L) \in \pm\partial\mathcal{N}_L.$$

By the ε-neck property, we have $|d(y_3, y_0) - L| \leq \varepsilon L$.

Suppose that y_1 lies in the bounded component of $\mathcal{O}^4\setminus\mathcal{N}_L$. Then $y_2, y_3 \in -\partial\mathcal{N}_L$. Thus, there exists a point $y_- \in -\partial\mathcal{N}_L$ such that

$$\begin{aligned}
r &= d(p, y_0) \\
&= d(p, y_-) + d(y_-, y_0) \\
&\geq d(p, y_2) - d(y_2, y_-) + d(y_3, y_0) - d(y_-, y_3) \\
&\geq r + (1-\varepsilon)L - 2\operatorname{diam}(-\partial\mathcal{N}_L) \\
&> r,
\end{aligned}$$

which is a contradiction. The case where y_1 lies in the unbounded component of $\mathcal{O}^4 \setminus \mathcal{N}_L$ similarly leads to a contradiction. This proves Claim 6.82. ☐

Now let $y \in \partial B(p,r)$. By (6.189) and Claim 6.82, we have

$$d_g(y, y_0) \leq \operatorname{diam}_{\text{extr}}(\Sigma_0) + 2L \leq o(r). \tag{6.191}$$

Since $|\nabla f| \leq 1$ and by (6.187) and (6.189), we have that

$$f(y) \leq f(y_0) + d_g(y, y_0) \leq -r + C\sqrt{r} + o(r) \leq -r + o(r).$$

Lemma 6.81 follows. ☐

LEMMA 6.83.

$$R d_p = \frac{n-1}{2} + o(1). \tag{6.192}$$

PROOF. This follows from Lemma I-8.13 (the proof applies to our situation) and (6.185). ☐

6.4.3.2.b. *Positive curvature operator with linear decay.*

Let $p \in \mathcal{O}^4$ and consider the level set $\Sigma(t) = \{x \in \mathcal{O}^4 : f(x) = t\}$, where $t = f(p)$, so that $p \in \Sigma(t)$. Without loss of generality, we may assume that $\Sigma(t)$ is a smooth hypersurface. Let $\{e_i\}_{i=1}^4$ be an orthonormal frame at p with e_1, e_2, e_3 tangent to $\Sigma(t)$ and with $e_4 = \frac{\nabla f}{|\nabla f|}$. Let α be a 2-form at p. Then we can write

$$\alpha = \sum_{i,j=1}^{3} a_{ij}\, e_i^* \wedge e_j^* + \sum_{\ell=1}^{3} b_\ell\, e_4^* \wedge e_\ell^*, \tag{6.193}$$

where $a_{ij}, b_\ell \in \mathbb{R}$ and $a_{ji} = -a_{ij}$.[20] Recall from (4.197) that the curvature operator $\mathrm{R} : \wedge^2(T^*M^n) \to \wedge^2(T^*M^n)$ is defined by

$$\mathrm{R}(\alpha)_{ij} = \sum_{k,\ell=1}^{4} R_{ijk\ell}\alpha_{\ell k} \quad \text{for all } \alpha \in \wedge^2(T^*M^n), \tag{6.194}$$

where $R_{ijk\ell} := \mathrm{Rm}(e_i, e_j, e_k, e_\ell) = \langle \mathrm{Rm}(e_i, e_j)e_k, e_\ell\rangle$. Recall that the inner product on $\wedge^2(T^*M^n)$ is defined by

$$\langle \alpha, \beta\rangle = \frac{1}{2}\sum_{i,j=1}^{4} \alpha_{ij}\beta_{ij}. \tag{6.195}$$

Define

$$\mathrm{R}(\alpha, \alpha) := \langle \mathrm{R}(\alpha), \alpha\rangle = \sum_{i,j,k,\ell=1}^{4} R_{ijk\ell}\alpha_{ij}\alpha_{\ell k}. \tag{6.196}$$

Now,

$$\mathrm{R}(e_i^* \wedge e_j^*, e_\ell^* \wedge e_k^*) = 2R_{ijk\ell}.$$

The curvature operator applied to α yields

$$\mathrm{R}(\alpha, \alpha) = 2\sum_{i,j,k,\ell=1}^{3} R_{ijk\ell}a_{ij}a_{\ell k} + 4\sum_{i,j,k=1}^{3} R_{ijk4}a_{ij}b_k + 2\sum_{j,k=1}^{3} R_{4jk4}b_j b_k. \tag{6.197}$$

Define

$$A_{jk} := \sum_{i,\ell=1}^{3} R_{ijk\ell}\nabla_i f\nabla_\ell f \tag{6.198}$$

for $1 \leq j, k \leq 3$, where $\nabla_i f := e_i(f)$. Since $\nabla_i f = |\nabla f|\delta_{i4}$, we have that

$$A_{jk} = |\nabla f|^2 R_{4jk4}.$$

By (1.49), we have that

$$A_{jk} = \Delta R_{jk} - \frac{1}{2}\nabla^2_{jk}R + 2\sum_{i,\ell=1}^{4} R_{ijk\ell}R_{i\ell} - \sum_{\ell=1}^{4} R_{j\ell}R_{k\ell}. \tag{6.199}$$

Since our steady gradient Ricci soliton is asymptotically cylindrical, we have that

$$|\nabla \mathrm{Ric}| = \mathrm{o}(R^{\frac{3}{2}}), \quad |\Delta \mathrm{Ric}| = \mathrm{o}(R^2), \quad |\nabla^2 R| = \mathrm{o}(R^2). \tag{6.200}$$

Thus,

$$A_{jk} = 2R_{ijk\ell}R_{i\ell} - R_{j\ell}R_{k\ell} + \mathrm{o}(R^2). \tag{6.201}$$

Let $(\mathbb{S}^3 \times \mathbb{R}, \overline{g})$ be an exact cylinder and let $\overline{p} = (x, y) \in \mathbb{S}^3 \times \mathbb{R}$. Let $\{\overline{e}_i\}_{i=1}^4$ be an orthonormal frame at $\overline{p}$ with $\overline{e}_1, \overline{e}_2, \overline{e}_3$ tangent to the slice $\mathbb{S}^3 \times \{y\}$ and $\overline{e}_4 = \frac{\partial}{\partial y}$. With respect to this frame, the components of $\overline{\mathrm{Rm}}$ for this geometry satisfy

$$\overline{R}_{ijk\ell} = \frac{\overline{R}}{6}(\overline{g}_{i\ell}\overline{g}_{jk} - \overline{g}_{ik}\overline{g}_{j\ell}) = \frac{\overline{R}}{6}(\delta_{i\ell}\delta_{jk} - \delta_{ik}\delta_{j\ell}) \quad \text{for all } 1 \leq i, j, k, \ell \leq 3 \tag{6.202}$$

and

$$\overline{R}_{ijk4} = 0 \quad \text{for all } 1 \leq i, j, k \leq 4.$$

---

[20]The wedge product is defined by $e_i^* \wedge e_j^* = e_i^* \otimes e_j^* - e_j^* \otimes e_i^*$.

This implies that the Ricci tensor of the cylinder is given by

$$\overline{R}_{jk} = \sum_{i,\ell=1}^{2} \overline{g}^{i\ell}\overline{R}_{ijk\ell} = \frac{\overline{R}}{3}\overline{g}_{jk} = \frac{\overline{R}}{3}\delta_{jk} \quad \text{for all } 1 \leq j,k \leq 3$$

and $\overline{R}_{j4} = 0$ for $1 \leq j \leq 4$. Thus,

$$\overline{A}_{jk} := 2\overline{R}_{ijk\ell}\overline{R}_{i\ell} - \overline{R}_{j\ell}\overline{R}_{k\ell} = \frac{\overline{R}}{9}\overline{g}_{jk} = \frac{\overline{R}}{9}\delta_{jk} \quad \text{for all } 1 \leq j,k \leq 3. \tag{6.203}$$

By (6.201), (6.203), and again using that our steady gradient Ricci soliton is asymptotically cylindrical, we obtain

$$\sum_{i,\ell=1}^{3} R_{ijk\ell}\nabla_i f \nabla_\ell f = A_{jk} = \frac{1}{9}R^2 g_{jk} + \mathrm{o}(R^2) \quad \text{for all } 1 \leq j,k \leq 3. \tag{6.204}$$

Likewise, in view of (6.202), we have that

$$R_{ijk\ell} = \frac{R}{6}(g_{i\ell}g_{jk} - g_{ik}g_{j\ell}) + \mathrm{o}(R) \quad \text{for all } 1 \leq i,j,k,\ell \leq 3. \tag{6.205}$$

Furthermore, by $R_{ijk\ell}\nabla_\ell f = \nabla_j R_{ik} - \nabla_i R_{jk}$, which is from (1.47), and since (6.200),

$$R_{ijk4} = \mathrm{o}(R^{\frac{3}{2}}). \tag{6.206}$$

By applying the approximations above to (6.197), we obtain

$$\begin{aligned} \mathrm{R}(\alpha,\alpha) &= 2\sum_{i,j,k,\ell=1}^{3} R_{ijk\ell}a_{ij}a_{\ell k} + 4\sum_{i,j,k=1}^{3} R_{ijk4}a_{ij}b_k + 2\sum_{j,k=1}^{3} R_{4jk4}b_j b_k \\ &= \frac{2}{3}R|a|^2(1 \pm \mathrm{o}(1)) \pm |a|\,|b|\,\mathrm{o}(R^{\frac{3}{2}}) + (1 \pm \mathrm{o}(1))\left(\frac{1}{9}R^2 \pm \mathrm{o}(R^2)\right)|b|^2, \end{aligned} \tag{6.207}$$

where $|a|^2 := \sum_{i,j,k,\ell=1}^{3} g^{i\ell}g^{jk}a_{ij}a_{\ell k}$ and $|b|^2 := g^{jk}b_j b_k$. Thus, outside a sufficiently large compact subset of $\mathcal{O}^4$, we have that

$$\mathrm{R}(\alpha,\alpha) \geq \frac{1}{3}R|a|^2 + \frac{1}{18}R^2|b|^2, \tag{6.208}$$

which is positive provided $\alpha \neq 0$. This proves that the steady gradient Ricci soliton has positive curvature operator outside a compact subset. Moreover, by (6.207) and Lemma 6.83, we obtain (6.166). This completes the proof of Theorem 6.77(2).

6.5. 4-Dimensional steady soliton singularity models have at most quadratic scalar curvature decay

In this section, we apply Bamler's theory to obtain scalar curvature decay estimates for 4-dimensional steady gradient Ricci solitons and certain n-dimensional singularity models.

6.5.1. Scalar curvature decay estimate for singularity models.

The scalar curvature lower bound (2.27) for shrinking gradient Ricci solitons can be extended to ancient singularity models in a certain sense. Namely, Chan, Ma, Zhang, and one of the authors [**CCMZ23**, Theorem 2.1] proved the following result regarding singularity models for the Ricci flow. Later in this section we will apply this result to 4-dimensional steady gradient Ricci soliton singularity models, in which case we will obtain a quadratic decay lower bound for the scalar curvature.

THEOREM 6.84 (Curvature decay estimate for singularity models satisfying 'Property R'). *Let $(\mathcal{O}^n, g(t))$, $t \in (-\infty, 0]$, be a complete ancient solution on an orbifold with isolated singularities that is also a singularity model of the Ricci flow. Let $x_0 \in \mathcal{O}^n$ and let f be the potential of the conjugate heat kernel based at $(x_0, 0)$ (defined by (6.15)). Define*

$$(6.209)\qquad a_0(r) := \inf \left\{ (|t|+1)R(x,t) : x \in B_{g(t)}\left(z_t, r\sqrt{|t|+1}\right),\ t \in (-\infty, 0) \right\},$$

where (z_t, t) is some choice of H_n-center for each $t \in (-\infty, 0)$. Assume ***Property R***, *which says that*

$$a_0(r) > 0 \quad \textit{for all } r > 0.$$

Then the scalar curvature R satisfies the following lower bound:

$$(6.210)\qquad (|t|+1)R(x,t) \geq \frac{c}{f(x,t) + C(n) - \mu_\infty},$$

where $\mu_\infty > -\infty$ is the asymptotic pointed Nash entropy defined by (6.155) and where c is some positive constant depending on n, μ_∞, x_0, and the function a_0.

Note that in Theorem 6.84, we do not assume that $g(t)$ has bounded curvature on compact time intervals. We remark that the curvature hypothesis of the theorem says that in the vicinity of the H_n-centers the scalar curvature satisfies a positive Type I lower bound. We also remark that Theorem 6.84 can also be used to rederive the scalar curvature lower bound (2.27) in the special case of shrinking gradient Ricci soliton singularity models.

REMARK 6.85. Suppose that $n = 4$ and that $(\mathcal{O}^4, g(t))$, $t \in (-\infty, 0]$, is a complete ancient solution that is also a singularity model of the Ricci flow. If Property R does not hold for this singularity model, then it has a tangent flow at infinity with $R \equiv 0$, which implies that this tangent flow at infinity is $\mathbb{R}^4/\Gamma$, where Γ is a nontrivial finite subgroup of $\mathrm{O}(4)$.

For singularity models with tangent flows at infinity that are smooth, we have the following (see [**CCMZ23**, Corollary 2.2]):

COROLLARY 6.86 (Curvature decay estimate for singularity models with smooth tangent flows). *Let $(\mathcal{O}^n, g(t), x_0)$, $t \in (-\infty, 0]$, be a complete pointed ancient solution on an orbifold with isolated singularities that is also a singularity model of the Ricci flow. If all tangent flows at infinity of $(\mathcal{O}^n, g(t))$ (see Definition 6.67) are smooth, i.e., their underlying spaces are smooth n-manifolds, then*

$$(6.211)\qquad (|t|+1)R(x,t) \geq \frac{c}{f(x,t) + C(n) - \mu_\infty}$$

for some positive constant $c > 0$ (depending on $(\mathcal{O}^n, g(t), x_0)$) and where f is the potential of the conjugate heat kernel based at $(x_0, 0)$.

PROOF OF COROLLARY 6.86. By Theorem 6.84, we just need to show that the solution $(\mathcal{O}^n, g(t))$ satisfies Property R. Assume, for a contradiction, that $(\mathcal{O}^n, g(t))$ does not satisfy Property R that $a_0(r) > 0$ for all $r > 0$. Then there would exist $r > 0$, $t_i \to -\infty$, and $x_i \in B_{g(t_i)}\big(z_{t_i}, r\sqrt{|t_i| + 1}\big)$ such that

$$(|t_i| + 1)R(x_i, t_i) \to 0.$$

Consider a tangent flow at infinity $(M^n_\infty, g_\infty(t))$ associated to a subsequence of the rescaling factors $\lambda_i := |t_i|^{-1} \to 0$, where, by our hypothesis, the underlying manifold M^n_∞ is smooth and $g_\infty(t)$ is a smooth solution to the Ricci flow.

The sequence of pointed smooth Riemannian orbifolds $\big(\mathcal{O}^n, \lambda_i g(-\lambda_i^{-1}), z_{t_i}\big)$ subconverges in the smooth pointed Cheeger–Gromov sense to $(M^n_\infty, g_\infty(-1), z_\infty)$ for some $z_\infty \in M^n_\infty$. By the conditions above satisfied by x_i, we have that with respect to this Cheeger–Gromov convergence, x_i subconverges to some point $x_\infty \in \overline{B}_{g_\infty(-1)}(z_\infty, r)$ with

$$R_{g_\infty(-1)}(x_\infty) = 0.$$

By the strong maximum principle applied to the evolution equation for $R_{g_\infty(t)}$, this implies that the metric $g_\infty(-1)$ is Ricci flat and hence is Euclidean since it is also a shrinking gradient Ricci soliton metric. This implies that the asymptotic pointed Nash entropy $\mu_\infty = 0$, which in turn implies by Perelman's monotonicity formula that $(\mathcal{O}^n, g(t))$ is static Euclidean space. This is a contradiction to the assumption that it is a singularity model. □

6.5.2. Scalar curvature decay estimate for steady solitons.

We next discuss an application of the result above to 4-dimensional steady gradient Ricci soliton singularity models (see [**CCMZ23**, Corollary 2.7]).

COROLLARY 6.87 (Quadratic curvature decay estimate for 4-dimensional steady solitons). *If $(\mathcal{O}^4, g, \overline{f})$ is a non-Ricci-flat 4-dimensional steady gradient Ricci soliton singularity model with bounded sectional curvature and if $o \in \mathcal{O}^4$, then there exists a positive constant c such that*

$$R(x) \geq c\big(d(x, o) + 1\big)^{-2}. \tag{6.212}$$

REMARK 6.88. If $\mathcal{O}^4$ is a smooth manifold, then by Theorem 5.10, we do not need to assume the sectional curvature bound.

PROOF OF COROLLARY 6.87. Let $(\mathcal{O}^4, g(t), \overline{f}(t))$ be the canonical form associated to $(\mathcal{O}^4, g, \overline{f})$ restricted to the time interval $t \in (-\infty, 0]$. By our hypothesis and by Theorem 6.72, the tangent flow at infinity of this steady gradient Ricci soliton singularity model is unique, smooth, and splits. Thus, we may apply Corollary 6.86 to obtain

$$2R(x, -1) \geq \frac{c}{f(x, -1) + C(n) - \mu_\infty}, \tag{6.213}$$

where f is the potential of the conjugate heat kernel based at $(o, 0)$.

On the other hand, since the solution has bounded curvature, by the Gaussian lower bound for the conjugate heat kernel K defined by Definition 6.2 (see [**CCG$^+$10**, Theorem 26.31]), we have that

$$(4\pi)^{-\frac{n}{2}} \mathrm{e}^{-f(x,-1)} = K(o, 0; x, -1) \geq \frac{c}{\sqrt{\operatorname{Vol} B_g(o, 1)}\sqrt{\operatorname{Vol} B_g(x, 1)}} \mathrm{e}^{-\frac{d_g^2(x,o)}{c}} \tag{6.214}$$

for some constant $c > 0$. Moreover, there is a constant C such that $\operatorname{Vol} B_g(p,1) \leq C$ for all $p \in \mathcal{O}^4$ by the Bishop–Gromov volume comparison theorem and the curvature bound for g. Thus,

$$f(x,-1) \leq \frac{d_g^2(x,o)}{c} + C. \tag{6.215}$$

Now recall that $g(t) = \varphi_t^* g$, where the 1-parameter group of diffeomorphisms $\varphi_t : \mathcal{O}^4 \to \mathcal{O}^4$ is defined by $\partial_t \varphi_t(x) = (\nabla_g \overline{f})(\varphi_t(x))$ and $\varphi_0 = \mathrm{id}$. So

$$R_g(x) = R(x,0) = R(\varphi_1(x),-1). \tag{6.216}$$

Moreover, since $\left|\nabla_g \overline{f}\right| \leq 1$, we have that $d(\varphi_1(x),x) \leq 1$. The corollary follows. □

We remark that for this scalar curvature lower bound for 4-dimensional steady gradient Ricci soliton singularity models, the conjugate heat kernels of the associated canonical form (solving the Ricci flow) have an advantage over the potential function of the gradient Ricci soliton.

6.5.3. Proof of the scalar curvature decay estimate for singularity models.

The idea of the proof of Theorem 6.84 is to mimic the proof of the quadratic decaying scalar curvature lower bound for shrinking gradient Ricci solitons (see Theorem 4.31 in Volume 1). We will use the potential of the conjugate heat kernel as a proxy for the potential function of a shrinking gradient Ricci soliton. For the case of shrinking gradient Ricci solitons, the elliptic maximum principle was used. Analogously, for ancient singularity models, we will use the parabolic maximum principle. As we will see, various issues arise when we employ this strategy. With this in mind, we proceed as follows:

Proof of Theorem 6.84. Denote $\tau = |t|$. Recall from (6.33) that, as a consequence of Perelman's differential Harnack estimate, we have:

$$\tau \Box f \geq f - \frac{n}{2}, \tag{6.217}$$

where f is the potential of the conjugate heat kernel based at $(x_0, 0)$ defined by (6.15). For convenience, we adjust the scaling of f and we define

$$\widetilde{f}(x,t) := \sqrt{\tau+1}\left(f(x,t) + C(n) - \mu_\infty\right), \tag{6.218}$$

where μ_∞ is given by (6.155). We will employ $\widetilde{f}$ to construct a barrier function for the scalar curvature R in order to estimate R. The $\sqrt{\tau+1}$ factor is to facilitate the application of the parabolic maximum principle. By the Gaussian upper bound (6.154) for the conjugate heat kernel and definition (6.218),

$$\widetilde{f}(x,t) \geq \sqrt{\tau+1}\left(\frac{d_t(x,z_t)^2}{9\tau} + 1\right). \tag{6.219}$$

We compute using (6.217) that

$$\begin{aligned}
\tau \Box \widetilde{f} &= \sqrt{\tau+1}\,\tau \Box f - \frac{\tau}{2(\tau+1)} \widetilde{f} \\
&\geq \sqrt{\tau+1}\left(f - \frac{n}{2}\right) - \frac{\tau}{2(\tau+1)} \widetilde{f} \\
&= \frac{\tau+2}{2(\tau+1)} \widetilde{f} - \frac{n}{2}\sqrt{\tau+1} + \sqrt{\tau+1}\left(-C(n) + \mu_\infty\right).
\end{aligned}$$

Thus,

$$\tau \square \widetilde{f} \geq \frac{1}{2}\widetilde{f} - C_1\sqrt{\tau+1}, \tag{6.220}$$

where $C_1 := C(n) + \frac{n}{2} - \mu_\infty$.

Our barrier function will be a linear combination of $\widetilde{f}^{-1}$ and $\widetilde{f}^{-2}$. We compute that

$$\begin{aligned}
\tau \square \widetilde{f}^{-1} &= -\tau \widetilde{f}^{-2} \square \widetilde{f} - 2\tau \widetilde{f}^{-3}|\nabla \widetilde{f}|^2 \\
&\leq -\widetilde{f}^{-2}\left(\frac{1}{2}\widetilde{f} - C_1\sqrt{\tau+1}\right) - 2\tau \widetilde{f}^{-3}|\nabla \widetilde{f}|^2 \\
&\leq -\frac{1}{2}\widetilde{f}^{-1} + C_1\sqrt{\tau+1}\,\widetilde{f}^{-2}.
\end{aligned}$$

Similarly, we compute that

$$\begin{aligned}
\tau \square \widetilde{f}^{-2} &= -2\tau \widetilde{f}^{-3} \square \widetilde{f} - 6\tau \widetilde{f}^{-4}|\nabla \widetilde{f}|^2 \\
&\leq -2\widetilde{f}^{-3}\left(\frac{1}{2}\widetilde{f} - C_1\sqrt{\tau+1}\right) - 6\tau \widetilde{f}^{-4}|\nabla \widetilde{f}|^2 \\
&\leq -\widetilde{f}^{-2} + 2C_1\sqrt{\tau+1}\,\widetilde{f}^{-3}.
\end{aligned}$$

For scaling reasons, we further compute that

$$\begin{aligned}
\tau \square \big(\sqrt{\tau+1}\,\widetilde{f}^{-2}\big) &= \sqrt{\tau+1}\,\tau \square \widetilde{f}^{-2} - \frac{\tau}{2\sqrt{\tau+1}}\widetilde{f}^{-2} \\
&\leq \sqrt{\tau+1}\left(-\widetilde{f}^{-2} + 2C_1\sqrt{\tau+1}\,\widetilde{f}^{-3}\right) - \frac{\tau}{2\sqrt{\tau+1}}\widetilde{f}^{-2} \\
&\leq -\sqrt{\tau+1}\,\widetilde{f}^{-2} + 2C_1(\tau+1)\widetilde{f}^{-3}.
\end{aligned}$$

By combining the calculations above, we obtain

$$\begin{aligned}
&\tau \square \big(\widetilde{f}^{-1} + 4C_1\sqrt{\tau+1}\,\widetilde{f}^{-2}\big) \\
&\quad\leq -\frac{1}{2}\widetilde{f}^{-1} + C_1\sqrt{\tau+1}\,\widetilde{f}^{-2} - 4C_1\sqrt{\tau+1}\,\widetilde{f}^{-2} + 8C_1^2(\tau+1)\widetilde{f}^{-3} \\
&\quad= -\frac{1}{2}\big(\widetilde{f}^{-1} + 4C_1\sqrt{\tau+1}\,\widetilde{f}^{-2}\big) - C_1\sqrt{\tau+1}\,\widetilde{f}^{-3}\big(\widetilde{f} - 8C_1\sqrt{\tau+1}\big).
\end{aligned} \tag{6.221}$$

Moreover, the standard formula $\square R = 2|\mathrm{Ric}|^2$ yields

$$\begin{aligned}
\tau \square \big(\sqrt{\tau+1}\,R\big) &= 2\tau\sqrt{\tau+1}\,|\mathrm{Ric}|^2 - \frac{\tau}{2\sqrt{\tau+1}}R \\
&\geq -\frac{1}{2}\sqrt{\tau+1}\,R.
\end{aligned}$$

Consider the quantity

$$\phi := \sqrt{\tau+1}\,R - c\big(\widetilde{f}^{-1} + 4C_1\sqrt{\tau+1}\,\widetilde{f}^{-2}\big), \tag{6.222}$$

where c is a suitable positive constant to be determined below. By (6.219), we have that

(6.223)
$$\begin{aligned}\phi &\geq \sqrt{\tau+1}R - \frac{c}{\sqrt{\tau+1}}\left(\left(\frac{d_t(x,z_t)^2}{9\tau}+1\right)^{-1} + 4C_1\left(\frac{d_t(x,z_t)^2}{9\tau}+1\right)^{-2}\right)\\ &\geq \sqrt{\tau+1}R - \frac{c(1+4C_1)}{\sqrt{\tau+1}}\left(\frac{d_t(x,z_t)^2}{9\tau}+1\right)^{-1}\\ &\geq \sqrt{\tau+1}R - \frac{c(1+4C_1)}{\sqrt{\tau+1}}.\end{aligned}$$

By the calculations above, we obtain

$$\begin{aligned}\tau\Box\phi &\geq -\frac{1}{2}\sqrt{\tau+1}R + \frac{c}{2}(\tilde{f}^{-1} + 4C_1\sqrt{\tau+1}\tilde{f}^{-2})\\ &\quad + cC_1\sqrt{\tau+1}\tilde{f}^{-3}(\tilde{f} - 8C_1\sqrt{\tau+1})\\ &= -\frac{1}{2}\phi + cC_1\sqrt{\tau+1}\tilde{f}^{-3}(\tilde{f} - 8C_1\sqrt{\tau+1}).\end{aligned} \tag{6.224}$$

Suppose, for a contradiction, that $\inf_{\mathcal{O}^n\times(-\infty,0)}\phi < 0$. By the properties that we will prove below regarding the limiting behavior of $\phi(x,t)$ as $x\to\infty$, $t\to-\infty$, or $t\to 0_-$, we would from this assumption obtain the following:

CLAIM 6.89. *If ϕ is not non-negative everywhere on $\mathcal{O}^n\times(-\infty,0)$, then the negative infimum of ϕ is attained at some point $(x_1,t_1)\in\mathcal{O}^n\times(-\infty,0)$.*

Assuming this claim and by applying the maximum principle to the parabolic inequality (6.224), we obtain that

$$\begin{aligned}0 &\geq |t_1|\Box\phi(x_1,t_1)\\ &\geq -\frac{1}{2}\phi(x_1,t_1) + cC_1\sqrt{|t_1|+1}\tilde{f}^{-3}(x_1,t_1)\left(\tilde{f}(x_1,t_1) - 8C_1\sqrt{|t_1|+1}\right).\end{aligned}$$

Thus, by this and (6.219), we have that

$$8C_1\sqrt{|t_1|+1} > \tilde{f}(x_1,t_1) \geq \sqrt{|t_1|+1}\left(\frac{d_{t_1}(x_1,z_{t_1})^2}{9|t_1|}+1\right) \geq \sqrt{|t_1|+1}. \tag{6.225}$$

This implies that the minimum point (x_1,t_1) of ϕ satisfies

$$d_{t_1}(x_1,z_{t_1})^2 < 72C_1|t_1|. \tag{6.226}$$

That is, x_1 has, after rescaling, bounded distance from the H_n-centers at time t_1.

Now, since $a_0(9\sqrt{C_1}) > 0$ and since $x_1 \in B_{g(t_1)}(z_{t_1}, 9\sqrt{C_1}\sqrt{|t_1|+1})$ by (6.226), we have by the Property R assumption (6.209) that

$$(|t_1|+1)R(x_1,t_1) \geq a_0(9\sqrt{C_1}). \tag{6.227}$$

Thus, by the definition (6.222) of ϕ and by the lower bound for $\widetilde{f}(x_1,t_1)$ given by (6.225), we conclude that

$$\begin{aligned}\phi(x_1,t_1) &= \sqrt{|t_1|+1}\,R(x_1,t_1) - c\left(\widetilde{f}^{-1}(x_1,t_1) + 4C_1\sqrt{|t_1|+1}\,\widetilde{f}^{-2}(x_1,t_1)\right)\\ &\geq \frac{a_0(9\sqrt{C_1})}{\sqrt{|t_1|+1}} - \frac{c\,(1+4C_1)}{\sqrt{|t_1|+1}}.\end{aligned}$$

Since $\phi(x_1,t_1) < 0$, we obtain a contradiction provided $c > 0$ is chosen so that $c < \frac{a_0(9\sqrt{C_1})}{1+4C_1}$. This proves that ϕ is non-negative and hence proves the theorem modulo Claim 6.89.

We now proceed to prove Claim 6.89. Assume that the ancient solution has bounded curvature on compact time intervals. In order to apply the parabolic maximum principle, we analyze the asymptotic behavior of the function ϕ as time approaches zero, as time tends to negative infinity, and as the distance from the point x to the base point o becomes unbounded.

SUBCLAIM 1.

$$\liminf_{t\to 0_-}\left(\inf_{x\in\mathcal{O}^n}\phi(x,t)\right) \geq 0. \tag{6.228}$$

PROOF OF SUBCLAIM 1. Let $t\in(-\infty,0)$ and let $x\in\mathcal{O}^n$.

CASE 1: *Suppose* $x\in B_t(z_t, D\sqrt{\tau+1})$. Fix any $D>0$. By the Property R assumption (6.209), we have that

$$\phi(x,t) \geq \frac{a_0(D)}{\sqrt{\tau+1}} - \frac{c(1+4C_1)}{\sqrt{\tau+1}}.$$

Thus, $\phi(x,t)\geq 0$ for all $x\in B_t(z_t, D\sqrt{\tau+1})$ provided $c>0$ is chosen so that $c\leq\frac{a_0(D)}{1+4C_1}$.

CASE 2: *Suppose* $x\notin B_t(z_t, D\sqrt{\tau+1})$. Then, by (6.219), we have that

$$\widetilde{f}(x,t) \geq \sqrt{\tau+1}\left(\frac{d_t(x,z_t)^2}{9\tau}+1\right) \geq \sqrt{\tau+1}\left(\frac{D^2(\tau+1)}{9\tau}+1\right) \geq \frac{D^2}{9\tau}.$$

Thus, by the definition of ϕ, we have that

$$\begin{aligned}\phi(x,t) &\geq -c\big(\widetilde{f}^{-1}(x,t) + 4C_1\sqrt{\tau+1}\,\widetilde{f}^{-2}(x,t)\big)\\ &\geq -c\left(\frac{9\tau}{D^2} + 4C_1\sqrt{\tau+1}\,\frac{81\tau^2}{D^4}\right).\end{aligned}$$

Subclaim 1 now follows directly from Cases 1 and 2. ○

SUBCLAIM 2.

$$\liminf_{t\to-\infty}\left(\inf_{x\in\mathcal{O}^n}\phi(x,t)\right)\geq 0. \tag{6.229}$$

PROOF OF SUBCLAIM 2. By (6.219), we have that

$$\widetilde{f}(x,t)\geq\sqrt{\tau+1}$$

for all $x\in\mathcal{O}^n$ and $t\in(-\infty,0)$. Since $R\geq 0$, we thus obtain

$$\phi(x,t)\geq -c\left(\frac{1}{\sqrt{\tau+1}}+4C_1\sqrt{\tau+1}\frac{1}{(\sqrt{\tau+1})^2}\right)=-\frac{c(1+4C_1)}{\sqrt{\tau+1}}.$$

Subclaim 2 follows directly from this inequality. ○

SUBCLAIM 3. *For each $\varepsilon>0$, we have that*

$$\liminf_{x\to\infty}\phi(x,t)\geq 0 \tag{6.230}$$

uniformly for $t\in[-\varepsilon^{-1},-\varepsilon]$. Equivalently, for each $\varepsilon>0$, there exists a positive constant $r(\varepsilon)$ such that

$$\phi(x,t)\geq-\varepsilon\quad\textit{for all } x\in\mathcal{O}^n\setminus B(x_0,r(\varepsilon)),\ t\in[-\varepsilon^{-1},-\varepsilon]. \tag{6.231}$$

PROOF OF SUBCLAIM 3. Let $\varepsilon>0$. There exists a constant $C(\varepsilon)$ such that

$$d_t(z_t,x_0)\leq C(\varepsilon) \tag{6.232}$$

for all H_n-centers (z_t,t) with $t\in[-\varepsilon^{-1},-\varepsilon]$. This is because we can estimate the distance of an ℓ-center to a fixed point and in turn we can estimate the distance between H_n and ℓ-centers; see §6.6 below for the details. Thus, by (6.219), there exists a constant $C'(\varepsilon)$ such that

$$\widetilde{f}(x,t)\geq\sqrt{\tau+1}\left(\frac{d_t(x,z_t)^2}{9\tau}+1\right)\geq\varepsilon^{-1} \tag{6.233}$$

for all $x\in\mathcal{O}^n\setminus B(x_0,C'(\varepsilon))$ and $t\in[-\varepsilon^{-1},-\varepsilon]$. From this we obtain that there exists a constant $C''(\varepsilon)$ such that

$$\phi(x,t)\geq-\varepsilon \tag{6.234}$$

for all $x\in\mathcal{O}^n\setminus B(x_0,C''(\varepsilon))$ and $t\in[-\varepsilon^{-1},-\varepsilon]$, proving Subclaim 3. ○

Finally, by Subclaims 1–3, if ϕ is negative somewhere, then it attains its infimum. Consequently, we can apply the maximum principle argument as carried out above to obtain a contradiction provided $c>0$ is chosen sufficiently small. The theorem is proved. □

6.6. Volume estimates for steady soliton singularity models

In this section, we discuss sharp volume growth estimates for noncollapsed steady gradient Ricci soliton singularity models in all dimensions, following the work of Bamler, Chan, Ma, and Zhang [**BCMZ**].

6.6.1. Statement of the theorem.

Recall that any steady gradient Ricci soliton must have at least linear volume growth by Theorem 5.2. Additionally, by Theorem 5.1, any steady gradient Ricci soliton has at most exponential square root volume growth.

In the case where the steady gradient Ricci soliton is also a singularity model, one can improve the volume estimates to the following qualitatively sharp result which holds in any dimension:

THEOREM 6.90. *Let (M^n, g, f) be a steady gradient Ricci soliton singularity model. Then there exist positive constants $c(n, \mu_\infty)$, $C(n)$, and $r_0(n, \mu_\infty)$ such that for any $o \in M^n$,*

$$c(n, \mu_\infty) r^{\frac{n+1}{2}} \leq \operatorname{Vol} B(o, r) \leq C(n) r^n, \tag{6.235}$$

where the lower bound holds for all $r \geq r_0(n, \mu_\infty)$ and the upper bound holds for all $r > 0$.

The upper volume bound is a special case of Proposition 6.65, which is Bamler's non-inflating result for singularity models. In this section we discuss the proof of the lower volume bound. The upper and lower estimates are qualitatively sharp on the Gaussian steady soliton (and Ricci-flat ALE spaces when in dimension 4) and the Bryant steady soliton, respectively. The key ingredients of the proof are Perelman's $\mathcal{L}$-geometry and the breakthrough by Bamler [**Bam20a**] on conjugate heat kernel geometry, including the pointed Nash entropy and conjugate heat kernel bounds as well as his improved noncollapsing result for Ricci flow.

In more detail, the idea of the proof is to first establish the existence of points that are "ℓ-central" at scale τ, and that are also at distance τ from a fixed base point.[21] On the other hand, by Bamler's no local collapsing theorem, at scale τ, we have the noncollapsing at radius $\sqrt{\tau}$ of balls centered at H_n centers. By using estimates for the conjugate heat kernel, we can bound the distance between an ℓ-central point and an H_n-center. Combining all of this, as well as stringing disjoint balls together, yields the lower volume estimate.

6.6.2. Perelman's reduced distance viewed as a function on a time-slice.

Let (M^n, g, f) be a (static) steady gradient Ricci soliton singularity model, as in the hypothesis of Theorem 6.90. Let $(M^n, g(t), f(t))$, $t \in (-\infty, \infty)$, be its associated (dynamic) canonical form as given by Proposition 1.73, which is a solution to the Ricci flow.[22] We may consider the reduced distance ℓ (defined in §6.1.1.4) for this Ricci flow as a family of functions on the static steady gradient Ricci soliton (M^n, g, f), parametrized by τ, as follows:

By the definition of the canonical form, we have that

$$g(t) = \varphi_t^* g(0) = \varphi_t^* g, \quad \text{so that} \quad g = \varphi_{-t}^* g(t), \tag{6.236}$$

where the 1-parameter group of diffeomorphisms $\{\varphi_t\}_{t \in \mathbb{R}}$ of M^n satisfies the ODE $\partial_t \varphi_t(x) = \nabla_g f(\varphi_t(x))$ with the initial condition $\varphi_0 = \mathrm{id}$. In view of this, Perelman's

[21] For convenience, using that we are on a steady gradient Ricci soliton, we consider the static equivalent of Perelman's ℓ-geometry.

[22] For clarification, the steady gradient Ricci soliton singularity model (M^n, g, f), by definition, corresponds to a time-slice of a Ricci flow singularity model. This model, as a limit of Ricci flows, possesses the property that it is diffeomorphism-equivalent to its associated canonical form.

$\mathcal{L}$-geometry on the space-time of the canonical form $(M^n, g(t))$ may be recast as being on (M^n, g), i.e., on the time-slice $t = 0$.

To this end, associated to a path $\gamma : [0, \overline{\tau}] \to M^n$, define $\beta : [0, \overline{\tau}] \to M^n$ by

$$\beta(\tau) = \varphi_{-\tau}(\gamma(\tau)). \tag{6.237}$$

By differentiating this formula, we have

$$\beta'(\tau) = d\varphi_{-\tau}(\gamma'(\tau)) - (\nabla_g f)(\beta(\tau)). \tag{6.238}$$

Thus, we may rewrite the $\mathcal{L}$-length of γ (see (6.18)) in terms of β as

$$\begin{aligned}\mathcal{L}(\gamma) &= \int_0^{\overline{\tau}} \sqrt{\tau}\big(R_{g(-\tau)}(\gamma(\tau)) + |\gamma'(\tau)|^2_{g(-\tau)}\big)d\tau \\ &= \int_0^{\overline{\tau}} \sqrt{\tau}\big(R_{\varphi_\tau^* g(-\tau)}\big(\varphi_{-\tau}(\gamma(\tau))\big) + \big|d\varphi_{-\tau}(\gamma'(\tau))\big|^2_{\varphi_\tau^* g(-\tau)}\big)d\tau \\ &= \int_0^{\overline{\tau}} \sqrt{\tau}\big(R_g(\beta(\tau)) + \big|\beta'(\tau) + (\nabla_g f)(\beta(\tau))\big|^2_g\big)d\tau \\ &=: \overline{\mathcal{L}}(\beta).\end{aligned} \tag{6.239}$$

We call $\overline{\mathcal{L}}(\beta)$ defined in this way the **$\overline{\mathcal{L}}$-length** of β. Notice that this $\overline{\mathcal{L}}$-length is defined using the static steady gradient Ricci soliton structure (M^n, g, f) and is equivalent to Perelman's $\mathcal{L}$-length by (6.239).

Given a point $y \in M^n$ and a scale $\overline{\tau} > 0$, define the **$\overline{L}$-distance** (to $o \in M^n$) by[23]

$$\overline{L}(y, \overline{\tau}) := \inf_\beta \overline{\mathcal{L}}(\beta), \tag{6.240}$$

where the infimum is taken over all $\beta : [0, \overline{\tau}] \to M^n$ satisfying $\beta(0) = o$ and $\beta(\overline{\tau}) = y$. If a path β minimizes the $\overline{\mathcal{L}}$-length among all paths having the same pair of endpoints, then β is a **minimal $\overline{\mathcal{L}}$-geodesic**.

The **static reduced distance** is defined by

$$\overline{\ell}(y, \overline{\tau}) := \frac{1}{2\sqrt{\overline{\tau}}}\overline{L}(y, \overline{\tau}). \tag{6.241}$$

Observe that this is related to Perelman's reduced distance ℓ by

$$\overline{\ell}(y, \overline{\tau}) = \ell\big(\varphi_{\overline{\tau}}(y), \overline{\tau}\big), \tag{6.242}$$

where the equality follows from (6.239); note that $\beta(\overline{\tau}) = y$ implies that $\gamma(\overline{\tau}) = \varphi_{\overline{\tau}}(y)$.

By (6.20), for any $\overline{\tau} > 0$, there exists a point $p_{\overline{\tau}} \in M^n$ such that

$$\overline{\ell}(p_{\overline{\tau}}, \overline{\tau}) = \ell\big(\varphi_{\overline{\tau}}(p_{\overline{\tau}}), \overline{\tau}\big) \le \frac{n}{2}. \tag{6.243}$$

The point $p_{\overline{\tau}}$ is called a **static $\overline{\ell}$-center** at time $-\overline{\tau}$. This corresponds to Perelman's ℓ-center $q_{\overline{\tau}} := \varphi_{\overline{\tau}}(p_{\overline{\tau}})$ at scale $\overline{\tau}$ for the space-time of the Ricci flow.

---

[23] This is a different quantity from Perelman's $\overline{L}$-distance.

Since we are on a steady gradient Ricci soliton, we may express $\overline{\mathcal{L}}$ as

$$\begin{aligned}(6.244)\quad \overline{\mathcal{L}}(\beta) &= \int_0^{\overline{\tau}} \sqrt{\tau}\big((R_g + |\nabla_g f|_g^2)(\beta(\tau)) + |\beta'(\tau)|_g^2 + 2(\nabla_g f)(\beta(\tau)) \cdot \beta'(\tau)\big)d\tau \\ &= \frac{2}{3}\overline{\tau}^{\frac{3}{2}} + \int_0^{\overline{\tau}} \sqrt{\tau}\big(|\beta'(\tau)|_g^2 + 2(f \circ \beta)'(\tau)\big)d\tau,\end{aligned}$$

where we used the steady gradient Ricci soliton equation $R + |\nabla f|^2 = 1$.

We may now prove the following lower bound for the static reduced distance $\overline{\ell}$ at the points $(o, \overline{\tau})$, which form the "traditional" world-line of the base point $(o, 0)$. Observe that this lower bound is linear in the scale $\overline{\tau}$.

LEMMA 6.91. *Let (M^n, g, f) be a steady gradient Ricci soliton and let $o \in M^n$. For any $\overline{\tau} > 0$, the static reduced distance at the base point o for the scale $\overline{\tau}$ satisfies the estimate*

$$(6.245)\qquad \overline{\ell}(o, \overline{\tau}) \geq \frac{1}{12}\overline{\tau}.$$

PROOF. Let $\beta : [0, \overline{\tau}] \to M^n$ satisfy $\beta(\overline{\tau}) = \beta(0) = o$. By integrating by parts on (6.244), we obtain

$$\begin{aligned}(6.246)\qquad \overline{\mathcal{L}}(\beta) &= \frac{2}{3}\overline{\tau}^{\frac{3}{2}} + \int_0^{\overline{\tau}} \sqrt{\tau}\big(|\beta'(\tau)|_g^2 + 2(f \circ \beta - f(o))'(\tau)\big)d\tau \\ &= \frac{2}{3}\overline{\tau}^{\frac{3}{2}} + \int_0^{\overline{\tau}} \sqrt{\tau}\left(|\beta'(\tau)|_g^2 - \frac{1}{\tau}\big(f(\beta(\tau)) - f(o)\big)\right)d\tau.\end{aligned}$$

Define

$$(6.247)\qquad \mathrm{D} := \sup_{\tau \in [0, \overline{\tau}]} d_g(o, \beta(\tau)).$$

Choose $\tau_1 \in (0, \overline{\tau}]$ so that

$$(6.248)\qquad d_g(o, \beta(\tau_1)) = \mathrm{D}.$$

Regarding the integral terms on the right-hand side of (6.246), we compute as follows:

Firstly, we have that (without loss of generality, we may assume that $\tau_1 < \overline{\tau}$)

$$\begin{aligned}\int_0^{\overline{\tau}} \sqrt{\tau}\,|\beta'(\tau)|_g^2 d\tau &= \int_0^{\tau_1} \sqrt{\tau}\,|\beta'(\tau)|_g^2 d\tau + \int_{\tau_1}^{\overline{\tau}} \sqrt{\tau}\,|\beta'(\tau)|_g^2 d\tau \\ &\geq \frac{1}{2\sqrt{\tau_1}}\left(\int_0^{\tau_1} |\beta'(\tau)|_g d\tau\right)^2 + \frac{1}{2(\sqrt{\overline{\tau}} - \sqrt{\tau_1})}\left(\int_{\tau_1}^{\overline{\tau}} |\beta'(\tau)|_g d\tau\right)^2 \\ &\geq \frac{\mathrm{D}^2}{2}\left(\frac{1}{\sqrt{\tau_1}} + \frac{1}{\sqrt{\overline{\tau}} - \sqrt{\tau_1}}\right) \\ &\geq \frac{2\,\mathrm{D}^2}{\sqrt{\overline{\tau}}},\end{aligned}$$

where for the last inequality we used that $\frac{1}{a} + \frac{1}{b} \geq \frac{4}{a+b}$ for $a, b > 0$.

Secondly, since $|\nabla f| \leq 1$, we have that

$$-\int_0^{\overline{\tau}} \frac{1}{\sqrt{\tau}}\big(f(\beta(\tau)) - f(o)\big)d\tau \geq -\int_0^{\overline{\tau}} \frac{1}{\sqrt{\tau}} d_g(o, \beta(\tau))\,d\tau \geq -2\sqrt{\overline{\tau}}\,\mathrm{D}.$$

Therefore, (6.246) implies that

$$\overline{\mathcal{L}}(\beta) \geq \frac{2}{3}\overline{\tau}^{\frac{3}{2}} + \frac{2\,\mathrm{D}^2}{\overline{\tau}^{\frac{1}{2}}} - 2D\overline{\tau}^{\frac{1}{2}} \geq \frac{1}{6}\overline{\tau}^{\frac{1}{2}}. \tag{6.249}$$

The lemma now follows from taking the infimum over β and from the definition of $\overline{\ell}$. □

The fact that for any static $\overline{\ell}$-center $p_{\overline{\tau}}$, $\overline{\ell}(p_{\overline{\tau}},\overline{\tau}) \leq \frac{n}{2}$, while $\overline{\ell}(o,\overline{\tau}) \geq \frac{1}{12}\overline{\tau}$, suggests that the point o is not close to $p_{\overline{\tau}}$, at least for large $\overline{\tau}$. To show this, we first compare the static reduced distance $\overline{\ell}$ at different points in space-time.

LEMMA 6.92. *For any points $x_1, x_2 \in M^n$, $\overline{\tau} > 0$, and $\delta \in (0,1)$, we have that*

$$\overline{\ell}\big(x_2,(1+\delta)^2\overline{\tau}\big) \leq \overline{\ell}(x_1,\overline{\tau}) + 3\delta\overline{\tau} + \frac{d_g(x_1,x_2)^2}{2\delta\overline{\tau}}.$$

PROOF. Define the path $\beta : [0,(1+\delta)^2\overline{\tau}] \to M^n$ as the concatenation of the following two paths β_1 and β_2. The path $\beta_1 : [0,\overline{\tau}] \to M^n$ is a minimal $\overline{\mathcal{L}}$-geodesic from o to x_1. The path $\beta_2 : [\overline{\tau},(1+\delta)^2\overline{\tau}] \to M^n$ is given by $\beta_2(\tau) = \alpha(2\sqrt{\tau})$, where $\alpha : \big[2\sqrt{\overline{\tau}}, 2(1+\delta)\sqrt{\overline{\tau}}\big] \to M^n$ is a constant speed minimal geodesic with respect to the metric g satisfying $\alpha\big(2\sqrt{\overline{\tau}}\big) = x_1$ and $\alpha\big(2(1+\delta)\sqrt{\overline{\tau}}\big) = x_2$.

Using β as a test path, we estimate the $\overline{L}$-distance by:

$$\begin{aligned}
\overline{L}\big(x_2,&(1+\delta)^2\overline{\tau}\big) \leq \overline{\mathcal{L}}(\beta) \\
&= \int_0^{\overline{\tau}} \sqrt{\tau}\big(R_g(\beta_1(\tau)) + \big|\beta_1'(\tau) + (\nabla_g f)(\beta_1(\tau))\big|_g^2\big)d\tau \\
&\quad + \int_{\overline{\tau}}^{(1+\delta)^2\overline{\tau}} \sqrt{\tau}\big(R_g(\beta_2(\tau)) + \big|\beta_2'(\tau) + (\nabla_g f)(\beta_2(\tau))\big|_g^2\big)d\tau \\
&= \int_{\overline{\tau}}^{(1+\delta)^2\overline{\tau}} \sqrt{\tau}\big((1 + |\beta_2'(\tau)|_g^2 + 2(\nabla_g f)(\beta_2(\tau))\cdot\beta_2'(\tau)\big)d\tau + \overline{L}(x_1,\overline{\tau}),
\end{aligned} \tag{6.250}$$

where we used (6.244) for the last equality. Now we estimate the right-hand side of (6.250) in terms of the Riemannian distance of g. Let $\sigma = 2\sqrt{\tau}$. We compute that

$$\begin{aligned}
\int_{\overline{\tau}}^{(1+\delta)^2\overline{\tau}} &\sqrt{\tau}\big((1 + |\beta_2'(\tau)|_g^2 + 2(\nabla_g f)(\beta_2(\tau))\cdot\beta_2'(\tau)\big)d\tau \\
&\leq \int_{\overline{\tau}}^{(1+\delta)^2\overline{\tau}} \sqrt{\tau}\big(2 + 2|\beta_2'(\tau)|_g^2\big)\,d\tau \\
&\leq \int_{2\sqrt{\overline{\tau}}}^{2(1+\delta)\sqrt{\overline{\tau}}} \left(\frac{1}{2}\sigma^2 + 2|\alpha'(\sigma)|_g^2\right)d\sigma \\
&= \frac{4}{3}\left((1+\delta)^3 - 1\right)\overline{\tau}^{\frac{3}{2}} + \frac{d_g(x_1,x_2)^2}{\delta\sqrt{\overline{\tau}}},
\end{aligned}$$

where we have used $|\nabla_g f|^2 \leq 1$ for the first inequality and where we used $d\tau = \frac{1}{2}\sigma d\sigma$. By combining the above and dividing by $2(1+\delta)\sqrt{\overline{\tau}}$, we obtain

$$\overline{\ell}\big(x_2,(1+\delta)^2\overline{\tau}\big) \leq \frac{1}{1+\delta}\overline{\ell}(x_1,\overline{\tau}) + \frac{2}{3}\frac{(1+\delta)^3-1}{1+\delta}\overline{\tau} + \frac{d_g(x_1,x_2)^2}{2(1+\delta)\delta\overline{\tau}}. \tag{6.251}$$

Since $\delta \in (0,1)$, we have that

$$u(\delta) := \frac{(1+\delta)^3 - 1}{1+\delta} = (1+\delta)^2 - \frac{1}{1+\delta} \leq (1+3\delta) - (1-\delta) \leq 4\delta.$$

Therefore,

$$\overline{\ell}(x_2, (1+\delta)^2\overline{\tau}) \leq \overline{\ell}(x_1, \overline{\tau}) + \frac{8}{3}\delta\overline{\tau} + \frac{d_g(x_1,x_2)^2}{2\delta\overline{\tau}},$$

where we used that $\overline{\ell} \geq 0$ since $R \geq 0$. □

By combining the previous two lemmas, we can obtain a linearly growing lower estimate for the distance to o of any static $\overline{\ell}$-center.

PROPOSITION 6.93 (Lower bound for the distance of any static $\overline{\ell}$-center). *For any static $\overline{\ell}$-center $p_{\overline{\tau}}$, where $\overline{\tau} \geq 20n$, we have that*

$$d_g(p_{\overline{\tau}}, o) \geq \frac{\sqrt{5}}{96}\overline{\tau} \geq \frac{1}{49}\overline{\tau}. \tag{6.252}$$

PROOF. By Lemma 6.92,

$$\overline{\ell}\big(o, (1+\delta)^2\overline{\tau}\big) \leq \overline{\ell}(p_{\overline{\tau}}, \overline{\tau}) + 3\delta\overline{\tau} + \frac{d_g(p_{\overline{\tau}},o)^2}{2\delta\overline{\tau}}.$$

By combining this with Lemma 6.91 and using the definition of a static $\overline{\ell}$-center, we obtain

$$\frac{1}{12}\overline{\tau} \leq \frac{1}{12}(1+\delta)^2\overline{\tau} \leq \overline{\ell}\big(o, (1+\delta)^2\overline{\tau}\big) \leq \frac{n}{2} + 3\delta\overline{\tau} + \frac{d_g(p_{\overline{\tau}},o)^2}{2\delta\overline{\tau}}.$$

Thus, by taking $\delta = \frac{1}{96}$ and since $\overline{\tau} \geq 20n \geq \frac{96n}{5}$, we obtain

$$\frac{48 d_g(p_{\overline{\tau}},o)^2}{\overline{\tau}} \geq \frac{5}{96}\overline{\tau} - \frac{n}{2} \geq \frac{5}{192}\overline{\tau},$$

which is equivalent to inequality (6.252). □

We next show that there exists a point $x_{\overline{\tau}}$ at the distance $\overline{\tau}$ from o with $\overline{\ell}$-value at some scale comparable to $\overline{\tau}$ bounded by a dimensional constant. We will call $x_{\overline{\tau}}$ an **$\overline{\ell}$-central point**.

LEMMA 6.94 (Existence of an $\overline{\ell}$-central point $x_{\overline{\tau}}$ at distance $\overline{\tau}$). *For any $\overline{\tau} \geq 392n$, there exists a point $x_{\overline{\tau}} \in M^n$ and a $\tau_0 \in \left[\frac{1}{16}\overline{\tau}, 49\overline{\tau}\right]$ satisfying*

$$d_g(x_{\overline{\tau}}, o) = \overline{\tau} \quad \text{and} \quad \overline{\ell}(x_{\overline{\tau}}, \tau_0) \leq 14n. \tag{6.253}$$

Note that τ_0 is "comparable" to $\overline{\tau}$.

PROOF. Let $\overline{\tau} \geq 392n$ and let $\gamma : [0, 49\overline{\tau}] \to M^n$ be a minimal $\overline{\mathcal{L}}$-geodesic from o to a static $\overline{\ell}$-center $p_{49\overline{\tau}}$, so that

$$\overline{\mathcal{L}}(\gamma) = \overline{L}(p_{49\overline{\tau}}, 49\overline{\tau}). \tag{6.254}$$

By Proposition 6.93,

$$d_g(p_{49\overline{\tau}}, o) \geq \overline{\tau}. \tag{6.255}$$

This allows us to define

$$\tau_0 := \sup\left\{\tau \in [0, 49\overline{\tau}] : d_g(\gamma(\tau), o) \leq \overline{\tau}\right\} \tag{6.256}$$

and define

$$x_{\overline{\tau}} := \gamma(\tau_0). \tag{6.257}$$

CLAIM 6.95. *We have that*

$$\tau_0 \geq \frac{1}{16}\overline{\tau}. \tag{6.258}$$

PROOF OF CLAIM 6.95. Define $\alpha : \left[0, 14\sqrt{\overline{\tau}}\right] \to M^n$ by $\gamma(\tau) =: \alpha(2\sqrt{\tau})$. Then $\alpha(0) = o$ and $\alpha\left(14\sqrt{\overline{\tau}}\right) = \gamma(49\overline{\tau}) = p_{49\overline{\tau}}$. By (6.244) and $R \geq 0$,

$$\overline{L}(p_{49\overline{\tau}}, 49\overline{\tau}) = \overline{\mathcal{L}}(\gamma) \geq \int_0^{\tau_0} \sqrt{\tau}\big(1 + |\gamma'(\tau)|_g^2 + 2(\nabla_g f)(\gamma(\tau)) \cdot \gamma'(\tau)\big)d\tau.$$

Thus, we have that

$$\begin{aligned}\int_0^{2\sqrt{\tau_0}} |\alpha'(\sigma)|_g^2 d\sigma &\leq \overline{L}(p_{49\overline{\tau}}, 49\overline{\tau}) + \int_0^{2\sqrt{\tau_0}} \langle \alpha'(\sigma), (\nabla f)_{\alpha(\sigma)} \rangle \sigma \, d\sigma \\ &\leq 2\sqrt{49\overline{\tau}}\,\frac{n}{2} + \frac{1}{2}\int_0^{2\sqrt{\tau_0}} |\alpha'(\sigma)|_g^2 d\sigma + 2\int_0^{2\sqrt{\tau_0}} |\nabla f|_g^2 \sigma^2 \, d\sigma,\end{aligned}$$

and so

$$\begin{aligned}\int_0^{2\sqrt{\tau_0}} |\alpha'(\sigma)|_g^2 d\sigma &\leq 14n\sqrt{\overline{\tau}} + 4\int_0^{2\sqrt{\tau_0}} \sigma^2 \, d\sigma \\ &= 14n\sqrt{\overline{\tau}} + \frac{32}{3}\tau_0^{\frac{3}{2}},\end{aligned}$$

where we have used $|\nabla f|_g^2 \leq 1$.

On the other hand,

$$\begin{aligned}\int_0^{2\sqrt{\tau_0}} |\alpha'(\sigma)|_g^2 d\sigma &\geq \frac{1}{2\sqrt{\tau_0}}\left(\int_0^{2\sqrt{\tau_0}} |\alpha'(\sigma)|_g \, d\sigma\right)^2 \\ &\geq \frac{1}{2\sqrt{\tau_0}} d_g(p_{49\overline{\tau}}, o)^2 \\ &\geq \frac{\overline{\tau}^2}{2\sqrt{\tau_0}},\end{aligned}$$

where the last inequality follows from (6.255). By combining the two displayed equations above, we obtain

$$\begin{aligned}\overline{\tau}^2 &\leq 28n(\overline{\tau}\tau_0)^{\frac{1}{2}} + \frac{64}{3}\tau_0^2 \\ &\leq 196n\overline{\tau} + \frac{64}{3}\tau_0^2 \\ &\leq \frac{1}{2}\overline{\tau}^2 + \frac{64}{3}\tau_0^2,\end{aligned}$$

where we used that $\tau_0^{\frac{1}{2}} \leq 7\overline{\tau}^{\frac{1}{2}}$ and since $\overline{\tau} \geq 392n$. Thus,

$$\overline{\tau} \leq \left(\frac{128}{3}\right)^{\frac{1}{2}} \tau_0 \leq 16\tau_0. \tag{6.259}$$

This proves Claim 6.95.

Finally, we conclude that

$$\overline{\ell}(x_{\overline{\tau}}, \tau_0) = \overline{\ell}(\gamma(\tau_0), \tau_0) \leq \left(\frac{49\overline{\tau}}{\tau_0}\right)^{\frac{1}{2}} \overline{\ell}(p_{49\overline{\tau}}, 49\overline{\tau}) \leq 14n, \tag{6.260}$$

where we used that $\overline{\ell}(p_{49\overline{\tau}}, 49\overline{\tau}) \leq \frac{n}{2}$ and where we used $R \geq 0$ for the first inequality in the displayed equation. This proves the lemma. □

6.6.3. Lower bounds for the volumes of balls.

The following result says that for any τ greater than a dimensional constant, there exists a point x_τ at distance τ from o such that a ball centered at x_τ of radius comparable to $\sqrt{\tau}$ is noncollapsed. Qualitatively, this is the same as the radius at which the Bryant steady soliton is noncollapsed.

PROPOSITION 6.96 (Volumes of balls at distance τ and radius $\sqrt{\tau}$ in steady solitons). *Let (M^n, g, f) be a steady gradient Ricci soliton singularity model. Then there exist positive constants $c(n)$ and $C(n)$ such that for any $\tau \geq 392n$, there exists a point $x_\tau \in M^n$ such that $d_g(x_\tau, o) = \tau$ and*

$$\operatorname{Vol}_g B\left(x_\tau, \sqrt{A\tau}\right) \geq c(n) \mathrm{e}^{\mu_\infty} \tau^{\frac{n}{2}}, \tag{6.261}$$

where $A := 196H_n + 882\big(C(n) - \mu_\infty\big)$ and μ_∞ is the asymptotic pointed Nash entropy defined by (6.65).

PROOF. Let $\tau \geq 392n$ and let x_τ and $\tau_0 \in \left[\frac{1}{16}\tau, 49\tau\right]$ be as in Lemma 6.94, with $\overline{\tau} := \tau$ therein. Then

$$d_g(x_\tau, o) = \tau \quad \text{and} \quad \overline{\ell}(x_\tau, \tau_0) \leq 14n. \tag{6.262}$$

Let A be a positive constant to be chosen later. Since $\tau_0 \leq 49\tau$ and $g_{-\tau_0} = \varphi^*_{-\tau_0} g$, we have that

$$\begin{aligned} \operatorname{Vol}_g B_g\left(x_\tau, \sqrt{A\tau}\right) &\geq \operatorname{Vol}_g B_g\left(x_\tau, \sqrt{A\tau_0/49}\right) \\ &= \operatorname{Vol}_{g_{-\tau_0}} B_{g_{-\tau_0}}\left(\varphi_{\tau_0}(x_\tau), \sqrt{A\tau_0/49}\right), \end{aligned} \tag{6.263}$$

where the equality follows from being on a steady gradient Ricci soliton.

Choose an H_n-center $(z_{-\tau_0}, -\tau_0)$ of $(o, 0)$. We proceed to bound the volume on the right-hand side of (6.263) from below using Bamler's no local collapsing result and estimating the distance between $\varphi_{\tau_0}(x_\tau)$ and the H_n-center $z_{-\tau_0}$.

On one hand, by Perelman's Harnack estimate for the conjugate heat kernel (6.28), we have that

$$K\big(o, 0; \varphi_{\tau_0}(x_\tau), -\tau_0\big) \geq (4\pi\tau_0)^{-\frac{n}{2}} \mathrm{e}^{-\ell(\varphi_{\tau_0}(x_\tau), \tau_0)} \geq (4\pi\tau_0)^{-\frac{n}{2}} \mathrm{e}^{-14n}, \tag{6.264}$$

where the last inequality is (6.262), which yields

$$\ell(\varphi_{\tau_0}(x_\tau), \tau_0) = \overline{\ell}(x_\tau, \tau_0) \leq 14n.$$

On the other hand, by the Gaussian upper bound (6.153), we also have that

$$K\big(o, 0; \varphi_{\tau_0}(x_\tau), -\tau_0\big) \leq C(n) \mathrm{e}^{-\mu_\infty} \tau_0^{-\frac{n}{2}} \exp\left(-\frac{d_{-\tau_0}(\varphi_{\tau_0}(x_\tau), z_{-\tau_0})^2}{9\tau_0}\right). \tag{6.265}$$

From combining the lower and upper estimates (6.264) and (6.265) for the conjugate heat kernel, we can estimate the distance between $\varphi_{\tau_0}(x_\tau)$ and $z_{-\tau_0}$ by

$$d_{-\tau_0}(\varphi_{\tau_0}(x_\tau), z_{-\tau_0})^2 \leq 9\big(C(n) - \mu_\infty\big)\tau_0 \tag{6.266}$$

for a different constant $C(n) > 0$. Hence, if

$$A \geq 882\big(C(n) - \mu_\infty\big), \tag{6.267}$$

then

$$d_{-\tau_0}(\varphi_{\tau_0}(x_\tau), z_{-\tau_0}) \leq \sqrt{A\tau_0/98}\,.$$

By this property, we have that

$$\mathrm{Vol}_{g_{-\tau_0}} B_{g_{-\tau_0}}\left(\varphi_{\tau_0}(x_\tau), \sqrt{A\tau_0/49}\right) \geq \mathrm{Vol}_{g_{-\tau_0}} B_{g_{-\tau_0}}\left(z_{-\tau_0}, \sqrt{A\tau_0/98}\right).$$

On the other hand, by Bamler's noncollapsing result, i.e., by (6.112) in Theorem 6.38, we have that

$$\mathrm{Vol}_{g_{-\tau_0}} B_{g_{-\tau_0}}\left(z_{-\tau_0}, \sqrt{2H_n\tau_0}\right) \geq c(n)\,\mathrm{e}^{\mu_\infty}\tau_0^{\frac{n}{2}}. \tag{6.268}$$

In view of this estimate and condition (6.267), we choose $A := 196H_n + 882(C(n) - \mu_\infty) \geq 196H_n$, so that

$$\mathrm{Vol}_{g_{-\tau_0}} B_{g_{-\tau_0}}\left(\varphi_{\tau_0}(x_\tau), \sqrt{A\tau_0/49}\right) \geq c(n)\,\mathrm{e}^{\mu_\infty}\tau_0^{\frac{n}{2}}.$$

Combining this with (6.263) completes the proof of the proposition. □

6.6.4. The proof of the main theorem.

We can now improve the estimate in Proposition 6.96 to prove Theorem 6.90 concerning a lower bound for the volume growth of steady gradient Ricci soliton singularity models. The idea is that we can string together a sequence of disjoint balls, each of which possesses a lower volume estimate.

PROOF OF THEOREM 6.90. Let A be as defined in Proposition 6.96 and let $B = 10$. Assume that the radius satisfies $r > BA + 392n$. Inductively, we define a decreasing sequence of scales starting at r by $\tau_1 := r$ and satisfying

$$\tau_i - \tau_{i+1} = \sqrt{A}\left(\sqrt{\tau_i} + \sqrt{\tau_{i+1}}\right). \tag{6.269}$$

We claim that as long as $\tau_i \geq \frac{1}{B}r$, the next term of the sequence $\tau_{i+1} > 0$ is well-defined. Firstly, (6.269) says that

$$\left(\sqrt{\tau_{i+1}} + \frac{1}{2}\sqrt{A}\right)^2 = \left(\sqrt{\tau_i} - \frac{1}{2}\sqrt{A}\right)^2.$$

Since $\tau_i > A$, this equation has the unique positive solution

$$\tau_{i+1} := \left(\sqrt{\tau_i} - \sqrt{A}\right)^2 < \tau_i. \tag{6.270}$$

Secondly, we have as long as $\tau_i \geq \frac{1}{B}r$ that

$$\tau_i - \tau_{i+1} = 2\sqrt{A\tau_i} - A > \sqrt{A\tau_i} \geq \sqrt{Ar/B}.$$

This ensures that there exists a smallest positive integer N such that the corresponding scale satisfies $\tau_N < \frac{1}{B}r$, at which point we stop our inductive definition to obtain the sequence of scales

$$r = \tau_1 > \tau_2 > \cdots > \tau_N > 0. \tag{6.271}$$

We can now apply Proposition 6.96 to obtain that for each $1 \leq i \leq N-1$ there exists a point $x_{\tau_i} \in M^n$ such that $d_g(x_{\tau_i}, o) = \tau_i$ and

$$\mathrm{Vol}_g B\left(x_{\tau_i}, \sqrt{A\tau_i}\right) \geq c(n)\mathrm{e}^{\mu_\infty}\tau_i^{\frac{n}{2}}, \tag{6.272}$$

since $\tau_i \geq \frac{1}{B}r > A \geq 196H_n > 392n$ (the last inequality is by (6.93)).

Now, for any $x \in B\left(x_{\tau_i}, \sqrt{A\tau_i}\right)$, we have that

$$\tau_i - \sqrt{A\tau_i} \leq d(x, o) \leq \tau_i + \sqrt{A\tau_i}.$$

Since $\tau_i + \sqrt{A\tau_i} = \tau_{i+1} - \sqrt{A\tau_{i+1}}$ for all $1 \leq i \leq N-1$ by (6.269), we have that the collection of open balls $\{B(x_{\tau_i}, \sqrt{A\tau_i})\}_{i=1}^N$ is pairwise disjoint.

For all $1 \leq i \leq N$, since $\tau_i + \sqrt{A\tau_i} \leq \tau_1 + \sqrt{A\tau_1} = r + \sqrt{Ar} \leq 2r$, we have that $B(x_{\tau_i}, \sqrt{A\tau_i}) \subset B(o, 2r)$. Therefore,

$$\begin{aligned}
\operatorname{Vol} B(o,2r) &\geq \sum_{i=1}^N \operatorname{Vol} B\left(x_{\tau_i}, \sqrt{A\tau_i}\right) \\
&\geq c(n) \mathrm{e}^{\mu_\infty} \sum_{i=1}^N \tau_i^{\frac{n}{2}} \\
&\geq \frac{c(n)}{\sqrt{A}} \mathrm{e}^{\mu_\infty} \sum_{i=1}^{N-1} \tau_i^{\frac{n-1}{2}} (\tau_i - \tau_{i+1}),
\end{aligned}$$

where the third inequality follows from (6.269). Since $\tau^{\frac{n-1}{2}} \leq \tau_i^{\frac{n-1}{2}}$ for $\tau \in [\tau_{i+1}, \tau_i]$, we have that

$$\begin{aligned}
\operatorname{Vol} B(o,2r) &\geq \frac{c(n)}{\sqrt{A}} \mathrm{e}^{\mu_\infty} \sum_{i=1}^{N-1} \int_{\tau_{i+1}}^{\tau_i} \tau_i^{\frac{n-1}{2}} d\tau \\
&\geq \frac{c(n)}{\sqrt{A}} \mathrm{e}^{\mu_\infty} \sum_{i=1}^{N-1} \int_{\tau_{i+1}}^{\tau_i} \tau^{\frac{n-1}{2}} d\tau \\
&= \frac{c(n)}{\sqrt{A}} \mathrm{e}^{\mu_\infty} \int_{\tau_N}^{r} \tau^{\frac{n-1}{2}} d\tau \\
&\geq \frac{2c(n)}{(n+1)\sqrt{A}} \mathrm{e}^{\mu_\infty} \left(r^{\frac{n+1}{2}} - \left(\frac{r}{B}\right)^{\frac{n+1}{2}} \right),
\end{aligned}$$

where the last inequality is because $\tau_N \leq \frac{r}{B}$. The theorem follows. □

6.7. Open problems for steady solitons

OPEN PROBLEM 6.1. *Is each Appleton cohomogeneity-one steady gradient Ricci soliton M_k^4, $k \geq 3$, (see* §1.3.3.2.c*) the singularity model of some 4-dimensional Ricci flow?*

OPEN PROBLEM 6.2. *Is each Kronheimer space (a.k.a. hyper-Kähler Ricci-flat ALE; see* §1.3.3*) the singularity model of some 4-dimensional Ricci flow?*

OPEN PROBLEM 6.3. *Can one prove that any non-Ricci-flat complete noncompact 4-dimensional steady gradient Ricci soliton singularity model has at most linear curvature decay, $R(x) \geq c(d(x,o)+1)^{-1}$?*

Corollary 6.87 *shows that it has at most quadratic curvature decay; i.e., $R(x) \geq c(d(x,o)+1)^{-2}$.*

OPEN PROBLEM 6.4. *Let (M^4, g, f) be a shrinking gradient Ricci soliton singularity model. Can a steady gradient Ricci soliton limit*

$$(M_\infty^4, g_\infty, f_\infty, x_\infty)$$

of such a shrinking gradient Ricci soliton, as in Proposition 2.57*, have tangent flow at infinity isometric to $\mathbb{R} \times \mathbb{S}^3/\Gamma$? For example, can one show that $(M_\infty^4, g_\infty, f_\infty)$ cannot be the Bryant steady soliton?*

OPEN PROBLEM 6.5. *Let* (M^4, g, f) *be a steady gradient Ricci soliton singularity model with tangent flow at infinity given by* $\mathbb{S}^2 \times \mathbb{R}^2$. *Is it possible that* (M^4, g, f) *is not* $\mathrm{O}(3) \times \mathbb{Z}_2$*-symmetric?*

OPEN PROBLEM 6.6. *In the paper "Uniqueness of asymptotically cylindrical steady gradient Ricci solitons" (arXiv:2505.20576), Law assumes that the tangent flow at infinity of a noncompact steady gradient Ricci soliton is* $\mathbb{S}^3/\Gamma \times \mathbb{R}$, *where* $\mathbb{S}^3/\Gamma$ *is homogeneous, and he concludes that—up to a possible* $\mathbb{Z}_2$*-quotient—every such soliton must be one of the cohomogeneity-one examples constructed by Appleton* [**App17**].

Do there exist steady solitons with tangent flow at infinity $\mathbb{S}^3/\Gamma \times \mathbb{R}$, *where* $\mathbb{S}^3/\Gamma$ *is not homogeneous? If so, can they be classified?*

APPENDIX A

Review of Riemannian Orbifolds

In the study of the Ricci flow on smooth closed 4-manifolds, surgeries can introduce isolated orbifold singularities. Furthermore, Bamler's higher-dimensional compactness and singularity theory should ensure that we stay in the category of 4-manifolds with isolated orbifold singularities (see Chapter 6). To accommodate this eventuality, we expand our category to include 4-dimensional smooth Riemannian orbifolds with isolated singularities. Rigorous treatments of the differential geometry of orbifolds are available in the literature, and we refer the reader to these sources for a detailed understanding. In this appendix, the discussion of orbifolds is more cursory, focusing on an intuitive and working understanding rather than a comprehensive or detailed exposition; references for proofs are given.

A.1. Smooth orbifolds

The foundational definitions and theory in the orbifold category are due to Satake [**Sat56**], Thurston [**Thu79**, **Thu97**], and Haefliger [**Hae90**]; various researchers (see references below) have since provided specific formulations or adaptations, which we follow.

A.1.1. Definition of smooth orbifold.

See Boileau, Maillot, and Porti [**BMP03**, §2.1] for the following definition of an orbifold (cf. Cooper, Hodgson, and Kerckhoff [**CHK00**, §2.1] and Kleiner and Lott [**KL14**]). More abstract definitions are given in terms of Lie groupoids and stacks (see Adem, Leida, and Ruan [**ALR07**, Definition 1.1] and Lerman [**Ler10**]).

Definition A.1 (Smooth orbifold). An n-dimensional **smooth orbifold** $\mathcal{O}^n$ (without boundary)[1] is a metrizable topological space O^n, called the **underlying topological space**, with the following structure. An **orbifold atlas** is a collection of **orbifold charts**:

$$\{(U_i, \widetilde{U}_i, \phi_i, \Gamma_i)\}_{i\in I},$$

where for each $i \in I$, U_i is an open subset of O^n with the induced topology, $\widetilde{U}_i$ (called a **Euclidean model**) is a connected open subset of $\mathbb{R}^n$ with the induced smooth structure, the map

$$\phi_i : \widetilde{U}_i \to U_i$$

is continuous, and Γ_i (called the **chart group**) is a finite group of self-diffeomorphisms of $\widetilde{U}_i$ (i.e., Γ_i acts smoothly and effectively on $\widetilde{U}_i$ on the left with group multiplication corresponding to compositions of diffeomorphisms) satisfying the

[1]These are also called **effective orbifolds** in the literature.

following conditions:

(1) $\{U_i\}_{i\in I}$ is an open cover of O^n.

(2) Let $\pi_i : \widetilde{U}_i \to \widetilde{U}_i/\Gamma_i$ denote the quotient map defined by $\pi_i(\widetilde{x}) := [\widetilde{x}]$, the orbit of $\widetilde{x}$ under the action of the chart group Γ_i on $\widetilde{U}_i$. There exists a homeomorphism

$$\eta_i : \widetilde{U}_i/\Gamma_i \to U_i \tag{A.1}$$

such that $\phi_i = \eta_i \circ \pi_i$; that is, the following diagram commutes:

$$\begin{array}{ccc} \widetilde{U}_i & & \\ {\scriptstyle \pi_i}\downarrow & \searrow^{\phi_i} & \\ \widetilde{U}_i/\Gamma_i & \xrightarrow[\eta_i]{\cong} & U_i \end{array} \tag{A.2}$$

We call $(\widetilde{U}_i, \Gamma_i)$ a **local model** (of $\mathcal{O}^n$).

(3) (Orbifold chart compatibility) If $\widetilde{x}_i \in \widetilde{U}_i$ and $\widetilde{x}_j \in \widetilde{U}_j$ satisfy $\phi_i(\widetilde{x}_i) = \phi_j(\widetilde{x}_j) =: x_{ij} \in U_i \cap U_j \subset O^n$, then there exist open neighborhoods $V_i \subset \widetilde{U}_i$ of $\widetilde{x}_i$ and $V_j \subset \widetilde{U}_j$ of $\widetilde{x}_j$ and a diffeomorphism

$$\psi_{ij} : V_i \to V_j$$

such that $\phi_j(\psi_{ij}(\widetilde{x})) = \phi_i(\widetilde{x})$ for all $\widetilde{x} \in V_i$, and $\psi_{ij}(\widetilde{x}_i) = \widetilde{x}_j$; that is, the following diagram commutes:

$$\begin{array}{ccc} (V_i, \widetilde{x}_i) & \xrightarrow[\cong]{\psi_{ij}} & (V_j, \widetilde{x}_j) \\ & {\scriptstyle \phi_i}\searrow \quad \swarrow{\scriptstyle \phi_j} & \\ & (U_i \cap U_j, x_{ij}) & \end{array} . \tag{A.3}$$

The orbifold charts $(U_i, \widetilde{U}_i, \phi_i, \Gamma_i)$ and $(U_j, \widetilde{U}_j, \phi_j, \Gamma_j)$ above are **compatible**.

Analogous to the case of smooth manifolds, given a smooth orbifold atlas $\mathcal{A} := \{(U_i, \widetilde{U}_i, \phi_i, \Gamma_i)\}_{i\in I}$ for $\mathcal{O}^n$, the associated **maximal orbifold atlas** is the collection $\mathcal{M}$ of all orbifold charts that are compatible with all charts in $\mathcal{A}$. We have that $\mathcal{M}$ is an orbifold atlas and contains any orbifold atlas that contains $\mathcal{A}$. We will assume that the orbifold atlas is maximal. In particular, if an orbifold is described by a non-maximal orbifold atlas, then we replace this atlas by the maximal one.

REMARK A.2. An **embedding** of a local model $(\widetilde{U}_i, \Gamma_i)$ into a local model $(\widetilde{U}_j, \Gamma_j)$ is an embedding $\widetilde{f} : \widetilde{U}_i \to \widetilde{U}_j$ and a homomorphism $\rho : \Gamma_i \to \Gamma_j$ such that

$$\widetilde{f}(\gamma(x)) = \rho(\gamma)\big(\widetilde{f}(x)\big)$$

for all $\gamma \in \Gamma_i$ and $x \in \widetilde{U}_i$. Since the group actions are effective, it follows that ρ is injective.

The orbifold chart compatibility condition (3) in Definition A.1 is equivalent to the condition that for any local models $(\widetilde{U}_i, \Gamma_i)$ and $(\widetilde{U}_j, \Gamma_j)$ and point $x \in U_i \cap U_j$, there exists a local model $(\widetilde{U}_k, \Gamma_k)$ with $x \in U_k$ and embeddings of $(\widetilde{U}_k, \Gamma_k)$ into $(\widetilde{U}_i, \Gamma_i)$ and of $(\widetilde{U}_k, \Gamma_k)$ into $(\widetilde{U}_j, \Gamma_j)$ (cf. [**KL14**, Definition 2.1]).

We refer to a smooth orbifold as an **orbifold**. We also denote $|\mathcal{O}^n| := O^n$ for the underlying topological space. If each Γ_i in the definition of an orbifold is the trivial group, then $\mathcal{O}^n$ is an n-dimensional smooth manifold.

An **orientation** on an orbifold $\mathcal{O}^n$ is an orientation on each Euclidean model $\widetilde{U}_i$ for $\mathcal{O}^n$ such that Γ_i acts by orientation-preserving self-diffeomorphisms and the diffeomorphisms $\psi_{ij} : V_i \to V_j$ in Definition A.1(3) are orientation-preserving. An **oriented orbifold** is an orbifold with an orientation.

Given $x \in O^n$, let $\phi_i : \widetilde{U}_i \to U_i$ be any orbifold chart with $x \in U_i$. The **local group** (or **isotropy group**) Γ_x of x is defined to be the stabilizer of any point $\widetilde{x}$ in $\phi_i^{-1}(x)$; that is,[2]

$$\Gamma_x := \{\gamma \in \Gamma_i : \gamma(\widetilde{x}) = \widetilde{x}\}. \tag{A.4}$$

One can show that the isomorphism class of Γ_x is independent of the orbifold chart $(U_i, \widetilde{U}_i, \phi_i, \Gamma_i)$ and the choice of $\widetilde{x}$ in $\phi_i^{-1}(x)$, and that the subgroup Γ_x of Γ_i is unique up to conjugacy within Γ_i, meaning that different choices of $\widetilde{x}$ may yield different but conjugate subgroups.

The **regular locus** O^n_{reg} of an orbifold $\mathcal{O}^n$ is the set of points $x \in O^n$ satisfying Γ_x is the trivial group. The **singular locus** O^n_{sing} is the set of points $x \in O^n$ satisfying that Γ_x is a non-trivial group. The regular locus is an open set and the singular locus is a closed set. Moreover, the regular locus is naturally an n-dimensional smooth manifold. A point in the regular locus is a **regular point**, while a point in the singular locus is a **singular point** or **singularity**.

DEFINITION A.3. An orbifold $\mathcal{O}^n$ is an **orbifold with isolated singularities** if its singular locus is a discrete subset of O^n.

In the case that $x_0 \in O^n$ is an isolated singularity and $\phi_i : \widetilde{U}_i \to U_i$ is an orbifold chart with $x_0 \in U_i$, we have that $\phi_i^{-1}(x_0) = \{\widetilde{x}_0\}$ consists of a single point. Given $\gamma \in \Gamma_{x_0}$, we have that $\gamma(\widetilde{x}_0) = \widetilde{x}_0$ and $d\gamma_{\widetilde{x}_0} : T_{\widetilde{x}_0}\widetilde{U}_i \to T_{\widetilde{x}_0}\widetilde{U}_i$ is a linear automorphism. Thus,

$$d\Gamma_{x_0} := \{d\gamma_{\widetilde{x}_0} : \gamma \in \Gamma_{x_0}\} \tag{A.5}$$

is a finite subgroup of $\mathrm{GL}(T_{\widetilde{x}_0}\widetilde{U}_i)$. It is easy to see that $d\Gamma_{x_0}$ is isomorphic to Γ_{x_0} since the derivative of the composition of two maps is equal to the composition of the derivatives of the two maps.

EXAMPLE A.4 (Teardrops and American footballs). Let k and ℓ be positive integers and define two orbifold charts by $(\mathbb{R}^2, \mathbb{R}^2/\mathbb{Z}_k, \phi_k, \mathbb{Z}_k)$ and $(\mathbb{R}^2, \mathbb{R}^2/\mathbb{Z}_\ell, \phi_\ell, \mathbb{Z}_\ell)$, where the group $\mathbb{Z}_m$ acts on $\mathbb{R}^2$ generated by the rotation $z \mapsto \mathrm{e}^{i2\pi/m}$ for $m = k, \ell$. Let $\pi_m : \mathbb{R}^2 \to \mathbb{R}^2/\mathbb{Z}_m$ be the quotient map defined by $\pi_m(z) = [z]$ for $m = k, \ell$. Define the topological space

$$O^2 := (\mathbb{R}^2/\mathbb{Z}_k \sqcup \mathbb{R}^2/\mathbb{Z}_\ell)/\sim,$$

where the equivalence relation $\sim$ is defined by the identification

$$(\mathbb{R}^2 \setminus \{0^2\})/\mathbb{Z}_k \ni [r\mathrm{e}^{i\theta}] \sim [r^{-1}\mathrm{e}^{i\theta k/\ell}] \in (\mathbb{R}^2 \setminus \{0^2\})/\mathbb{Z}_\ell.$$

[2]The group Γ_x is also called the **isotropy group** or **stabilizer group** of x.

Let $i_m : \mathbb{R}^2/\mathbb{Z}_m \to O^2$ be the inclusion map and define $\phi_m : \mathbb{R}^2 \to O^2$ by $\phi_m := i_m \circ \pi_m$ for $m = k, \ell$. These two orbifold coordinate charts define a closed 2-dimensional orbifold $\mathcal{O}^2_{k,\ell}$ with its underlying topological space homeomorphic to $\mathbb{S}^2$.

If $k = \ell = 1$, then $\mathcal{O}^2_{k,\ell}$ is a smooth 2-dimensional manifold diffeomorphic to the 2-sphere. If $k = 1$ and $\ell > 1$, then $\mathcal{O}^2_{k,\ell}$ is called a **teardrop**. If $k > 1$ and $\ell > 1$, then $\mathcal{O}^2_{k,\ell}$ is called an **American football**.

If U is an open subset of the underlying topological space O^n of an orbifold $\mathcal{O}^n$, then the orbifold structure of $\mathcal{O}^n$ restricts to an orbifold with its underlying topological space U, which we denote by $\mathcal{O}^n|_U$. Note that the local groups Γ_x of $\mathcal{O}^n|_U$ coincide with those of $\mathcal{O}^n$ at each point $x \in U$. The regular and singular loci of $\mathcal{O}^n|_U$ are equal to the intersections of the regular and singular loci of $\mathcal{O}^n$ with U, respectively.

Let Γ be a finite group acting by diffeomorphisms on an open subset U of $\mathbb{R}^n$. Suppose that the stabilizer of $\widetilde{x}_0 \in U$ is equal to Γ and that the stabilizers of all points $\widetilde{x} \in U \setminus \{\widetilde{x}_0\}$ are equal to 1. Then Γ is locally conjugate (by a diffeomorphism) to an orthogonal action. Consequently, for any orbifold with isolated singularities, every local group is isomorphic to a finite subgroup of $\mathrm{O}(n)$, and there exists an orbifold atlas $\{(U_i, \widetilde{U}_i, \phi_i, \Gamma_i)\}_{i \in I}$ such that each $\widetilde{U}_i$ is a Euclidean n-ball[3] centered at the origin 0^n and Γ_i acts as a subgroup of $\mathrm{O}(n)$, acting orthogonally on $\widetilde{U}_i$ with respect to the Euclidean metric g_{Euc}. Such orbifold charts **canonical**. Observe that Γ_i acts freely on $\widetilde{U}_i \setminus \{0^n\}$. In particular, denoting $x_0 := \phi_i(0^n)$, we have that $U_i \setminus \{x_0\}$ is an n-manifold diffeomorphic to $(\widetilde{U}_i \setminus \{0^n\})/\Gamma_i$; see Figure A.1. We note that the local groups are

$$\Gamma_{x_0} \cong \Gamma_i \quad \text{and} \quad \Gamma_x \cong 1 \text{ for } x \in U_i \setminus \{x_0\}.$$

The notion of a canonical orbifold chart is not essential but we find it useful for discussing examples.

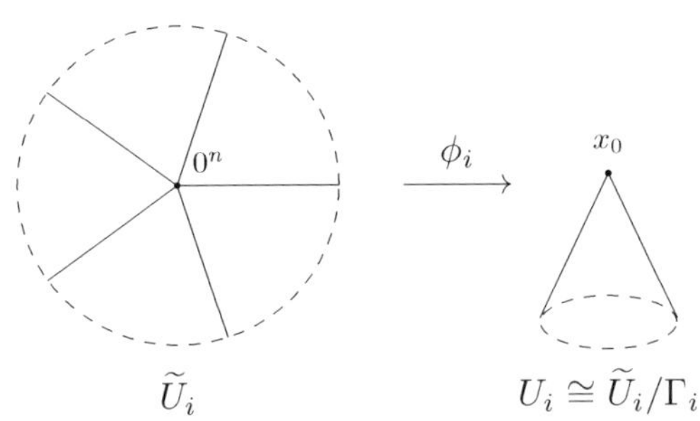

FIGURE A.1. A canonical orbifold chart with a Euclidean model projecting to an open set in an orbifold with isolated singularities.

The **product** of two orbifolds $\mathcal{O}_1^{n_1}$ and $\mathcal{O}_2^{n_2}$ is an orbifold, denoted by $\mathcal{O}_1^{n_1} \times \mathcal{O}_2^{n_2}$. If $(U_i, \widetilde{U}_i, \phi_i, \Gamma_i)$ is an orbifold chart for $\mathcal{O}_1^{n_1}$ and if $(U_j, \widetilde{U}_j, \phi_j, \Gamma_j)$ is an orbifold chart for $\mathcal{O}_2^{n_2}$, then $(U_i \times U_j, \widetilde{U}_i \times \widetilde{U}_j, \phi_i \times \phi_j, \Gamma_i \times \Gamma_j)$ is an orbifold chart for $\mathcal{O}_1^{n_1} \times \mathcal{O}_2^{n_2}$. Here, $\phi_i \times \phi_j : \widetilde{U}_i \times \widetilde{U}_j \to U_i \times U_j$ denotes the product map defined by

[3]We allow the ball to have infinite radius; i.e., we include all of $\mathbb{R}^n$ as a ball.

$\phi_i \times \phi_j(\widetilde{x}_i, \widetilde{x}_j) := \big(\phi_i(\widetilde{x}_i), \phi_j(\widetilde{x}_j)\big)$, and $\Gamma_i \times \Gamma_j$ is the product group acting by the product action on $\widetilde{U}_i \times \widetilde{U}_j$. The singular set of such a product is

$$(\mathcal{O}_1^{n_1} \times \mathcal{O}_2^{n_2})_{\mathrm{sing}} = \big((\mathcal{O}_1^{n_1})_{\mathrm{sing}} \times \mathcal{O}_2^{n_2}\big) \cup \big(\mathcal{O}_1^{n_1} \times (\mathcal{O}_2^{n_2})_{\mathrm{sing}}\big).$$

For example, for any 3-dimensional orbifold $\mathcal{O}^3$,

$$(\mathcal{O}^3 \times \mathbb{R})_{\mathrm{sing}} = \mathcal{O}^3_{\mathrm{sing}} \times \mathbb{R}. \tag{A.6}$$

In particular, any 4-dimensional orbifold with isolated singularities and that splits as the product of $\mathbb{R}$ with a 3-dimensional orbifold must be a smooth 4-manifold.

A.2. Orbifold maps, bundles, and tensors

The definition of a smooth structure on an orbifold is based on local models, where the smooth structure is manifest on the Euclidean model. Likewise, for maps between orbifolds, the notion of smoothness is manifest via the local models on the domain and codomain. This leads to restrictions on the nature of smooth orbifold maps near singular points.

A.2.1. Orbifold maps.

Let $\mathcal{O}_1^{n_1}$ and $\mathcal{O}_2^{n_2}$ be smooth orbifolds with underlying topological spaces $O_1^{n_1}$ and $O_2^{n_2}$, respectively. A map $f : O_1^{n_1} \to O_2^{n_2}$ is a **smooth orbifold map** if for any $x \in O_1^{n_1}$ there exist orbifold charts $(U_{i,1}, \widetilde{U}_{i,1}, \phi_{i,1}, \Gamma_{i,1})$ for $O_1^{n_1}$ with $x \in U_{i,1}$ and $(U_{i,2}, \widetilde{U}_{i,2}, \phi_{i,2}, \Gamma_{i,2})$ for $O_2^{n_2}$ with $f(x) \in U_{i,2}$ such that $f(U_{i,1}) \subset U_{i,2}$ and there exists a smooth (in the Euclidean sense) map $\widetilde{f} : \widetilde{U}_{i,1} \to \widetilde{U}_{i,2}$ satisfying

$$\phi_{i,2} \circ \widetilde{f} = f \circ \phi_{i,1};$$

that is, the following diagram commutes:

$$\begin{array}{ccc} \widetilde{U}_{i,1} & \xrightarrow{\ \widetilde{f}\ } & \widetilde{U}_{i,2} \\ {\scriptstyle \phi_{i,1}}\big\downarrow & & \big\downarrow{\scriptstyle \phi_{i,2}} \\ U_{i,1} & \xrightarrow{\ f\ } & U_{i,2} \end{array} . \tag{A.7}$$

See [**KL14**, §2.1]. In particular, if $\mathcal{O}^n$ is a smooth orbifold, then a function $f : O^n \to \mathbb{R}$ is **smooth** if for every $x \in O^n$ there exists an orbifold chart $(U_i, \widetilde{U}_i, \phi_i, \Gamma_i)$ for $\mathcal{O}^n$ with $x \in U_i$ and if there exists a smooth function $\widetilde{f} : \widetilde{U}_i \to \mathbb{R}$ such that $\widetilde{f} = f \circ \phi_i$; that is, the following diagram commutes:

$$\begin{array}{ccc} \widetilde{U}_i & & \\ {\scriptstyle \phi_i}\big\downarrow & \searrow^{\widetilde{f}} & \\ U_i & \xrightarrow[\ f\ ]{} & \mathbb{R} \end{array} . \tag{A.8}$$

Let $\mathcal{O}^n$ and $\mathcal{P}^n$ be smooth orbifolds of the same dimension and with underlying topological spaces O^n and P^n, respectively. We say that a bijective map $f : O^n \to P^n$ is an **orbifold diffeomorphism** if f and its inverse f^{-1} are both smooth orbifold maps. In this case, $f|_{O^n_{\mathrm{reg}}}$ is a diffeomorphism between the smooth manifolds O^n_{reg} and P^n_{reg}, and $f|_{O^n_{\mathrm{sing}}}$ is a bijection between the sets O^n_{sing} and P^n_{sing}. Moreover, we have that $\Gamma_{f(x)}$ is isomorphic to Γ_x for all $x \in O^n$. Two orbifolds are **orbifold diffeomorphic** if there exists an orbifold diffeomorphism between them.

A smooth orbifold map $f : O^n \to P^m$ is an **orbifold immersion** if, for any $x \in O^n$, the local models around x and $f(x)$ in the definition of smooth orbifold map can be chosen in such a way that the local lift $\widetilde{f}$ is an immersion.

EXAMPLE A.5. Let $f : O^n \to P^n$ be an orbifold immersion between orbifolds with isolated singularities of the same dimension. Then for each singular point x_0 in O^n, $f(x_0)$ is a singular point of P^n. Furthermore, there exist canonical local models $(U_1, \mathbb{R}^n, \phi_1, \Gamma_1)$ and $(U_2, \mathbb{R}^n, \phi_2, \Gamma_2)$ around x and $f(x)$,[4] respectively, such that the following diagram commutes:

$$\begin{array}{ccc} \mathbb{R}^n & \xrightarrow{\mathrm{id}_{\mathbb{R}^n}} & \mathbb{R}^n \\ \pi_1\downarrow & & \downarrow\pi_2 \\ \mathbb{R}^n/\Gamma_1 & \xrightarrow{\widetilde{f}} & \mathbb{R}^n/\Gamma_2 \\ \eta_1\downarrow & & \downarrow\eta_2 \\ U_1 & \xrightarrow{f|_{U_1}} & U_2 \end{array} \quad . \tag{A.9}$$

Here, Γ_1 and Γ_2 are finite subgroups of $\mathrm{O}(n)$, acting by orthogonal transformations on $\mathbb{R}^n$, with $\Gamma_1 \cong \Gamma_{x_0}$ and $\Gamma_2 \cong \Gamma_{f(x_0)}$. In particular, $\widetilde{f}(\pi_1(z)) = \pi_2(z)$. For any $\gamma \in \Gamma_1$ and $\widetilde{x} \in \mathbb{R}^n$, we have

$$\pi_2(\widetilde{x}) = \widetilde{f}(\pi_1(\widetilde{x})) = \widetilde{f}\big(\pi_1(\gamma(\widetilde{x}))\big) = \pi_2(\gamma(\widetilde{x})).$$

Thus, each $\gamma \in \Gamma_1$ corresponds to an element $\rho(\gamma) \in \Gamma_2$ and we have an injective homomorphism $\rho : \Gamma_1 \to \Gamma_2$. Therefore, the local group Γ_{x_0} is isomorphic to a subgroup of the local group $\Gamma_{f(x_0)}$.

For example, let $\phi_m = \pi_m : \mathbb{R}^2 \to \mathbb{R}^2/\mathbb{Z}_m$ be the quotient map and let $x_m = [0^2] \in \mathbb{R}^2/\mathbb{Z}_m$ for $m = k, l$. Let $\mathcal{O}^2_m$ be the 2-dimensional orbifold with its orbifold atlas consisting of the single orbifold coordinate chart $(\mathbb{R}^2, \mathbb{R}^2/\mathbb{Z}_m, \phi_m, \mathbb{Z}_m)$ for $m = k, l$. We have that there exists a smooth orbifold immersion $f : \mathcal{O}^2_k \to \mathcal{O}^2_\ell$ if and only if k divides ℓ. An example of such a map is $f(\pi_k(z)) = \pi_\ell(z)$ for $z \in \mathbb{R}^2$. This map is well-defined since k divides ℓ. Observe that we have the commutative diagram:

$$\begin{array}{ccc} \mathbb{R}^2 & \xrightarrow{\mathrm{id}_{\mathbb{R}^2}} & \mathbb{R}^2 \\ \pi_k\downarrow & & \downarrow\pi_\ell \\ \mathbb{R}^2/\mathbb{Z}_k & \xrightarrow{f} & \mathbb{R}^2/\mathbb{Z}_\ell \end{array} \quad . \tag{A.10}$$

A.2.2. Orbifiber bundles and covering spaces.

Let $\mathcal{E}$, $\mathcal{B}$, and $\mathcal{F}$ be orbifolds (called the **entire space**, **base**, and **fiber**, respectively) and let E, B, and F denote their underlying topological spaces. An **orbifiber bundle** consists of a surjective smooth orbifold map

$$p : E \to B$$

such that for every $x \in \mathcal{B}$ there exist a local model $(\widetilde{U}_i, \Gamma_i)$ of $\mathcal{B}$ with $x \in U_i \subset B$ and $\Gamma_i \cong \Gamma_x$ (the local group of x), an action of Γ_i on F, and an orbifold diffeomorphism

[4]Alternatively, one can replace $\mathbb{R}^n$ by the open unit n-ball.

$\psi : (F \times \widetilde{U}_i)/\Gamma_i \to p^{-1}(U_i)$ from $(\mathcal{F} \times \widetilde{U}_i)/\Gamma_i$ to $\mathcal{E}|_{p^{-1}(U_i)}$ such that the following diagram commutes:

$$\begin{array}{ccc} (F \times \widetilde{U}_i)/\Gamma_i & \xrightarrow{\psi} & p^{-1}(U_i) \subset E \\ {\scriptstyle \mathrm{proj}_2}\downarrow & & \downarrow{\scriptstyle p} \\ \widetilde{U}_i/\Gamma_i & \xrightarrow{\eta_i} & U_i \subset B \end{array}, \tag{A.11}$$

where proj_2 is the map induced by the projection onto the second factor and where η_i is defined by (A.1). Here, the group Γ_i acts diagonally on $F \times \widetilde{U}_i$ by $\gamma(f, \widetilde{x}) = (\gamma(f), \gamma(\widetilde{x}))$.

An orbifiber bundle is a **orbifold covering space** if the fiber is a 0-dimensional orbifold.

A.2.3. Orbifold universal cover.

For a manifold, the fundamental group is defined as its fundamental group of its underlying topological space. The universal cover of a topological space may be defined as the simply-connected covering space, which is unique up to equivalence. This definition depends on the notion of simply-connectedness, which in turn depends on the definition of the fundamental group. On the other hand, there is a standard construction of the universal cover of a topological space using homotopy classes of based paths. This construction can be modified to suit the orbifold category, and we can define the orbifold universal cover. The orbifold fundamental group is then defined as the group of deck transformations of the orbifold universal cover. Generally, the definitions are due to Thurston, with various later specific formulations.

Before we formally define the orbifold universal covering space, we consider an example that will illustrate some of the key features of this concept.

EXAMPLE A.6. Let $k \geq 2$ be an integer and let $\mathcal{O}_k^2$ be the orbifold, with its underlying topological space $\mathbb{R}^2/\mathbb{Z}_k$, defined in Example A.5. Consider the orbifold covering space $\pi_k : \mathbb{R}^2 \to \mathbb{R}^2/\mathbb{Z}_k = O_k^2$ given by the quotient map $\pi_k(z) := [z]$. Its group of orbifold deck transformations is generated by the rotation $z \mapsto e^{i2\pi/k}z$ and is isomorphic to $\mathbb{Z}_k$. Correspondingly, for any $r > 0$ we have the free loop $\gamma : [0,1] \to O_k^2$ defined by

$$\gamma(t) := \left[re^{i\frac{2\pi}{k}t}\right].$$

In defining the orbifold fundamental group below, the orbifold free homotopy class $[\gamma]$ of γ will be non-trivial, whereas the orbifold free homotopy class $[\gamma]^k$ will be trivial. The latter corresponds to the path $\gamma^k : [0,1] \to O_k^2$ defined by $\gamma^k(t) := \left[re^{i2\pi t}\right]$ lifting to a (homotopically trivial) free loop in $\mathbb{R}^2$. In particular, we should have $\pi_1^{\mathrm{orb}}(\mathcal{O}_k^2) \cong \mathbb{Z}_k$, which is isomorphic to the group of orbifold deck transformations.

Let $x_0 \in O_{\mathrm{reg}}^n$. An **orbifold path**[5] in $\mathcal{O}^n$ based at x_0 is a path $\gamma : [0,1] \to O^n$ such that $\gamma(0) = x_0$ and the preimage of the singular locus $\gamma^{-1}(O_{\mathrm{sing}}^n)$ consists of a finite number of points.[6]

[5]Kleiner and Lott [**KL14**, §2.2] call this a *special curve*.

[6]Hong Huang has observed that this definition of orbifold path is not sufficient to define the notion of holonomy. We would like to thank John Lott for pointing this out to us.

We now define the notion of "orbifold homotopy" between two orbifold paths in $\mathcal{O}^n$ through "elementary short homotopies" in the local lifts. Let $\gamma_0, \gamma_1 : [0,1] \to O^n$ be orbifold paths with $\gamma_0(0) = \gamma_1(0) =: x_0$ and $\gamma_0(1) = \gamma_1(1) =: x_1$.

A **short homotopy** between $\gamma_0(t)$ and $\gamma_1(t)$ consists of a partition $0 = t_0 < t_1 < \cdots < t_k = 1$, local models $(U_i, \widetilde{U}_i, \phi_i, \Gamma_i)$, $1 \leq i \leq k$, and smooth homotopies $\widetilde{H}_i(t,s)$, where $t, s \in [t_{i-1}, t_i]$, in $\widetilde{U}_i$ (called **elementary short homotopies**) in $\widetilde{U}_i$ between lifts $(\widetilde{\gamma}_0)_i(t)$ and $(\widetilde{\gamma}_1)_i(t)$ of $\gamma_0|_{[t_{i-1},t_i]}(t)$ and $\gamma_1|_{[t_{i-1},t_i]}(t)$, respectively, such that

(1) $\phi_i\big(\widetilde{H}_i(t_i, s)\big) = \phi_{i+1}\big(\widetilde{H}_{i+1}(t_i, s)\big)$ for $1 \leq i < k$ and $s \in [0,1]$.

(2) $\phi_1\big(\widetilde{H}_1(0, s)\big) = x_0$ and $\phi_k\big(\widetilde{H}_k(1, s)\big) = x_1$ for $s \in [0,1]$.

Two orbifold paths in $\mathcal{O}^n$ are **orbifold homotopic** if there exists a sequence of short homotopies connecting one orbifold path to the other. Being orbifold homotopic is an equivalence relation. For the following definition, see [**BMP03**, §2.2.1] and cf. [**KL14**, §2.1]:

DEFINITION A.7. The underlying set of the **orbifold universal cover** of $\mathcal{O}^n$ is the set of orbifold paths based at x_0 modulo the equivalence relation of being orbifold homotopic.

THEOREM A.8 (Thurston [**Thu79**]). *The orbifold universal cover of an orbifold has the structure of an orbifold. If an orbifold has isolated singularities, then its orbifold universal cover has isolated singularities.*

A.2.4. Orbifold fundamental group.

We follow Thurston [**Thu02**] and define the **orbifold fundamental group** $\pi_1^{\mathrm{orb}}(\mathcal{O}^n)$ of an orbifold $\mathcal{O}^n$ as the group of deck transformations of the orbifold universal cover of $\mathcal{O}^n$.

REMARK A.9. The orbifold fundamental group $\pi_1^{\mathrm{orb}}(\mathcal{O}^n)$ of an orbifold $\mathcal{O}^n$ can equivalently be defined as the fundamental group of the classifying space $B\mathcal{G}$ of the Lie groupoid $\mathcal{G}$ representing the orbifold structure of $\mathcal{O}^n$ (see [**ALR07**, Definition 1.50]).

EXAMPLE A.10.

(1) Let $k \geq 2$ be an integer and consider the action of $\mathbb{Z}_k$ on $\mathbb{S}^2$ given by rotation by the angle $2\pi/k$ about the z-axis. Then $\mathbb{S}^2/\mathbb{Z}_k$ is an orbifold with two singular points, at the north and south poles, each with its local group isomorphic to $\mathbb{Z}_k$. The universal orbifold covering space is the quotient map $p : \mathbb{S}^2 \to \mathbb{S}^2/\mathbb{Z}_k$ and the orbifold fundamental group is $\pi_1^{\mathrm{orb}}(\mathbb{S}^2/\mathbb{Z}_k) \cong \mathbb{Z}_k$.

(2) Let $\mathcal{O}^2_{k,\ell}$ be a closed 2-dimensional orbifold with its underlying topological space homeomorphic to $\mathbb{S}^2$ and with two singular points, one of order $k \geq 2$ and one of order $\ell \geq 2$. Denote $\mathrm{g} := \gcd(k, \ell)$. The universal orbifold covering space is the quotient map $p : \mathcal{O}^2_{k/\mathrm{g},\ell/\mathrm{g}} \to \mathcal{O}^2_{k,\ell}$ and the orbifold fundamental group is $\pi_1^{\mathrm{orb}}(\mathcal{O}^2_{k,\ell}) \cong \mathbb{Z}_{\mathrm{g}}$.

We end this subsection with a question that hopefully is not difficult to answer: Does Wylie's Theorem 1.78 that shrinkers have finite fundamental groups extend to orbifolds with isolated singularities?

A.2.5. Tangent and tensor orbibundles.

We will now define the **tangent orbibundle**

$$p : T\mathcal{O} \to \mathcal{O}^n$$

of an orbifold $\mathcal{O}^n$ in a way so that $T\mathcal{O}$ is a $2n$-dimensional smooth orbifold and the projection map p is a smooth orbifold map.

We first describe the tangent orbibundle locally, over the open subset $U_i \subset O^n$ of each orbifold chart. Given an orbifold chart $(U_i, \widetilde{U}_i, \phi_i, \Gamma_i)$, the group Γ_i acts on the tangent bundle $T\widetilde{U}_i$ of $\widetilde{U}_i$ by

$$\gamma(\widetilde{x}, v) := \big(\gamma(\widetilde{x}), d\gamma_{\widetilde{x}}(v)\big) \quad \text{for } \gamma \in \Gamma_i, \tag{A.12}$$

where $(\widetilde{x}, v)$ denotes an element of $T\widetilde{U}_i$ so that v is a tangent vector to $\widetilde{U}_i$ at $\widetilde{x} \in \widetilde{U}_i$. Let $[\widetilde{x}, v] \in T\widetilde{U}_i/\Gamma_i$ denote the orbit of $(\widetilde{x}, v)$ under the Γ_i action on $T\widetilde{U}_i$, and let

$$p_i([\widetilde{x}, v]) := \phi_i(\widetilde{x}).$$

The tangent orbibundle of $\mathcal{O}^n$ restricted to U_i is given (up to equivalence) by

$$p_i : T\widetilde{U}_i/\Gamma_i \to U_i. \tag{A.13}$$

In the case that $(U_i, \widetilde{U}_i, \phi_i, \Gamma_i)$ is a canonical orbifold chart, we can identify $T\widetilde{U}_i/\Gamma_i$ with the disjoint union of the tangent bundle $T(U_i \setminus \{x_0\})$ and the quotient tangent space $(T_{0^n}\widetilde{U}_i)/\Gamma_i \cong \mathbb{R}^n/\Gamma_i$.

We now patch together the local tangent orbibundles described in (A.13). As before, let $\{(U_i, \widetilde{U}_i, \phi_i, \Gamma_i)\}_{i \in I}$ be an atlas for $\mathcal{O}^n$. The underlying topological space of the tangent orbibundle $T\mathcal{O}$ is the quotient space

$$TO := \bigsqcup_{i \in I} (T\widetilde{U}_i/\Gamma_i)\Big/ \sim, \tag{A.14}$$

where the equivalence relation $\sim$ is defined as follows: Let $\widetilde{x}_i \in \widetilde{U}_i$, $\widetilde{x}_j \in \widetilde{U}_j$, and $\psi_{ij} : V_i \to V_j$ be as in the chart compatibility property of the definition of orbifold, so that $x_{ij} := \phi_i(\widetilde{x}_i) = \phi_j(\widetilde{x}_j) \in U_i \cap U_j$ and $\psi_{ij}(\widetilde{x}_i) = \widetilde{x}_j$. On the disjoint union $\bigsqcup_{i \in I} T\widetilde{U}_i/\Gamma_i$, we define an equivalence relation by identifying $[\widetilde{x}, v] \in T\widetilde{U}_i/\Gamma_i$ with $[\psi_{ij}(\widetilde{x}), d\psi_{ij}(v)] \in T\widetilde{U}_j/\Gamma_j$.

The projection map $p : TO \to O^n$ is defined by

$$p\big([[\widetilde{x}, v]]\big) := \phi_i(\widetilde{x}) \tag{A.15}$$

for $[\widetilde{x}, v] \in T\widetilde{U}_i/\Gamma_i$, where $\big[[\widetilde{x}, v]\big]$ denotes the equivalence class of $[\widetilde{x}, v]$ in TO. Note that each **tangent orbispace**

$$T_x O := p^{-1}(x) \quad \text{where } x \in O^n,$$

is isomorphic to

$$T_x O \cong T_{\widetilde{x}}\widetilde{U}_i/\Gamma_x \cong \mathbb{R}^n/\Gamma_x,$$

where Γ_x is the local group of x and where $\widetilde{x} \in \phi_i^{-1}(x)$. Note that this isomorphism class is independent of the choice of $\widetilde{x} \in \phi_i^{-1}(x)$; see [**ALR07**, Proposition 1.21].

Let U be an open subset of the orbifold $\mathcal{O}^n$, and let $\mathcal{U}$ denote the orbifold corresponding to U. Then the restriction $T\mathcal{O}|_{\mathcal{U}}$ of the tangent bundle of $\mathcal{O}^n$ to $\mathcal{U}$ has underlying topological space

$$TO|_U := \bigsqcup_{x \in U} T_x O.$$

We have that $TO|_U$ is equal to the underlying topological space TU of the tangent bundle of the orbifold $\mathcal{U}$. To express this, we write $T\mathcal{O}|_{\mathcal{U}} = T\mathcal{U}$.

A **section of the tangent orbibundle**, also known as a **vector field**, is defined as a map $V : O^n \to TO$ such that $p \circ V = \mathrm{id}_{O^n}$. A vector field V is **smooth** if for each $i \in I$, there exists a smooth vector field $\widetilde{V}_i$ on $\widetilde{U}_i$ such that the following diagram commutes:

$$\begin{array}{ccc} \widetilde{U}_i & \xrightarrow{\widetilde{V}_i} & T\widetilde{U}_i \\ {\scriptstyle \phi_i}\downarrow & & \downarrow{\scriptstyle \pi_i} \\ U_i & \xrightarrow{V} & TO|_{U_i} \cong T\widetilde{U}_i/\Gamma_i \end{array}, \tag{A.16}$$

where

$$\pi_i : T\widetilde{U}_i \to T\widetilde{U}_i/\Gamma_i$$

is the quotient map and where we identified $TO|_{U_i}$ with $T\widetilde{U}_i/\Gamma_i$. This condition on the vector field $V : O^n \to TO$ is equivalent to there existing a Γ_i-invariant smooth vector field $\widetilde{V}_i$ on $\widetilde{U}_i$ such that

$$V_x = \pi_i\big((\widetilde{V}_i)_{\widetilde{x}}\big) \quad \text{for all } \widetilde{x} \in \phi_i^{-1}(x).$$

This is because for $\gamma \in \Gamma_i$ and $\widetilde{x} \in \widetilde{U}_i$,

$$\pi_i\big((\widetilde{V}_i)_{\widetilde{x}}\big) = V_{\phi_i(\widetilde{x})} = V_{\phi_i(\gamma(\widetilde{x}))} = \pi_i\big((\widetilde{V}_i)_{\gamma(\widetilde{x})}\big).$$

All the vector fields that we will consider will be smooth either by necessity or by assumption.

If V is a smooth vector field on $\mathcal{O}^n$, then for any orbifold coordinate chart $(U_i, \widetilde{U}_i, \phi_i, \Gamma_i)$, V lifts to a smooth vector field $\widetilde{V}_i$ on $\widetilde{U}_i$ which is invariant under Γ_i. In particular, since the local group of an isolated singular point is non-trivial, every smooth vector field on $\mathcal{O}^n$ must vanish at all isolated singular points of $\mathcal{O}^n$. In more detail, let $x_0 \in U_i \subset O^n$ be an isolated singularity and let $\{\widetilde{x}_0\} = \phi_i^{-1}(x_0)$. Recall that $\{d\gamma_{\widetilde{x}_0} : \gamma \in \Gamma_{x_0}\} \cong \Gamma_{x_0}$. We have that $d\gamma_{\widetilde{x}_0}\big((\widetilde{V}_i)_{\widetilde{x}_0}\big) = (\widetilde{V}_i)_{\widetilde{x}_0}$ for all $\gamma \in \Gamma_{x_0}$. Since x_0 is an isolated singularity, for any $\gamma \in \Gamma_{x_0} \setminus \{1\}$ we have that the only fixed point of $d\gamma_{\widetilde{x}_0}$ is 0. Thus, $(\widetilde{V}_i)_{\widetilde{x}_0} = 0$. In summary, a vector field V on $\mathcal{O}^n$ is considered smooth if it can be locally lifted to a smooth vector field on $\widetilde{U}_i$ for each orbifold chart. This forces the vector field to be zero at any point where the local group is nontrivial.

We now describe the **cotangent orbibundle**

$$p : T^*\mathcal{O} \to \mathcal{O}^n.$$

Given an orbifold chart $(U_i, \widetilde{U}_i, \phi_i, \Gamma_i)$, the group Γ_i acts on the cotangent bundle $T^*\widetilde{U}_i$ of $\widetilde{U}_i$ by

$$\gamma(\widetilde{x}, \alpha) := \big(\gamma(\widetilde{x}), (\gamma^{-1})^*(\alpha)\big) \quad \text{for } \gamma \in \Gamma_i, \tag{A.17}$$

where $(\widetilde{x}, \alpha)$ denotes an element of $T^*\widetilde{U}_i$ so that α is a cotangent vector to $\widetilde{U}_i$ at $\widetilde{x}$. The underlying topological space of the cotangent orbibundle $T^*\mathcal{O}$ is the quotient space

$$T^*O := \bigsqcup_{i \in I} (T^*\widetilde{U}_i/\Gamma_i) \Big/ \sim, \tag{A.18}$$

where the equivalence relation $\sim$ is defined in a way analogous to that for the tangent orbibundle. For each point $x \in O^n$, the cotangent orbispace $T_x^*O := p^{-1}(x)$ is isomorphic to $T_{\widetilde{x}}^*\widetilde{U}_i/\Gamma_x$, where Γ_x is the local group of x. In the case that $(U_i, \widetilde{U}_i, \phi_i, \Gamma_i)$ is a canonical orbifold chart, we can identify $T^*\widetilde{U}_i/\Gamma_i$ with the disjoint union of the cotangent bundle $T^*(U_i \setminus \{x_0\})$ and the quotient cotangent space $(T_{0^n}^*\widetilde{U}_i)/\Gamma_i \cong \mathbb{R}^n/\Gamma_i$.

A section of the cotangent is called a **1-form**. The definition of a **smooth** 1-form is analogous to that of a smooth vector field, and we leave it to the reader to define this. Any smooth 1-form vanishes at each singular point of $\mathcal{O}^n$. If f is a smooth function on $\mathcal{O}^n$, then its **exterior derivative** df is defined in the expected way. Namely, if $x \in O^n$ and $(U_i, \widetilde{U}_i, \phi_i, \Gamma_i)$ is an orbifold chart with $x \in U_i$, then

$$df_x := \big[[d\widetilde{f}_{\widetilde{x}}]\big] \in T_x^*O,$$

where $[d\widetilde{f}_{\widetilde{x}}] \in T^*\widetilde{U}_i/\Gamma_i$. We have that df is a smooth 1-form. In particular, df vanishes at each singular point of $\mathcal{O}^n$.

We define the **tensor orbibundles** $\bigotimes^r T^*\mathcal{O} \otimes \bigotimes^s T\mathcal{O}$ in a similar way to tangent and cotangent orbibundles. Given an orbifold chart $(U_i, \widetilde{U}_i, \phi_i, \Gamma_i)$ of $\mathcal{O}^n$, we have that the group Γ_i acts on the local model tensor bundle $\bigotimes^r T^*\widetilde{U}_i \otimes \bigotimes^s T\widetilde{U}_i$, so we may define in a similar way to (A.14) and (A.18) the underlying topological space to be the quotient space

$$\bigotimes^r T^*O \otimes \bigotimes^s TO := \bigsqcup_{i \in I} \Big(\bigotimes^r T^*\widetilde{U}_i \otimes \bigotimes^s T\widetilde{U}_i\Big)/\Gamma_i \Big/ \sim.$$

Here, the action of $\gamma \in \Gamma_i$ is given by

$$\gamma\big(v_1 \otimes \cdots \otimes v_r \otimes \alpha_1 \otimes \cdots \otimes \alpha_s\big) := \gamma(v_1) \otimes \cdots \otimes \gamma(v_r) \otimes \gamma(\alpha_1) \otimes \cdots \otimes \gamma(\alpha_s)$$

and then by linear extension. The equivalence relation $\sim$ is defined in a way analogous to that for the tangent and cotangent orbibundles. An (r, s)**-tensor** is a section of the tensor orbibundle $\bigotimes^r T^*\mathcal{O} \otimes \bigotimes^s T\mathcal{O}$. An (r, s)-tensor on $\mathcal{O}^n$ is considered **smooth** if it can be locally lifted to a smooth (r, s)-tensor on $\widetilde{U}_i$ for each orbifold chart.

The local operations of tensor product, Lie bracket, Lie derivative, and the pullback of tensors extend in a natural way from manifolds to orbifolds (we leave it to the reader to work out the details of why they are well-defined).

A.2.6. Differential forms on orbifolds.

Given the earlier definition of tensors on orbifolds, we define the **bundle of** p**-forms** $\Lambda^p T^*\mathcal{O}$ as the set of *alternating* p-tensors on $\mathcal{O}^n$, which is an orbivector bundle over $\mathcal{O}^n$. A p**-form** is a section of $\Lambda^p T^*\mathcal{O}$; we will assume that all p-form are smooth. The space of smooth p-forms is denoted by $\Omega^p(\mathcal{O}^n)$.

The definition of the **exterior derivative**

$$d : \Omega^p(\mathcal{O}^n) \to \Omega^{p+1}(\mathcal{O}^n),$$

extends from manifolds to orbifolds. The reason for this is that if a differential form on $\widetilde{U}_i$ is Γ_i-invariant, then its exterior derivative is also Γ_i-invariant. Similarly, the wedge product and interior product extend in a natural way from manifolds to orbifolds.

A.2.7. Integration and Stokes's theorem on orbifolds.

Using the idea of passing to local lifts, the theory of integration of differential forms extends from manifolds to orbifolds (cf. [**Fau19**, §2.1.1]). Let $\mathcal{O}^n$ be an orbifold. Let $\{(U_i, \widetilde{U}_i, \phi_i, \Gamma_i)\}_{i\in I}$ be a locally finite orbifold atlas. Given an n-form ω with compact support in an orbifold chart U_i, we define

$$\int_{\mathcal{O}^n} \omega := \frac{1}{|\Gamma_i|} \int_{\widetilde{U}_i} \widetilde{\omega}_i, \tag{A.19}$$

where $\widetilde{\omega}_i$ is a smooth lift of ω to an n-form on $\widetilde{U}_i$.

In general, given an n-form ω with compact support in $\mathcal{O}^n$, we choose a partition of unity $\{\eta_i\}_{i\in I}$ subordinate to the orbifold open cover $\{U_i\}_{i\in I}$. For a proof of the existence of partitions of unity on orbifolds, see Barletta, Dragomir, and Duggal [**BDD07**, Theorem B.12]. We define

$$\int_{\mathcal{O}^n} \omega := \sum_{i\in I} \int_{\mathcal{O}^n} \eta_i \omega. \tag{A.20}$$

This definition is independent of the orbifold open cover and subordinate partition of unity.

We have **Stokes's theorem** on orbifolds (see e.g. Faulk [**Fau19**, Lemma 2.7]).

THEOREM A.11. *Let $\mathcal{O}^n$ be an orbifold with isolated singularities. If α is an $(n-1)$-form with compact support on $\mathcal{O}^n$, then*

$$\int_{\mathcal{O}^n} d\alpha = 0. \tag{A.21}$$

To verify this theorem, we compute that

$$\int_{\mathcal{O}^n} d\alpha = \int_{\mathcal{O}^n} d\left(\sum_{i\in I} \eta_i \alpha\right) = \sum_{i\in I} \int_{\mathcal{O}^n} d(\eta_i \alpha) = \sum_{i\in I} \frac{1}{|\Gamma_i|} \int_{\widetilde{U}_i} d(\widetilde{\eta}_i \widetilde{\alpha}_i) = 0,$$

where $\widetilde{\eta}_i$ and $\widetilde{\alpha}_i$ are the smooth local lifts to $\widetilde{U}_i$ of η_i and α, respectively, and where the vanishing on the far right-hand side follows from Stokes's theorem in the manifold case.

We can also define the **de Rham cohomology** of orbifolds in the usual way.

A.3. Basic methods of geometric analysis on orbifolds

Geometric analysis on orbifolds is for the most part the same as on manifolds.

A.3.1. Riemannian orbifolds.

DEFINITION A.12 (Riemannian orbifold). Let $\mathcal{O}^n$ be an orbifold with orbifold atlas $\{(U_i, \widetilde{U}_i, \phi_i, \Gamma_i)\}_{i\in I}$. An (orbifold) **Riemannian metric** g on $\mathcal{O}^n$ is a smooth $(2,0)$-tensor on $\mathcal{O}^n$, that is, a smooth section of the tensor orbibundle $T^*\mathcal{O} \otimes T^*\mathcal{O}$, whose local lifts to the Γ_i-invariant sections $\widetilde{g}_i$ of $T\widetilde{U}_i \otimes T\widetilde{U}_i$ are symmetric and positive-definite at each point. That is, the lift $\widetilde{g}_i$ is a Γ_i-invariant Riemannian metric on $\widetilde{U}_i$, with respect to the action of Γ_i on $\widetilde{U}_i$. We call $(\mathcal{O}^n, g)$ a **Riemannian orbifold**.

Recall that the action of Γ_i on $\widetilde{U}_i$ induces an action of Γ_i on $T\widetilde{U}_i$ which is defined by (A.12). In particular, if $x_0 = \phi_i(0^n)$ in a canonical orbifold chart, then Γ_i acts on $T_{0^n}\widetilde{U}_i$ by orthogonal transformations with respect to the inner product $(g_i)_{0^n}$ on $T_{0^n}\widetilde{U}_i$.

EXAMPLE A.13. Let Γ be a finite subgroup of $\mathrm{SO}(n)$ acting freely on $\mathbb{R}^n$. Since the Euclidean metric $g_{\mathbb{E}}$ on $\mathbb{R}^n$ is $\mathrm{SO}(n)$-invariant, it induces an orbifold Riemannian metric on the oriented orbifold $\mathbb{R}^n/\Gamma$.

Given a Riemannian metric g on an orbifold $\mathcal{O}^n$, we can define its **Levi-Civita connection** ∇ and **Riemann curvature tensor** Rm in an analogous way to that for Riemannian metrics on manifolds. In particular, ∇ is the unique torsion-free connection compatible with the metric g, and ∇ satisfies the equation[7]

$$\begin{aligned}2g(\nabla_X Y, Z) &= X(g(Y,Z)) + Y(g(X,Z)) - Z(g(X,Y)) \\ &\quad + g([X,Y],Z) - g([X,Z],Y) - g([Y,Z],X).\end{aligned}$$

This formula holds on orbifolds since it holds on the local lifts. The Riemann curvature tensor of $(\mathcal{O}^n, g)$ is defined in the usual way by

$$\mathrm{Rm}(X,Y)Z := \nabla_X(\nabla_Y Z) - \nabla_Y(\nabla_X Z) - \nabla_{[X,Y]}Z.$$

This is a smooth $(3,1)$-tensor on $\mathcal{O}^n$. By tracing, we can define the **Ricci tensor** $\mathrm{Ric}(X,W) := \sum_{i=1}^n \langle \mathrm{Rm}(X,e_i)e_i, W\rangle$ and the **scalar curvature** $R := \sum_{i=1}^n \mathrm{Ric}(e_i,e_i)$ in the usual way, which is a smooth $(2,0)$-tensor and a function, respectively.

Given a smooth function f on a Riemannian orbifold $(\mathcal{O}^n, g)$, its **Hessian** and (**rough**) **Laplacian** are defined in the usual way by

$$\nabla^2 f := \nabla df \quad \text{and} \quad \Delta f := \mathrm{tr}_g(\nabla^2 f), \tag{A.22}$$

respectively, where tr_g denotes the trace with respect to g.

Given a 1-form α on $(\mathcal{O}^n, g)$, we define its **divergence** by

$$\mathrm{div}(\alpha) := \mathrm{tr}_g(\nabla\alpha). \tag{A.23}$$

Observe that $\Delta f = \mathrm{div}(df)$.

A.3.2. Orthonormal frame orbibundle.

Given a Riemannian manifold (M^n, g), its orthonormal frame bundle $\mathrm{OF}(M^n)$ is a principal $\mathrm{O}(n)$-bundle over M^n. Given a Riemannian orbifold $(\mathcal{O}^n, g)$ with isolated singularities, we can define its **orthonormal frame orbibundle** $\mathrm{OF}(\mathcal{O}^n)$ as follows: Let $(U_i, \widetilde{U}_i, \phi_i, \Gamma_i)$ be an orbifold chart and let $\widetilde{g}_i$ be the lifted, Γ_i-invariant Riemannian metric on $\widetilde{U}_i$. The orthonormal frame bundle of $\widetilde{U}_i$ is

$$\mathrm{OF}(\widetilde{U}_i) := \bigsqcup_{\widetilde{x} \in \widetilde{U}_i} \mathrm{OF}(\widetilde{x}),$$

where

$$\mathrm{OF}(\widetilde{x}) := \left\{ \{e_k\}_{k=1}^n : (\widetilde{g}_i)_{\widetilde{x}}(e_k, e_\ell) = \delta_{k\ell} \right\}.$$

[7] The notion of connection and the definitions of torsion-free and compatible with the metric extend from manifolds to orbifolds in a natural way.

The orthonormal frame orbibundle has its underlying topological space given by

$$\mathrm{OF}(\mathcal{O}^n) := \bigsqcup_{i\in I} (\mathrm{OF}(\widetilde{U}_i)/\Gamma_i)\Big/\sim, \tag{A.24}$$

where the equivalence relation $\sim$ is defined similarly to that for the definition of the tangent orbibundle.

The orbibundle projection map

$$p : \mathrm{OF}(\mathcal{O}^n) \to \mathcal{O}^n$$

is defined by

$$p\big([[\{e_k\}_{k=1}^n]]\big) := \pi_i(\widetilde{x}),$$

where $\{e_k\}_{k=1}^n$ is an orthonormal frame at $\widetilde{x}$. The fibers are defined by $\mathrm{OF}_x(\mathcal{O}^n) := p^{-1}(x)$. Observe that if $x \in O^n_{\mathrm{reg}}$, then we may choose an orbifold chart

$$(U_i, \widetilde{U}_i, \phi_i, \Gamma_i)$$

around x with $\Gamma_i = 1$ (so that $\widetilde{U}_i = U_i$), and so the associated local orbibundle trivialization is $\pi_i : \mathrm{OF}(U_i) \to U_i$, which is the usual orthonormal frame bundle over the Riemannian manifold $(U_i, g|_{U_i})$. Since the action of the finite group Γ_i on $\mathrm{OF}(\widetilde{U}_i)/\Gamma_i$ is free for each $i \in I$, we have that $\mathrm{OF}(\mathcal{O}^n)$ is a smooth manifold of dimension $n + \dim(\mathrm{O}(n))$. In a canonical orbifold chart, we have that $\mathrm{OF}_x(\mathcal{O}^n) \cong \mathrm{O}(n)$ for $x \in U_i \setminus \{x_0\}$ and $\mathrm{OF}_x(\mathcal{O}^n) \cong \mathrm{O}(n)/\Gamma_i$, where the action of Γ_i on $\mathrm{O}(n)$ is free.

The orbifold Riemannian metric on $\mathcal{O}^n$ induces a Riemannian metric $\overline{g}$ on $\mathrm{OF}(\mathcal{O}^n)$, where the orbibundle projection map $p : \big(\mathrm{OF}(\mathcal{O}^n), \overline{g}\big) \to (\mathcal{O}^n, g)$ is an orbifold Riemannian submersion.

EXAMPLE A.14. Let $k \geq 2$ be an integer and consider the orbifold $\mathcal{O}^2_k$ with orbifold chart $(\mathbb{R}^2/\mathbb{Z}_k, \mathbb{R}^2, \phi_k, \mathbb{Z}_k)$, where $\phi_k : \mathbb{R}^2 \to \mathbb{R}^2/\mathbb{Z}_k$ is the quotient map. Let $g_{\mathbb{E}}$ be the Euclidean metric on $\mathbb{R}^2$. Let g be the smooth orbifold metric on $\mathcal{O}^2_k$ whose local lift is $g_{\mathbb{E}}$. The underlying topological space of the orthonormal frame orbibundle of $\mathcal{O}^2_k$ is $\mathrm{OF}(O^2_k) = \mathrm{O}(2) \times_{\mathbb{Z}_k} \mathbb{R}^2$. Since the diagonal action of $\mathbb{Z}_k$ on $\mathrm{O}(2) \times \mathbb{R}^2$, generated by the map $(A, z) \mapsto (\mathrm{e}^{i2\pi/k}A, \mathrm{e}^{i2\pi/k}z)$ for $A \in \mathrm{O}(2)$ and $z \in \mathbb{R}^2$, is free, we have that $\mathrm{OF}(O^2_k)$ is a smooth 3-manifold. Observe that for $[z] \in \mathbb{R}^2/\mathbb{Z}_k \setminus \{[0^2]\}$, we have $\mathrm{OF}(\mathbb{R}^2/\mathbb{Z}_k)_{[z]} = [\mathrm{O}(2) \times \{z\}] \cong \mathrm{O}(2)$, whereas $\mathrm{OF}(\mathbb{R}^2/\mathbb{Z}_k)_{[0^2]} = [\mathrm{O}(2) \times \{0^2\}] \cong \mathrm{O}(2)/\mathbb{Z}_k \cong \mathrm{O}(2)$.

The orthonormal frame orbibundle $\mathcal{O}^2_k$ is an example of a closed **Seifert fiber space**,[8] the latter of which may be characterized as a circle bundle over a closed 2-dimensional orbifold with isolated singularities (see Scott [**Sco83**, §3]).

We have a smooth, effective, and almost free[9] action of $\mathrm{O}(n)$ on the smooth manifold $\mathrm{OF}(\mathcal{O}^n)$ and the quotient space $\mathrm{OF}(\mathcal{O}^n)/\,\mathrm{O}(n)$ is orbifold diffeomorphic to $\mathcal{O}^n$; cf. [**ALR07**, Theorem 1.23].

[8] Epstein [**Eps72**] characterized Seifert fiber spaces as those 3-manifolds that can be foliated by circles.

[9] An action is **almost free** if its stabilizers are all finite.

A.3.3. The elliptic maximum principle on orbifolds.

For the elliptic maximum principle, we consider a smooth function f defined on an orbifold $\mathcal{O}^n$ with isolated singularities, and we suppose that $x_0 \in O^n$ is a maximum point of f.

If x_0 is in the regular locus, then the same results as in the manifold case apply: we have that

$$\nabla f(x_0) = 0, \quad \nabla^2 f(x_0) \leq 0, \quad \Delta f(x_0) \leq 0. \tag{A.25}$$

On the other hand, suppose that x_0 is in the singular locus. Then there exists an orbifold chart $\phi_i : \widetilde{U}_i \to U_i$ with $x_0 \in U_i$ such that $\widetilde{U}_i$ is a Euclidean ball centered at the origin 0^n and $\phi_i^{-1}(x_0) = \{0^n\}$ and Γ_i acts orthogonally on $\widetilde{U}_i$. Let $\widetilde{g}_i := \phi_i^*(g)$, which is a smooth Riemannian metric on $\widetilde{U}_i$. Define $\widetilde{f} : \widetilde{U}_i \to \mathbb{R}$ by $\widetilde{f} := f \circ \phi_i$. Then $\widetilde{f}$ attains its maximum at 0^n. Thus, by the elliptic maximum principle,

$$\widetilde{\nabla} \widetilde{f}(0^n) = 0, \quad \widetilde{\nabla^2} \widetilde{f}(0^n) \leq 0, \quad \widetilde{\Delta} \widetilde{f}(0^n) \leq 0, \tag{A.26}$$

where $\widetilde{\nabla}$, $\widetilde{\nabla^2}$, and $\widetilde{\Delta}$ are the gradient, Hessian, and Laplacian with respect to $\widetilde{g}_i$, respectively. Therefore, we again have that

$$\nabla f(x_0) = 0, \quad \nabla^2 f(x_0) \leq 0, \quad \Delta f(x_0) \leq 0. \tag{A.27}$$

A.3.4. Integration by parts on Riemannian orbifolds.

Let $(\mathcal{O}^n, g)$ be an oriented Riemannian orbifold. The **volume form** dv, which is an n-form, and the **Hodge star operator** $* : \Omega^p(\mathcal{O}^n) \to \Omega^{n-p}(\mathcal{O}^n)$ of $(\mathcal{O}^n, g)$ are defined in the usual ways. We have that the Laplace operator Δ satisfies

$$\Delta f \, dv = (-1)^n d * df := (-1)^n d(*(df)). \tag{A.28}$$

As a consequence of Theorem A.11, we have the following **integration by parts** formula:

$$\int_{\mathcal{O}^n} h \Delta f \, dv = -\int_{\mathcal{O}^n} \langle dh, df \rangle \, dv = -\int_{\mathcal{O}^n} \langle \nabla h, \nabla f \rangle \, dv, \tag{A.29}$$

where h and f are smooth functions with compact support on $\mathcal{O}^n$. In particular, for any smooth function f with compact support on $\mathcal{O}^n$,

$$\int_{\mathcal{O}^n} \Delta f \, dv = 0. \tag{A.30}$$

A.3.5. Hodge theory on orbifolds.

Let $(\mathcal{O}^n, g)$ be a Riemannian orbifold. We have the **codifferential** $\delta : \Omega^{p+1}(\mathcal{O}^n) \to \Omega^p(\mathcal{O}^n)$, which is the L^2-adjoint of the exterior derivative d. The **Hodge Laplacian** acting on p-forms is defined by

$$\Delta_d \alpha := (d\delta + \delta d)\alpha. \tag{A.31}$$

A differential form α on $\mathcal{O}^n$ is said to be **harmonic** if

$$\Delta_d \alpha = 0. \tag{A.32}$$

The **space of harmonic** p**-forms** on $\mathcal{O}^n$ is denoted by

$$\mathbf{H}^p(\mathcal{O}^n) = \{\alpha \in \Omega^p(\mathcal{O}^n) : \Delta_d \alpha = 0\}. \tag{A.33}$$

The following **Hodge decomposition theorem** holds for orbifolds (see Baily [**Bai56**]):

THEOREM A.15. *Let $(\mathcal{O}^n, g)$ be a closed Riemannian orbifold.*

(1) *Given $\gamma \in \Omega^p(\mathcal{O}^n)$, the equation*

$$\Delta_d \alpha = \gamma \tag{A.34}$$

has a solution $\alpha \in \Omega^p(\mathcal{O}^n)$ if and only if

$$\langle\langle \gamma, \beta \rangle\rangle_{L^2} = 0 \tag{A.35}$$

for all $\beta \in \mathbf{H}^p(\mathcal{O}^n)$.

(2) *Consequently, we have the following orthogonal decomposition (with respect to the L^2-inner product) of the space of p-forms:*

$$\begin{aligned} \Omega^p(\mathcal{O}^n) &= \Delta_d\big(\Omega^p(\mathcal{O}^n)\big) \oplus \mathbf{H}^p(\mathcal{O}^n) \\ &= d\delta\big(\Omega^p(\mathcal{O}^n)\big) \oplus \delta d\big(\Omega^p(\mathcal{O}^n)\big) \oplus \mathbf{H}^p(\mathcal{O}^n). \end{aligned} \tag{A.36}$$

Moreover, the space $\mathbf{H}^p(\mathcal{O}^n)$ is finite-dimensional.

As a consequence, we have the following existence theorem for solutions to the Poisson equation:

COROLLARY A.16. *Let $(\mathcal{O}^n, g)$ be a closed Riemannian orbifold. Let $\phi : O^n \to \mathbb{R}$ be a smooth function satisfying $\int_{O^n} \phi dv = 0$. Then there exists a smooth function $f : O^n \to \mathbb{R}$ satisfying the Poisson equation*

$$\Delta f = \phi. \tag{A.37}$$

The function f is unique up to an additive constant.

A.4. Short-time existence of the Ricci flow on orbifolds

We sketch a proof of the short-time existence and uniqueness Theorem 1.4 for the Ricci flow on a closed orbifold $\mathcal{O}^n$ with isolated singularities. For details in the manifold case, see for example the book by Knopf and one of the authors [**CK04**, Chapter 3].

Let g_0 be a smooth Riemannian metric on $\mathcal{O}^n$. The **Ricci–DeTurck flow** is

$$\partial_t g_t = -2\mathrm{Ric}_{g_t} + \mathcal{L}_{W_t} g_t,$$

where the vector field W_t is defined by

$$W_t^k := g_t^{ij}\big((\Gamma_t)_{ij}^k - (\Gamma_0)_{ij}^k\big).$$

This is a second-order strictly parabolic system of equations which is diffeomorphism equivalent to the Ricci flow. By the strict parabolicity, there exists $\varepsilon > 0$ and a unique solution $\widetilde{g}(t)$, $t \in [0, \varepsilon)$, to the Ricci–DeTurck flow with initial metric $\widetilde{g}(0) = g_0$ (see e.g. Taylor [**Tay11**, Proposition 8.2 in Chapter 15]).

Let $\varphi_t : \mathcal{O}^n \to \mathcal{O}^n$ be the solution to the ODE

$$\partial_t \varphi_t = -W_t$$

with initial condition $\varphi_0 = \mathrm{id}_{\mathcal{O}^n}$. Then $g(t) := \varphi_t^* \widetilde{g}(t)$, $t \in [0, \varepsilon)$, is a solution to the Ricci flow with $g(0) = g_0$.

To show that this solution is unique, suppose that $g(t)$, $t \in [0, \varepsilon)$, is a solution to the Ricci flow with $g(0) = g_0$. Let $\psi_t : \mathcal{O}^n \to \mathcal{O}^n$ be the unique solution to the **harmonic map heat flow**

$$\partial_t \psi_t = \Delta_{g(t),g_0} \psi_t$$

with $\psi_0 = \mathrm{id}_{\mathcal{O}^n}$. Here, $\Delta_{g(t),g_0}$ denotes the **map Laplacian** for smooth maps from $(\mathcal{O}^n, g(t))$ to $(\mathcal{O}^n, g_0)$; see [**CC24**, §12.8]. Then $\widetilde{g}(t) := (\psi_t^{-1})^* g(t)$ is the unique solution to the Ricci–DeTurck flow with $\widetilde{g}(0) = g_0$. The uniqueness of solutions to the Ricci flow follows easily.

Regarding this proof of existence and uniqueness, the main issue is to justify that it works on orbifolds with isolated singularities. In this regard, no essential difficulties are encountered.

A.5. Compactness theorem for orbifolds

The compactness theorem for orbifolds with isolated singularities, Theorem 1.6, is due to Lu [**Lu01**] (cf. Kleiner and Lott [**KL14**, Proposition 4.1] and [**CCG$^+$07**, Theorems 3.24 and 3.25]). In [**KL14**] the proof of this compactness theorem is simplified by passing to the orthonormal frame orbibundle of the orbifold, which is a smooth manifold. Namely, suppose that $(\mathcal{O}_i^n, g_i, x_i)$ is a sequence of pointed Riemannian orbifolds with uniformly bounded curvatures (and their derivatives) on $\mathcal{O}_i^n$ and uniformly (volume) noncollapsed at x_i. Then $(\mathrm{OF}(\mathcal{O}_i^n), \overline{g}_i, F_i)$, where F_i is an orthonormal frame at x_i, is a sequence of pointed Riemannian manifolds with uniformly bounded curvature on $\mathrm{OF}(\mathcal{O}_i^n)$ and uniformly noncollapsed at F_i. We can now apply the pointed C^∞ Cheeger–Gromov compactness theorem to the orthonormal frame bundles to obtain a smooth limit $(\mathrm{OF}_\infty, \overline{g}_\infty, F_\infty)$ of $(\mathrm{OF}(\mathcal{O}_i^n), \overline{g}_i, F_i)$. One can then deduce that $(\mathrm{OF}_\infty, \overline{g}_\infty, F_\infty)$ is the orthonormal frame orbibundle of a pointed Riemannian orbifold $(\mathcal{O}_\infty^n, g_\infty, x_\infty)$. From this one can conclude that $(\mathcal{O}_\infty^n, g_\infty, x_\infty)$ is the pointed C^∞ Cheeger–Gromov limit of $(\mathcal{O}_i^n, g_i, x_i)$.

Bibliography

[AC23] Vestislav Apostolov and Charles Cifarelli, *Hamiltonian 2-forms and new explicit Calabi–Yau metrics and gradient steady Kähler–Ricci solitons on* $\mathbb{C}^n$, J. Differential Geom. **130** (2025), no. 3, 517–570, DOI 10.4310/jdg/1749492078. MR4918214

[AIK15] Sigurd B. Angenent, James Isenberg, and Dan Knopf, *Degenerate neckpinches in Ricci flow*, J. Reine Angew. Math. **709** (2015), 81–117, DOI 10.1515/crelle-2013-0105. MR3430876

[AK22] Sigurd B. Angenent and Dan Knopf, *Ricci solitons, conical singularities, and nonuniqueness*, Geom. Funct. Anal. **32** (2022), no. 3, 411–489, DOI 10.1007/s00039-022-00601-y. MR4431121

[Ale09] Spyros Alexakis, *Unique continuation for the vacuum Einstein equations*, arXiv:0902.1131, 2009.

[ALR07] Alejandro Adem, Johann Leida, and Yongbin Ruan, *Orbifolds and stringy topology*, Cambridge Tracts in Mathematics, vol. 171, Cambridge University Press, Cambridge, 2007, DOI 10.1017/CBO9780511543081. MR2359514

[And89] Michael T. Anderson, *Ricci curvature bounds and Einstein metrics on compact manifolds*, J. Amer. Math. Soc. **2** (1989), no. 3, 455–490, DOI 10.2307/1990939. MR999661

[And08] Michael T. Anderson, *Einstein metrics with prescribed conformal infinity on 4-manifolds*, Geom. Funct. Anal. **18** (2008), no. 2, 305–366, DOI 10.1007/s00039-008-0668-5. MR2421542

[App17] Alexander Appleton, *A family of non-collapsed steady Ricci solitons in even dimensions greater or equal to four*, arXiv:1708.00161, 2017.

[App23] Alexander Appleton, *Eguchi-Hanson singularities in* $U(2)$*-invariant Ricci flow*, Peking Math. J. **6** (2023), no. 1, 1–141, DOI 10.1007/s42543-022-00048-y. MR4552641

[APS75a] M. F. Atiyah, V. K. Patodi, and I. M. Singer, *Spectral asymmetry and Riemannian geometry. I*, Math. Proc. Cambridge Philos. Soc. **77** (1975), 43–69, DOI 10.1017/S0305004100049410. MR397797

[APS75b] M. F. Atiyah, V. K. Patodi, and I. M. Singer, *Spectral asymmetry and Riemannian geometry. II*, Math. Proc. Cambridge Philos. Soc. **78** (1975), 405–432, DOI 10.1017/S0305004100051872. MR397798

[Art25] Emil Artin, *Zur Isotopie zweidimensionaler Flächen im* R_4 (German), Abh. Math. Sem. Univ. Hamburg **4** (1925), no. 1, 174–177, DOI 10.1007/BF02950724. MR3069446

[Aub82] Thierry Aubin, *Nonlinear analysis on manifolds. Monge-Ampère equations*, Grundlehren der mathematischen Wissenschaften [Fundamental Principles of Mathematical Sciences], vol. 252, Springer-Verlag, New York, 1982, DOI 10.1007/978-1-4612-5734-9. MR681859

[Bai56] Walter L. Baily Jr., *The decomposition theorem for* V*-manifolds*, Amer. J. Math. **78** (1956), 862–888, DOI 10.2307/2372472. MR100103

[Bam] Richard H. Bamler, *Personal communication*, 2025.

[Bam18a] Richard H. Bamler, *Long-time behavior of 3-dimensional Ricci flow: introduction*, Geom. Topol. **22** (2018), no. 2, 757–774, DOI 10.2140/gt.2018.22.757. MR3748679

[Bam18b] Richard H. Bamler, *Long-time behavior of 3-dimensional Ricci flow: A: Generalizations of Perelman's long-time estimates*, Geom. Topol. **22** (2018), no. 2, 775–844, DOI 10.2140/gt.2018.22.775. MR3748680

[Bam18c] Richard H. Bamler, *Long-time behavior of 3-dimensional Ricci flow: B: Evolution of the minimal area of simplicial complexes under Ricci flow*, Geom. Topol. **22** (2018), no. 2, 845–892, DOI 10.2140/gt.2018.22.845. MR3748681

[Bam18d] Richard H. Bamler, *Long-time behavior of 3-dimensional Ricci flow: C: 3-manifold topology and combinatorics of simplicial complexes in 3-manifolds*, Geom. Topol. **22** (2018), no. 2, 893–948, DOI 10.2140/gt.2018.22.893. MR3748682

[Bam18e] Richard H. Bamler, *Long-time behavior of 3-dimensional Ricci flow: D: Proof of the main results*, Geom. Topol. **22** (2018), no. 2, 949–1068, DOI 10.2140/gt.2018.22.949. MR3748683

[Bam20a] Richard H. Bamler, *Entropy and heat kernel bounds on a Ricci flow background*, `arXiv:2008.07093v3`, 2020.

[Bam20b] Richard H. Bamler, *Structure theory of non-collapsed limits of Ricci flows*, `arXiv:2009.03243v2`, 2020.

[Bam21a] Richard H. Bamler, *On the fundamental group of non-collapsed ancient Ricci flows*, `arXiv:2110.02254`, 2021.

[Bam21b] Richard H. Bamler, *Recent developments in Ricci flows*, Notices Amer. Math. Soc. **68** (2021), no. 9, 1486–1498, DOI 10.1090/noti2343. MR4323820

[Bam22] Richard H. Bamler, *New developments in Ricci flow with surgery*, Surveys in differential geometry 2020. Surveys in 3-manifold topology and geometry, Surv. Differ. Geom., vol. 25, Int. Press, Boston, MA, 2022, pp. 1–31. MR4479748

[Bam23] Richard H. Bamler, *Compactness theory of the space of super Ricci flows*, Invent. Math. **233** (2023), no. 3, 1121–1277, DOI 10.1007/s00222-023-01196-3. MR4623543

[Ban87] Shigetoshi Bando, *Real analyticity of solutions of Hamilton's equation*, Math. Z. **195** (1987), no. 1, 93–97, DOI 10.1007/BF01161602. MR888130

[BBI01] Dmitri Burago, Yuri Burago, and Sergei Ivanov, *A course in metric geometry*, Graduate Studies in Mathematics, vol. 33, American Mathematical Society, Providence, RI, 2001, DOI 10.1090/gsm/033. MR1835418

[BC23] Richard H. Bamler and Eric Chen, *Degree theory for 4-dimensional asymptotically conical gradient expanding solitons*, `arXiv:2305.03154`, 2023.

[BCCD24] Richard H. Bamler, Charles Cifarelli, Ronan J. Conlon, and Alix Deruelle, *A new complete two-dimensional shrinking gradient Kähler-Ricci soliton*, Geom. Funct. Anal. **34** (2024), no. 2, 377–392, DOI 10.1007/s00039-024-00668-9. MR4715366

[BCD+22] Richard H. Bamler, Bennett Chow, Yuxing Deng, Zilu Ma, and Yongjia Zhang, *Four-dimensional steady gradient Ricci solitons with 3-cylindrical tangent flows at infinity*, Adv. Math. **401** (2022), Paper No. 108285, 21, DOI 10.1016/j.aim.2022.108285. MR4392220

[BCJN] Richard Bamler, Bennett Chow, Wenshuai Jiang, and Aaron Naber, *Lectures on Ricci curvature*, In preparation.

[BCMZ] Richard H. Bamler, Pak-Yeung Chan, Zilu Ma, and Yongjia Zhang, *An optimal volume growth estimate for noncollapsed steady gradient Ricci solitons*, Peking Math. J. **6** (2023), no. 2, 353–364, DOI 10.1007/s42543-023-00060-w. MR4619596

[BDD07] Elisabetta Barletta, Sorin Dragomir, and Krishan L. Duggal, *Foliations in Cauchy-Riemann geometry*, Mathematical Surveys and Monographs, vol. 140, American Mathematical Society, Providence, RI, 2007, DOI 10.1090/surv/140. MR2319199

[BE85] D. Bakry and Michel Émery, *Diffusions hypercontractives* (French), Séminaire de probabilités, XIX, 1983/84, Lecture Notes in Math., vol. 1123, Springer, Berlin, 1985, pp. 177–206, DOI 10.1007/BFb0075847. MR889476

[Bes08] Arthur L. Besse, *Einstein manifolds*, Classics in Mathematics, Springer-Verlag, Berlin, 2008. Reprint of the 1987 edition. MR2371700

[BGR07] Florin Alexandru Belgun, Nicolas Ginoux, and Hans-Bert Rademacher, *A singularity theorem for twistor spinors* (English, with English and French summaries), Ann. Inst. Fourier (Grenoble) **57** (2007), no. 4, 1135–1159, DOI 10.5802/aif.2289. MR2339323

[BK19] Richard H. Bamler and Bruce Kleiner, *Ricci flow and contractibility of spaces of metrics*, `arXiv:1909.08710`, 2019.

[BK21] Richard H. Bamler and Bruce Kleiner, *On the rotational symmetry of 3-dimensional κ-solutions*, J. Reine Angew. Math. **779** (2021), 37–55, DOI 10.1515/crelle-2021-0037. MR4319063

[BK22] Richard H. Bamler and Bruce Kleiner, *Uniqueness and stability of Ricci flow through singularities*, Acta Math. **228** (2022), no. 1, 1–215, DOI 10.4310/acta.2022.v228.n1.a1. MR4448680

[BK23] Richard H. Bamler and Bruce Kleiner, *Ricci flow and diffeomorphism groups of 3-manifolds*, J. Amer. Math. Soc. **36** (2023), no. 2, 563–589, DOI 10.1090/jams/1003. MR4536904

[BK24] Richard H. Bamler and Bruce Kleiner, *Diffeomorphism groups of prime 3-manifolds*, J. Reine Angew. Math. **806** (2024), 23–35, DOI 10.1515/crelle-2023-0069. MR4685084

[BKK+21] Stefan Behrens, Boldizsár Kalmár, Min Hoon Kim, Mark Powell, and Arunima Ray (eds.), *The disc embedding theorem*, Oxford University Press, Oxford, 2021. MR4519498

[BKN89] Shigetoshi Bando, Atsushi Kasue, and Hiraku Nakajima, *On a construction of coordinates at infinity on manifolds with fast curvature decay and maximal volume growth*, Invent. Math. **97** (1989), no. 2, 313–349, DOI 10.1007/BF01389045. MR1001844

[BL02] Lev A. Borisov and Anatoly Libgober, *Elliptic genera of singular varieties, orbifold elliptic genus and chiral de Rham complex*, Mirror symmetry, IV (Montreal, QC, 2000), AMS/IP Stud. Adv. Math., vol. 33, Amer. Math. Soc., Providence, RI, 2002, pp. 325–342, DOI 10.1090/amsip/033/22. MR1969036

[BL03] Lev Borisov and Anatoly Libgober, *Elliptic genera of singular varieties*, Duke Math. J. **116** (2003), no. 2, 319–351, DOI 10.1215/S0012-7094-03-11625-7. MR1953295

[BLP05] Michel Boileau, Bernhard Leeb, and Joan Porti, *Geometrization of 3-dimensional orbifolds*, Ann. of Math. (2) **162** (2005), no. 1, 195–290, DOI 10.4007/annals.2005.162.195. MR2178962

[BM15] Jacob Bernstein and Thomas Mettler, *Two-dimensional gradient Ricci solitons revisited*, Int. Math. Res. Not. IMRN **1** (2015), 78–98, DOI 10.1093/imrn/rnt177. MR3340295

[BMP03] Michel Boileau, Sylvain Maillot, and Joan Porti, *Three-dimensional orbifolds and their geometric structures* (English, with English and French summaries), Panoramas et Synthèses [Panoramas and Syntheses], vol. 15, Société Mathématique de France, Paris, 2003. MR2060653

[Boc56] S. Bochner, *Weak solutions of linear partial differential equations*, J. Math. Pures Appl. (9) **35** (1956), 193–202. MR81446

[Bre10] Simon Brendle, *Ricci flow and the sphere theorem*, Graduate Studies in Mathematics, vol. 111, American Mathematical Society, Providence, RI, 2010, DOI 10.1090/gsm/111. MR2583938

[Bre13] Simon Brendle, *Rotational symmetry of self-similar solutions to the Ricci flow*, Invent. Math. **194** (2013), no. 3, 731–764, DOI 10.1007/s00222-013-0457-0. MR3127066

[Bre14] Simon Brendle, *Rotational symmetry of Ricci solitons in higher dimensions*, J. Differential Geom. **97** (2014), no. 2, 191–214. MR3231974

[Bre20] Simon Brendle, *Ancient solutions to the Ricci flow in dimension 3*, Acta Math. **225** (2020), no. 1, 1–102, DOI 10.4310/acta.2020.v225.n1.a1. MR4176064

[Bre22] Simon Brendle, *Singularity models in the three-dimensional Ricci flow*, Recent progress in mathematics, KIAS Springer Ser. Math., vol. 1, Springer, Singapore, 2022, pp. 87–118, DOI 10.1007/978-981-19-3708-8_3. MR4510955

[Bry] Robert L. Bryant, *Ricci flow solitons in dimension three with $SO(3)$-symmetries*, `http://www.math.duke.edu/bryant/3DRotSymRicciSolitons.pdf`.

[BS11] Simon Brendle and Richard Schoen, *Curvature, sphere theorems, and the Ricci flow*, Bull. Amer. Math. Soc. (N.S.) **48** (2011), no. 1, 1–32, DOI 10.1090/S0273-0979-2010-01312-4. MR2738904

[But21] Timothy Buttsworth, *$su(2)$-invariant steady gradient Ricci solitons on four-manifolds*, `arXiv:2111.12807`, 2021.

[Cao96] Huai-Dong Cao, *Existence of gradient Kähler-Ricci solitons*, Elliptic and parabolic methods in geometry (Minneapolis, MN, 1994), A K Peters, Wellesley, MA, 1996, pp. 1–16. MR1417944

[Cao10] Huai-Dong Cao, *Recent progress on Ricci solitons*, Recent advances in geometric analysis, Adv. Lect. Math. (ALM), vol. 11, Int. Press, Somerville, MA, 2010, pp. 1–38. MR2648937

[Cao11] Huai-Dong Cao, *Geometry of complete gradient shrinking Ricci solitons*, Geometry and analysis. No. 1, Adv. Lect. Math. (ALM), vol. 17, Int. Press, Somerville, MA, 2011, pp. 227–246. MR2882424

[Car39] T. Carleman, *Sur un problème d'unicité pur les systèmes d'équations aux dérivées partielles à deux variables indépendantes* (French), Ark. Mat. Astr. Fys. **26** (1939), no. 17, 9. MR334

[CC95] Bennett Chow and Sun-Chin Chu, *A geometric interpretation of Hamilton's Harnack inequality for the Ricci flow*, Math. Res. Lett. **2** (1995), no. 6, 701–718, DOI 10.4310/MRL.1995.v2.n6.a4. MR1362964

[CC96] Bennett Chow and Sun-Chin Chu, *A geometric approach to the linear trace Harnack inequality for the Ricci flow*, Math. Res. Lett. **3** (1996), no. 4, 549–568, DOI 10.4310/MRL.1996.v3.n4.a13. MR1406020

[CC13] Huai-Dong Cao and Qiang Chen, *On Bach-flat gradient shrinking Ricci solitons*, Duke Math. J. **162** (2013), no. 6, 1149–1169, DOI 10.1215/00127094-2147649. MR3053567

[CC20] Huai-Dong Cao and Xin Cui, *Curvature estimates for four-dimensional gradient steady Ricci solitons*, J. Geom. Anal. **30** (2020), no. 1, 511–525, DOI 10.1007/s12220-019-00152-z. MR4058524

[CC24] Bennett Chow and Yutze Chow, *Lectures on differential geometry*, Graduate Studies in Mathematics, vol. 245, American Mathematical Society, Providence, RI, 2024. MR4809877

[CCD22] Charles Cifarelli, Ronan J. Conlon, and Alix Deruelle, *On finite time type I singularities of the Kähler–Ricci flow on compact Kähler surfaces*, arXiv:2203.04380, to appear in J. Eur. Math. Soc., 2022.

[CCG+07] Bennett Chow, Sun-Chin Chu, David Glickenstein, Christine Guenther, James Isenberg, Tom Ivey, Dan Knopf, Peng Lu, Feng Luo, and Lei Ni, *The Ricci flow: techniques and applications. Part I*, Mathematical Surveys and Monographs, vol. 135, American Mathematical Society, Providence, RI, 2007. Geometric aspects, DOI 10.1090/surv/135. MR2302600

[CCG+08] Bennett Chow, Sun-Chin Chu, David Glickenstein, Christine Guenther, James Isenberg, Tom Ivey, Dan Knopf, Peng Lu, Feng Luo, and Lei Ni, *The Ricci flow: techniques and applications. Part II*, Mathematical Surveys and Monographs, vol. 144, American Mathematical Society, Providence, RI, 2008. Analytic aspects, DOI 10.1090/surv/144. MR2365237

[CCG+10] Bennett Chow, Sun-Chin Chu, David Glickenstein, Christine Guenther, James Isenberg, Tom Ivey, Dan Knopf, Peng Lu, Feng Luo, and Lei Ni, *The Ricci flow: techniques and applications. Part III. Geometric-analytic aspects*, Mathematical Surveys and Monographs, vol. 163, American Mathematical Society, Providence, RI, 2010, DOI 10.1090/surv/163. MR2604955

[CCG+15] Bennett Chow, Sun-Chin Chu, David Glickenstein, Christine Guenther, James Isenberg, Tom Ivey, Dan Knopf, Peng Lu, Feng Luo, and Lei Ni, *The Ricci flow: techniques and applications. Part IV*, Mathematical Surveys and Monographs, vol. 206, American Mathematical Society, Providence, RI, 2015. Long-time solutions and related topics, DOI 10.1090/surv/206. MR3409114

[CCL24] Pak-Yeung Chan, Ronan Conlon, and Yi Lai, *A family of Kähler flying wing steady Ricci solitons*, arXiv:2403.04089, 2024.

[CCMZ23] Pak-Yeung Chan, Bennett Chow, Zilu Ma, and Yongjia Zhang, *Lower bounds for the scalar curvatures of Ricci flow singularity models*, J. Reine Angew. Math. **794** (2023), 253–265, DOI 10.1515/crelle-2022-0086. MR4529416

[CCZ08] Huai-Dong Cao, Bing-Long Chen, and Xi-Ping Zhu, *Recent developments on Hamilton's Ricci flow*, Surveys in differential geometry. Vol. XII. Geometric flows, Surv. Differ. Geom., vol. 12, Int. Press, Somerville, MA, 2008, pp. 47–112, DOI 10.4310/SDG.2007.v12.n1.a3. MR2488948

[CD15] Chih-Wei Chen and Alix Deruelle, *Structure at infinity of expanding gradient Ricci soliton*, Asian J. Math. **19** (2015), no. 5, 933–950, DOI 10.4310/AJM.2015.v19.n5.a6. MR3431684

[CD20] Ronan J. Conlon and Alix Deruelle, *Expanding Kähler-Ricci solitons coming out of Kähler cones*, J. Differential Geom. **115** (2020), no. 2, 303–365, DOI 10.4310/jdg/1589853627. MR4100705

[CDM22] Bennett Chow, Yuxing Deng, and Zilu Ma, *On four-dimensional steady gradient Ricci solitons that dimension reduce*, Adv. Math. **403** (2022), Paper No. 108367, 61, DOI 10.1016/j.aim.2022.108367. MR4405369

[CDS24] Ronan J. Conlon, Alix Deruelle, and Song Sun, *Classification results for expanding and shrinking gradient Kähler-Ricci solitons*, Geom. Topol. **28** (2024), no. 1, 267–351, DOI 10.2140/gt.2024.28.267. MR4711837

[CE08] Jeff Cheeger and David G. Ebin, *Comparison theorems in Riemannian geometry*, AMS Chelsea Publishing, Providence, RI, 2008. Revised reprint of the 1975 original, DOI 10.1090/chel/365. MR2394158

[CFG92] Jeff Cheeger, Kenji Fukaya, and Mikhael Gromov, *Nilpotent structures and invariant metrics on collapsed manifolds*, J. Amer. Math. Soc. **5** (1992), no. 2, 327–372, DOI 10.2307/2152771. MR1126118

[CFSZ20] Bennett Chow, Michael Freedman, Henry Shin, and Yongjia Zhang, *Curvature growth of some 4-dimensional gradient Ricci soliton singularity models*, Adv. Math. **372** (2020), 107303, 17, DOI 10.1016/j.aim.2020.107303. MR4128573

[CFSZ25] Bennett Chow, Michael Freedman, Henry Shin, and Yongjia Zhang, *Correction to: Curvature growth of some* 4*-dimensional gradient Ricci soliton singularity models*, `arXiv:2505.03823`, 2025.

[CG86] Jeff Cheeger and Mikhael Gromov, *Collapsing Riemannian manifolds while keeping their curvature bounded. I*, J. Differential Geom. **23** (1986), no. 3, 309–346. MR852159

[CG90] Jeff Cheeger and Mikhael Gromov, *Collapsing Riemannian manifolds while keeping their curvature bounded. II*, J. Differential Geom. **32** (1990), no. 1, 269–298. MR1064875

[Cha19] Pak-Yeung Chan, *Curvature estimates for steady Ricci solitons*, Trans. Amer. Math. Soc. **372** (2019), no. 12, 8985–9008, DOI 10.1090/tran/7954. MR4029719

[Cha23] Pak-Yeung Chan, *Curvature estimates and gap theorems for expanding Ricci solitons*, Int. Math. Res. Not. IMRN **1** (2023), 406–454, DOI 10.1093/imrn/rnab257. MR4530113

[Che44] Shiing-shen Chern, *A simple intrinsic proof of the Gauss-Bonnet formula for closed Riemannian manifolds*, Ann. of Math. (2) **45** (1944), 747–752, DOI 10.2307/1969302. MR11027

[Che75] Shiu Yuen Cheng, *Eigenvalue comparison theorems and its geometric applications*, Math. Z. **143** (1975), no. 3, 289–297, DOI 10.1007/BF01214381. MR378001

[Che09] Bing-Long Chen, *Strong uniqueness of the Ricci flow*, J. Differential Geom. **82** (2009), no. 2, 363–382. MR2520796

[Chi90] Yuan-Jen Chiang, *Harmonic maps of V-manifolds*, Ann. Global Anal. Geom. **8** (1990), no. 3, 315–344, DOI 10.1007/BF00127941. MR1089240

[CHK00] Daryl Cooper, Craig D. Hodgson, and Steven P. Kerckhoff, *Three-dimensional orbifolds and cone-manifolds*, MSJ Memoirs, vol. 5, Mathematical Society of Japan, Tokyo, 2000. With a postface by Sadayoshi Kojima. MR1778789

[CHM25] Ronan Conlon, Max Hallgren, and Zilu Ma, *Non-collapsed finite time singularities of the Ricci flow on compact Kähler surfaces are of Type i*, `arXiv:2502.19804`, 2025.

[Cho91a] Bennett Chow, *On the entropy estimate for the Ricci flow on compact* 2*-orbifolds*, J. Differential Geom. **33** (1991), no. 2, 597–600. MR1094471

[Cho91b] Bennett Chow, *The Ricci flow on the* 2*-sphere*, J. Differential Geom. **33** (1991), no. 2, 325–334. MR1094458

[Cho14] Otis Chodosh, *Expanding Ricci solitons asymptotic to cones*, Calc. Var. Partial Differential Equations **51** (2014), no. 1-2, 1–15, DOI 10.1007/s00526-013-0664-y. MR3247379

[Cho23] Bennett Chow, *Ricci solitons in low dimensions*, Graduate Studies in Mathematics, vol. 235, American Mathematical Society, Providence, RI, 2023, DOI 10.1090/gsm/235. MR4624811

[Cho25] Bennett Chow, *Small improvements for a trio of estimates for Ricci solitons, Proc. Amer. Math. Soc.* **153** (2025), no. 7, 3153–3157, DOI 10.1090/proc/16545. MR4912143

[CK04] Bennett Chow and Dan Knopf, *The Ricci flow: an introduction*, Mathematical Surveys and Monographs, vol. 110, American Mathematical Society, Providence, RI, 2004, DOI 10.1090/surv/110. MR2061425

[CK20] Mary Cook and Brett Kotschwar, *The full holonomy group under the Ricci flow*, Proc. Amer. Math. Soc. **148** (2020), no. 12, 5313–5321, DOI 10.1090/proc/15156. MR4163843

[CL22] Huai-Dong Cao and Tianbo Liu, *Curvature estimates for 4-dimensional complete gradient expanding Ricci solitons*, J. Reine Angew. Math. **790** (2022), 115–135, DOI 10.1515/crelle-2022-0039. MR4472870

[CLN06] Bennett Chow, Peng Lu, and Lei Ni, *Hamilton's Ricci flow*, Graduate Studies in Mathematics, vol. 77, American Mathematical Society, Providence, RI; Science Press Beijing, New York, 2006, DOI 10.1090/gsm/077. MR2274812

[CLT06] Xiuxiong Chen, Peng Lu, and Gang Tian, *A note on uniformization of Riemann surfaces by Ricci flow*, Proc. Amer. Math. Soc. **134** (2006), no. 11, 3391–3393, DOI 10.1090/S0002-9939-06-08360-2. MR2231924

[CLY11] Bennett Chow, Peng Lu, and Bo Yang, *Lower bounds for the scalar curvatures of noncompact gradient Ricci solitons* (English, with English and French summaries), C. R. Math. Acad. Sci. Paris **349** (2011), no. 23-24, 1265–1267, DOI 10.1016/j.crma.2011.11.004. MR2861997

[CM21] Tobias Holck Colding and William P. Minicozzi, II, *Singularities of Ricci flow and diffeomorphisms*, arXiv:2109.06240, 2021.

[CMZ21] Pak-Yeung Chan, Zilu Ma, and Yongjia Zhang, *Hamilton-Ivey estimates for gradient Ricci solitons*, arXiv:2112.11025, 2021.

[CMZ22] Pak-Yeung Chan, Zilu Ma, and Yongjia Zhang, *The rate of* $\mathbb{F}$*-convergence for Ricci flows with closed and smooth tangent flows*, J. Geom. Anal. **32** (2022), no. 12, Paper No. 293, 28, DOI 10.1007/s12220-022-01023-w. MR4487757

[CMZ24] Pak-Yeung Chan, Zilu Ma, and Yongjia Zhang, *On noncollapsed* $\mathbb{F}$*-limit metric solitons*, arXiv:2401.03387, 2024.

[CN15] Jeff Cheeger and Aaron Naber, *Regularity of Einstein manifolds and the codimension 4 conjecture*, Ann. of Math. (2) **182** (2015), no. 3, 1093–1165, DOI 10.4007/annals.2015.182.3.5. MR3418535

[CRZ21] Huai-Dong Cao, Ernani Ribeiro Jr., and Detang Zhou, *Four-dimensional complete gradient shrinking Ricci solitons*, J. Reine Angew. Math. **778** (2021), 127–144, DOI 10.1515/crelle-2021-0026. MR4308617

[CT24] Xiaodong Cao and Hung Tran, *Geometry and analysis of gradient Ricci solitons in dimension four*, Surveys in differential geometry 2022. Essays on geometric flows—celebrating 40 years of Ricci flow, Surv. Differ. Geom., vol. 27, Int. Press, Somerville, MA, 2024, pp. 213–233. MR4862087

[CW91] Bennett Chow and Lang-Fang Wu, *The Ricci flow on compact 2-orbifolds with curvature negative somewhere*, Comm. Pure Appl. Math. **44** (1991), no. 3, 275–286, DOI 10.1002/cpa.3160440302. MR1090433

[CW13] Xiuxiong Chen and Bing Wang, *On the conditions to extend Ricci flow(III)*, Int. Math. Res. Not. IMRN **10** (2013), 2349–2367, DOI 10.1093/imrn/rns117. MR3061942

[CW17] Xiuxiong Chen and Bing Wang, *Space of Ricci flows (II)—Part A: Moduli of singular Calabi-Yau spaces*, Forum Math. Sigma **5** (2017), Paper No. e32, 103. MR3739253

[CXZ13] Bing-Long Chen, Guoyi Xu, and Zhuhong Zhang, *Local pinching estimates in 3-dim Ricci flow*, Math. Res. Lett. **20** (2013), no. 5, 845–855, DOI 10.4310/MRL.2013.v20.n5.a3. MR3207356

[CY75] S. Y. Cheng and S. T. Yau, *Differential equations on Riemannian manifolds and their geometric applications*, Comm. Pure Appl. Math. **28** (1975), no. 3, 333–354, DOI 10.1002/cpa.3160280303. MR385749

[CZ06] Bing-Long Chen and Xi-Ping Zhu, *Uniqueness of the Ricci flow on complete noncompact manifolds*, J. Differential Geom. **74** (2006), no. 1, 119–154. MR2260930

[CZ10] Huai-Dong Cao and Detang Zhou, *On complete gradient shrinking Ricci solitons*, J. Differential Geom. **85** (2010), no. 2, 175–185. MR2732975

[CZ12] Qiang Chen and Meng Zhu, *On rigidity of gradient Kähler-Ricci solitons with harmonic Bochner tensor*, Proc. Amer. Math. Soc. **140** (2012), no. 11, 4017–4025, DOI 10.1090/S0002-9939-2012-11648-X. MR2944741

[Der16] Alix Deruelle, *Smoothing out positively curved metric cones by Ricci expanders*, Geom. Funct. Anal. **26** (2016), no. 1, 188–249, DOI 10.1007/s00039-016-0360-0. MR3494489

[Der17a] Alix Deruelle, *Asymptotic estimates and compactness of expanding gradient Ricci solitons*, Ann. Sc. Norm. Super. Pisa Cl. Sci. (5) **17** (2017), no. 2, 485–530. MR3700376

[Der17b] Alix Deruelle, *Unique continuation at infinity for conical Ricci expanders*, Int. Math. Res. Not. IMRN (2017), no. 10, 3107–3147. MR3658133

[DeT83] Dennis M. DeTurck, *Deforming metrics in the direction of their Ricci tensors*, J. Differential Geom. **18** (1983), no. 1, 157–162. MR697987

[DeT03] Dennis M. DeTurck, *Deforming metrics in the direction of their Ricci tensors (improved version)*, Collected papers on the Ricci flow, Series in Geometry and Topology, vol. 37, Int. Press, Somerville, MA, 2003, pp. 163–165.

[DK90] S. K. Donaldson and P. B. Kronheimer, *The geometry of four-manifolds*, Oxford Mathematical Monographs, The Clarendon Press, Oxford University Press, New York, 1990. Oxford Science Publications. MR1079726

[DLM02] Chongying Dong, Kefeng Liu, and Xiaonan Ma, *On orbifold elliptic genus*, Orbifolds in mathematics and physics (Madison, WI, 2001), Contemp. Math., vol. 310, Amer. Math. Soc., Providence, RI, 2002, pp. 87–105, DOI 10.1090/conm/310/05399. MR1950942

[Don83] S. K. Donaldson, *An application of gauge theory to four-dimensional topology*, J. Differential Geom. **18** (1983), no. 2, 279–315. MR710056

[Don87] S. K. Donaldson, *Irrationality and the h-cobordism conjecture*, J. Differential Geom. **26** (1987), no. 1, 141–168. MR892034

[dP87] Pasquale del Pezzo, *Sulle superficie dell'n^{mo} ordine immerse nello spazio din dimensioni*, Rendiconti del Circolo Matematico di Palermo **1** (1887), 241–271.

[DS23] Alix Deruelle and Felix Schulze, *A relative entropy and a unique continuation result for Ricci expanders*, Comm. Pure Appl. Math. **76** (2023), no. 10, 2613–2692, DOI 10.1002/cpa.22086. MR4630599

[DW07] Xianzhe Dai and Guofang Wei, *Hitchin-Thorpe inequality for noncompact Einstein 4-manifolds*, Adv. Math. **214** (2007), no. 2, 551–570, DOI 10.1016/j.aim.2007.02.010. MR2349712

[DW09] Andrew S. Dancer and McKenzie Y. Wang, *Some new examples of non-Kähler Ricci solitons*, Math. Res. Lett. **16** (2009), no. 2, 349–363, DOI 10.4310/MRL.2009.v16.n2.a11. MR2496749

[DW11] Andrew S. Dancer and McKenzie Y. Wang, *On Ricci solitons of cohomogeneity one*, Ann. Global Anal. Geom. **39** (2011), no. 3, 259–292, DOI 10.1007/s10455-010-9233-1. MR2769300

[DZ20a] Yuxing Deng and Xiaohua Zhu, *Classification of gradient steady Ricci solitons with linear curvature decay*, Sci. China Math. **63** (2020), no. 1, 135–154, DOI 10.1007/s11425-019-1548-0. MR4050577

[DZ20b] Yuxing Deng and Xiaohua Zhu, *Higher dimensional steady Ricci solitons with linear curvature decay*, J. Eur. Math. Soc. (JEMS) **22** (2020), no. 12, 4097–4120, DOI 10.4171/jems/1003. MR4176787

[EF03] Luis Escauriaza and Francisco Javier Fernández, *Unique continuation for parabolic operators*, Ark. Mat. **41** (2003), no. 1, 35–60, DOI 10.1007/BF02384566. MR1971939

[EFV06] L. Escauriaza, F. J. Fernández, and S. Vessella, *Doubling properties of caloric functions*, Appl. Anal. **85** (2006), no. 1-3, 205–223, DOI 10.1080/00036810500277082. MR2198840

[EKPV06] L. Escauriaza, C. E. Kenig, G. Ponce, and L. Vega, *Decay at infinity of caloric functions within characteristic hyperplanes*, Math. Res. Lett. **13** (2006), no. 2-3, 441–453, DOI 10.4310/MRL.2006.v13.n3.a8. MR2231129

[ELNM08] Manolo Eminenti, Gabriele La Nave, and Carlo Mantegazza, *Ricci solitons: the equation point of view*, Manuscripta Math. **127** (2008), no. 3, 345–367, DOI 10.1007/s00229-008-0210-y. MR2448435

[EMT11] Joerg Enders, Reto Müller, and Peter M. Topping, *On type-I singularities in Ricci flow*, Comm. Anal. Geom. **19** (2011), no. 5, 905–922, DOI 10.4310/CAG.2011.v19.n5.a4. MR2886712

[Eps60] D. B. A. Epstein, *Linking spheres*, Proc. Cambridge Philos. Soc. **56** (1960), 215–219, DOI 10.1017/s0305004100034496. MR117740

[Eps72] D. B. A. Epstein, *Periodic flows on three-manifolds*, Ann. of Math. (2) **95** (1972), 66–82, DOI 10.2307/1970854. MR288785

[Esc00] Luis Escauriaza, *Carleman inequalities and the heat operator*, Duke Math. J. **104** (2000), no. 1, 113–127, DOI 10.1215/S0012-7094-00-10415-2. MR1769727

[Esp25] Carlos Esparza, *Shrinking gradient Kähler-Ricci solitons are simply-connected*, `arXiv:2503.05838`, 2025.

[ESŠ03a] L. Escauriaza, G. Seregin, and V. Šverák, *Backward uniqueness for parabolic equations*, Arch. Ration. Mech. Anal. **169** (2003), no. 2, 147–157, DOI 10.1007/s00205-003-0263-8. MR2005639

[ESŠ03b] L. Escauriaza, G. Seregin, and V. Šverák, *Backward uniqueness for the heat operator in half-space*, Algebra i Analiz **15** (2003), no. 1, 201–214. MR1979722

[Fau19] Mitchell Faulk, *Some canonical metrics on Kähler orbifolds*, Thesis (Ph.D.)—Columbia University, 2019.

[FIK03] Mikhail Feldman, Tom Ilmanen, and Dan Knopf, *Rotationally symmetric shrinking and expanding gradient Kähler-Ricci solitons*, J. Differential Geom. **65** (2003), no. 2, 169–209. MR2058261

[FM88a] Robert Friedman and John W. Morgan, *On the diffeomorphism types of certain algebraic surfaces. I*, J. Differential Geom. **27** (1988), no. 2, 297–369. MR925124

[FM88b] Robert Friedman and John W. Morgan, *On the diffeomorphism types of certain algebraic surfaces. II*, J. Differential Geom. **27** (1988), no. 3, 371–398. MR940111

[FMZ08] Fu-quan Fang, Jian-wen Man, and Zhen-lei Zhang, *Complete gradient shrinking Ricci solitons have finite topological type* (English, with English and French summaries), C. R. Math. Acad. Sci. Paris **346** (2008), no. 11-12, 653–656, DOI 10.1016/j.crma.2008.03.021. MR2423272

[FQ90] Michael H. Freedman and Frank Quinn, *Topology of 4-manifolds*, Princeton Mathematical Series, vol. 39, Princeton University Press, Princeton, NJ, 1990. MR1201584

[Fre82] Michael Hartley Freedman, *The topology of four-dimensional manifolds*, J. Differential Geometry **17** (1982), no. 3, 357–453. MR679066

[Fri80] D. Friedan, *Nonlinear models in* $2+\varepsilon$ *dimensions*, Phys. Rev. Lett. **45** (1980), no. 13, 1057–1060, DOI 10.1103/PhysRevLett.45.1057. MR584365

[FS09] Ronald Fintushel and Ronald J. Stern, *Six lectures on four 4-manifolds*, Low dimensional topology, IAS/Park City Math. Ser., vol. 15, Amer. Math. Soc., Providence, RI, 2009, pp. 265–315, DOI 10.1090/pcms/015/09. MR2503498

[FT86] Michael H. Freedman and Laurence R. Taylor, *A universal smoothing of four-space*, J. Differential Geom. **24** (1986), no. 1, 69–78. MR857376

[FU91] Daniel S. Freed and Karen K. Uhlenbeck, *Instantons and four-manifolds*, 2nd ed., Mathematical Sciences Research Institute Publications, vol. 1, Springer-Verlag, New York, 1991, DOI 10.1007/978-1-4613-9703-8. MR1081321

[Fur01] M. Furuta, *Monopole equation and the* $\frac{11}{8}$*-conjecture*, Math. Res. Lett. **8** (2001), no. 3, 279–291, DOI 10.4310/MRL.2001.v8.n3.a5. MR1839478

[GH94] Phillip Griffiths and Joseph Harris, *Principles of algebraic geometry*, Wiley Classics Library, John Wiley & Sons, Inc., New York, 1994. Reprint of the 1978 original, DOI 10.1002/9781118032527. MR1288523

[GHMS03] Michael Gutperle, Matthew Headrick, Shiraz Minwalla, and Volker Schomerus, *Space-time energy decreases under world-sheet RG flow*, J. High Energy Phys. **1** (2003), 073, 20, DOI 10.1088/1126-6708/2003/01/073. MR1969853

[Gom85] Robert E. Gompf, *An infinite set of exotic* $\mathbf{R}^4$*'s*, J. Differential Geom. **21** (1985), no. 2, 283–300. MR816673

[Gri85] A. A. Grigor′yan, *The existence of positive fundamental solutions of the Laplace equation on Riemannian manifolds* (Russian), Mat. Sb. (N.S.) **128(170)** (1985), no. 3, 354–363, 446. MR815269

[Gri06] Alexander Grigor′yan, *Heat kernels on weighted manifolds and applications*, The ubiquitous heat kernel, Contemp. Math., vol. 398, Amer. Math. Soc., Providence, RI, 2006, pp. 93–191, DOI 10.1090/conm/398/07486. MR2218016

[GS99] Robert E. Gompf and András I. Stipsicz, *4-manifolds and Kirby calculus*, Graduate Studies in Mathematics, vol. 20, American Mathematical Society, Providence, RI, 1999, DOI 10.1090/gsm/020. MR1707327

[GT01] David Gilbarg and Neil S. Trudinger, *Elliptic partial differential equations of second order*, Classics in Mathematics, Springer-Verlag, Berlin, 2001. Reprint of the 1998 edition. MR1814364

[Guo18] Siao-Hao Guo, *Uniqueness of self-shrinkers to the degree-one curvature flow with a tangent cone at infinity*, Calc. Var. Partial Differential Equations **57** (2018), no. 3, Paper No. 81, 43, DOI 10.1007/s00526-018-1356-4. MR3795213

[GZ08] Hui-Ling Gu and Xi-Ping Zhu, *The existence of type II singularities for the Ricci flow on* S^{n+1}, Comm. Anal. Geom. **16** (2008), no. 3, 467–494, DOI 10.4310/cag.2008.v16.n3.a1. MR2429966

[Hae90] André Haefliger, *Orbi-espaces* (French), Sur les groupes hyperboliques d'après Mikhael Gromov (Bern, 1988), Progr. Math., vol. 83, Birkhäuser Boston, Boston, MA, 1990, pp. 203–213, DOI 10.1007/978-1-4684-9167-8_11. MR1086659

[Ham82] Richard S. Hamilton, *Three-manifolds with positive Ricci curvature*, J. Differential Geometry **17** (1982), no. 2, 255–306. MR664497

[Ham86] Richard S. Hamilton, *Four-manifolds with positive curvature operator*, J. Differential Geom. **24** (1986), no. 2, 153–179. MR862046

[Ham88] Richard S. Hamilton, *The Ricci flow on surfaces*, Mathematics and general relativity (Santa Cruz, CA, 1986), Contemp. Math., vol. 71, Amer. Math. Soc., Providence, RI, 1988, pp. 237–262, DOI 10.1090/conm/071/954419. MR954419

[Ham93a] Richard S. Hamilton, *Eternal solutions to the Ricci flow*, J. Differential Geom. **38** (1993), no. 1, 1–11. MR1231700

[Ham93b] Richard S. Hamilton, *The Harnack estimate for the Ricci flow*, J. Differential Geom. **37** (1993), no. 1, 225–243. MR1198607

[Ham95a] Richard S. Hamilton, *A compactness property for solutions of the Ricci flow*, Amer. J. Math. **117** (1995), no. 3, 545–572. MR1333936

[Ham95b] Richard S. Hamilton, *The formation of singularities in the Ricci flow*, Surveys in differential geometry, Vol. II (Cambridge, MA, 1993), Int. Press, Cambridge, MA, 1995, pp. 7–136. MR1375255

[Ham95c] Richard S. Hamilton, *An isoperimetric estimate for the Ricci flow on the two-sphere*, Modern methods in complex analysis (Princeton, NJ, 1992), Ann. of Math. Stud., vol. 137, Princeton Univ. Press, Princeton, NJ, 1995, pp. 191–200, DOI 10.1080/09502389500490321. MR1369139

[Ham97] Richard S. Hamilton, *Four-manifolds with positive isotropic curvature*, Comm. Anal. Geom. **5** (1997), no. 1, 1–92, DOI 10.4310/CAG.1997.v5.n1.a1. MR1456308

[Ham99] Richard S. Hamilton, *Non-singular solutions of the Ricci flow on three-manifolds*, Comm. Anal. Geom. **7** (1999), no. 4, 695–729, DOI 10.4310/CAG.1999.v7.n4.a2. MR1714939

[Hat83] Allen E. Hatcher, *A proof of the Smale conjecture,* $\mathrm{Diff}(S^3) \simeq \mathrm{O}(4)$, Ann. of Math. (2) **117** (1983), no. 3, 553–607, DOI 10.2307/2007035. MR701256

[Hat02] Allen Hatcher, *Algebraic topology*, Cambridge University Press, Cambridge, 2002. MR1867354

[Hir95] Friedrich Hirzebruch, *Topological methods in algebraic geometry*, Classics in Mathematics, Springer-Verlag, Berlin, 1995. Translated from the German and Appendix One by R. L. E. Schwarzenberger; With a preface to the third English edition by the author and Schwarzenberger; Appendix Two by A. Borel; Reprint of the 1978 edition. MR1335917

[Hit74] Nigel Hitchin, *Compact four-dimensional Einstein manifolds*, J. Differential Geometry **9** (1974), 435–441. MR350657

[HKMR12] Sungbok Hong, John Kalliongis, Darryl McCullough, and J. Hyam Rubinstein, *Diffeomorphisms of elliptic 3-manifolds*, Lecture Notes in Mathematics, vol. 2055, Springer, Heidelberg, 2012, DOI 10.1007/978-3-642-31564-0. MR2976322

[HKP25] Mark Haskins, Ilyas Khan, and Alec Payne, *Uniqueness of asymptotically conical gradient shrinking solitons in* G_2*-Laplacian flow*, Math. Ann. **391** (2025), no. 4, 5033–5116, DOI 10.1007/s00208-024-03049-7. MR4884543

[HL97] Qing Han and Fanghua Lin, *Elliptic partial differential equations*, Courant Lecture Notes in Mathematics, vol. 1, New York University, Courant Institute of Mathematical Sciences, New York; American Mathematical Society, Providence, RI, 1997. MR1669352

[HL23] Bobo Hua and Zhiqin Lu, *Graphs with positive spectrum*, J. Lond. Math. Soc. (2) **107** (2023), no. 6, 2054–2078, DOI 10.1112/jlms.12733. MR4598179

[HLSX22] Michael J. Hopkins, Jianfeng Lin, XiaoLin Danny Shi, and Zhouli Xu, *Intersection forms of spin 4-manifolds and the* $\mathrm{Pin}(2)$*-equivariant Mahowald invariant*, Commun. Am. Math. Soc. **2** (2022), 22–132, DOI 10.1090/cams/4. MR4385297

[HM11] Robert Haslhofer and Reto Müller, *A compactness theorem for complete Ricci shrinkers*, Geom. Funct. Anal. **21** (2011), no. 5, 1091–1116, DOI 10.1007/s00039-011-0137-4. MR2846384

[HN13] Wolfhard Hansen and Ivan Netuka, *On the existence of Evans potentials*, Math. Ann. **356** (2013), no. 4, 1283–1302, DOI 10.1007/s00208-012-0873-2. MR3072801

[HN14] Hans-Joachim Hein and Aaron Naber, *New logarithmic Sobolev inequalities and an ϵ-regularity theorem for the Ricci flow*, Comm. Pure Appl. Math. **67** (2014), no. 9, 1543–1561, DOI 10.1002/cpa.21474. MR3245102

[IKŠ19] James Isenberg, Dan Knopf, and Nataša Šešum, *Non-Kähler Ricci flow singularities modeled on Kähler-Ricci solitons*, Pure Appl. Math. Q. **15** (2019), no. 2, 749–784, DOI 10.4310/pamq.2019.v15.n2.a5. MR4047391

[IMŠ11] James Isenberg, Rafe Mazzeo, and Natasa Sesum, *Ricci flow in two dimensions*, Surveys in geometric analysis and relativity, Adv. Lect. Math. (ALM), vol. 20, Int. Press, Somerville, MA, 2011, pp. 259–280. MR2906929

[Ive93] Thomas Ivey, *Ricci solitons on compact three-manifolds*, Differential Geom. Appl. **3** (1993), no. 4, 301–307, DOI 10.1016/0926-2245(93)90008-O. MR1249376

[Ive96] Thomas A. Ivey, *Local existence of Ricci solitons*, Manuscripta Math. **91** (1996), no. 2, 151–162, DOI 10.1007/BF02567946. MR1411650

[Joh79] Klaus Johannson, *Homotopy equivalences of 3-manifolds with boundaries*, Lecture Notes in Mathematics, vol. 761, Springer, Berlin, 1979. MR551744

[JS79] William H. Jaco and Peter B. Shalen, *Seifert fibered spaces in 3-manifolds*, Mem. Amer. Math. Soc. **21** (1979), no. 220, viii+192, DOI 10.1090/memo/0220. MR539411

[Kaw81] Tetsuro Kawasaki, *The index of elliptic operators over V-manifolds*, Nagoya Math. J. **84** (1981), 135–157. MR641150

[Kha23] Ilyas Khan, *Uniqueness of asymptotically conical higher codimension self-shrinkers and self-expanders*, J. Reine Angew. Math. **805** (2023), 55–99, DOI 10.1515/crelle-2023-0063. MR4669035

[Khi57] A. I. Khinchin, *Mathematical foundations of information theory*, Dover Publications, Inc., New York, 1957. Translated by R. A. Silverman and M. D. Friedman. MR92709

[Kin93] J. R. King, *Exact polynomial solutions to some nonlinear diffusion equations*, Phys. D **64** (1993), no. 1-3, 35–65, DOI 10.1016/0167-2789(93)90248-Y. MR1214546

[Kir89] Robion C. Kirby, *The topology of 4-manifolds*, Lecture Notes in Mathematics, vol. 1374, Springer-Verlag, Berlin, 1989, DOI 10.1007/BFb0089031. MR1001966

[KL08] Bruce Kleiner and John Lott, *Notes on Perelman's papers*, Geom. Topol. **12** (2008), no. 5, 2587–2855, DOI 10.2140/gt.2008.12.2587. MR2460872

[KL14] Bruce Kleiner and John Lott, *Geometrization of three-dimensional orbifolds via Ricci flow*, Astérisque (2014), no. 365, 101–177. MR3244330

[KL17] Bruce Kleiner and John Lott, *Singular Ricci flows I*, Acta Math. **219** (2017), no. 1, 65–134, DOI 10.4310/ACTA.2017.v219.n1.a4. MR3765659

[KL20] Bruce Kleiner and John Lott, *Singular Ricci flows II*, Geometric analysis—in honor of Gang Tian's 60th birthday, Progr. Math., vol. 333, Birkhäuser/Springer, Cham, 2020, pp. 137–155. MR4181000

[KM63] Michel A. Kervaire and John W. Milnor, *Groups of homotopy spheres. I*, Ann. of Math. (2) **77** (1963), 504–537, DOI 10.2307/1970128. MR148075

[KN96] Shoshichi Kobayashi and Katsumi Nomizu, *Foundations of differential geometry. Vol. I*, Wiley Classics Library, John Wiley & Sons, Inc., New York, 1996. Reprint of the 1963 original; A Wiley-Interscience Publication. MR1393940

[Kne29] Hellmuth Kneser, *Geschlossene flächen in dreidimensionalen mannigfaltikeiten*, Jahresbericht der Deutschen Mathematiker-Vereinigung **38** (1929), 248–260.

[Koi90] Norihito Koiso, *On rotationally symmetric Hamilton's equation for Kähler-Einstein metrics*, Recent topics in differential and analytic geometry, Adv. Stud. Pure Math., vol. 18, Academic Press, Boston, MA, 1990, pp. 327–337, DOI 10.2969/aspm/01810327. MR1145263

[Kot98] Dieter Kotschick, *On the Gromov-Hitchin-Thorpe inequality* (English, with English and French summaries), C. R. Acad. Sci. Paris Sér. I Math. **326** (1998), no. 6, 727–731, DOI 10.1016/S0764-4442(98)80039-6. MR1641774

[Kot08] Brett Kotschwar, *On rotationally invariant shrinking Ricci solitons*, Pacific J. Math. **236** (2008), no. 1, 73–88, DOI 10.2140/pjm.2008.236.73. MR2398988

[Kot10] Brett Kotschwar, *Backwards uniqueness for the Ricci flow*, Int. Math. Res. Not. IMRN (2010), no. 21, 4064–4097. MR2738351

[Kot13] Brett L. Kotschwar, *A local version of Bando's theorem on the real-analyticity of solutions to the Ricci flow*, Bull. Lond. Math. Soc. **45** (2013), no. 1, 153–158, DOI 10.1112/blms/bds074. MR3033963

[Kot14] Brett Kotschwar, *Ricci flow and the holonomy group*, J. Reine Angew. Math. **690** (2014), 133–161, DOI 10.1515/crelle-2012-0023. MR3200338

[Kot16a] Brett Kotschwar, *An energy approach to uniqueness for higher-order geometric flows*, J. Geom. Anal. **26** (2016), no. 4, 3344–3368. MR3544962

[Kot16b] Brett Kotschwar, *A short proof of backward uniqueness for some geometric evolution equations*, Internat. J. Math. **27** (2016), no. 12, 1650102, 17. MR3575926

[Kot17] Brett Kotschwar, *Short-time persistence of bounded curvature under the Ricci flow*, Math. Res. Lett. **24** (2017), no. 2, 427–447, DOI 10.4310/MRL.2017.v24.n2.a9. MR3685279

[Kot18] Brett Kotschwar, *Kählerity of shrinking gradient Ricci solitons asymptotic to Kähler cones*, J. Geom. Anal. **28** (2018), no. 3, 2609–2623, DOI 10.1007/s12220-017-9922-0. MR3833809

[Kot24a] Brett Kotschwar, *Backward propagation of warped product structures and asymptotically conical shrinkers*, `arXiv:2401.00606`, 2024.

[Kot24b] Brett Kotschwar, *Ricci flows which terminate in cones*, `arXiv:2401.00607`, Math. Annalen, to appear.

[Kro89] P. B. Kronheimer, *The construction of ALE spaces as hyper-Kähler quotients*, J. Differential Geom. **29** (1989), no. 3, 665–683. MR992334

[KS77] Robion C. Kirby and Laurence C. Siebenmann, *Foundational essays on topological manifolds, smoothings, and triangulations*, Annals of Mathematics Studies, No. 88, Princeton University Press, Princeton, NJ; University of Tokyo Press, Tokyo, 1977. With notes by John Milnor and Michael Atiyah. MR645390

[KT87] Ryoichi Kobayashi and Andrey N. Todorov, *Polarized period map for generalized $K3$ surfaces and the moduli of Einstein metrics*, Tohoku Math. J. (2) **39** (1987), no. 3, 341–363, DOI 10.2748/tmj/1178228282. MR902574

[KW15] Brett Kotschwar and Lu Wang, *Rigidity of asymptotically conical shrinking gradient Ricci solitons*, J. Differential Geom. **100** (2015), no. 1, 55–108. MR3326574

[KW24] Brett Kotschwar and Lu Wang, *Isometries of asymptotically conical shrinking Ricci solitons*, Comm. Anal. Geom. **32** (2024), no. 1, 1–20. MR4806761

[Lai24] Yi Lai, *A family of 3D steady gradient solitons that are flying wings*, J. Differential Geom. **126** (2024), no. 1, 297–328, DOI 10.4310/jdg/1707767339. MR4704550

[Lai25] Yi Lai, *$O(2)$-symmetry of 3D steady gradient Ricci solitons*, Geom. Topol. **29** (2025), no. 2, 687–789, DOI 10.2140/gt.2025.29.687. MR4879366

[Law80] H. Blaine Lawson Jr., *Lectures on minimal submanifolds. Vol. I*, 2nd ed., Mathematics Lecture Series, vol. 9, Publish or Perish, Inc., Wilmington, DE, 1980. MR576752

[Ler10] Eugene Lerman, *Orbifolds as stacks?*, Enseign. Math. (2) **56** (2010), no. 3-4, 315–363, DOI 10.4171/LEM/56-3-4. MR2778793

[Li90] Peter Li, *On the structure of complete Kähler manifolds with nonnegative curvature near infinity*, Invent. Math. **99** (1990), no. 3, 579–600, DOI 10.1007/BF01234432. MR1032881

[Li12] Peter Li, *Geometric analysis*, Cambridge Studies in Advanced Mathematics, vol. 134, Cambridge University Press, Cambridge, 2012, DOI 10.1017/CBO9781139105798. MR2962229

[Lic63] André Lichnerowicz, *Spineurs harmoniques* (French), C. R. Acad. Sci. Paris **257** (1963), 7–9. MR156292

[LM89] H. Blaine Lawson Jr. and Marie-Louise Michelsohn, *Spin geometry*, Princeton Mathematical Series, vol. 38, Princeton University Press, Princeton, NJ, 1989. MR1031992

[LP61] M. Lees and M. H. Protter, *Unique continuation for parabolic differential equations and inequalities*, Duke Math. J. **28** (1961), 369–382. MR140840

[LS84] Terry Lyons and Dennis Sullivan, *Function theory, random paths and covering spaces*, J. Differential Geom. **19** (1984), no. 2, 299–323. MR755228

[LŠ12] Lu Li and Vladimír Šverák, *Backward uniqueness for the heat equation in cones*, Comm. Partial Differential Equations **37** (2012), no. 8, 1414–1429, DOI 10.1080/03605302.2011.635323. MR2957545

[LT92] Peter Li and Luen-Fai Tam, *Harmonic functions and the structure of complete manifolds*, J. Differential Geom. **35** (1992), no. 2, 359–383. MR1158340

[Lu01] Peng Lu, *A compactness property for solutions of the Ricci flow on orbifolds*, Amer. J. Math. **123** (2001), no. 6, 1103–1134. MR1867313

[LV16] Michael T. Lock and Jeff A. Viaclovsky, *Quotient singularities, eta invariants, and self-dual metrics*, Geom. Topol. **20** (2016), no. 3, 1773–1806, DOI 10.2140/gt.2016.20.1773. MR3523069

[LW01] Peter Li and Jiaping Wang, *Complete manifolds with positive spectrum*, J. Differential Geom. **58** (2001), no. 3, 501–534. MR1906784

[LW02] Peter Li and Jiaping Wang, *Complete manifolds with positive spectrum. II*, J. Differential Geom. **62** (2002), no. 1, 143–162. MR1987380

[LW06] Peter Li and Jiaping Wang, *Weighted Poincaré inequality and rigidity of complete manifolds* (English, with English and French summaries), Ann. Sci. École Norm. Sup. (4) **39** (2006), no. 6, 921–982, DOI 10.1016/j.ansens.2006.11.001. MR2316978

[LW20] Yu Li and Bing Wang, *Heat kernel on Ricci shrinkers*, Calc. Var. Partial Differential Equations **59** (2020), no. 6, Paper No. 194, 84, DOI 10.1007/s00526-020-01861-y. MR4169048

[LW23] Yu Li and Bing Wang, *On Kähler Ricci shrinker surfaces*, `arXiv:2301.09784`, 2023.

[LW24a] Yu Li and Bing Wang, *Heat kernel on Ricci shrinkers (II)*, Acta Math. Sci. Ser. B (Engl. Ed.) **44** (2024), no. 5, 1639–1695, DOI 10.1007/s10473-024-0502-7. MR4791450

[LW24b] Yu Li and Bing Wang, *Rigidity of the round cylinders in Ricci shrinkers*, J. Differential Geom. **127** (2024), no. 2, 817–897, DOI 10.4310/jdg/1717772425. MR4756093

[LY02] Fanghua Lin and Xiaoping Yang, *Geometric measure theory—an introduction*, Advanced Mathematics (Beijing/Boston), vol. 1, Science Press Beijing, Beijing; International Press, Boston, MA, 2002. MR2030862

[Mat57] Yozô Matsushima, *Sur la structure du groupe d'homéomorphismes analytiques d'une certaine variété kählérienne* (French), Nagoya Math. J. **11** (1957), 145–150. MR94478

[Mat82] Yukio Matsumoto, *On the bounding genus of homology 3-spheres*, J. Fac. Sci. Univ. Tokyo Sect. IA Math. **29** (1982), no. 2, 287–318. MR672065

[May99] J. P. May, *A concise course in algebraic topology*, Chicago Lectures in Mathematics, University of Chicago Press, Chicago, IL, 1999. MR1702278

[MH73] John Milnor and Dale Husemoller, *Symmetric bilinear forms*, Ergebnisse der Mathematik und ihrer Grenzgebiete [Results in Mathematics and Related Areas], Band 73, Springer-Verlag, New York-Heidelberg, 1973. MR506372

[Mil61] John Milnor, *A procedure for killing homotopy groups of differentiable manifolds*, Proc. Sympos. Pure Math., Vol. III, Amer. Math. Soc., Providence, RI, 1961, pp. 39–55. MR130696

[Mil62] J. Milnor, *A unique decomposition theorem for 3-manifolds*, Amer. J. Math. **84** (1962), 1–7, DOI 10.2307/2372800. MR142125

[Mil65] John Milnor, *Lectures on the h-cobordism theorem*, Princeton University Press, Princeton, NJ, 1965. Notes by L. Siebenmann and J. Sondow. MR190942

[Miz58] Sigeru Mizohata, *Unicité du prolongement des solutions des équations elliptiques du quatrième ordre* (French), Proc. Japan Acad. **34** (1958), 687–692. MR105553

[MMŠ] Zilu Ma, Hamidreza Mahmoudian, and Nataša Šešum, *Unique asymptotics of steady Ricci solitons with symmetry*, `arXiv:2311.09405`.

[Mor38] L.J. Mordell, *The definite quadratic forms in eight variables with determinant unity*, J. Math. Pures Appl. **17** (1938), 41–46.

[MS39] S. B. Myers and N. E. Steenrod, *The group of isometries of a Riemannian manifold*, Ann. of Math. (2) **40** (1939), no. 2, 400–416, DOI 10.2307/1968928. MR1503467

[MS74] John W. Milnor and James D. Stasheff, *Characteristic classes*, Annals of Mathematics Studies, No. 76, Princeton University Press, Princeton, NJ; University of Tokyo Press, Tokyo, 1974. MR440554

[MŠ13] Ovidiu Munteanu and Natasa Sesum, *On gradient Ricci solitons*, J. Geom. Anal. **23** (2013), no. 2, 539–561, DOI 10.1007/s12220-011-9252-6. MR3023848

[MSW19] Ovidiu Munteanu, Chiung-Jue Anna Sung, and Jiaping Wang, *Poisson equation on complete manifolds*, Adv. Math. **348** (2019), 81–145, DOI 10.1016/j.aim.2019.03.019. MR3925015

[MSW21] Ovidiu Munteanu, Felix Schulze, and Jiaping Wang, *Positive solutions to Schrödinger equations and geometric applications*, J. Reine Angew. Math. **774** (2021), 185–217, DOI 10.1515/crelle-2020-0046. MR4250475

[MT07] John Morgan and Gang Tian, *Ricci flow and the Poincaré conjecture*, Clay Mathematics Monographs, vol. 3, American Mathematical Society, Providence, RI; Clay Mathematics Institute, Cambridge, MA, 2007. MR2334563

[MT14] John Morgan and Gang Tian, *The geometrization conjecture*, Clay Mathematics Monographs, vol. 5, American Mathematical Society, Providence, RI; Clay Mathematics Institute, Cambridge, MA, 2014. MR3186136

[MW] Ovidiu Munteanu and Jiaping Wang, *Remarks on Ricci solitons*, Unpublished.

[MW11] Ovidiu Munteanu and Jiaping Wang, *Smooth metric measure spaces with nonnegative curvature*, Comm. Anal. Geom. **19** (2011), no. 3, 451–486, DOI 10.4310/CAG.2011.v19.n3.a1. MR2843238

[MW12] Ovidiu Munteanu and Jiaping Wang, *Analysis of weighted Laplacian and applications to Ricci solitons*, Comm. Anal. Geom. **20** (2012), no. 1, 55–94, DOI 10.4310/CAG.2012.v20.n1.a3. MR2903101

[MW14] Ovidiu Munteanu and Jiaping Wang, *Geometry of manifolds with densities*, Adv. Math. **259** (2014), 269–305, DOI 10.1016/j.aim.2014.03.023. MR3197658

[MW15a] Ovidiu Munteanu and Jiaping Wang, *Geometry of shrinking Ricci solitons*, Compos. Math. **151** (2015), no. 12, 2273–2300, DOI 10.1112/S0010437X15007496. MR3433887

[MW15b] Ovidiu Munteanu and Jiaping Wang, *Kähler manifolds with real holomorphic vector fields*, Math. Ann. **363** (2015), no. 3-4, 893–911, DOI 10.1007/s00208-015-1192-1. MR3412346

[MW15c] Ovidiu Munteanu and Jiaping Wang, *Topology of Kähler Ricci solitons*, J. Differential Geom. **100** (2015), no. 1, 109–128. MR3326575

[MW17] Ovidiu Munteanu and Jiaping Wang, *Conical structure for shrinking Ricci solitons*, J. Eur. Math. Soc. (JEMS) **19** (2017), no. 11, 3377–3390, DOI 10.4171/JEMS/741. MR3713043

[MW19] Ovidiu Munteanu and Jiaping Wang, *Structure at infinity for shrinking Ricci solitons* (English, with English and French summaries), Ann. Sci. Éc. Norm. Supér. (4) **52** (2019), no. 4, 891–925, DOI 10.11650/tjm/180806. MR4038455

[MW22] Ovidiu Munteanu and Jiaping Wang, *Ends of gradient Ricci solitons*, J. Geom. Anal. **32** (2022), no. 12, Paper No. 303, 26. MR4487767

[Mye35] Sumner Byron Myers, *Riemannian manifolds in the large*, Duke Math. J. **1** (1935), no. 1, 39–49, DOI 10.1215/S0012-7094-35-00105-3. MR1545863

[MZ21] Zilu Ma and Yongjia Zhang, *Perelman's entropy on ancient Ricci flows*, J. Funct. Anal. **281** (2021), no. 9, Paper No. 109195, 31, DOI 10.1016/j.jfa.2021.109195. MR4290285

[Nab10] Aaron Naber, *Noncompact shrinking four solitons with nonnegative curvature*, J. Reine Angew. Math. **645** (2010), 125–153, DOI 10.1515/CRELLE.2010.062. MR2673425

[Nak62] Mitsuru Nakai, *On Evans potential*, Proc. Japan Acad. **38** (1962), 624–629. MR150296

[Nak90] Hiraku Nakajima, *Self-duality of ALE Ricci-flat 4-manifolds and positive mass theorem*, Recent topics in differential and analytic geometry, Adv. Stud. Pure Math., vol. 18, Academic Press, Boston, MA, 1990, pp. 385–396, DOI 10.2969/aspm/01810385. MR1145266

[Nas58] J. Nash, *Continuity of solutions of parabolic and elliptic equations*, Amer. J. Math. **80** (1958), 931–954, DOI 10.2307/2372841. MR100158

[Nom60] Katsumi Nomizu, *On local and global existence of Killing vector fields*, Ann. of Math. (2) **72** (1960), 105–120, DOI 10.2307/1970148. MR119172

[NW08] Lei Ni and Nolan Wallach, *On a classification of gradient shrinking solitons*, Math. Res. Lett. **15** (2008), no. 5, 941–955, DOI 10.4310/MRL.2008.v15.n5.a9. MR2443993

[NW24] Jan Nienhaus and Matthias Wink, *New expanding Ricci solitons starting in dimension four*, J. Geom. Anal. **34** (2024), no. 11, Paper No. 327, 18, DOI 10.1007/s12220-024-01778-4. MR4792856

[Omo67] Hideki Omori, *Isometric immersions of Riemannian manifolds*, J. Math. Soc. Japan **19** (1967), 205–214, DOI 10.2969/jmsj/01920205. MR215259

[O'N66] Barrett O'Neill, *The fundamental equations of a submersion*, Michigan Math. J. **13** (1966), 459–469. MR200865

[OVdV86] C. Okonek and A. Van de Ven, *Stable bundles and differentiable structures on certain elliptic surfaces*, Invent. Math. **86** (1986), no. 2, 357–370, DOI 10.1007/BF01389075. MR856849

[Pag78] Don N. Page, *Taub-nut instanton with an horizon*, Phys. Lett. B **78** (1978), 249–251.

[Per02] Grisha Perelman, *The entropy formula for the Ricci flow and its geometric applications*, arXiv:math.DG/0211159, 2002.

[Per03a] Grisha Perelman, *Ricci flow with surgery on three-manifolds*, arXiv:math.DG/0303109, 2003.

[Per03b] Grisha Perelman, *Finite extinction time for the solutions to the Ricci flow on certain three-manifolds*, arXiv:math.DG/0307245, 2003.

[Pet16] Peter Petersen, *Riemannian geometry*, 3rd ed., Graduate Texts in Mathematics, vol. 171, Springer, Cham, 2016, DOI 10.1007/978-3-319-26654-1. MR3469435

[PRS11] Stefano Pigola, Michele Rimoldi, and Alberto G. Setti, *Remarks on non-compact gradient Ricci solitons*, Math. Z. **268** (2011), no. 3-4, 777–790, DOI 10.1007/s00209-010-0695-4. MR2818729

[PT25] Luke T. Peachey and Peter M. Topping, *All two-dimensional expanding Ricci solitons*, J. Lond. Math. Soc. (2) **111** (2025), no. 3, Paper No. e70072, DOI 10.1112/jlms.70072. MR4877592

[PW10] Peter Petersen and William Wylie, *On the classification of gradient Ricci solitons*, Geom. Topol. **14** (2010), no. 4, 2277–2300, DOI 10.2140/gt.2010.14.2277. MR2740647

[Ram13] Daniel Ramos, *Gradient Ricci solitons on surfaces*, arXiv:1304.6391, 2013.

[Rok52] V. A. Rohlin, *New results in the theory of four-dimensional manifolds* (Russian), Doklady Akad. Nauk SSSR (N.S.) **84** (1952), 221–224. MR52101

[Ros95] Philip Rosenau, *On fast and super-fast diffusion*, Phys. Rev. Lett. **74** (1995), 1056–1059.

[Rud76] Walter Rudin, *Principles of mathematical analysis*, 3rd ed., International Series in Pure and Applied Mathematics, McGraw-Hill Book Co., New York-Auckland-Düsseldorf, 1976. MR385023

[Sat56] I. Satake, *On a generalization of the notion of manifold*, Proc. Nat. Acad. Sci. U.S.A. **42** (1956), 359–363, DOI 10.1073/pnas.42.6.359. MR79769

[Sch91] Richard M. Schoen, *A report on some recent progress on nonlinear problems in geometry*, Surveys in differential geometry (Cambridge, MA, 1990), Lehigh Univ., Bethlehem, PA, 1991, pp. 201–241. MR1144528

[Sco83] Peter Scott, *The geometries of* 3-*manifolds*, Bull. London Math. Soc. **15** (1983), no. 5, 401–487, DOI 10.1112/blms/15.5.401. MR705527

[Sco05] Alexandru Scorpan, *The wild world of 4-manifolds*, American Mathematical Society, Providence, RI, 2005. MR2136212

[Ser73] J.-P. Serre, *A course in arithmetic*, Graduate Texts in Mathematics, No. 7, Springer-Verlag, New York-Heidelberg, 1973. Translated from the French. MR344216

[Shi89] Wan-Xiong Shi, *Deforming the metric on complete Riemannian manifolds*, J. Differential Geom. **30** (1989), no. 1, 223–301. MR1001277

[Sma61] Stephen Smale, *Generalized Poincaré's conjecture in dimensions greater than four*, Ann. of Math. (2) **74** (1961), 391–406, DOI 10.2307/1970239. MR137124

[Sog90] C. D. Sogge, *A unique continuation theorem for second order parabolic differential operators*, Ark. Mat. **28** (1990), no. 1, 159–182, DOI 10.1007/BF02387373. MR1049649

[SS87] Jean-Claude Saut and Bruno Scheurer, *Unique continuation for some evolution equations*, J. Differential Equations **66** (1987), no. 1, 118–139, DOI 10.1016/0022-0396(87)90043-X. MR871574

[Sta62] John Stallings, *The piecewise-linear structure of Euclidean space*, Proc. Cambridge Philos. Soc. **58** (1962), 481–488. MR149457

[Sto22] Maxwell Stolarski, *Closed Ricci flows with singularities modeled on asymptotically conical shrinkers*, 2022, arXiv:2202.03386.

[STW00] Chiung-Jue Sung, Luen-Fai Tam, and Jiaping Wang, *Spaces of harmonic functions*, J. London Math. Soc. (2) **61** (2000), no. 3, 789–806, DOI 10.1112/S0024610700008759. MR1766105

[SY05] Takashi Shioya and Takao Yamaguchi, *Volume collapsed three-manifolds with a lower curvature bound*, Math. Ann. **333** (2005), no. 1, 131–155, DOI 10.1007/s00208-005-0667-x. MR2169831

[Tat97] Daniel Tataru, *Carleman estimates, unique continuation and controllability for anizotropic PDEs*, Optimization methods in partial differential equations (South Hadley, MA, 1996), Contemp. Math., vol. 209, Amer. Math. Soc., Providence, RI, 1997, pp. 267–279, DOI 10.1090/conm/209/02771. MR1472300

[Tau87] Clifford Henry Taubes, *Gauge theory on asymptotically periodic 4-manifolds*, J. Differential Geom. **25** (1987), no. 3, 363–430. MR882829

[Tay11] Michael E. Taylor, *Partial differential equations III. Nonlinear equations*, 2nd ed., Applied Mathematical Sciences, vol. 117, Springer, New York, 2011, DOI 10.1007/978-1-4419-7049-7. MR2744149

[Tho69] John A. Thorpe, *Some remarks on the Gauss-Bonnet integral*, J. Math. Mech. **18** (1969), 779–786. MR256307

[Thu79] W. P. Thurston, *Geometry and topology of three-manifolds*, Princeton Math. Dept. Lecture Notes, 1979.

[Thu82] William P. Thurston, *Three-dimensional manifolds, Kleinian groups and hyperbolic geometry*, Bull. Amer. Math. Soc. (N.S.) **6** (1982), no. 3, 357–381, DOI 10.1090/S0273-0979-1982-15003-0. MR648524

[Thu97] William P. Thurston, *Three-dimensional geometry and topology. Vol. 1*, Princeton Mathematical Series, vol. 35, Princeton University Press, Princeton, NJ, 1997. Edited by Silvio Levy. MR1435975

[Thu02] William P. Thurston, *Geometry and topology of three-manifolds*, Electronic version 1.1 — March 2002, 13 chapters, 2002, https://library.slmath.org/nonmsri/gt3m/.

[Tia90] G. Tian, *On Calabi's conjecture for complex surfaces with positive first Chern class*, Invent. Math. **101** (1990), no. 1, 101–172, DOI 10.1007/BF01231499. MR1055713

[Top06] Peter Topping, *Lectures on the Ricci flow*, London Mathematical Society Lecture Note Series, vol. 325, Cambridge University Press, Cambridge, 2006, DOI 10.1017/CBO9780511721465. MR2265040

[Tsu81] Kenji Tsuboi, *Eta invariants and conformal immersions*, Publ. Res. Inst. Math. Sci. **17** (1981), no. 1, 201–214, DOI 10.2977/prims/1195186712. MR613942

[TY24] Peter M. Topping and Hao Yin, *Smoothing a measure on a Riemann surface using Ricci flow*, Ars Inven. Anal. (2024), Paper No. 4, 28. MR4845225

[vdB59] F. van der Blij, *An invariant of quadratic forms mod* 8, Indag. Math. **21** (1959), 291–293. Nederl. Akad. Wetensch. Proc. Ser. A **62**. MR108467

[Vil03] Cédric Villani, *Topics in optimal transportation*, Graduate Studies in Mathematics, vol. 58, American Mathematical Society, Providence, RI, 2003, DOI 10.1090/gsm/058. MR1964483

[Voi07] Claire Voisin, *Hodge theory and complex algebraic geometry. I*, Reprint of the 2002 English edition, Cambridge Studies in Advanced Mathematics, vol. 76, Cambridge University Press, Cambridge, 2007. Translated from the French by Leila Schneps. MR2451566

[Š05] Nataša Šešum, *Curvature tensor under the Ricci flow*, Amer. J. Math. **127** (2005), no. 6, 1315–1324. MR2183526

[Š06] Natasa Sesum, *Convergence of the Ricci flow toward a soliton*, Comm. Anal. Geom. **14** (2006), no. 2, 283–343, DOI 10.4310/cag.2006.v14.n2.a4. MR2255013

[Wal60] Andrew H. Wallace, *Modifications and cobounding manifolds*, Canadian J. Math. **12** (1960), 503–528, DOI 10.4153/CJM-1960-045-7. MR125588

[Wal16] C. T. C. Wall, *Differential topology*, Cambridge Studies in Advanced Mathematics, vol. 156, Cambridge University Press, Cambridge, 2016, DOI 10.1017/CBO9781316597835. MR3558600

[Wan14] Lu Wang, *Uniqueness of self-similar shrinkers with asymptotically conical ends*, J. Amer. Math. Soc. **27** (2014), no. 3, 613–638, DOI 10.1090/S0894-0347-2014-00792-X. MR3194490

[Wol11] Joseph A. Wolf, *Spaces of constant curvature*, 6th ed., AMS Chelsea Publishing, Providence, RI, 2011, DOI 10.1090/chel/372. MR2742530

[Wu91] Lang-Fang Wu, *The Ricci flow on 2-orbifolds with positive curvature*, J. Differential Geom. **33** (1991), no. 2, 575–596. MR1094470

[WW09] Guofang Wei and Will Wylie, *Comparison geometry for the Bakry-Emery Ricci tensor*, J. Differential Geom. **83** (2009), no. 2, 377–405, DOI 10.4310/jdg/1261495336. MR2577473

[WW25] Jie Wang and Youde Wang, *Rigidity and ε-regularity theorems of Ricci shrinkers*, Calc. Var. Partial Differential Equations **64** (2025), no. 2, Paper No. 42, 27, DOI 10.1007/s00526-024-02903-5. MR4846770

[WY12] Willie Wai-Yeung Wong and Pin Yu, *On strong unique continuation of coupled Einstein metrics*, Int. Math. Res. Not. IMRN **3** (2012), 544–560, DOI 10.1093/imrn/rnr038. MR2885982

[Wyl08] William Wylie, *Complete shrinking Ricci solitons have finite fundamental group*, Proc. Amer. Math. Soc. **136** (2008), no. 5, 1803–1806, DOI 10.1090/S0002-9939-07-09174-5. MR2373611

[WZ04] Xu-Jia Wang and Xiaohua Zhu, *Kähler-Ricci solitons on toric manifolds with positive first Chern class*, Adv. Math. **188** (2004), no. 1, 87–103, DOI 10.1016/j.aim.2003.09.009. MR2084775

[Yan12] Bo Yang, *A characterization of noncompact Koiso-type solitons*, Internat. J. Math. **23** (2012), no. 5, 1250054, 13, DOI 10.1142/S0129167X12500541. MR2914656

[Yau75] Shing Tung Yau, *Harmonic functions on complete Riemannian manifolds*, Comm. Pure Appl. Math. **28** (1975), 201–228, DOI 10.1002/cpa.3160280203. MR431040

[Yau76] Shing Tung Yau, *Some function-theoretic properties of complete Riemannian manifold and their applications to geometry*, Indiana Univ. Math. J. **25** (1976), no. 7, 659–670, DOI 10.1512/iumj.1976.25.25051. MR417452

[Yau78] Shing Tung Yau, *On the Ricci curvature of a compact Kähler manifold and the complex Monge-Ampère equation. I*, Comm. Pure Appl. Math. **31** (1978), no. 3, 339–411, DOI 10.1002/cpa.3160310304. MR480350

[Yok09] Takumi Yokota, *Perelman's reduced volume and a gap theorem for the Ricci flow*, Comm. Anal. Geom. **17** (2009), no. 2, 227–263, DOI 10.4310/CAG.2009.v17.n2.a3. MR2520908

[Yok12] Takumi Yokota, *Addendum to 'Perelman's reduced volume and a gap theorem for the Ricci flow' [MR2520908]*, Comm. Anal. Geom. **20** (2012), no. 5, 949–955, DOI 10.4310/CAG.2012.v20.n5.a2. MR3053617

[Zee65] E. C. Zeeman, *Twisting spun knots*, Trans. Amer. Math. Soc. **115** (1965), 471–495, DOI 10.2307/1994281. MR195085

[Zha07] Zhei-lei Zhang, *Compact blow-up limits of finite time singularities of Ricci flow are shrinking Ricci solitons* (English, with English and French summaries), C. R. Math. Acad. Sci. Paris **345** (2007), no. 9, 503–506, DOI 10.1016/j.crma.2007.09.017. MR2375111

[Zha09a] Zhu-Hong Zhang, *Gradient shrinking solitons with vanishing Weyl tensor*, Pacific J. Math. **242** (2009), no. 1, 189–200, DOI 10.2140/pjm.2009.242.189. MR2525510

[Zha09b] Zhu-Hong Zhang, *On the completeness of gradient Ricci solitons*, Proc. Amer. Math. Soc. **137** (2009), no. 8, 2755–2759, DOI 10.1090/S0002-9939-09-09866-9. MR2497489

[Zha12] Qi S. Zhang, *Bounds on volume growth of geodesic balls under Ricci flow*, Math. Res. Lett. **19** (2012), no. 1, 245–253, DOI 10.4310/MRL.2012.v19.n1.a19. MR2923189

[Zha19] Yongjia Zhang, *Compactness theorems for 4-dimensional gradient Ricci solitons*, Pacific J. Math. **303** (2019), no. 1, 361–384, DOI 10.2140/pjm.2019.303.361. MR4045370

Index

Selected Published Titles in This Series

293 **Bennett Chow, Brett Kotschwar, and Ovidiu Munteanu,** Ricci Solitons in Dimensions 4 and Higher, 2025

291 **Tadashi Ochiai,** Iwasawa Theory and Its Perspective, Volume 3, 2025

287 **Jacques Sauloy,** Basic Modern Theory of Linear Complex Analytic q-Difference Equations, 2024

286 **Zalman Balanov, Wieslaw Krawcewicz, Dmitrii Rachinskii, Hao-Pin Wu, and Jianshe Yu,** Degree Theory and Symmetric Equations Assisted by GAP System, 2025

285 **Niles Johnson and Donald Yau,** Bimonoidal Categories, E_n-Monoidal Categories, and Algebraic K-Theory, 2024

284 **Donald Yau,** Bimonoidal Categories, E_n-Monoidal Categories, and Algebraic K-Theory, 2024

283 **Donald Yau,** Bimonoidal Categories, E_n-Monoidal Categories, and Algebraic K-Theory, 2024

282 **Michael Kapovich and Pranab Sardar,** Trees of Hyperbolic Spaces, 2024

281 **Horst R. Thieme,** Discrete-Time Dynamics of Structured Populations and Homogeneous Order-Preserving Operators, 2024

280 **Tadashi Ochiai,** Iwasawa Theory and Its Perspective, Volume 2, 2024

279 **Ellen Elizabeth Eischen, Wee Teck Gan, Aaron Pollack, and Zhiwei Yun,** Automorphic Forms Beyond GL_2, 2024

278 **Delaram Kahrobaei, Ramón Flores, Marialaura Noce, Maggie E. Habeeb, and Christopher Battarbee,** Applications of Group Theory in Cryptography, 2024

277 **George Metcalfe, Francesco Paoli, and Constantine Tsinakis,** Residuated Structures in Algebra and Logic, 2023

276 **Balázs Bárány, Károly Simon, and Boris Solomyak,** Self-similar and Self-affine Sets and Measures, 2023

275 **Alekos Vidras and Alain Yger,** Multidimensional Residue Theory and Applications, 2023

274 **Willi Freeden and M. Zuhair Nashed,** Recovery Methodologies: Regularization and Sampling, 2023

273 **Gennadiy Feldman,** Characterization of Probability Distributions on Locally Compact Abelian Groups, 2023

272 **Tadashi Ochiai,** Iwasawa Theory and Its Perspective, Volume 1, 2023

271 **Hideto Asashiba,** Categories and Representation Theory, 2022

270 **Leslie Hogben, Jephian C.-H. Lin, and Bryan L. Shader,** Inverse Problems and Zero Forcing for Graphs, 2022

269 **Ralph S. Freese, Ralph N. McKenzie, George F. McNulty, and Walter F. Taylor,** Algebras, Lattices, Varieties, 2022

268 **Ralph S. Freese, Ralph N. McKenzie, George F. McNulty, and Walter F. Taylor,** Algebras, Lattices, Varieties, 2022

267 **Dragana S. Cvetković Ilić,** Completion Problems on Operator Matrices, 2022

266 **Kate Juschenko,** Amenability of Discrete Groups by Examples, 2022

265 **Nabil H. Mustafa,** Sampling in Combinatorial and Geometric Set Systems, 2022

264 **J. Scott Carter and Seiichi Kamada,** Diagrammatic Algebra, 2021

263 **Vugar E. Ismailov,** Ridge Functions and Applications in Neural Networks, 2021

262 **Ragnar-Olaf Buchweitz,** Maximal Cohen–Macaulay Modules and Tate Cohomology, 2021

For a complete list of titles in this series, visit the
AMS Bookstore at **www.ams.org/bookstore/survseries/**.

Kolibri 360 GmbH
Ferdinand-Jühlke-Straße 7
99095 Erfurt, Deutschland
produktsicherheit@kolibri360.de